Key Technologies on New Energy Vehicles

Key Technologies on New Energy Vehicles publishes the latest developments in new energy vehicles - quickly, informally and with high quality. The intent is to cover all the main branches of new energy vehicles, both theoretical and applied, including but not limited:

- Control Technology of Driving System
- Hybrid Electric Vehicle Coupling Technology
- Cross Disciplinary design optimization technology
- Single and Group Battery Technology
- Energy Management Technology
- Lightweight Technology
- New Energy Materials and Device
- Internet of Things (IoT)
- Cloud Computing
- 3D Printing
- Virtual Reality Technologies

Within the scopes of the series are monographs, professional books or graduate textbooks, edited volumes, and reference works purposely devoted to support education in related areas at the graduate and post-graduate levels.

Jianqiang Wang · Bingbing Nie · Hong Wang

The Intelligent Safety of Automobile

人民交通出版社股份有限公司
China Communications Press Co.,Ltd.

Springer

Jianqiang Wang
School of Vehicle and Mobility
Tsinghua University
Beijing, China

Bingbing Nie
School of Vehicle and Mobility
Tsinghua University
Beijing, China

Hong Wang
School of Vehicle and Mobility
Tsinghua University
Beijing, China

ISSN 2662-2920 ISSN 2662-2939 (electronic)
Key Technologies on New Energy Vehicles
ISBN 978-981-99-6398-0 ISBN 978-981-99-6399-7 (eBook)
https://doi.org/10.1007/978-981-99-6399-7

Jointly published with China Communications Press Co., Ltd.
The print edition is not for sale in China (Mainland). Customers from China (Mainland) please order the print book from: China Communications Press Co., Ltd.
ISBN of the Co-Publisher's edition: 978-7-114-17579-4

This Springer imprint is published by the registered company Springer Nature Singapore Pte Ltd.
The registered company address is: 152 Beach Road, #21-01/04 Gateway East, Singapore 189721, Singapore

Contents

Chapter 1
Introduction

1.1 Preface

1.1.1 Concept and Connotation of Intelligent Safety of Vehicle

Intelligent safety of vehicle (ISV) refers to the whole process of safety assurance for reliable safety performance, controllable operation risk and effective collision protection of intelligent vehicle (IV) employing advanced theoretical approaches and technologies. With IV in the driver-vehicle–road (DVE) micro traffic system as the core, it establishes and comprehensively evaluates the systematized fundamental safety theory, core methods, key technologies and application systems of vehicles in the design, development and application stages based on the system idea, cooperative theory and intelligent technology, to realize the system safety, operation safety and protection safety of vehicles.

With the development of vehicle intelligent and connected technology, the coupling relationship among traffic objects is continuously enhancing, and the behavior characteristics of traffic participants and the DVE interaction mechanism are changing, which puts forward higher requirements for vehicle driving safety. The active and passive safety technologies for traditional vehicles are difficult to meet driving safety requirements under the strong interaction of traffic elements. It is necessary to study vehicle safety in line with the idea of systematization, cooperation and intelligence. In this context, ISV is developed with a set of theoretical approaches and technical systems for driving safety. It takes the micro traffic system of DVE as the object, takes the IV under the interaction of DVE as the core element, and aims to ensure driving safety and protect traffic participants (including drivers, passengers and pedestrians) from injury. Based on the risk generation and accident causation mechanism in the driving process, it integrates the system design and development,

J. Wang et al., *The Intelligent Safety of Automobile*, Key Technologies on New Energy Vehicles, https://doi.org/10.1007/978-981-99-6399-7_1

operation risk identification, collision damage protection, safety performance evaluation and other related theories, comprehensively makes use of advanced technical means and safeguard measures such as artificial intelligence, big data and network communication to establish the ISV system with common structure, shared information, integrated perception, unified decision and coordinated control, and thus realizes the safety capability of reliable performance in vehicle design, controllable risk in application and effective protection in collision.

1.1.2 Significance of ISV

The development of ISV is in line with the national vehicle industry development plan, and is of social and economic significance to the reduction of road traffic accidents and improvement of road traffic safety. In 2019, the Ministry of Industry and Information Technology issued the *Development Plan of New Energy Vehicle Industry (2021–2035)*, which pointed out the principle of adhering to the development concept of innovation, coordination, green, openness and sharing, as well as the development direction toward electrification, networking, and intelligence. The gradual popularization of intelligent and connected technologies has laid the foundation for the theoretical development and technical application of ISV.

Compared with traditional vehicle active and passive safety, ISV is remarkably characterized in multiple aspects, mainly including:

(1) The vehicle safety performance in intelligent and connected road traffic is not limited to simple collision protection, but includes quite a few factors in the whole driving process, such as functional safety (FuSa), safety of the intended functionality (SOTIF), cybersecurity and intelligent protection. The traditional active and passive safety technology has been difficult to meet the new safety performance requirements. Based on ISV theory, method and technology, a new safety assurance system is developed and applied, to reduce traffic accidents, implement intelligent protection, and protect lives and property more effectively.
(2) The research of ISV has a complete and self-reinforcing system architecture, including system safety, driving safety, intelligent protection, and safety testing and evaluation, all of which are related and promotive to each other. And subsequently a more perfect system with comprehensive theories and approaches can be established, facilitating the rapid and mature development of various sub-technologies.
(3) The R&D and deployment of systems and products related to ISV technology have positively promoted the development and application of vehicle intelligent connected technology, and can drive the development and application of achievements in related industries, including intelligent manufacturing and processing, vehicle system design and intelligent road construction, and ultimately serve the travel demand and the construction of a transportation power.

1.1.3 Research Object and Overall Framework of ISV

The DVE micro traffic system consists of driver system, vehicle system and road system. ISV takes DVE system as the research and application object, and takes IV involved in DVE interaction as the core element, as shown in Fig. 1.1. With the development trend of "new four modernizations" (networking, intelligence, electrification and sharing), traffic travel ways and vehicle forms are changed in multiple ways, the coupling relationship between traffic objects is continuously enhanced, and the action mechanism is dynamically evolved. The three single objects in the DVE system have different features, which brings corresponding challenges for research. As the core of intelligent safety research, the "vehicle" is possessed with such properties as dynamic complexity, nonlinearity, and strong coupling, which will lead to research challenges such as the difficulties in balancing both the high accuracy and real-time of the model, as well as adapting to the vehicle dynamics under complicated and ultimate conditions. In addition, the application of perception, cognition, decision, control, execution and other technologies in the process of vehicle intellectualization will bring changes in element attributes, as well as numerous safety problems; The "driver" is characterized by randomness, time variability and diversity, which will lead to the research difficulties in identification of the interaction between the cognition mechanism and personality features, as well as individual specific demand and behavior mechanism; The "road" mainly includes traffic participants (such as pedestrians) and the environment. It is characterized by uncertainty, dynamics, and interactivity, and is composed of diversified elements. Hence, it is difficult to accurately assess its risks. Meanwhile, due to the interaction and deep coupling between driver and vehicle, vehicle and road, as well as driver and road, new requirements for vehicle safety in complicated DVE system have been put forward, together with the derivation of updated theories, approaches, and technologies.

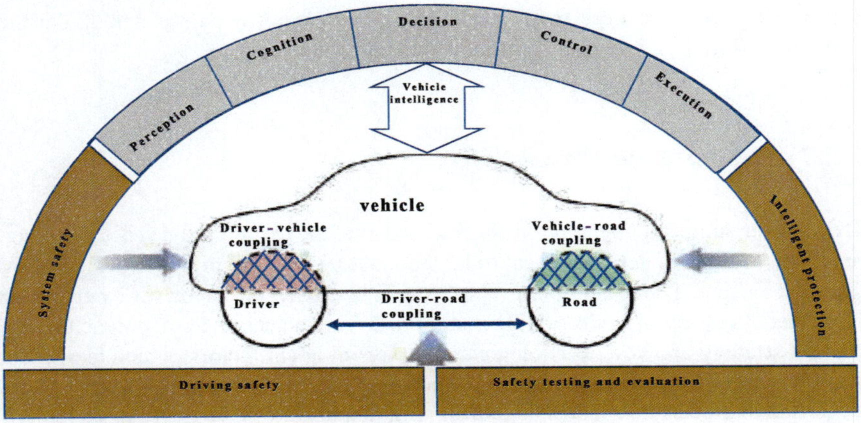

Fig. 1.1 Coupling relationship among various elements of DVE system

According to different dimensions such as the action object, action time, safety objectives and technical means of the DVE micro traffic system, the research field of ISV can be divided into four aspects, i.e., system safety, driving safety, intelligent protection, and safety testing and evaluation, each of which includes fundamental theories, core methods, key technologies, and typical systems. The overall structure is shown in Fig. 1.2. In terms of system development, based on the safety development theory of system engineering, the three technologies of FuSa, SOTIF and cybersecurity are applied to establish a typical safety system for ensuring safety functions in the vehicle design and development stage. In terms of the driving process, based on the driving safety theory, the risk identification and quantitative evaluation models are established, and a complete intelligent driving safety system is developed with key technologies in single-vehicle and group intelligent safety decision and control. In terms of intelligent protection, the injury risk quantification and cooperative protection methods and technologies are constructed through in-depth mining of collision accident data and identification of the injury mechanism of traffic participants, and an adaptive and integrated protection system is designed and developed. In terms of safety testing and evaluation, based on the safety performance evaluation theory, a multi-dimensional and comprehensive evaluation framework is established, and the intelligent safety evaluation system is constructed for technical improvement. ISV can effectively ensure the travel safety of traffic participants through full-stage, multi-dimensional and systematic theories and approaches, technology development and system application.

1.2 Status of Vehicle Safety Research

Vehicle safety research and related technologies have been developed for many years. Regarding DVE system, the current vehicle safety research is conducted in terms of driver safety, vehicle safety, road safety, and DVE cooperative safety. And ISV is the frontier of vehicle safety research in the context of intelligence.

1.2.1 Research on Driver Safety

Driver is the subject for perception and decision-making in traditional vehicle driving, and is also the considerable object to be learned and referred to in the design of IV decision system. The existing relevant research is carried out on driver's perception characteristics, driving style, driving intention, driving behavior, human–machine co-driving strategy and key psychological and physiological factors, which is of great significance for analyzing the causes of accidents, establishing driver's perception, cognition and decision models, and designing human-like driving strategies, etc.

The safety research on drivers mainly includes statistical indexes and driver models, as shown in Fig. 1.3. Research on statistical indexes is to analyze the potential

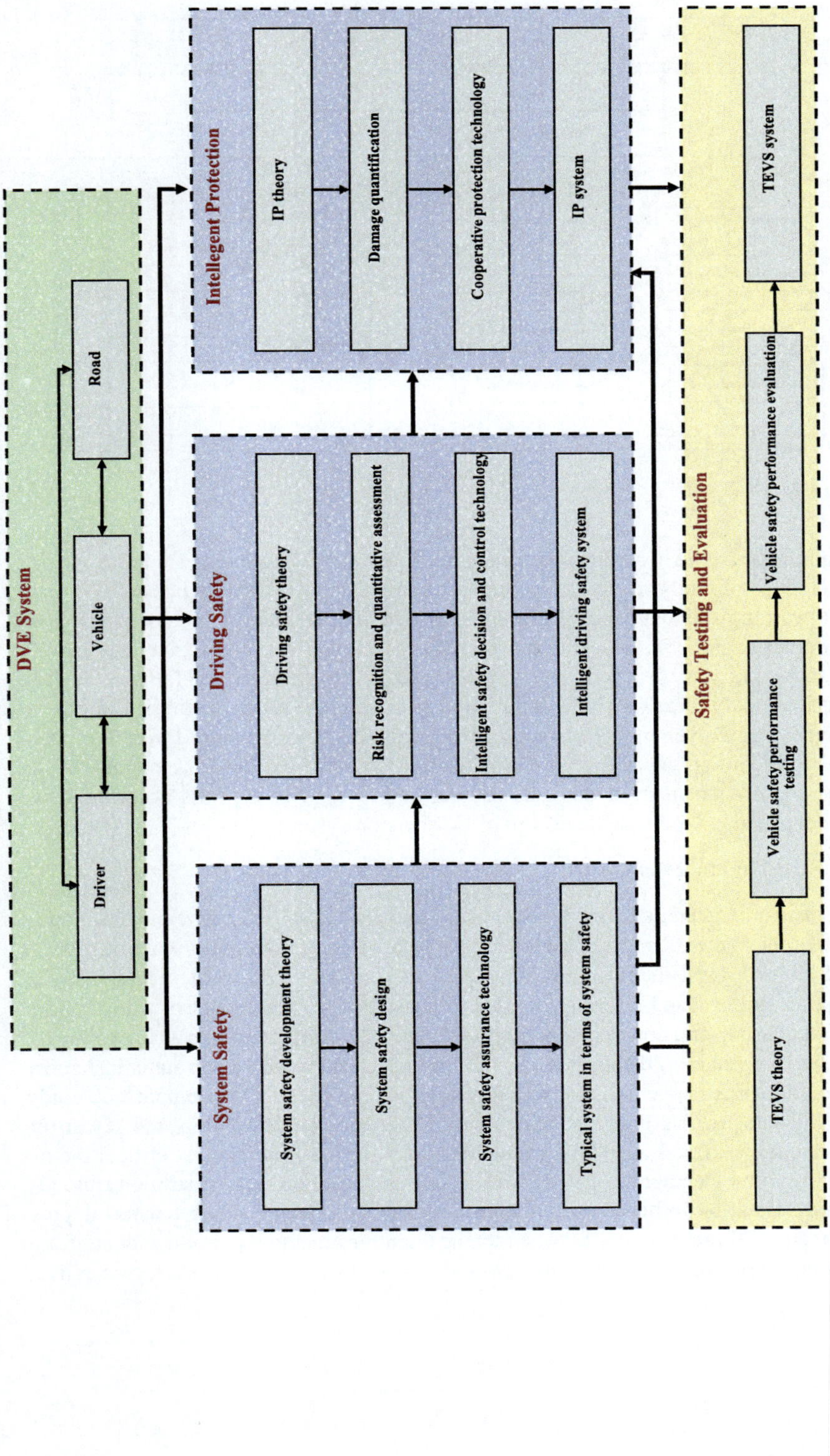

Fig. 1.2 Overall architecture of ISV

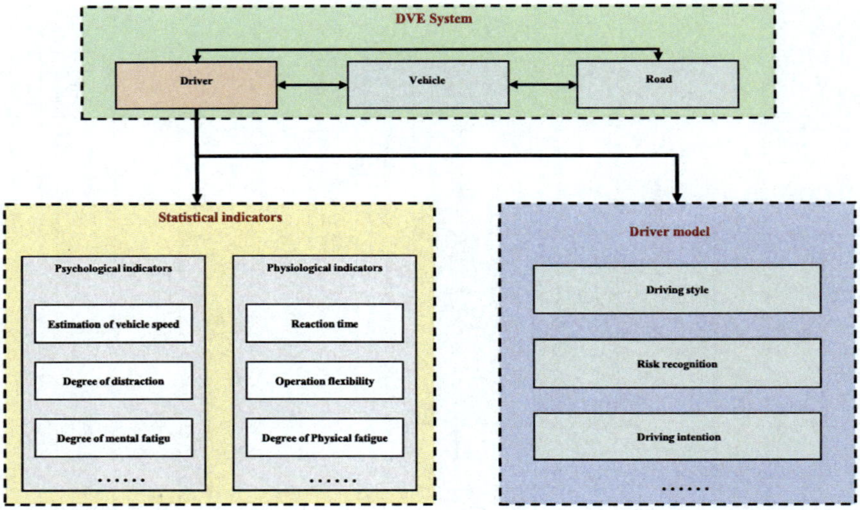

Fig. 1.3 Driver safety research

relationship between certain factors and traffic accidents employing statistics, extract key indexes directly related to accidents, and then carry out safety research on accident warning and avoidance. Relevant statistical indexes mainly include psychological indexes and physiological indexes. The psychological indexes include driver's estimation of vehicle speed, degree of distraction, degree of mental fatigue, etc. The physiological indexes include driver's reaction time, operation flexibility, degree of fatigue, etc. Driver model research is to establish driver's model through characteristic identification and style classification, so as to explore risk cognition and prediction mechanism or describe driving intention. A proper and accurate driver model can also facilitate the design of IV decision system.

1. Statistical Indexes

The research on statistical indexes at home and abroad started early, and has made a lot of achievements, most of which are data statistics on both psychological indexes and physiological indexes, revealing their relationship with traffic accidents. The existing studies can be further divided into two types. One is control study, that is, accident group and non-accident group are set up for comparison, so as to explore the causes of accident. The other is correlation study, which focuses on statistical correlation of various psychological and physiological indexes. In the comparative study, driving data are classified by accident group and non-accident group, and key difference indexes in each category are analyzed for the identification of critical factors leading to the accident. Such indexes as driver's reaction time, operation time [1], speed estimation ability [2], negative emotions (anger, fatigue, panic, tension, depression, etc.) [3] are different between the accident group and the non-accident group. Among them, driver's reaction time and operation time are positively correlated with the tendency of traffic accidents. The speed estimation ability of non-accident group is better than that of accident group, so it can be taken as the key factor for accident representation. In addition, these key factors are also related to their own vehicle speed, driving condition, age, driving age, etc. Taking truck drivers for example, the speed estimation accuracy of non-accident group is better than that of accident group. The difference between the two data is not remarkable under general low-speed conditions, but remarkable under high-speed and low-speed long-distance conditions; The drivers in accident group overestimate their speed more frequently than those in non-accident group, and show great differences under high-speed conditions. Driver's age is not obviously correlated with speed estimation accuracy [4], yet remarkably related to energy and spirit; Driving age has no significant effect on the driver's emotional state [3].

In terms of correlation research, typical indexes for analysis include dynamic visual field, dynamic visual acuity, dark adaptation, listening, masking listening, short-term memory, judgment ability, attention, reaction time, manipulation ability, etc. Drivers are grouped according to gender, age and driving mileage, and engaged in continuous simulated driving test for subsequent analysis. The results show that the three indexes of reaction time, attention and judgment ability have significant differences among groups, so that they can be regarded as the main factors for driving behavior risk identification [5]. There is also a certain correlation between drivers' psychological and physiological quality and their accident records. Through clustering data processing, the driver detection system suitable for China's national conditions can be established with relevant detection standards [6]. In addition, the existing correlation studies have also found that drivers with insufficient driving skills and experience are prone to operational errors; Although it is easy to assess and control such errors, it is extremely difficult to predict random errors that seriously affect traffic safety [7]; Drivers' insufficient identification of risk factors in traffic system and overestimation of their driving skills lead to decrease of safety margin of driving behavior. And the occurrence of driving errors is also strongly related to the low self-recovery ability of drivers in the error state [8]; The changes of drivers' psychological and physiological states during driving are related to their fatigue degree. And inertia is the precursor signal of fatigue in psychological adaptation mechanism [9].

2. Driver Model

Driver model research is conducted substantially for studying driver's driving style, risk cognition characteristics and driving intention. Compared with statistical indexes, it can show driver's subjective factors more concretely, and play an important role in predicting driving behavior and evolution law of driving risk. The current driving style evaluation mainly includes subjective evaluation and objective evaluation. Subjective evaluation is mainly for driving style by means of driver-filled questionnaire [10] and expert scoring [11], which highly relies on correct and effective evaluation questionnaire and scoring experience of experts, and is vulnerable to subjective factors and uncontrollable external factors. Objective evaluation is mainly to obtain a large amount of driving data through driving simulator [12] or real vehicle road experiment [13] for analyzing driving style, which relies on reliable data and can directly reflect driving style. In addition to the research with a single factor, studies that combine subjective scoring and objective driving simulator evaluation are also conducted for revealing the impact of driving style on driving behavior [14]. For example, according to the results of driving style questionnaire, the tested drivers are divided into three categories: cautious, normal, and aggressive. Then, the driving simulator equipped with an eye tracker is used to simulate the driving experiment of the tested drivers in the highway driving environment, and collect the visual characteristic parameters of the perception layer and the driving performance parameters of the operation layer. And the relevant parameters of gaze, saccade, lateral, and longitudinal control behavior are comprehensively evaluated. The results show that the more aggressive the driving style is, the less attention the driver pays to the surrounding environment, the worse the lateral control stability is, and the higher the frequency of rapid acceleration and deceleration is; Such parameters as the number of rear-view mirror gaze, steering wheel angle entropy, accelerator pedal opening and brake pedal opening are apparently different among divers with different driving styles, so that they can be used as characteristic parameters for driving styles. Existing studies on driving styles also show that the frequency of such driving modes as frequent lane changing, reckless lane changing and close-up car-following is a significant characterization of aggressive drivers [15]; The application of driving mode transition probability, together with the analysis of driver's eye movement behavior, would better facilitate the evaluation for driving style [16].

Drivers' risk perception and prediction mechanism for risk evolution are one of the key factors determining driving behavior. Key factors for drivers' perception of traffic risk can be screened by means of questionnaire. Driver risk cognition mechanism is significantly affected by their own accident records, driving history, personality traits, license holding time, violation records, etc. [17], and is remarkably different between experienced and inexperienced drivers. Through the virtual driving simulation experiment, the reaction difference between drivers in experienced group and inexperienced group under hazard scenarios is compared, and the results show that the reaction time of experienced group is shorter [17]. In accordance with basic objects of risk cognition, driver risk cognition mechanism can be divided into "risk cognition based on potential collision trajectory in the future" and "risk cognition based on abnormal driving behavior" [18]. In the study of risk cognition mechanism based on the collision trajectory, the psychological interaction among traffic participants should also be taken into account [19], and the potential benefits of driving behavior and collision risk should be balanced [20]. Abnormal driving behaviors are more frequently present in conditions such as bad weather, fatigue, visual occlusion, and boredom. Therefore, the risk assessment involving these semantic information can quantify risk more accurately in complex environment [21].

Drivers tend to dive unstably, such as changing lane and overtaking. Therefore, accurate driving intention recognition is of great significance for the design of driving assistance system and auto drive system. The important idea that "the driver's intention as an inner state can be inferred from external performance such as driving behavior" was originally put forward in 1997 [22]. In the subsequent research, the mainstream idea is to explore the impact of different factors on the accuracy of driving intention recognition by establishing the interaction model of various factors, among which lane changing intention and braking intention are the focus. The factors that often considered in the research of lane changing intention recognition are steering wheel angle and angle speed, accelerator pedal position, braking pressure, vehicle acceleration or vehicle speed, etc. Steering wheel angle is the most effective parameter for representing lane changing and turning intention [23]. When the combination of steering wheel angle and angular speed is taken as the observation sequence, the intention recognition accuracy based on multi-modal lane changing models such as lane keeping, emergency lane changing and normal lane changing is the highest [24]. The driver's visual characteristics also have a critical impact on the intention of changing lanes. By collecting driving data of drivers with multiple types, analyzing their visual characteristics in different driving stages and dividing their areas of interest, the driver's lane changing intention recognition model can be finally established by means of support vector machine, etc. [25].

Brake pedal force and vehicle speed are the critical factors in the research of braking intention recognition. By studying the influence of different parameters and their combinations on the recognition effect of braking intention, and comprehensively considering various factors such as driving habits, it can be found that the brake pedal displacement is the most effective key parameter for driver's braking intention [26]. The research on braking intention plays an important role in the optimization of vehicle braking control strategy [27]. There are also studies on joint recognition of various lateral and longitudinal driving intentions. The typical way is to establish intention recognition model with multi-input and single output from vehicle speed, accelerator pedal opening, accelerator pedal opening change rate to driver's intention, which can effectively identify five driving intentions of parking, deceleration, vehicle speed maintenance, acceleration, and rapid acceleration. The quantified driving intention results can be applied to the selection and correction of automatic transmission shift strategy [28].

In addition to description and prediction of driving intention with key parameters, a driver's driving decision mechanism can also be established for inference of driving intention in a specific environment. The theory that "driver's driving decision mechanism follows the principle of least action" was put forward in 2020. Inspired by numerous extreme phenomena of physical and biological behaviors in nature, the decision model based on least action is established by a physical model that abstractly describes driving process. The research results show that there is no significant difference between driver's actual action and least action during driving, which proves that driver's driving intention follows the principle of least action [29].

1.2.2 Research on Vehicle Safety

The birth of vehicle brings about the improvement of daily travel efficiency, as well as the progress and development of society. Since the first internal combustion engine vehicle came into being, technological progress has been promoting the rapid development and gradual maturity of vehicles and the whole industry. From the initial manual small-batch production, to the first assembly line production of Ford Model T, then to the diversification of vehicle types after the first World War and the diversification of products after the 1950s, vehicles have been popularized all over the world, and the industry has become the pillar in many countries. During the development toward the new four modernizations, vehicle industry has ushered in new transforms. Different from the traditional vehicle system, IV is a comprehensive system that integrates environment perception, decision planning, multi-level driving assistance and other functions. AV (autonomous vehicle, or self-driving vehicle) is an advanced form of IV, and realizes driverless by computer system.

With the popularity of vehicles and frequent occurrence of traffic accidents, the safety problems brought by the vehicle development have attracted the attention of various countries and major vehicle manufacturers. The research on vehicle safety is mainly conducted from three aspects: traditional passive/active safety, advanced driving assistance and intelligent safety, as shown in Fig. 1.4. The traditional passive and active safety technologies are relatively mature, mainly covering seatbelt, airbag, anti-locked braking system, electronic stability program, etc. Since the wide application of sensors in vehicles, advanced driving assistance has been rapidly developed and applied, with technologies such as forward collision warning, lane departure warning, automatic emergency braking and blind spot detection. In the context of the new four modernizations, intelligent safety technology is becoming the mainstream development direction currently, involving a number of specific technologies such as system safety, driving safety, intelligent protection, and safety testing and evaluation.

1. Traditional Passive/Active Safety

Passive safety-related research is early and covers more specific technologies. It mainly focuses on the occupant restraint system, to alleviate direct injuries of occupants during the collision and effectively reduce the loss in the accident. As one of the most typical devices for occupant protection, seatbelt restrains occupants in collision, so as to prevent them from secondary collision with steering wheel, instrument panel and interior trim panel, etc., or even rushing out of the vehicle. Its development process is shown in Fig. 1.5. In 1885, before the invention of vehicle, the prototype of seatbelt appeared on the carriage for preventing passengers from falling. In 1902, a racing driver was inspired by the cloth in the children's go-cart for preventing babies from falling, and installed a rope in his racing car, which protected him well in an accident. In 1922, seatbelts were widely set in racing cars for protecting racing drivers, while were not popularized in civilian cars. After the Second World War, the vehicle industry ushered in a period of rapid development. Due to the pursuit for vehicle speed, how to better ensure the safety of drivers and passengers has become a concerned research topic. In 1955, the seatbelt technique was applied by Ford and other vehicle manufacturers, yet with insufficiently scientific design in structure. For example, in a frontal collision, the traditional two-point seatbelt was probably incapable to prevent the upper body of passengers from leaning forward. In

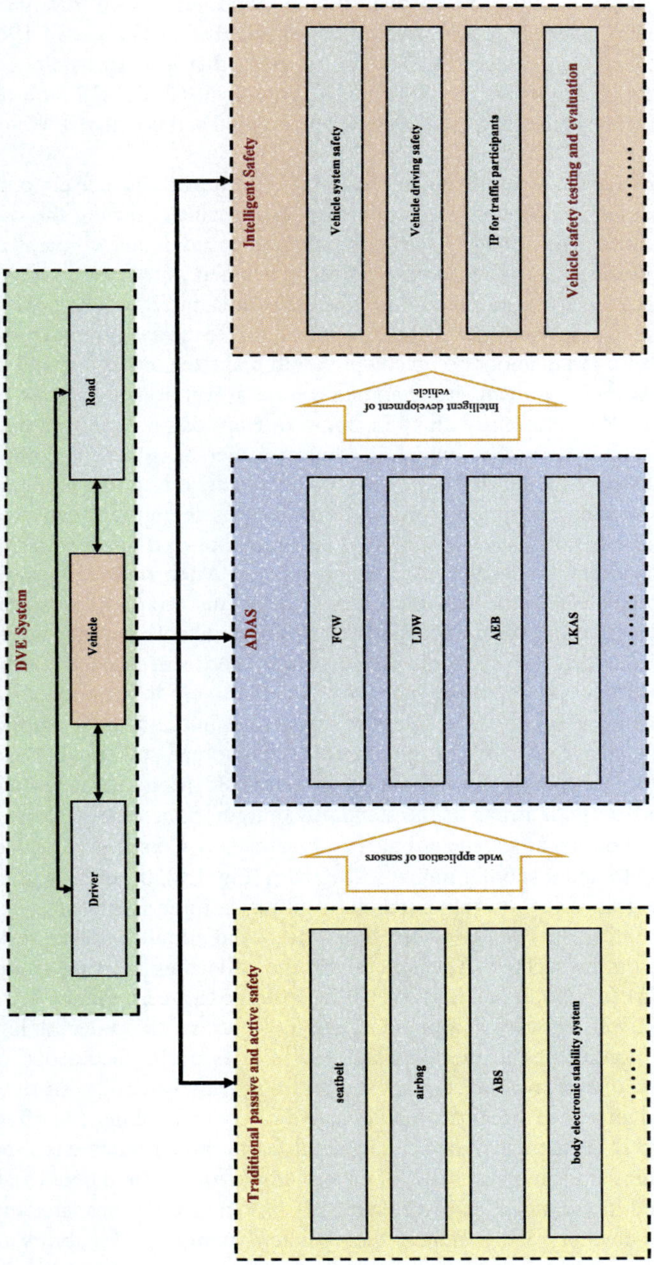

Fig. 1.4 Research on vehicle safety

1958, engineers of Volvo car invented the three-point seatbelt which overcame the defects of the above-mentioned two-point seatbelt, and announced that the patent for the three-point safety belt was open to the public free of charge. In 1967, the U.S. Traffic Safety Conference released data proving that the three-point seatbelt can reduce occupant injury by 50~60% in accidents. In 1968, the US requires cars to be equipped with seatbelts, which were widely applied and popularized from then on.

Airbags, also known as supplemental restraint system (SRS), are used to restrain and cushion passengers' bodies in conjunction with seatbelts during the collision for safety assurance. The airbag system is mainly composed of anti-collision airbag, sensor and electronic controller. During collision accident, the sensor receives the impact signal and sends it to the electronic controller when the signal reaches a certain strength. The electronic controller will send a start signal to the gas generator in airbag for generation of a large amount of gas, which will make the airbag expand rapidly and provide passengers with an elastic air cushion for absorption of collision energy, so as to avoid or alleviate injury. In 1953, John W. Hetrick, an American engineer, invented the vehicle air cushion, which is a buffer device based on the compressed air system for alleviating injuries of passengers in vehicle to a certain extent in case of emergency braking or frontal collision. However, as the system can not make the airbag expand rapidly, it is not sufficient for protection and thus not adopted by vehicle manufacturers. Until 1967, American engineer Allen K. Breed invented an airbag controlled by electromechanical sensors, which could sense the occurrence of collision accidents, and solved the problem that airbag could not expand rapidly with a small explosive device. It is considered to be the embryonic form of modern airbag. As an auxiliary restraint system, airbag can not work effectively for protection without seatbelt, yet can even aggravate accident injury. At the initial stage of application, airbag was not well accepted by consumers, some of whom mistakenly believed that airbag could work independently without seatbelt. In 1980, Mercedes Benz launched the world's first car with airbag as the standard configuration, which safety system was composed of both seatbelt and airbag. In 1995, the U.S. Congress passed a bill to make the airbag a standard configuration for vehicles (Fig. 1.6). In addition, the driver may hit the steering wheel during collision, resulting in injury. The safety steering system can optimize the rigidity of steering wheel, and simultaneously soften the plastic coating on the surface. Assembled with the collapsing steering column for displacement in collision, it can limit the collision force to a certain range.

With the gradual maturity and popularization of passive safety technology, the expectation no longer lies in the alleviation of injuries during accidents, but the further reduction or even avoidance of accidents through timely monitoring of vehicle state and improvement of stability. Vehicle active safety technology can effectively facilitate drivers in accident avoidance. In general, its working principle is to provide necessary warning information in face of danger employing radar, camera and other sensors installed in various parts of vehicle body or take certain measures to effectively prevent accidents. For instance, the vehicle dynamics and stability control technology is for monitoring the driving state, reasonably distributing the longitudinal and lateral tire adhesion, precisely controlling the vehicle dynamics behavior

Fig. 1.5 Development process of seatbelts

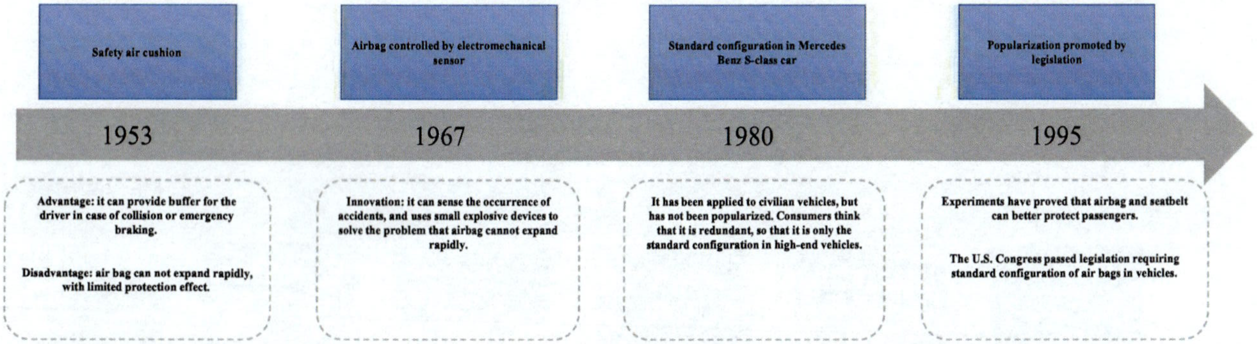

Fig. 1.6 Development process of airbag

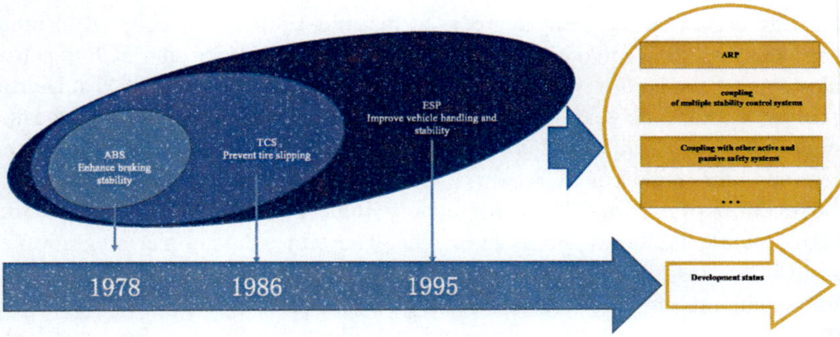

Fig. 1.7 Development process of vehicle dynamics and stability control technology

under the limit adhesion, and consequently improving the stability during driving [30]. The specific development process is shown in Fig. 1.7.

One of the most typical applications of active safety technology is the anti-locked braking system (ABS), which automatically adjusts the force during braking for preventing wheel locking and enhancing braking stability. ABS first appeared in the aviation field. As early as in 1920, Gabriel Voisin, a French aircraft designer, tried to adjust the hydraulic braking pressure of aircraft brakes to reduce risk of tire slip. In 1950, the Maxaret anti-skid system invented by Dunlop company was widely used in the aviation industry, with the capacity of shortening braking distance by 30% and extending service life of tires Since then, ABS has been gradually applied in the vehicle industry. For example, Lincoln vehicle was equipped with ABS device of Air France in 1954. Modern ABS was invented by Mario Palazzetti of Fiat Research Center. Later, the patent was sold to Bosch, who named it ABS. In 1978, Bosch and Mercedes Benz developed ABS with three channels, four wheels and a digital controller, laying the foundation for the basic mode of ABS in a breakthrough. Since then, ABS has been popularized and applied rapidly, with increasing market share. In the mid-1980s, with the progress of electronic control technology, ABS was possessed with more perfect function and lower cost. More and more small and medium-sized household vehicles were also installed with ABS. In 1987, European regulations stipulated that all new models must be equipped with ABS from 1991.

Along with ABS, traction control system (TCS), also known as anti-slip regulation (ASR), is widely used. With the increasing power of vehicle engine, the force between tire and ground is easy to break through the critical value, the excessive slip of driving wheels occurs from time to time, and the loss of road grip will affect the steering control and stability of vehicle. TCS limits the engine output power by controlling the engine throttle opening, or adjusts the braking force by the braking system, to prevent driving wheels from slipping. In 1972, Japan registered a system to control ignition of engine cylinders for the first time. When the wheel acceleration exceeds a certain threshold, the cylinder ignition is stopped. In 1978, Germany registered a device to decrease cylinder oil supply for driving anti-skid function. Since then, a series of devices have appeared, such as applying braking torque to driving wheels or adjusting clutch engagement degrees. In the mid-1980s, TCS ushered in rapid development. In 1985, Volvo cars took the lead in applying electronic TCS to actual vehicles in mass production. In 1986, GM exhibited the Corvette with TCS at Detroit Auto Show. In the same year, Bosch combined ABS and TCS for the first time and assembled them on Mercedes Benz S-class cars, marking the gradual maturity and application in the market of TCS.

The functions of ABS and TCS are relatively single. For example, ABS only works in braking, while TCS only works in driving. Therefore, the simple integration of ABS and TCS is insufficient to deal with vehicle lateral instability caused by ultimate conditions such as braking steering and high-speed steering. In 1992, Bosch and BMW cooperated to develop a new generation of Electronic Stability Program (ESP) integrating two technologies, which further expanded the functions of anti-locked braking and traction control. In 1995, ESP system was first applied to Mercedes Benz S-class car. The composition of ESP is shown in Fig. 1.8. Each sensor collects signals, communicates with the engine management system through the CAN bus, and then analyzes the signals by the electronic control unit (ECU) for judging the vehicle driving state. The stability and controllability of the vehicle when entering hazardous conditions are improved through the regulation of braking and steering systems [30]. The popularity of ESP has effectively improved driving safety performance. According to the investigation of the National Highway Traffic Safety Administration (NHTSA), ESP can reduce 35% of collision accidents and 43% of the possibility of fatal collisions [31, 32]. In 2011, the 126th Regulation of Federal Motor Vehicle Safety Standards (FMVSS) of US stipulated that vehicles produced and sold in North America must be equipped with ESP. In recent years, the research on vehicle stability control systems has still been advancing. For example, the active rollover protection (ARP) [33] based on ESC system further considers the force and roll of vehicle in the vertical direction, and avoids rollover by braking and reducing the engine output torque.

2. Advanced Driver Assistance

Most road traffic accidents are closely related to driver factors [34]. In order to reduce accidents caused by drivers' low skills, bad habits, careless observation, wrong decisions, improper handling and other factors, advanced driver assistance systems (ADAS) can be applied to sense the road traffic environment and provide driving information for drivers with the help of modern electronic information and sensing technology, identify risks from the comprehensive information of road and vehicle conditions, send warning or driving assistance to drivers in case of danger, prevent traffic accidents or reduce accident losses [30], as well as minimize man-made errors and reduce casualties [35].

In the late 1970s, vehicle passive safety technology which can better protect passengers in accidents has gradually matured and popularized. In order to further avoid accidents, the research focus has shifted to the development of ADAS. Compared with ABS, ESP and other systems, which mostly rely on the vehicle's state data, ADAS focuses on the perception of driving environment and the judgment of relative motion state between the vehicle and the environment. It makes use of

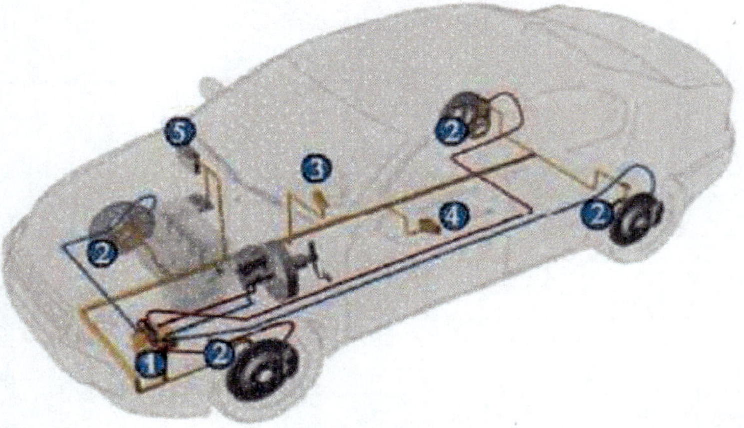

Fig. 1.8 ESP composition [30]. 1-ESP electronic control hydraulic unit and ECU; 2-active wheel speed sensor; 3-steering wheel angle sensor; 4-yaw rate sensor and lateral acceleration sensor; 5-communication with engine management system via CAN bus

sensors including camera, laser radar, millimeter wave radar, etc., or directly obtains information from other vehicles or road infrastructure with wireless communication technology. Based on perceived information, ADAS can give warnings to drivers in hazardous situations or directly take measures to avoid accidents. Common ADAS technology applications include forward collision warning (FCW), autonomous emergency braking (AEB), adaptive cruise control (ACC), lane departure warning (LDW), lane keeping assist (LKA), etc., as shown in Fig. 1.9.

The FCW system can timely obtain the position information of the vehicle or object in front through sensors such as radar, judge and analyze the relative distance and speed between the ego-vehicle and the object in front, and warn the driver by flashing a warning light or sound in case of a collision risk. The traditional FCW algorithm usually judges the existence of danger by a risk assessment model with fixed parameters. In order to adapt to different driver characteristics, the warning threshold can be adjusted in real time according to behavior changes (including behavior fluctuations and individual differences) for better improve the adaptability to drivers and reduce the system false alarm rate [36]. The FCW system can enhance driver's risk perception ability (about 23%) in the face of collision scenario, and shorten risk response time by 36.4% within the speed range of 45-60 km/h [37]. It can monitor the vehicle or object in front, and automatically brake to avoid or mitigate the collision, when there is no braking taken or too small braking force. The research on active collision avoidance began early in 1960. Due to the limitations of microwave technology at that time and high cost, no substantive results were achieved. Later, the development and popularization of radar technology promoted R&D and breakthrough of active anti-collision technology. In 1986, more than 10 European vehicle manufacturers, led by Mercedes Benz, intended to develop an

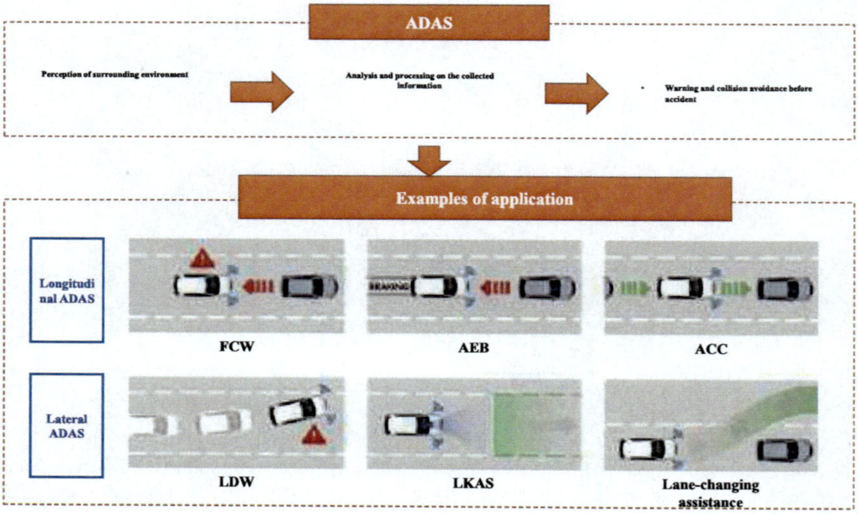

Fig. 1.9 ADAS

intelligent active anti-collision system for commercial use. In 1999, they successfully launched the Distronic system, which has the functions of cruising, distance keeping and collision avoidance. With the technical development and market demand, collision avoidance is expected not only to realize early warning, but also to assist the brake in an emergency. Therefore, research institutions and major manufacturers have been dedicated a lot to the development of AEB functions. When drivers fail to respond to FCW warning and take measures, AEB can make emergency braking in hazardous situations for avoiding collision. In 1991, the Japanese Ministry of Transport implemented the Advanced Safe Vehicle (ASV) program. With the support and promotion of the program, Toyota designed and developed an active preventive safety system, which can realize the auxiliary braking function. In the twentieth century, major vehicle manufacturers have developed and launched their own mature AEB products, such as Honda's Collision Mitigation Brake System (CMBS) and Volvo's City Safety System. A study based on European and Australian data in 2015 showed that AEB system can reduce rear-end collision accidents by 38% [38].

ACC system is a common configuration of current mass-produced vehicles, and is an automatic control system developed based on early cruise control technology. It senses the driving environment through on-board sensors, obtains the position and speed information of the vehicle ahead, and realizes automatic following to a certain extent in combination with constant speed cruise and safe distance maintenance control. For example, when the vehicle is too close to its front vehicle, ACC control unit can coordinate with ABS and the engine control system to make the wheels brake properly and control the output power of the engine to decrease, so as to maintain a safe and appropriate following distance. ACC system can reduce driver's operation task to a certain extent and improve driving safety. Its working principle was first proposed by the University of Michigan in the 1960s. With the progress of sensor and computer technology, ACC system ushered in rapid development in the 1990s. At first, it mainly aimed at highway scenarios and worked within a certain speed range. In 1992, Mitsubishi took the lead in applying a laser-based distance detection system Debonair in the Japanese market, which can warn drivers without affecting acceleration, braking or gear shifting. In 1997, Toyota proposed the lidar-based ACC system, which controls the speed through accelerator pedal or downshift. In 1999, Mercedes Benz S-class car was equipped with the first ACC system based on millimeter wave radar. In 2000, Toyota's ACC system was added with brake control, which can control vehicle speed by the brake system. Since the beginning of the twenty-first century, ACC system has been continuously improved and developed, with the capability of working in the full speed range, coping with urban road scenarios, as well as start-stop following and low-speed following.

LDW system can assist the driver to avoid lane departure in case of misoperation, distraction and sleepiness. It generally monitors the lateral position of the vehicle in the lane based on vision. When the vehicle has crossed the lane line without turn signal on, the system will send a warning signal to the driver. In 2000, the first LDW system was unveiled for the first time. Developed by American company Iteris for Mercedes Benz Actros commercial trucks, it can warn drivers of unintentional deviation through the sound generated by vehicles departing from the lane. If the driver sends an active turn signal, no warning will be generated. In 2012, Mobileye developed LDW system for detecting lane markings and alerting drivers to focus by issuing visual and audible warnings. In 2014, Tesla Model S was released with an advanced LDW system, which can read the traffic speed limit through the front camera with computer vision character recognition technology, and then transmits it to the vehicle. When the vehicle leaves the lane at a speed of more than 48 km per hour, the system will give a warning and the steering wheel will vibrate to remind driver of an unexpected lane changing. From 2009 to 2015, police data from 25 states in the US showed that the occurrence of sideslip and frontal collision of vehicles with LDW system as an optional configuration decreased by 11%, and the injuries in these collisions reduced by 21% [40]. Similar to LDW system, LKA system can also remind driver of the detected driving lane departure. In addition, it can control the steering wheel, slightly correct the heading, and actively keep the vehicle in the driving lane. To improve system reliability, avoid frequent steering correction and reduce driving burden, the LKA system with two switchable modes of LDW and LKA has been proposed [39]. In 2001, Nissan's fourth generation CIMA was first equipped with LKA system. In 2006, Lexus introduced the multi-mode LKA system in the LS 460 model, which uses stereo cameras and more complex pattern recognition processors to issue audio-visual warnings or use electric power steering to guide the vehicle in the lane. In 2012, Ford applied LKA in several high-end models, and its technology was supported by Mobileye. At present, Mobileye's safety technology has been integrated into hundreds of new models from the world's major vehicle manufacturers. The development process of the above-mentioned typical ADAS applications is shown in Fig. 1.10.

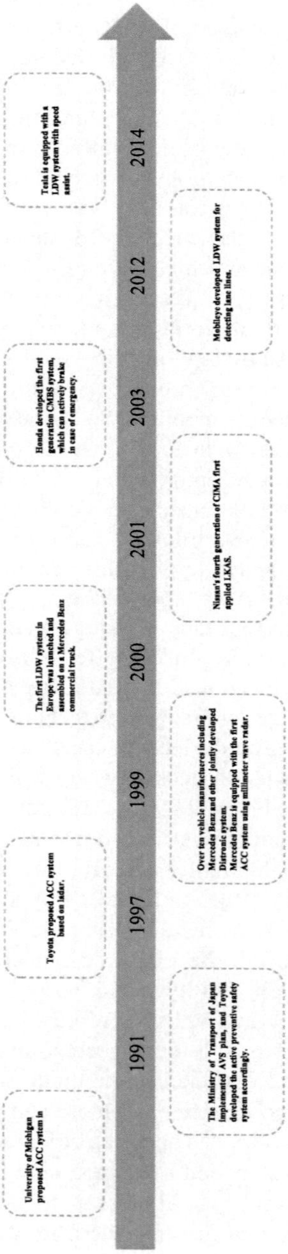

Fig. 1.10 Development process of typical ADAS applications

Since 2011, the US New Car Assessment Program (US-NCAP) has included some ADAS applications into vehicle safety scoring, and introduced FCW and LDW as testing bonus items. In 2014, the European New Car Assessment Program (E-NCAP) incorporated ADAS into the safety scoring system and introduced the evaluation of AEB and LDW. Since 2014, the configuration of ADAS has become a necessary condition for vehicles to obtain E-NCAP five stars. Nowadays, China's New Car Evaluation Program (C-NCAP) has also brought some ADAS applications into the evaluation system. In addition to enhancing the importance of ADAS in the new car assessment system, countries have issued regulations to promote the mandatory installation of ADAS. For example, in Europe, AEB and LDW are mandatory for newly-produced heavy commercial vehicles.

3. Intelligent Safety of Vehicles

With the rapid development of advanced electronics, communication and information technology, a new generation of vehicle technology innovation toward intelligence and networking is being carried out all over the world. Intelligent safety has become the mainstream technology in vehicle safety, covering system safety, driving safety, intelligent protection, and safety testing and evaluation.

In terms of system safety, the research focuses on the improvement of FuSa, SOTIF and cybersecurity employing appropriate assurance technology. There are such standards as ISO 26262 [41], ISO 21448 [42], SAE J3061 [43] and ISO 21434 [44] issued for these three types of system safety, which stipulate the basic design and development process, and involve the application of safety analysis, verification, and confirmation. FuSa technology aims at the risk induced by abnormal function of the electronic and electrical system, detects, and isolates the fault, and further controls the impact of the fault on the system; The existing technical system is relatively complete, involving typical technologies in fault detection and diagnosis and fault-tolerant control, etc. The fault detection and diagnosis are based on an analytical model, signal processing and intelligent fault detection and diagnosis. Taking point-to-point communication for example [45], the technology developed based on AUTOSAR (automotive open system architecture) can effectively detect repeated transmission, verification, and transmission sequence errors in the communication process employing E2EProfile2, so as to ensure the safe communication of data between different cores in ECU. The fault-tolerant control includes passive fault-tolerant control and active fault-tolerant control technologies, such as redundancy mechanisms. The redundant system [46] can provide a multi-threaded framework as an important operating system for hardware error detection and recovery, and minimize the increased cost of redundant replication. SOTIF technology aims at the risk induced by intended function of the system or its insufficient action. The relevant research is in the initial stage, and a perfect technical system has not been established. Its technologies mainly include SOTIF state monitoring, perception-based SOTIF assurance, decision-based SOTIF assurance, HMI-based SOTIF design, etc. Among them, the SOTIF state monitoring is conducted employing MEMS (micro-electro-mechanical system) sensors, three-dimensional acceleration sensors, and 12 accelerometers on four vertices of the cube, which can detect and configure vehicle motion attitude parameters (MAP) [47]. Through the estimation on state of vehicles, environment and drivers and passengers, it provides a reference for risk protection, and inputs information for the real-time assurance of perception, decision and HMI during driving; Through hardware improvement such as sensor anti-interference and network communication, as well as software improvement such as data repair and multi-source information fusion, the SOTIF based on perception is ensured; Through the improvement of decision algorithm, hybrid decision as well as application of risk monitoring and protection technologies based on upstream and downstream algorithms, the SOTIF based on decision is ensured; Through the analysis on misoperation behavior and the research on protection technology, the SOTIF based on HMI is designed. For example, when the perception and decision functions are realized by neural network technology, both ensuring the robustness of neural network against external interference by resisting samples [48] and enhancing the fault recovery capability of neural network by fault-tolerant strategy [49] are conducive to the improvement of the perception-decision based SOTIF. Aiming at the risk posed by cyberattack threats, cybersecurity technology is developed based on artificial intelligence (AI) algorithm, control theory, infection theory, game theory and graph theory, etc. Typical technology includes cybersecurity network topology design, cybersecurity intrusion detection and defense, as well as cybersecurity encryption and communication, etc. The network topology comprises of traditional distributed architecture, functional domain-based architecture, and central computing architecture, and supports the fundamental architecture for other cybersecurity technologies. As an active defense technique, intrusion detection and defense are applied to warn and respond to intrusion attacks, and provide feedback to passive defense technologies such as encryption and communication, so as to effectively ensure cybersecurity. By design of the identity authentication scheme under the structure of IOV, as well as application of digital certificates, challenge response, multi-factor authentication and other technical means, the reliability and security of information transmission of IV and cloud service platforms can be ensured in face of password attack, replay attack, man-in-the-middle attack and other intrusion means [50].

In terms of driving safety, the research focuses on the safety of single vehicle or multi-vehicle group driving through accurate driving risk assessment and reasonable decision and control. The single vehicle intelligent safety decision technology is based on the vehicle kinematics model and dynamics model. In view of dynamic risks in the driving process, it can realize global path planning, behavior decision and trajectory planning by hierarchical decision and safety decision algorithms, so as to ensure trajectory tracking control and critical instability control. In terms of intelligent safety decision and control of a single vehicle, cutting-edge decision and control technology include safety decision based on driving safety field, and multi-objective cooperative decision technology based on least action [29, 51], etc. Other typical technologies include driving trajectory planning based on multi-cluster polynomial curves and Bessel curves [52], vehicle lateral and longitudinal control based on model predictive control (MPC) [53], and intelligent decision based on machine learning [54]. In the single vehicle intelligent safety decision, the state of other vehicles around has a great impact on the ego-vehicle safety. Therefore, the intention identification and trajectory prediction towards other vehicles as well as the driving risk assessment towards the ego-vehicle are the main research focuses. For vehicle behavior intention identification, an algorithm combining hidden Markov model (HMM) and Bayesian filtering (BF) can be used to identify driver's lane-changing intention. The experiment shows that the identification accuracy of left and right Lane changing can reach 93.5% and 90.3% respectively [55]. In terms of vehicle trajectory prediction, an interactive multi-model trajectory prediction can be conducted by trajectory prediction based on both physical model and behavior, for more accurate trajectory prediction in a long prediction period [56]. In terms of risk assessment, researchers propose a driving risk field model in a complex dynamic environment based on the DVE interactive coupling, for effective quantification in comprehensive driving risks of multi-factor coupling and assistance in safe driving of IV [57]. Group intelligent safety decision technology can realize string formation, intersection traffic optimization and cooperative collision avoidance through sub-scenario decision and safety decision algorithm. The typical representative of multi-vehicle group intelligent safety decision and control technology is the multi-vehicle string cooperation [58]. With the help of vehicle-vehicle communication and vehicle–road-cloud cooperative control technology, the information of multi-vehicle system is collected and broadcast to the vehicles in the string. Each vehicle obtains information of other vehicles based on its sensor perception and communication means, and comprehensively determines the current safe following strategy, such as appropriate following distance and speed. In addition to applying to single-lane road segments, multi-vehicle string cooperation technology can also be applied in such scenarios as multi-lane road segments and intersections for safety assurance of multi-vehicle systems. For example, in the road segment scenario, in order to minimize the risk of rear-end collision, a multi-vehicle coordinated braking control strategy based on MPC framework is adopted, for reducing the possibility of collision between vehicle groups and mitigating the impact of the inevitable collision [59]. In the intersection scenario, the vehicle is accurately positioned by multi-source information fusion technology, and the intersection traffic light phase and its expected change are obtained by vehicle- road interaction technology. Combined with the vehicle movement status, driving risk is judged in real time through speed threshold algorithm that reflects driver's deceleration characteristics. In case of danger, the system can send warning to driver, or realize the automatic braking by the adaptive braking control based on deceleration feedback control [60].

In terms of intelligent protection, the research focuses on prediction of hazardous scenarios in advance and effective protection for traffic participants and key vehicle system by means of collision risk warning and safety protection technology. From the perspective of structure and function, it tries to design and develop a new type of constraint system, which is expected to meet the protection demand in the current and future complex and changeable traffic scenarios. For example, the seatbelt system that can provide uniform restraint can disperse the concentrated load, effectively reduce the amount of chest compression, and serve as a new means for the protection of occupants in a frontal collision [61]. The pretensioned seatbelt is used together with the on-board warning system, and activated by the data of active safety systems such as electronic stability control and brake assist device, or triggered by the data of environmental sensors such as radar. It can realize reversible seatbelt retraction pretension in case of emergency, and reduce the risk of occupant injury during high-speed recovery of seatbelt with extended pretension time. For vulnerable traffic groups such as pedestrians, on-board cameras and sensors are applied to sense their appearance and predict their behaviors, with the help of the control system for vehicle manipulation and collision avoidance; If the collision cannot be avoided, pedestrian airbag, pop-up bonnet and other system components are triggered for alleviating injury severity of pedestrians. With the development of new energy vehicles, another hot research on intelligent protection is the safety protection of power system. Relevant technologies shall not only consider the external impact load of collision accident, but also the safety of new energy power system itself. For instance, the study on thermal model and electrochemical model of power battery [62] is conducted from perspectives of energy storage properties, electrochemical properties of power battery, thermal management system and structural mechanics of battery box. The battery management system can collect the voltage, temperature, gas, and other information of key parts in the battery through sensors based on advanced materials, so as to monitor and even warn the battery pack.

In terms of safety testing and evaluation, the research focuses on comprehensive, accurate, efficient, and safe testing and evaluation of vehicle performance. With testing scenario as the input data, the test is realized from virtual to real, from components to complete vehicle, and from specific functions to comprehensive performance, utilizing various in-loop testing, closed site testing and open road-testing technology. For testing scenario generation, common ways include automatic generation of typical scenarios [63] and accelerated generation of high-risk scenarios [64]. For vehicle testing, classic technology includes virtual testing based on various simulation software [65–67], radar in-loop testing based on simulation of RF environment in an anechoic chamber [68], and vehicle in-loop testing based on closed-site [69] or rotating hub platform [70]. Cutting-edge technology includes digital network testing site construction based on modular design idea [71], and end-to-end test based on both virtual and real environment [72], etc. In accordance with the data obtained in the testing process, vehicle performance evaluation is conducted for qualitative or quantitative results of a certain function or vehicle performance.

For vehicle performance evaluation, the basic principles of safety evaluation shall be followed. Through the screening, optimization and grading of evaluation indexes, the vehicle safety performance evaluation index system [73] shall be established with classified evaluation focus and key indexes. For selection of evaluation methods, the subjective evaluation [74] which reflects subjective feeling of driving and riding and objective evaluation [75] which reflects the objective performance of vehicle can be used together, so as to comprehensively represent vehicle safety level in theory. Meanwhile, the single function evaluation technology and the vehicle system evaluation technology can explore the safety degree of vehicle functions and whole system in the actual testing environment, so as to more truly reflect vehicle safety level. The conclusion of analysis based on evaluation results can not only lay a theoretical basis for the optimization of software and hardware of each module of vehicle [76], reduce high-risk behaviors and scenarios, but also guide the improvement of testing and evaluation for more accurate and reasonable evaluation results.

1.2.3 Research on Road Safety

The research on road is closely related to the continuous emergence of new structures, new processes, and new materials. The research at home and abroad mainly focuses on four aspects: subgrade engineering, pavement engineering, highway retaining structure and road geometric design [77]. Subgrade engineering mainly studies the features of subgrade settlement deformation, settlement control, slope stability, etc.; Pavement engineering pays more attention to the influence of pavement materials and pavement properties on slip ratio; Highway retaining structure focuses on the design of the most reasonable retaining ways, such as gravity type and anchor bolt type.; Road geometric design shall consider all aspects of the road, including load, speed, geographical factors and so on. With the concept of smart city and smart transportation, the research on road has also introduced new technologies such as roadside perception, roadside communication, edge computing and cloud control interconnection, which puts forward higher requirements for new road design.

The research on road safety includes two aspects, i.e., road safety design and road intelligent safety, as shown in Fig. 1.11. Road safety design mainly refers to the road alignment, subgrade and pavement, special sections, ancillary facilities, and other elements that shall be considered in the design of road and related traffic facilities. Road intelligent safety mainly refers to the way to maintain and improve road safety with intelligent safety technology and equipment after road safety design, mainly including roadside perception, information service, emergency dispatching and intelligent highway.

1. Road Safety Design

With the attention increasingly given to road safety, the existence of concentrated road segments promotes the study of road alignment. As early as in 1985, foreign researchers proposed a plane alignment accident prediction model [78] applicable for expressways, and clarified the positive correlation between the total amount of traffic accidents and road curvature. In 1991, some scholars extracted the statistical law that "traffic accidents are positively correlated with the length of plane alignment curve, and negatively correlated with the setting of transition curve and the width of driveway" according to the data [79]. The research on road alignment in China started relatively late. Until 2009, the comprehensive indexes of road alignment are put forward for the first time in China [80], and relevant evaluation models were established. The model evaluates the quality of highway alignment with such indexes as curvature and its change rate, curve corner, longitudinal slope, lane width and number, left and right shoulder width, etc. In the same year, the rationality of the road alignment design standard for long downgrade sections was also verified

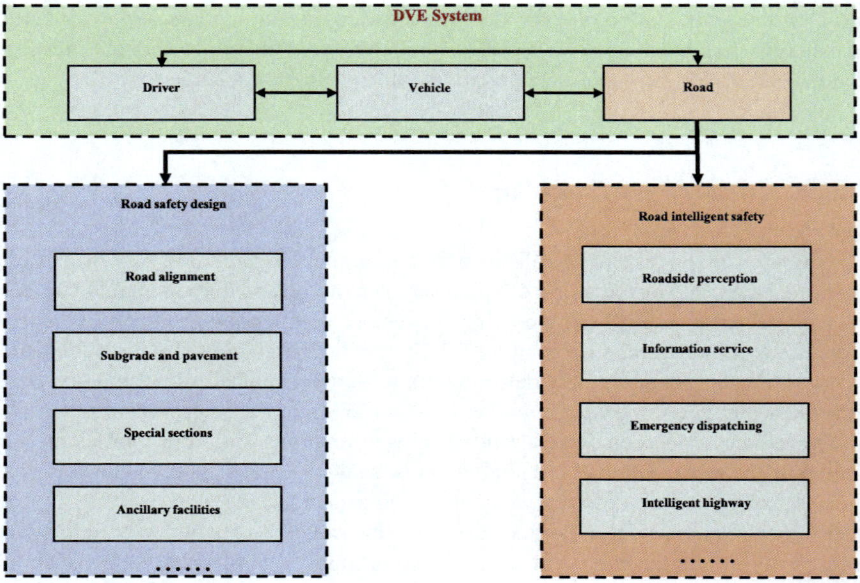

Fig. 1.11 Research on road safety

with the help of the evaluation model [81]. The quality of road alignment design directly affects road traffic safety. Studies at home and abroad show that [82, 83], some traffic accidents are directly caused by poor road conditions, and most traffic accidents caused by driver factors are also affected by road traffic conditions to a certain extent. Appropriate road alignment design is helpful to the reduction of traffic accidents.

The subgrade and pavement also have an important impact on the occurrence of traffic accidents. Pavement conditions usually include pavement strength, stability, flatness, diseases, etc. Internal factors affecting subgrade and pavement mainly include subgrade height and slope, pavement width and flatness, etc. External factors mainly include weather and maintenance management, etc. China has put forward a variety of technical standards and established a complete safety inspection system for highway subgrade and pavement, mainly including the safety evaluation guide for domestic highway projects, technical standards for highway engineering, highway subgrade design specifications, highway pavement design specifications, etc. The design of subgrade and pavement shall be in accordance with environmental terrain conditions and national specifications, as well as driver behavior and safety psychology [84]. For example, the high-filled subgrade shall be less in urban roads, for prevention of rollover and falling accidents; The pavement width shall be between 3.4–3.7 m, for minimizing the accident rate [85]. Based on new polymer concrete materials, self-luminous pavement with high brightness and duration can ensure driving safety at night [86]. New technologies such as PCC (large diameter pipe pile by using cast-in-place concrete) for foundation reinforcement have greatly facilitated the stability and safety of subgrade [87]. Appropriate subgrade and pavement design is conducive to the decreasing occurrence of ultimate conditions such as vehicle sideslip, as well as the reduction of accident rate.

After the completion of main road design, safety inspection and design shall be further carried out for special road segments, such as ramps, tunnels, and intersections, to ensure road safety. Taking the tunnel for example, since 2002, based on the local road design situation, many scholars have put forward models in accordance with the characteristics of tunnel traffic flow in China to reproduce the operation of tunnel traffic, and studied the tunnel traffic control and guidance measures under various traffic conditions [88, 89]. In 2006, the safety warning for extra-long highway tunnels was discussed, and the methods for tunnel traffic congestion warning and traffic accident warning were put forward together [90]. In addition, the Ministries of Transport at home and abroad have issued relevant laws and regulations for tunnel safety and established relevant evaluation systems. Such laws and regulations provide important guidance for road safety evaluation, management, maintenance, and reinforcement. Due to the peculiarity of function, environment, geographical location and other factors, special road segments are usually constrained in the design of the alignment, subgrade, and pavement. Therefore, efficient supervision, management and maintenance means are required to ensure safety.

Supporting road ancillary facilities are indispensable in modern roads for driving safety and economy. Road safety facilities are the general term of safety-related road ancillary facilities, including safety guardrails, traffic signs and markings, anti-dazzle facilities, isolation and closure facilities, sight guidance facilities, video monitoring facilities, etc., which can provide guidance, instructions, warnings, prohibitions, and other information for traffic participants, and maximize the safety of traffic participants after accidents [91]. Among them, the safety guardrails can be mainly divided into roadside guardrail, median guardrail, bridge guardrail, movable guardrail, transition section guardrail, gradual transition section guardrail, end guardrail, etc., with the function of absorbing collision energy by elastic (plastic) deformation. Traffic signs and markings mainly include warning signs, prohibition signs, indication signs, road guidance signs, etc., which can be queried in the national standard of *Road Traffic Signs and Markings* implemented in 2009. Anti-dazzle facilities are used to prevent night driving from being affected by the glare of headlights, including anti-dazzle panel, anti-dazzle net, anti-dazzle shed, etc. And roadside trees can also be regarded as a kind of anti-dazzle facilities. Closure and isolation facilities are the general term of the structures that isolate and close the driveway, mainly including overpass and protective net, etc. Sight guidance facilities refer to the settings on both sides of the driveway for indicating the road alignment, direction, boundary, and other information, mainly including four types of contour sign, diversion and merging guidance sign, warning guidance sign and protruding road sign, which are of great help to driver's navigation under adverse weather conditions. Pavement vibration marking is a new structure of road facilities. With high pollution resistance and whiteness, it serves for deceleration, skid resistance and reflection of light at rainy nights. Due to the focused usage and low cost, it is widely applied. Video monitoring facilities mainly include roadside cameras and video transmission devices, etc., which are used to detect violations and collect accident information. The improvement of road ancillary facilities is conducive to enhancing driving safety, and lays corresponding hardware foundation and support for subsequent road intelligent safety.

The road safety design shall be considered and completed before the actual application of the road, so as to maximize the expected road safety. However, in the practical use, due to the complicated behavior modes of drivers and other traffic participants and the induction of accidental factors, it is still necessary to supervise and regulate the road with real-time road intelligent safety means for complex DVE hybrid traffic scenarios.

2. Intelligent Road Safety

As the basis of intelligent road safety, roadside perception is to obtain real-time road information through the cooperative sensing of roadside sensors such as cameras and radars on the roadside. It mainly has functions of multi-target detection, multi-target tracking, intention identification, and trajectory prediction, etc. In 2018, with the help of the input information from the questionnaire survey and network communication, researchers analyzed the interaction mechanism between pedestrians and other traffic participants in the pedestrian-bicycle hybrid section by phase field theory, and established a pedestrian intention identification model by fuzzy logic [92]. Experiments show that the results of pedestrian motion intention identification model are basically consistent with the actual situation. However, the roadside perception system is difficult to obtain psychological questionnaire data of pedestrians in the application, and shall obtain the pedestrian intention according to objective information instead. In 2021, some studies used variance analysis to obtain the observed variables and potential variables that affect pedestrian intention, collected relevant data through roadside perception system, and modeled the pedestrian intention identification process with Bayesian network, which showed excellent intention identification effect in many typical scenarios [93]. In terms of short-term trajectory prediction, motion prediction methods such as Kalman filter and particle filter [93] are mainly used to predict trajectory at present. Motion prediction is based on the historical observation position of the target. When the target motion state is unchanged, the prediction error is small. To predict long-term trajectory of the target more accurately, some researchers tend to first predict the change of the target motion state, and then predict the target trajectory. For example, in 2020, researchers [94] applied the dynamic Bayesian network and probabilistic reasoning to infer pedestrian's motion intention according to environmental characteristics and kinematic information from roadside monitoring, and then used the sequence-to-sequence model to generate future trajectory according to motion intention information and past trajectory by regarding the trajectory prediction as a sequence generation task. Rich roadside perception information is the basis of subsequent decision control, and makes a variety of applications possible.

Roadside perception has rich information, as well as diverse usage. In the early stage, it is mainly conducted by manual intervention and engagement, and supplemented by sensor cooperation. For example, the road monitoring system shall acquire abnormal events on the current road through manual watching of video information, so that it is unable to respond in time. In the later stage, the cooperation of sensors is expected to complete the task. For example, the roadside sensing device is combined with the varistor to make the vehicle weight detection and prevent road from being damaged due to overload. The roadside sensing function plays a great role in the safety supervision of the traffic system at all stages, and is of great significance in maintaining road safety.

With the popularization of the idea of Internet of things, road also plays a role in providing information for traffic safety improvement. As shown in Fig. 1.12, road information service takes roadside communication facilities as the carrier to transmit objective information such as roadside perception information, vehicle perception information, road history information, and current environmental information (such as weather and road conditions) to other agents in the road area by wireless or wired means, so as to realize information sharing and dissemination in the road area [95]. This way helps to avoid the problems that are difficult to be solved by single vehicle intelligence, such as driver's visual blind spot, and makes vehicle with over distance perception and thus improves driving safety. In addition, road information service can store subjective information such as driver's driving style, driver's intention, and environmental risk, and assist safer decisions of the overall traffic environment based on this information. It can also upload information to the cloud service center for subsequent processing. For example, during a traffic accident, the alarm signal and accident information can be quickly uploaded to the cloud service center through roadside communication facilities. Then, the service consultant will immediately contact accident participants and comprehend their safety status, to facilitate the rescue and effectively reduce the accident mortality.

In addition to providing information for traffic participants, the road itself can also realize emergency dispatching and improve traffic safety based on decision and control technology. Firstly, the road fuses the information from multiple roadside sensors with information fusion technology and obtains the most real road environment information. Secondly, the road will conduct a situation assessment on the environment to confirm the existence of hazardous areas with possible occurrence of traffic accidents in the environment, and prevent traffic accidents by means of signal light control and lane control, etc. If a traffic accident occurrs, rescue vehicle shall reach the accident site as soon as possible. At this time, the road will assess the traffic situation of the roads where rescue vehicle drives, and realize emergency dispatching through the adjustment of traffic signs and other information, to avoid the congestion of safe passage and ensure the rescue efficiency as much as possible.

With the layout of roadside equipment, the development of Internet of things and the development of big data-based technologies such as edge computing, road will gradually break away from the traditional definition and become a carrier that can undertake the tasks of perception, communication, computing, decision-making and provide information for traffic participants, that is, intelligent road. An intelligent highway shall realize the data-driven information service mode and traffic management mode based on the hardware support of roadside equipment, and provide better analysis, judgment, and decision support than manual driving. The luminous road surface can improve driving environment, transmit road warning information, and

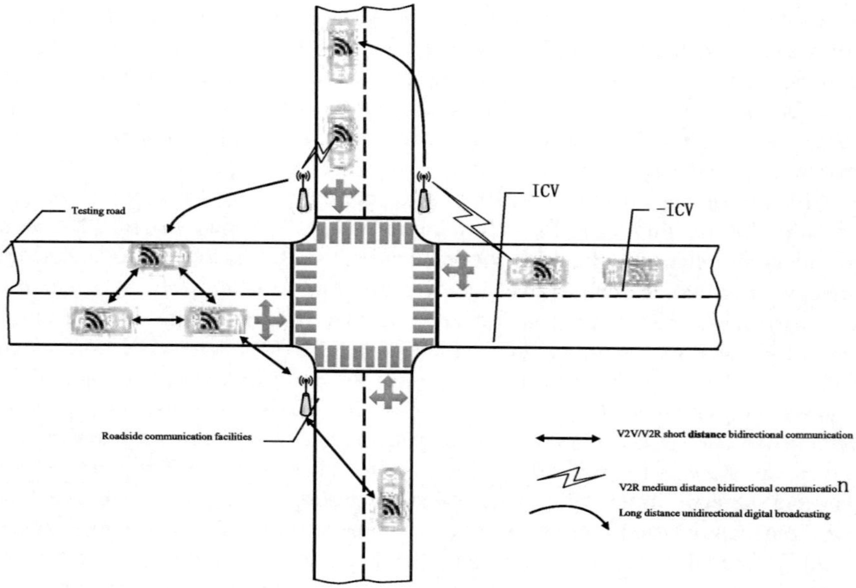

Fig. 1.12 Road information service

reduce accident rate at night. Temperature response dynamic paint can enable driver to clearly identify ice surface of the slippery road in cold weather. Interactive induction lamp can automatically adjust the light according to the distance between vehicle and index lamp for the best indication effect. Intelligent highway shall combine road safety design means and road intelligent safety functions to form an optimal scheme for traffic organization, management, and control by means of intelligent algorithms, so as to create a safe and orderly traffic environment. For the traffic accident problems due to imperfect system, intelligent highway shall improve the accident edge analysis mechanism, enhance the accident handling, and rescue capability, and iteratively optimize the system itself. In a word, intelligent highway is one of the ideal integration means of road safety technology, an important part of smart city, and is of great significance to the realization of vehicle intelligent safety.

1.2.4 Research on DVE Cooperative Safety

DVE cooperative safety comprehensively takes into consideration various factors including driver, vehicle, road, etc., and realize the cooperative safety of multi-factor coupling by means of network communication, etc., which is of great significance for ensuring traffic safety, improving traffic efficiency, reducing energy consumption, optimizing travel experience and so on. Typical representatives of existing research and application include vehicle–road/vehicle-vehicle cooperative safety and the further DVE cooperative safety in view of driver characteristics.

Vehicle–road/vehicle-vehicle cooperative safety is one of the early development directions. Based on wireless communication, sensor detection and other technologies, it obtains the information of road and their users, and thus realize the intelligent coordination and cooperation between vehicle and roadside facilities as well as between vehicles, so as to improve road traffic safety. The representative projects of vehicle–road/vehicle-vehicle cooperation at the national level mainly include the Vehicle Infrastructure Integration (VII) of the US, the Smartway of Japan and the eSafety of the EU, which were carried out around 2000. Since 2003, based on road facilities, the typical representative VII of the vehicle–road cooperation system in the US has realized the integration of vehicles and road infrastructure by information and communication technology, obtained real-time traffic data, perceived driving risks, provided safety assistance for drivers, and ensured driving safety [96]. The Smartway project in Japan, which began at the end of the twentieth century, has established a unified platform for vehicle on-board units, detected traffic jams or obstacles at the turning of the road ahead or in the blind area of the sight through roadside sensing, and provided real-time road information to drivers through the vehicle–road communication system for reducing traffic accidents [97]. The EU eSafety project, which was launched in 2003, provided a more comprehensive safety solution for road traffic by communication technology. In addition to autonomous on-board safety devices, it also considered the vehicle–road/vehicle-vehicle cooperation modes, that is, to obtain road environment information with communication technology, effectively assess potential driving hazards and optimize functions of on-board safety system. The subsystems developed for vehicle accident prevention and protection include on-board intelligent terminal system, pre-accident safe driving assistance system, in-accident occupant protection system, post-accident emergency rescue system, etc. [98].

Typical applications of vehicle–road/vehicle-vehicle cooperation technology include cooperative adaptive cruise control (CACC), road hazard warning, red light-running warning, intersection cooperation control, etc. The traditional ACC technology obtains the position and speed information of the front vehicle through radar and other sensors of the ego-vehicle, with limited information accuracy and types. On this basis, CACC technology can obtain the control state information such as acceleration of other vehicles that are difficult to be perceived by radar and other sensors through vehicle-vehicle and vehicle–road communication, and obtain more accurate vehicle position information through roadside auxiliary positioning [99, 100]. With wider range of obtained information, larger string scale and better string stability can be achieved [53]. In addition, CACC technology can also support quantitative analysis of the impact of non-fixed communication topology and non- ideal communication conditions such as time delay and packet loss on the driving process of multi-vehicle string, which is also of great significance to improvement of string driving safety [101, 102]. Road hazard warning technology can send the road warning information to ICV within a certain range in the form of broadcast based on vehicle–road communication when roadside equipment detects the slippery, icy, windy, and foggy roads or hazardous road segments, so as to reduce possible traffic accidents. Red light-running warning technology can send the hazard information to all ICV within the intersection when roadside equipment detects vehicles running through the red light at the intersection, to provide timely information supplement for other vehicles that cannot directly observe the situation, and reduce the occurrence of intersection collision accidents [98]. Intersection cooperative control can obtain the phase information of ICV and traffic lights within the intersection through integrated application of wireless communication and other means, comprehensively optimize the phase timing of traffic lights and vehicle speed trajectory, judge driving danger degree in real time, and provide warning to drivers or carry out emergency braking control on vehicles in hazardous situations [103]. In addition, the coordinated optimal scheduling of vehicle groups and roadside units in the unsignalized intersections can realize the conflict-free traffic of multiple vehicles, which can not only ensure driving safety, but also eliminate energy consumption of vehicle stop-and-go led by the phase switching of signal lights [104].

DVE cooperative safety is the research and application based on vehicle–road/vehicle-vehicle cooperation, which further considers the characteristics and requirements of drivers, and ensures safety of vehicles and drivers in an all-around way. Traffic data statistics show that accurate quantification and assessment of the risk of DVE coupling in dynamic traffic systems can effectively avoid or reduce possible collision accidents [105]. The research on "driving risk field" of DVE closed-loop system starts from the idea of field theory and comprehensively considers the degree of risks posed by various elements of DVE to vehicle driving safety, including the "kinetic energy field" determined by motor vehicles, non-motor vehicles and other moving objects in the road, the "potential energy field" determined by road environment elements and the "behavior field" determined by individual characteristics of driver, and thus facilitates quantitative assessment of driving risk in complex environment and vehicle safety decision control [106]. For example, the collision warning algorithm based on driving risk field theory comprehensively considers the characteristic parameters that may lead to collision in the DVE elements, so as to make a unified risk field description and provide important input for safety decision and control [107]. Another typical research idea is to determine the safe vehicle speed in accordance with various factors of DVE, that is, to modify the controller parameters in real time based on road parameters and driver's utilization of road adhesion coefficient, and make the vehicle speed conform to driver's individual speed preference and risk level. The experimental results of driving simulator show that this way can timely and effectively provide drivers with a speed assistance mode consistent with their driving habits, and improve driving safety and comfort [108]. In addition, the research reveals the instability mechanism of vehicle running state under the comprehensive action of DVE, constructs a multi-agent steady-state safety simulation framework including driver, vehicle, road, coordination center and human–machine interface, and thus establishes driver preview control and follow-up model, vehicle multi-body dynamics model, and three-dimensional road space simulation model for analyzing the steady-state safety performance from such perspectives as vehicle stability and driver's operating load [109].

1.2.5 Summary of Vehicle Safety Research Status

Safety assurance is one of the most important goals in the process of vehicle development. In order to ensure safety of drivers, other passengers, traffic participants and vehicles, research institutions and vehicle enterprises around the world have carried out a lot of theoretical research and technological development.

The research on driver safety starts from the perspectives of statistical indexes and driver model, extracts key psychological and physiological indexes from a large number of accident data, and establishes a driver model for describing driving characteristics, risk cognition characteristics and driving intention, so as to help researchers better understand drivers and design a more reasonable human-like driving algorithm. The sorting of key accident indexes is also conducive to the development of accident warnings and other technologies.

The research on vehicle safety has evolved from traditional passive and active safety to advanced driving assistance, and gradually promotes the emergence of vehicle intelligent safety. Passive safety is used to reduce occupant injury and accident loss. Active safety and advanced driving assistance judges and warns possible accidents according to sensor information, and reminds drivers or directly controls the safety of vehicle, so as to reduce accident rate or severity. With the help of intelligent and connected technology, the intelligent safety of vehicles will continue to ensure vehicle safety from the aspects of system safety, driving safety, intelligent protection, and safety testing and evaluation.

The research on road safety mainly includes road safety design and road intelligent safety. The research on road safety design relies on the design of the alignment, subgrade, and pavement of the road for reducing the occurrence of accidents. Road intelligent safety further ensures the safety of traffic participants with the help of advanced technologies such as sensing, information communication and cloud computing.

Based on the vehicle–road/vehicle-vehicle cooperation, the research on DVE cooperative safety has been gradually developed and improved. Meanwhile, it considers and realizes the coupling and cooperation of multiple elements in DVE, and more comprehensively ensures road traffic safety.

Overall, by applying advanced intelligent and connected technologies, and considering multiple factors of DVE, the intelligent safety technology shows great advantages in the comprehensiveness of protection objects, the continuity of action time and the reliability of protection performance, and is the main development direction of vehicle safety.

1.3 Intelligent Safety Technology System

Vehicle intelligent safety technology consists of four types, i.e., system safety assurance, intelligent safety decision and control, intelligent cooperative protection, and safety performance evaluation. Its architecture is shown in Fig. 1.13.

The system safety assurance technology designs the overall system from the system point of view, and defines functional requirements and application opportunities of all systems. It specifically includes FuSa, SOTIF, cybersecurity assurance, etc. Among them, FuSa technology obtains information by means of fault detection and diagnosis, and applies fault-tolerant control to eliminate faults. SOTIF obtains the system state, external environment, and driver state through condition monitoring, comprehends environment from the data by perception-based SOTIF, outputs the perceived results to decision-based SOTIF, and adopts non-learning or learning-based decision algorithm optimization to realize hybrid decision-making. The SOTIF design of HMI carries out the analysis of misoperation behavior and behavior protection according to the obtained driver state. Based on the vehicle cybersecurity topology, cybersecurity technology designs the in-depth defense system architecture, and collects, analyzes, detects, and defends the intrusion information through vehicle cybersecurity intrusion detection and defense. Cybersecurity encryption and communication technology can ensure the cybersecurity inside and outside vehicle.

Intelligent safety decision and control technology ensures safety during driving, reduces accident risks and severity while judging or predicting accidents, and provides input for cooperative protection technology. It specially involves two levels, i.e., single vehicle and group vehicles. Single vehicle intelligent safety decision technology realizes global path planning, behavior decision and trajectory planning through hierarchical decision and safety decision algorithm. Single vehicle intelligent safety control technology is based on vehicle kinematics model and dynamics model, and designs safety control algorithm in accordance with single vehicle safety control indexes for vehicle trajectory tracking control and critical instability control, etc. Group intelligent safety decision technology can realize the functions of string formation, intersection traffic optimization and cooperative collision avoidance employing sub-scenario decision and safety decision algorithm. Group safety control technology is based on four-element model and other vehicle group models, and designs group cooperative control under ideal communication and vehicle group control under non-ideal conditions in accordance with group safety control indexes.

Intelligent cooperative protection technology serves for reducing damage risk and severity of vehicle, occupants of the vehicle and other traffic participants in the process of accident, as well as providing accident damage feedback for decision and control technology for together improving system functions. It specifically includes technologies for prediction of collision risk and severity of driver-vehicle system, vehicle protection based on advanced materials and sensors, and cooperative protection of traffic participants. Among them, the prediction technology can establish damage indexes according to damage mechanism and data of traffic participants,

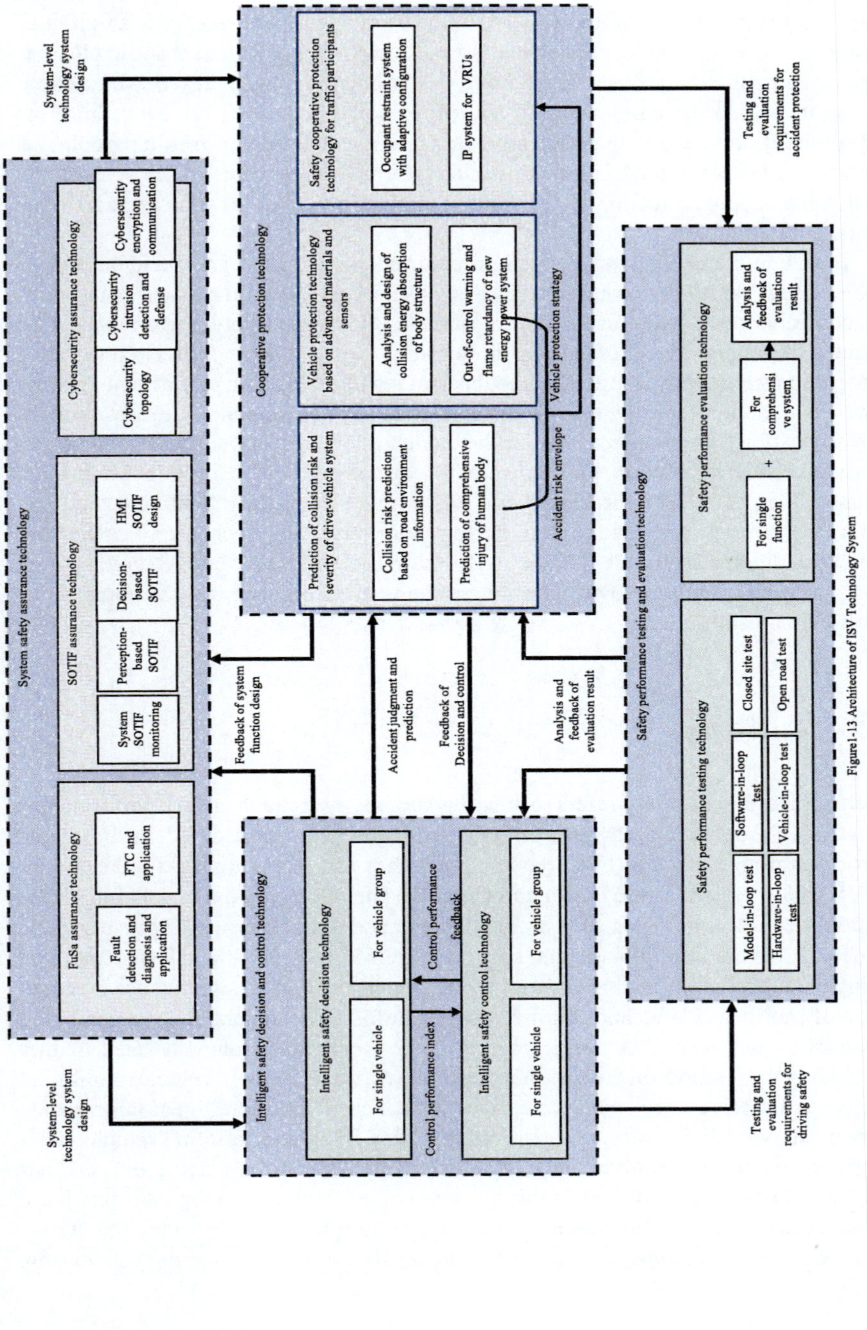

Fig. 1.13 Architecture of vehicle intelligent safety technology system

and realize pre-accident collision risk prediction and post-accident comprehensive damage prediction according to road environment information. Vehicle protection technology based on advanced materials and sensors formulates structural collision resistance indexes through the analysis and design of collision energy absorption of body structure, the optimization of battery pack layout, flame retardant and explosion-proof, and collision protection structure design for the new energy power system, so as to realize the early warning and protection for thermal runaway. In terms of different objects, safety cooperative protection technology of traffic participants includes adaptive configuration constraint system for occupants in the vehicle and intelligent safety protection system for vulnerable traffic groups such as pedestrians outside the vehicle.

Based on the testing and evaluation requirements of safety and accident protection system, safety performance testing and evaluation technology carries out multi-dimensional testing and evaluation of vehicle intelligent safety performance, and guides the improvement of intelligent decision and control as well as cooperative protection technology through the feedback of results, including performance testing and evaluation. According to different test efficiency and authenticity, safety performance testing technology can realize the model in-loop test and software in-loop test through virtual simulation, the hardware in-loop test and vehicle in-loop test through some physical objects, the closed site test through the testing field, and the open road test through real road. Safety performance evaluation technology carries out safety evaluation on the single function of the vehicle and the whole vehicle system, and infers risks and improve technical problems in accordance with comprehensive evaluation results.

1.4 Summary

Vehicle intelligent safety refers to safety guarantee for reliable safety performance, controllable operation risk and effective collision protection of IV in the whole process through advanced theoretical approaches and technologies. It takes micro DVE traffic system as the object and IV as the core element, and makes full use of cutting-edge technologies such as intelligence and networking to improve vehicle safety. It covers multiple aspects, such as system safety, driving safety, intelligent protection, and safety testing and evaluation, which are of great significance in reduction of traffic accidents and improvement of traffic environment. The research and technology development on vehicle safety has been conducted for many years, mainly including the research on driver (statistical indexes, driver model), vehicle (traditional active and passive safety, advanced driver assistance, vehicle intelligent safety), road (road safety design, road intelligent safety), DVE integrated system (vehicle–road/vehicle-vehicle cooperative safety, DVE integrated cooperative safety), etc., and has given birth to a series of vehicle safety technologies. Vehicle intelligent safety has a comprehensive protection range, long protection time and reliable protection performance. It integrates with advanced technologies such as intelligence and networking, and establishes a technical system composed of system safety assurance, intelligent safety decision and control, intelligent cooperative protection, and safety performance testing and evaluation. It can ensure the safety of DVE system in an all-round way, and is a critical development direction in the field of vehicle safety.

References

1. Wang Shengchang, Li Xinyao. Study on the correlation between driver's action, reaction characteristics and traffic accidents [J]. Journal of Xi'an Highway University, 1995 (04): 59-65. [in Chinese]
2. Mao Enrong, Zhou Yiming. The influence of field dependence and speed estimation ability of motor-vehicle drivers on driving safety [J]. Journal of China Agricultural University, 1997 (02): 114-118. [in Chinese]
3. He Cundao, Tang Zhendong, Zheng Ninghua. Research on the emotional state of truck drivers [J]. Chinese Journal of Ergonomics, 2001 (02): 18–21+70. [in Chinese]
4. Li Xiaohua, He Cundao, Peng Chuqiao, Guo Weili. Research on speed estimation of truck drivers [J]. Psychological Science, 1997 (06): 525–529+575–576. [in Chinese]
5. Guo Zizheng. Theory and method of dangerous state identification of driving behavior [D]. Southwest Jiaotong University, 2009. [in Chinese]
6. Li Baichuan, Sun Jianhong, Xiao Lijun. Research on the establishment of suitability testing standards for professional car drivers in China[J]. Journal of Safety and Environment, 2001 (03): 7-10. [in Chinese]
7. Brehmer B. Variable errors set a limit to adaptation[J]. Ergonomics, 1990, 33(10-11): 1231-1239.
8. Brown I D. Drivers' margins of safety considered as a focus for research on error[J]. Ergonomics, 1990, 33(10-11): 1307-1314.
9. Raggatt P T F, Morrissey S A. A field study of stress and fatigue in long-distance bus drivers[J]. Behavioral medicine, 1997, 23(3): 122-129.
10. Zhang Lei, Wang Jianqiang, Yang Furui, Li Keqiang. Factor analysis and fuzzy clustering of driver behavior modes [J]. Journal of Traffic and Transportation Engineering, 2009, 9 (05): 121-126. [in Chinese]
11. Holman A C, Havârneanu C E. The Romanian version of the multidimensional driving style inventory: Psychometric properties and cultural specificities[J]. Transportation research part F: traffic psychology and behaviour, 2015, 35: 45-59.
12. Chen S W, Fang C Y, Tien C T. Driving behaviour modelling system based on graph construction[J]. Transportation research part C: emerging technologies, 2013, 26: 314-330.
13. Johnson D A, Trivedi M M. Driving style recognition using a smartphone as a sensor platform[]//2011 14th International IEEE Conference on Intelligent Transportation Systems (ITSC). IEEE, 2011: 1609–1615.
14. Hou Haijing, Jin Lisheng, Guan Zhiwei, Du Haixing, Li Jingjun. The influence of driving style on driving behavior [J]. China Journal of Highway and Transport, 2018, 31 (04): 18-27. [in Chinese]
15. Berdoulat E, Vavassori D, Sastre M T M. Driving anger, emotional and instrumental aggressiveness, and impulsiveness in the prediction of aggressive and transgressive driving[J]. Accident Analysis & Prevention, 2013, 50: 758-767.
16. Muñoz M, Reimer B, Mehler B. Exploring new qualitative methods to support a quantitative analysis of glance behavior[]//Proceedings of the 7th International Conference on Automotive User Interfaces and Interactive Vehicular Applications. 2015: 125–132.
17. Zheng Dongpeng. Research on drivers' risk perception and influencing factors [D]. Shanghai Jiaotong University, 2013. [in Chinese]

18. Lefèvre S, Vasquez D, Laugier C. A survey on motion prediction and risk assessment for intelligent vehicles[J]. ROBOMECH journal, 2014, 1(1): 1-14.
19. Lawitzky A, Althoff D, Passenberg C F, et al. Interactive scene prediction for automotive applications[]//2013 IEEE Intelligent Vehicles Symposium (IV). IEEE, 2013: 1028–1033.
20. Hillenbrand J, Spieker A M, Kroschel K. A multilevel collision mitigation approach—Its situation assessment, decision making, and performance tradeoffs[J]. IEEE Transactions on intelligent transportation systems, 2006, 7(4): 528-540.
21. Worrall S, Orchansky D, Masson F, et al. Improving vehicle safety using context based detection of risk[]//13th International IEEE Conference on Intelligent Transportation Systems. IEEE, 2010: 379–385.
22. Liu A, Pentland A. Towards real-time recognition of driver intentions[]//Proceedings of Conference on Intelligent Transportation Systems. IEEE, 1997: 236–241.
23. Berndt H, Emmert J, Dietmayer K. Continuous driver intention recognition with hidden markov models[]//2008 11th International IEEE Conference on Intelligent Transportation Systems. IEEE, 2008: 1189–1194.
24. Kuge N, Yamamura T, Shimoyama O, et al. A driver behavior recognition method based on a driver model framework[J]. SAE transactions, 2000: 469–476.
25. Hou Haijing. Research on recognition way of drivers' intention for changing roads on expressway [D]. Jilin University, 2013. [in Chinese]
26. Wang Yingfan, Ning Guobao, Yu Zhuoping. Selection of recognition parameters for passenger-vehicle driver's braking intention [J]. Automotive Engineering, 2011, 33 (03): 213–216+230. [in Chinese]
27. Wang Qingnian, Tang Xianzhi, Wang Pengyu, Sun Lei. Hybrid vehicle control strategy based on driving intention recognition [J]. Journal of Jilin University (Engineering Edition), 2012, 42 (04): 789-795. [in Chinese]
28. Wang Yuhai, Song Jian, Li Xingkun. Unified recognition and real-time algorithm of driver's intention and driving environment [J]. Journal of Mechanical Engineering, 2006 (04): 206-212. [in Chinese]
29. Wang Jianqiang, Zheng Xunjia, Huang Heye. The principle of least action followed by driver's driving decision-making mechanism [J]. China Journal of Highway and Transport, 2020,33 (04): 155-168. [in Chinese]
30. Song Jian, Wang Weiwei, Li Liang, et al. Research status and prospect of vehicle safety technology [J] Journal of Automotive Safety and Energy, 2010,1 (2): 98–106. [in Chinese]
31. FERGUSON S A. The Effectiveness of Electronic Stability Control in Reducing Real-World Crashes: A Literature Review[J]. Journal of Crash Prevention & Injury Control, 2007, 8(4):329-338.
32. Insurance Institute for Highway Safety (IIHS). Electronic stability control could prevent nearly one-third of all fatal crashes and reduce rollover risk by as much as 80%; effect is found on single- and multiple-vehicle crashes [EB/OL]. News Release (2006–06–13)[2021–07–03]. http://www.iihs.org/news/rss/pr061306.html.
33. Volkswagen. Active roll-over protection system. Retrieved 28 August 2012. https://www.volkswagen.co.uk/technology/car-safety/crumple-zones.
34. HAMID A, ZAKIR U; et al. Autonomous emergency braking system with potential field risk assessment for frontal collision mitigation[J]. 2017 IEEE Conference on Systems, Process and Control (ICSPC). 2017:71–76.
35. JOAN L. Studies: Automated safety systems are preventing car crashes [EB/OL]. AP News.(2017–08–27)[2020–11–23]. https://phys.org/news/2017-08-automated-safety-car.html.
36. Jianqiang Wang, Chenfei Yu, Shengbo Eben Li*, Likun Wang. A Forward Collision Warning Algorithm With Adaptation to Driver Behaviors[J]. IEEE Transactions On Intelligent Transportation Systems. 2016, 17(4), 1157–1167.
37. Li Y, Zheng Y, Wang J, et al. Evaluation of Forward Collision Avoidance system using driver's hazard perception[]. Intelligent Transportation Systems (ITSC), 2016 IEEE 19th International Conference on. IEEE, 2016: 2273–2278.

38. "New study confirms real-world safety benefits of autonomous emergency braking". European Transport Safety Council. 11 July 2015. Retrieved 8 June 2019.
39. Yougang Bian, Jieyun Ding, Manjiang Hu, Qing Xu, Jianqiang Wang, and Keqiang Li. An Advanced Lane-Keeping Assistance System With Switchable Assistance Modes[J]. IEEE Transactions on Intelligent Transportation Systems, (IF=5.744) .2020, VOL: 21 NO: 1 PP. 385–396.
40. Lowy, Joan. "Studies: Automated safety systems are preventing car crashes". AP News. Archived from the original on 2017–08–27.
41. Road vehicles—Functional safety: ISO 26262[S]. Geneva, Switzerland: ISO: 2018.
42. Road Vehicles—Safety of the Intended Functionality: ISO/DIS 21448: 2021[S]. Geneva, Switzerland: ISO: 2021.
43. Cybersecurity Guidebook for Cyber-Physical Vehicle Systems: SAE J3061[S].Geneva:ISO,2016.
44. Road Vehicles-Cybersecurity Engineering: ISO/SAE DIS 21434[S]. Geneva:ISO,2020.
45. Zhong Zaimin, Huang Xi, Zhang Hongbin. Implementation of point to point secure communication based on AUTOSAR [J]. Computer Measurement and Control, 2017, 25 (10): 5. [in Chinese]
46. B Döbel. Operating system support for redundant multithreading[J]. 2012.
47. Gao Yi. Research on optimal dispatching of motor vehicle operation safety monitoring system [D]. South China University of Technology, 2012. [in Chinese]
48. Zheng S , Song Y , Leung T , et al. Improving the Robustness of Deep Neural Networks via Stability Training[J]. Computer Vision & Pattern Recognition, 2016:4480–4488.
49. Santos F , Draghetti L , Weigel L , et al. Evaluation and Mitigation of Soft-Errors in Neural Network-Based Object Detection in Three GPU Architectures[]// 2017 47th Annual IEEE/IFIP International Conference on Dependable Systems and Networks Workshop (DSN-W). IEEE, 2017.
50. Liu Nan. Design and prototype development of identity authentication mechanism in cybersecurity of intelligent Internet of vehicles [D]. Beijing University of Posts and Telecommunications, 2020. DOI: https://doi.org/10.26969/d.cnki.gbydu.2020.001589. [in Chinese]
51. Zheng X, Huang H, Wang J, et al. Behavioral decision-making model of the intelligent vehicle based on driving risk assessment[J]. Computer-Aided Civil and Infrastructure Engineering, 2021, 36(7): 820-837.
52. URMSON C, ANHALT J, BAGNELL D, et al. Autonomous driving in urban environments: Boss and the urban challenge[J]. Journal of Field Robotics, 2008, 25(8): 425-466.
53. LI S, LI K, RAJAMANI R, et al. Model predictive multi-objective vehicular adaptive cruise control[J]. IEEE Transactions on control systems technology, 2010, 19(3): 556-566.
54. SCHWARTING W, ALONSO-MORA J, RUS D. Planning and decision-making for autonomous vehicles[J]. Annual Review of Control, Robotics, and Autonomous Systems, 2018:357–366.
55. Keqiang Li, Xiao Wang, Youchun Xu, Jianqiang Wang*. Lane changing intention recognition based on speech recognition models[J]. Transportation Research Part C: Emerging Technologies, 2016, 69: 497–514.
56. Xie, G., Gao, H., Qian, L., Huang, B., Li, K., & Wang, J*. Vehicle trajectory prediction by integrating physics-and maneuver-based approaches using interactive multiple models[J]. IEEE Transactions on Industrial Electronics, 2018, 65(7): 5999–6008.
57. Wang J, Wu J, Li Y. The driving safety field based on driver–vehicle–road interactions[J]. IEEE Transactions on Intelligent Transportation Systems, 2015, 16(4): 2203-2214.
58. Zheng Yang. Dynamics modeling and distributed control of vehicle platoon based on four-element architecture [D]. Beijing: Tsinghua University, 2015. [in Chinese]
59. Jianqiang Wang, Shengbo Eben Li*, Yang Zheng and Xiaoyun Lu. Longitudinal collision mitigation via coordinated braking of multiple vehicles using model predictive control [J]. Integrated Computer-Aided Engineering, 2015:171–185.
60. Wang Jianqiang, Wang Haipeng, Liu Jiaxi, et al. Vehicle driving assistance system at intersections based on vehicle-road integration [J]. Journal of China Highway Engineering, 2013, 26 (4): 169-175. [in Chinese]

61. Editorial Department of China Journal of Highway and Transport. Summary of China's academic research on vehicle engineering. 2017 [J]. China Journal of Highway and Transport, 2017,30 (06): 1–197. [in Chinese]
62. Chen Ningchuan. Design and control of thermal management system for vehicle power battery pack [D] Beijing: Tsinghua University, 2016. [in Chinese]
63. Chen Junyi, Feng Tianyue, Liu Lihao, et al. Automatic generation of specific scenarios for decision-planning system testing [J]. Automobile Technology, 2020 (10): 45-50. [in Chinese]
64. HALLERBACH S,XIA Y,EBERLE U,et al.Simulation-based identification of critical scenarios for cooperative and automated vehicles[J].SAE International Journal of Connected and Automated Vehicles,2018,1(2018–01–1066): 93–106.
65. Li Lin, Zhu Xichan, Dong Xiaofei, et al. Research on collision avoidance strategy of autonomous emergency braking system [J] Automotive Engineering, 2015, 37 (2): 168–174. [in Chinese]
66. Zhu B, Yan S, Zhao J, et al. Personalized lane-change assistance system with driver behavior identification[J]. IEEE Transactions on Vehicular Technology, 2018, 67(11): 10293-10306.
67. Sundaravadivelu K, Shantharam G, Prabaharan P, et al. Analysis of vehicle dynamics using co-simulation of AVL-CRUISE and CarMaker in ETAS RT environment[]//2014 International Conference on Advances in Electrical Engineering (ICAEE). IEEE, 2014: 1–4.
68. Huang H, Pan M, Lu Z. Hardware-in-the-loop simulation technology of wide-band radar targets based on scattering center model[J]. Chinese Journal of Aeronautics, 2015, 28(5): 1476-1484.
69. Tettamanti T, Szalai M, Vass S, et al. Vehicle-in-the-loop test environment for autonomous driving with microscopic traffic simulation[]//2018 IEEE International Conference on Vehicular Electronics and Safety (ICVES). IEEE, 2018: 1–6.
70. Xu Z, Wang M, Zhang F, et al. PaTAVTT: A hardware-in-the-loop scaled platform for testing autonomous vehicle trajectory tracking[J]. Journal of Advanced Transportation, 2017, 2017.
71. Li Xiaochi, Zhao Xiangmo, Xu Zhigang, et al. Modular flexible testing ground for intelligent connected transport system [J]. China Journal of Highway and Transport, 2019, 32 (06): 137-146. [in Chinese]
72. Wang J, Shao Y, Ge Y, et al. A survey of vehicle to everything (V2X) testing[J]. Sensors, 2019, 19(2): 334.
73. Sun Yang, Xiong Guangming, Chen Huiyan. Evaluation of unmanned vehicle's intelligent behavior based on Fuzzy-EAHP [J]. Automotive Engineering, 2014, 36 (01): 22-27. [in Chinese]
74. Guo Zhangyong. Research on evaluation of driver acceptance of adaptive cruise control system [D]. Changchun: Jilin University, 2014. [in Chinese]
75. Xiong Guangming, Gao Li, Wu Shaobin, et al. Intelligent behavior of driverless vehicles and its testing and evaluation [M] Beijing: Beijing University of Technology Press, 2015. [in Chinese]
76. Yu Rongjie, Tian Ye, Sun Jian. Virtual testing of high-level automated driving vehicle: research progress and frontier [J]. China Journal of Highway and Transport, 2020, 33 (11): 125-138. [in Chinese]
77. Editorial Department of China Journal of Highway and Transport. Summary of academic research on China's road engineering. 2013 [J]. China Journal of Highway and Transport, 2013, 26 (03): 1–36. [in Chinese]
78. Glennon J C, Neuman T R, Leisch J E. SAFETY AND OPERATIONAL CONSIDERATIONS FOR DESIGN OF RURAL HIGHWAY CURVES. FINAL REPORT[R]. 1983.
79. Vogt A, Bared J. Accident models for two-lane rural segments and intersections[J]. Transportation Research Record, 1998, 1635(1): 18-29.
80. Guo Zhongyin, Yang Yang, Cao Jiwei, et al. Safety evaluation model based on comprehensive index of expressway alignment [J] Journal of Tongji University: Natural Science Edition, 2009 (11): 1472–1476. [in Chinese]
81. Shi Yang, Chen Yongsheng. Safety evaluation for road alignment design of downhill sections [J]. Western Transportation Science and Technology, 2008 (5): 43-46. [in Chinese]

82. Hou Tao. Research on safety evaluation of trunk roads based on running speed [D] [D]. Chongqing Jiaotong University, 2009. [in Chinese]
83. Pei Yulong, Ma Ji. Analysis on the causes of road traffic accidents and research on prevention countermeasures [J]. China Journal of Highway and Transport, 2003, 16 (4): 77-82. [in Chinese]
84. Xu Zhongyang. Research on safety inspection system of highway subgrade and pavement design [D]. Chang'an University, 2006. [in Chinese]
85. Guo Zhongyin, Fang Shou'en, et al. Road safety engineering [M]. Beijing: People's Communications Press, September 2003. [in Chinese]
86. Zhi Fan, Guo Runhua, Liu Yingqiang. Research on the properties of self-luminous pavement materials based on polymer concrete [J]. Thermosetting Resin, 2022,37 (01): 38-43. DOI: https://doi.org/10.13650/j.cnki.rgxsz.2022.01.017. [in Chinese]
87. Liu Hanlong, Fei Kang, Ma Xiaohui, Gao Yufeng. Vibratory die sinking technology for large-diameter cast-in-situ thin-walled pipe piles and its application (I): Development and design [J]. Geotechnical Mechanics, 2003 (02): 164-168. DOI: https://doi.org/10.16285/j.rsm.2003.02.005. [in Chinese]
88. Chen Jinfeng. Research on traffic guidance and control of long tunnel [D]. Chang'an University, 2002. [in Chinese]
89. Shan Yongxin. Research on traffic guidance and control strategy of highway tunnel [D]. Chang'an University, 2004. [in Chinese]
90. Han Zhi, Peng Jianguo, Zheng Hao. Research on anomaly classification and early warning technology of Xuefeng Moutain super long highway tunnel [J]. Journal of Central South Highway Engineering, 2006 (01): 76–78+119. [in Chinese]
91. Yi Wei. Research on safety evaluation of expressway road safety facilities [D]. South China University of Technology, 2012. [in Chinese]
92. Liu Z, Wang X, Wang J, et al. Pedestrian movement intention identification model in mixed pedestrian-bicycle sections based on phase-field coupling theory[J]. Advances in Mechanical Engineering, 2018, 10(2): 1687814017746515.
93. Xu Q, Wu H, Wang J, et al. Roadside pedestrian motion prediction using Bayesian methods and particle filter[J]. IET Intelligent Transport Systems, 2021, 15(9): 1167-1182.
94. Yang Li, Xiao-Yun Lu, Jianqiang Wang*, Keqiang Li. Pedestrian Trajectory Prediction Combining Probabilistic Reasoning and Sequence Learning[J]. IEEE transactions on intelligent vehicles, 2020,5 (3):461–474. EI:20200308051536.
95. Zhang Changlong, Bao Haixing, Du Xiantong, Xie Pengcheng. Application and future of roadside sensing in ICVs [J]. Artificial Intelligence, 2019 (01): 58-66. DOI: https://doi.org/10.16453/j.cnki.issn2096-5036.2019.01.007. [in Chinese]
96. Wang Xiaojing. Intelligent transportation and road traffic safety—Development trends and suggestions []//Proceedings of the Fourth China Intelligent Transportation Annual Conference 2008: 21–32. [in Chinese]
97. Makino H. Smartway project[J]. Development, 2005.
98. Chen Chao, Lv Zhiyong, Fu Shanshan, Peng Qi. Overview of the development of vehicle- road collaborative system at home and abroad [J]. Journal of Transport Information and Safety, 2011, 29 (01): 102–105+109. [in Chinese]
99. Van Arem B, Van Driel C J G, Visser R. The impact of cooperative adaptive cruise control on traffic-flow characteristics[J]. IEEE Transactions on intelligent transportation systems, 2006, 7(4): 429-436.
100. Naus G J L, Vugts R P A, Ploeg J, et al. String-stable CACC design and experimental validation: A frequency-domain approach[J]. IEEE Transactions on vehicular technology, 2010, 59(9): 4268-4279.
101. Zheng Y, Li S E, Wang J, et al. Stability and scalability of homogeneous vehicular platoon: Study on the influence of information flow topologies[J]. IEEE Transactions on intelligent transportation systems, 2015, 17(1): 14-26.
102. Li K, Bian Y, Li S E, et al. Distributed model predictive control of multi-vehicle systems with switching communication topologies[J]. Transportation Research Part C: Emerging Technologies, 2020, 118: 102717.

103. Wang Jianqiang, Wang Haipeng, Liu Jiaxi, Li Keqiang. Vehicle driving assistance system at intersections based on vehicle-road integration [J]. China Journal of Highway and Transport, 2013, 26 (04): 169–175+183. [in Chinese]
104. Xu B, Li S E, Bian Y, et al. Distributed conflict-free cooperation for multiple connected vehicles at unsignalized intersections[J]. Transportation Research Part C: Emerging Technologies, 2018, 93: 322-334.
105. Wang J, Huang H, Li Y, et al. Driving risk assessment based on naturalistic driving study and driver attitude questionnaire analysis[J]. Accident Analysis & Prevention, 2020, 145: 105680.
106. Wang Jianqiang, Wu Jian, Li Yang. Concept, principle and modeling of driving risk field based on DVE cooperation [J]. China Journal of Highway and Transport, 2016,29 (01): 105-114. [in Chinese]
107. Wang J, Wu J, Zheng X, et al. Driving safety field theory modeling and its application in pre-collision warning system[J]. Transportation research part C: emerging technologies, 2016, 72: 306-324.
108. Jia Xingli, Yang Hongzhi, Liu Chen. Vehicle steady state safety simulation in DVE coupling environment [J]. China Safety Science Journal , 2015, 25 (01): 40-45. [in Chinese]
109. Chen Xiaolei. Safety speed assistance system for curve based on DVE cooperation [D]. China Agricultural University, 2014. [in Chinese]

Chapter 2
System Safety

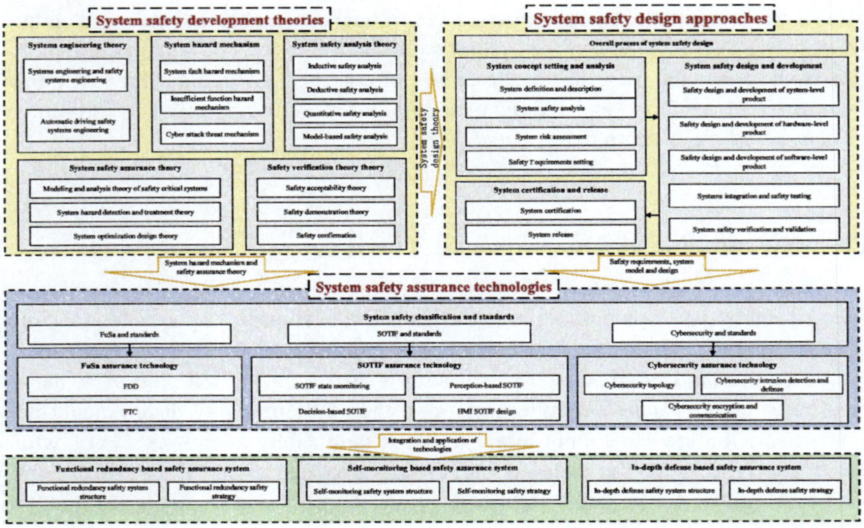

System Safety of Vehicles (SSV) aims to identify potential hazards of the whole vehicle, components, and algorithms with the application of relevant theories on safety systems engineering and approaches to system safety design, reduce system risks to an acceptable level with effective safety assurance measres and technologies, and subsequently achieve the best safety state within a specified range of performance, time and cost.

Intelligent vehicle (IV) is a typical complex system with in-depth integration of complicatedly interlinked constituent elements such as vehicles, transportation, and information. As a combination of systems engineering and safety engineering,

J. Wang et al., *The Intelligent Safety of Automobile*, Key Technologies on New Energy Vehicles, https://doi.org/10.1007/978-981-99-6399-7_2

system safety study is mainly conducted to solve the safety problems of complicated systems. Since activities related to safety cover the whole life cycle of the system, the study is an essential guarantee for the development of IVs with high reliability and safety. It involves a basic theory of safety development, system safety design flow, a key technology of system safety assurance and a typical system.

The framework of Chapter II is shown in Fig. 2.0. Based on systems engineering theory, system safety development theory explores three main hazard mechanisms of fault, insufficient function, and cyber attack of complex systemsand studies theories of guarantee and safety verification as the basis of design and assurance; System safety design is the core to system safety study and assurance. Based on the overall design process of "V", it successively carries out system concept setting and analysis, system safety design and development, and system certification and release, to provide safety requirements, system, model and design basis for system safety assurance technology; Aiming at the three types of hazard mechanisms, system safety assurance technology derives three key technologies in terms of functional safety (FuSa), safety of the intended functionality (SOTIF), and cybersecurity, and establishes corresponding implementation approaches, characteristics, and standard frameworks; Typical system for system safety is the carrier of integrated application of the above theories, approaches, and technologies, including safety assurance system based on functional redundancy and self-monitoring, and a cybersecurity system based on in-depth defense, etc.

2.1 System Safety Development Theory

System safety development theory is the basis of development approach and assurance technology. The overall framework is V-shaped model, as shown in Fig. 2.1. Based on system design input, system hazard theory on the left side of V model is used to identify safety requirements, and then system safety development and implementation are carried out by means of system safety assurance theory, while verification and validation methods and related safety acceptance theory on the right side of V model verify safety implementation. If safety requirements are met, the product safety release can be launched.

2.1.1 Systems Engineering Theory

As the foundation of system safety development, systems engineering theory facilitates system safety analysis, safety assurance and verification techniques. As shown in Fig. 2.1, targeted at the application of IV, systems engineering theory includes systems engineering, safety systems engineering and IV safety systems engineering.

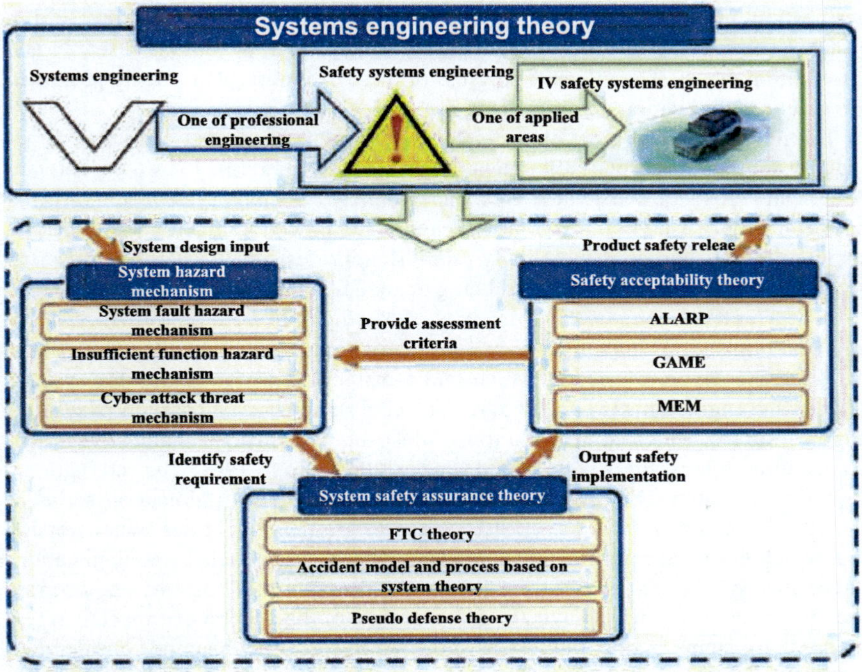

Fig. 2.1 Theoretical framework of system safety development

1. Systems Engineering and Safety Systems Engineering

Systems engineering is the premise of realizing system safety, with safety systems engineering as one of its professional engineering.

(1) Systems Engineering

In the long-term production practice, human beings have gradually formed a simple systematic thought, i.e., connecting the various components of things and making a comprehensive analysis as a whole. Systems thought has existed since ancient times, but the concept of "systems engineering" has been generated for less than 100 years, mainly from the practical application of quantitative system approach since the 1940s [1]. As an emerging specialty, systems engineering is in the process of rapid development, with relevant approaches and techniques being present and improved, as well as application fields expanded continuously.

Under the guidance of systems thought, systems engineering comprehensively applies the relevant theories, approaches and tools in natural science and social science, so as to achieve optimal planning, design and control management, analyze and deal with practical problems such as system function, composition and information flow. It is not only a science of overall planning and coordinated research on the system, but also an engineering technique for system development, design, testing, verification, implementation and application [2]. The basic disciplines involved in systems engineering include systems theory, cybernetics, operations research and information theory. The research progress of these disciplines has greatly promoted the practical application of systems engineering.

The main research object of systems engineering is a relatively complex large-scale system. In the early development, it has played a great role in pushing forward American aerospace and related weapons. In today's society, systems engineering has been widely used in transportation, energy and power, military, agriculture and social economy, and correspondingly established professional systems engineering [3], with new application fields still being opened up and expanded [4].

In 1957, HH. Goode of the University of Michigan first introduced such mathematical branches as queuing theory, linear programming and decision theory into his book *Systems Engineering*, which laid a mathematical foundation for systems science and engineering [5]. In 1965, R.E. machol summarized all aspects of systems engineering [6], which contributed to the most successful early practice of systems engineering, that is, the Apollo Moon Landing Program in 1969. The International Council on Systems Engineering (INCOSE) has started the publication series of *Systems Engineering Handbook* (SEH) since 1994. Till now, it has issued version 4.0, which systematically illustrates relevant theories, approaches and applications in the field of systems engineering [7]. In the discussion of systems engineering, the "V" model is often employed to give a picture of the decomposition-integration process of systems engineering [8], as shown in Fig. 2.2, in which the consideration on safety and reliability is described.

With the development of information technology, the interaction and coupling between systems enhances the system complexity. In the late 1990s, System of Systems (SoS, also named as systematic system or decentralized complex system) guided by complex adaptive theory came into being. SoS and its corresponding

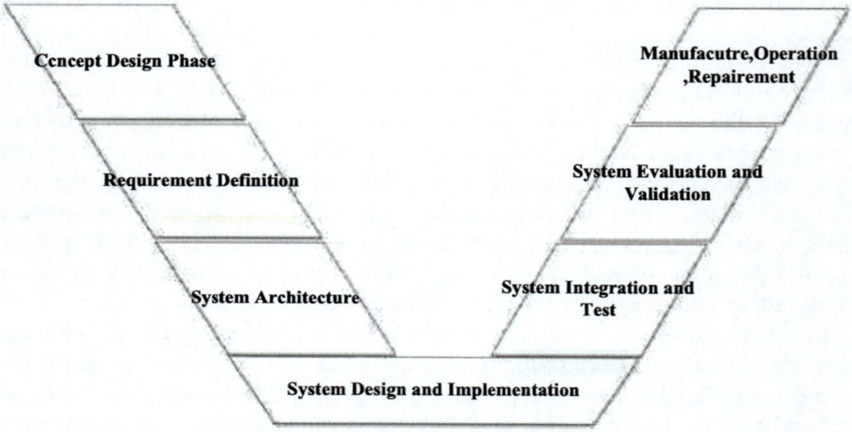

Fig. 2.2 "V" model of systems engineering

systems engineering have become the key research field of systems engineering. Typical SoS includes military system, computer system, global ground observation system, intelligent transportation system, social system, etc. At present, the rapid developing IV, IOV and intelligent transportation can also be regarded as a large-scale and complicated system integrated by multiple systems.

(2) Safety Systems Engineering

Systems Engineering involves two areas from the perspective of knowledge dimension, including traditional discipline engineering and professional engineering. Traditional discipline engineering indicates that specific techniques in such fields as mechanics, thermotics and electrics take into account the common laws that artificial engineering system must obey, mostly by extension from natural science. Based on relations and contradictions between the system and its various stakeholders, professional engineering usually applies special knowledge and theoretical approaches across disciplines to improve functional characteristics of the system [9]. According to brief introduction to various professional engineering in *Systems Engineering Handbook* of INCOSE, it can be divided into 6 categories in terms of specific connotation, as shown in Table 2.1.

Safety Systems Engineering is an essential systems engineering in line with safety. With the guidance of systematic ideology and application of advanced systems engineering theories and approaches, it analyzes and evaluates safety and its influencing factors, so as to establish a comprehensive safety prevention and control system and realize continuous and effective operation of the system. The scientific basis of

Table 2.1 Professional engineering classification in systems engineering handbook of INCOSE

Order	Category	Professional engineering
1	Professional engineering related to economy and value	Economic affordability; Value engineering
2	Professional engineering related to properties of the system	Mass property engineering; Electromagnetic compatibility
3	Professional engineering related to the environment of model system	Environmental engineering/impact analysis; Interoperability analysis
4	Professional engineering related to manufacturing	Manufacturing and productibility analysis
5	Features related to usage and manipulation	Usability analysis/human-systems integration; Training needs analysis; System safety engineering
6	Professional engineering related to availability	Reliability, availability, and maintainability, RAM; Safety systems engineering; Logistics engineering; Resilience engineering; System security engineering

Safety Systems Engineering includes system theory, cybernetics, information theory, optimization theory of operations research, reliability engineering, man–machine engineering, behavioral science, engineering psychology, etc.

Safety Systems Engineering mainly covers the following aspects as identifying and analyzing hazard, and predicting accident; eliminating and controlling the risk of accidents; analyzing the relationship and mutual influence among different units that constitute the safety system, coordinating the relationship among the units, and obtaining the best design of system safety; proving the satisfaction of corresponding safety requirements by the verification and validation on safety design and measures, and finally realizing system design and operation safety with reduction of risk to an acceptable level. The research and application of safety systems engineering has been conducted internationally for a long time, mainly in safety–critical systems of military industry, aerospace, chemical petroleum, railway and highway transportation, etc.

2. IV Safety Systems Engineering

IV Safety Systems Engineering is the specific application of safety systems engineering for IV. The overall development architecture [10] is shown in Fig. 2.3. The key steps include: ① Conduct safety analysis based on the establishment of system model, including Failure Mode and Effect Analysis (FMEA) on comprehensive utilization of systems engineering, Fault Tree Analysis (FTA) and Systems-Theoretical Process Analysis (STPA), etc.; ② Based on the results of safety analysis, put forward technical safety requirements, including software safety requirements and hardware safety requirements, so as to guide and track the specific safety development and implementation; ③ Develop safety mechanism based on safety requirements and make the system safety architecture design; ④ Generate test cases based on safety analysis and requirements to guide test and verification evaluation; ⑤ Apply such approaches as Goal Structure Notation (GSN) and Claims-Arguments-Evidence (CAE) to conduct safety argument.

(1) System Modelling

Based on systems engineering and model development approach, the construction of comprehensive systems architecture model for IV by various development tools and best practices of the industry is the foundation of safety analysis, safety requirements, safety architecture design and safety verification, etc.

Mapping all safety requirements into the functional architecture of IV system can obtain the tracking relationship between requirements and functional architecture elements. In addition, the traceability relationship among a series of safety activities such as "stakeholder requirements → safety requirements → safety architecture → test cases" is established based on system model. Among them, safety requirements are derived from stakeholder requirements, met by safety architecture, and finally verified by test cases, which is also the framework for demonstration of SSV with high intelligence.

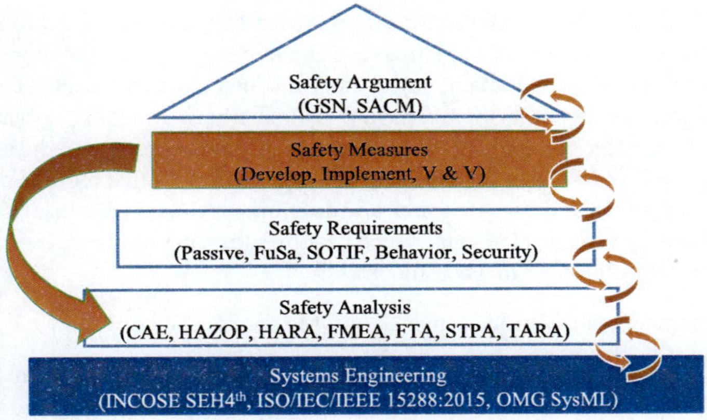

Fig. 2.3 IV safety systems engineering

(2) Safety Analysis

Based on system model, the safety analysis on IV and its safety–critical systems can be carried out. Due to safety challenges at all levels, its safety analysis also involves many different angels. It is necessary to comprehensively apply a variety of safety analysis approaches to obtain systematic and complete safety requirements.

1) Hazard Analysis and Risk Assessment (HARA): comprehensively utilize Hazard and Operability Study (HAZOP), FMEA and FTA to analyze the vehicle-level hazards caused by system failure and hardware fault, and assess their risk levels to obtain safety goals.
2) STPA: discover the hazards caused by component failure, and further explore the hazards induced by non-functional interaction among more components and human misuse.
3) Responsibility Sensitive Safety (RSS): model and analyze the interaction behavior between IV and other traffic participants around to ensure that traffic accidents will not be caused by the decision-making mistakes of IV itself.
4) Threat Analysis and Risk Assessment (TARA): find possible attacks on the system, and evaluate and assess their risks.

(3) Safety Requirements

Based on comprehensive safety analysis, the safety requirements of IV at all levels are derived, i.e., passive safety, functional safety (FuSa), safety of the intended functionality (SOTIF), behavior safety, cybersecurity etc., which constitute a very complex systems engineering [11].

It should be noted that classification of the above safety analysis is only a form adopted by the "inside out" safety system, and there can be other methods to classify it. For example, Automated Driving Systems 2.0 A Vision for Safety released by USDOT in 2017 has proposed 12 safety design elements that need to be prioritized in the development of autonomous vehicle development (including system safety, operational design domain, object and event detection and response, minimal risk condition, verification approach, human–machine interaction, cybersecurity, passive safety, after-collision behavior, data recording, consumer education and training, as well as laws and regulations). It is recommended that all the enterprises carrying out automated driving research and development should disclose their safety design according to these elements, which becomes the top safety requirements [14] for IV safety development to a certain extent.

(4) Development and Verification of Safety Measures

The core step in safety development is to promote development and verification of corresponding safety measures based on relatively complete safety requirements, cover corresponding safety requirements with safety functions, and finally form an overall safety architecture. As safety requirements involve many different types, it is necessary to carry out a large number of safety tests to verify safety measures, so as to ensure these requirements are fulfilled. For example:

1) Test and verification for FuSa includes: ① Performance test of vehicles, systems, subsystems and components; ② Verification of systems, subsystems and components based on requirements; ③ Fault injection test of safety critical control on input, output, calculation and communication; ④ Verification of fault transition (transition to auxiliary control path in case of main path fault) to safe state within fault tolerance time interval; ⑤ Intrusive tests, such as electromagnetic interference and electromagnetic compatibility tests, as well as exposure tests of other environmental elements (including temperature, humidity, radio frequency and light energy); ⑥ Durability test; ⑦ Software verification based on regression and simulation.
2) Test and verification for SOTIF includes: ① Exposure degree of IV within performance boundary of Operational Design Domain; ② Identification and iterative test of driving scenarios and boundaries; ③ The ability of detecting and responding to vehicle targets and events, as well as identifying environmental objects and situations with requirement for safe behavior; ④ Qualitative and quantitative evaluation on whether the vehicle behavior meets the standards.
3) Test and verification for cybersecurity includes: ① Demand-based test; ② Interface test; ③ Resource utilization evaluation test; ④ Control flow and data flow verification; ⑤ Penetration test, vulnerability scanning, fuzzy test, etc.

    During IV design, the above process will be repeated to verify and implement various safety performance. The safety requirements will be continuously met, and system architecture in systems engineering will be continuously improved to form a more complete systems safety architecture.

(5) Safety Argument

Complete safety files are formed through system safety activities, and finally the safety argument methods such as GSN and structured assurance case metamodel

(SACM) are used to demonstrate the safety of automated driving systems. It should be pointed out that in order to ensure IV safety, multidisciplinary approaches across multiple functional levels are needed, including hardware fault tolerant design, machine learning algorithms, harmonious interaction with drivers in traditional vehicles, verification of normal operation of the system in a highly unstructured environment and applicable legal supervision, which is far beyond the research scope of safety engineering itself. Multidisciplinary research and cooperation should also be implemented in hardware, software, robot, safety, testing, HMI, social acceptance, legal supervision, etc.

2.1.2 System Hazard Mechanism

Based on physical compositions, working principles and operational properties of the system, the hazard mechanism mainly analyzes the formation causes and evolution process of system hazard. According to the order of causes from inside out, the mechanism can be divided into systems fault hazard mechanism led by system's own fault, functional deficiency hazard mechanism by system's internal and external interaction, and cyber attack hazard mechanism absolutely by external malice.

1. System Fault Hazard Mechanism

The hazard mechanism for system fault mainly focuses on causal inference process of system hazard led by system's own physical principles and design defects. Generally speaking, fault refers to the event that the whole system's function deteriorates due to functional failure of some components in the system, as well as abnormal conditions that can cause the failure of elements or related items. According to different hazard mechanisms, faults can be categorized into systematic faults and random hardware faults [13].

Systematic fault is caused by insufficient design specifications for design or production processes, operating procedures, documents and other relevant factors, including software fault and some hardware fault. Systematic fault will occur in a certain way, that is, definitely along with easily-reproduced systematic failure. Therefore, it is likely to be avoided by updating and standardizing the design or production processes.

Random hardware fault is induced by physical processes, such as fatigue, physical degradation or environmental stress. It occurs unexpectedly and obeys the objective law of probability distribution in the lifecycle of hardware elements. For example, some hardware does not work as expected due to oxidation, which eventually leads to component failure. Since not all random hardware faults will result in hazards, the focus shall be whether they violate vehicle safety objective. Accordingly, the random hardware faults are classified, as shown in Fig. 2.4.

When studying random hardware failure, it is necessary to distinguish differences and connections between the three concepts of "fault", "error" and "failure", which is the basis for understanding numerously related concepts of random hardware failure.

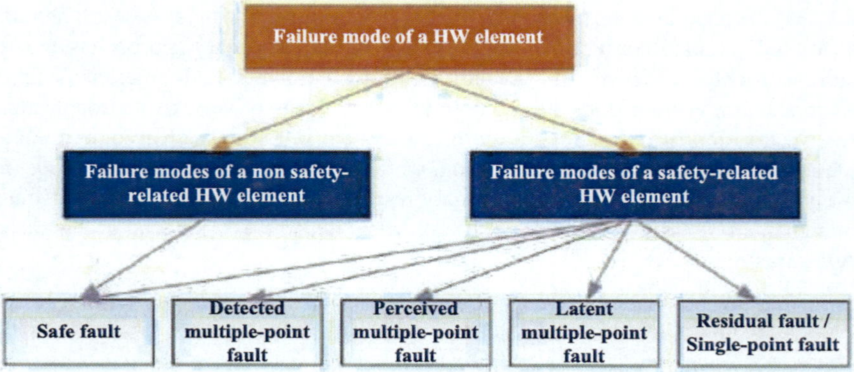

Fig. 2.4 Classification of random hardware faults

1) Fault: the abnormal condition that can result in the failure of elements or related items.
2) Error: the difference between the calculated, observed, measured value or condition and the real, specified and theoretically correct value or condition.
3) Failure: the termination of an element's capability to perform a function as required.

In the same level, fault is the cause of failure, failure is the result of fault, and error is the manifestation of fault. Within different levels, component-level failure is system-level fault.

The relationship of fault, error and failure is shown in Fig. 2.5. The development process from fault to error and from error to failure are described from three different types of causes (system software problem, random hardware problem and system hardware problem).

2. Hazard Mechanism of Insufficient Function

With increasing complexity of IV functional architecture, only considering the systems safety problems caused by systems fault can not meet the safety requirements of highly complex systems. The safety risks led by no fault or insufficient functions have attracted more and more attention. Based on physical principle of the system itself and the induction of the external trigger environment, the hazard mechanism induced by system functional deficiency under internal and external interactive coupling of the system is clarified.

As a means for IV obtaining internal and external information, sensor has functional deficiencies in terms of two aspects: First, extreme weather (such as rain, snow, fog and adverse light) affects its surrounding perception capability; Second, subject to principles, sensors for detecting some specific targets are limited. For example, when lidar detects strongly reflective mirrors (such as puddles on the ground, and smooth iron plates) and millimeter wave radar detects specific materials (such as

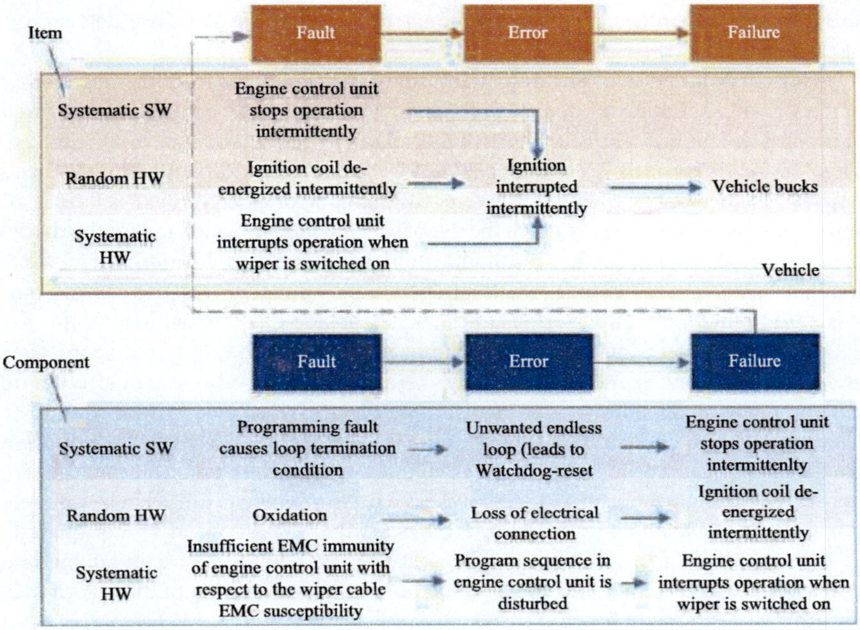

Fig. 2.5 Example of relationship among fault, error and failure

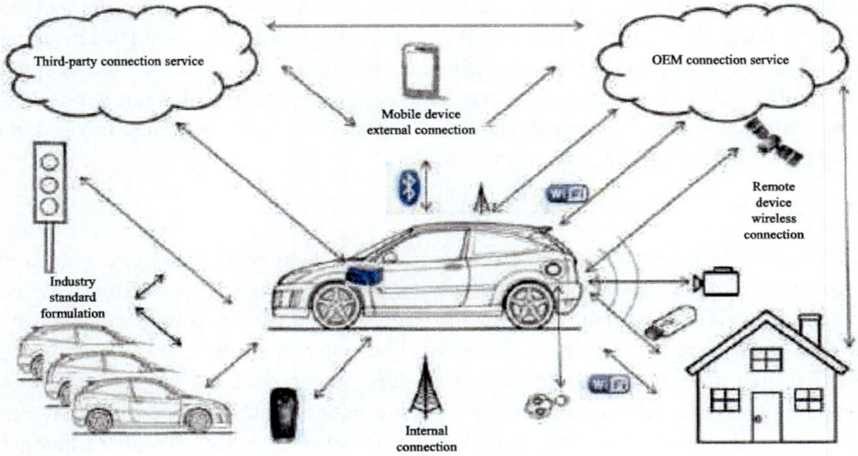

Fig. 2.6 Information interaction around ICV

strongly reflective objects, non-metallic materials), the missed or false detection is likely to occur.

Insufficient function of perceptual algorithm mainly comes from the uncertainty of the algorithm. Taking the deep learning model as an example, errors are propagated backward by labeling a large amount of data, and then the parameters are optimized. This learning approach is compared to a "black box". It is difficult to understand impact of the number of hidden layers, the number of neurons and the form of activation function on the results in the deep learning model, which makes the model hard to interpret and the output result uncertain. In practical application, "black box" problem leads to poor output results of deep learning model in some cases. Since it is hard to find the reason, serious traffic accidents are likely to happen.

Planning and decision-making algorithms mainly include rule-based approaches and learning-based approaches. Among them, the rule-based ones usually ignore environmental details that lead to decision-making mistakes, which makes it unable to cover all the potential hazard scenarios for IV on real road; In addition, the division conditions for state cutting and triggering conditions for the behavior rule database are easy to overlap, and the scenario depth traversal is insufficient, resulting in insufficient functions of these approaches. Learning-based algorithms depend on the quantity and quality of data. Insufficient number of samples, data quality overlapping and unreasonable network will lead to over learning or under learning of the algorithms, resulting in insufficient functions as well.

The lack of control function mainly comes from vehicle dynamics. The vehicle dynamic model is not sufficient to demonstrate vehicle dynamic properties under working conditions of large curvature curve, strong crosswind and wet pavement. At present, the transverse and longitudinal control of IV is generally based on the linear model with two degrees of freedom, which is an approximate model in case of small curvature. When the tracked target has large curvature, the model is not applicable, for it will lead to large tracking deviation. Besides, under the condition of low ground adhesion, vehicle control is restrained as well.

3. Threat Mechanism of Cyber Attack

Due to the rapid development of electrification, intellectualization, networking, and sharing, ICV needs to defend against cyber attacks such as hacker intrusion. When connected to IOV, the private information of ICV and its owner is probably exposed to danger anytime and anywhere, and easy to be stolen, disturbed, or even modified. At the same time, cyber attacks are also related to personal and property security. Once a private vehicle is attacked by a person with an ulterior motive, it may not only cause the loss of property in the vehicle or the vehicle itself but also endanger the lives of drivers and passengers. In addition, the cyber vulnerability of ICV will also lead to systematic risks and social security problems. For example, criminals may control ICV to make terrorist attacks.

In terms of privacy, property, and social security, the main targets of cyber attacks include stealing sensitive information or records, suspending service, interrupting communication, damaging the function of embedded network physical systems, and so on. Table 2.2 lists some common cyber attack risks. Among them, the network application attack is targeted at the communication process between the client and the server. The attack is launched during the communication between ICV and the cloud or terminal by sniffing, to easily obtain accounts and passwords of various network applications [14]; Targeted at the client, the social engineering attack deceives customers into providing privacy information by phishing emails and other means; Targeted at the server, denial of service attack makes connected vehicles service suspended by consuming the network bandwidth of the server.

According to the attack subject or intrusion interface, cyber attack threats for IOV are divided into the following categories [15]. First of all, ICV itself can be the subject of attack. Potential threats of cyber attack exist in the weak parts of data, code, and functional design. The intrusion interface includes key components or functions such as the onboard gateway, sensor, ECU, and Over-the-Air (OTA). Especially the In-Vehicle Infotainment (IVI) and Telematics-BOX (TBOX) are vulnerable to intrusion, which exerts a great influence on the safety of the whole vehicle system. For example, for TBOX communicating with background and mobile app, it is easy to carry out the denial of service attack; For IVI, it is easy to obtain permissions and steal data, resulting in high-risk threats such as remote malicious control and sensitive information disclosure. Secondly, communication in IOV can also be attacked. Figure 2.6 shows information interaction of ICV and OEM with the third-party service platform and mobile devices, etc. According to the access boundary, complex information interaction in IOV can be divided into the in-vehicle domain, Vehicle-to-Everything (V2X) domain, and infrastructure equipment domain, with different major safety threats. For example, the threats to the gateway, ECU, and CAN of the in-vehicle domain mainly include distortion of communication data packets and software systems; communications such as Vehicle to vehicle (V2V) and Vehicle-to-Infrastructure (V2I) in the V2X domain are vulnerable to threats of eavesdropping, interference and forged identity, etc.; Traffic infrastructure and communication infrastructures such as signal lights and base stations are vulnerable to threats such as denial of service and data theft.

Table 2.2 Classification of common cyber attack risks

Cybersecurity risk	Common attack means
Cyber application attack	Sniffing, code injection, cross-site scripting, intermediary attack, WiFi penetration, etc.
Malware attack	Advertising software, attack software, criminal software, spyware, etc.
Social engineering attack	Face-to-face, deception, phishing, social media, etc.
Hacker attack	Access control vulnerabilities, cloud side-channel attacks, domain name server redirection, cryptographic attacks, etc.
Denial of service attack	Distributed Denial of Service (DDoS), Ransom DDoS, etc.
Advanced persistent attack	Botnets, industrial worms, malicious networks, rootkits techniques, etc.

The Telematics Service Provider (TSP) platform, which provides navigation, entertainment, control, and other functions for ICV, is also easy to be attacked. Related threats include cross-site script attack, Structured Query Language (SQL) injection, information disclosure, denial of service, and so on. Due to the openness of the IOV service platform, attackers and ordinary users have the same permissions and can access and share various resources provided by the service platform. Therefore, this kind of cyber attack has low cost, less physical contact, and serious harm. The mobile application of ICV also faces various hidden dangers from cyber attacks. Attackers can intervene in the mobile terminal to communicate with the Bluetooth or WiFi module of IV and realize attacks such as interface hijacking and override control. Finally, the threat to the data safety of IOV can not be underestimated. Its collection, transmission, storage, use, migration, destruction, and other processes are faced with risks such as tampering, counterfeiting, and leakage, among which the recovery of incompletely destroyed data is particularly common [16]. In addition, according to regulations on the cybersecurity Management System (CSMS) of UNECE R155 (WP. 29), security process design and supervision should be improved to reduce human misoperation [17]. To manage these risks, vehicle manufacturers must reinforce and standardize their safety design process, supervise suppliers and partners, and find customers with potential weaknesses, to protect themselves and suppliers' products and services.

ITo formulate proper guidelines to deal with various cyber attacks, it is necessary to conduct TARA by taking into account its occurrence and consequence and then defining countermeasures. Table 2.3 is a simple example of qualitative analysis, showing risk quantification levels. Judgment on the possibility of a cyber attack is based on experience on the one hand and the analysis of system attacks on the other hand. The Data Resource Center of China Automotive Technology and Research Center released the first ranking of the top 10 factors for vehicle cyber attack risks [18] in May 2019. Figure 2.7 presents statistics of the occurrence probability of each risk. The threat degree corresponding to a certain cyber attack depends on the cyber attack motivation, place, threat degree of system safety, etc. The higher the possibility and threat of cyber attacks are, the higher the risk level is, for which more prioritized protection is needed.

The ranking of the top 10 factors for vehicle cyber attack risk is determined by various items such as risk occurrence probability, utilization difficulty, and harm degree. As shown in Fig. 2.7, "vehicle network without security isolation" is most

Table 2.3 Hierarchy table of risk quantification level in terms of possibility and consequences of cyber attack

Threat degree of consequence	Possibility		
	Extremely possible	Possible	Unlikely
High	Extremely high	High	Medium
Medium	High	Medium	Low
Low	Medium	Low	Extremely low

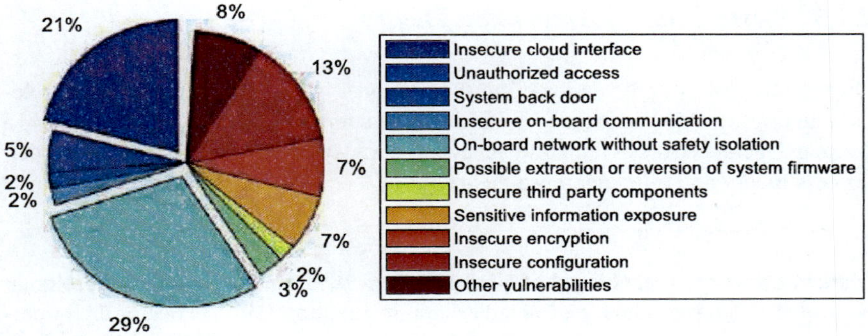

Fig. 2.7 Occurrence and comprehensive ranking of the top 10 factors for vehicle cyber attack risk (from top to bottom)

likely to occur, with 28.9%, which is the most common design error. However, owing to its local attacks and difficulty in use, it is finally ranked fifth. The "insecure cloud interface" has a high incidence and great threat, which may lead to the disclosure of private information stored in the database, the tampering of data tables, the destruction of hard disk data, and even the collapse and downtime of the server. Therefore, it ranks first in the comprehensive ranking.

To sum up, the risk level of a cyber attack mainly depends on the source, object, motivation, and means of attack, as well as the risk degree of ICV being threatened in each attack stage.

The intrusion of cyber attack generally includes several stages, such as data leakage, software installation, horizontal movement, attack maintenance, voucher acquisition, system penetration, and target reconnaissance [14]. Firstly, the attacker can extract and tamper with the attacked data and install malware on the target system or network; then, the attacker can move laterally from one access point in the system or network to another and keep it for a while. At the same time, the attacker may install a back door on the damaged system or network for repeated access in the future. Finally, after obtaining the permission of the root user or administrator, the attacker can deeply access the target system or network, observe the target and formulate a further attack plan until the plan is completed or stopped. Among them, various network attacks can pose different threats to the information in the connected vehicles. For instance, the disclosure of unauthorized information threatens the confidentiality of cybersecurity; truncation, tampering, and destruction of data threaten the integrity of cybersecurity; and denial of service attack threatens the availability of cybersecurity.

2.1.3 System Safety Assurance Theory

On the basis of systems engineering theory and system hazard mechanism, relevant theories to safety assurance are illustrated in terms of analyzed systems hazard from such perspectives as fault tolerant cybernetics, system-theoretic accident model, process theory and mimic defense theory.

1. Fault Tolerant Cybernetics

Marked by the Integral Control of Niederlinski in 1971, the concept of fault tolerant control (FTC) has gradually attracted academic attention [19]. As one of the important foundations for hazard treatment technique of system fault, FTC is an interdisciplinary and comprehensive theory, with theoretical basis as modern cybernetics, signal processing, optimization approach, decision theory and statistical mathematics, etc. During system operation, it is inevitable that some parts will fail and lead to functional degradation. However, people can usually accept the function degradation that does not affect the system stability. The goal of FTC is to make the system stable in case of component fault during operation, and realize small or acceptable function degradation.

FTC of complex systems usually includes two parts: hardware redundancy and analytical redundancy. The hardware redundancy FTC originates from the reliability analysis theory. According to the theory, the parallel system can effectively ensure system safety. As any component in the parallel system fails, the other parallel components can complete the function of the failed component [20]. If the parallel components fail, the normal operation of the system will not be affected. However, due to space and cost, some system components are not suitable for hardware redundant FTC, such as actuators.

In case of main component failure, the system cannot achieve intended performance due to functional degradation. At this time, the objective is to maintain the system within an acceptable range. Analytical redundancy realizes FTC of the system passively or actively by control algorithms. Among them, the passive analytical redundancy mainly takes the system uncertainty as well as possible disturbances and faults into account in the design of control algorithm, in order to reduce system dependence on the operation of a single component and improve system robustness to faults. However, it is necessary to know the disturbances and faults affecting system stability in advance. The active analytical redundancy is to identify and locate faults through detection and diagnosis, and further reconstruct and update the controller online according to the specific fault, so as to maintain system stability. Therefore, fault detection and diagnosis is vital for active analytical redundancy. Any error from fault detection and diagnosis module will lead to the decline of FTC effect. Since the active approach is not limited to the prior knowledge of faults, its effect toward on-line processing of complex fault system is better than the passive one.

2. Systems-Theoretic Accident Modeling and Process

System safety analysis is an important part of the "V" model of IV system safety development. Based on the initial design input of systems, it adopts an appropriate accident causality model for the three types of systems hazard mechanism to analyze various systems safety problems and derives requirements. The establishment of causality model provides a theoretical basis for systems safety analysis.

The traditional causality model mainly refers to the event chain model, which models the accident as a series of independent events over time, in which the former event is the direct inducement of the latter one. Traditional model mainly deals with the system by means of analysis and decomposition. By decomposing physical or functional components into smaller sub-parts and assuming their interaction in a direct or known way, each part is checked and analyzed separately and then combined, so as to tackle the complexity. Based on traditional causal model, a series of safety analysis approaches have been derived, such as FTA and FMEA. However, such models are grounded on assuming the independence of various parts of the system and the direct correlation among hazardous events.

IV system adopts advanced communication, computer, network, and control technologies, and realizes the functions of intelligent perception, prediction, decision-making, control and HMI. It brings changes in system attributes such as the change of accident essence, the increase of system complexity and coupling, and the complexity of HMI, which pose greater challenges to the in-depth analysis of system hazards such as specific faults, insufficient functions, and cyber attacks. Therefore, the traditional causality model is not fully applicable, and requires to be combined with a more effective model as theoretical basis. In the early twentieth century, Nancy G. Leveson of Massachusetts Institute of Technology proposed system-theoretical accident modeling and process (STAMP) [21], which is a new accident model based on system theory. System theory regards the system as a whole rather than a simple sum of all parts. Its main focus is the emergency attribute, which comes from the interaction and synthesis among system components. From the perspective of system theory, STAMP extends the traditional causality model to a series of directly related system hazard events or component faults, in order to cover more complex processes and unsafe interactions among system components.

STAMP takes safety as a control issue and manages it by the control structure embedded in the system, which goal is to constrain system development and system operation leading to safety behavior. Accidents will occur, when external interference, component fault, or dysfunctional interaction among system components are not fully handled by the control system. Therefore, in order to understand the causes of accidents, it is necessary to determine the causes of the ineffective control structure in this framework. With regard to prevention of potential accidents, it is necessary to design a control structure to enforce essential restrictions. The basic concepts of STAMP include constraint, control cycle, control model and control hierarchy. Constraint plays a central role in system safety. Instead of being understood from a series of events, causes of accidents should be regarded as the lack of constraints

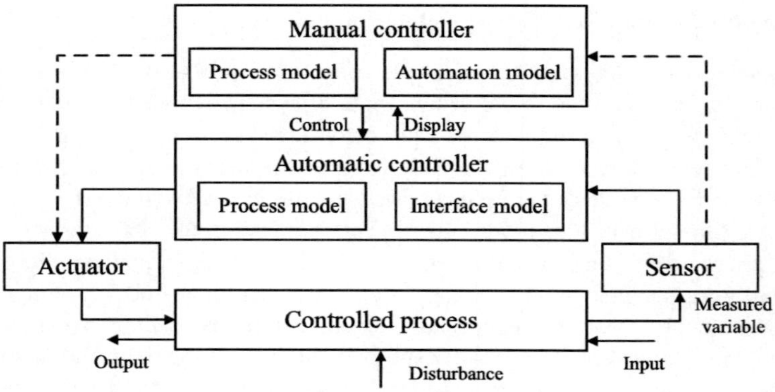

Fig. 2.8 Typical control loop and its process model

on system design and operation. The role of system safety engineer is to determine required design constraints for safety maintenance and ensure the capability of system design for implementation.

Control loop and process model is the modeling form of STAMP. Different from the traditional event-based model decomposing the system and accident into structural components and event flow, STAMP describes system and accident according to the control level of the adaptive feedback mechanism. Within it, open systems are regarded as interrelated components, which are kept in dynamic equilibrium through the feedback loop of information and control. Figure 2.8 shows a typical control loop and its process model, in which the automatic controller is supervised by the manual controller and jointly monitors and controls the controlled process based on their respective process models. This loop corresponds to the process of joint control over the vehicle by the driver and IV system. The dotted line indicates that the driver can directly access system status information and control the controlled process through manual operation.

In addition, when modeling complex organizations or systems, reasonable hierarchies should be built, with each layer imposing constraints over the next layer. Effective communication channels are needed between layers of each control structure. The downward reference channel provides information necessary to impose constraints on subordinates, and the upward measurement channel provides feedback on an effective way of implementing constraints. At each level, inadequate control can be induced by a lack of constraints, inadequate transmission of constraints, or incorrect enforcement of constraints at lower levels.

Based on fundamental assumption of STAMP on the cause of accident—insufficient control, it can be further classified as the following problems: (1) insufficient implementation of safety constraints, including unidentified hazards (and related constraints) and insufficient implementation of constraints by control measures. The latter one may be caused by defects in control algorithms, inconsistent use of control algorithms, incorrect process model, inadequate coordination among multiple controllers and decision-makers, etc. (2) insufficient execution of control actions, such as problems occurring during transmission or execution of control commands. (3) insufficient or missing feedback. Specific reasons include feedback not involved in system design, defects in monitoring or feedback communication channels, untimely feedback or sensor operation problems, etc.

Adopting a top-down working mechanism, STAMP is more suitable for complex systems than traditional models. It embodies software, human and other factors as the cause of accidents or other losses, without dealing with a variety of factors in different ways. It is applicable for analyzing various types of system hazards including faults, insufficient functions, cyber attacks, etc.; In addition, more powerful analysis tools such as STPA can be further created based on STAMP.

3. Mimic Defense Theory

In view of cyber attack threat, traditional assurance theory applies feature base and rule base to detect and deal with typical cyber attacks such as denial of service and sniffing. This kind of defense approach takes a certain time to respond and can not deal with endless new threats. Cyber Mimic Defense (CMD) innovatively provides defense theories and approaches of universal significance, to cope with various potential threats such as unknown vulnerabilities, backdoors, and viruses in the network [22].

The core of Mimic Defense is a closed-loop dynamic heterogeneous redundant architecture, as shown in Fig. 2.9. Its dynamic, heterogeneous, and redundant properties make it difficult for threats to be successfully applied. The target information system can be reconstructed based on this framework to possess endogenous safety defense capability in the system structure. Instead of relying on prior knowledge for matching detection, the framework perceives the risk as a disturbance and cleans and schedules abnormal processing units through feedback and scheduling mechanism, to ensure the implementation of real-time security of the system in the face of unknown threat attacks. The redundancy idea of Mimic Defense is reflected, by utilizing multiple heterogeneous actuators with the same function but different operating conditions to make a multi-mode judge on the processing results of the same input data. The abnormality of a certain actuator indicates that its vulnerabilities have been exploited by attackers, and the system can perceive the threats and risks. The negative feedback controller is used to replace the abnormal actuator with the reconfigurable one, which presents the dynamic convergence property of mimic defense. The threat is perceived and processed by being normalized to disturbance, to ensure that system safety does not decrease over time.

Traditional cybersecurity theory focuses on detecting the known threats in information transmission links outside the system, while the innovative Mimic Defense theory mainly focuses on the unknown threats that make use of vulnerabilities to invade inside the system. The two theories are complementary. In practical application, the unknown threat perception function of Mimic Defense can transform unknown threats into a priori knowledge through traceability learning together with traditional traceability means, to serve traditional cybersecurity theory. Therefore,

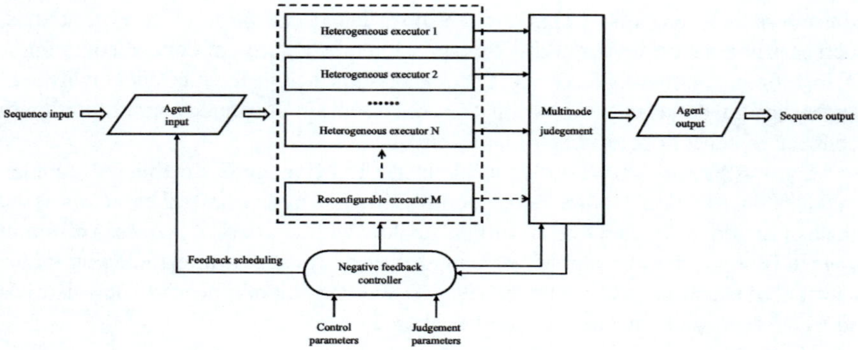

Fig. 2.9 Dynamic heterogeneous redundancy architecture of mimic defense

mimic defense and traditional defense techniques can jointly construct a complete cybersecurity solution.

2.1.4 Safety Acceptability Theory

As an important part on the right side of the "V" model of Safety Systems Engineering, safety acceptability theory is applied to confirm whether the system can reach the appropriate safety level and then realize the final product safety release. Like any complex systems, IV cannot be absolutely safe and free from all accidents. On the contrary, due to the complexity of its own system and the unknown driving environment, it is also quite difficult to judge its safety acceptability level.

Generally speaking, the real traffic accident data, test field data and relevant statistical data can be of reference to the acceptable level of vehicle system. However, due to the lack of the above data in the field of IVs, the safety acceptability level of IV and its main systems can be evaluated by appropriate acceptability theory. At present, the acceptability theories that can be referred to include As Low As Reasonably Practicable (ALARP), Globalement Au Moins Equivalent (GAMEB) and Minimum Endogenous Mortality (MEM).

ALARP means that residual risk needs to be reduced to the lowest level within a reasonable and practical range. ALARP divides risks into three categories: High risk. In accordance with social rules, high risk is intolerable at the current stage, and must be mitigated by safety measures; Low risk. Referred to as ignorable risk, it is acceptable in social architecture where the system is deployed, and the system can be put into use directly. However, once the risk is identified, it needs to be recorded and monitored in case of any increase; ALARP risk. Between high risk and low risk, it can be tolerated on the premise of "return-cost" balance.

As shown in Fig. 2.10, risk area is divided into three parts, in which ALARP area has different boundaries in different projects. At the initial stage of risk analysis of a

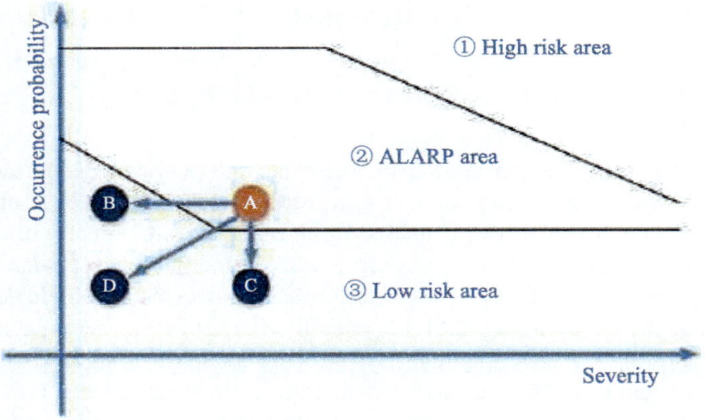

Fig. 2.10 Risk area division based on ALARP

specific project, the boundaries of ALARP area should be consistent. Assuming that risk A is marked within the ALARP area shown in the Figure, it can be transferred to risk B by reducing injury severity caused by the risk, or to risk C by decreasing occurrence probability of the risk. At the same time, risk A can be transferred to risk D by the above two approaches. In the process of reducing ALARP area risk, it is necessary to consider the balance between the cost of risk reduction and the benefits obtained. For the risks in ALARP area, appropriate approaches need to be taken to lessen risks, until it can be proved that the cost of further reduction is far higher than the benefits obtained.

Introduced from France, GAMEB acceptability criterion indicates that any new system or any modification to the existing system should be more safety than the existing recognized reference system on a global scale. Without considering safety by specific risk level, it studies the global safety level of the system in accordance with global requirements, which means that designers are free to assign risks to subsystems with similar risk levels in the system. In terms of different products, the global safety indexes can be the number of casualties per year, the number of injuries per hour using the system or other appropriate indexes.

MEM is an approach of obtaining absolute value of risk acceptance based on human natural mortality in a specific age group. For individuals, the total risks of all systems shall be no more than natural mortality of human beings. For a single system, the fatal risks to individuals shall be less than 10^{-5} times/year, which can be further decomposed into subsystems [23]. MEM is mainly used in railway, and its utilization rate in other technical fields is lower than the above two approaches.

2.2 Systems Safety Design Approach

2.2.1 Overall Process of Systems Safety Design

Safety is the basic requirement and key index for all kinds of system design and R & D, including IV design. Standard system design processes and principles can effectively reduce the probability of software and hardware failure, insufficiencies of specification or performance limitations. At present, a series of international standards such as ISO 26262 and ISO 21448 clarify the safety-related processes in vehicle design and development. As shown in Fig. 2.11, the overall process of IV systems safety design is divided into three stages, including system concept setting and analysis, system safety design and development, and system certification and release. Based on the system safety concept, potential system risks and safety requirements are clarified through the setting and analysis of relevant items of IV. Safety design and development are carried out toward the software and hardware respectively, simultaneously together with system-level design, development and integration. The final step is to conduct certification in accordance with the relevant safety standards of IV products and products-release, so that system safety design process is completed.

Defining system concept setting and analysis includes three sub-stages: project safety definition, hazard analysis and risk assessment, and safety concept determination.

1) Project safety definition: comprehensively describe the development project of IV (at least one system), clarify relevant item's definition, and initialize safety life cycle of product;
2) Hazard analysis and risk assessment: identify potential hazards of IV system, and quantify and classify risks according to certain principles, so as to set specific safety objectives for different levels of risks;
3) Safety concept determination: based on the safety objectives and IV basic architecture, refine safety requirements specified by each system, so as to form a complete safety concept.

System safety design and development involves with system, hardware, and software product development, all of which should follow the "V" development model.

1) Carry out system-level IV safety development planning, extract technical requirements & specifications, and conduct system design containing system architecture based on the specifications;
2) Execute R & D of software and hardware products at the same time of system design. The software and hardware design and specifications-extracting are in the left branch of the V-model, while the testing, verification and confirmation of software and hardware are in the right branch;

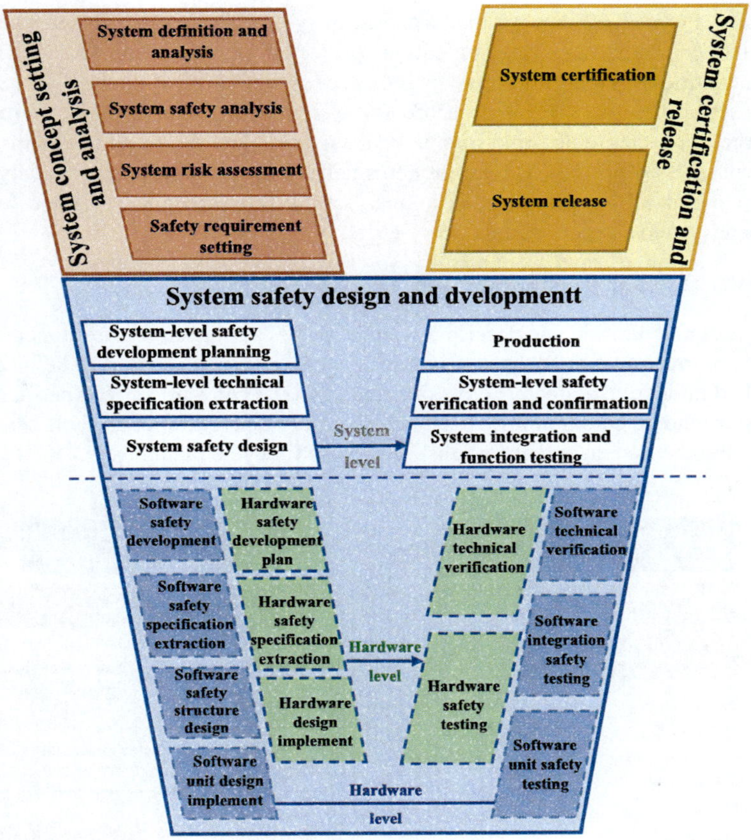

Fig. 2.11 Overall process of IV systems safety design

3) Implement system-level functional integration and testing, which verify and confirm the system bsafety, and then try to produce protptype.

System certification and release is the last stage of IV safety design. Before the release, it is necessary to verify production according to the previously specified IV criteria, and then provide relevant documents to obtain the license through relevant review. Finally, mass production, sales and application can be permitted.

2.2.2 System Concept Setting and Analysis

System concept is the basis of IV system safety design and development. Based on setting and analysis, system and corresponding safety requirements can be fully explained. Figure 2.12 shows the logical flow of system concept design and analysis. The first part determines key contents of project safety definition from terms of relevant entities of IV, and clarifies key project of system safety lifecycle; The second part describes the classical approaches for system safety analysis; The third part show the important tools for system risk assessment; The fourth part explains how to assign safety objectives to relevant items defined by the project successively, and match corresponding safety levels, so as to satisfy safety requirements and form a complete system safety concept.

1. System Definition and Description

In the system definition and description stage, the IV systems to be developed (such as decision-making system, chassis control system or vehicle) should be defined and described in detail from perspective of safety, and the concepts of relevant items of the system should be futhur initialized, which ensure that relevant designers can fully understand the system and accomplish the safety lifecycle planning.

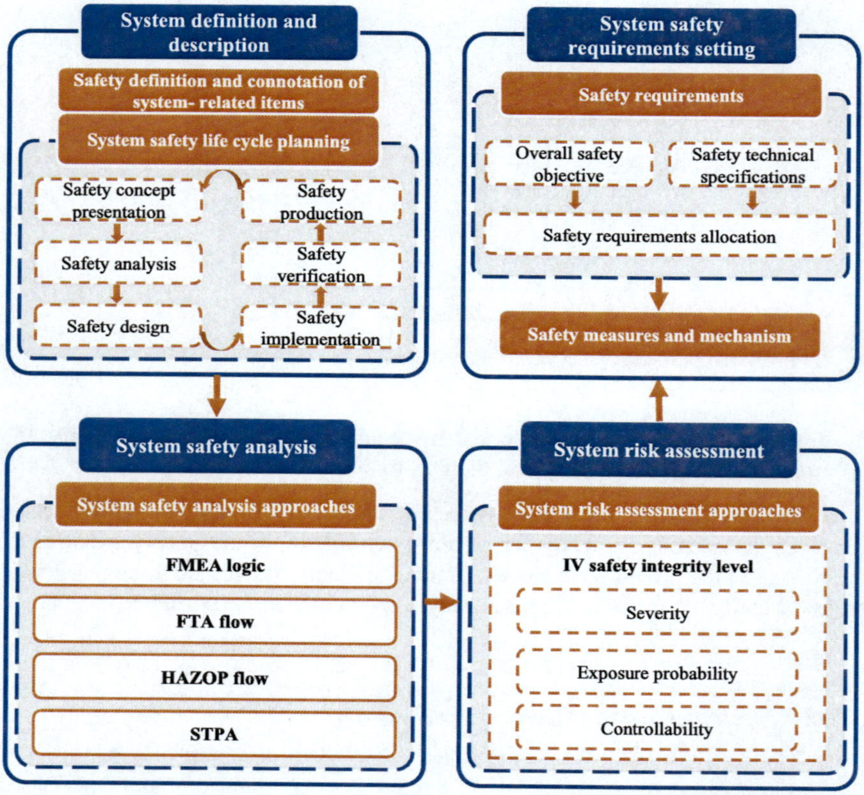

Fig. 2.12 Systems concept setting and analysis process

(1) Safety Definition and Connotation of Systems-Related Items

System-related items refer to systems or system combinations that realize specific functions of IV, including architecture, component, and unit, etc. In the traditional architecture, the main IV systems generally comprise of perception, location, communication, decision, control and human–machine interface (HMI) system. Furthermore, system-related items also include electrical architecture, chassis, sensor, audio, on-board computer, etc.

In the initial substage of system definition and description, it is necessary to define the safety of relevant items, including functional requirements and non-functional requirements (such as operation requirements and environmental restrictions), regulatory requirements of relevant items, national and international standards, definition of the system intended behaviors, potential impact resulted from known failure modes and functional limitations, etc. Also, the safety definition of related items shall consider the project boundary conditions and the interface between related items, such as interaction and constraints between related elements or environmental elements, functional planning and resource allocation among various elements, and impact of various elements interaction on the functional operation.

Before formal IV-development and design, the safety definition of system related items gives a complete and comprehensive description of system concept, which ensure that designers can fully understand the design requirements and boundaries. And then a reasonable safety lifecycle for R & D is planned, laying the foundation for subsequent safety analysis, design and development.

(2) Systems Safety Lifecycle Planning

In the system safety design substage, safety lifecycle is an essential concept, which refers to all the safety related process from product concept proposal to manufacturing. Lifecycle usually includes six conceptual stages: safety concept proposal, safety analysis, safety design, safety implementation, safety verification and safety production. In the system definition and description stage, it is necessary to plan the safety lifecycle of IV, that is, to implement and refine the specific process of relevant items. In the refinement process, the designer should distinguish wheather the system is newly designed or transformated from the existing product. if it is a new creation, it needs to start from safety definition of relevant items, and implement the complete safety process for all sub-stages of the lifecycle; If the system is transformated from the existing production, it is feasible to adjust and customize the items' safety lifecycle process with current audit and certification documents in the lifecycle planning which can omit unnecessary repeated processes. For example, the upgradation or update of software is only necessary to arrange safety process for changing part and implement part of the software safety lifecycle planning, without re-executing the hardware process.

During system upgrading, it is necessary to firstly focus on comprehensive safety investigation of system and application, so as to clarify the issue of upgrading and changing, and further evaluate the impact on IV safety. The second step is to clearly describe the differences of the conditions before and after the system transformation (such as operating condition, operation mode, environmental interface, and installation requirement), so as to determine the safety process that needs to be adjusted or supplemented. In addition, it is also necessary to confirm the system changes and their impact on safety. If the impact of system changes on safety of IV relevant elements and their operating environment cannot be clarified, relevant safety test data should be recorded to support further analysis. In the above process, the items associated with the changes are also required to meet IV safety requirements.

2. System Safety Analysis

System safety analysis is an important stage to identify, classify and analyze IV safety issues for preventing or alleviating relevant hazards by avoiding unreasonable risks. Figure 2.13 shows the logical flow of system safety analysis. Based on definition and description of system related items, IV safety issues is analysed, potential sources of hazards are investigated, and the corresponding risks are quantitatively evaluated. Furthermore, hazard control approaches are implemented in the subsequent stage to assess whether the risks after hazard control can be accepted. If the risks are not acceptable, IV design will be modified and operation conditions restricted; If acceptable, the safety R & D will be followed up, and periodic system evaluation will be continuously conducted during manufacturing and application. If hazards not found before are detected in process of manufacturing or application, corresponding risk management measures shall be executed for the renewal of systems safety analysis process.

(1) Hazard Analysis

Hazard is the potential source of harm induced by abnormal function of system related items. System safety analysis is to analyze and investigate IV systems and components by certain means, so as to find potential causes of harm (such as failure and functional limitation). Common hazard risk analysis include FMEA, Event Tree Analysis (ETA), FTA, Model-Based Safety Analysis (MBSA), HAZOP, and STPA, etc.

FMEA is an inductive safety analysis proposed by NASA in the 1960s based on Domino Theory. The logical flow is shown in Fig. 2.14. In the development process, FMEA first lists all potential failure modes according to system definition and deduces their consequences and causes. Then, the influence degrees of failure modes are determined and ranked. Finally, the issues should be improved with the effect evaluated. During IV development, manufacturing and operation, it is necessary to record, analyze and manage new failures, collect and record users' experience and suggestions, and execute corresponding safety control on the issues with the greatest impact on safety.

FMEA can be further catogarized into Design Failure Mode and Effects Analysis (DFMEA) and Potential Failure Mode and Effects Analysis (PFMEA). DFMEA mainly applies in IV design process, which analyzes the failure of software, hardware, algorithm, function system and vehicle. DFMEA needs checking whether the system design meets all safety requirements, (except failure in IV manufacturing process).

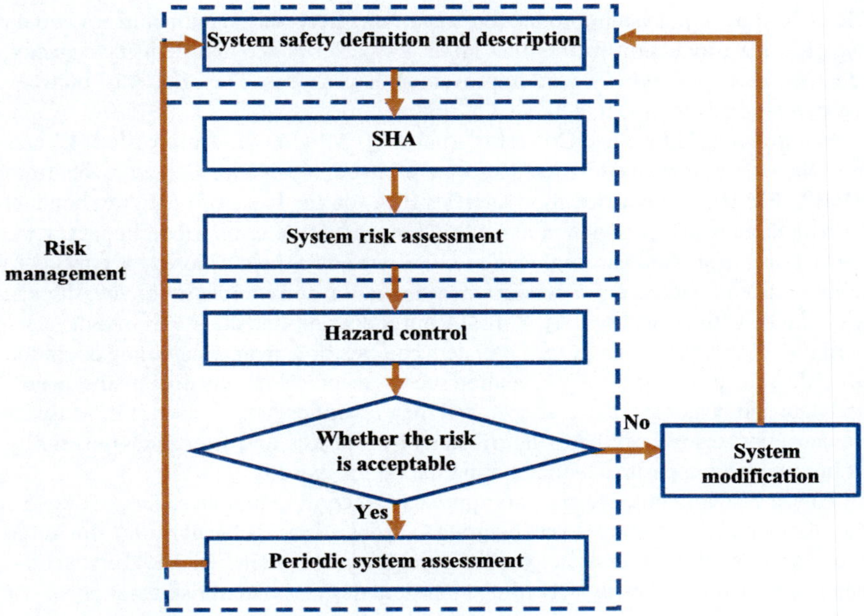

Fig. 2.13 Logic flow of systems safety analysis

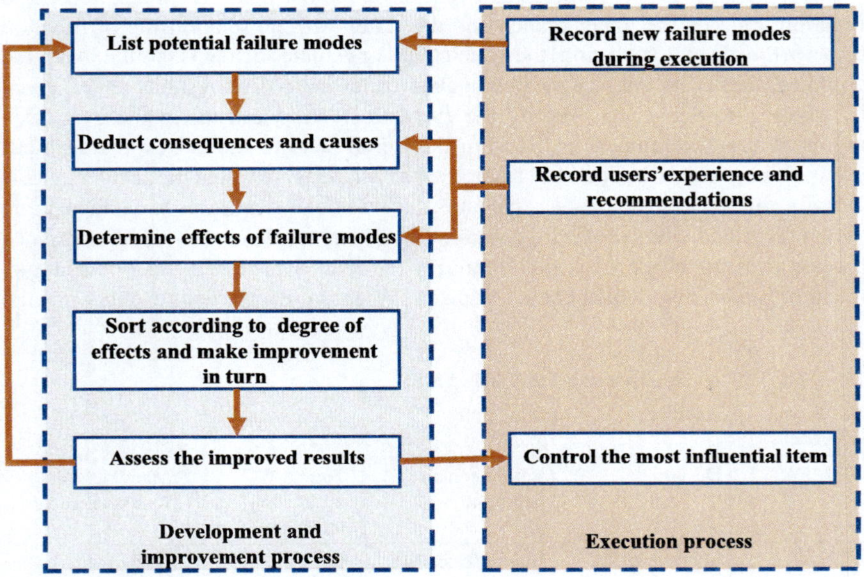

Fig. 2.14 Logic flow of FMEA

PFMEA applies in system production stage (assembly and manufacturing), which analyzes the production factors that affect IV performance and delivery capacity, identifies risks and puts forward improvement approaches. The difference between the two methods is shown in Table 2.4.

Failure Mode Effects and Criticality Analysis (FMECA) and Failure Mode Effects and Diagnostic Analysis (FMEDA) are quantitative analysis approaches derived from FMEA. FMECA adds Criticality Analysis (CA) on the basis of FMEA to conduct fault mode analysis for all potential faults of the product, determine their impact on the normal operation, find out single faults and determinate their criticality according to failure mode impact, and select relevant optimization and compensation measures in accordance with criticality [24]. FMEDA introduces on-line diagnosis into the standard FMEA analysis. It analyzes IV system composition, maintenance and operation according to comprehensive information such as technical quality documents, historical data and customer requirements, and then identifies hazard scenarios, analyzes the causes of hazards, estimates the criticality by statistics, and determines the priority of improvement measures by comparing the criticality [25].

ETA is also an inductive analysis approach. According to time sequence of event development, ETA uses tree event diagram to deduce the possible hazards from initial event, so as to built safety strategy. ETA logic flow is shown in Fig. 2.15. Firstly, ETA determines the initial event according to system design, system risk assessment, IV operation condition and traditional vehicle accident information. Starting from the initial event, the event tree is draw from left to right in line to exhibite safety event process, with branches representing event development path. The states related to function draw on the upper branch and the states with no relationship of function draw on the lower branch, until the system fault or vehicle accident is found. After the completion of event tree, the event branches irrelevant to system safety are removed by expert knowledge, to simplify the event tree. The branches in the event tree represent the development paths from the initial state to the corresponding fault event, among which the branches leading to the accident are called the accident paths. Usually, there is more than one accident path. Based on the analysis of accident path, which provide the tips of hazard control. To prevent accidents, the safety assurance paths is an essential approach which could be introduced into event tree. For example, the introduction of redundant power paths in IV can avoid hazardous events when the

Table 2.4 Difference between DFMEA and PFMEA

Analytical approach	Apartment	Start node	Completion node	Assumption
DFMEA	Design	Before concept definition	Formal production releasing	Manufacturing with no failure
PFMEA	Manufacturing	Before feasibility analysis	Before completion of production manufacturing	Design with no error

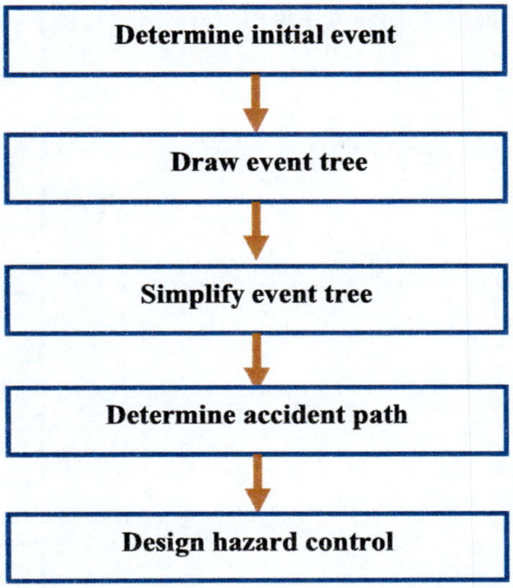

Fig. 2.15 Logic flow of ETA

main battery system entering the failure accident path. Furthermore, the occurrence of the accident can be calculated quantitatively based on the event tree, that is, the probability sum of all accident path. ETA can apply in IV fault diagnosis, weak sector analysis, traffic accident probability calculation, etc., so as to prevent accidents, increase the continuous safe operation time of the system and improve the design of IV.

For instance, the wrong identification on rollover vehicle by IV perception system is defined as a hazardous event. Starting from structural composition of the perception system, the perception process is divided into a function series from the camera to the algorithm. All possible event branches of the perception system are drawn, and the branches that do not affect the accident path such as wind and road width are omitted. After completion of the event tree, the engineers could search all potential paths that lead to the perception error, such as abnormal traffic participant appearance or irregular observation angle, and improve the sensors or supplement the training set for the potential cases, so as to control error identification hazard of perception system.

FTA is a deductive safety analysis approach, proposed by Watson and Mearns from AT&T Bell Laboratories in 1961. FTA describes accidents as discrete event that occur in a specific time sequence. The logic flow of FTA is shown in Fig. 2.16. Firstly, FTA determine the hazard event according to system definition and description, which is regarded as the root event of the failure tree. Then, it conducts system safety analysis on the causes and records of all root events, find the correlation and occurrence, and finally draws he corresponding fault tree. The fault tree is generally established with event symbols and logic gates structually from top to bottom, which display the system state (top), original component state (basic) and marking path intersections

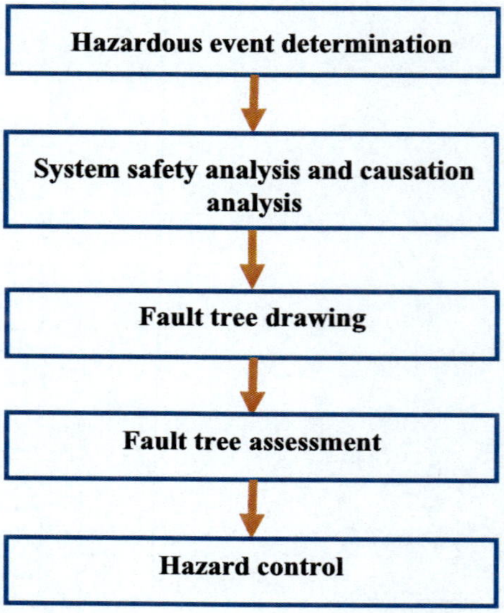

Fig. 2.16 Logic flow of FTA

with logic gate symbols. After completion of the fault tree, engineers analyze and evaluate the effect of the system-improvement, control the identified risks and reduce the exposure of hazardous events.

FTA is intuitive and logical method, which can be used for qualitative analysis, or quantitative analysis when the fault rate of each component is known.

As the mathematical description of the fault tree, FTA quantitative analysis can calculate the occurrence of top event by the occurrence of root event or the minimum cut set, and can also evaluate the importance of root event [26]. The occurrence of root event (i.e. hazardous event) can be divided into several formular in terms of different logic structures:

(1) The occurrence of root event with AND gate structure:

$$F_s(t) = F_1(t)F_2(t)\dots F_n(t) \tag{2.1}$$

(2) The occurrence of root event with OR gate structure:

$$F_s(t) = 1 - [1 - F_1(t)][1 - F_2(t)]\dots[1 - F_n(t)] \tag{2.2}$$

(3) The occurrence of root event with random gate structure:

$$F_s(t) = \sum_{j=1}^{r}\left[\prod_{i\in k_j} F_i(t)\right] \tag{2.3}$$

where r represents the minimum number of disjoint cut sets, $F_i(t)$ is the occurrence of the i-th branch failure in the j-th minimum cut set at time t.

In addition, in order to determine the effect of branch events on the occurrence of root events, the importance of branch events can be calculated. The importance is defined as the variation in the occurrence of root event when branch event changes from normal state to fault state. The calculation formula for $I_p(i)$, i.e. importance, is:

$$I_p(i) = \frac{\partial g(Q)}{\partial q_i} = g(1_u, Q) - g(0_u, Q) \tag{2.4}$$

where $g(1_u, Q)$ represents the occurrence of root event when the u-th branch event must occur, and $g(0_u, Q)$ indicates the occurrence of root event when the u-th branch event must not occur.

With the introduction of algorithm and model of IV, Model-Based Safety Analysis (MBSA) has been developed from traditional safety analysis. On the basis of modeling constituent elements of the system, MBSA further model system safety elements (such as fault mode and fault impact), abstracts physical model into a formal model composed of fault behaviors of the system with its components, and describes system properties in case of one or more faults in combination with system properties obtained from the mutual connection and logical relationship of various related items. Based on system model and properties, MBSA can automatically generate FTA and FMEA. MBSA promote the cooperation of different engineers in IV development, save resources and improve IV safety by applying automatic tools for safety analysis.

Based on the system modeling language SysML, Object Management Group (OMG) also launched SafeML for safety modeling, which built a system safety model including safety component model. In addition, relevant institutions have also introduced safety modeling or analysis approaches, such as safety analysis based on Architecture Analysis and Design Language (AADL) and Hierarchically Performed Hazard Origin and Propagation Studies (HiP-HOPS), and AltaRica, a formal language for system safety behavior description and analysis.

The logic flow of HAZOP is shown in Fig. 2.17. Firstly, HAZOP establishes description model for system development, and decomposes the system into basic logic units. In the same time, one or more units combine into a component or system. And then HAZOP determines design requirements and relevant parameters for each component or system, and allocates several guide words for each parameter. Finally, according to the combination of parameters and guide words, engineers checks and confirms the possible deviation of each system unit successively, analyzes causes and consequences of the deviation, formulates corresponding control strategies, and summarizes and fills them in the HAZOP analysis table in detail without omission.

Guide words refer to simple words used in HAZOP for qualitatively or quantitatively describing R & D indexes, which guide and identify dangerous situations in the process of system development. Deviation refers to the difference between the determined index and setting value after analysis of the parameters at each analysis

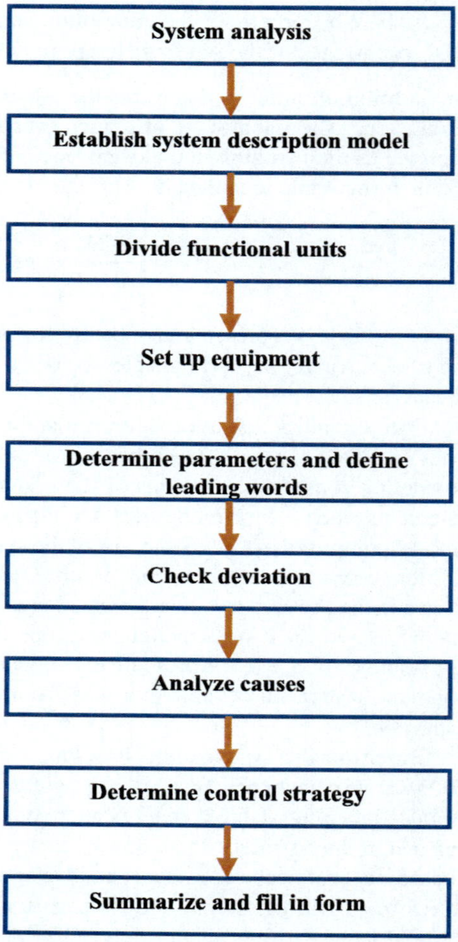

Fig. 2.17 Logic flow of HAZOP

node by guide words. Names and meanings of common guide words is listed in Table 2.5.

Taking IV automatic cruise control system as an example, if the distance monitoring module does not transmit data to the algorithm unit, the guide word is recorded as "no"; If the following distance exceeds the recommended value given by the algorithm unit, the guide word is recorded as "more", otherwise it is recorded as "less"; If the acceleration control signal appears when the vehicle is too close and should decelerate, the guide word is recorded as "reverse". Then, taking the autonomous pedestrian identification module as an example, if the camera is damaged and there is a lot of noise in the data collected, the guide word is recorded as "as well as"; If an unknown object cannot be recognized and only image warning appears when sound

Table 2.5 Common guide words and their meanings

Guide words	Deviation	Meaning	Explanation
No	No	Negation of design intention	Indexes/events required by design or operation do not occur at all
Less	Less	Reduction in numbers	The value is smaller than the standard value
More	More	Increase in numbers	The value is larger than the standard value
Part of	Part of	Only part of operation is completed	Complete only part of the established functions
As well as	As well as	Noise or extra operation	Complete the established functions with redundant events
Reverse	Reverse	Logical opposite to design intention	Occur things completely opposite to design requirements
Other than	Other than	Different from design intention	Occur things different from design requirements

and image synchronization warning is required, the guide word is recorded as "part of".

The steps of STPA are shown in Fig. 2.18. Firstly, STPA confirms corresponding high-level safety constraints for the known high-level hazards, and then establishes hierarchical control system structure to clarify the relationship between system related items and layers. Based on the control structure, it deduces the control behavior leading to system hazard, analyzes the behavior and refines safety constraints on the original system safety requirements. Finally, it analyzes the causal scenarios of hazard control behavior, and finds out all the possible reasons for hazard behavior in the control loop. The final requirements obtained from STPA is described in formal constraint model with formal language, and further verify the correctness of model design.

3. System Risk Assessment

System Risk refers to the combination of harm occurrence, controllability and severity at the system level. Risk in IV hazard events is usually related to the frequency of system hazard events (f), the capability to avoid harm through specific behaviors (C), and the severity of system hazard events (S).

Automotive Safety Integrity Level (ASIL) is an important tool for level quantification of IV system risk analysis, which describes the probability that system can achieve the specified safety objectives [13]. Exposure refers to the probability that the system is exposed to hazards under given environmental and operating conditions, and can be used as an index to measure the frequency of system hazard events; Controllability measures the capability of the participants involved in the event (including the driver, passengers or other people around the vehicle) to avoid specific injuries or losses by timely response; Severity is used to evaluate the degree of harm

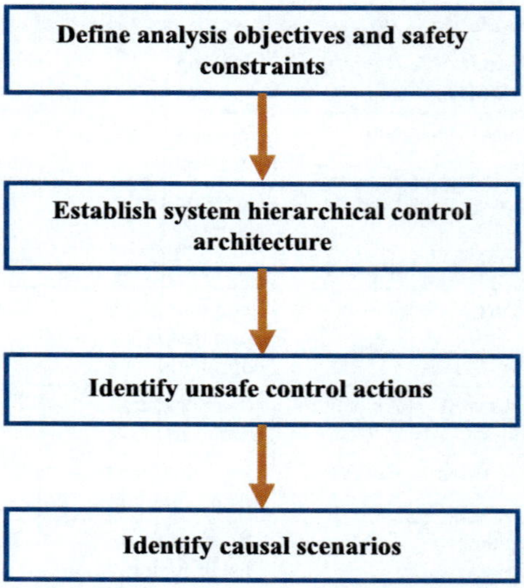

Fig. 2.18 Specific steps of STPA

that may affect one or more related targets in potential system hazard events. System Risk can quantitatively evaluate the above three indexes by certain approaches, and finally grade the risk. One of the quantitative evaluation and grading approaches is shown in Table 2.6.

Based on exposure, controllability and severity, the corresponding ASIL can be determined according to the actual design operation conditions and operation modes, and further transformated into safety objectives. ASIL is divided into four levels: A, B, C and D, where ASIL A is the lowest and ASIL D is the highest. In addition to these four levels, a QM level means it does not affect safety. In the process of risk analysis, similar safety objectives can be merged with the highest level among the merged items. Accordingly, the safety objectives can also be gradually decomposed

Table 2.6 Index classification of system risk assessment

Exposure probability	E0	E1	E2	E3	E4
	Hardly happen	Extremely low occurrence	Low occurrence	Medium occurrence	High occurrence
Controllability	C0	C1	C2	C3	
	Usually controllable	Simply controllable	Normally controllable	Difficult or uncontrollable	
Severity	S0	S1	S2	S3	
	No harm	Minor harm	Serious or life-threatening harm (surviving)	Life-threatening or fatal harm	

into specific requirements for IV operating environment. Relationship among S, E. C and ASIL is shown in Table 2.7.

Besides the above-mentioned ASIL for risk quantification, other approaches such as historical data, expert scoring, theoretical analysis, extrapolation, sensitivity analysis, Decision Tree Analysis and simulation can also be adopted to quantitatively evaluate and accurately describe the system risk.

4. Safety Requirement Setting

Systems concept setting and analysis is to obtain safety objectives of the whole IV through hazard analysis and risk assessment, determine the specific technical safety requirements and specifications of systems or software and hardware, further allocate the above safety requirements to the preliminary design architecture, and finally put forward necessary safety measures and mechanisms to meet relevant safety requirements and realize a complete safety concept.

(1) Contents of Safety Requirements

Safety requirements define safety behaviors or safety measures independent from specific implementation means, which includes concepts of safety objectives and safety technical specifications, determines safety-related attribute requirements, and finally allocates them to specific systems, software and hardware. Safety objectives are requirements at the highest level of hazard analysis and risk assessment results in the whole vehicle. Safety technical specifications are technical constraints based on specified safety requirements. For instance, the overall safety goal of IV is to achieve less accident rate than human drivers in the design of operating conditions, the safety requirement assigned to the perception system is to correctly identify all entities,

Table 2.7 ASIL level

Severity level	Exposure level	Controllability level		
		C1	C2	C3
S1	E1	QM	QM	QM
	E2	QM	QM	QM
	E3	QM	QM	A
	E4	QM	A	B
S2	E1	QM	QM	QM
	E2	QM	QM	A
	E3	QM	A	B
	E4	A	B	C
S3	E1	QM	QM	A
	E2	QM	A	B
	E3	A	B	C
	E4	B	C	D

and the safety requirement assigned to the decision-making system is able to handle various behavior trajectories to avoid collision. Accordingly, the system-level safety requirements can be further disassembled into safety technical specifications such as calculation speed of calculation unit, camera resolution, and number of training set cases. The purpose of setting safety requirements is to promote system related items into or maintain safe operation by considering definite hazardous events.

Based on specific safety requirements from the analysis, and combination with expected architecture, function, operation mode and system state, safety requirements are appropriately distributed, and the ASIL requirement is reasonably allocated to subsystems. For all elements with certain relevance, safety requirements shall be allocated according to safety importance and element relevance. During distribution, the ASIL and safety requirements need to be inherited in the transmission process, that is, the sub-element must reach the same or higher than safety requirements of upper-level. If multiple safety requirements are assigned to the same system element, the system element shall be subject to the ASIL with the highest safety requirements. IV are usually composed of multiple systems, so the safety requirements of each independent system and interrelated interface need to be obtained from the preset system safety requirements, which should be also allocated to corresponding systems and functions. For the fixed elements and designs, their ASIL requirements are no longer considered, and their safety requirements can be realized through redundancy or other approaches. For the safety requirements related to the external risk reduction, technical safety specifications should also be obtained from corresponding external risk reduction approaches and further allocated to corresponding components. In addition, it is necessary to clarify interface safety requirements related to external risk reduction and ensure the proper implementation of external risk reduction approaches.

Starting from the safety goal of the whole vehicle, safety requirements shall be decomposed and inherited whin the development process. ASIL, as a key attribute of safety goal, shall be inherited by each subsequent safety requirement. From the initial basic vehicle structure to the final software and hardware elements, functional purpose and technical safety requirements should be allocated to each element. In the process of ASIL decomposition, if an ASIL Z is decomposed into new ASIL X and ASIL Y, it should be marked as ASIL X (Z) and ASIL Y (Z). If ASIL X (Z) is further decomposed into new ASIL X1 and ASIL X2, the original mark should be retained and shown as ASIL X1 (Z) and ASIL X2 (Z). The specific decomposition strategy is shown in Fig. 2.19.

During decomposition of ASIL, each initial safety requirement shall be considered separately and each decomposed safety requirement shall also follow its own initial safety requirements. If ASIL decomposition is carried out at the software and hardware level, sufficient independence among software and hardware elements should be checked at the system level. If it is not fully independent, appropriate measures should be taken at the software and hardware level as well as system level to achieve full independence. If the decomposition of an ASIL with an initial safety requirement will result in redistribution of functions and relevant safety mechanisms, the relevant safety mechanisms should be allocated to a higher level in ASIL. In the process of decomposition and inheritance at each level, integration of elements and subsequent

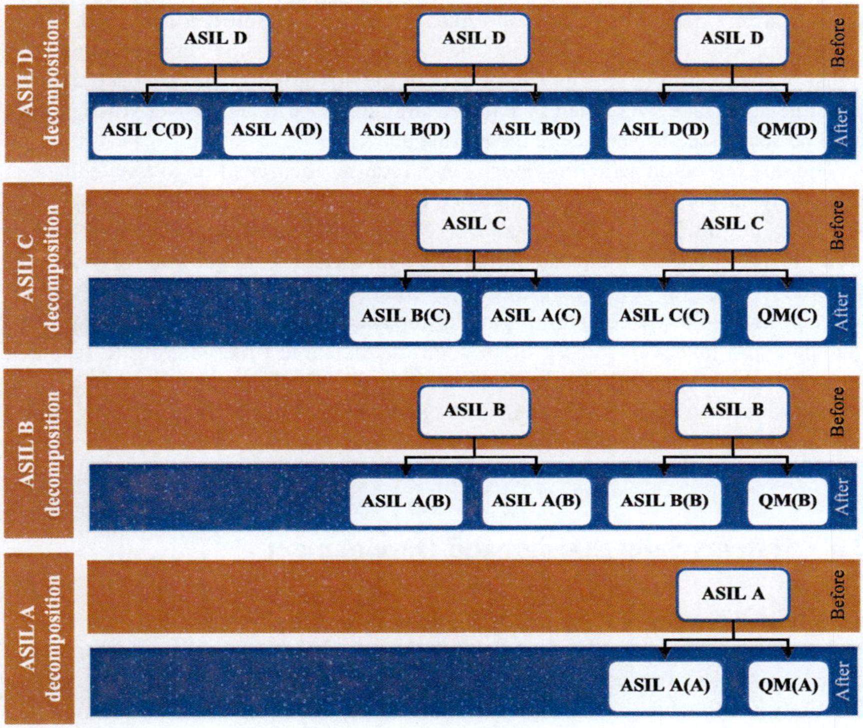

Fig. 2.19 ASIL decomposition strategy

safety activities should be carried out step by step based on ASIL requirements before decomposition.

(2) Safety Measures and Mechanisms

Compared with traditional vehicles, IV involves a wider development range and more complex software & hardware, whose safety requirements must be guaranteed by certain measures and mechanisms. By means of basic safety mechanisms and measures, the safety requirements can be better satisfied in the development and design of IV system.

As a common safety assurance mechanism, fault tolerant mechanism makes the system still maintain safe operation whin certain range when a fault occurs, so as to ensure that the fault will not directly result in events that violate safety goal (such as degradation). For example, power redundancy system is added, in order to make reliable protection in case of electrification faults.

Fault detection and failure mitigation mechanism requires that when the system has identified corresponding fault signals, certain measures should be taken to reduce or eliminate the risk caused by failure. For instance, when rainstorm or skid has been detected, IV shall slow down to prevent potential collision. Fault detection and driver warning device is a common typical measure for fault detection and failure mitigation, which requires that when the system recognizes the corresponding fault signal, a clear warning prompt can be sent to the driver before hazard occurs. If the driver needs to take over when IV enters a strange scenario, a takeover voice prompt shall be sent to the driver before the entrance.

The safe operation switching mechanism requires the transition to a state that can effectively ensure IV safe operation when it encounters possible system harm, such as emergency brake or remote takeover. When different system functions trigger multiple task requests at the same time, the appropriate sequence control process is selected through the logical arbitration mechanism to avoid potential risks such as task conflict and resource preemption. For instance, when music playing, map voice guidance and takeover voice prompt occur at the same time, leading to resource preemption of IV sound system, it is necessary to firstly conduct takeover voice prompt, then map voice guidance in order, and finally return to music playing, based on the priority of task and safety.

2.3 Systems Safety Design and Development

System-level product development shall determine and plan the safety activities in each sub-stage of system development, illustrate technical safety requirements, and list safety specifications. In the process of systems design, the first step is to establish IV safety architecture and assign technical safety requirements to corresponding software and hardware component and their interfaces, then refine specific items of technical safety requirements according to architecture complexity. After completion of software and hardware design, it is necessary to further integrate and test software and hardware, system and project afterwards to provide relevant records and evidence by verification and confirmation process, which could prove IV designed already achieves the vehicle-level safety and meets relevant safety requirements.

1. Safety Design and Development of System Products

The first goal of system-level safety design and development is to carry out system design and determines functions that meet technical safety requirements and specifications of the project. The second is to verify system design and functional requirements. The design and development of IV system architecture shall take into account the safety requirements and functional requirements. Therefore, both requirements should be dealt with during this stage.

(1) System Safety Development Plan

The system safety development plan shall consider the safety assigned to the system design elements, i.e., system design shall meet specific technical safety specifications. In order to meet the specifications, verifiability of system design, technical realizability of software and hardware, and execution & test capability in system integration process should be considered in system design. In addition, relevant scenarios should be preset as the representative of IV operation conditions, for the implementation of system test planning.

The system and subsystem architecture shall meet safety objectives and technical safety requirements under their ASIL. Besides, it is necessary to clearly define safety-related software and hardware interfaces, and standardize interfacing property, interaction mode, resource allocation, operation mechanism, configuration and constraints according to certain standards, so as to avoid the impact on safety of software and hardware operation and interaction due to interface matching problems.

(2) System Technical Requirements and Specifications

In addition to functional specifications and design criteria, system technical specifications shall also specify relevant mechanisms for systems or elements to achieve safety objectives according to IV architecture design and safety goal, including safety-related operation modes, failure conditions defined by the system and system elements combination leading to failure. The safety mechanism required by safety technical specifications is usually realized by relevant measures, such as system self-detection, indication and fault control measures (including the detection of random hardware faults by the system or components, and the self-monitoring measures of system faults), external fault detection interacting with the system, fault indication and fault control measures, measures to keep or maintain the system safe operation, refinement and implementation measures of warning and degradation concepts, and measures to prevent faults from being hidden.

Due to the complexity of IV systems, the time sequence of fault detection and resolution is particularly critical. The time interval related concepts of fault are the basis for solving, as follows: Fault Detection Time Interval is the time interval from fault occurring to being detected; Fault Reaction Time Interval refers to the time interval from detecting fault to entering safe operation or emergency operation; Fault Handling Time Interval is the sum of Fault Detection Time Interval and Fault Reaction Time Interval; Fault Tolerant Time Interval is the shortest tolerant time interval from internal faults occurrence of related items to possible hazardous events when safety mechanism is not activated; Diagnostic Test Time Interval is the time interval for online diagnostic of safety problem. The corresponding time sequence of fault time interval is shown in Fig. 2.20.

In the mechanism for realization or maintenance of IV safe-operation, the switching of safe operation state, fault tolerant time interval and measures to maintain safe operation should be specified. Within Fault Tolerant Time Interval, if system related items maintain safe operation, or in the process of transition to safe operation state and emergency operation mode, it indicates that safety mechanism handles fault in time. Generally speaking, Fault Tolerant Time Interval is defined at the system level, and the longest fault handling time interval and the required safe operation state after fault handling, which are defined at the component level to meet safety requirements. When Diagnostic Test Time Interval is much shorter than Fault Detection Time Interval, which means Fault Detection Time Interval can accommodate multiple Diagnostic Test Time Intervals and the edge of error disturbance can be further eliminated.

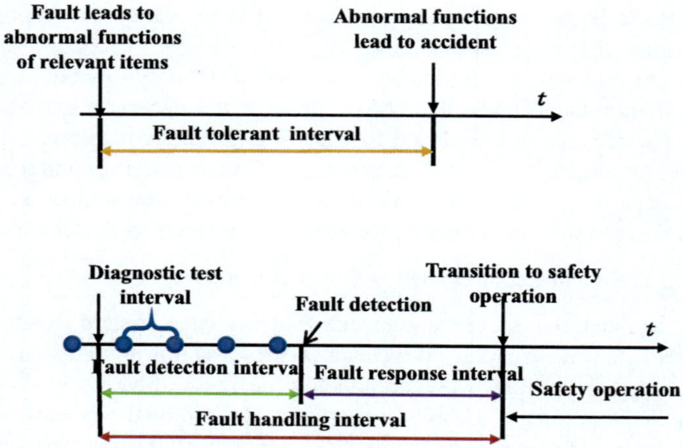

Fig. 2.20 Corresponding time sequence of fault time interval

If safe operation cannot be reached immediately, IV shall enter emergency operation mode. Emergency operation mode refers to functional operation during system transition from failure occuring to safe operation. A common emergency operation mode is the remote takeover of IV. Emergency Operation Time Interval refers to the time interval when emergency operation mode is lasted. Emergency Operation Tolerance Time Interval refers to the specific time interval that can maintain emergency operation mode without unreasonable risk, that is, the maximum time interval allowed to keep emergency operation mode. Within emergency operation tolerance time, IV can operate safely in emergency mode. The corresponding time sequence of emergency operation is shown in Fig. 2.21.

Within it, Warning and Degradation are two common safety operations based on the concept of emergency operation. When percepting unexpected situation (out of operation design condition), it will remind (warn) the driver in a certain way and conduct reasonable function reduction (degradation) to achieve safe operation. For example, when IV encounters a rainstorm leading to abnormal performance of the

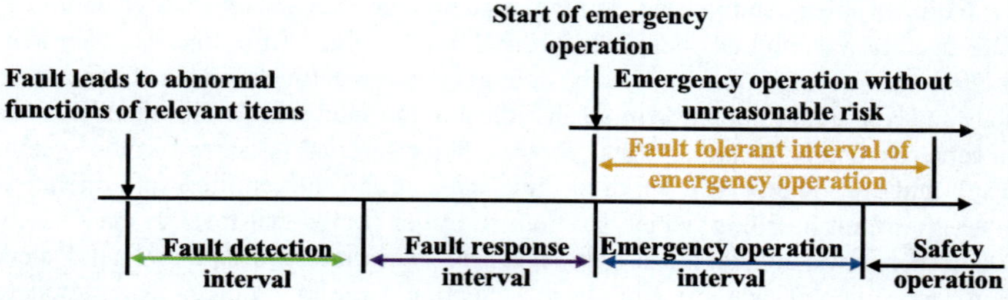

Fig. 2.21 Corresponding time sequence of emergency operation

camera, it will give a takeover voice reminder to the driver and automatically slow down to the side of the road until being taken over safely by the driver. Warning and Degradation strategy remind the driver with tactile, sound or visual prompts. The reminder includes the upcoming degradation, description of current safe operation state, operation for transition to safe operation, condition for recovery from safe operation state, emergency operation under applicable conditions, corresponding Tolerance Time Interval during emergency operation, etc.

2. Safety Design and Development of Hardware Products

The safety design and development of hardware products first need to arrange safety activities of hardware development and extract corresponding safety requirements, so as to implement relevant hardware design and development. After completion of hardware design, safety-related conditions need to be tested to verify technical requirements. At the hardware level, necessary product development activities and processes include hardware implementation of technical safety concept, potential hardware failure and impact analysis, and coordination with software development. The first goal for the hardware safety requirements specification extraction is to specify hardware safety requirements with reference to technical safety requirements and system safety specifications; The second is to verify the consistency of hardware safety requirements with technical safety requirements and system safety specifications; The third is to describe safety requirements for hardware interface in detail.

According to safety requirements specification of hardware design, the attributes of hardware and its related safety mechanisms are required to control internal faults of hardware units, match safety requirements of external units, and deal with potential faults. In addition, it is required to be able to detect and prompt faults of hardware-related units. For example, in IV energy system, the temperature of the battery can be detected by hardware, and the battery explosion in the case of too high temperature can be prevented by physical measures.

The first goal of hardware design is to depict the hardware according to system design specifications and hardware safety requirements, and the second is to verify the design. Hardware design includes hardware architecture design and hardware unit design. Hardware architecture design shall show all hardware units and their connections. Hardware unit design refers to the principle, structure, time sequence, electrical and other designs of related units. Hardware architecture design shall follow safety design principles including hierarchical design and redundancy reduction, etc.

In addition to the above six design principles, non-functional terms in hardware design process shall also consider the impacts of temperature, vibration, humidity, dust, electromagnetic interference and other hardware or environment factors, in view of potential faults of safety-related hardware components.

At the same time, single point fault, multi-point fault, safety fault or residual fault should be considered for hardware safety analysis. In most cases, the analysis can be limited to two-point fault. Recognization of two-point fault requires the identification of potential fault towards each hardware combination, which includes fault analysis of a single hardware and fault analysis after the combination.

3. Safety Design and Development of Software Products

The safety design and development of software products shall firstly arrange safety activities of software development, and extract corresponding safety requirements and specifications, so as to implement relevant design and development of software. After completion of the design, it is necessary to test safety-related conditions of software unit and integration respectively, and verify and confirm software technical requirements. For IV, the software part is not only an important embodiment of its intelligence, but also the core for its safety assurance. IV software includes a series of development projects and sub-stages at different levels such as network, algorithm and system. Based on development level and complexity of the project, it is required to extract corresponding specifications by certain means and make them comply with relevant regulations and their ASIL evaluation.

The first goal of software safety specification extraction is to specify software safety requirements with reference to technical safety requirements and system safety specifications; the second is to verify the consistency of software safety requirements with technical safety requirements and system safety specifications; and the third is to describe safety requirements of software interface in detail.

The first goal of software architecture design is to depict software architecture for software safety requirements; the second is to verify software architecture. The first goal of software unit design is to extract technical safety requirements of software unit according to software system design and relevant software safety requirements; the second is to realize the function of specified software unit; and the third is to verify design and implementation of software units in terms of safety. The realization of software unit includes generation of source code and compilation of object code.

(1) Safety Design Specification for Software Architecture

For the safety of design and implementation process, less complicated language should be applied in design and modeling language or programming language. The software design and implementation process shall actively use language subsets, cyber attack and defense technology, clear graphical representation and uniform naming, etc., and adopt established design principles and guidelines to gurantee safety.

At the same time, in order to avoid the faults caused by high complexity, the software architecture design shall be modular, encapsulated and simplified, which is also beneficial for rapidly iterative update of IV system or algorithm. System architecture design shall follow unit design of hierarchical structure, consider such factors as size limit of software unit and interface, coupling limit between units, suspended usage limit and high cohesion of software unit, and also adopt appropriate scheduling attributes. In particular, mechanisms should be built to avoid mutual interference and meet their ASIL and safety requirements, for resource sharing in software partitions.

According to International Standard ISO 26262-9:2018 of ADAS, IV system development process shall include safety analysis in software architecture, so as to identify or verify software safety-related parts and prove the validity of software safety mechanism. In order to obtain the results of software architecture safety analysis, the input and output data range detection, authenticity detection, data error detection, external detection, control flow monitoring and diversified approaches can be used for safety detection and analysis.

According to safety analysis in software architecture, software safety mechanisms should be specified for software architecture, such as static recovery mechanism, fault softening, independent parallel redundancy, data error correction code, etc.

(3) Safety Design Specification for Software Unit

The safety design of software unit shall consider all involving relevant elements, which is mainly divided into static design and dynamic design. The static design includes software structure, logical sequence of data processing, data type and its characteristics, internal and external interface, constraint condition, external dependent unit, etc. The dynamic design includes unit function, behavior, control flow, concurrency, interactive data flow, external interface, data transmission, time sequence constraint, etc.

Each safety-related software unit can be classified into new development unit, modified reuse unit or non-modified reuse unit. These three types of units shall meet corresponding safety standards during design. IV system or algorithm update is integrated by the above three types of units. Among them, the highest software safety requirements should be gradually decomposed and distributed to each software unit. Therefore, each software unit should be carefully designed, so as to meet the requirements of the highest assigned ASIL. For the update of the new version of IV system or algorithm, it is necessary to carefully confirm the changes before and after the update as well as the safety goal of the whole vehicle after the update. If necessary, the vehicle safety can be monitored periodically after the update.

The software unit design specification requires a detailed description of software functional behavior and internal design, so as to facilitate the verification of relevant development. At the same time, the source code of software unit should adhere to relevant specifications for time sequence, interface and information flow during the design and implementation, with such properties as readability, robustness and simplicity.

4. Systems Integration and Safety Testing

Systems integration includes three stages and two verification goals. The first stage is component integration of hardware and software; the second is integration of system elements into a complete vehicle; and the third is integration of system verification in the project. The first goal of integration verification is to verify that all related items comply with the assigned ASIL and corresponding safety requirements; and the second is to verify that safety measures proposed in system design can be properly implemented to meet safety requirements.

After completion of software and hardware development, systems integration shall be carried out. Starting from software and hardware units, system integration combines relevant systems and finally forms products (such as the whole vehicle system). Safety testing is likely to be conducted at all stages of system development, in order to prove that related items meet safety requirements and work properly. Safety testing shall focus on system safety requirement, including accuracy of functional characteristics, correctness of safety mechanism time sequence, interface consistency, robustness, and coverage of safety mechanism diagnosis and fault detection. For technical safety requirements of components, methods such as demand-oriented test, fault injection test or back-to-back test can be adopted; For performance, execution accuracy and time sequence correctness, methods such as back-to-back test and performance test can be adopted; For diagnosis coverage (DC) of safety mechanism, fault injection and Error Guessing test can be applied; For robustness, methods such as resource utilization and stress testing can be used; And for interface consistency and correctness, the interface internal test, external test and consistency check should be carried out respectively.

With reference to the expert experience of IV development, process specification for system integration and safety test is usually further divided into the integration and test of software, hardware and even components on the basis of integration and test plan of the project. Project integration and testing requirements shall be specific to the system level and vehicle level, ensuring that there are corresponding verification means for solving the existing open safety problems, and focusing on IV internal interface, subsystem (including internal and external related items) and operation conditions. In addition to the above testing approaches, long-term test, interactive test or communication test, real user test and field test shall also be applied for vehicle-level test.

5. System Safety Verification and Validation

The purpose of verification activities (such as safety verification, safety analysis, integration and validation of hardware, software and system) is to provide evidence that the result of each specific activity meets the specified safety requirements. In the vehicle integration project, the purpose of the validation is to provide appropriate evidence that safety measures fully meet safety requirements within the intended operating range.

System safety verification is required in different stages. In the concept definition stage, the verification shall ensure that the concept is correct, complete and consistent with boundary conditions of the project, and all safety concepts can be realized. In the design stage, system design evaluation is executed to ensure that system design maintains the correctness, integrity and consistency of the previously established safety requirements. In the testing stage, it is to evaluate products under specified test environment to ensure that they meet safety requirements. In production and operation stage, it is to ensure that corresponding public safety requirements can be realized during production, repair and maintenance, and these information should be reflected in the user manual. In addition, it is also necessary to meet project safety requirements through control measures in the production process.

The first goal of safety validation is to provide evidence that is satisfactory to safety objectives and applicable to safety concept proposed by the project from the perspective of safety; and the second is to prove that safety objectives are proper, complete and achievable at the vehicle level.

Similarly, system safety verification and validation activities are to provide evidence for system safety to be in line with the definition of IV "safety". At the same time, safety verification and validation prove system safety from different perspectives. Safety verification shall prove that the activities in each stage of system design meet relevant safety specifications and requirements, and gurantee system safety from the process. Safety validation shall prove that the developed system and its corresponding functions are safe, and can meet the requirements of users and relevant personnel.

The evaluation on safety objectives shall include controllability, effectiveness of safety measures for random faults and system faults, effectiveness of external measures, effectiveness of relevant safety techniques, etc. At the vehicle level, safety objectives and intended function shall be tested and confirmed. For each safety objective, the inspection procedure and test cases shall clarify the configuration, sector, driving, operation and application of IV, and formulate detailed pass or fail criteria. For safety validation, in addition to the specified inspection process, it can also be confirmed together by analysis, long-term test, user test, expert test, anonymous review, retrial and other means. At the same time, IV is dependent on scenarios, comprehensive tests in different dimensions such as simulation, hardware in the loop and field comparison can be carried out based on scenarios to obtain the recognition of the public and regulators.

2.3.1 Systems Certification and Release

1. Systems Certification

Systems certification is to demonstrate the functionality and safety in a certain way, and provide evidence to support the safety goal. The certification on systems or components can be carried out in two different ways of testing or analysis. Related approaches can be applied alone or in combination according to their properties. During the analysis, it depends on the applied analysis approach and assumptions, and infers safety according to test data. During the test, the system or component shall be placed in the preset environment and operating conditions for evaluation of its functional performance and compatibility. For IV, the two ways are usually conducted together. By setting a certain safety-oriented challenge scenario and test indexes, the evaluation data are collected and analyzed, so as to obtain the characterization of safety as certification evidence. When the unit that already has passed the test needs to be changed, it is necessary to analyze the corresponding impact brought by the change. For instance, regarding the update iteration of IV operation interface, it is necessary to analyze whether it exerts an impact on HMI of the driver and leads to relevant potential risks subsequently.

Even if different evaluation approaches are used, the final results should be displayed in the certification report (the certification report can be composed of a group of documents, including report, evidence, note, annotation, etc.), and provide system or component assumptions, restrained, opeation conditions, test cases and supportting evidences. It is better to develop a comprehensive way to conduct inspection, which usually includes performance data, certification process, results and basic principles.

Materials should be provided in accordance with relevant certification specifications, for determining whether the system or component can pass the certification. These materials usually cover safety conditions of the system or component, including requirement, behavior under failure, configuration, interface, function, component integration description, functional response under abnormal operation conditions, dependency between system or components and other units or components, emergency treatment measures under abnormal conditions, etc. In addition, the system or component shall also indicate the coverage of safety mechanism, offer the operation or conditions to return to normal operation state in case of failure of relevant components, as well as the behavior leading to violation of safety objectives and its results. For the IV, it is necessary to provide applicable intended behaviors and safety certificates, and describe corresponding national or international standards that should be met in development of IV-related systems or components.

In particular, the newly developed products with the same definition and operation conditions as the released products or with high universality can be applied to products related to the original products, which demonstrates that the products can be reused or replaced. However, unless it is proved that the new product meets the standard of the designated product used by the upper-level product, it cannot be considered as replaceable when the new product meets the original function. Units, components and systems can be regarded as products for demonstration of product reusability or substitutability. The units, components and systems should be demonstrated according to technical safety concept, algorithm, model, source code, object code, configuration or calibration data, etc.

After arrangement of analysis, certification and verification plan, the qualification test shall be carried out according to standardized test specifications. The development and test plan shall include system or component functional description, test scheme, requirements for assembly and connection, the operating conditions of the system or component, simulated operating and environmental description and measurement calibration of environmental parameters, pass/fail criteria, etc.

2. Systems Release

Systems release refers to the issue of system standard description after completion of specified project development, with safety requirements at the vehicle level. Systems release is a prerequisite for subsequent production and mass production. It is required that the verification and validation of software and hardware units, components, systems and products shall be conducted during the development, and the overall evaluation release document shall be signed by the government. And the system safety document shall include the name and signature of the person in charge, version of released project, configuration of released project, as well as relevant references, release date and other information.

Based on the setting and analysis of system concept, system safety design establishes the basic understanding and requirements of IV system safety. It designs, develops and integrates IV hardware, software and system at hierarchical levels, and further ensures IV safety by means of testing and verification. Finally, the systematic certification is carried out in accordance with relevant safety standards of IV products, and the project is completed in the system release stage. The system safety design process with V-architecture ensures the reliability of design and development by standardized and complete engineering experience, as well as the safety of the final product by implementation of safety principles and details in each stages.

2.4 Systems Safety Assurance Technology

Given the three types of systems hazard mechanisms, requirements for the three major systems safety assurance technologies (FuSa, SOTIF, and cybersecurity) are derived. At present, corresponding standards have been issued or being formulated by the industry, to realize the concretization and standardization of system safety design approaches. And safety assurance technology systems have also been established respectively.

2.4.1 System Safety Classification and Standard

According to different sources of system hazards, system safety can be divided into three categories: FuSa problems due to system faults, SOTIF problems due to insufficient functions, and information safety problems due to cyber attacks. The solutions to the three types of safety problems together constitute the current system safety research. In addition, there are relevant standards as industry norms. The definitions and standards for the three types of problems are illustrated as follows.

1. Functional Safety and Standards

Aimed at system fault hazards, Functional Safety (FuSa) [13] specifically refers to the avoidance of unreasonable risks caused by hazards in case of abnormal functional performance of Electrical and Electronic (E/E) system. The IV power battery management system, anti-lock braking system and body stability system may face FuSa challenges at the levels of system, hardware and software, which are manifested in braking failure, steering failure, airbag ejection failure, and other problems during operation, resulting in serious system hazards. With the increasing complexity of the system as well as the extensive use and rapid iteration of software and electromechanical equipment, the IV FuSa issues show a trend of complexity and difficulty.

FuSa standard specifies the realization and requirements of system safety design in terms of FuSa. The first version of the FuSa standard in the field of road vehicles—ISO 26262:2011 [27] was released by International Organization for Standardization (ISO) on November 11, 2011, as the specific application of IEC 61,508 for FuSa requirements of E/E System, mainly for mass-produced passenger vehicles with a maximum net mass of 3500 kg. And the second edition of FuSa standard-ISO 26262:2018 [13] was released in December 2018, extending the coverage from passenger vehicles to all road vehicles except mopeds.

ISO 26262:2018 includes ten normative parts and two guideline parts, as shown in Fig. 2.22, emphasizing the development and assurance of FuSa in the whole system life cycle. The ten normative parts consist of parts 1–9 and 12. In addition to defining and standardizing activities of the whole life cycle such as concept stage, system product development, hardware product development, software product development, and post-development production, operation, service, and scrapping, it also includes FuSa management, support process, criteria, ASIL, safety-oriented analysis, etc. The two guideline parts refer to the ISO 26262 guide in part 10 and the semiconductor application guide in part 11. The FuSa development of the whole product as well as that of hardware and software products all follow the three "V" models, as shown in the Figure.

2. SOTIF and Standards

Aimed at hazards of insufficient function, SOTIF [28] specifically refers to the avoidance of unreasonable risks caused by hazards in case of insufficient intended function or its realized function. The specific manifestations of insufficient function include: ① insufficient specifications for intended function at the vehicle level; ② insufficient specifications or performance limitations for the implementation of system E/E. With practical application and problem exposure of IV products in recent years, SOTIF becomes an extended system safety concept based on FuSa.

The evolution of hazard events related to SOTIF is shown in Fig. 2.23. In the scenario with trigger conditions, the lack of function will lead to hazard. If factor that may lead to harm exists in the scenario, it will further evolve into hazard events and result in harm in the case of insufficient controllability. Typical trigger conditions include bad weather, bad road condition, unexpected behavior of traffic participants, etc. And drivers' and passengers' reasonably foreseeable misoperation over the intended function or the whole IV system is a potential trigger as well.

The already-known IV operation scenarios can be summarized and deduced concerning driving experience, driving data, and expert analysis. However, due to the openness of traffic, there will always be some unknown scenarios. According to conditions whether the scenario is known and dangerous or not, it can be divided into four categories: known safety scenario, known dangerous scenario, unknown safety scenario, and unknown dangerous scenario, as shown in Fig. 2.24. The SOTIF objectives can be illustrated through the transformation between scenarios. For area 1 (known safety scenario), it is required to maximize or maintain the known scenario set, and realize the transformation from other areas to this area at the same time; For area 2 (known dangerous scenario), it is to evaluate its potential risks, and take

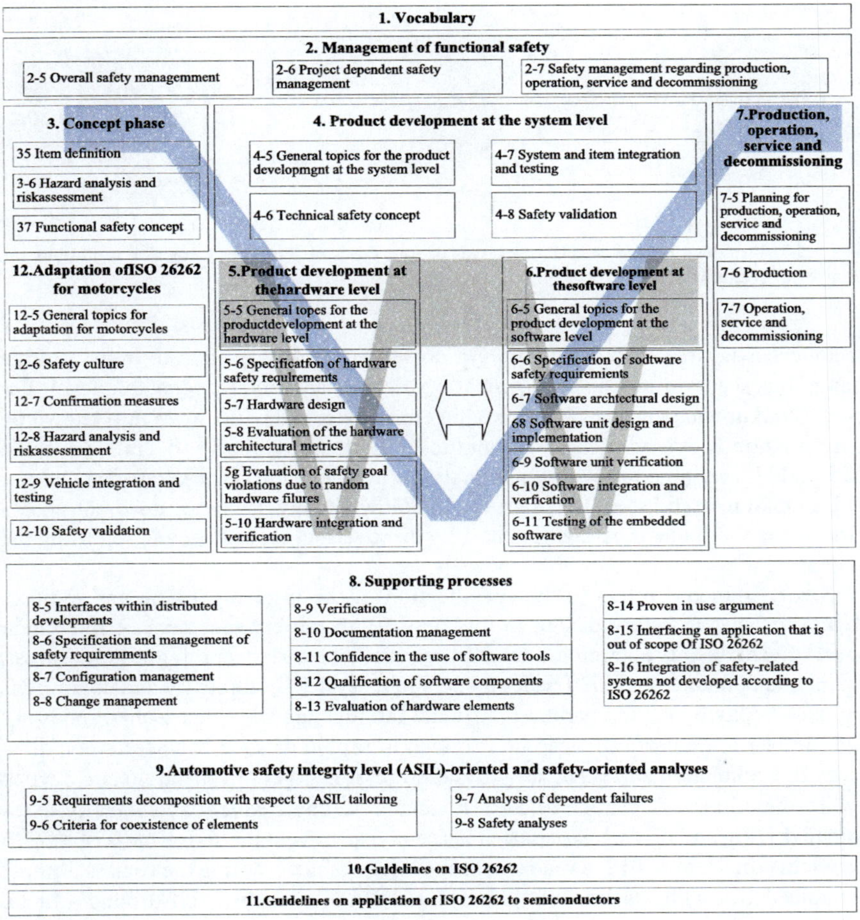

Fig. 2.22 FuSa standard architecture [13]

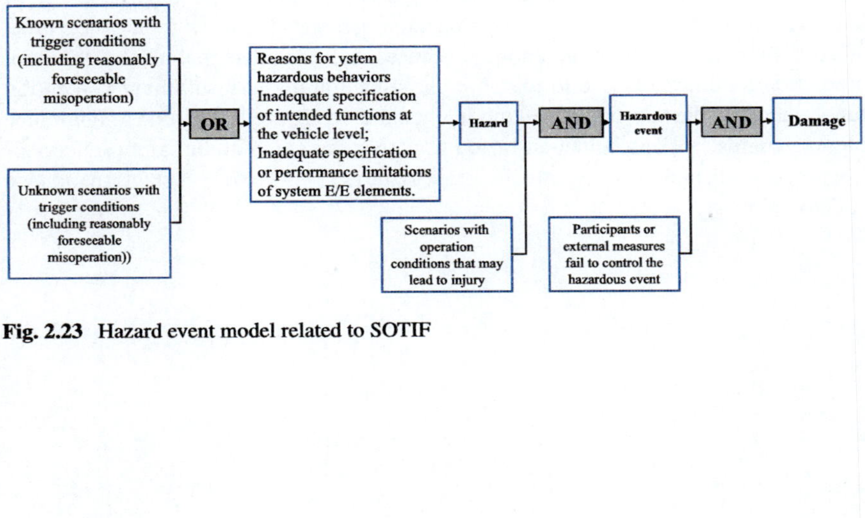

Fig. 2.23 Hazard event model related to SOTIF

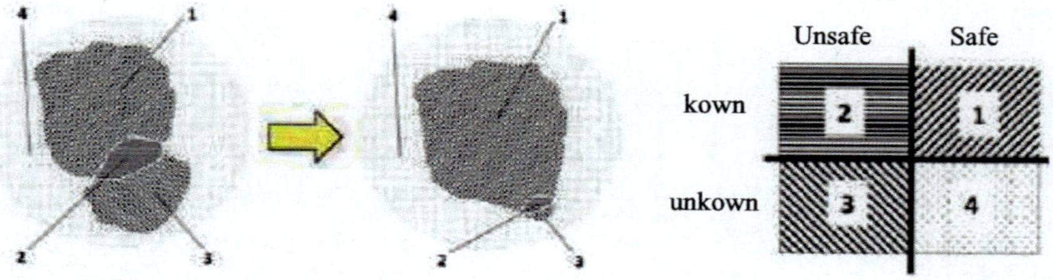

Fig. 2.24 Visualization and SOTIF objectives of four types of scenarios (area size represents the number of scenarios)

technical measures such as function improvement and restriction to transform the dangerous scenario into a safe scene, to reduce the area to an acceptable level; For area 3 (unknown dangerous scenario), it is to transform the unknown into known by confirmation activities such as random test, and then reduce the overall risk to an acceptable level together with functional improvement and other measures; Despite being unknown, the scenario in area 4 (unknown safety scenario) does not pose a risk. It can facilitate maximizing area 1 by discovering, recording, and reporting the scenario in this area.

SOTIF standard refers to the specification of the implementation and requirements for system safety design in this field. With the enhancement of intelligent level of vehicles, its operation scenario presents the trend toward high-dimensional, open, and dynamic random development, and its system complexity and coupling is increasing day by day. In addition, artificial intelligence and other technologies not only improve the performance but also lead to problems such as uncertainty, difficulty in explanation and verification. Therefore, it is urgent to speed up and deepen the formulation and practice of SOTIF standards. The frequent accidents of automated driving and driving assistance in recent years reveal the seriousness of SOTIF problems. In March 2018, a road test drone hit a woman crossing the road in Tempe, Arizona, USA, which is the world's first pedestrian accident by a self-driving vehicle. The vehicle has two prediction modules. The main prediction module works based on the historical trajectory of the object and other information. When the historical trajectory cannot be obtained long enough, the sub-prediction module is enabled for path prediction, which presets the possible moving targets in line with the object type. According to the investigation report, the causes of the accident include that the main prediction module fails due to unstable classification under insufficient perception function, while the sub-prediction module makes wrong prediction due to ignorance of pedestrian's crossing behavior. In addition, the distraction of the safety supervisor (regarded as misoperation) eventually leads to the delay in vehicle obstacle avoidance and collision.

Since February 2016, the SOTIF working group (ISO/TC22/SC32/WG8) under ISO has started the research and formulation of the SOTIF standard-ISO 21448. The participating members include experts from China, France, Germany, and other countries. It has formed versions of ISO/PAS 21,448 [29], ISO/CD 21448 [30], ISO/DIS 21448 [28] and solicited suggestions from different countries. The final standard is planned to be released in March 2022. The existing draft ISO 21448 standard identifies nine activities in SOTIF development: specification and design, hazard identification and assessment, potential functional deficiency, and trigger condition identification and assessment, function improvement, verification and validation strategy definition, known hazard scenario assessment, unknown hazard scenario assessment, SOTIF release criteria and operation phase activities. The logical relationship of SOTIF development activities is shown in Fig. 2.25. The number in the circle is the number of the corresponding chapter in the standard.

3. Cybersecurity and Standards

Aimed at the threat of cyber attack, cybersecurity [15, 31] specifically refers to the protection of the vehicle and its functions from unacceptable risks due to cyber attack threats on its E/E components. The protected objects of cybersecurity are mainly divided into three parts: in-vehicle system (such as software system, E/E hardware, in-vehicle data, and communication), out-of-vehicle communication (such as out-of-vehicle long-distance communication, and out-of-vehicle short-range communication) and out-of-vehicle system (such as shared sensing system and back-end server). IOV and intelligent trend lead to the requirement for the system to deal with more

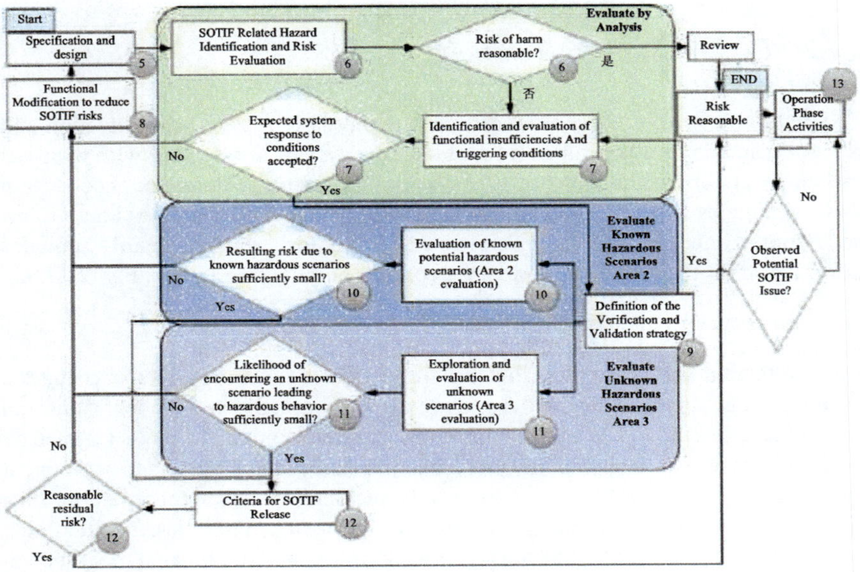

Fig. 2.25 Logical relationship of SOTIF development activities [28]

diversified and targeted cyber attack threats, and the safety consequences after being attacked become more serious. Therefore, higher requirements are put forward for cybersecurity.

Cybersecurity Standard specifies the implementation and requirements of system safety design in its related field. On January 14, 2016, Society of Automotive Engineers (SAE International) released SAE j3061 (*Cybersecurity Guidebook for Cyber-Physical Vehicle Systems)* [15], which refers to the "V" development model and integrates cybersecurity into the definition, production, operation, service and scrapping stage. On February 12, 2020, SAE International and ISO jointly issued ISO/SAE DIS 21434 (*Road Vehicles-Cybersecurity Engineering*) [31], based on and aiming to replace the original SAE J3061.

The standard provides key terms, objectives, requirements, guidelines, etc. (shown in Fig. 2.26), which can guide cybersecurity policies and processes, manage cybersecurity risks and cultivate cybersecurity culture. For the complete life cycle of vehicles, the cybersecurity system development is described in terms of the stages of definition, product development, cybersecurity verification, production, operation, and maintenance, as well as equipment scrapping. In addition, to ensure effective activities in each stage, the standard also stipulates the contents of cybersecurity management, continuous cybersecurity activities, and distributed cybersecurity activities.

Standards in the three system safety fields refine and standardize the systems safety design process and requirements in terms of different problems. In the actual product development, it is also necessary to improve FuSa, SOTIF, and Cybersecurity respectively with specific safety techniques.

2.4.2 FuSa Assurance Technology

The main objective of FuSa assurance technology is to realize fault detection and isolation, and control the impact of faults on the system, for dealing with the risks caused by abnormal functions of E/E system. The mature FuSa assurance system usually includes two parts, i.e., fault detection and diagnosis module and fault tolerance control module. The fault detection and diagnosis module mainly provides information for the active fault tolerance control module, as shown in Fig. 2.27.

1. Fault Detection and Diagnosis

Fault Detection and Diagnosis (FDD) mainly realizes functions of supervising the overall system safety, dealing with various system abnormalities, and identifying and separating system faults. From the perspective of fault location, the faults targeted by FuSa assurance technology mainly include sensor fault, actuator fault and component fault [32]. A complete FDD system usually includes system representation, data acquisition and preprocessing, signal processing method, fault classification, etc., as shown in Fig. 2.28 [33]. Specifically, the approaches based on analytical model, signal processing and intelligent FDD are three common types of FDD technologies.

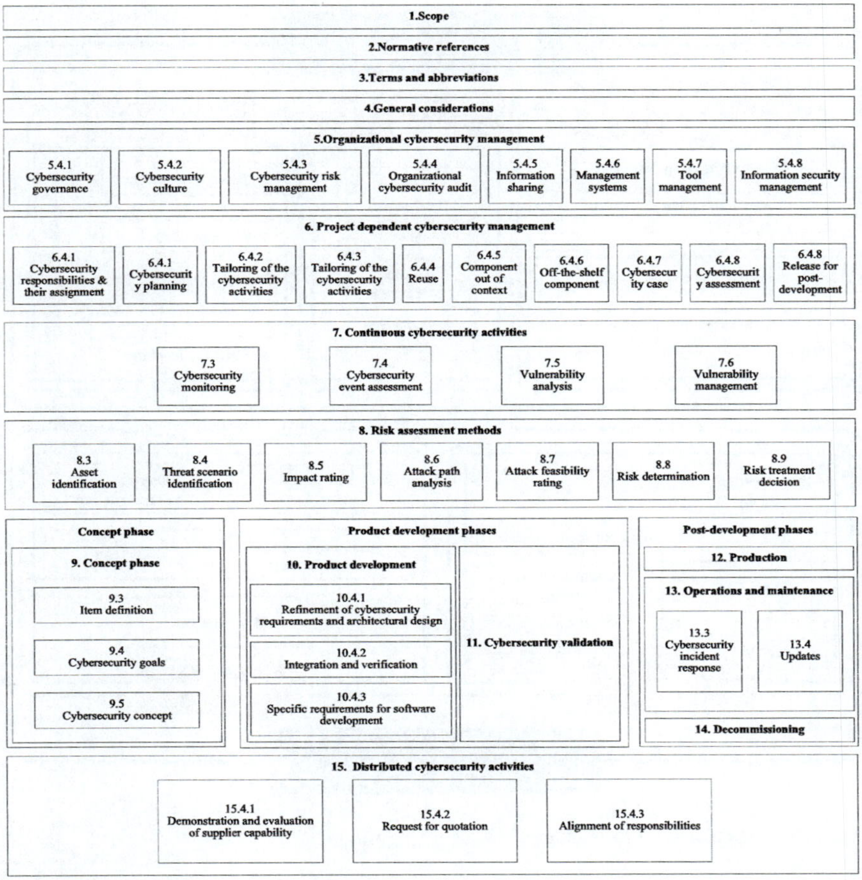

Fig. 2.26 Cybersecurity standard architecture [31]

(1) Approach Based on Analytical Model

The analytical model approach is usually on the basis of modern cybernetics and modern optimization method. Such means as mathematical modeling of the system, estimation and identification of parameter model and Kalman filter are used to generate residual sequence, which is further evaluated with certain threshold or criteria for fault diagnosis [34]. Here, analytical model refers to that initial conditions, model parameters and input–output relationship in the model are established by mathematical formulas. It is required to accurately establish mathematical model for the diagnosed object, which is mainly applicable for small-scale systems with less input and output and clear mathematical model. Generally speaking, it can be subdivided into state estimation diagnosis and parameter estimation diagnosis, etc. The premise of using state estimation diagnosis is that the system can be observed or

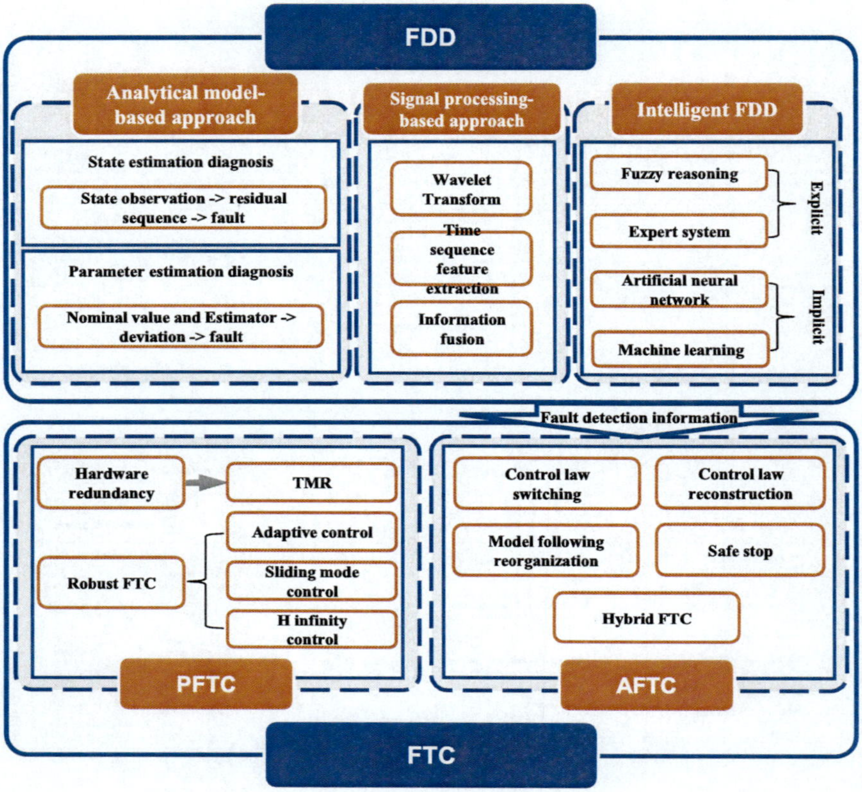

Fig. 2.27 FuSa assurance technology

partially observed, and various state observers or filters can be used for state estimation. Firstly, by comparing the reconstructed state variables of the controlled system with the measurable variables of the actual system, the residual sequence of state variables can be obtained. Secondly, by combining a suitable model construction and statistical test, the residual sequence is evaluated for fault detection.

Different from state estimation diagnosis, parameter estimation diagnosis mainly detects faults according to statistical characteristics of parameter changes. When the system fault causes significant changes in specific parameters, the parameter estimation diagnosis can be applied to detect fault information, i.e., to detect fault in accordance with the deviation between nominal value of the parameter and the estimator.

Any fault from the sensor will affect the actual control system. Usually, the diagnosis based on state estimation is used for sensor fault detection and diagnosis. Taking the semi-active suspension system as an example, the fault of the sensor will affect the ride comfort of IV, so it is very necessary to detect and diagnose the fault. Firstly,

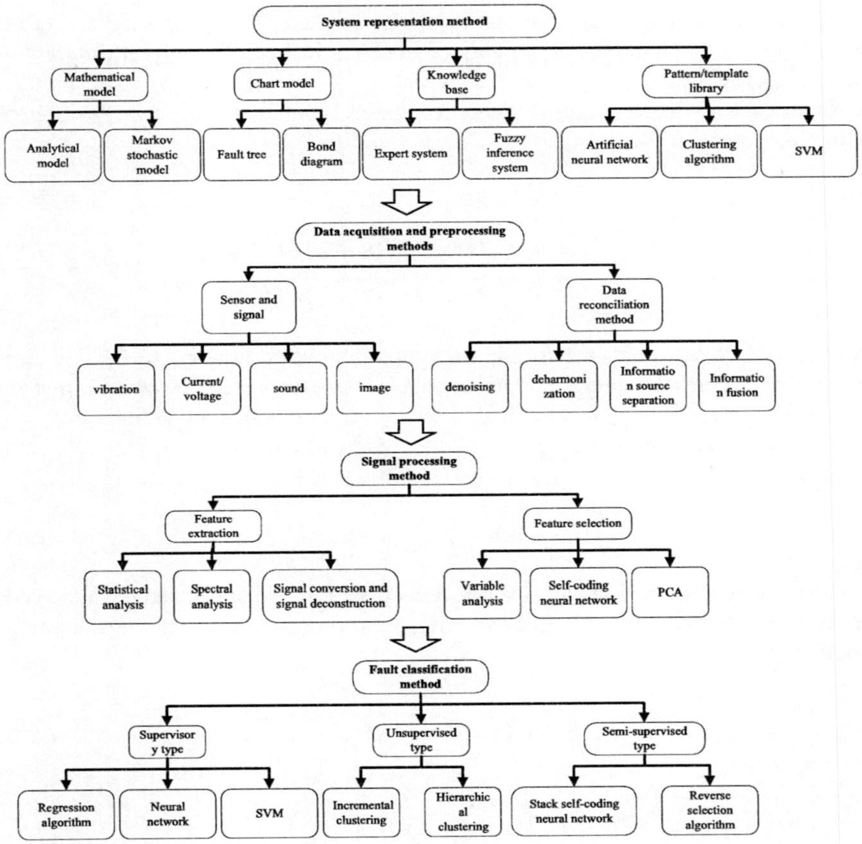

Fig. 2.28 FDD system

sensor faults are classified, including sensor jamming, sensor offset and sensor noise faults, as well as the combined faults of the two. Secondly, according to the properties of target system, appropriate state estimation is selected to estimate system state and compare it with sensor data, so as to realize fault detection. According to the nonlinear property of semi-active suspension system, the traceless Kalman filter is selected for state estimation, in which 7-DOF suspension model can be expressed as:

$$\begin{cases} X(k+1) = f[X(k), u(k)] + W(k) \\ Z(k) = h[X(k)] + V(k) \end{cases} \tag{2.5}$$

where, $u(k)$ is determinant control term; f is nonlinear state function; h is nonlinear observation function.

Assuming that the covariance matrix of *V(k)* is *R* and the covariance matrix of *W(k)* is *Q*, the implementation process of traceless Kalman filtering at different *k* times is as follows:

Firstly, the nonlinear mapping point set at time *k* is obtained, that is, 2n + 1 Sigma sampling points.

$$X^i(k|k) = \begin{bmatrix} \widehat{X}(k|k) \\ \widehat{X}(k|k) + \sqrt[\square]{(n+\lambda)P(k|k)} \\ \widehat{X}(k|k) - \sqrt[\square]{(n+\lambda)P(k|k)} \end{bmatrix} \tag{2.6}$$

where, $X^i(k|k)$ represents the *i*th Sigma sampling point at time *k*, $i = 0,1,2,\ldots,2n$; $X^i(k|k)$ represents the state prediction at time *k*; $P(k|k)$ represents covariance matrix at time *k*; $\lambda = \theta^2 (n + \kappa)$-*n* is the scaling factor used to reduce prediction error;

Calculate the one-step prediction of $2n + 1$ Sigma sampling point set, $i = 0, 1, 2, \ldots, 2n$.

$$X^i(k+1|k) = f[X^i(k|k), u(k|k)] \tag{2.7}$$

The one-step prediction and covariance matrix of the system state are calculated from the predicted value of Sigma sampling point set combined with the weighting coefficient.

$$\widehat{X}(k+1|k) = \sum_{i=0}^{2n} \omega^i X^i(k+1|k) \tag{2.8}$$

$$P(k+1|k) = \sum_{i=0}^{2n} \omega^i \left[X^i(k+1|k) - \widehat{X}(k+1|k)\right] \cdot \left[X^i(k+1|k)\right] \tag{2.9}$$

Combined with the one-step prediction results, a new Sigma sampling point set is generated by traceless transformation.

$$X^i(k+1|k) = \begin{bmatrix} \widehat{X}(k+1|k) \\ \widehat{X}(k+1|k) + \sqrt[\square]{(n+\lambda)P(k+1|k)} \\ \widehat{X}(k+1|k) - \sqrt[\square]{(n+\lambda)P(k+1|k)} \end{bmatrix}^{\mathrm{T}} \tag{2.10}$$

The new Sigma sampling point set is substituted into the observation equation to produce new observation results.

$$Z^i(k+1|k) = h\left[X^i(k+1|k)\right] \tag{2.11}$$

The new observation results are weighted to obtain the prediction mean and corresponding covariance of the system.

$$\widehat{Z}(k+1|k) = \sum_{i=0}^{2n} \omega^i Z^i(k+1|k) \tag{2.12}$$

$$P_{Z_k Z_k} = \sum_{i=0}^{2n} \omega^i \left[Z^i(k+1|k) - \widehat{Z}(k+1|k) \right] \cdot \left[Z^i(k+1|k) - \widehat{Z}(k+1|k) \right]^{\mathrm{T}} + R \tag{2.13}$$

$$P_{Z_k X_k} = \sum_{i=0}^{2n} \omega^i \left[X^i(k+1|k) - \widehat{Z}(k+1|k) \right] \cdot \left[Z^i(k+1|k) - \widehat{Z}(k+1|k) \right]^{\mathrm{T}} \tag{2.14}$$

Calculate the Kalman filter gain matrix:

$$K(k+1) = P_{Z_k X_k} P_{Z_k Z_k}^{-1} \tag{2.15}$$

Status update and covariance update:

$$\widehat{X}(k+1|k+1) = \widehat{X}(k+1|k) + K(k+1)\left[Z(k+1) - \widehat{Z}(k+1|k) \right] \tag{2.16}$$

$$P(k+1|k+1) = P(k+1|k) - K(k+1) P_{Z_K Z_K} K^T(k+1) \tag{2.17}$$

Together with 7-DOF suspension model of the vehicle, the above traceless Kalman filter is designed to obtain the residual sequence formed by the difference between the output value of the filter and the data collected by the sensor, which are then compared with the preset threshold for fault determination [35].

However, the approach based on analytical model is easy to be affected by system interference and uncertain factors that are not established in the reference model, resulting in poor fault detection effect. And there are some systems that can not establish analytical model for the tested object in practice.

(2) Approach Based on Signal Processing

The signal processing-based approach directly applies information processing and feature extraction means such as spectrum analysis, adaptive signal processing and time series feature extraction to diagnose the fault, which avoids the difficulty of establishing mathematical model of the tested object. Wavelet Transform, time series feature extraction and information fusion are the main approaches based on signal processing.

As a time–frequency domain analysis for observation signals, Wavelet Transform has the characteristics of multi-resolution. Utilizing the singularity of observation signal and the change of frequency structure of observation signal are the two main ways of fault diagnosis based on wavelet transform.

The time sequence feature extraction constructs the time series process model by selecting state variables directly related to the fault, and then uses model parameters as the feature vector to evaluate the type of fault.

Information fusion mainly solves the problems of low signal-to-noise ratio and low diagnostic reliability of single information source in fault diagnosis. It carries out fault diagnosis by fusing multi-source information.

As one of the main actuators of IV, permanent magnet synchronous motor directly affects IV operation safety in terms of safety and reliability. Due to its poor heat dissipation conditions, small working space and mechanical factors, various faults will inevitably occur. Among them, one of the common faults is inter turn short circuit, which is highly destructive and has great joint effect. In order to avoid the functional failure of permanent magnet synchronous motor, it is very important to detect the initial fault of inter turn short circuit. Usually, the diagnosis based on Discrete Wavelet Transform (DWT) is adopted to process the signal of stator current or zero sequence voltage, eliminate the interference of noise and other harmonics, and accurately extract fundamental component with small amplitude for effective detection of the initial fault. The specific process is shown in Fig. 2.29 [36].

(3) Intelligent Fault Detection and Diagnosis

Intelligent Fault Detection and Diagnosis (IFDD) is a new fault diagnosis approach based on computer technology and artificial intelligence technology. Its advantage is that it can obtain fault knowledge base from a large amount of data of the tested

Fig. 2.29 Fault diagnosis process based on discrete wavelet transform

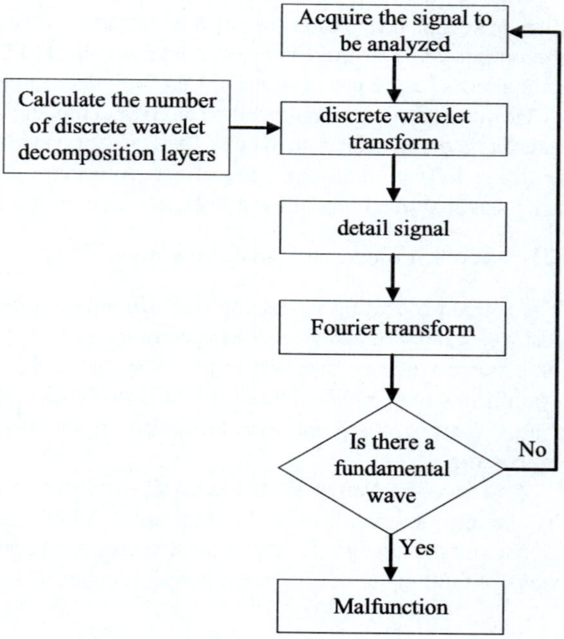

object, and then cluster and detect faults based on knowledge [37]. Two common forms of knowledge are explicit knowledge and implicit knowledge. Among them, IFDD based on explicit knowledge include fuzzy reasoning and expert system, etc. And IFDD based on implicit knowledge include artificial neural network method and machine learning, etc.

The diagnosis based on expert system is to establish corresponding storage mode and operation rules in the computer according to a large number of expert experience and formed knowledge laws in the field, and simulate the reasoning and decision-making process of experts, so as to detect and diagnose faults. Figure 2.30 is the composition diagram of expert system, in which the human–machine interface is responsible for the function of inputting expert rules and displaying corresponding status and diagnosis suggestions to users. And the inference engine is divided into interpretation module and rule reasoning module. The interpretation module is responsible for outputting results of fault diagnosis and maintenance opinions. The rule reasoning module is to realize fault diagnosis in the way of forward reasoning or reverse reasoning by combining the rules in the knowledge base and the currently obtained data and information. The knowledge base mainly stores relevant expert rules, including knowledge acquisition module and knowledge base management module. And the database contains records of various performance parameters and working states obtained by signal detection and data acquisition module [38].

Artificial neural network has the functions of dealing with complex nonlinear mapping problems, association, reasoning and memory, etc. Neural network approach is quite suitable for fault detection and diagnosis of tested system with modeling difficulty. With capability of self-organization and self-learning, it can overcome limitations of traditional expert system reasoning rules. There are two main diagnosis modes based on artificial neural network, namely off-line diagnosis and on-line diagnosis. Figure 2.31 shows the neural network model commonly used in fault diagnosis [34].

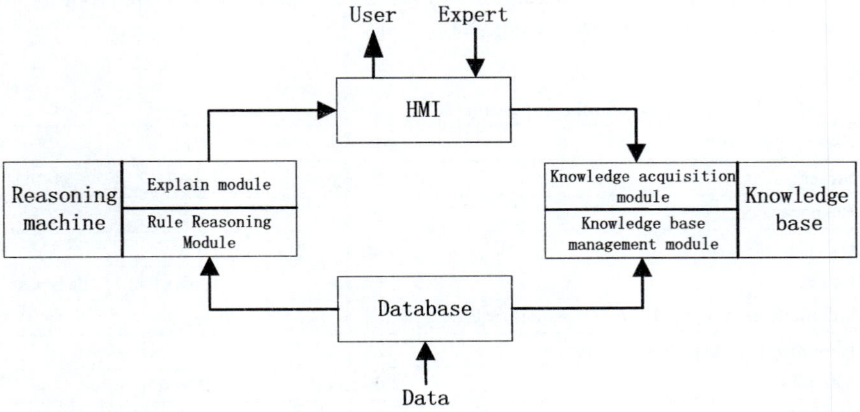

Fig. 2.30 Composition of expert system

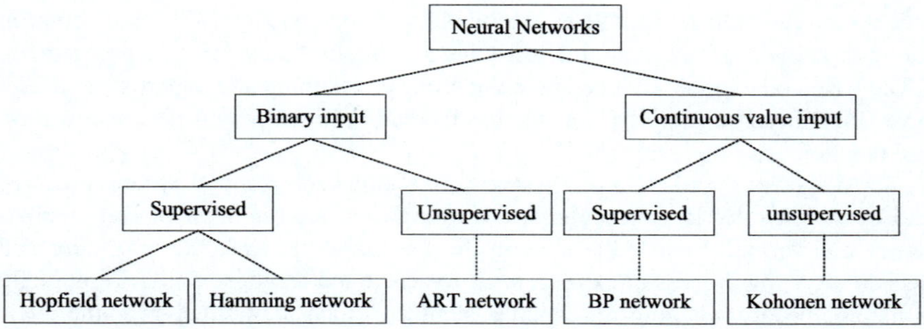

Fig. 2.31 Neural network model commonly used in fault diagnosis

2.5 Fault Tolerant Control

According to different principles, Fault Tolerant Control is divided into Active Fault Tolerant Control (AFTC) and Passive Fault Tolerant Control (PFTC). The main differences between them are shown in Table 2.8.

(1) PFTC

Whether the system has faults or not, PFTC can maintain system stability without changing the structure of the controller. Because there is no need to detect and separate faults and change the controller structure, PFTC does not need fault detection and diagnosis module, and its response is fast. At present, the main PFTC include hardware redundancy and robust FTC, etc.

Hardware redundancy FTC is to connect multiple components with the same function in parallel to form functional redundancy. At present, Triple Modular Redundancy (TMR) is widely applied. The main idea is that three parallel devices work at the same time, and then the majority of results are taken as the system output by voting mechanism, as shown in Fig. 2.32.

Table 2.8 Fault tolerant control comparison [20]

System properties	AFTC	PFTC
Structure	Complex	Simple
Response speed	Slow	Fast
Calculation quantity	Large	Less
Fault type	Various	Preset fixed fault types
Is fault detection and diagonisis model required?	Yes	No
Is controller reconfiguration required	Yes	No
Noise Impact	Disturbed	Robust to noise
Time Delay	Possible existing	No time delay

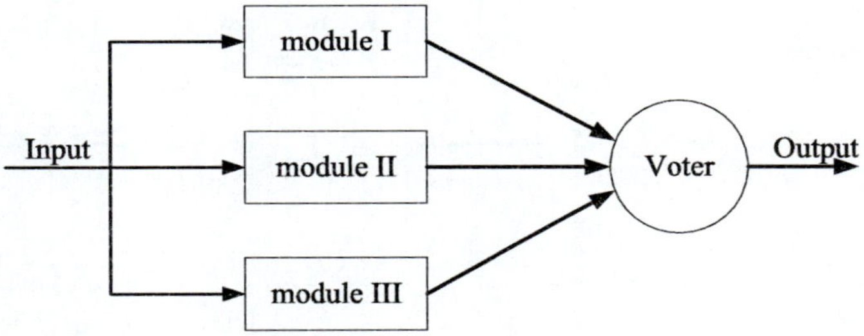

Fig. 2.32 TMR work principle

The robust FTC mainly takes the system uncertainty and possible disturbances and faults into account in the design stage, and then formulates a certain performance index and designs corresponding robust controller, so that the system can maintain a certain performance when there is a fault. Robust FTC can be roughly divided into three types: reliable stabilization, integrity and simultaneous stabilization, mainly including such approaches as adaptive control, sliding mode control and H ∞ control.

As an important part of the chassis by wire, the steering by wire system is vital for IV driving safety. At present, the steering by wire system usually has a set of mechanical steering system as a kind of hardware redundancy, in order to prevent the occurrence of system fault and failure. In addition, robust FTC, such as adaptive control and H ∞ control, is also applied to FTC of steering by wire system [39]. However, due to off-line design, PFTC is difficult to deal with system faults that are not considered. For a wide variety of faults with uncertain time sequence, the functional limitations of PFTC are very prominent.

(2) AFTC

AFTC can reconstruct and update the controller on line according to specific faults, in order to maintain the stability of the system. Different from PFTC, it considers a variety of faults at the same time. It usually covers three parts, i.e., fault detection and diagnosis module, reconfiguration mechanism and reconfigurable active controller. Its work flow is as follows: the fault detection and diagnosis module identifies and locates the fault, sends the fault information to AFTC system, which accordingly updates and reconstructs the controller, so as to compensate the output. In terms of different fault compensation, AFTC can be divided into control law switching, control law reconstruction, model following reorganization, safe stop, hybrid FTC, etc. [40].

The basic idea of control law switching is to design corresponding candidate control laws offline for various faults. After the fault detection and diagnosis module identifies and locates the fault, the AFTC system will select the appropriate control law for control [32], as shown in Fig. 2.33.

The control law reconstruction is to reconstruct the control law online after the fault is diagnosed by the fault detection and diagnosis module. The reconstruction of the controller include adaptive control strategy and rolling time domain control

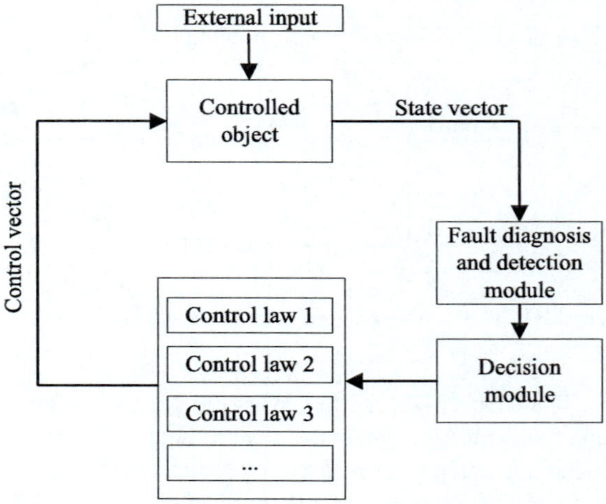

Fig. 2.33 FTC based on control law switching

strategy [31]. The model following reorganization mainly adopts the idea of model reference adaptive control, and its FTC is realized by keeping the state quantity of the controlled object to adaptively follow the state quantity of the reference model [32]. The safe stop aims to solve system imbalance caused by transient faults that cannot be handled by redundant control or FTC. When the fault occurs, it should be ensured firstly that the controlled object is maintained at the safe stop until the AFTC takes over the system and changes the system state to the nominal equilibrium state [41].

The hybrid FTC includes both AFTC module and PFTC module. When a fault occurs, the passive fault tolerant controller is first responsible for the control of the system. After the fault diagnosis and detection module identifies and locates the fault, the active fault tolerant controller takes over the system. It can not only overcome the time delay of AFTC and ensure system stability after the fault occurs by PFTC, but also make use of the advantages of AFTC in controlling corresponding fault.

Taking IV driving system as an example, its control quantity is much larger than the motion dimension. It is a typical overdrive Multi-input Multi-output (MIMO) system. Its FTC problems mostly are treated with control law reconstruction. The fault tolerant algorithm of wire control system based on control reconfiguration is used for FTC of IV drive system [42].

Introduction of tire load rate η_{ij}, used to evaluate vehicle stability:

$$\eta ij = \frac{\sqrt{F_{xij}^2 + F_{yij}^2}}{\mu F_{zij}} (0 < \eta ij < 1) \tag{2.18}$$

where, F_{xij}, F_{yij}, F_{zij} respectively represent the longitudinal force, transverse force and vertical force of the corresponding wheel; μ represents the friction coefficient of pavement.

At the same time, the failure factor CIJ c_{ij} introduced to characterize the execution capability of the drive by wire system:

$$cij = \frac{F_{xij}^{*}}{F_{xij} + \zeta}(0 < cij < 1) \tag{2.19}$$

where, F_{xij}, F_{xij}^{*} represents the target value and actual value of tire longitudinal force respectively; ζ Represents the compensation coefficient to prevent the failure factor denominator from being 0 or the failure factor more than 1 due to driving current fluctuation.

During normal driving, the optimal distribution of control force is adopted, with the minimization of tire load rate as the optimization index. And its objective function is defined as:

$$JCA = \min = Bu_{CA} =_2^2 \tag{2.20}$$

$$uCA = [F_{xfl}\ F_{xfr}\ F_{xrl}\ F_{xrr}\ F_{yfl}\ F_{yfr}\ F_{yrl}\ F_{yrr}]^{T}$$

$$B = \left[\frac{1}{\mu F_{zfl}}\ \frac{1}{\mu F_{zfr}}\ \frac{1}{\mu F_{zrl}}\ \frac{1}{\mu F_{zrr}}\ \frac{1}{\mu F_{zfl}}\ \frac{1}{\mu F_{zfr}}\ \frac{1}{\mu F_{zrl}}\ \frac{1}{\mu F_{zrr}}\right]$$

In case of driving failure, the reconfiguration and distribution of control force based on failure factor and failure constraint is adopted, and the optimization index is to maximize the satisfaction of vehicle longitudinal force and transverse torque requirements, so as to take into account vehicle stability. The objective function is defined as:

$$JRCA = \min = W_{RCA}(B^{*}u_{CA} - y_{d}) =_2^2 \tag{2.21}$$

$$W_{RCA} = diag\{112/b\}$$

$$yd = [F_{xd}\ F_{yd}\ M_{zd}]^{T}$$

$$B* = \begin{bmatrix} c_{fl} & c_{fr} & c_{rl} & c_{rr} & 0\ 0\ 0 & 0 \\ 0 & 0 & 0 & 0 & 1\ 1\ 1 & 1 \\ -c_{fl}\frac{b}{2} & -c_{fr}\frac{b}{2} & -c_{rl}\frac{b}{2} & -c_{rr}\frac{b}{2} & l_f\ l_f\ -l_r & -l_r \end{bmatrix}$$

where W_{RCA} represents the weight coefficient of balancing the total body motion force and the total yaw motion moment M_z; b represents the trajectory width; l_f, and l_r respectively represent the distance from the center of mass to the front and rear axles; F_{xd}, F_{yd}, andM_{zd} respectively represent the expected longitudinal force, transverse force and total yaw moment of the whole vehicle.

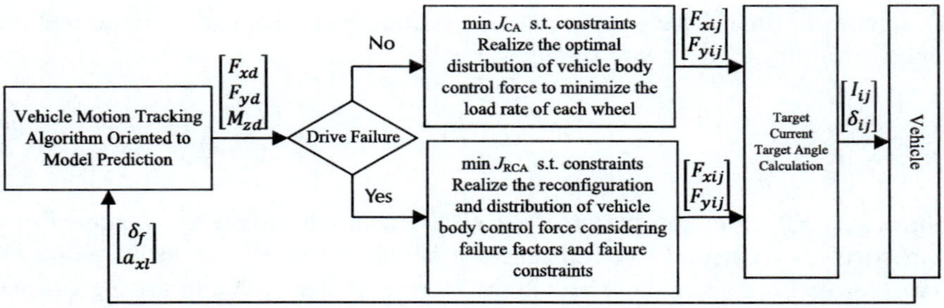

Fig. 2.34 Block diagram of control reconfiguration allocation [42]

As shown in Fig. 2.34, firstly, the desired vehicle control quantity is provided based on the vehicle motion tracking algorithm. Further, the fault detection and diagnosis module detects whether the drive fails. Then, the control output of four-wheel drive vehicle is realized by the above fault tolerant algorithm of wire control system based on control reconfiguration.

2.5.1 SOTIF Assurance Technology

The SOTIF design improvement of IV mainly considers three principles, i.e., system improvement, function limitation, taking over by drivers and passengers, and handling of misoperation. In terms of inherent problems such as performance limitations and insufficient specifications, operation risks can be reduced through system design improvement and performance optimization; In terms of the upper limit of system improvement and openness of road scenario, safety can be guaranteed by limiting the intended function or applied scenario; In terms of safety takeover ability and misoperation risk of drivers and passengers, the driving right can be properly handed over when the system function is insufficient, and reasonable and foreseeable misoperation can be prevented and handled.

Based on the above three principles and the characteristics of each IV functional subsystem, the SOTIF assurance technology is illustrated, as shown in Fig. 2.35: potential risks can be found online by SOTIF status monitoring technology, to provide information input for risk protection based on system adjustment, function restriction and driver takeover; In IV system layer, it can be divided into perception, decision-making, control, and other modules according to different function realization. Among them, the insufficient function of perception and decision-making is the most obvious, and SOTIF shall be guaranteed in combination with corresponding technologies, such as the function improvement technology of hardware and software required for the realization of perception function, and the performance optimization technology of different decision-making algorithms; In addition, the SOTIF based on

HMI can realize the takeover and misoperation of drivers and passengers, to further ensure SOTIF of the system.

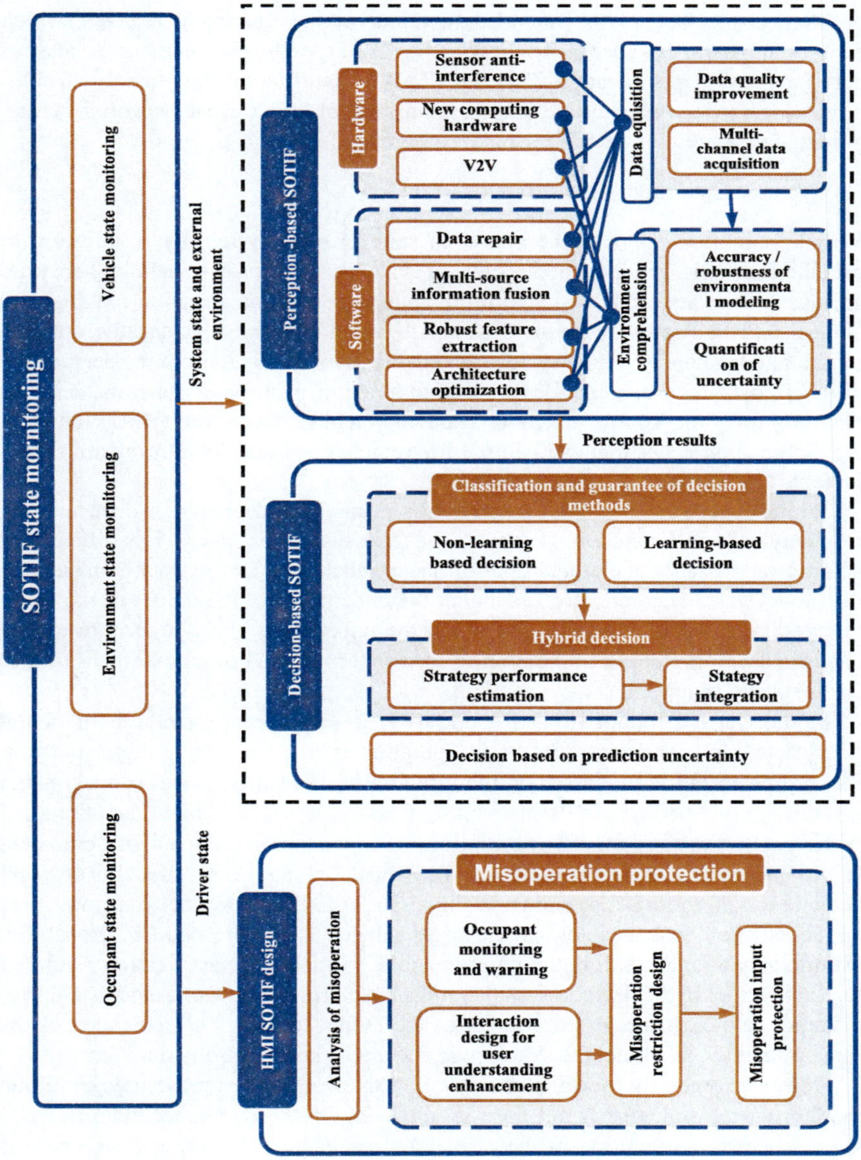

Fig. 2.35 SOTIF assurance technology

1. SOTIF Condition Monitoring

Real-time, dynamic and effective system condition monitoring is the premise of realizing prevention and treatment of SOTIF problems. The essence of the SOTIF problem is the lack of the intended function or the function it realizes, which evolves into hazards under the trigger conditions of specific environments or reasonably foreseeable misoperation. Therefore, SOTIF status monitoring mainly involves three aspects, i.e., vehicle state, trigger conditions of the external environment, and potential misoperation of drivers and passengers.

(1) Vehicle Condition Monitoring

Vehicle condition refers to the operation state of each system from hardware to specific functions. From the perspective of system architecture, it includes hardware such as sensors and actuators, algorithms such as perception, decision-making, and control, and their software means. Through the monitoring of vehicle status, possible functional deficiencies of the system can be directly identified. From the perspective of functional attribute, it includes perception function, positioning function, decision planning function, control function, Human–Machine Interaction (HMI) function, etc. Some approaches that can be used for vehicle condition monitoring are shown in Table 2.9.

Taking uncertainty evaluation as an example, the AI model based on deep learning is widely used in the system. However, due to its black-box nature, it is difficult for common approaches to predict its performance and failure. Uncertainty is an index of the model state. Theoretically, the higher the uncertainty of the model is, the lower the credibility of its output is. Therefore, the uncertainty of the algorithm can be quantified into system state monitoring for the formulation of countermeasures, to ensure safety requirements.

Monte Carlo Dropout (MC dropout) [49] is an inference method for neural network model uncertainty. Its basic operation is to carry out multiple network forward propagation processes on the same input, randomly cancel some connections in the middle layer of the network (called dropout), and obtain the outputs of "different network structures" by sampling. With statistics of these outputs, predicted results and uncertainty of the model can be gained. Taking the classification of target detection as an example, as shown in Fig. 2.36, Yolo is a neural network model for target detection, which can output the target category and corresponding probability estimation value. Regarding the input image, if probability classification results of the same target in the detection results of multiple dropout network models are very different, it indicates high uncertainty of the model for target classification of the input image. At this time, the system can take some backup protection measures.

However, generally speaking, the mature framework of current vehicle condition monitoring technology has not been established. On the one hand, there is still a large space for exploration of theoretical approaches; On the other hand, there is no sufficient transformation of practical application. With a certain research value, approaches such as anomaly detection in the field of machine learning probably can

Table 2.9 Approaches available for vehicle condition monitoring

Approach type	Approach description	Advantages	Shortcomings
Mutual inspection of multi-source information	Use mutual detection of multi-source information (such as spatial and temporal dimensions) [43] to identify the possible functional deficiencies of the corresponding module of single-source information	• Multi-source information provides mutual reference and redundancy, which can improve the reliability of detection; • It can deal with the problems caused by different trigger conditions of single-source information	• High cost in redundant design and calculation during operation; • Priority problem when there is a contradiction among multi-source information;
Uncertainty evaluation	Obtain the uncertainty output by using the Bayesian model and other methods, and monitor the performance of corresponding functional modules on the current input by evaluating the uncertainty [44]	• Uncertainty can be obtained without truth-value comparison; • Uncertainty represents the randomness of the real world and the reliability of the algorithm	• High modeling difficulty and calculation cost for obtaining uncertainty; • Existence of uncertain calibration problems; • Subjective uncertainty without unified standards

(continued)

Table 2.9 (continued)

Approach type	Approach description	Advantages	Shortcomings
Anomaly detection	Through distribution modeling of the model input data or output data, identify the "unknown" (e.g. Training samples are not covered) input. Specific technologies includes novelty detection [45], domain mismatch estimation [46], etc. ·	• It has certain application potential for identifying "unknown" scenarios; • Modeling from the perspective of data has certain universality for different functional models	• Difficult to accurately model distribution migration; • Difficult to accurately quantify the influence mechanism of distribution offset on algorithm performance
Decision safety evaluation	Evaluate safety of decision results, with Time-to-collision (TTC) [47], safety potential field [48], and other indexes	• It is the most direct evaluation for decision safety; • It has certain generality for different decision models	• Difficult to verify rationality and representativeness of the evaluation index itself; • Should avoid using the same model or index for decision making and evaluation

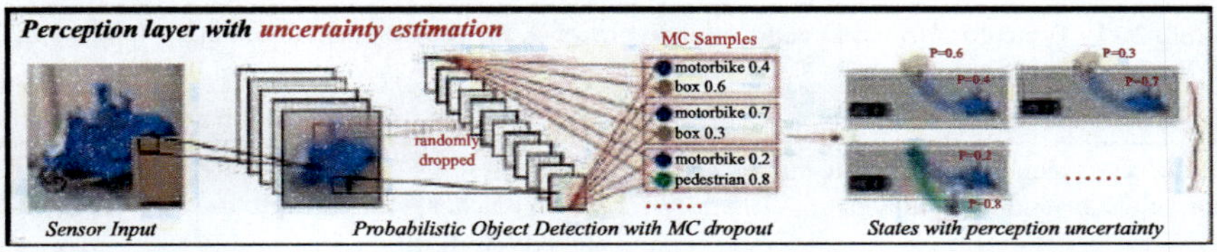

Fig. 2.36 Estimation of uncertainty in AI target detection with MC-dropout [49]

be further applied to realize the real-time monitoring technology of the system SOTIF state.

(2) Environmental Condition Monitoring

The performance limitations or insufficient specifications of various functional components of the system may lead to system hazards under corresponding external environment triggering conditions. Table 2.10 summarizes typical environmental conditions that may trigger the SOTIF problems of different functional components. Table 2.11 summarizes typical information acquisition approaches for environmental condition monitoring.

Table 2.10 External environmental conditions possibly leading to system SOTIF problems and their descriptions

Functional components	External environmental conditions and description (example)
Lidar	Rain, snow, and fog weather (macromolecular substances block and interfere with the laser)
Camera	Rain, snow, and fog (the image is blurred due to rainfall, snowfall, and heavy fog, for the rain and snowflakes may attach to the camera lens); Special lighting conditions (such as direct sunlight, night, entering and leaving the tunnel, resulting in changes in brightness)
Radar	Special road conditions (such as curves, ramps, and depressions); Material properties of special road facilities (such as high transmittance or absorption)
Global Navigation Satellite System (GNSS)	Signal blocking or reflection scenarios (such as viaduct, tunnel, underground garage, urban high-rise building, and mountainous area)
Decision algorithm	Complex road conditions (such as complex intersections), temporary traffic events, uncertain or extreme behaviors of traffic participants, etc
Controller	Road slope, curve, pavement defect, and pavement adhesion coefficient (such as mountain road, rain, and snow road); Weather conditions (such as strong wind)

Table 2.11 Typical information acquisition approaches for environmental condition monitoring [50]

Environmental conditions	Typical information acquisition approaches
Road type (such as intersection, tunnel, and curve); Surrounding buildings	HD map; Management system background (such as cloud platform); Environmental model (on-board sensor fusion)
Friction coefficient	Management system background (obtaining the friction characteristic diagram); Environmental model (measuring friction coefficient)
Lateral wind	Deviation from the predetermined trajectory; Management system background
Temperature	Temperature sensor
Road surface and geometry (such as slopes, cross slopes, and potholes)	Environmental model; HD map
Rain, snow, and fog	Rain sensor; Environmental model
Extreme weather conditions	Management system background

Taking the application of rainfall monitoring technology as an example, rainfall sensors can be used for direct measurement, which mainly includes optical, capacitive, resistive, piezoelectric vibrator, charge-coupled device imaging, and other types. Among them, the optical rainfall sensor is made according to the principle of light refraction and total reflection. Owing to its advantages of accuracy and convenience, it has been widely used. It includes a transmitting module, receiving module, glass prism, and so on. Combined with the principle of photoelectric conversion, the change of light intensity is transformed into the change of current signal. Other studies try to use the original data from cameras and other sensors to design rainfall recognition algorithms, to establish a more realistic environmental model for the estimation of rainfall levels.

In addition, through theoretical research and experimental verification, it is necessary to deeply explore the impact mechanism of external environmental factors on the insufficient function of the system, sort out more complete potential trigger conditions, and provide clear and executable objects for environmental state monitoring. At the same time, the research on specific monitoring indexes and monitoring technology shall also be further carried out to improve the accuracy and efficiency of environmental state monitoring.

(3) Monitoring for Drivers and Passengers Status

In IV of L3 and below, the driver is the direct user, who undertakes the responsibility of taking over in time in dangerous situations, and has the highest authority over the system. Taking driver's fatigue monitoring as an example, a study by Transportation Research Board (TRB) shows that about 7% of traffic accidents are related to driver's fatigue. To ensure safe interaction of drivers and passengers with the system, it is necessary to monitor their status and find out in time the conditions that may lead to reasonable and predictable misoperation, including fatigue status, distraction status, posture status (such as normal/abnormal driving posture, and in/out of the driving position), extreme abnormal status (such as driving under the influence of alcohol and drug), seatbelt status, etc. Passengers are directly affected by the system operation and may interfere with the operation. Therefore, it is also required to monitor their basic status, including posture status (such as there is/is not the behavior of grabbing driving equipment), extreme abnormal status (such as angry or irritable behavior), child restraint system status, seatbelt status, etc.

Different sensors and related techniques can support the monitoring of various states of drivers and passengers. Table 2.12 summarizes some advanced techniques in the industry.

Typical monitoring technology is shown in Fig. 2.37. The fatigue symptoms perceived by the sensor (such as pupil response, fixation mode, steering response, and EEG) can be used to monitor the driver's fatigue. In addition, the system can predict the subsequent fatigue behavior and generate an approximate behavior model for each driver with more personalized intervention.

2. Perceptual SOTIF Assurance

The perception system obtains information of IV's own state and surrounding environment as the input to subsequent decision-making and control modules. Its key requirement is the accuracy and robustness of driving environment modeling. As a whole, perception function can be divided into two parts, i.e., data acquisition and environment understanding in terms of the process, which mainly involves

Table 2.12 Typical sensors used for monitoring status of drivers and passengers [51]

Status of drivers and passengers	Driver monitoring camera	Seat position sensor	Sensor on steering wheel	Occupant sensor	Brake/accelerator pedal sensor	Shift lever sensor	Steering wheel torque sensor
User intention	+		++		++	+	++
User takeover capability	+	+	+		+		+
User distraction	++			+			
Mode mixture	++	+	++		++		++
Driver's absence	++			++			

Note More "+" indicates that the sensor signal has a higher potential to correctly infer the status attribute of the left driver and passenger

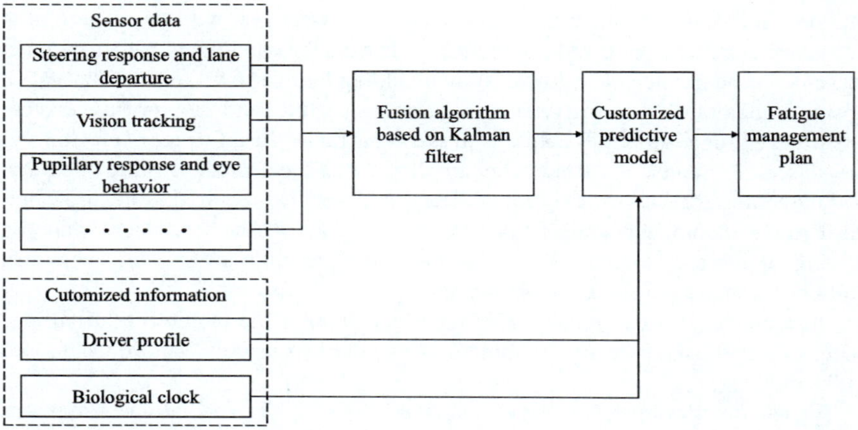

Fig. 2.37 Driver fatigue monitoring and prediction model [52]

such specific functional deficiencies as limited coverage of environment perception, poor real-time performance, and limited performance of perception algorithm. Regarding the perception function, some guarantee means at the hardware and software levels are conducted for the improvement of data quality and the strengthening of environmental understanding, as shown in Table 2.13.

In Table 2.13, the general idea of ensuring SOTIF from the perspective of perception is to improve the accuracy and robustness of environmental understanding, that is, the improvement of perception performance itself. For example, image restoration is widely used to improve the quality of image information. Multi-sensor information fusion can significantly improve the redundancy and fault tolerance of the system, and the enabling and empowering fusion perception based on intelligent IOV can cope with more complex and challenging scenarios. Comprehensive assurance for perception function safety will be the development direction in the future. In actual situation, due to the limitations of hardware conditions, scenarios, algorithm performance and other factors, it is difficult to completely eliminate the problem of insufficient sensing function from the sensing end. Monitoring the state of perception model by the above uncertainty quantification technology and taking it into account as reference in the decision-making process is also an effective means for solving the problem of insufficient sensing function in the whole system.

(1) Image restoration

In the case that sensors are generally disturbed by the environment, targeted data restoration is often carried out in combination with the interference process and the nature of sensor data, which is called image restoration technology in terms of image data. For image degradation caused by atmospheric particles, scattering of imaging light source and other factors, image restoration technology restores the image by establishing mathematical model of degradation process and using prior knowledge

Table 2.13 Typical perceptual SOTIF technology

Technology type	Technical means	Technical description	Specific technology
Hardware	Sensor anti-interference	Based on senor property, the accuracy and coverage should be ensured, for the interference reduction of environmental factors	• Select sensors with higher accuracy and wider coverage; • Combination of sensor kits; • Use anti-jamming sensor technology (such as multiple echo or area array lidar technology); • Parameter real-time dynamic correction technology
	V2X	It realizes information interaction and sharing between IV and its external nodes. The perception system can obtain external environmental information and improve the integrity, accuracy and robustness of environmental modeling	• V2x technology; • High-speed wireless communication technology
	New computing hardware	The huge amount of computation of perceptual tasks requires high computing power. In order to achieve the purpose of low latency, targeted design should be made in terms of different computing requirements for significant performance improvement	• AI chip design technology (combination of CPU, GPU, FPGA and ASIC); • Edge computing hardware technology

(continued)

Table 2.13 (continued)

Technology type	Technical means	Technical description	Specific technology
Software	Data repair	The data obtained by sensors often has noise, environmental interference and partial missing. Before extraction of environmental information, it should be repaired to improve the reliability of environmental modeling of perception algorithm	• Noise reduction technology (point cloud filtering, etc.); • Sampling technology on point cloud (encryption and completion); • Image enhancement technology (improve image visual effect); • Image restoration technology (image restoration based on image degradation model)
	Robust feature extraction	As a key step of environmental understanding, feature extraction should take into account accuracy, rapidity and robustness; For the increasing application of machine learning algorithms, corresponding measures should be taken to ensure the above performance, and the uncertainty can be quantified to serve the safety monitoring	• Training sample enhancement technology (data expansion, countermeasure sample generation, etc.); • Model optimization technology (model pruning, model quantification, knowledge distillation, etc.); • Uncertainty quantification technology (for detection [48], segmentation, etc.)

(continued)

Table 2.13 (continued)

Technology type	Technical means	Technical description	Specific technology
	Multi-source information fusion	It analyzes and synthesizes the information from different sources under certain criteria for the consistent description of environmental objects, and obtains better perceptual performance than a single information source	• Multisensor information fusion technology; • Integrated positioning and navigation technology; • Enabling and empowering fusion perception technology
	Architecture optimization	The huge data from sensors and other sources need to be processed in real time to implement the perception of security. The software and hardware architecture is designed and optimized by system approach to meet the functional requirements and achieve the balance between safety and cost	• Database management technology; • Computing framework technology; • Hybrid computing architecture technology

of degradation process. For example, the atmospheric degradation model is often utilized for image degradation due to fog [53]. The imaging expression is:

$$I(x) = J(x)t(x) + A[1 - t(x)] \tag{2.22}$$

Where, $I(x)$ represents the observed foggy image; $J(x)$ represents the fog-free image to be restored.

In this model, the imaging is divided into reflection attenuation part $J(x)\ t(x)$ and ambient light part $A\ [1 - t(x)]$. Therefore, the core of the image defogging algorithm based on this model is to estimate the atmospheric light intensity$A$ and transmittance $t(x)$, which mainly depends on some artificial assumptions (such as classical dark channel a priori [54]) or deep learning-based algorithms[55, 56]. The restored image can be better applied for subsequent tasks such as target detection.

(2) Multi-sensor Information Fusion

In the perception system, the interference and information integrity problems faced by a single type of sensor in signal reception and data processing are difficult to be completely solved. Multi-Sensor Information Fusion (MSIF) can not only correct the measurement error, but also give full play to the advantages of various types of sensors and significantly enhance the accuracy and robustness of environmental information modeling. It is an information processing in which the information from multi-sensors is analyzed and synthesized under certain criteria to complete the required decision-making and estimation. Multi-sensor fusion is classified mainly in accordance with the abstract level of data fusion, as shown in Table 2.14.

Data level fusion based on such statistical approaches as weighted average can reduce accidental error and interference and retain as much information as possible.

Table 2.14 Data fusion levels, characteristics and fusion approaches

Fusion levels	Characteristics	Common approaches
Data level	Carry out data fusion of the same type of sensor data at the data level	Adaptive weighted average, neural network, etc.
Feature level	Extract the feature vector of the original information from the sensor (the feature can be the target edge, direction, speed, etc.) and fuse the feature information from different sources to perform the subsequent perception task	Feature compression and clustering algorithm, parameter template, and deep learning-based algorithm, etc.
Decision level	On the basis of feature extraction, each sensor component independently completes certain perceptual tasks such as recognition, classification and tracking, and then coordinates with each other according to certain criteria and the reliability between sensors, so as to make the optimal decision	Bayesian inference, D-S evidence theory, fuzzy logic theory, etc.

However, it has large burden of data transmission and processing, and only can carry out the integration within the same type of data.

Feature level fusion is to extract common feature information (object pose, speed, edge, direction, etc.) from the data collected by sensors with different modes (point cloud, image, etc.), match the corresponding feature information, and then perform subsequent specific perception tasks on this basis. Feature fusion can be subdivided into target state information fusion and target feature information fusion. The former fusion uses simultaneous interpreting of target state information acquired by different sensors to jointly estimate state for uncertainty reduction. The latter fusion is mainly to aggregate the features of the lower layer to improve the accuracy and robustness. In fact, it is also classified as pattern recognition issue. The feature level fusion has advantage in retaining the main environmental information and significantly reducing the amount of data, yet still with some information lost and difficulty in selection of appropriate features. In recent years, the deep learning approach has become increasingly popular in the field of feature fusion. It mainly relies on the powerfully automatic feature extraction capability of neural network for feature information fusion. For instance, it utilizes convolution network to fuse laser point cloud and RGB image features, and output the three-dimensional trajectory of the target end to end [57].

Decision level fusion is to respectively establish models based on the data of each modal sensor, accordingly describe the target (such as position, speed and other state variables), independently perform "decision" level tasks (identification and tracking, etc.), and then fuse these local "decisions". As the highest level fusion, it loses a lot of relevant information in data and features and reduces the integrity of environmental information, but has strong anti-interference capability due to small amount of communication.

Different levels of fusion approaches have their own advantages and disadvantages. In practical application, the specific fusion scheme is often formulated based on the task type, sensor property, speed, accuracy, reliability and other requirements.

For multiple object tracking (MOT), target state information fusion is commonly applied. The fusion of lidar and camera is shown as an example in Fig. 2.38. It uses image and point cloud to detect the target separately, establishes state models for the detection results respectively, and uses Kalman filter to predict and estimate the motion state of the target at the next moment. Since motion state of the target detected by point cloud is nonlinear due to rotation, the extended Kalman filter is used. Firstly, the fusion shall transform the targets detected from visual and point cloud data to the same coordinate system, so as to correlate the targets. Usually, according to the parameter relationship among sensors, it projects the three-dimensional boundary frame of the target detected from the point cloud onto the camera plane, calculates the overlap rate between projection frame and original detection target frame in the camera view, and uses this index as the basis for target association to determine the same target by threshold. After the target correlation of the corresponding frame, it calculates the feature correlation degree of the target of the front and rear frames for the image and point cloud detection results and state estimation results respectively, obtain the correlation matrix by fusion, then correlates the correlation matrix between

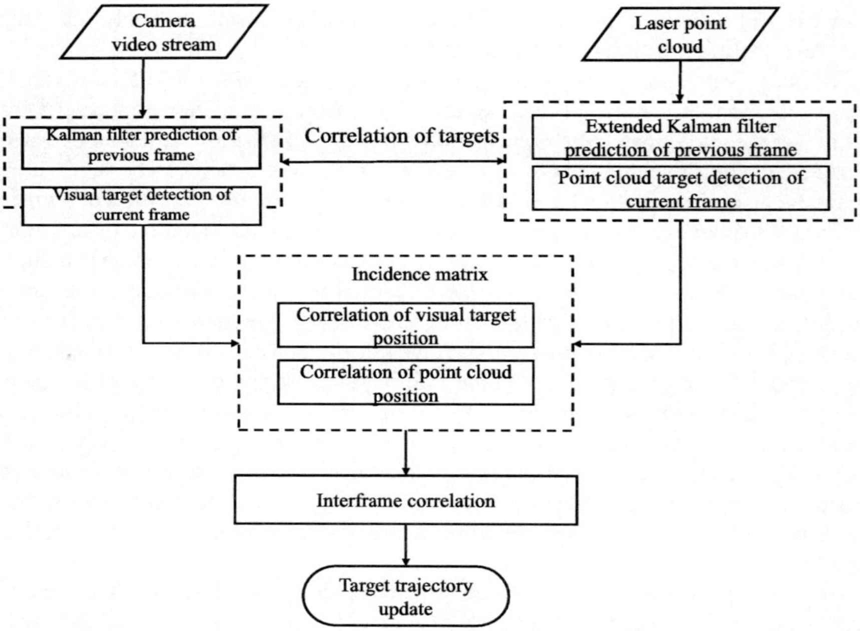

Fig. 2.38 Detection and tracking process of lidar and camera fusion

frames by the matching algorithm, and updates the tracking trajectory of Kalman filter and the target. The commonly-used matching algorithms include Hungarian algorithm and Kuhn-Munkres (KM) algorithm, etc.

The core step in the whole fusion process is obtaining the target feature correlation degree and correlation matrix of the front and rear frames:

The visual target detection algorithm obtains the center point position, length and width of the target boundary box in the pixel coordinate system, which is used as the visual state vector of target i, represented by D_i^{C}:

$$D_i^{\mathrm{C}} = (x_i^{\mathrm{C}}, y_i^{\mathrm{C}}, w_i^{\mathrm{C}}, h_i^{\mathrm{C}})^{\mathrm{T}} \tag{2.23}$$

where, x_i^{C} and y_i^{C} respectively represent the abscissa and ordinate of the center of the visual boundary box of target i; w_i^{C} and h_i^{C} represent the pixel width and height of the target boundary box respectively.

At the previous time, the state model is used to predict the position of multiple target center points at the current time. Within the target detection results at the current time, the correlation degree of visual target position is measured by Mahalanobis distance, represented by $C_{i,j}^{(1)}$:

$$C_{i,j}^{(1)} = (D_j^{\mathrm{C}} - Y_i^{\mathrm{C}})^{\mathrm{T}} P_{\mathrm{C}}^{-1} (D_j^{\mathrm{C}} - Y_i^{\mathrm{C}}) \tag{2.24}$$

where, i and j respectively represent the detection targets in the front and rear frames; $D^{C}{}_{j}$ represents the state variable of the target j detected and calculated by vision at the current time; $Y^{C}{}_{i}$ represents the prediction of the current time state variable of target i at the previous time (assuming constant speed, transverse acceleration, etc.); p_{C} represents the covariance matrix of the observed variables at the current time updated by Kalman filter. It is believed that the greater the degree of correlation is, the higher the probability of belonging to the same objective is.

Similarly, the point cloud target detection algorithm obtains the position, length, width and height of the three-dimensional frame of the target as the state, represented by the vector D_i^{l}:

$$D_i^{\mathrm{l}} = (x_i^{\mathrm{l}}, y_i^{\mathrm{l}}, w_i^{\mathrm{l}}, h_i^{\mathrm{l}}, l_i^{\mathrm{l}})^{\mathrm{T}} \tag{2.25}$$

where, x_i^{l} and y_i^{l} represents the x-axis and y-axis coordinates of the central point of the point cloud target (ignoring the displacement of the target in the z-axis direction); w_i^{l}, h_i^{l}, and l_i^{l} represents the length, width and height of the object respectively.

$C_{i,j}^{(2)}$, the correlation degree of point cloud target position in front and rear frames, is also measured by Mahalanobis distance:

$$C_{i,j}^{(2)} = (D_j^{\mathrm{l}} - Y_i^{\mathrm{l}})^{\mathrm{T}} P_{\mathrm{l}}^{-1} (D_j^{\mathrm{l}} - Y_i^{\mathrm{l}}) \tag{2.26}$$

where, D_j^{l} represents the state variable of target j detected by point cloud at the current moment; Y_i^{l} represents the state variable of target i predicted at the last moment; P_{l} represents the covariance matrix of the observation space at the current moment predicted by the extended Kalman filter.

The correlation degree fusion of point cloud position and visual position is conducted next:

$$\begin{cases} C_{i,j} = \lambda_1 C_{i,j}^{(1)} + \lambda_2 C_{i,j}^{(2)} \\ \lambda_1 + \lambda_2 = 1 \end{cases} \tag{2.27}$$

where, $C_{i,j}$ represents the inter-frame target feature correlation value after fusion; λ_1 and λ_2 indicate the weight factor of correlation degree (adjusted manually or by adaptive approach).

In addition to the visual target position correlation degree and point cloud target position correlation degree, it can also integrate the correlation degree from other target state models. After the correlation value of the recognition target between the front and rear frames is obtained by the above calculation, the subsequent matching algorithm is applied on the basis of the correlation matrix composed of the correlation value. This method can integrate visual and laser point cloud information, better adaptable to diversified scenarios and conducive to ensuring driving safety.

(3) Enabling and Enpowering Fusion Perception

Compared with the increasing expectation on IV safety, the perception information obtained by on-board sensors is difficult to fully meet requirements of safe operation in various environments, and the enabling and enpowering fusion perception based on intelligent IOV will be an effective means to overcome this kind of problem. The traditional perception system mainly relies on the sensor data of IV to perform the perception, which is called on-board enabling perception. Using information from other sources other than the vehicle, such as roadside infrastructure, to complete the perception task is called enpowering perception. Introduced by IOV technology, the enpowering perception information is fused with the enabling perception information at the same vehicle end for perception realization, which is called enabling and enpowering fusion perception. It comprehensively applies V2X technology, high-speed communication technology, high-precision map and other technologies, as shown in Fig. 2.39.

Through the enabling and enpowering fusion of vehicle-end, roadside and cloud information, it can effectively deal with performance limitations owing to constrained sensor coverage, interference and complex scenarios that cannot be solved or inconvenient to be solved by traditional vehicle-end enabling perception, such as occlusion, blind area of vision, complex environment perception and insufficient

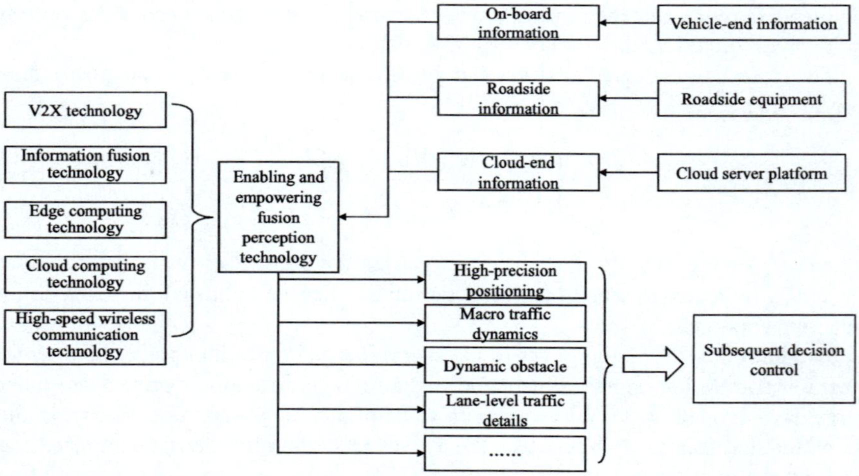

Fig. 2.39 Enabling and enpowering fusion perception technology

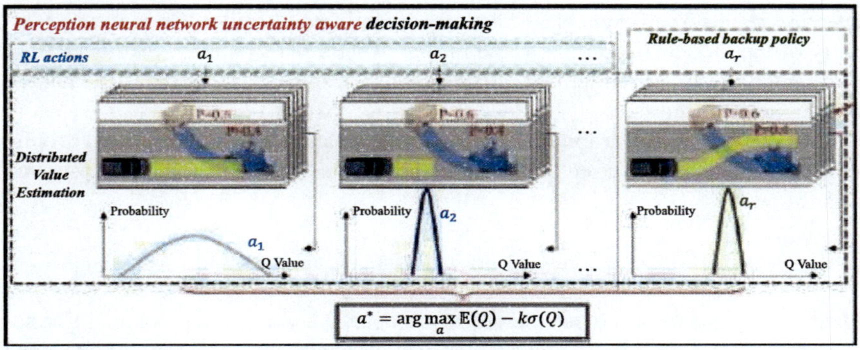

Fig. 2.40 AI decision framework of IV based on perceptual uncertainty

computing performance, and greatly improve safe operation capability of IV in open environment.

(4) Decision-making Based on Perceptual Uncertainty

By incorporating perceptual uncertainty into the decision-making end, the impact of insufficient perception can be made up to a certain extent. A framework of reinforcement learning safety decision-making approach given the uncertainty of target classification is shown in Fig. 2.40:

Decision-making approach based on deep reinforcement learning is widely studied and applied in the field of autonomous vehicles because of its potential to explore the optimal strategy in complex environments. A reinforcement learning strategy based on the action value function estimates the value of different actions in each environmental state and selects the action with the highest value. However, one defect of this approach is that it completely depends on the perceived result. When the perceived result is wrong, its strategy will choose the optimal action in the wrong environmental state, resulting in the degradation of system performance and SOTIF risk. The central idea of this approach is to estimate the uncertainty of the perceived result at the perception level. At the decision level, based on the action value function of the reinforcement learning approach, the perceptual uncertainty is considered in its value function estimation, so that the decision-making system can obtain the perceived uncertainty information while obtaining the perceived result, make optimal decisions in the case of perceived uncertainty, and improve the intelligent safety of autonomous vehicles.

As shown on the left side of Fig. 2.40, the target detection algorithm inputs the image and outputs the probability vector to which the target belongs:

$$c = [p_1, p_2, ..p_N]^T$$

where, $p_k, k = 1, 2 \ldots, N$ represents the probability of belonging to the target of class k. Multiple MC dropout samples are taken for the target detection model, and detection results of different classification probabilities can reflect the current uncertainty of the model. According to the results of multiple sampling, the uncertainty representation of the environment can be established and used in the decision-making of integrating uncertainty.

Suppose T times of dropout sampling are conducted, and the uncertain driving scenario is constructed based on the sampling results. The mathematical expression is as follows:

$$\{s_1, s_2, \ldots, s_T\} = \mathcal{L}(I, e) \tag{2.28}$$

where, I represents the input picture to be detected, e represents other information required for planning, such as self-vehicle state, and s represents the state of the driving environment constructed on each sampling result. Here, s includes classification and other non-semantic information, expressed as:

$$s_i = [s_r, \boldsymbol{c}] \tag{2.29}$$

where, $i = 1, 2, \ldots, T$. s_r represents other non-semantic information besides the target category. Suppose s_r is accurate, only the classification uncertainty of the target is considered. The set of possible states represented by all target classifications is recorded as S: $S = \{s^k | \forall k, \mathbf{c}[k]\}$

After the establishment of uncertainty state, the selection to realize the current optimal action can be expressed as value optimization:

$$a^* = \mathcal{R}(\mathcal{L}(I, e)) = \underset{a \in A}{\operatorname{argmax}} Q_u(a, \mathcal{L}(I, e)) \tag{2.30}$$

where, A is a discrete set, which can be a complete set of discrete action space. For continuous or high-dimensional action space, $A = \bigcup_{\forall s^k \in S} \{\pi(s^k)\}$ can be expressed; π is the AI strategy without considering uncertainty, Q_u represents a value function based on uncertainty.

The construction of Q_u needs to take into account environmental representation based on classification uncertainty. In the reinforcement learning decision-making process based on the action value function, the strategy is selected by maximizing the value function, and the best action in the action space can be selected as a result. The value function (called ordinary value function, for distinction in the follow-up) can be estimated by the depth Q network:

$$Q = \tilde{Q}(s^k, a) \tag{2.31}$$

where, $\tilde{Q}$ represents the value estimation network, which can be solved by deep Q-learning.

For the sampling result of a certain state represented by s, the accidental uncertainty, i.e., the mixed value function of classification probability represented by Q_p is:

$$Q_p(s,a) = \mathbb{E}_{s^k \sim P(s^k|s)}\left[\tilde{Q}(s^k,a)\right] = \sum_k \mathbf{c}[k]\tilde{Q}(s^k,a) \tag{2.32}$$

T-test results obtained by T times of sampling with MC dropout, represented by $\mathbf{c}$, constitute T environmental states s_i. The integration of cognitive uncertainty is achieved by using the standard deviation of the approximate value function based on the results of T sampling:

$$\sigma\left(Q_p(s,a)\right) \approx \sqrt{\frac{1}{T}\sum_i^T \left(Q_p(s_i,a) - \mathbb{E}\left(Q_p(s,a)\right)\right)^2} \tag{2.33}$$

$$\mathbb{E}\left(Q_p(s,a)\right) \approx \frac{1}{T}\sum_i^T Q_p(s_i,a) \tag{2.34}$$

The higher $\sigma\left(Q_p(s,a)\right)$ means that the higher the uncertainty of the model during detection is, the higher the inaccuracy of motion estimation is. Therefore, the value function considering uncertainty is expressed as:

$$Q_u(a|I,e) = \mathbb{E}\left(Q_p(s,a)\right) - \theta\sigma\left(Q_p(s,a)\right) \tag{2.35}$$

where, δ is a parameter that determines the proportion of uncertainty in decision-making. The decision-making process considering uncertainty can be expressed as:

$$a^* = \underset{a \in A}{\operatorname{argmax}}\, Q_u(a|I,e) \tag{2.36}$$

If δ is higher, IV will behave more conservatively. With $\sigma\left(Q_p(s,a)\right)$ as a penalty, when the perceived uncertainty is higher, the planner tends to choose safer actions in most cases. Accordingly, in the case of low perceived uncertainty, the planner will give an action close to making the ordinary value function optimal, that is, generally determining the optimal in the environment.

In the case of high perceived uncertainty, the real state s may not be in the detected set of possible states S. At this time, even the most conservative actions in A may not be able to ensure safety. The idea to solve this problem is to use other non-semantic information with low uncertainty besides classification information, represented by s_r, for designing backup policy π_r and generating actions to ensure safety $a_r = \pi_r(s_r)$. In terms of safety decision efficiency, a_r is designed based on rules.

In the final mixed action space $A_h = A \cup \{a_r\}$, the decision-making process of perceived uncertainty modified by backup policy can be described as:

$$a_h^* = \underset{a \in A_h}{\operatorname{argmax}}\, Q_u(a|I,e) \tag{2.37}$$

In general, this approach forms the representation of an uncertain environment by quantifying the classification uncertainty and establishes the reinforcement learning decision-making under uncertainty, which balances the decision-making performance and safety. For a high uncertainty environment, the backup policy is designed to ensure the minimum safety requirements under high uncertainty conditions. With high environmental adaptability, this approach can be extended to a wider scope of application and can improve the actual problem of insufficient perception function.

3. SOTIF Assurance Based on Decision-making

The decision-making system obtains the perceived result and performs tasks such as behavior decision making, trajectory planning, and emergency obstacle avoidance through reasonable prediction over traffic participants. However, the complexity and variability of scenarios and the limitations of the algorithm itself lead to SOTIF problems for decision-making. Typical environmental factors that may trigger the risk of decision SOTIF include highly complex traffic scenarios, high randomness and unexpected behavior of traffic participants, and "unknown hazard" cases not considered in the algorithm design, etc.; In addition, from the perspective of decision algorithm, the problems in the stages of model/rule design, parameter training/calibration, and real-time operation may result in the insufficient performance of various forms of decision-making functions.

To guarantee SOTIF for decision-making, it is necessary to first carry out targeted assurance according to the characteristics and differences of different decision-making approaches, and complement their advantages as well. In addition, taking into account possible function deficiency in the prediction process can further improve the safety of decision-making results.

(1) Classification and Assurance for Decision-making Approaches

The decision-making approaches are divided into two categories, i.e., non-learning decisions and learning-based decisions. The former is based on rules, search and optimization solutions, etc., and the latter mainly refers to machine learning driven by data. Table 2.15 lists typical potential function deficiency problems and assurance technologies of the two decision types.

Some problems in the functional deficiencies described in Table 2.15 can be solved using the listed technologies, but a considerable part of the problems can only be initially alleviated.

(2) Hybrid Decision-making

In recent years, hybrid decision-making integrating the two kinds of approaches has attracted more and more attention. It can overcome or alleviate the function shortcomings of single type of decision-making by applying the principle of complementary advantages and redundancy. For example, non-learning decision-making is difficult to model the environment with high-dimensional uncertainty, but reliable and conservative rules or models can be used to formulate the underlying safety and alleviate the risk caused by the lack of generalization of learning-based decision-making in some

Table 2.15 Examples for potential function deficiency and assurance technology of two types of decision-making approaches

Decision type	Potential functional deficiency	Problem description	Assurance/ mitigation technology (example)
Non-learning decision	Insufficient specification	The decision-making rules or models are not comprehensive enough, such as the lack of corresponding rules for decision-making scenarios including individual road conditions (e.g., ramps), traffic participants' behavior (i.e., red-light running of pedestrians), weather conditions (e.g., the reduction of visual distance caused by rain, snow, and fog)	• Decision logic optimization (such as extracting universal safety indexes of collision time and headway as rule conditions); • Design of decision system based on system theory and process analysis (such as STPA)
	Insufficient accuracy of modeling and adaptability and expansibility to complex differentiated scenarios	Due to the complexity, interactivity, and uncertainty of real traffic scenarios, it is difficult to establish rules, models, or methods that can effectively deal with dynamic complex scenarios; Due to the openness and differences of scenarios, the set rules and models may be difficult to cover all operation scenarios, thus triggering dangers in unknown scenarios; In addition, models or rules are likely to be redesigned manually each time for new scenarios	• Adopt new modeling theory (artificial potential field, partially observable Markov model, game theory, etc.); • Introduce new modeling information (human vehicle/ vehicle interaction information, driver confidence index, etc.); • Scenario template technology (extracting common and different features of scenarios); • Calibration of model parameters based on big data samples

(continued)

Table 2.15 (continued)

Decision type	Potential functional deficiency	Problem description	Assurance/ mitigation technology (example)
	Difficult to solve the model	The complex model objective function/constraint, nonlinear, nonconvex, and multiple local optima, and other problems may lead to difficulty in solving the model or time-consuming	• Linearization technology of nonlinear model; • Adoption of specific solutions (particle swarm optimization algorithm, genetic algorithm, simulated annealing algorithm, subgradient method, near point method, etc.)
Learning-based decision	Insufficient interpretability	Due to the high complexity of models (such as neural networks), learning-based decision algorithms are difficult to explain, which poses challenges to safety verification, driver trust establishment, accident analysis, algorithm improvement, etc	• Active/passive interpretation; • Interpretation based on example/ attribute/hidden semantic/rule; • Local/semi-local/ global interpretation; • Pre-model/ in-model/ post-model interpretation
	Insufficient training samples	Insufficient scenario coverage, unreasonable and imbalanced distribution, and omission of edge dangerous cases of training samples may lead to the decline in the decision-making safety performance of the training algorithm	• Data augmentation technology; • Data disturbance technology; • Countermeasure sample generation technology

(continued)

Table 2.15 (continued)

Decision type	Potential functional deficiency	Problem description	Assurance/ mitigation technology (example)
	Training process problems	Due to the complexity and high non-convexity of the model, many problems are likely to occur in the training process, such as local optimization, over fitting/under fitting, gradient explosion/ disappearance, and convergence difficulty/divergence	• Optimization solution technology; • Weight regularization technology; • Dropout, batch normalization technologies, etc
	Complex computation	Due to a large amount of data and complex model, AI algorithm has the problem of high computational complexity, which affects the real-time decision-making response and emergency response	• Hardware upgrade and computing power improvement technology; • Parallel computing technology; • Model simplification technology (such as neural network pruning)

scenarios. In addition, due to its dependence on the perception and prediction effect, the decision algorithm can be monitored and integrated by uncertainty quantification in the case of insufficient performance, which can strengthen the adaptability of the system environment and ensure safety.

A typical hybrid decision-making [58] architecture is shown in Fig. 2.41. The two types of decision-making approaches receive environmental condition information from the current traffic scenario for parallel decision-making, and fuse strategies given strategy performance estimation. Specifically, the estimation of strategy performance can be realized by the calculation of the strategy activation function:

$$C(\pi_{\mathrm{AI}}, \pi_{\mathrm{b}}, s) = E_{\pi_{\mathrm{AI}}}\left(\sum_{k=h}^{h+H} \gamma^{k-h} r_k \middle| s_h = s\right) - E_{\pi_{\mathrm{b}}}\left(\sum_{k=h}^{h+H} \gamma^{k-h} r_k \middle| s_h = s\right) \quad (2.38)$$

where, h represents the current time; s_h indicates the corresponding state; H is the decision length; r_k represents the amount of decision-making return in different states; γ represents the loss constant of return in time sequence; E indicates expectation; π_{b} and π_{AI} represent the strategies determined by rule/model decision and AI decision respectively.

The two items on the right of the equal sign in Formula (2.38) are the performance estimation results of the two strategies respectively, and the indexes that directly

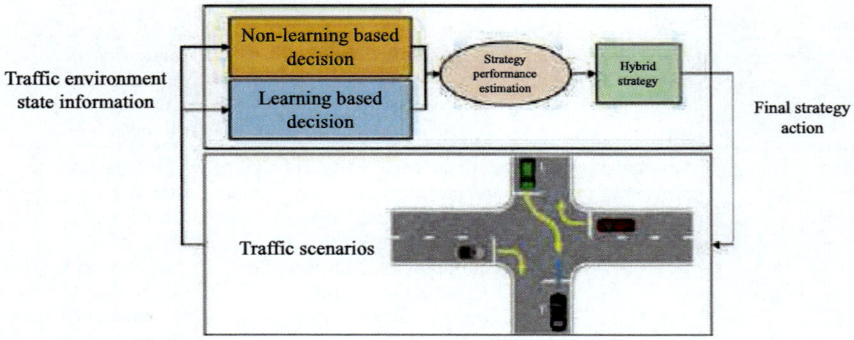

Fig. 2.41 Hybrid decision-making

reflect the safety, such as the shortest distance and the number of collisions, can be used as the reward function; Uncertainty evaluation, anomaly detection, and other technologies can also be used to evaluate the current state of the decision model. This formula can be intuitively understood as: when $C(\pi_{\mathrm{AI}},\pi_{\mathrm{b}},s) > 0$, it is considered that the performance of the learning-based decision strategy is better than that of the non-learning decision strategy; When $C(\pi_{\mathrm{AI}},\pi_{\mathrm{b}},s) < 0$, the result is the opposite.

In the final strategy formulation stage, strategies can be switched or mixed. Strategy switching refers to the selection of a comprehensive strategy for better decision-making performance according to the actual operation. And strategy mixing refers to the integration and smooth transition of the two strategies through weighting and other ways, as shown in the following formula:

$$\pi_{\mathrm{h}} = \pi_{\mathrm{b}} + \frac{\pi_{\mathrm{AI}} - \pi_{\mathrm{b}}}{1 + \mathrm{e}[-wC(\pi_{\mathrm{AI}},\pi_{\mathrm{b}},s)]} \tag{2.39}$$

where π_{h} represents the final hybrid strategy; w indicates that the influence factor determines the influence degree of strategy activation function on the evolution trend of hybrid strategy.

(3) Decision-making in View of Prediction Uncertainty

The decision-making process, especially the local motion planning, largely depends on the prediction of the future trajectory of surrounding moving objects. However, on the one hand, the future trajectory of the object itself is uncertain. On the other hand, the deep learning algorithm applied for prediction has the nature of black box and uncertainty. The existence of uncertainty will affect the safety of decision-making to a certain extent. As shown in Fig. 2.42, the possible cutting behavior of the vehicle in front of the right makes the future trajectory uncertain. In this case, if the prediction module fails to judge the future trajectory of the target vehicle in time and accurately, there is a potential risk of collision.

The quantification of prediction uncertainty and the decision-making mechanism integrating uncertainty are the ways to solve the above problems. On the one hand,

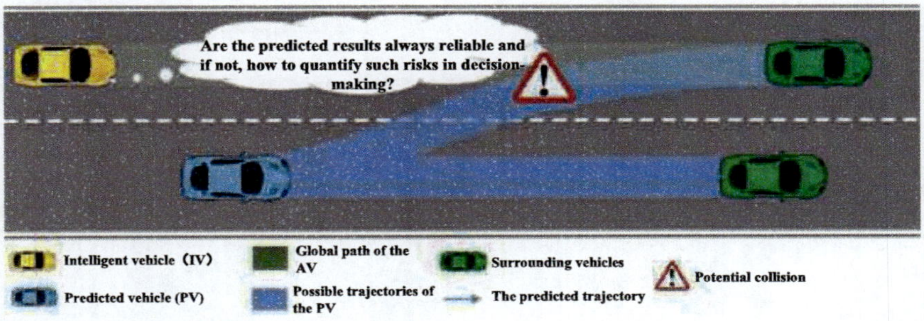

Fig. 2.42 Uncertain trajectory scenario

the future trajectory of the vehicle itself is uncertain, and the trajectory points in the future time can be modeled as mixed Gaussian distribution. On the other hand, based on the neural network [59], the existing prediction algorithms can characterize the insufficient function of the prediction model by the above uncertainty quantification technology. And Deep Ensemble [60] can be adopted in practice. The main idea of Deep Ensemble is to use multiple randomly-initialized models to predict the same input, take the mean value of multiple predictions as the final prediction result, and the variance as the uncertainty. Its advantage lies in its ease of use, adaptability to distributed computing, and excellent uncertainty extraction effect [61].

As shown in the upper part of Fig. 2.43, there are M complete prediction models with random initialization. For the target vehicle to be predicted, the model based on LSTM is used for prediction. Convolution Social-pooling module [62] is added to each model for considering the interaction with other surrounding vehicles. The output of each model m is the prediction of the future position of the target vehicle. The position of future time t is modeled as binary Gaussian distribution, which can be expressed as:

$$R(\mathcal{X})_m = \left[\boldsymbol{E}^t_m, \boldsymbol{cov}^t_m\right]. \tag{2.40}$$

where, $\boldsymbol{E}^t_m = \left[E^t_{xm}, E^t_{ym}\right]$, $\boldsymbol{cov}^t_m = \begin{bmatrix} {\sigma^t_{xm}}^2 & \rho^t_m \sigma^t_{xm} \sigma^t_{ym} \\ \rho^t_m \sigma^t_{xm} \sigma^t_{ym} & {\sigma^t_{ym}}^2 \end{bmatrix}_m$

$X \, and \, Y$ represent the longitudinal and transverse directions of the road respectively; E_x and E_y as the output from the model are the predicted mean value in the $XandY$ directions; σ_x and σ_y are the variances; ρ is the correlation coefficient, and the subscripts m and t refer to the output value of the m-th model corresponding to time t (since each variable of the calculation belongs to the same time t, it is omitted in subsequent writing). M Gaussian distributions are synthesized into a hybrid Gaussian distribution, and the prediction of the overall uncertainty is expressed as:

$$\boldsymbol{E}_{y*}(\mathcal{X}) = \left[E_{x*}, E_{y*}\right] = \frac{1}{M}\sum_{m=1}^{M} \boldsymbol{E}_m \tag{2.41}$$

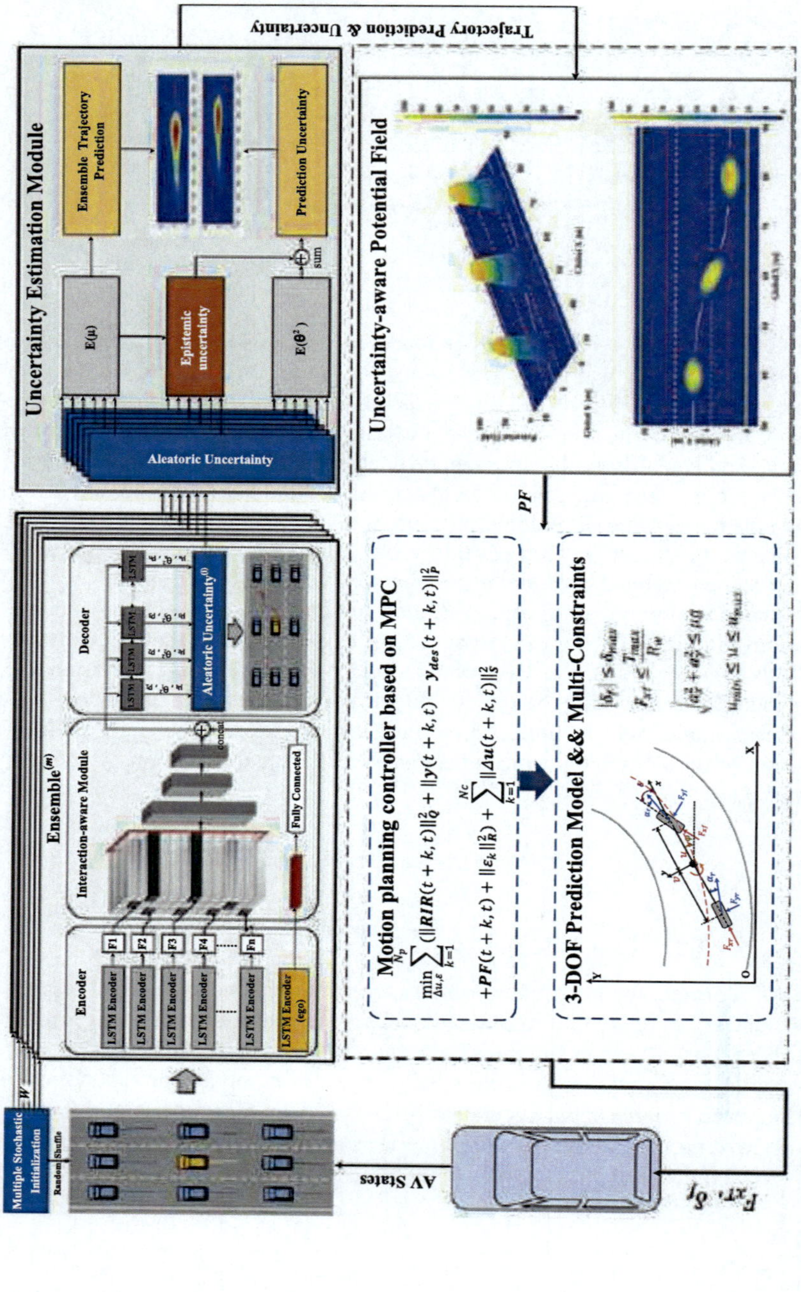

Fig. 2.43 Control framework of uncertain trajectory prediction and planning

$$\boldsymbol{cov}_{y*}(\mathcal{X}) = \frac{1}{M}\sum_{m=1}^{M}\begin{bmatrix} \sigma_x^2 + E_x^2 & \rho\sigma_x\sigma_y + E_xE_y \\ \rho\sigma_x\sigma_y + E_xE_y & \sigma_y^2 + E_y^2 \end{bmatrix}_m - \begin{bmatrix} E_{x*}^2 & E_{x*}E_{y*} \\ E_{x*}E_{y*} & E_{y*}^2 \end{bmatrix} \tag{2.42}$$

In terms of the uncertainty prediction results, the design of decision-making considering uncertainty can improve the safety in the case of inaccurate prediction. Take a local planning based on potential field and model predictive control as an example, as shown in Fig. 2.43.

Taking the components of the covariance matrix of Gaussian mixture distribution in the X and Y directions as the parameters to characterize the uncertainty of the surrounding vehicle trajectory, an artificial potential field function incorporating the prediction uncertainty is constructed:

$$UPF_i(X,Y) = a_i e^{\beta_i} \tag{2.43}$$

$$\beta_i = -\left\{\frac{\left((X - E_{x*})\cos\vartheta + (Y - E_{y*})\sin\vartheta\right)^2}{2\lambda_{x_i}^2} + \frac{\left(-(X - E_{x*})\sin\vartheta + (Y - E_{y*})\cos\vartheta\right)^2}{2\lambda_{y_i}^2}\right\}^{b_i}$$

$$\lambda_{xi} = L_{xi} + M^{-1}\sum_{m=1}^{M}\left(\sigma_x^2 + E_x^2\right)_m - E_{x*}^2$$

$$\lambda_{yi} = L_{yi} + M^{-1}\sum_{m=1}^{M}\left(\sigma_y^2 + E_y^2\right)_m - E_{y*}^2$$

where, i is the vehicle number on the road; $UPF_i(X,Y)$ represents the artificial potential field generated by the i-th vehicle; a_i and b_i represent the strength and shape parameters of the potential field respectively; X and Y are the vertical and horizontal global coordinates of IV respectively; ϑ Is the heading angle of the vehicle, and L_{xi} and L_{yi} are the parameters related to the shape of IV.

As shown in the lower right of Fig. 2.43, when the target vehicle i is in the process of lane change (e.g. [58 m, 4 m], UPF_i will adaptively adjust the intensity according to the uncertainty. For the target vehicle with large uncertainty, it offers a higher potential field peak and a larger high-value area.

The artificial potential field with sensitive uncertainty considers the requirement for safe obstacle avoidance of surrounding vehicles. The planner based on model predictive control operates the vehicle for safe obstacle avoidance and path following, by the globally planned trajectory and uncertain artificial potential field, as well as the control constraints and vehicle dynamics constraints. Compared to prediction with certainty, this approach can realize better safety planning and control, and improve the adaptability of the system in an uncertain environment.

4. SOTIF Design for HMI

Reasonable and predictable misoperation is an important trigger condition of SOTIF. For example, in the urban environment, if the driver mistakenly activates an autonomous function only applied in the expressway, the system may have insufficient functions in identifying and responding to the movement of pedestrians or cyclists, resulting in risks. The premise for misoperation avoidance of drivers and passengers is to systematically deduce and analyze such potential scenarios, as shown in Fig. 2.44. Regarding the potential misoperation behavior obtained from the analysis, it is necessary to carry out systematic protection. The existing misoperation protection technologies mainly include four categories: monitoring and reminding for drivers and passengers, interactive design for users' understanding improvement, misoperation restriction design, and misoperation input protection.

By monitoring the state of drivers and passengers with SOTIF state monitoring technology, misoperation behavior or risk can be identified in advance for the in-time reminder. In view of the reaction time and attention of drivers and passengers, the reminder information should last for a certain time (e.g. more than 3 s) in complex road environment. The first objective of the reminder is warning, that is, to attract the attention of drivers and passengers. And the second is to action suggestion, that is, to inform the drivers and passengers of the specific operation way (e.g., necessary to hold the steering wheel). The typical techniques for the reminder function can be realized using vision, hearing, touch, etc. The specific techniques are shown in Table 2.16.

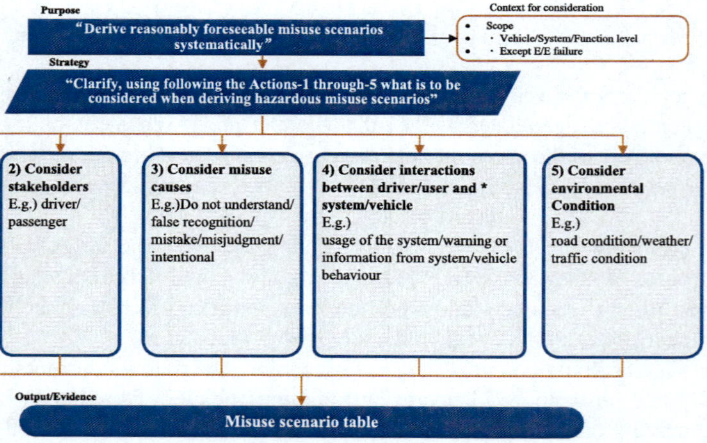

Fig. 2.44 System derivation and analysis process of SOTIF misoperation scenarios of drivers and passengers [28]

Table 2.16 Techniques for driver and passenger reminder with HMI

Type	Techniques
Vision	Warning light: different positions (instrument panel, steering wheel, door, etc.), different colors, and different flashing modes
	Display screen: warn drivers and passengers by dangerous color and image flashing, and convey action suggestions in the form of text and image
Hearing	Alarm sound (warning function), voice prompt (action suggestion)
Touch	Seat vibration, steering wheel vibration, seatbelt vibration

The unfamiliarity with systems and the pressure in an emergency may lead to the driver's and passenger's incomprehension or wrong understanding of the system function, resulting in misoperation. To solve these problems, improving the intuitiveness and understandability of the HMI function is also an important technical means, in addition to increasing user training and improving the HMI instruction manual. The existing technology often carries out HMI design through multi-channel information interaction, enhances the driver's and passenger's understanding of functions with redundant information, and adopts different standards (such as shape, size, position, and color) for different function switches to make a distinction. In the specific design, the following information should be transmitted to the drivers and passengers, including the current driving mode (such as manual driving, autonomous activation, and partial auxiliary function activation), the current allowed/disabled operation (such as the prohibition of urban functions on the highway), the current system status (such as partial function fault or failure), the system takeover request, etc. Besides, interactive information can be enhanced, such as providing visual, auditory, and tactile feedback when drivers and passengers turn on the automated driving function.

In some cases, drivers and passengers may still take misoperation actions due to not receiving the reminder, tension, subconscious action, or unconscious behavior. Therefore, safety design for limiting such actions needs to be considered in HMI design. Typical techniques are shown in Table 2.17.

Finally, although the above SOTIF design technology based on HMI can avoid misoperation of drivers and passengers to a large extent, it is still necessary to consider the misoperation in special cases to ensure that the system has certain protection capability against misoperation input. On the one hand, it can restrain some mismatched functions in specific scenarios, such as the activation prohibition of automatic parking function in the expressway. On the other hand, the misoperation input can be monitored and corrected. For example, in the Traffic Jam Pilot (TJP) system of Honda L3, when the driver does not respond to the handover request after repeated reminders, the system will assist IV to slow down and stop and flash the hazard warning light, and whistle to remind surrounding vehicles.

Table 2.17 Safety design for misoperation

Technical type	Explanation
Safety design of the position	Place the critical safety operation button at the position with a low contact probability
Safety design of operation mode	Special confirmation or unlocking mechanism, such as multiple confirmation operations or multiple simultaneous operations; Special activation actions, such as low probability activation gestures designed according to human operation habits
Restrictions on drivers and passengers	Driver restrictions when the advanced autonomous function is turned on, such as automatic seat reverse, pedal acceleration or braking, and steering wheel locking

2.5.2 Cybersecurity Assurance Technology

With the development of autonomous technology and IOV technology, there are more and more functional modules of IV cyber-physical systems, with the attack surface larger and malicious cyber attacks more likely to happen. At the same time, traditional cyber attack tools can be easily and directly applied to IOV, which poses a great threat to the cybersecurity of the IV system. The number, speed, and ways of these attacks make it difficult for effective defense. To resist the cybersecurity threat to IV cyber-physical systems, experts and scholars have studied the cybersecurity assurance technology from the perspectives of AI algorithms, cybernetics, infection theory, game theory, graph theory, etc. [63, 64] In terms of AI algorithms, agents are constructed and trained through artificial neural networks, evolutionary algorithms, fuzzy sets, genetic algorithms, deep learning, etc., which can analyze cyber attacks and start appropriate countermeasures in most cases; Cybernetics is generally applied to the control analysis and design in the case of data delay and packet loss; In terms of infection theory, its dynamic model is often used in the design of cybersecurity risk transmission control approaches and devices; Game theory can be applied in the decision-making design of multi-module interaction; And graph theory can be used to design the topology of complex networks.

Cybersecurity assurance technology can be summarized into three parts, as shown in Fig. 2.45. The cyber topology is divided into traditional distributed architecture, functional domain-based architecture, and central computing architecture, which provides a basic framework for the research of other cybersecurity technologies. Intrusion detection and defense system is an active defense technology to ensure cybersecurity. It resolves intrusion attacks through three stages, including status monitoring, intrusion detection, and intrusion defense. In the status monitoring stage, the communication data of IOV is used to monitor the cyber status. In the intrusion detection stage, the historical data and the knowledge base stored in the vehicle are used to identify intrusion attacks, and the intrusion defense results can be transmitted to the surrounding IV, and then the cloud end through V2V communication. Encryption and communication is a passive protection technology for cybersecurity. The encryption system designs algorithms such as symmetric password, public key password, digital digest, and digital signature applicable for the IOV environment, with Public Key Infrastructure (PKI) serving for key management. The communication system considers information interaction of different distances such as in-vehicle communication, near-end communication, V2X communication, and satellite communication, with PKI serving for identity authentication.

1. Network Topology for Cybersecurity

With the intelligent interconnection of IV and mobile devices, cellular networks, and road infrastructure, the interaction between the out-of-vehicle environment and the in-vehicle system is gradually increasing, and more and more data exchange among ECUs are also expected to be faster, more reliable, more convenient and more economical, which drives the great change of the on-board network from distributed to hybrid and then to central computing architecture. Its characteristics are shown in Table 2.18. The traditional distributed architecture is a flat system with a simple combination and low cost, but requires information synchronization among all nodes,

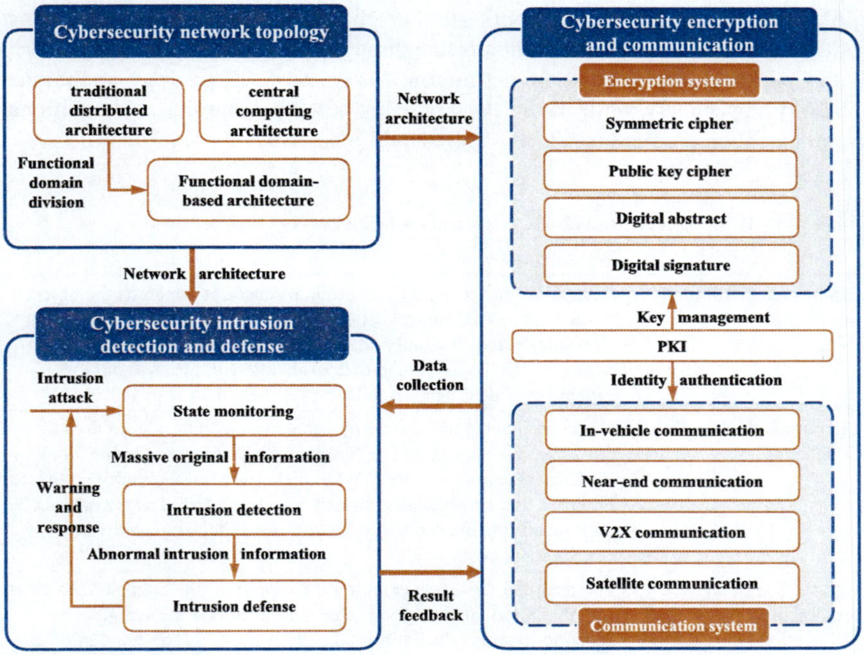

Fig. 2.45 Cybersecurity assurance technology

without cybersecurity design in it. Hybrid architecture is the most common architecture at present. Adopting the technical concept of separation of concerns, the system is divided into five block domains according to functions, including the connection domain, driver replacement product domain, transmission and power system domain, body and comfort system domain, and onboard experience domain. The gateway is introduced for isolation, focus, and efficiency in different domains. In the central computing architecture, all original data processing and decision-making are completed in the central processor, without data tampering and distortion. However, it has high requirements for data transmission frequency and processor computing power, only applied in some small start-up autonomous vehicles now [65]. Controller Area Network (CAN), Local Interconnect Network (LIN), and FlexRay are three bus systems customized for on-board networks, among which CAN is one of the most widely used systems. The bus topology of the onboard network architecture mainly includes forms of line, star, and ring, with cost, deployment difficulty, and robustness against cable fault enhanced in sequence.

Based on the functional domain architecture, data forwarding and transmission are conducted among different domains through the vehicle gateway. Serving as a firewall, it can not only control the access from the external interface to the internal network but also ensure the isolation of internal functional domains. Vehicle gateways are generally divided into CAN gateways, Ethernet gateways, and new generation hybrid gateways. The typical CAN gateway topology is shown in Fig. 2.46 [66]. Within the in-vehicle network structure based on CAN or Ethernet, most ECUs and domain processors communicate through CAN or Ethernet. The gateway with CAN bus interface or Ethernet interface is called CAN gateway or Ethernet Gateway respectively, while the gateway with both Ethernet interface and traditional communication protocol interface is called hybrid gateway.

Table 2.18 Working mode and characteristics of on-board network architecture

Network architecture	Working modes and characteristics
Traditional distributed architecture	After data processing by the terminal sensor, it transmits the object or metadata to the central processing unit, which makes a decision after integration; It is only adaptable for networks with a small number of ECUs and small data exchange volume, without the capability of dealing with radar and camera data
Functional domain-based architecture	The centralized data processing can be carried out in each domain by the domain processor or the distributed architecture; On the whole, the latest domain controller and transmission bus technology, such as Ethernet, can be gradually applied. The traditional framework can be used locally At the same time to improve reliability and reduce development costs
Central computing architecture	The terminal sensor does not modify and filter the data, and the CPU can obtain the most original information for operation and decision-making; Each terminal node is aware of the trend of other nodes, but the data transmission reaches GB at this time and needs to be processed in time

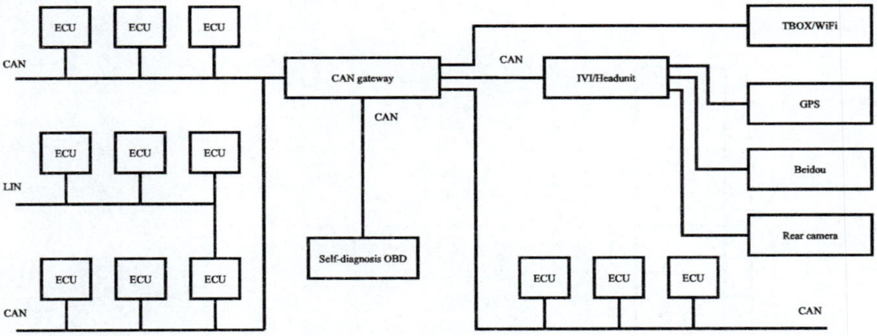

Fig. 2.46 Example of vehicle CAN gateway topology

As the cybersecurity issue of the CAN bus is not considered enough in the initial design, more and more research focuses on improving the cybersecurity attributes such as system confidentiality, authenticity, reliability, integrity, and non-repudiation, with continuous occurrence of related accidents. Among the attributes, non-repudiation ensures that all participants cannot deny or repudiate the completed operations and commitments in the process of information interaction. Specifically, information source evidence can be used to prevent the sender from denying the sent information, and submission and reception evidence can be applied to prevent the recipient from denying the received information.

Cybersecurity can be assured by optimizing the safety design of network topology. When the number of ECUs is small and the bus load rate is less than 50%, the vehicle usually has only one CAN and adopts distributed architecture; When there are a large number of ECUs or the bus load rate is greater than 50%, there are usually multi-channel CAN lines. The hybrid architecture is adopted to communicate with ECUs directly connected to the same bus through the direct bus, and communicate with ECUs directly connected to different buses through the gateway. Generally speaking, the topology will divide the network according to the distance of the equipment controlled by ECUs, so that the power-related ECU, body-related ECU, and chassis-related ECU will be distributed into different networks. Once TBOX and CAN gateway are attacked, these ECUs may not be able to communicate normally, resulting in abnormal basic functions of the vehicle, such as failure of braking, and posing serious safety threats. To solve this problem, different vehicle manufacturers have made different designs. For example, the network of electronic controllers such as steering power and body stability can be divided according to the correlation between handling and driving. In this way, even if TBOX and CAN gateway are attacked, the vehicle can still ensure the basic driving function, for the communication between electronic controllers related to handling and driving does not need to be forwarded by the gateway, as shown in Fig. 2.47 [67].

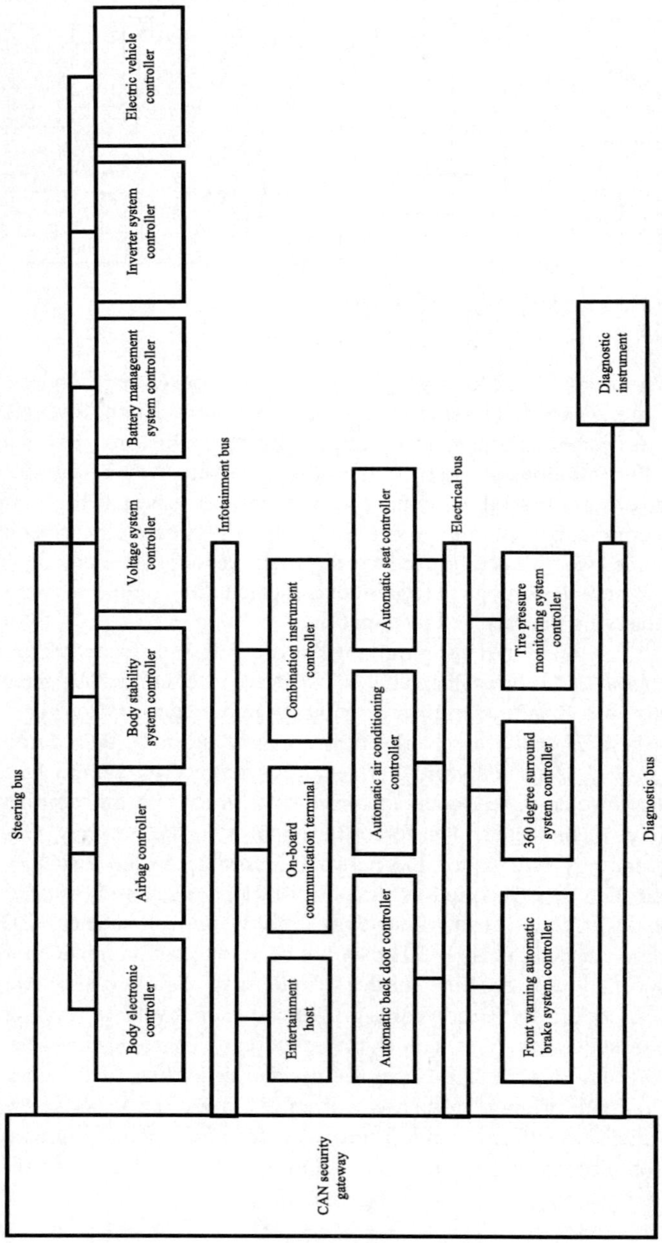

Fig. 2.47 IV Cybersecurity topology with network divided in terms of control driving correlation

2. Cybersecurity Intrusion Detection and Defense

The key component of IV is one of the main objects of cyber attacks. The attacker looks for and attacks the vulnerabilities of the network, equipment, and application, and then obtains relevant sensitive information, system configuration, and other resources. Due to the openness and dynamics of the IOV environment, once it is invaded interiorly by the attacker in some way, serious cybersecurity problems will be generated. Therefore, intrusion detection and defense is one of the key issues faced by automobile manufacturers, original equipment manufacturers (OEM), and suppliers. Equivalent to a network real-time monitoring system, the intrusion detection system can judge whether the data source information and the user behavior mode are abnormal. As the extension of the intrusion detection system, the intrusion prevention system provides protection measures and tries to prevent the possible intrusion identified and recorded by the detection system. The two systems together constitute the active defense technology for IV cybersecurity.

There are many types of intrusion attacks in IOV communication, mainly including replay, tampering, interruption, camouflage, etc., as shown in Table 2.19 [68]. Various intrusion attacks will make the user system behave abnormally. For example, multiple password input errors are suspicious abnormal behavior; Different times and different places login is a suspicious abnormal use of I/O interface; The rapid rewriting or storage of executable files caused by Trojan virus is a suspicious abnormal use of CPU. These suspicious abnormal behaviors make intrusion detection and defense possible.

The operation of intrusion detection and defense system mainly has three stages, including status monitoring, intrusion detection, and intrusion defense stage. The status monitoring stage reflects the monitoring feature that the system needs to collect network status, system status, user behavior, and other information of IV and some key nodes in IOV. The system often can not directly find the problem from the massive original information collected. It needs to analyze the information in the intrusion detection stage to filter out the abnormal intrusion information. The methods used in this stage include statistical analysis, pattern matching, integrity analysis, and neural network, for extraction of important data related to intrusion. If a cyber attack is

Table 2.19 Some common intrusion types in IOV communication

Type of cyber attack	Attack means
Replay	When the vehicle communicates with the outside, the attacker repeatedly sends useless information and occupies vehicle resources
Tampering	When the vehicle communicates with the outside world, the attacker illegally modifies the outside information and sends it to the vehicle
Interruption	The attacker interrupts the communication of the vehicle through dos and other means, making it unable to receive information
Camouflage	The attacker, disguised as a legitimate device, communicates with the vehicle and obtains information or misleads the vehicle

confirmed, it will carry out an attack alarm, lock the components or subsystems that the intruder is trying to attack, and determine the risk and severity of the attack at the same time [69, 70]. In the intrusion prevention stage, the system will take measures to prevent the intrusion. Several common warning and response countermeasures include: ① the system itself directly cuts off the intrusion connection, such as terminating the network connection or user dialogue being attacked by the intrusion; ② Change the security environment, such as reconfiguring firewall, router, and switch and other network equipment to prevent the access of intruders; ③ Filter the security information, such as deleting the infected attachments in the email and then allowing the cleaned email to reach the inbox normally; ④ Send warning information to the administrator, and record the information and response related to intrusion behavior such as attack time, type and target in the log [71].

Intrusion detection and defense can be realized generally in two approaches. One approach transforms the detection problem into pattern matching and indicates that the system has been invaded by monitoring the specific system state transformation. It needs to build a rule base on a priori knowledge and is widely applied at present. But in theory, the artificially defined specifications are usually not enough to cover all possible intrusion scenarios. Another approach is to build an artificial neural network, which can detect new attacks. However, it is prone to missing and false reports, and its effectiveness should be improved by further research [72].

The approach on rule base makes full use of abnormal performance of each node of IOV in the face of intrusion, and defines each possible anomaly measure as a binary variable A_i, with the value of 1 indicating abnormal behavior or state. H represents the assumption that IOV is being invaded, and $\overline{H}$ represents its opposite event. Then the reliability and sensitivity of each anomaly measure are expressed by the following formula:

$$\begin{cases} \text{reliability} = p(A_i = 1|H) \\ \text{sensitivity} = p(A_i = 1|\overline{H}) \end{cases} \tag{2.44}$$

According to the Bayes theorem, the prior knowledge of the abnormal performance when the intrusion occurs can be used to deduce the probability of intrusion given the abnormal performance:

$$p(H|A_1, A_2, \ldots, A_n) = \frac{p(A_1, A_2, \ldots, A_n|H)p(H)}{p(A_1, A_2, \ldots, A_n|H)p(H) + p\big(A_1, A_2, \ldots, A_n|\bar{H}\big)p\big(\bar{H}\big)} \tag{2.45}$$

where, it should be noted that the anomaly measure A_i and A_j are likely to be related. After encoding all anomaly measures into vector A, it is also necessary to model its covariance matrix [73].

The detection and defense of unknown vulnerabilities is an urgent field to be developed, which needs to be extended by artificial neural networks and machine learning. The training and testing of artificial neural networks need to collect a large amount of data and extract feature entities, including the number of packets transmitted between vehicles, the delay of packet transmission, the number of lost packets, and so on. AI system for intrusion detection generally uses a feedforward neural network to describe as follows: input extracted features, pass through several hidden layers, output the final classification decision results, and finally warn or accept the abnormal behavior based on the threshold [74]. Network evaluation often makes use of some statistical data, including detection rate (the ratio of the number of correctly detected intrusions to the number of attacks), false alarm rate (the ratio of the number of incorrectly detected intrusions to the number of non-attacks), accuracy rate (the ratio of the number of correctly classified times to the total amount of data), etc.

According to the type of objects monitored by the intrusion detection and defense system and the way of deployment, several categories are classified as shown in Table 2.20. The host-based system takes the onboard system log and application log as the information source, which can identify the attack intention and has excellent real-time performance. However, due to the limited information provided by the system log, the secret intrusion is likely to be ignored. The network-based system takes each network packet as the information source and can detect the relevant information of multiple vehicle nodes, but can not detect the pre-encrypted packet. For potential intrusion attacks on IOV, WLAN technology is also one of the important technologies in the intrusion prevention system. Most intrusion prevention systems can specify the prevention configuration for each type of alarm, including enabling or disabling prevention and specifying the type of prevention.

The distributed intrusion detection and defense system is designed according to the characteristics of VANET and is composed of various nodes in the network. Figure 2.48 shows the system structure and information transmission process between the two nodes [75]. Each vehicle node uses the improved Naive Bayesian classification algorithm to start local detection and adopts the update algorithm to continuously renew the feature library so that the abnormal data features are rich enough for improvement of the efficiency and capability of local detection. When detecting an

Table 2.20 Types of intrusion detection and defense system

System type	Operation mode
Host-based system	Monitor the characteristics of a single host as well as suspicious activities and events on the host
Network-based system	Monitor network data of specific network segments or devices, and analyze network and application protocol activities
Distributed detection system	Based on the above two systems and combined with the characteristics of the vehicular ad-hoc network (VANET)
Network behavior analysis system	Check network traffic and identify threats that generate abnormal traffic, such as DoS attacks
Wireless LAN (WLAN) monitoring system	Monitor and analyze wireless network traffic to identify suspicious activities involving wireless network protocols

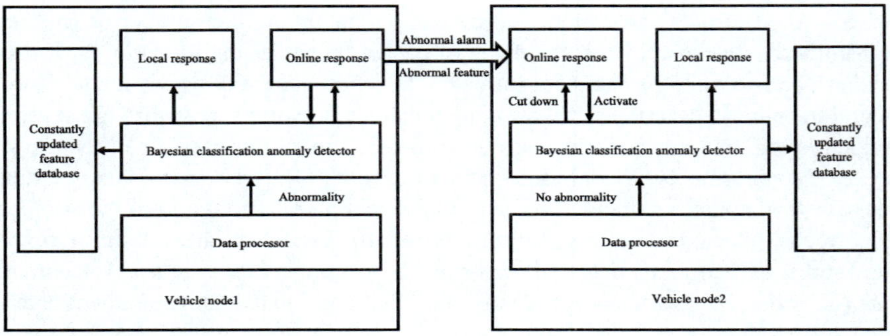

Fig. 2.48 Structure of distributed intrusion detection system

abnormality, a node will store it in the local feature library, and send its alarm and feature to the adjacent nodes. The adjacent nodes can also use their feature library for independent local detection and disconnect from the source node, to enhance the network analysis and detection capability and reduce the overall hazard brought by attacks.

3. Cybersecurity Encryption and Communication

Compared with the intrusion detection and defense system which actively monitors, analyzes, and protects the communication environment and content, the security encryption and communication technology develop passively in the process when information is consistently intercepted, cracked, and forged. In the traditional field of computer communication, the basic cryptography such as symmetric cryptography and public-key cryptography and the related theories and technologies of Applied Cryptography such as digital signature and access control have a very long history, which is of certain reference for IOV cybersecurity. However, due to the frequent switching of communication networks, effective links are relatively short and real-time requirements are quite strict. Therefore, traditional encryption measures are not fully applicable to ICV. It is necessary to design faster and safer cybersecurity technology according to the special safety requirements of IOV.

Generally speaking, basic encryption and decryption algorithms are mainly divided into the symmetric cryptosystem and the public-key cryptosystem (asymmetric encryption and decryption algorithm). The symmetric cryptosystem has the same encryption and decryption key, fast computing speed, and complete dependence on the key. It mainly includes stream cipher and block cipher. Block encryption is the idea that the well-known Advanced Encryption Standard (AES) adopts. The calculation of the public-key cryptosystem is slow but relatively more secure. The more popular ones are asymmetric encryption RSA[1] system based on large integer factorization, the public-key system of Differ-Hellman based on discrete logarithm problem over a finite field, the public-key system of Elliptic Curve Cryptography (ECC) based on discrete logarithm problem on the elliptic curve, etc. [76]

As a standard public-key encryption algorithm recommended by ISO, RSA can resist most cryptographic attacks. Its basic idea is that it is easy to multiply two large prime numbers, but difficult to factorize the product. It is estimated that the RSA key with a length of 2048 bits will not be cracked until 2030. The relevant theories of Rivest Shamir Adleman (RSA) are as follows:

Any given positive integer *n* in the positive integer less than *n*, the number of coprime with *n* is recorded as the value of Euler function $\varphi(n)$, such as $\varphi(7) = 6$ and $\varphi(8) = 4$. Later, these lemmas related to Euler functions will be used: if *n* is a prime number, then $\varphi(\mathrm{n}) = n - 1$; If *n* can be decomposed into the product of two prime numbers, then $\varphi(n) = \varphi(p \cdot q) = \varphi(p) \cdot \varphi(q)$ Euler's theorem, that is, if two positive integers *m* and *n* are coprime, then the power of $\varphi(n)$ of *m* subtracts 1 can be divided by *n*, as follows,

$$m^{\varphi(n)} \mathrm{mod} n \equiv 1 \tag{2.46}$$

K power operation is performed on both sides of Formula (2.46) and then multiplied by *m*. Following the law of modular operation, it can be deduced,

$$m^{k \cdot \varphi(n)+1} \bmod n \equiv m \tag{2.47}$$

Another definition is that if two positive integers *e* and *x* are coprime, the integer *d* must be found so that *e*·*d* minus 1 can be divided by *x*. At this time, *d* can be regarded to be the modular inverse element of *e* relative to *x*.

$$(e \cdot d) \bmod x \equiv 1 \tag{2.48}$$

If $\varphi(n)$ is regarded as *x*, that is, $e \cdot d = k \cdot \varphi(n) + 1$, it can be gained that

$$m^{e \cdot d} \bmod n \equiv m \tag{2.49}$$

The condition of this formula is that *m* and *n* are coprime, and *e* and φ(n) and are coprime. The intermediate quantity *c* is introduced, so that,

$$m^{e} \mathrm{mod} n = c \tag{2.50}$$

It can be deduced,

$$c^{d} \mathrm{mod} n = m \tag{2.51}$$

The above two formulas indicate the operation principle of key exchange in the RSA algorithm, where *m* represents the plaintext to be encrypted, *c* represents the sent ciphertext, (*n*, *e*) is the open public key, and (*n*, *d*) is the private key. The selection approach of parameters *n*, *e*, and *d* involved in the algorithm is as follows: ① *n* is the product of two large prime numbers *p* and *q*, that is, $n = p \cdot q$. At this time, the number of bits occupied by the binary of *n* is the key length; ② it is required e and $\varphi(n)$

are coprime, and $\varphi(n) = \varphi(p) \cdot \varphi(q) = (p-1) \cdot (q-1)$. Therefore, taking *e* only requires to be coprime with $(p-1)(q-1)$; ③ The selection of *d* is related to *e*. *d* is the modular inverse element of $\varphi(n)$ relative to *e*. The safety of the RSA algorithm depends on large number decomposition. If an attacker wants to deduce the private key (*n*, *d*) in the case of having obtained the public key (*n*, *e*), it is necessary to know the value of the Euler function $\varphi(n)$ and calculate the modular inverse element of *e* relative to it, which is difficult when *n* is very large.

Asymmetric encryption and decryption algorithms are often used in conjunction with digital digest algorithms to generate digital signatures. The digital digest algorithm, also known as the Hash Algorithm, only performs one-way encryption without a key. The same digest can be obtained only by the same plaintext input and the same digital digest algorithm. Therefore, the sender can encrypt the digest of the message with the private key to generate a digital signature and send it together with the message. Then the receiver decrypts the digital signature with the public key and processes the received message with the same digital digest algorithm. If the results are consistent, it can be proved that the message has not been tampered with.

Due to the outstanding advantages of asymmetric encryption and decryption algorithms and the rapid application of public-key technology, a series of cybersecurity management systems have been gradually derived. Among them, Public Key Infrastructure (PKI) is regarded as the basic authentication system. The public key cryptosystem is used to transparently provide key management and certificate management services for encryption, signature, and other operations in network applications. PKI generally includes certification Authority (CA), registration authority, policy management, key management, certificate management, key backup, key recovery, revocation system, and other functional modules. In the public-key encryption system, CA signs the digital certificate to prove the relationship between the public key and the user and provides users with identity authentication or signature authentication and other services. The recipient decrypts the sender's digital certificate using the CA's public key to get the sender's real public key and then can process the received message and digital signature. This idea of encryption and authentication can be adopted for reference in the IOV environment. However, the key generation speed of the RSA algorithm is slow, taking 2–3 s when the dominant frequency is 5 GHz, which is unacceptable in IOV [77]. For a long time, international common cryptographic algorithm systems and relevant standards such as Triple Data Encryption Algorithm (3DES), Secure Hash Algorithm 1 (SHA-1), and RSA have been applied in significant national fields. However, with the continuous change in the international situation and the breakthrough development of new technology, it is urgent to realize domestic development and independent control of the cryptographic algorithm of security products. In recent years, China has independently developed a series of SM[3] algorithms and the Zuchongzhi algorithm (ZUC), etc. Table 2.21 summarizes the applicable national commercial cipher algorithms and corresponding international algorithms in IOV. With the same security strength and related software and hardware performance as AES, the SM1 algorithm is not disclosed and only stored in the chip in the form of Intellectual Property Core (IPC). As a kind of ECC, the SM2 algorithm is more advanced and secure than the RSA algorithm. It has high password complexity, small storage space, and fast processing speed used to replace the RSA algorithm in China's commercial cryptosystem. Improved and implemented based on the Secure Hash Algorithm of SHA-256, the SM3 algorithm is suitable for digital signature and verification as well as random number generation in commercial cryptographic applications. Being a block symmetric cipher algorithm specially designed for WLAN products in China, the SM4 algorithm is faster than 3DES and is used to ensure the confidentiality of data and information. Utilizing SM2 and SM3 algorithms, various independent innovative technologies and approaches such as V2X information signature, short certificate, key derivation, and reconstruction can be realized in China's 5G-V2X communication security and trust system [78]. As shown in Fig. 2.49, the Intelligent Transportation Certificate Authentication System (ITS-CA) based on PKI technology and V2X security chip constructs a unified vehicle-environment-cloud identity authentication system, builds the cybersecurity support platform of IOV for auto enterprises, and provides end-to-end safety protection products for vehicle-environment cooperation [79].

The communication between ICV and the outside environment includes in-vehicle communication, near-end communication, V2X communication, and satellite communication according to the distance from near to far. Among them, in-vehicle

Table 2.21 Applicable national secret algorithms and corresponding international algorithms in IOV

National secret algorithm	International algorithm	Algorithm type
SM1	AES	Block cipher algorithm
SM2	RSA	Public key algorithm
SM3	SHA-256	Digital digest algorithm
SM4	3DES	Block cipher algorithm

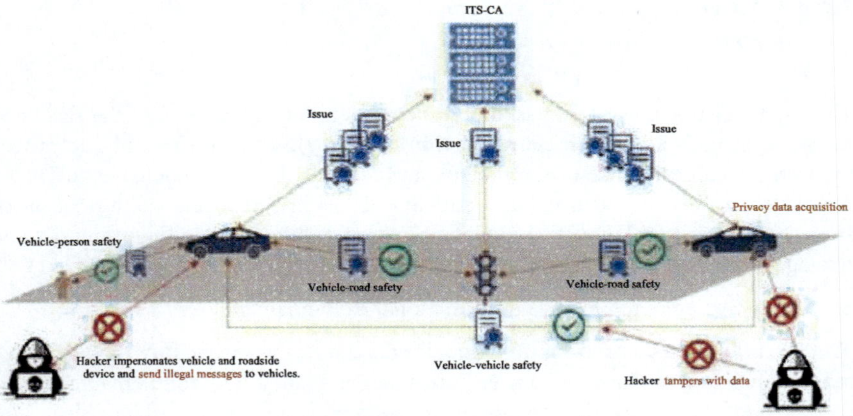

Fig. 2.49 ITS-CA safety authentication system

communication mainly involves information interaction among various domains, which is related to the design of isolation domain and gateway. Near-end communication includes the near-end wireless communication security techniques such as WiFi and Bluetooth and the physical contact devices such as USB and optical drive. For example, the cybersecurity hidden dangers in suspicious open hotspots are analyzed through the wireless communication device defense system; V2X communication includes vehicle-vehicle, vehicle-human, vehicle-environment, vehicle-cloud, and other communications, involving technologies such as physical interface security between vehicle terminal and roadside equipment and network access security in the scenario of the cellular network and direct communication. For example, remote updates of the vehicle system can be carried out employing OTA technology in vehicle-cloud communication; Satellite communication mainly involves safety techniques such as positioning and navigation. For instance, anomaly detection, Open Service Navigation Message Authentication (OS-NMA), and other technologies can be applied to mitigate interference and prevent deception.

2.6 Typical Systems in Terms of System Safety

In the process of IV development, it is necessary to integrate system safety technology into the specific design based on system safety development theory and design approach, to develop a vehicle system that can ensure system safety. According to the main principles of design basis, there are three types of typical systems, including safety assurance system based on functional redundancy, safety assurance system based on self-monitoring, and cybersecurity system based on in-depth defense.

2.6.1 Safety Assurance System Based on Functional Redundancy

The safety assurance system based on functional redundancy reduces the risks due to potential fault and insufficient function of a single function module, improves the overall fault tolerance of the system, and realizes the comprehensive guarantee for system safety, especially FuSa and SOTIF, by the repeated configuration of some components in IV system and combination with corresponding monitors and managers.

1. Safety System Architecture for Functional Redundancy

The overall architecture of the functional redundancy system mainly includes such main functional modules as power module redundancy, environment perception module redundancy, positioning module redundancy, decision control module redundancy, execution module redundancy, and architecture redundancy, as shown in Fig. 2.50.

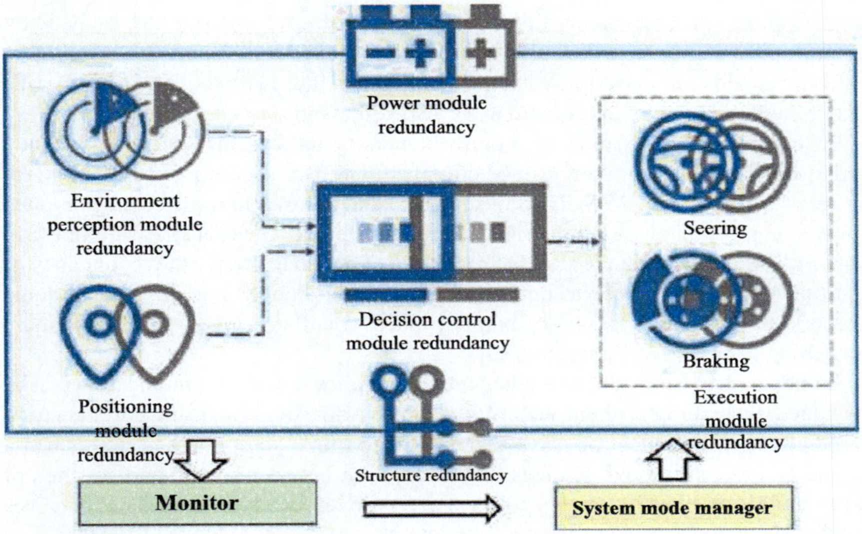

Fig. 2.50 Overall architecture of functional redundancy system

In terms of the power supply system, the standby power supply can be set to continue power support for some time when the main power supply fails, to ensure safe parking of IV. By setting an independent power supply in each key drive system, it can ensure that these systems can still operate safely for a period of time in case of single circuit fault or circuit interruption.

In terms of environmental perception, the redundancy scheme of multiple sensors such as ultrasonic sensor + millimeter wave radar + camera + laser radar is adopted. By making full use of complementary advantages of different sensors and sensor status monitoring techniques, the adaptive adjustment of multi-sensor fusion mode and the treatment of individual sensor's faults or insufficient functions can be carried out, in view of the applicability of each sensor to different application scenarios. Thus, the environmental information modeling is more reliable.

In terms of IV positioning, a redundant positioning scheme composed of absolute positioning based on satellite signals and relative positioning based on road features can be adopted. The former can be based on satellite positioning intelligent sensor, which integrates GNSS signal receiver, high-precision Inertial Measurement Unit (IMU), and signal deviation correction service, to ensure the positioning stability under various working conditions; The latter can collect, upload and compare road features by various on-board sensors such as camera and radar, to achieve high-precision positioning. The combination of the two schemes realizes further redundancy and can check each other under adverse conditions such as bad weather or occlusion, to ensure the reliability of the positioning system.

In terms of decision control, two sets of core computing units can be configured to ensure that the other system can take over in time when a single system fails so that IV can still run safely or stop at a degraded level. In addition, by backing up multiple collision detection and avoidance systems, the surrounding environment can be scanned continuously. When the main system fails to effectively identify risks due to faults or insufficient functions, IV can still avoid obstacles safely.

In terms of execution, braking system redundancy and steering system redundancy are involved. Vehicle braking can be equipped with two sets of main and auxiliary systems. For example, BOSCH adopts ESP 9.3 and ibooster to form a dual guarantee for braking. As an electric braking system, ibooster can facilitate the safe brake in case of ESP failure; Vehicle steering can also be applied to realize the redundancy of multiple components, such as double winding motor, double axle drive, and double sensors, to establish two sets of parallel independent steering systems for timely switching in case of system fault [80].

In terms of architecture, redundant architecture for secure communication can be designed. For example, Great Wall Motor adopts a dual communication architecture, builds three independent physical communication links to connect in time when a single link is interrupted, realizes the "distinction between public and private" of communication lines by establishing ten software logic control paths, and achieves transmission and sharing of information among more than 30 kinds of sensors.

To ensure reasonable deployment of the above redundant systems, monitors need to be designed to monitor the status and behavior of system elements in terms of performance as well as safety events and faults, such as whether the system reaches specific nominal performance, whether system failure occurs, the availability of failure degradation capability and cybersecurity events. Hence, system hazards can be timely identified, with specific causes analyzed. If the monitor recognizes insufficient function or failure of the system, it will send this information to the system mode manager. After receiving the above failure, nominal performance indication, degraded performance indication, vehicle, and user status, and other information, the mode manager can further control the vehicle and adopt reasonable safety strategies to ensure the safe operation of the system.

2. Safety Strategy for Functional Redundancy

In view of different risk conditions led by a system failure, the Minimal Risk Condition (MRC) is defined as a safe operation state with the least risk when the function of the system is not fully available due to system failures (main system failure and insufficient function caused by trigger conditions, etc.), such as driver taking over, restricted operation, and end of the operation; The minimal risk maneuver (MRM) is defined as a series of emergency operations to reach the MRC, including timely enabling the backup system, takeover request, effective function status, gentle/emergent available parking, etc. Its purpose is to make the vehicle reach a tolerable risk level at the functional level it can support. The relationship between MRM and MRC is shown in Fig. 2.51 [51], where nominal operation refers to the operation state when the system function is fully available. If the conditions of the nominal operation state cannot be met due to system failure, different MRMs can be implemented according to the actual situation to realize degraded operation and achieve

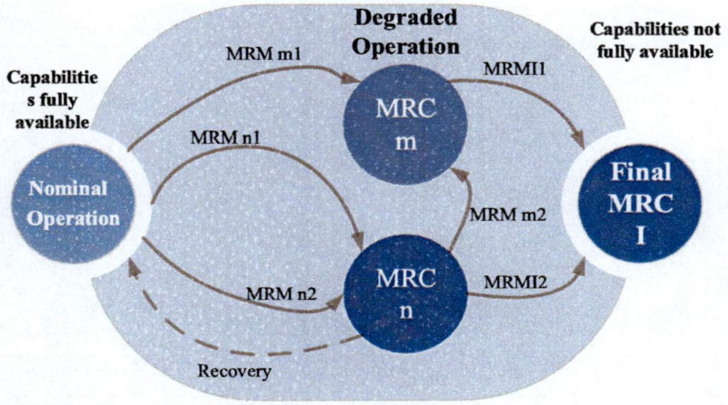

Fig. 2.51 Relationship between MRM and MRC [51]

appropriate MRC. Regarding the complexity of the system and influencing factors, multiple MRCs and MRMs can be experienced continuously.

In terms of different lowest risk states, the scenarios can be divided into three categories [50]: Fail Operational (FO). For instance, even with a sensor fault in the system, redundant sensors can still retain the safe operation of the system; Fail Degraded (FD). Degraded operation is conducted to ensure that the system can still operate in case of fault, yet possibly without all functions available; Fail-Safe (FS). Even if the system is no longer running, the failure will no longer lead to unsafe consequences, such as a shutdown. Based on the requirements of different scenarios, essential capabilities for the IV system include FS capability and FD capability [50], which relationship is shown in Fig. 2.52. FS capability can be interrupted or covered by FD capability because the correlation between its unavailability and safety is low enough. FS capability mainly includes seven categories: perceiving relevant objects, determining location, predicting the future behavior of relevant objects, formulating collision-free and legal driving planning, correctly implementing driving planning, communicating and interacting with other vulnerable traffic participants, and determining whether specific nominal performance is achieved. FD capability guarantees stable performance at a certain level in case of system failure, to realize the safe transformation to the lowest risk situation. It mainly includes six categories: ensuring controllability of the driver, detecting the availability of degradation mode, ensuring safe mode transformation and user cognition, dealing with nominal performance deficiency and other failures, degrading system performance in case of failure, and executing degradation mode within the reduced system limit.

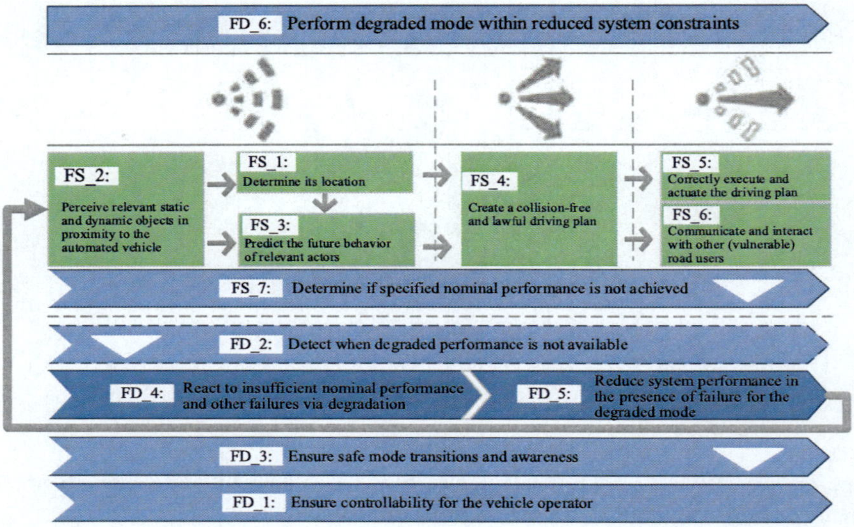

Fig. 2.52 Relationship between FS capability and FD capability [51]

2.6.2 *Self-monitoring-Based Safety Assurance System*

The safety assurance system based on self-monitoring focuses on the promotion and utilization of "self-awareness" of the system. The "self-awareness" of the system refers to its capability to identify its state, possible behaviors, and their impact on itself and the environment, and to respond safely [81]. IV is a complex safety–critical system. If "self-awareness" of such layers as perception, decision-making, and control conflicts with each other, disastrous consequences are likely to occur. Therefore, cross-layer solutions must be considered to ensure that indexes from different levels can be aggregated into consistent self-expression of the system, to realize the overall safety monitoring.

1. Safety System Architecture for Self-Monitoring

The self-monitoring safety system is usually integrated into the IV system in the form of safety modules. A typical implementation architecture is shown in Fig. 2.53. The Brunswick University of Technology integrates relevant safety modules (shown in red box) into the Guidance and Control System (GCS) of its autonomous vehicle Leonie. The data acquisition part of the system obtains information on the environment, traffic participants, vehicle status, and its location, inputs it to the safety monitoring module, and then carries out adaptive safety control by the monitored system status.

Safety monitoring system collects data from functional modules and sensors, calculates performance indexes, and judges the execution of degradation actions. Its core elements are controlling the watchdog and safety unit. The controlling watchdog can monitor horizontal and vertical controllers and other modules, collect heartbeat

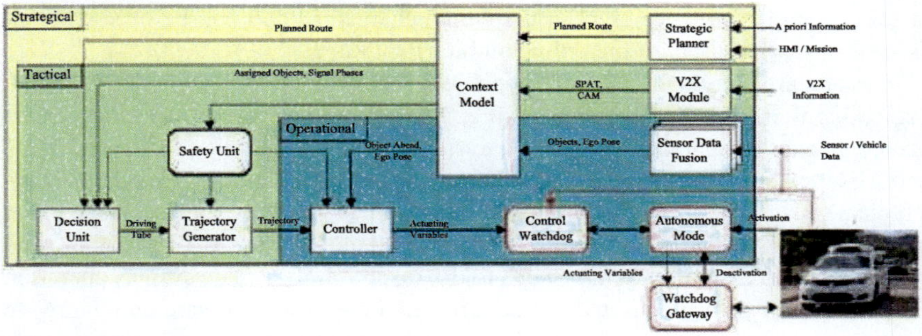

Fig. 2.53 Safety system architecture for self-monitoring [82]

and other signals from the associated hardware, and calculate performance indexes including position accuracy, visible area, system operation status, response time, and so on. The safety unit combines kinetic data from IV sensors and GPS/INS, estimates grip index according to road and weather conditions, and dynamically adjusts driving strategy to prevent accidents under severe road and weather conditions.

The watchdog gateway and autonomous mode module are jointly responsible for activating and stopping automated driving mode. As the interface between IV and its system, the watchdog gateway is responsible for: monitoring CAN signal and providing important vehicle parameters; Verifying the control value of the vehicle actuator to ensure safe actuator control; Limiting the steering angle of the vehicle at the current speed; and monitoring driver interference during the test on IV systems. The autonomous mode module receives inputs from the vehicle, HMI, and controlling watchdog. The module requests activation from the watchdog gateway only when all doors are closed, ESP is activated, the watchdog gateway and controlling watchdog send signals, the system and vehicle are in good condition, and the driver requests activation of automated driving.

2. Safety Strategy for Self-Monitoring

Self-monitoring-based system safety shall be realized together with activities in system safety design. Specifically, it can be assisted by a capability diagram and skill diagram [83], as shown in Fig. 2.54. Capability refers to all conditions necessary for a certain part of the driving task. In the conceptual stage of system safety design, the design of the capability diagram can be parallel to the project definition and used to support further safety development. Its main components include capability node and quality demand edges; Skill is defined as the description of specific task-related activities, such as performance level. As a specific technical solution after the instantiation of abstract concepts in the capability diagram, the skill diagram mainly includes skill nodes and performance impact edges and should focus on the implementation of the redundancy mechanism. Skill chart is essential for system safety self-monitoring. If the performance of each skill can be measured and there is a sufficiently complete mathematical model to express the propagation of the above performance level in the whole chart, the chart can be of significant assistance for the safety of the online monitoring system.

The actual safety self-monitoring process mainly includes two stages: performance monitoring and problem handling. The former can rely on specific performance indexes to characterize possible system problems regarding FuSa or SOTIF, and the latter can ensure system safety by functional degradation. Figure 2.55 shows a typical self-monitoring safety system strategy, which measures performance index by monitoring system parameters. The system response time is the waiting time from the occurrence of external events to the response of IV; The system operation state includes signal runout and periodic information of all software and hardware modules involved; The visual area represents the space covered by sensors around; The grip reflects the vehicle kinetic performance, which changes dynamically in line with road, weather, and other environmental conditions; Position accuracy is the evaluation of the accuracy of current IV position estimation. Appropriate function degradation can be further taken by calculating results of performance indexes, including modifying driving parameters (e.g., reducing vehicle speed, and increasing safety time-distance), modifying driving strategies (e.g., lane changing and turning), enforcing safety strategies, or prohibiting driving strategies (e.g., disabling invalid sensors or actuator components), etc.

Taking vehicle adhesion performance (grip) as an example [84], such indexes can be monitored to adapt to changing road and weather conditions, and to adjust appropriate safety strategies in real-time according to system performance constraints. The system status data can be collected by IMU (provide current traction angle), Lidar, digital map (store traffic signal information), rainfall sensor, temperature sensor, wheel speed sensor, as well as status monitors for partial functions (e.g., ESC, Automatic Traction Control-ATC, and ABS). And corresponding performance indexes can be calculated accordingly, as shown in Fig. 2.56.

Where, the initial adhesion performance index $g_{n\text{-}1}$ conducts condition determination and correction according to the interference of ESC, ATC, ABS, and other functions identified by the sensor, and the parameters including difference between the average values of front and rear wheel speedsδ_{v}, sideslip angles_{c}, rainfall r,, temperature$T$, etc., and finally obtain the current adhesion performance index determine and correct the conditions of such as rainfall R and temperature T, and finally obtain the current adhesion performance index. x_i in the Figure is the adhesion performance index changed according to the functional interference,t_{v},t_{s}, and $T_{\max}$ are the judgment thresholds,c_{v},c_{s}, and c_{t} are the scale factors.

The performance index value can be further combined with the vehicle longitudinal control strategy. Taking the intelligent driver model as an example, safe and stable car-following behavior can be achieved by the output of the longitudinal expected acceleration $\dot{v}$.

$$\dot{v} = a \cdot \left[1 - \left(\frac{v}{v_0}\right)^d - \left(\frac{S^*}{S_a}\right)^2\right] \tag{2.52}$$

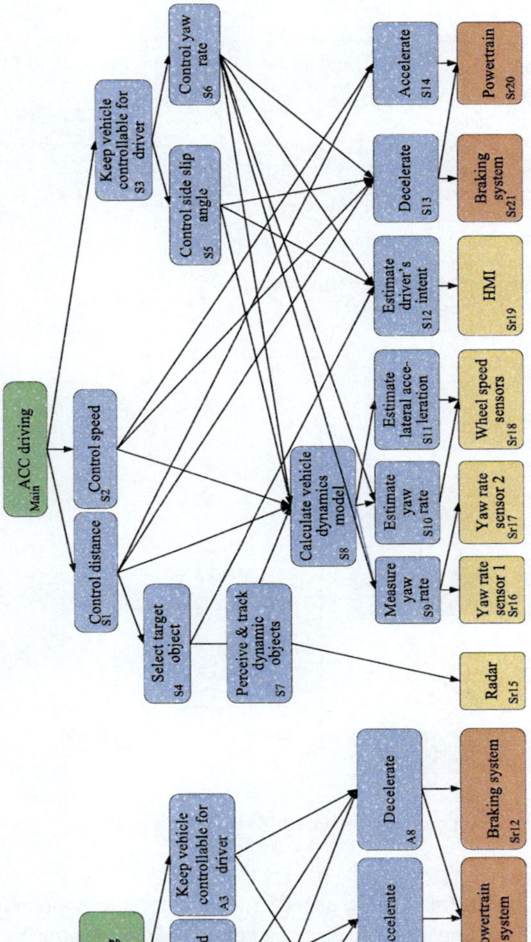

(a) Simplified ACC ability graph.

(b) Further detailed ACC skill graph with redundancies and exemplary metrics.

Fig. 2.54 Example of capability diagram and skill diagram [83]

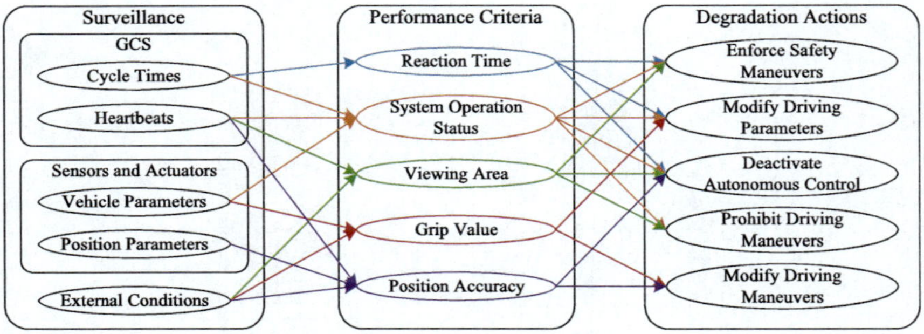

Fig. 2.55 A typical self-monitoring safety systems strategy [82]

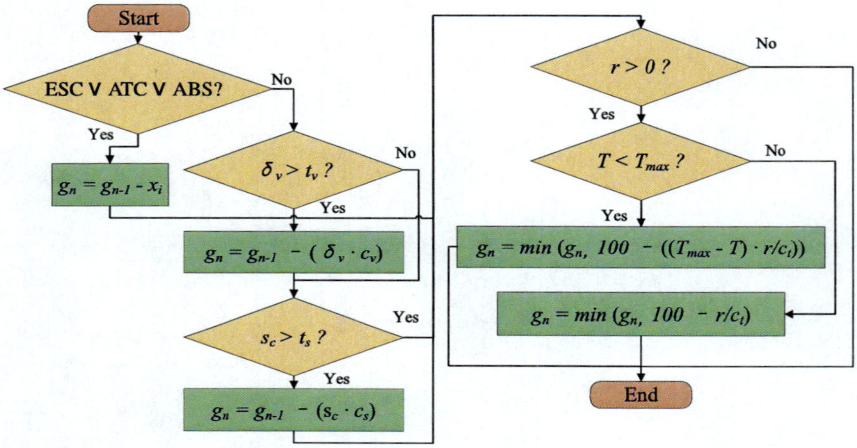

Fig. 2.56 Calculation of adhesion performance index [84]

where, v represents the current vehicle speed; v_0 indicates the expected speed; s_a represents the distance between the front and rear vehicles; a represents the maximum acceleration; d represents the acceleration index; s^* indicates the safe distance.

The calculation relationship between s^* and such parameters as congestion spacing s_0, safe time headway T, the speed difference between the ego-vehicle and front vehicle Δv, and the expected deceleration b is as follows:

$$s^* = s_0 + T \cdot v + \frac{v \cdot \Delta v}{2\sqrt{a \cdot b}} \tag{2.53}$$

Reduction of the expected speed v_0 can facilitate a more conservative and safer drive under severe road and weather conditions. Specifically, the speed coefficient c_{vf} integrated by the grip index and longitudinal controller can be used:

$$v_0' = v_0 \cdot \left[\left(1 - c_{vf}\right) + \left(\frac{g_n \cdot c_{vf}}{100}\right)\right] \tag{2.54}$$

where, v_0 and $v_0\prime$ respectively indicate the expected speed before and after adjustment.

If the grip index value is lower than 100, the safe time headway T can be appropriately extended, to ensure car-following safety by earlier braking, as well as the safety when encountering obstacles and traffic lights:

$$T' = T \cdot \left[2 - \left(\frac{g_n}{100}\right)\right] \tag{2.55}$$

where, T and $T\prime$ respectively represent the safe time headway before and after adjustment.

In addition, the acceleration value a_n can be reduced directly to achieve an adaptive safety strategy for grip.

$$a_n' = \min\left[a_n, c_f + a_n \cdot g_n \cdot \frac{\left(1 - \frac{c_f}{100}\right)}{100}\right] \tag{2.56}$$

where, parameter c_f is used to determine the lower bound and influence factor of g_n; a_n and a_n' respectively represent the acceleration value before and after adjustment.

2.6.3 Cybersecurity System Based on In-Depth Defense

This system aims to introduce the concept of "in-depth defense" into the solution of ICV cybersecurity system, establish as many levels of protection as possible between threats of the external world and IOV, and protect key ICV applications of communication and control, etc., by deploying multi-level and different targeted security measures.

1. Safety System Architecture for In-depth Defense

The cybersecurity system of ICV generally adopts an architecture of in-depth defense safety system. Even if the shallow defense is broken, the deep defense can effectively resolve cyber attacks. In addition, the in-depth defense can ensure that the problem can be effectively controlled once it occurs and will not quickly spread to other IV systems. The development of the IOV cybersecurity infrastructure platform requires the application of key technologies such as end-pipe-cloud cybersecurity protection technology, autonomous and controllable on-board password technology, and intelligent detection technology for the security vulnerability, to build on-board security architecture and end-pipe-cloud system for in-depth defense of ICV.

The on-board cybersecurity architecture for in-depth defense can be summarized into four general levels, namely, secure interface, secure gateway, secure network, and secure processing, which are respectively represented by serial numbers ①–④ in Fig. 2.57. In addition, there are various access control and anti-theft systems as a supplement [66]. A secure interface realizes hardware encryption. Its functions mainly include secure Machine-to-Machine/Man (M2M) authentication and secure key storage, as well as secure access rights of Telematics Control Unit (TCU) and On-Board Diagnostics (OBD) ports. Secure gateway realizes the separation of concerns of on-board network systems and ensures the wireless update of safety. Its functions mainly include isolation domain, firewall, centralized intrusion detection, and so on. Secure network realizes secure vehicle network communication. Its functions mainly include message information authentication, dangerous message blocking, intrusion distributed detection, and so on. Secure processing is deployed on each control unit and processor for vehicle interconnection. Its functions mainly include the safe start of ECU and operation integrity guarantee, etc. Safe vehicle access control includes the smartphone-based remote switch and remote vehicle monitoring. This multi-level architecture as a whole ensures IV cybersecurity, posing many difficulties for Hackers' invasion.

DENSO released its in-depth defense system architecture for onboard cybersecurity of single vehicles in 2016, which includes four protection layers of External Connection, Gateway, LAN, and ECU, for the protection of people and vehicles from various network attacks, as shown in Fig. 2.58. The external interface of DENSO has the function of resisting attacks such as external abnormal data packets; the gateway has the function of keeping attacks away from the system; Identity authentication plays an important role in vehicle and LAN safety. As shown by the red line in Fig. 2.58, after a cyber attack against the ADAS system breaks through the V2X interface and gateway, it was disintegrated in the LAN layer due to identity authentication failure, which ensures LAN security; ECU protection measures can also protect ECU from damage during charging by Power Line Communication (PLC) technology [85]. The in-depth defense system deployed by DENSO shows the implementation scheme of general four-layer architecture and sets a benchmark in the industry. Since then, Continental of Germany, Baidu Apollo of China, and AutoCryp released cybersecurity solutions for in-depth defense respectively in 2017, 2018, and 2020. In addition to the deployment of the above layers, some of their characteristics are summarized in Table 2.22.

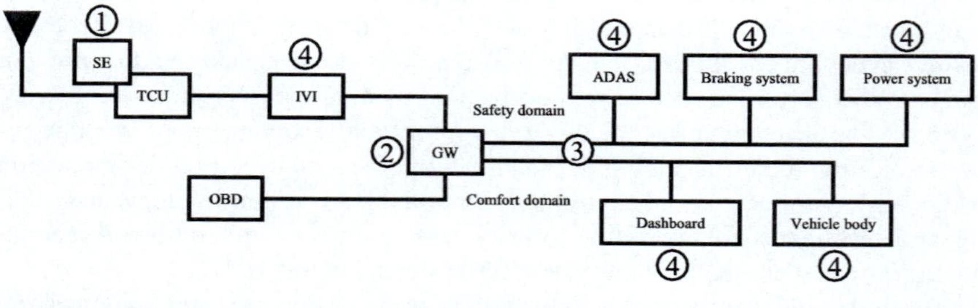

Fig. 2.57 General cybersecurity system architecture with four layers based on in-depth defense

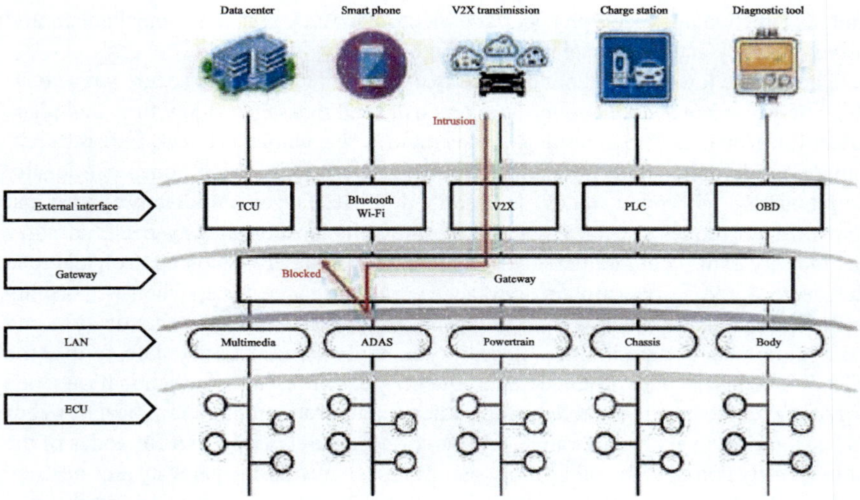

Fig. 2.58 Cybersecurity system architecture of DENSO based on In-depth defense

Table 2.22 Cybersecurity solutions for in-depth defense released by different enterprises

Enterprises	Scheme characteristics
Continental	It is end-to-end and modular; the safety operation center provides vehicle condition monitoring, OTA, and other services; provides real-time big data analysis and monitoring, and threat response for the fleet [86]
Baidu Apollo	Implement cybersecurity regulation with three key levels in external communication, access gateway, and in-vehicle applications; deploy firewalls and other tools in the cloud platform, for ensuring cloud-end cybersecurity [87]
AutoCrypt	It is comprehensive, safe, and 5G supportive. The verification technology can almost perfectly realize 5G communication in the intelligent transportation system. The fleet management system can also ensure the safe collection and analysis of fleet data [88]

2. Safety Strategy Based on In-depth Defense

The first layer of in-depth defense system architecture is the secure interface. If the interface is not safe enough, hackers can obtain access rights of TCU or OBD port, t carry out forgery attacks, and obtain the control right of key safety elements such as braking and power system by sending fake CAN messages. Therefore, to set the first barrier, it is necessary to prevent illegal reading and tampering of data by securely encrypting data and verifying exchange messages. Specifically speaking, a safety unit can be installed to maximize the security of TCU. The security unit can apply Transport Layer Security (TLS) on ordinary cellular connections or Wi-Fi connections. At the same time, it is also equipped with an advanced encryption accelerator and an effective advanced physical and electrical attack resistance mechanism and serves as a key and certificate safe with high security.

The second layer of in-depth defense system architecture is the secure gateway. If the gateway is not secure enough, hackers can send messages at any time and place after the invasion. The central gateway divides the whole network into different functional domains, and the gateway firewall determines which nodes can legally communicate with other nodes, as described above. For cross-domain control and transmission requests, even from outside the vehicle, the gateway needs to carry out identity identification, message authentication, and permission control, and then decide to agree to forward or deny access. At the same time, when performing security-related tasks such as encryption and decryption, message authentication, and random number generation, it is necessary to apply some sensitive data, with some cache of sensitive data produced subsequently, e.g., Message Authentication Code (MAC) with the key generated during authentication for comparison. The data needs to be stored in a very secure range, without direct access permission for nodes of the non-security domain. To realize these security mechanisms, the gateway requires safe and reliable symmetric and asymmetric encryption and decryption algorithms, true and pseudo-random number generators, including AES-128 based packet encryption and decryption algorithm, and corresponding message authentication code (Cipher-based MAC, CMAC) in the Security Hardware Extension (SHE), asymmetric RSA, ECC based signature algorithm, etc. In addition, the real-time risk monitoring mechanism should also be deployed based on the central gateway. Once AI finds any abnormality from in-vehicle network communication, it should notify the gateway to take early warning and other response measures.

The third layer of in-depth defense system architecture is the secure network. If the network is not secure enough, although the attack surface is greatly reduced after being decomposed into functional domains, the sub-network is still vulnerable to attack and intrusion, and the vulnerabilities may be maliciously exploited by malware, Trojan horse, etc. To ensure cybersecurity, the adopted strategies include: ① add a message authentication mechanism, expand each message with encryption code, ensure the authenticity of the sender, and confirm that the message has not been tampered with when it is received; ② Prevent the attacker from stealing data and identity by encrypting the messages exchanged between different ECUs in the vehicle; ③ The network-based intrusion detection system detects abnormal conditions in network traffic with pattern recognition and rule base, to prevent forged or tampered malicious packets from reaching the microcontroller, and limit message-sending rate for DoS attack prevention.

The fourth layer of in-depth defense system architecture is the secure process. If the process is not secure enough, hackers may use system defects to carry out cyber attacks such as code injection, so that the software functions can not operate normally, or even be replaced by malicious attacks. To ensure that the software running on the processor is authentic and unmodified, the microcontroller is equipped with a safe startup and real-time integrity check mechanism. At the same time, the controlled locking mechanism of MCU and ECU should be adopted in the production for debugging prohibition. In addition, when defects or security vulnerabilities of the system are detected, on-board software should be updated quickly, seamlessly, and safely through OTA technology.

2.7 Summary

The development theory, design approach, assurance technology, and typical system of IV system safety can ensure the scientificity, reliability, and comprehensiveness of system development. Within it, FuSa, SOTIF, and cybersecurity are three key system safety technologies, which not only show common features but also correspond to different hazard mechanisms, technical properties, and design assurance ideas:

1) The basic theories of systems engineering, safety systems engineering, and IV safety systems engineering are the guidance of system safety design. In the design process, it is necessary to study hazard mechanisms of the three types of system safety problems (system fault, insufficient function, and cyber attack); Fault tolerant cybernetics, accident model and process based on system theory, and mimic defense theory lays the foundation for system safety; In addition, relevant theories of safety verification such as safety acceptability, demonstration, and validation are the theoretical basis to prove whether system safety is adequately guaranteed.
2) The common structure of system safety design is in the shape of "V", and the overall process is divided into three stages: system concept setting and analysis is used to define the system, carry out safety analysis and risk assessment, and set safety requirements; Safety design and development is carried out at the system, hardware, and software level, and safety requirements shall be met through system integration, safety test, verification and validation; And system authentication and release is conducted finally.
3) The existing standards refine and standardize the common architecture of system safety design in FuSa, SOTIF, and cybersecurity. In addition, their respective characteristics need to be considered for targeted assurance technologies. Research in FuSa is relatively mature, and two key support technologies of fault detection and diagnosis and fault tolerant control can better solve relevant problems; the concept of SOTIF is relatively new. There are preliminary studies on SOTIF monitoring, assurance of SOTIF based on perception and decision-making, and design of SOTIF based on HMI, but a perfect technical system for safety assurance has not been established; The cybersecurity inside and outside the vehicle has a specific assurance technology system, in which safety topology design, intrusion detection and defense, and safety encryption and communication are the three typical key techniques.
4) Based on systems safety development theory and design universality, together with technologies in view of systems safety assurance properties, we can further develop intelligent systems to ensure system safety, such as safety system based

on functional redundancy, safety system based on self-monitoring and cybersecurity systems based on in-depth defense, and implement safety strategy based on safety systems architecture.

References

1. Qian Xuesen. System engineering. New century edition [M] Shanghai: Shanghai Jiaotong University Press, 2007 [in Chines].
2. Guo Baozhu, Wang Guoxin, Zheng Xinhua. System engineering: research and practice based on international standard process [M] Beijing: China Machine Press, 2020 [in Chinese].
3. Guo Baozhu. Research on system management of large and complex engineering projects [D]. Beijing: Beijing University of Aeronautics and Astronautics, 2006 [in Chinese].
4. Zheng Xinhua, Qu Xiaodong. Development course of Qian Xuesen's system engineering thought [J]. Science and Technology Review, 2018, 36 (20): 6–9 [in Chinese].
5. GOODE H H, MACHOL R E, TEICHMANN T. systems engineering [J]. Physics Today, 1957, 10(9): 34.
6. HALL A D. Systems engineering from an engineering viewpoint [J]. IEEE Transactions on Systems Science and Cybernetics, 1965, 1(1): 4–8.
7. HASKINS C, FORSBERG K, KRUEGER M, et al. Systems engineering handbook [M], INCOSE, 2006.
8. MOOZ H, FORSBERG K. 4.4.3 A Visual Explanation of Development Approaches and Strategies including the Waterfall, Spiral, Vee, Vee+, and Vee++ Models [J]. INCOSE International Symposium, 2001.
9. International Council on Systems Engineering. Systems engineering manual: System life cycle process and activity guide, [M]. Beijing: China Machine Press, 2017 [in Chinese].
10. Zhang Yuxin. Engineering method and application of automated driving safety system [R]. Changchun: Jilin University, 2019 [in Chinese].
11. Zhang Yuxin, He Wenqin, Chen Hong, et al. Overview of patent analysis on safety technology of automated driving vehicle [J] Science China: Information Sciences, 2020, 50 (11): 1732–1755 [in Chinese].
12. National Highway Traffic Safety Administration. Automated driving systems 2.0: A vision for safety [J]. Washington, DC: US Department of Transportation, DOT HS, 2017, 812442.
13. International Standardization Organization. Road vehicles — Functional safety: ISO 26262: 2018[S]. Geneva, Switzerland: ISO: 2018.
14. JOHNSON M. Cyber Crime, Security and Digital Intelligence [M]. New York: Routledge, 2016.
15. Vehicle Cybersecurity Systems Engineering Committee. Cybersecurity Guidebook for Cyber-Physical Vehicle Systems: SAE J3061 [J/OL]. SAE International, 2016, 128 [2016–01–14]. https://doi.org/10.4271/J3061_201601.
16. China Automobile Data Center. White paper on cybersecurity of Internet of vehicles (2020) [EB/OL] [2020–12]. https://www.secrss.com/articles/26174 [in Chinese].
17. The United Nations Economic Commission for Europe. Proposal for a new UN Regulation on uniform provisions concerning the approval of vehicles with regards to cybersecurity and cybersecurity management system [EB/OL] [2020–04–02]. https://undocs.org/ECE/TRANS/WP.29/2020/79.
18. China Automobile Data Center. The data resource center released the first TOP 10 of cybersecurity risks in the vehicle industry [EB/OL] [2019–05–24]. https://www.auto-testing.net/news/show-101871.html [in Chinese].
19. Zhou Donghua, et al. Modern fault diagnosis and fault tolerant control [M]. Beijing: Tsinghua University Press, 2000 [in Chinese].

20. ARSLAN A A, KHALID M H. A review of Fault Tolerant Control Systems: Advancements and applications [J]. Measurement, 2019, 143.
21. LEVESON N. A new accident model for engineering safer systems [J]. Safety science, 2004, 42(4): 237–270.
22. Wu Jiangxing. Cyberspace pseudo security defense [J] Secrecy Science and Technology, 2014 (10): 7 [in Chinese].
23. Yan Fei, Tang Tao. Development and research status of safety technology of rail transit signal system [J] China Safety Science Journal, 2005, 15 (6): 94–99 [in Chinese].
24. Kang Rui, Shi Rongde. FMECA technology and its application [M] Beijing: National Defense Industry Press, 2006 [in Chinese].
25. Shao Weigui. Research on the application of FMECA and FTA in landing gear system fault analysis of a certain aircraft [D]. Chengdu: Xihua University, 2019 [in Chinese].
26. IEC 61025 Technical Committee. Fault tree analysis [S]. Geneva, Switzerland: International Electrotechnical Commission: 2006.
27. International Standardization Organization. Road vehicles—Functional safety: ISO 26262: 2011[S]. Geneva: ISO, 2011.
28. International Standardization Organization. Road Vehicles—Safety of the Intended Functionality: ISO/DIS 21448: 2021 [S]. Geneva: ISO, 2021.
29. International Standardization Organization. Road Vehicles—Safety of the Intended Functionality: ISO/PAS 21448: 2019 [S]. Geneva: ISO, 2019.
30. International Standardization Organization. Road Vehicles—Safety of the Intended Functionality: ISO/CD 21448: 2019 [S]. Geneva: ISO, 2019.
31. International Standardization Organization. Road Vehicles—Cybersecurity Engineering: ISO/SAE DIS 21434 [S]. Geneva: ISO, 2020.
32. Zhou Donghua, et al. Modern fault diagnosis and fault tolerant control [M] Beijing: Tsinghua University Press, 2000 [in Chinese].
33. ANAM A, MUHAMMAD T K, JAVAID I. A review on fault detection and diagnosis techniques: basics and beyond [J]. Artificial Intelligence Review, 2020 (prepublish).
34. Qian Huaming. Fault diagnosis and fault tolerant technology and its application in integrated navigation system [D]. Harbin: Harbin Engineering University, 2004 [in Chinese].
35. Deng Xuecai. Research on active fault tolerant control way and prototype development of vehicle magnetorheological semi-active suspension system [D]. Hefei: Hefei University of Technology, 2020 [in Chinese].
36. Fang Jie. Research on early fault detection for short circuit of permanent magnet synchronous motor interturn based on wavelet analysis [D]. Hefei: Anhui University, 2020 [in Chinese].
37. Wang Zhongsheng. Intelligent fault diagnosis and fault tolerant control [M]. Xi'an: Northwest University of Technology Press, 2005 [in Chinese].
38. Xiong Qiufen. Research and implementation of signal detection and fault diagnosis system for fuel cell engines [D]. Wuhan: Wuhan University of Technology, 2018 [in Chinese].
39. CHAO H, FAZEL N, HAIPING D, et al. Fault tolerant steer-by-wire systems: An overview [J]. Annual Reviews in Control, 2019, 47.
40. ABBASPOUR A, MOKHTARI S, SARGOLZAEI A, et al. A Survey on Active Fault-Tolerant Control Systems [J]. Electronics, 2020, 9(9).
41. RAHUL G, PRASHANT M. A safe-parking framework for plant-wide fault-tolerant control [J].Chemical Engineering Science, 2009, 64(13).
42. Li Chunshan. Research on active fault tolerant control strategy for driving system of four-wheel independent electric vehicle by wire [D]. Changchun: Jilin University, 2018 [in Chinese].
43. SRINIVASAN RAMANAGOPAL M, ANDERSON C, VASUDEVAN R, et al. Failing to Learn: Autonomously Identifying Perception Failures for Self-Driving Cars [J]. IEEE Robotics and Automation Letters, 2018, 3(4): 3860–3867.
44. KENDALL A, GAL Y. What uncertainties do we need in bayesian deep learning for computer vision?[J]. arXiv preprint arXiv:1703.04977, 2017.
45. CHALAPATHY R,CHAWLA S. Deep learning for anomaly detection: A survey [J]. arXiv preprint arXiv:1901.03407, 2019.

46. LOHDEFINK J, FEHRLING J, KLINGNER M, et al. Self-Supervised Domain Mismatch Estimation for Autonomous Perception [C] // Proceedings of the IEEE/CVF Conference on Computer Vision and Pattern Recognition Workshops. 2020: 334–335.
47. HAYWARD J C. Near-miss determination through use of a scale of danger [J]. Highway Research Record, 1972.
48. WANG H, HUANG Y, KHAJEPOUR A, et al. Crash Mitigation in Motion Planning for Autonomous Vehicles [J]. IEEE Transactions on Intelligent Transportation Systems, 2019, 20(9): 3313–3323.
49. Gal Y, Ghahramani Z. Dropout as a Bayesian Approximation: Representing Model Uncertainty in Deep Learning [J]. JMLR.org, 2015.
50. GROUP B. BMW Safety Assessment Report [R/OL] [2020–11–06]. https://www.bmwusa.com/content/dam/bmwusa/innovation-campaign/autonomous/BMW-Safety-Assessment-Report.pdf.
51. WOOD M, ROBBEL P, MAASS M, et al. Safety first for automated driving [R/OL] [2019–10–25]. https://wwwdaimler.com/documents/innovation/other/safetyfirstforautomateddriving.Pdf.
52. ABBOOD H, AL-NUAIMY W, AL-ATABY A, et al. Prediction of driver fatigue: Approaches and open challenges [C] // Computational Intelligence. IEEE, 2014.
53. MCCARTNEY E J. Optics of the Atmosphere: Scattering by Molecules and Particles [J]. Phys. Today, 1977.
54. HE K, SUN J, TANG X. Single image haze removal using dark channel prior [J]. IEEE transactions on pattern analysis and machine intelligence, 2010, 33(12): 2341–2353.
55. LI B,PENG X,WANG Z,et al. AOD-Net: All-in-One Dehazing Network[C] // 2017 IEEE International Conference on Computer Vision (ICCV). IEEE,2017.
56. ZHANG H, PATEL V M. Densely Connected Pyramid Dehazing Network [C]// 2018 IEEE/CVF Conference on Computer Vision and Pattern Recognition (CVPR). IEEE, 2018.
57. FROSSARD D, URTASUN R. End-to-end Learning of Multi-sensor 3D Tracking by Detection [J]. 2018:s635–642.
58. Cao Chong. Research on hybrid decision-making of driverless vehicle based on rules and self-learning [D]. Beijing: Tsinghua University, 2020 [in Chinese].
59. Mozaffari S, Al-Jarrah O Y, Dianati M, et al. Deep Learning-Based Vehicle Behavior Prediction for Autonomous Driving Applications: A Review [J]. IEEE Transactions on Intelligent Transportation Systems, 2020, PP(99):1–15.
60. Lakshminarayanan B, Pritzel A, Blundell C. Simple and scalable predictive uncertainty estimation using deep ensembles [J]. Advances in neural information processing systems, 2017, 30.
61. Ovadia Y, Fertig E, Ren J, et al. Can you trust your model's uncertainty? evaluating predictive uncertainty under dataset shift [J]. Advances in neural information processing systems, 2019, 32.
62. Deo N, Trivedi M M. Convolutional social pooling for vehicle trajectory prediction [C] // Proceedings of the IEEE Conference on Computer Vision and Pattern Recognition Workshops. 2018: 1468–1476.
63. MÖLLER D P F, HAAS R E. Guide to Automotive Connectivity and Cybersecurity [M]. New York: Springer, 2019.
64. Qingji Safety Public Welfare Group. Panorama of vehicle cyberecurity knowledge and capability, Version 1 [EB/OL] [220–11–06]. http://www.hackdig.com/11/hack-188561.htm [in Chinese].
65. Talent Exchange Center of the Ministry of Industry and Information Technology. Cyberecurity technology and application of ICVs [M]. Beijing: Publishing House of Electronics Industry, 2018 [in Chinese].
66. National Technical Committee of Auto Standardization. Technical requirements for vehicle gateway cyberecurity (2020 Standard for comments) [EB/OL] [202–04–26]. https://www.doc88.com/p-19416996817839.html [in Chinese].

67. Zheng Yue, Zhao Xuguang, Zhong Zhijie. Vehicle network topology: China, CN109167712A [P]. 2019–01–08 [in Chinese].
68. Liu Yanbing. Analysis of key technologies for Internet of vehicles safety [M]. Beijing: Science Press, 2019 [in Chinese].
69. KONRARDY B, CHRISTENSEN S T, HAYWARD G, et al. Autonomous vehicle component malfunction impact assessment: U.S. Patent 10,168,703 [P]. 2019–01–01.
70. LANDRUM R, PACE S, HU F. Cyber-Physical System Security—Smart Grid Example [J]. Cyber-Physical Systems: Integrated Computing and Engineering Design, 2013: 1121.
71. SCARFONE K, MELL P. Guide to intrusion detection and prevention systems (idps) [J]. NIST special publication, 2007, 800(2007): 94.
72. KUMAR S, SPAFFORD E H. An application of pattern matching in intrusion detection: The COAST Project, Department of Computer Sciences, Purdue University, West Lafayette [R]. USA, Technical Report CSD-TR-94–013, 1994.
73. SMAHA S E. Haystack: An intrusion detection system. Proceedings of the 4th Aerospace Computer Security Applications Conference [C]. USA: Orlando, 1988, 44.
74. KARIM M E, PHOHA V V. Cyber-physical systems security [M]// Applied Cyber-Physical Systems. New York: Springer,2014:75–83.
75. Liu Xingwei, Wang Li, Huang Yuan, et al. A distributed intrusion detection way for on-board ad hoc networks: China, CN103237308A [P]. 2013–08–07 [in Chinese].
76. STINSON D R, PATERSON M. Cryptography: theory and practice [M]. Los Angeles: CRC press, 2018.
77. Wang Liangmin, Li Tingting, Chen Long. Structure and safety of Internet of vehicles based on vehicle identity [J] Chinese Journal of Network and Information Security, 2016, 2(2): 41 [in Chinese].
78. CICV. 5G-V2X cybersecurity trusted system of Internet of vehicles based on China's commercial cypher algorithm [EB/OL] [220–03–17]. https://www.auto-testing.net/news/show-106138.html [in Chinese].
79. Guangdong Commercial Cypher Association. Vehicle-road cooperative safety solution based on national cypher algorithm [EB/OL] [2020–08–24]. https://www.gdcca.net.cn/log/67 [in Chinese].
80. BOSCH. Automated Driving Redundancy For Sure! [EB/OL] [2021–7–19]. https://www.bosch.com/stories/redundant-systems-automated-driving.
81. SCHLATOW J, MOOSTL M, ERNST R, et al. Self-awareness in autonomous automotive systems [C]//Design, Automation & Test in Europe Conference & Exhibition (DATE), 2017. IEEE, 2017: 1050–1055.
82. RESCHKA A, BÖHMER J R, NOTHDURFT T, et al. A surveillance and safety system based on performance criteria and functional degradation for an autonomous vehicle [C]// 2012 15th International IEEE Conference on Intelligent Transportation Systems. IEEE, 2012: 237–242.
83. RESCHKA A, BAGSCHIK G, ULBRICH S, et al. Ability and skill graphs for system modeling,online monitoring,and decision support for vehicle guidance systems [C]// 2015 Ieee IVs symposium (Iv). IEEE, 2015: 933–939.
84. RESCHKA A, BÖHMER J R, SAUST F, et al. Safe, dynamic and comfortable longitudinal control for an autonomous vehicle [C]// 2012 IEEE IVs Symposium. IEEE, 2012: 346–351.
85. DENSO Official Channel. Vehicle cybersecurity [EB/OL] [2016–10–10]. https://www.youtube.com/watchv=3Kb_QQzeTR8.
86. Continental Automotive Global. cybersecurity @ automotive connectivity by Continental [EB/OL].[2017–10–23]. https://www.youtube.com/watchv=WEq6SC0Rcb0.
87. Baidu.Apollo Pilot Safety Report [EB/OL] [2018–07–11]. https://apollo-homepage.bj.bcebos.com/Apollo-Pilot-Safety-Report-2018.pdf [in Chinese].
88. AUTOCRYPT. Introducing AUTOCRYPT—Automotive Cybersecurity Solutions [EB/OL] [2020–08–13]. https://www.autocrypt.io/automotive-cybersecurity.

Chapter 3
Driving Safety

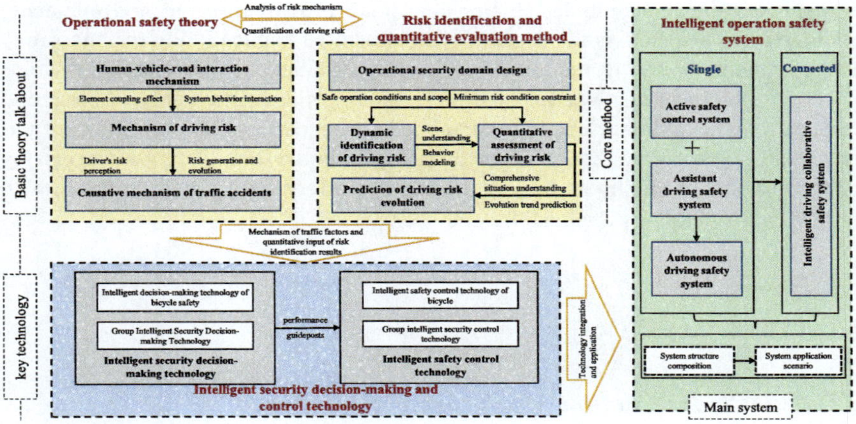

Driving safety of vehicle (DSV) refers to the realization of IV safety decision and control by means of quantitative evaluation and prediction over comprehensive driving risks, for preventing traffic participants inside and outside the vehicle from unacceptable risks.

DSV involves four aspects, i.e., fundamental theory for factor coupling and accident causation, core methods for risk identification and quantitative evaluation, key techniques for intelligent safety decision-making and control, and intelligent driving safety system. It aims at pre-collision accident warning and safety prevention during driving, comprehensively applies relevant theories and methods such as risk identification, system dynamics and optimal control, takes the micro traffic system of DVE as the object, makes the quantitative evaluation and prediction of driving risk

J. Wang et al., *The Intelligent Safety of Automobile*, Key Technologies on New Energy Vehicles, https://doi.org/10.1007/978-981-99-6399-7_3

by studying the interactive behavior characteristics of traffic participants, and adopts intelligent decision and control technology to ensure the timely safety response of the vehicle in the face of complicated and dynamic traffic environment, so as to comprehensively improve DSV.

The structure of this chapter is shown in the diagram on the previous page, in which the driving safety theory mainly starts from compositing objects of traffic system to explore the DVE interaction mechanism, driving risk generation mechanism and driving accident causation mechanism, so as to provide theoretical foundation for DSV; Risk identification and quantitative evaluation are the core means for DSV assurance. The operational safety conditions and scopes of IV are defined through the design of operational safety domain. Under the constraint of minimum risk conditions, the dynamic cognition and quantitative evaluation of driving risk are carried out, and then the risk evolution prediction in time and space domain is conducted. As the key technology for DSV, intelligent safety decision and control focuses on independent decision-making of single vehicle and cooperative decision of multiple vehicles, as well as trajectory tracking control of single vehicle and cooperative control of multiple vehicles, contributing to driving safety and efficiency; Intelligent DSV system is the carrier of the integrated application of driving safety theories, methods and techniques. By covering intelligent driving safety systems of single vehicle and networked multiple vehicles, this chapter analyzes structural compositions, working principles and application scenarios of key systems such as active safety control system, assistance driving safety system, autonomous driving safety system and intelligent driving cooperative safety system in accordance with engineering technology application.

3.1 Theory of DSV

3.1.1 DVE Interaction Mechanism

In the traffic scenario, the driver, vehicle and surrounding environment jointly constitute the traffic situation by interactions, with each referred to as "human", "vehicle" and "road" respectively. In accordance to relevant research, the current approaches for studying DVE interaction includes four categories, i.e., psychological motivation, double-layer interaction, control theory and virtual mechanics. This section will first outline the basic modeling ideas, advantages and disadvantages of the four approaches. Furthermore, the mechanism of DVE interaction is illustrated from such three levels as element property, element interaction and coupling mechanism. Starting from the analysis on the properties of the three types of elements, this section deeply explores the interaction between each two elements, so as to reveal the constraint and response mechanism for the interaction between human and vehicle, human and road, as well as vehicle and road. Finally, the DVE interaction mechanism and operational law are summarized, and the coupling mechanism is systematically described (Fig. 3.1).

1. Overview of DVE Interaction Analysis

According to the existing analysis on the interaction and coupling relationship of DVE elements, the current research approaches are divided into four categories, with each connotation, extension and research ideas as well as the advantages, disadvantages and future development trend summarized and analyzed as follows.

(1) Cybernetics

Cybernetics is adopted to establish the closed-loop feedback function for DVE system with reference to the transfer function in automatic control theory, as well as the modeling of driver and vehicle dynamics, which are applied to the coupling and interaction analysis on various elements of DVE system.

In the process of driving, the driver undertakes all the responsibilities of vehicle control, by which the environment state is directly transformed with the change of vehicle state. This interaction relationship can be expressed by the closed-loop structure of the three elements, as shown in Fig. 3.2. From the perspective of cybernetics, the core of DVE system is the driver. The driver interacts with the environment through perception of environmental information, recognizes the environment and ego-vehicle by comprehensively taking the environmental information and vehicle state into account, and then controls the vehicle in real time with the steering wheel and pedal. The main controlled object in the DVE system is the vehicle, on which the environment acts mainly through the media such as tire force and air resistance, and the driver acts by steering. For IV, their movement will be controlled by intelligent driving system. The environment in the DVE system can not only interact with the vehicle directly through the mechanical action of road surface with wheel, air and

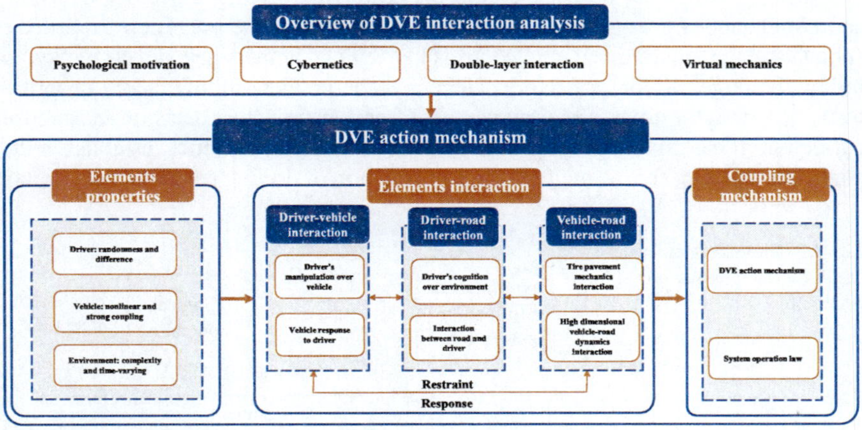

Fig. 3.1 Logical framework of DVE interaction mechanism

vehicle body, but also interact with the vehicle indirectly by affecting human's cognition towards such environmental factors as weather conditions, traffic conditions and traffic regulations (traffic lights, road signs, etc.). The driver steers to alter vehicle state according to environmental change and driving demand, then affects behaviors of traffic participants due to their coupling relationship with the ego-vehicle, and subsequently changes the state of traffic environment, that is, traffic situation. As a result, an integrated DVE system with mutual coupling and interaction is established.

It is essential to mathematically describe the characteristics of the driver and driver-vehicle interaction, for establishing DVE closed-loop system based on cybernetics. The driver's steering behavior model based on optimal control, fuzzy control and model predictive control can realize the objective description of the driver's steering behavior [1]. Preview-follow theory is one of the classical means for studying DVE closed-loop system based on cybernetics, and its framework towards the feedback control of the system is shown in Fig. 3.3. The overall framework can be divided into two parts: preview P(s) and follower F(s), with steering behavior of skilled drivers meeting the requirements of P(s)F(s) = 1. Where, y_0 is the expected trajectory and y is the actual output trajectory. Through the two means of preview and follow, the modeling for three steps of environmental perception, trajectory decision-making and control execution is conducted objectively. The ideal preview and tracking can realize the real-time and accurate tracking of the expected trajectory and the actual trajectory, and can basically achieve the control goal $y_0 = y$ within the acceptable range of error with favorable anti-interference and error-correction capability. In Fig. 3.3, the transfer function H(s) is used to describe the driver's control strategy, the feedback link B(s) is to describe the driver's prediction behavior, and G(s) is to describe the vehicle dynamics model. The environmental disturbance can affect the closed-loop control system by the vehicle dynamics model G(s), or be indirectly reflected in the driver's control strategy H(s) by the impact of the driver's environmental cognition change on control behavior.

During driving, the driver uses the previewer for driving trajectory decision, and simultaneously uses the follower for integration of the cognition toward the vehicle and the environment, and controls the vehicle by feedback and correction to follow the target driving trajectory. Therefore, the preview-follow theory quantitatively describes the driver's environmental perception, trajectory decision and control correction in the driving process based on the classical cybernetics, establishes the control framework for the mathematical abstraction of driving behavior, so as to lay

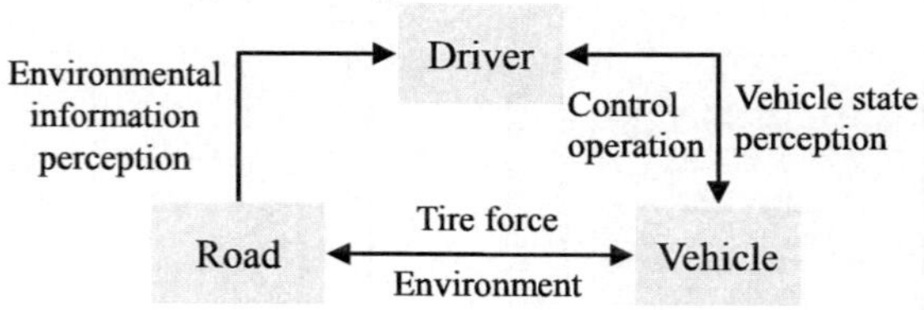

Fig. 3.2 Internal interaction of DVE system based on cybernetics

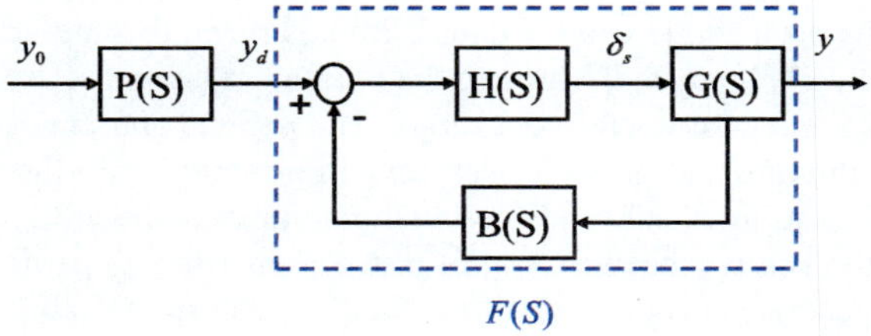

Fig. 3.3 Closed-loop feedback control on DVE system based on preview-follow theory [2]

a theoretical basis for the subsequent IV decision control technology and human-like automated driving.

The difficulty for establishing the interaction model of DVE system based on cybernetics lies in the modeling of each transfer function and the accurate calibration of model parameters. In order to improve the rationality of trajectory decision, such factors as weather, road and driver's personality are properly and comprehensively integrated into the original control system, which enhances the modeling accuracy of transfer function while ensuring the modeling efficiency, and has become a challenge and research focus in cybernetics-based driver modeling.

(2) Double-layer Interaction

According to the double-layer interaction analysis, the DVE system is divided into intelligent layer and physical layer, and the connection of various elements between the two levels is defined. This method can be applied to analyze the interaction of elements in IV-involved DVE system.

With the development of automated driving technology, there will be vehicles with different intelligent levels in the traffic environment. The interaction among human, vehicle and road includes not only the transmission of original information, but also some output judgments and decisions by intelligent processing. The three element closed-loop analysis framework under cybernetics system can no longer clearly explain the internal interaction of DVE system. In this context, Ref. [3] proposed a two-level interaction analysis, which divides the DVE system into intelligent layer and physical layer. Within it, the function of intelligent layer is defined independently from the physical execution carrier (Fig. 3.4). The intelligent layer not only endows the driver with the intelligence of information processing and decision making, but also facilitates the vehicle and environment in preliminary information processing with a certain intelligence. Therefore, the intelligent layer includes three units: human, vehicle and road, and correspondingly the physical layer also includes the same three units, which are mainly used to describe the physical properties of these three elements, such as driver's operation delay, vehicle dynamics and tire dynamics. Meanwhile, since the environment is actually composed of all traffic participants, including not only the ego-vehicle, but also surrounding vehicles and traffic lights, etc. The mutual interaction between the elements of DVE system is represented by connecting lines, and the overall relationship among the elements forms a "bus", so it is more appropriate to use the "bus" to represent the environment. The bus can collect the information of all traffic participants and release the information required by each single vehicle unit. With this form, the interaction of vehicle groups can be clearly analyzed. Through the intelligent layer, the driver comprehends traffic environment and makes decisions. The vehicle and environment can also judge traffic situation and provide necessary information with the assistance of on-board intelligent units and roadside intelligent units respectively; Through the physical layer, the vehicle receives control instructions from all parties, updates its own status, and generates original data to describe the system (Table 3.1). The intelligent layer is based on perception feature and decision mechanism, while the physical layer is based on driver's physical feature (or physiological feature) and vehicle dynamics mechanism.

The framework shown in Fig. 3.4 can be applied to analyze manual driving, human–machine co-driving or unmanned driving scenarios at the same time. The intuitive difference of the three scenarios is reflected in the connection between the elements. Within it, the environment unit of the intelligent layer only interacts with the driver and vehicle unit of the same layer, not with the driver and vehicle unit of the physical layer. If the environment unit of the intelligent layer directly provides real-time control instructions for the vehicle, the requirements for the real-time and reliability of the remote communication between the vehicle and infrastructure would be higher. At this stage, the direct interaction between the environment unit of the intelligent layer and the vehicle unit of the physical layer is not listed in the framework. In the future intelligent and connected vehicles environment, if the communication technology makes a breakthrough in assuring the real-time and reliability of information, the above interaction relationship is feasible. In order to ensure safe driving, the environment unit of the intelligent layer usually does not provide real-time control instructions for the vehicle. On the whole, compared with the original simple three-element analysis framework, the new one can more intuitively and clearly explain the interaction within the DVE system in the automated driving environment. In addition, it is proved that automated driving does not simply reduce the driver's cognitive tasks, but also introduces other cognitive tasks while reducing driving burdens in certain aspects.

The double-layer interaction is an emerging analysis for DVE system interaction. With the architecture of intelligent layer and physical layer, the DVE system interaction analysis can be conducted under three types of driving conditions, i.e., manual driving, human–machine hybrid driving and fully automated driving, but the quantitative expression for the information interaction process of each element shall

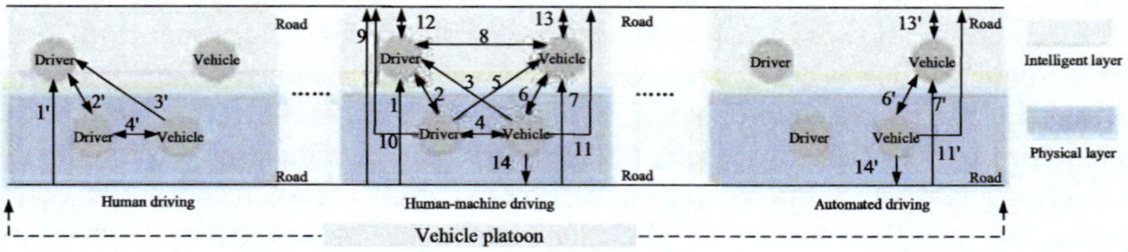

Fig. 3.4 DVE interaction coupling analysis based on double-layer interaction [3]

Table 3.1 Content of DVE interaction based on double-layer interaction [3]

Line	Content
Line 1′, 1, 2′ (↑), 2 (↑), 3′, 3	Drivers receive information about the environment, themselves and vehicle status to make decisions
Line 2′ (↓), 2 (↓)	Driver issues vehicle control instructions
Line 4′ (←), 4 (←)	The response of vehicle affects driver's feeling
Line 4′ (→), 4 (→)	Driver controls the steering wheel, pedal, etc.
Line 5, 6 (↑), 6′ (↑), 7 (↑), 7′ (↑)	Vehicle obtains the status information of the driver, itself and the environment through on-board sensors or Internet connection to make decisions
Line 6 (↓), 6′ (↓)	Vehicle executes commands or responds through some basic functional units such as path tracking controller [3]
Line 8 (←)	Vehicle provides prompt or warning information for the driver, such as driver fatigue warning (DFW) [4] and lane departure warning (LDW) [5]
Line 8 (→)	Driver participates in driving and cooperates with the vehicle
Line 9, 10, 11, 11′	Environment obtains the information of environment, driver and vehicle status respectively by means of sensors or network connection for decision-making
Line 12 (↓), 13 (↓), 13′ (↓)	Environment provides prompt or planning assistance for drivers and vehicles respectively, such as speed and traffic light phase planning [6, 7]
Line 12 (↑), 13 (↑), 13′ (↑)	Driver and vehicle provide assistance information to roadside or cloud intelligent units in the environment
Line 14, 14′	Change of vehicle status updates the environmental status

be further improved. In particular, key scientific issues such as the information interaction mechanism between AV and manual vehicle in the intelligent and connected vehicles environment shall be deeply explored.

(3) Psychological Motivation

By referring to the probabilistic analysis in psychology, the psychological motivation approach quantifies the subjective and objective factors affecting driving behavior, establishes the mathematical expression between the driving steering representation parameters and the elements of the DVE system, and objectively describes the system elements and steering behavior.

In the "Adaptive Integrated Driver-vehicle Interface (AIDE)" [8] project developed by Volvo Car Corporation of Sweden, a driver-vehicle-environment (DVE) system model is designed to support the development of a safer and more reliable HMI interface for vehicles with driving assistance function or other rich information functions. Based on the DVE system data of AIDE Research Institute, Ref. [9] proposed a psychological motivation method to analyze the interactive coupling mechanism of human, vehicle and road. The model includes five psychological concepts commonly applied in driver psychological motivation modeling, i.e., experience, attitude, task demand, driver state and situation awareness, all of which can be used together to describe the driver's state at any time (Fig. 3.5). And the quantitative expression model of the coupling relationship between the concepts is established based on fuzzy rule, as shown in Formula (3.1), and the influencing factors involved in DVE system in the model can be quantitatively described through the sub-factors listed in Table 3.2.

$$\begin{aligned} \text{Speed}_{\text{intended}} = [1 + 0.2 \cdot (\text{Attitude} - \text{Experience} + 1 + \text{Task}_{\text{demand}}) \\ + 0.2 \cdot \varphi(\text{Road}_{\text{condition}})] \cdot \text{Speed}_{\text{max\_allowed}} \end{aligned} \tag{3.1}$$

In Formula (3.1), experience refers to driving experience, with values 0, 1 and 2 corresponding to novice, moderately experienced and experienced drivers respectively. Attitude refers to the type of driving risk tendency, with values of 0, 1 and 2 respectively corresponding to those with low, medium and high risk tendency. $\text{Task}_{\text{demand}}$ indicates the task load, with the values -1, 0 and 1 corresponding to high, medium and low task loads respectively. $\varphi(\text{Road}_{\text{condition}})$ indicates the road condition, with values -1, 0 and 1 corresponding to poor, moderate and favorable road conditions respectively. Other factors, such as traffic environment properties (road characteristics, weather conditions and traffic state) and the driver's visual scanning strategy, will jointly influence the expression toward environment information. The vehicle model in the system is based on the vehicle dynamics equation. For instance, the lateral vehicle model can adopt the bicycle model. The quantitative basis of driver's psychological concept parameters is shown in Table 3.2.

The DVE system interaction based on psychological motivation model focuses on establishing the underlying cognitive behavior law of driver's driving behavior with psychological means, and quantifying driver's cognition on vehicle and environment with mathematical language. The advantage of this method is that it can thoroughly elaborate the sociological motivation behind different drivers' driving behaviors. The difficulty is how to obtain the subjective and objective psychological data in the real driving environment, so as to overcome the limitations of the existing simulator data and improve the reliability of the analysis results, and subsequently lay a theoretical

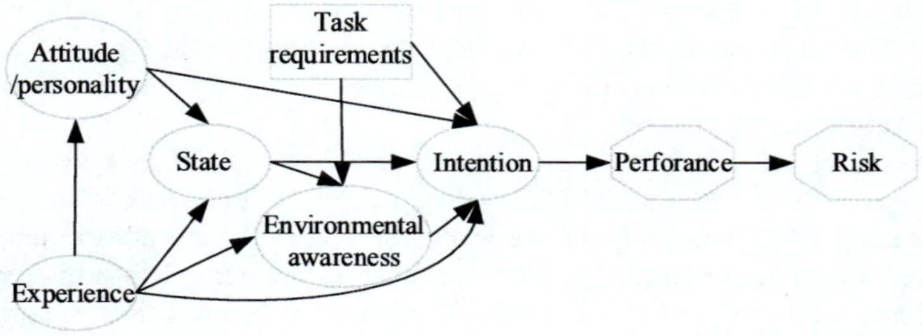

Fig. 3.5 DVE system interaction based on psychological motivation model [9]

Table 3.2 Quantitative basis of driver's psychological concept parameters [9]

Driver parameters	Definition	Measurable parameters
Experience	Knowledge and skills accumulated by drivers during driving	(1) Mileage per year (2) Years with driver's license
Attitude	Driver's attitude and concept towards some driving behaviors	(1) Vehicle speed level (2) Lane keeping frequency (3) Overtaking frequency (4) THW with the vehicle in front
Task demand	Efforts required for certain mission objectives	(1) Complexity of traffic environment (2) Weather (3) Lamps (4) Speed limit (5) Driving direction
Driver state	Physical and mental state of the driver, such as fatigue and drowsiness	(1) Control performance of lane keeping and distance keeping (2) Continuous driving time (3) Weather and road conditions (4) Complexity of traffic environment (5) Speed
Situation awareness	Drivers' perception and prediction of the state of various elements in a certain space–time range	(1) Distraction (2) Driver status (3) Task demand

basis for a more scientific and precise description of the cognitive process of driving behavior.

(4) Virtual Mechanics

By referring to the ideas of virtual displacement and virtual work equivalence in physics, the virtual mechanics method establishes the equal force or virtual force between the elements of DVE, so as to elaborate driving behavior and driving control mechanism, and provide reference for the human-like IV decision-making.

Reference [10] proposed the concept of social force and successfully applied it to the study of pedestrian traffic flow system. The research idea of human–human interaction elaborated by social force strongly promotes and inspires the development of other similar virtual mechanics methods in traffic modeling. In the DVE system, the interaction between various traffic elements can be regarded as mechanical action in a broad sense. With the idea of equivalent modeling, Ref. [11] adopted equivalent driving force to describe the force of driver's steering behavior on the DVE closed-loop system, inertial force to describe the force of vehicle inertial motion on the DVE system, and equivalent resistance to describe the force of road surface, traffic rules and other vehicle motion on the ego-vehicle. And based on the principle of least action, it proved that the sum of the above three forces meets the force balance of virtual mechanical system of DVE, which realizes the balance between driver's subjective driving force and equivalent resistance. When subjective driving force is larger than equivalent resistance, driving efficiency is high, whereas when equivalent resistance is larger than the subjective driving force, the driving risk is small. Therefore, the force balance is the optimization of driver's pursuit for driving efficiency and safety in theory. But in practice, the driver always tends to take risk within a certain threshold in order to pursue higher driving efficiency. Therefore, in actual driving, driver's subjective driving force is usually larger than the equivalent resistance. Reference [12] referred to the concept of "field" and realized the quantitative assessment over the risk of each element of DVE system with the concept of "driving safety field (DSF)". Through the qualitative analysis of potential risks generated by each element, the quantitative representation function is constructed for the quantitative assessment of driving risk by means of field strength. On this basis, the equivalent field force of DVE system is defined (Fig. 3.6). The interaction among the three elements of DVE is described based on field force. The application of field force for interaction modeling puts forward a new way for the equivalent force modeling of virtual mechanical system.

The DVE system model based on virtual mechanics simplifies driver's cognition and operation mechanism on the environment and vehicle into a generalized mechanical action on the basis of the original mechanical mechanics, and increases the generalized virtual forces such as social force, equal force and field force. This type of modeling has the advantages of clear physical meaning and strong interpretability, yet has the challenge in the modeling and parameter identification of generalized virtual forces by comprehensively taking road conditions, traffic flow, weather conditions, vehicle dynamics and driver characteristics into account.

2. Action Mechanism of DVE

Starting from the properties of the elements, this section analyzes the special attributes and essential features of each component of DVE system, and then analyzes the interaction relationship between the elements. It systematically summarizes the interaction relationship from the three perspectives of driver-vehicle interaction, driver-road interaction and vehicle–road interaction. On this basis, it illustrates the

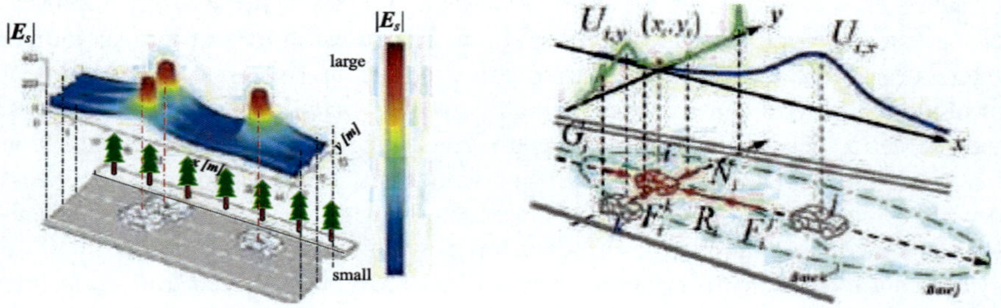

Fig. 3.6 Interaction among various traffic elements based on DSF [13, 14]

operation law and action mechanism and reveals the coupling mechanism of DVE system.

(1) Element Characteristics

1) Driver Characteristics

Driver characteristics mainly refer to the physiological and psychological characteristics shown by the driver during driving, which can be divided into visual and auditory characteristics, cognitive characteristics and controlling characteristics. The three aspects of characteristics respectively corresponding to the scientific laws in three procedures, including the driver obtaining environmental information with eyes and ears (audio-visual system), processing environmental information with brain (cognitive system) and forming driving strategies, and finally making real-time dynamic adjustment of driving strategy with hands and feet (control system). Driver characteristics are the common laws summarized from drivers' driving behaviors. In addition, there are certain personality laws, which can not exist alone, but still need to work with the common laws. Driver's personality is mainly reflected in the randomness of driver's visual and auditory information acquisition, the difference of cognitive characteristics and the time-varying driving strategies based on the first two characteristics. The description of these three characteristics are the difficulties for driver modeling at present.

With regard to the randomness of driver's visual and auditory information acquisition, driver's perception on the scope and object of environment has definite uncertainties in terms of different driving conditions, different weather and different individuality, and driver's sensitivity to the information also has certain fluctuations, resulting in the mathematical distribution of perception and impact on driver's cognitive results. For example, in the process of car following, there is some randomness in the driver's perception on the speed of the front vehicle and the distance from it. The fluctuation of the perception under different conditions will directly affect driver's judgment on the motion state of the front vehicle as well as driving strategy in the next step.

With regard to the difference of drivers' cognitive characteristics, there are differences in drivers' cognitive ability toward the environment due to the difference of age, gender, psychological state and driving experience, which is more prominent for complex driving tasks (such as passing through non-information controlled intersections with dense traffic flow, lane changing under urban congestion, and rush in and out of highway ramps). Under these conditions, driver's wrong cognition will easily induce traffic congestion and danger, resulting in the decrease of traffic efficiency and the intensification of traffic risk.

With regard to the time variability of drivers' driving strategies, there are differences in driving styles among different groups of people, and the driving strategies made by the same driver under different conditions are also different. The reason for the time variability of driving strategies is the combination of the randomness of drivers' visual and auditory information acquisition and the difference of drivers' cognitive characteristics. In the current research, in order to quantitatively describe the time variability, the clustering approach is usually used to divide driving strategies into three types, i.e., radical, moderate and conservative. Through naturalistic driving, the clustering center of various drivers' driving data is found, which further provides a reference for the formulation of control strategies of advanced driving assistance system and human–machine co-driving system as well as the decision-making of human-like IV.

The study of driver characteristics has always been the essential part of the study of DVE closed-loop system. The in-depth study can not only help to find the root cause of traffic accidents, but also lay a theoretical basis for IV decision-making.

2) Vehicle Characteristics

Vehicle characteristics mainly refer to the dynamic complexity, nonlinearity and strong coupling characteristics of vehicle system dynamics. It will directly affect vehicle power performance, braking performance, control stability, ride comfort, fuel economy, trafficability, etc. Among them, power performance and braking performance describe the longitudinal dynamics of vehicle, power performance reflects the longitudinal acceleration capacity, and braking performance reflects the longitudinal deceleration capacity; Control stability describes the response of vehicle lateral dynamics; Ride comfort describes the vertical dynamic characteristics; fuel economy reflects the energy consumption; And trafficability represents the off-road capability of vehicle. The above six types of vehicle performance can be quantified by establishment of mathematical models through differential equations or transfer functions.

In terms of the impact of dynamic complexity on vehicle performance, the dynamics characteristics of vehicle under different conditions are quite different. Vehicle dynamics system is a high-dimensional dynamic electromechanical system. Its dynamics performance will dynamically change in line with different time and space and user control characteristic. In other words, the vehicle dynamics system will be presented by different transfer functions due to the differences of service environment, service years of life and driver's controlling characteristics.

In terms of the impact of nonlinear characteristics on vehicle performance, the longitudinal driving and braking models, lateral steering models, vertical vibration models and even fuel consumption models have nonlinear characteristics. With the increase of vehicle speed and tire sideslip angle, the possibility of dynamics model entering the nonlinear region and degree of nonlinearity will rise sharply, which will further bring about the increasing system control difficulty as well as the system instability and other problems.

In terms of the influence of coupling on vehicle performance, two levels are involved, i.e., mechanical coupling and electro-hydraulic coupling. Firstly, mechanical coupling refers to the lateral-longitudinal-vertical three-dimensional coupling of vehicle dynamics. Accordingly, an integrated chassis control model is established to form an advanced driving assistance system with "shared structure and cooperative control", which has become an effective way to promote the coupling control of vehicle dynamics system with more and more consensus in the industry. Secondly, electro-hydraulic coupling refers to the correlation among the vehicle's original mechanical system, hydraulic system, and electronic and electrical system. For example, the vehicle's anti-brake lock system (ABS) is a typical electro-hydraulic coupling system. Coupling leads to certain contradictions in vehicle performance in different space–time scales and system dimensions. It is necessary to set an appropriate objective function to comprehensively optimize the overall performance.

In DVE system, the dynamic complexity, nonlinearity and strong coupling of the vehicle is likely to make the dynamics of the closed-loop system chaotic. It is an important topic in vehicle dynamics to enhance the prediction capability of the model, by improving the vehicle structure and control algorithm in consideration of the dynamic complexity, the nonlinearity within the predictable range, as well as the influence of coupling characteristics on the overall vehicle performance.

3) Environment Characteristics

Environment characteristics include not only road characteristics such as road adhesion characteristics, geometric characteristics (lateral and longitudinal slope and curvature, etc.), mechanical characteristics of road surface (have a more significant impact to the system on soft ground such as soil and snow), but also weather conditions, lighting conditions, traffic flow conditions and traffic rule constraints (intersection traffic signal, road speed limit, driving direction, etc.). The environment in DVE system has the characteristics of uncertainty, dynamics and interactivity.

In terms of uncertainty, the road attachment conditions of different roads and under different weather conditions are different, the geometric characteristics of different roads are also different, and the traffic rules in different regions and different periods are complicated. For example, on ice and snow roads, the road adhesion coefficient is low, which may lead to vehicle slip, steering deviation and tail flick and influence DSV. To abstract and simplify the uncertain characteristics of environment, establish a mathematical model, and then objectively describe the environment attributes has become one of the hot research topics with more attention.

In terms of dynamics, the weather, lighting, traffic flow and traffic rules of the same road at different times are possibly different. For example, the traffic flow in the same road segment has congested traffic flow in the morning and evening peak, as well as sparse and smooth traffic flow in non-peak hours. In the DVE model, the interaction behavior in congested traffic flow is more frequent, and the influence range of abnormal driving behavior on the disturbance and propagation of system stability constant is larger. In modeling, the dynamics of environment parameters should be fully considered for improving the universality of the model.

In terms of interactivity, other traffic participants in the traffic environment may interact with the ego-vehicle, which lead to more complicated influence mechanism of environment on vehicle decision. In the current research, these interactive behaviors are often described by means of game theory, partial state observable Markov process and reinforcement learning. The further exploration of environment interaction mechanism will facilitate the more accurate and objective description of DVE system, and lay a foundation for the next research on risk cognition and control decision-making.

The uncertainty, dynamics and interactivity of road environment can directly act on the vehicle and change its dynamics characteristics, and also change vehicle state through indirect vehicle control by driver's stress response to the environment.

(2) Element Interaction

1) Driver-Vehicle System Interaction

Driver-vehicle system interaction refers to the interaction between driver and vehicle. The specific interaction contents, modes and functions are illustrated in details as follows.

i. Somatosensory interaction between driver and vehicle. The driver perceives the state of vehicle, such as speed and stability, through body feeling. In this regard, driver's multi-perception channels work in a complementary way. Some studies have found that the front and surrounding vision as well as noise intensity can help the driver judge vehicle speed. The surrounding vision is more effective than the front vision, and the reduction of noise intensity easily lead to driver's underestimation over vehicle speed [15]. In addition, driver's perceived result is usually biased. Taking the perceived pedal force for example, as shown in Fig. 3.7, when pedal force is small, driver's perceived result is relatively large; when the pedal force is large, the driver's perception result is small; at the same time, the result changes with different pedal structures [16]. It should be noted that in Fig. 3.7, R^2 is the coefficient of determination representing the ratio between the sum of regression squares and the total sum of squares, which reflects the interpretable proportion of the independent variable to the dependent variable. In the right of Fig. 3.7, the value of R^2 is 0.9023, indicating that independent variable can explain about 90% of the variation of dependent variable. Steven's Power Law ($\psi = k\varphi^n$, where k is a scalar constant related to the current condition) [17] is often applied to determine the rate at which the perceived result increases with the real value. Based on this law, taking the perceived steering wheel force for example, the observed results of steering wheel force in steady state increases exponentially by 1.39 with the increase of force; and the observed results of the angle in steady state increases exponentially by 0.93 with the increase of angle [18]. The research on driver-vehicle somatosensory interaction is conducive to the improvement of the human-likeness and comfort of driving assistance system, as well as the human-like design of IV.
ii. Audio-visual and interface interaction between driver and vehicle. The driver can perceive the behavior characteristics of surrounding vehicles through audio-visual systems and obtain vehicle information through human–machine interface (HMI) for understanding intelligent behaviors of vehicles, and can also cooperate with vehicle intelligent unit in making complementary decisions. Driver's situation awareness can be effectively improved by favorable HMI design, such as adding uncertain information prompt in the interface [19]. The way of light reminder is more effective than that of graphical mark reminder [20], as group C is more effective than group B shown in Fig. 3.8. In addition, IV is expected to realize cooperative decision-making and shared control right switching with

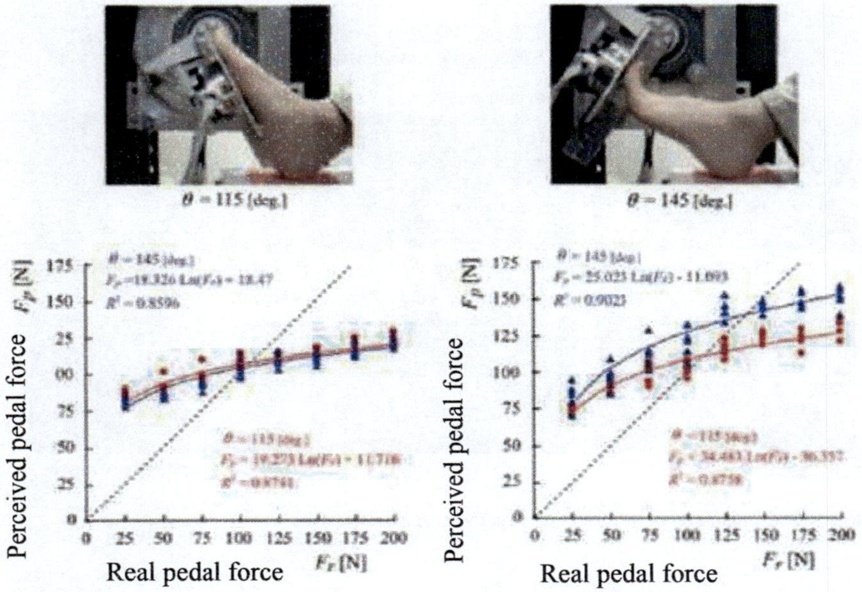

Fig. 3.7 Relationship between driver's perceived pedal force and real pedal force under different pedal structures [16]

intelligent units in the form of interface interaction in the future. Therefore, how to avoid control right conflict has become the key. Current studies have explored schemes such as control quantity weighted sum [21, 22] or game theory [23]. Further research on scientific, rational and comfortable human–machine control switching will facilitate the improvement of IV acceptance.

iii. Mechanical interaction between driver and vehicle. Mechanical interaction process refers to the driver changing the driving state by manipulating the steering wheel, pedal, shift handle and other devices of the vehicle. In the driver-vehicle mechanical interaction, the most critical factor is time delay, and its adverse effects are usually reflected in the control system. Time delay includes driver's time delay and mechanical system time delay. Among them, the mechanical system time delay is less affected by driving conditions and external environment, while driver's time delay is more dynamic and time-varying. Related research of crossover model divides the driver's delay into two parts, i.e., brain response delay and behavior delay [24]. In the nonlinear driver model with steering control and speed control as the main parts proposed in Ref. [25], the preview time also increases with the increase of driver's time delay [25]. As shown in Fig. 3.9, if the preview time is insufficient, it is difficult to ensure the stability of driving direction. Mechanical interaction is the focus of driver modeling. In order to improve the accuracy of driver model, it is necessary

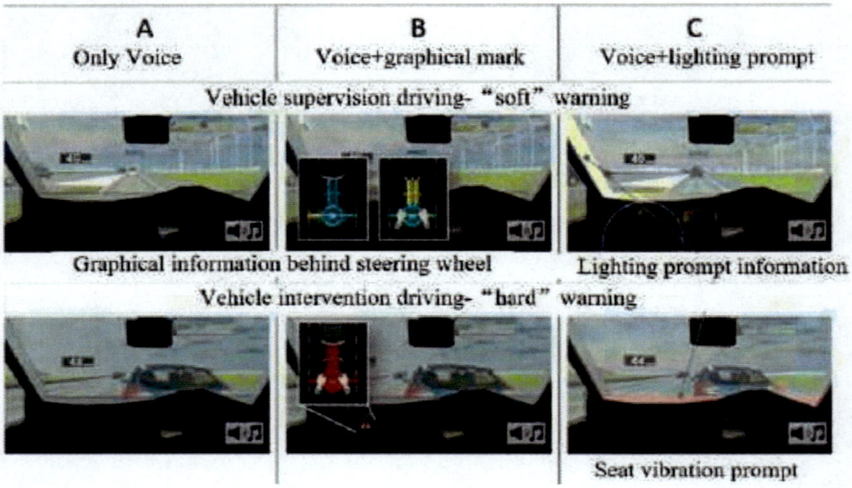

Fig. 3.8 Comparison of different information prompt modes [20]

to further expand the research on mechanical interaction between driver and vehicle.

The two difficulties in the driver-vehicle system interaction exist in the process of interaction between driver and intelligent driving system. On the one hand, due to the complex and diverse traffic environment, the coupling of a variety of interactive tasks as well as driver's personalized demand, it is still unclear about the type and quantity of effective information that the HMI is required to provide. Therefore, analyzing the characteristics and mechanism of driver information interaction is the key for establishment of an efficient and adaptive driver-vehicle information interaction system in complex environment. On the other hand, due to the difficulty of understanding driver's intention and the instability of driver's attention state, how to

Fig. 3.9 Relationship between driver's preview time and time delay [25]

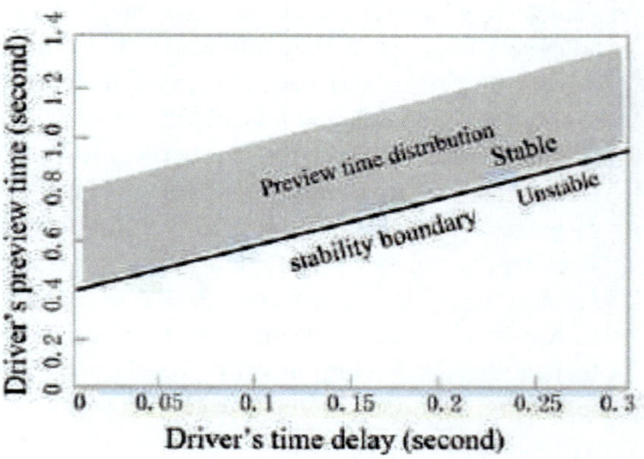

safely switch control during human–machine driving is quite challenging. Therefore, analyzing the driver's attention distribution model, response decision-making time and behavior performance is essential for safe and reliable driver-vehicle cooperative control.

2) Driver-Road System Interaction

Driver-road system interaction refers to the interaction between driver and surrounding traffic environment. The surrounding traffic environment includes traffic users, traffic regulations, road structure, weather, pavement, etc. At present, the roadside intelligent units or cloud intelligent platform relying on the Internet technology can facilitate the driver in intelligent decision-making or control information, but the relevant applications are not mature enough and shall be further explored and improved. The following focuses on the interaction between driver and the environment under the condition of manual driving.

i. Interaction between driver and surrounding traffic participants. Surrounding traffic participants include surrounding vehicles, pedestrians, non-motor vehicles, etc. When the potential motion trajectory of surrounding traffic participants conflicts with the driver's target trajectory, the driver can predict the trajectory trend of surrounding traffic participants and judge the risk, and then make reasonable decision to drive comfortably and efficiently without collision. In addition, the surrounding traffic participants can also predict the possible behavior of the driver and determine the trajectory of the ego-vehicle. The overall traffic situation is reflected in this dynamic interaction process. In terms of trajectory prediction on vehicle groups, the application of such approaches as neural networks for simulating the interaction between drivers and vehicles has become a research hot topic [13, 14, 26–28]. In the future, the more extensive and in-depth application of machine learning, reinforcement learning and other algorithms in this field is expected to lay the foundation for accurate and real-time trajectory prediction and risk assessment.
ii. Interaction between driver and traffic regulations. Existing research shows that traffic regulations definitely influence driver's decision, but the difficulty in the interaction is to clarify the effectiveness of traffic regulations on driver's scientific decision-making and reasonable guidance. Some traffic signs affect the driver's trajectory selection, such as no U-turn sign; Some traffic signs affect the driver's speed selection, such as signal lights, road speed limit and other traffic information. In order to quantitatively evaluate the possible impact of different traffic rules on driving behavior and improve the scientificity of traffic rule formulation, it is necessary to deeply explore the interaction mechanism between drivers and traffic regulations in terms of rationality, regionality and dynamics, so as to improve drivers' subjective recognition and compliance awareness on traffic regulations, and serve the traffic management department in scientific and reasonable revision of regulations for reducing traffic hazards induced by drivers' illegal driving and improving traffic efficiency and safety.

iii. Interaction between driver and environment. Road structure, weather, road adhesion conditions and other factors can influence the driver's decision-making, especially the driver's speed choice. Research shows that the driver usually tend to slow down on curves with large curvature [29] (Fig. 3.10), leaving sufficient time for steering operation adjustment. Regarding the influence of weather on driver's decision-making, fog, rain and snow and other weather result in driver's limited vision, which enhances the uncertainty of driver's cognition of the driving environment. In order to reserve sufficient reaction time and ensure safe driving, driver also tends to reduce the speed. Regarding the influence of road adhesion conditions on driver's decision, driver also tends to reduce speed on slippery road or uneven road in order to ensure the stability or comfort of the vehicle. To summarize the scientific decision-making mechanism from the operation of experienced drivers in extreme conditions and quantitatively evaluate the impact of environment factors on driving decision-making will help to clarify the accident causation mechanism and scientific operation mechanism in ultimate and extreme environment, and improve the DSV.

In the driver-road system interaction, driver's cognition on environment situation is subjective, and driver's behaviors are time-varying and uncertain, which increase the difficulty of analyzing driver's environment cognition and the formation mechanism of driving strategy. In the highly interactive traffic scenarios, the explicit and implicit information transmitted in the interaction process coexist, and human behavior is not completely rational. All these pose a great challenge to the clarification of driver-road user interaction mechanism. The human-like decision based on the concept of "learning from human, simulating human, surpassing human and serving human" provides a feasible way for capturing and exploring the driver-road interaction mechanism in the hybrid traffic environment.

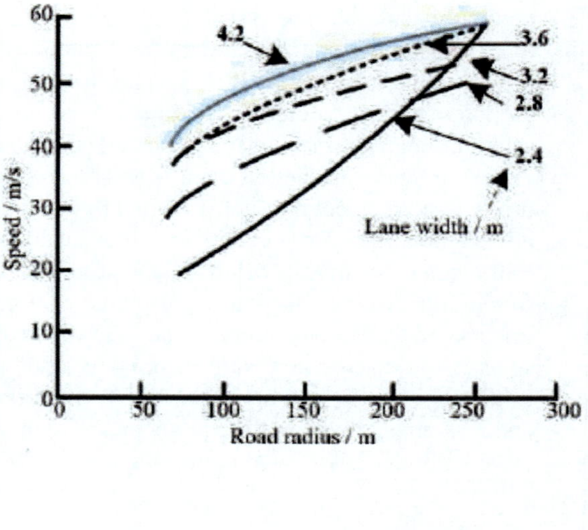

Fig. 3.10 Relationship between vehicle speed and road radius [29]

3) Vehicle–Road System Interaction

Vehicle–road system interaction refers to the interaction between the vehicle and surrounding traffic environment. Vehicle–road interaction can be regarded as the interaction between the vehicle and road in terms of cybernetics. From a broader perspective, the interaction between vehicle and environment can be included in the scope of vehicle–road interaction. For instance, the air resistance generated by the interaction between vehicle and air can also be regarded as generalized vehicle–road interaction. In the traditional sense, the interaction between vehicle and road is mainly conducted through tires in three dimensions, i.e., lateral, longitudinal and vertical [30]. The interaction between vehicle and road in the longitudinal direction is affected by the rolling resistance coefficient, adhesion performance and other properties of tires; The lateral interaction between the two is affected by the cornering property of tires; In the vertical direction, the basic modules such as damping, spring and mass block can be used to establish the vibration model of wheels, and analyze stiffness property, vibration mode and natural frequency of tires. In terms of longitudinal and lateral coupling, the attached ellipse represents the coupling relationship between the lateral and longitudinal directions of the vehicle and the road surface, that is, the envelope of the curve of the driving force and braking force with different cornering angles. Accordingly, the extreme value of the resultant force of tangential force and cornering force can be determined (Fig. 3.11). In addition, it should be noted that the installation of some active safety systems (such as ABS and ESP) can improve the response properties of vehicle lateral and longitudinal dynamics.

In the vehicle–road system interaction, it is difficult to measure the mechanical parameters of tires with high uncertainty in real time. In addition, the tire mechanical properties under extreme conditions have strong nonlinearity. Establishing a high-precision and real-time tire-road mechanical model available for all conditions

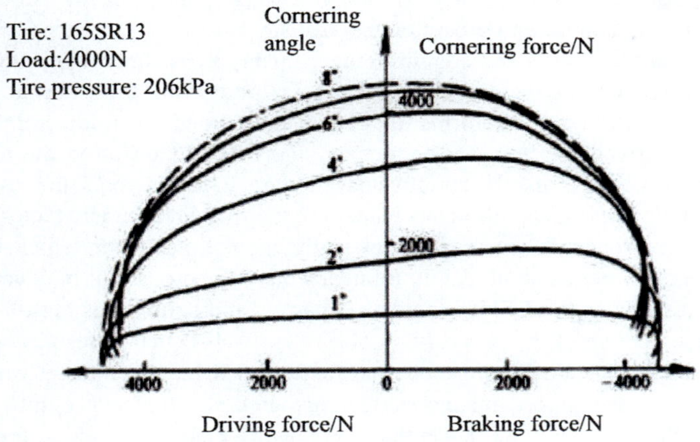

Fig. 3.11 Attached ellipse [30]

has become a critical issue for improvement of the control performance in complicated environment. Besides, the vehicle–road dynamics model has complex structure, strong nonlinearity and great uncertainty. The establishment of high-precision and high-dimensional vehicle–road dynamics model has become another key point for improving the safety and comfort of vehicle control. A feasible scheme proposed in the current research is to learn from the vehicle trajectory coupling dynamics [31] modeling of rail vehicles, establish two-dimensional and three-dimensional vehicle–road coupling systems by analyzing the dynamic response of the road under the load of moving vehicles, and then study the vertical vehicle–road coupling dynamics [32].

With the development of intelligent technology, the connotation of vehicle–road interaction in the traditional sense has changed, and the vehicle–road interaction in the intelligent layer has emerged. The intelligent vehicle–road interaction refers to the application of roadside intelligent units or cloud intelligent platforms for vehicle information release and guidance, which is conducive to the safety improvement of vehicle and the traffic efficiency of transportation system.

(3) Coupling Mechanism

1) DVE Interaction Mechanism

Based on the analysis of DVE element properties and element interaction, the DVE function can be summarized as a closed-loop feedback system mechanism centered on "human", which comprehensively takes the impact of dynamic environment and real-time nonlinear vehicle dynamics into account. Furthermore, the DVE interaction mechanism can truly and objectively reflect the driver's cognition on the environment and the operation on the vehicle in the actual driving, control the vehicle with correction and operation according to the disturbing and real-time dynamic environment information and vehicle state feedback, and finally appropriately implement the driving task. Specifically, the DVE interaction mechanism can be divided into three aspects: perception and cognition mechanism, intelligent decision-making mechanism, and control and execution mechanism.

In terms of perception and cognition mechanism, it is mainly manifested in the interaction between driver and environment. By perceiving the external environment information, the driver forms the understanding and cognition over environment safety, drivable area and vehicle state, and makes decision in the next step. The theory involved is mainly cognitive psychology. In recent years, the research on driver's perception and cognition mechanism shows that environment factors such as road curvature, road adhesion coefficient, traffic regulations, other vehicle behavior and weather conditions significantly influence driver's perception: in adverse environment, driver's cognitive load increases sharply, and continuous high-load work leads to the decline of driver's driving ability and subsequently results in various traffic accidents. Meanwhile, the internal factors such as driver's age, gender and driving age are also important factors affecting their perceptual and cognitive ability. The research shows that the perceptual and cognitive ability of older, female and inexperienced drivers is obviously lower than the average, and the possibility of road traffic accidents caused by them is rising evidently.

In terms of intelligent decision-making mechanism, it mainly explores the internal mechanism followed by the driver's decision-making response based on environment perception. Typical decision-making behaviors include straight ahead, left turn, right turn, U-turn, parking, etc., which are the driving behavior strategies of the driver in the face of driving environment. Current studies have shown that the driver's decision-making behavior is fuzzy and random to a certain extent, but the driver can follow the safety and efficiency in a uniform way. The coordination for threshold is closely related to the driver's personality. On the other hand, the research on IV decision-making mostly focuses on mechanism-driven and data-driven (also known as model-based and learning-based) modes, which have apparently complementary advantages in generalization capability, interpretability and robustness. Therefore, mechanism-data hybrid decision-making has become a significant trend in the development of IV decision-making technology. With the rapid development of brain science and technology, the cognitive mechanisms of brain-like decision and driver decision are expected to be further revealed, which will effectively facilitate intelligent decision-making under the DVE coupling mechanism.

In terms of control and execution mechanism, it is mainly reflected in the interaction between driver and vehicle. Under the decision response of the driver or intelligent driving system, the steering wheel and accelerator/brake pedal are directly controlled by the planning and tracking module or the driver in response to the driving intention. The research shows that the control delay and dead time effect inevitably appear in the actual vehicle control, which is bound to affect the driving and riding experience. Most of the existing studies eliminate them through compensation strategy and state feedback, but it is still difficult to meet its high real-time requirements for the vehicle control and execution under extreme conditions. The development of control by wire technology provides new technical support for accurate control and efficient execution over vehicle, mainly including steering by wire, braking by wire, drive by wire, etc. The control and execution efficiency is greatly improved by transmitting the control instructions of on-board sensing and control unit to each actuator. Meanwhile, the introduction of domain controller serves as a strong software architecture for chassis by wire technology, all of which lay the software and hardware foundation for the improvement of control and execution efficiency.

2) System Operation Law

The DVE system is a typical closed-loop feedback system. The control input of the driver (intelligent driving system) affects the dynamic response of the vehicle and is also affected by the vehicle and road environment. The three interact and couple with each other and form a complex micro road traffic system. Similar to other closed-loop systems, there are significant features and laws in terms of stability, observability and controllability, as follows:

In terms of stability, whether from the perspective of cybernetics or virtual mechanics, it can be proved that the DVE system is objectively stable, which meets the Lyapunov stability and the balance of virtual mechanics system. The stability of DVE system is normal, yet the environmental disturbance of system, vehicle fault or driver's improper operation and other factors are likely to lead to the abnormal state of system instability and result in accidents. Under the action mechanism of DVE system centered on "human", driver's environmental cognition and control ensure that the system is in a stable state, and realize the comprehensively cooperative optimization control of various performance of the system. The research shows that the DVE system under normal working conditions is locally and gradually stable. If the external interference (such as slippery road surface, tire burst and collision) reaches a certain degree, the system may be unstable, resulting in traffic congestion or traffic accidents. The commonly used stability analysis are phase plane method and Lyapunov temperature coefficient method.

In terms of observability, in order to estimate vehicle state parameters such as vehicle mass, centroid sideslip angle and yaw rate in real time, as well as road parameters such as road slope and road adhesion coefficient, indirect state estimation is usually conducted by designing state observer and Kalman filter. The research shows that the DVE system can be approximately transformed into a linear closed-loop system. The development of intelligent transportation technology greatly facilitates the state estimation and observation of key system parameters. The real-time monitoring of state parameters can lay a scientific basis for macro traffic management and control, and help the IV to realize a higher level of automated driving. Road adhesion coefficient, vehicle centroid sideslip angle and yaw rate are the key parameters of IV control, whereas traffic flow, traffic flow density and average speed of traffic flow are the key parameters to be obtained by intelligent traffic management.

In terms of controllability, the DVE system in stable state is linear and observably controllable. The operation state of the system can be changed through driver's operation, and the disturbance of external environment can be effectively dealt with. However, the vehicle dynamics equation under the tire mechanical extreme conditions such as large sideslip angle and large slip rate is nonlinear, so the controllability of the system is difficult to be guaranteed. Therefore, the intelligent chassis control system shall monitor the vehicle state parameters in real time and control the vehicle tire within the linear working range for the real-time controllability of the system. In the DVE system, driver realizes the motion response of the vehicle by manipulating the steering wheel and pedal, realizes the lateral control by steering wheel, and realizes the longitudinal control by accelerator pedal, brake pedal and gear. The coupling of the lateral and longitudinal controls shall take the influence of random interference in the vehicle and environment into account, and make real-time dynamic correction accordingly. The random interference factors probably include other vehicle driving behavior, road adhesion conditions, weather conditions, vehicle dynamics, etc. The existing chassis active control and driving assistance technology can assist the driver to control the ego-vehicle under emergent conditions and maintain the vehicle motion state under cruise conditions.

In addition, as the central control unit of the closed-loop system, the driver or intelligent driving system is vital for ensuring and improving the safety, comfort, energy saving and efficiency of the system. At present, the research on the operation law of closed-loop system mostly focuses on the above performance. Research on vehicle motion safety under the action of DVE closed-loop system shows that about

80% of road traffic accidents are caused by driver. The main reason is that it is difficult for driver to respond accurately and timely under the sudden change of road traffic environment and vehicle state; Comfort is closely related to driver's personality. Reducing the impact of driving environment on driving and riding sensitivity through DVE coordination has become an important way to improve the comfort of the system; Different driving styles and driving habits notably influence vehicle energy consumption. With system power parameter matching, the optimal configuration of power and energy consumption is a common means to improve energy efficiency of closed-loop system; The assurance for the efficient road traffic system with safety, comfort and energy saving is a new and urgent issue to be settled in the closed-loop road traffic system.

To sum up, the characteristics of various elements of the complex system composed of driver, vehicle and environment are quite complicated. Specifically speaking, drivers have randomness, time variability and differences, vehicles have complexity, nonlinearity and strong coupling, and the environment has dynamics, interactivity and uncertainty, which make the system not only possess the properties of mechanical system and information system, but also possess the properties of physiology and psychology. In the research of system modeling, to deeply explore the interaction of each subsystem and clarify its action mechanism and operation law by making full use of biomechanics, cognitive psychology, generalized dynamics, automatic control theory and machine learning represented by deep learning and reinforcement learning is the objective requirement of more comprehensive and objective DVE system modeling, performance analysis and optimal control.

3.1.2 Generation Mechanism of Driving Risk

The driving risk is influenced by quite a number of factors in the DVE system. In the system, the driver plays the role of decision-maker. The driver's perception-decision-operation performance directly affects the vehicle control stability and driving safety, and significantly affects the driving risk. The realization of assistance driving or automated driving technology with the concept of "learning from human, simulating human and surpassing human" is based on the deep understanding of the driver risk cognition and response mechanism in complex traffic environment. Therefore, this part first reveals how the objective driving risk is generated in terms of the influencing factors, generation mechanism and evolution law of driving risk; Secondly, from the perspective of driver's subjective understanding on potential objective risks in the traffic situation, driver's common cognition mechanism and individual control law for potential risks in the traffic environment is explored; Finally, by summarizing driver's risk response and decision, this part illustrates driver's operation behavior law and its motivation, and discusses fundamental compliance of driver's risk response. The logical structure of driving risk generation mechanism is shown in Fig. 3.12.

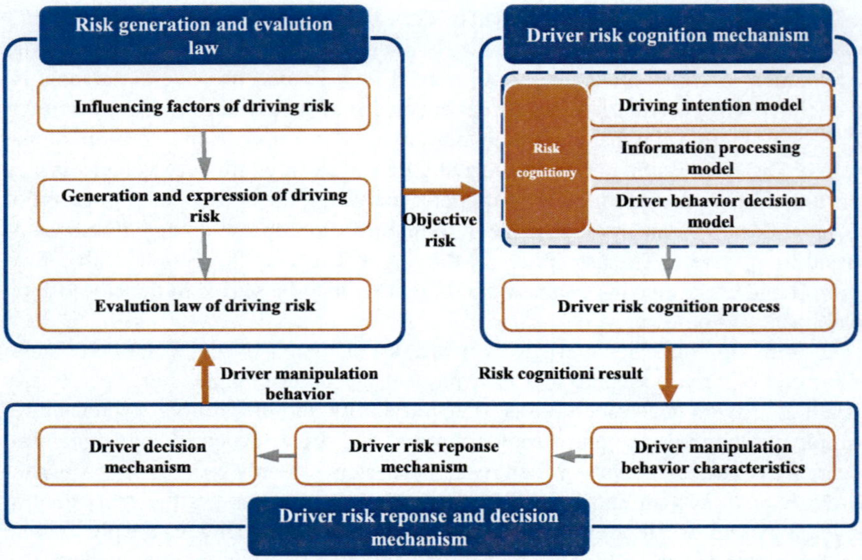

Fig. 3.12 Generation mechanism of driving risk

1. Risk Generation and Evolution Law

In order to clarify the mechanism of driving risk, the micro traffic system of DVE is taken as the research object. Firstly, the definition of driving risk is given, the traffic elements that generate driving risk are classified, and their impact on driving risk is analyzed; Then, the characteristics of driving risk and its generation process are analyzed, and the driving risk is quantified based on the theory of driving safety field; Finally, the evolution of driving risk with time is analyzed to reveal the evolution law of travel risk.

(1) Definition and Influencing Factors of Driving Risk

1) Definition of Driving Risk

The definition of risk in current research can be divided into two aspects. One is to regard risk as a measure of the possibility of harmful events or abnormal events in the system; The other is to define the risk as the product of the consequence of an accident and the probability of the accident. The research on system safety engineering believes that the root of the accident lies in hazard sources. Accidents can be prevented by controlling and eliminating hazard sources in the system. Generally speaking, hazard sources involve two categories: one is the energy or dangerous substances that may be accidentally released in the system; the second refers to various unsafe factors that lead to the failure of energy or hazardous substance constraints, mainly including system fault, human error and environment factors. In this book, the uncertainty of loss is defined as risk, which is comprised of two levels: one is the possibility of loss; the second is the uncertainty of loss. The uncertainty of loss is embodied in many aspects, such as its occurrence possibility, time, space, process and result.

2) Influencing Factors of Driving Risk

The factors affecting driving risk mainly include driver (human) factors, road user (vehicle) factors and road environment (road) factors. Figure 3.13 shows the main factors affecting driving risk in the traffic system composed of human, vehicle and road, as well as various traffic factor states that lead to sudden changes in driving risk and traffic accidents [33]. Among them, the reasons for the evolution of driving risk mainly include: ① driver's careless and inattentive observation of the driving environment; ② incorrect decision-making due to wrong estimation of current vehicle state, driving environment state and change trend; ③ operation errors due to immature driving skills and panic in the case of sudden situations; ④ traffic violations such as driving in violation of traffic regulations and other traffic safety regulations. The reasons for the evolution of driving risk made by relevant factors of road users mainly include: ① poor technical performance of vehicles, such as low or lost braking performance, failure of steering system, and failure of lighting system; ② pedestrians' road crossing without adhering to traffic regulations, such as running red lights, and walking on motor vehicle lanes. The reasons for the evolution of driving risk made by road environmental factors mainly include: ① poor road conditions, such as low road adhesion coefficient, too large slope, and too small radius of curvature; ② lack of road safety measures, such as no warning signs at sharp turns, narrow roads, steep slopes, intersections and railway crossings, and no clear restriction signs at the places of weight and height limits; ③ deterioration of road conditions and poor vision due to wind, snow, fog and other adverse weather conditions.

i. Driver-related Factors

Table 3.3 shows the survey results of International Drivers' Behavior Research Association (IDBRA) on 1500–2000 drivers in different countries, from which it can be found that traffic accidents are mainly related to drivers. Therefore, the impact of driver factors on driving risk has always been the focus of research, mainly including four aspects: physiological and psychological factors, driver cognitive ability, driving experience and driver violations.

Physiological and psychological factors include age, gender, personality, mood, etc. Young drivers are usually more prone to traffic accidents than older drivers, and are more prone to speeding, no seat belt wearing, too close vehicle—following and aggressive driving (including drunk driving) [35]. In terms of gender, young male drivers are more likely to have dangerous driving behavior than young female drivers [36]. In terms of driver's subjective risk assessment, researchers explored the impact of driver's gender, age and living environment (first-tier cities, second-tier cities and rural areas) on the subjective prediction of driving risk [37], and drew the following conclusions: under the same scenario, rural drivers and female drivers are more sensitive to risk than other types of drivers; compared with other drivers,

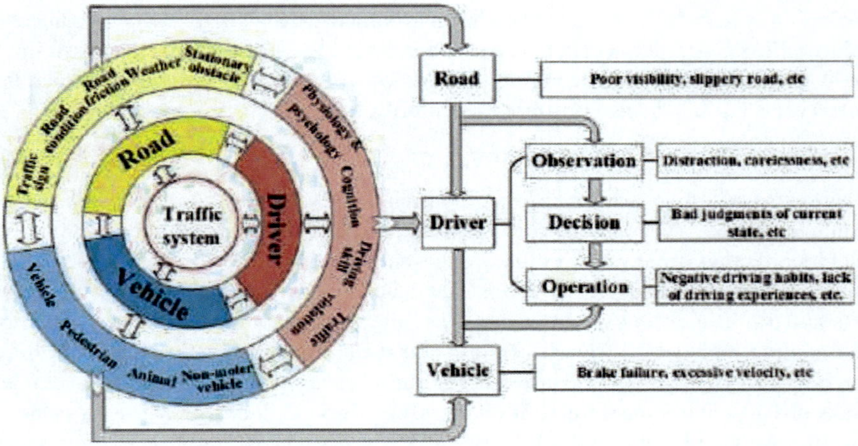

Fig. 3.13 Classification of influencing factors of driving risk

Table 3.3 Proportion of driver-related factors in traffic accidents in different countries [34] (unit: %)

Country	Proportion of driver-related factors	Country	Proportion of driver-related factors
Britain	56.1	Russia	52.7
Spain	92.0	Japan	40.0
France	85.5	South America	85.7
Sweden	81.1	Yugoslavia	69.7

young drivers have lower subjective risk assessment in low-risk scenarios; young male drivers in the survey are more likely to have violation and aggressive behaviors; compared with female drivers, male drivers have a higher proportion of illegal and aggressive driving behaviors, and the occurrence of illegal and aggressive driving behaviors decreases significantly with age. In terms of driver's personality, aggressive driving behavior has higher risk and is more likely to lead to traffic accidents [38]. The driver's character of pursuing stimulation is more related to a variety of dangerous driving behaviors such as drunk driving, speeding, racing, and illegal pass in no-go areas [39].

Cognitive level mainly represents driver's information processing ability. The related research at present can only reflect the relationship between driver's overall driving behavior and cognitive process, and pay less attention to the differences of drivers' cognitive levels. Different drivers have diverse personal attributes, driving habits and ways of thinking, which lead to different decisions in the face of the same scenario and even results in accidents. The influencing factors of driver's cognitive level mainly include driving skill, risk perception ability, environmental factors and so on.

In terms of driving experience, the wrong operation due to unskilled driving experience is one of the reasons for traffic accidents. Driving age is one of the indicators of driving experience. According to the causation analysis on 5664 passenger car accident cases in the database of China In-Depth Accident Study (CIDAS), 6967 accident causes are summarized. Among them, there are 4488 accidents induced by driver factors, accounting for about 81.5%. About 20.1% of the accidents are made by passenger car drivers with limited ability. Due to irregular driving behavior or unskilled driving, it is difficult for driver to make correct action response in the face of potential risks. Within it, accidents due to failure to effectively notice accident participants with potential conflicts account for the highest proportion, followed by failure to maintain a safe distance from other traffic participants.

In terms of violations, according to the causation analysis of passenger car drivers' accidents in CIDAS database, about 79.9% are induced by drivers' subjective errors. Due to the existence of subjective errors, drivers are unable to identify and perceive dangers in advance. Among them, the accidents induced by failure to yield in accordance with regulations account for the highest proportion, up to 43.4%, followed by excessive speed, illegal use of lanes, drunk driving, violation of traffic lights and fatigue driving. In recent years, China has begun to formulate corresponding laws and traffic regulations for motor vehicles to yield to pedestrians. In CIDAS database, the accidents caused by failure to yield to pedestrians account for 27% of the total number of accidents that fail to yield according to regulations. The accidents usually occur at intersections or signal indicators, etc.

ii. Factors Related to Traffic Participants

Traffic participants mainly include motor vehicles, non-motor vehicles, pedestrians, and even animals on the road. The risk factors related to motor vehicles and non-motor vehicles mainly include vehicle type, vehicle quality, vehicle technical conditions and motion state parameters, while the risk factors related to pedestrians and animals mainly include their location and motion state. For the impact of motor vehicles on driving risk of traffic participants, most evaluation models are based on vehicle kinematics and dynamics theory, and estimate the driving risk according to vehicle state information (speed, acceleration, yaw rate, etc.) and relative motion relationship information of two vehicles (relative speed, relative distance, etc.). In addition, some researchers use the artificial potential field (APF) theory in the field of robotics to analyze the impact of traffic participants on driving risk, use the potential field to describe the driving risk degree, and establish the artificial potential function.

iii. Factors Related to Road Environment

The factors related to road environment are divided into two types: one is the static obstacles on the road or roadside, such as roadside ditch and isolation belt in the middle of the road; and the other can be summarized as road attribute properties, including road adhesion, environmental visibility, road slope, curve curvature, lane width, etc. The influence of road conditional factors on driving risk is reflected in two aspects: one is to restrict driving, and the other is to affect driver's subjective judgment. For example, drivers are easy to overestimate the risk caused by horizontal curves and narrow lanes, and tend to underestimate the risk caused by intersections and roadside obstacles. And nearly 80% of the differences in drivers' subjective evaluation of road-related driving risks can be illustrated in terms of five major features of the road: road curvature, road width, shoulder width, road slope and obstacles [40].

(2) Generation and Expression of Driving Risk

1) Driving Risk Generation

Each constituent factor of transportation system has the possibility of driving risk. The risk caused by vehicle driving on the road is objective and inevitable. In the process of driving, vehicle safety is affected by many dynamic factors (such as driver's behavior, vehicle driving state and road traffic environment), so the possibility, risk degree and risk category of driving risk are also dynamic and uncertain. For vehicles traveling on the road, correctly identifying the essential feature of driving risk, that is, the macro expression of the generation and evolution law of driving risk, is of great guiding significance for establishing and improving vehicle decision-making and control and traffic control measures based on quantitative evaluation of driving risk.

The driver is the center of the transportation system composed of human, vehicle and road environment. The driving process of the vehicle is influenced by driver's "perception-decision-execution" process. The behavior characteristics of different drivers (radical or cautious) determine the driving modes and driving behaviors adopted by drivers. Meanwhile, the interaction of all vehicles on the road will lead to dynamic change of the traffic environment, which will lead to the change of driver's visual observation information. The driver's current observation information will become the basis for the next decision-making [33]. The impact of driver (human), traffic participant (vehicle) and road environment (road) on driving risk is shown in Fig. 3.14. Driver-related factors such as driver's biological characteristics, driving experience and driving proficiency directly affect driving control ability. Such factors as vehicle performance, other traffic participants and road traffic environment determine the driving task requirements. When the driving control ability is sufficient to meet the driving task requirements, the vehicle is in a controllable state, and the driving risk faced by the vehicle will also be within a safe range; When the control ability is not sufficient to fulfil the driving task, the vehicle movement state will be out of control, resulting in increased driving risk. If the driving risk exceeds the safety threshold, it is likely to evolve into traffic accident. If the surrounding vehicles take appropriate risk-avoidance operations, the driving risk faced by the ego-vehicle will be smaller and traffic accident can be avoided.

2) Driving Risk Expression

Accurate expression of driving risk is the basis for assistance or automated driving function of IV. A correct understanding on the features of driving risk, that is, the essence of driving risk and the external expression of its development law, is of

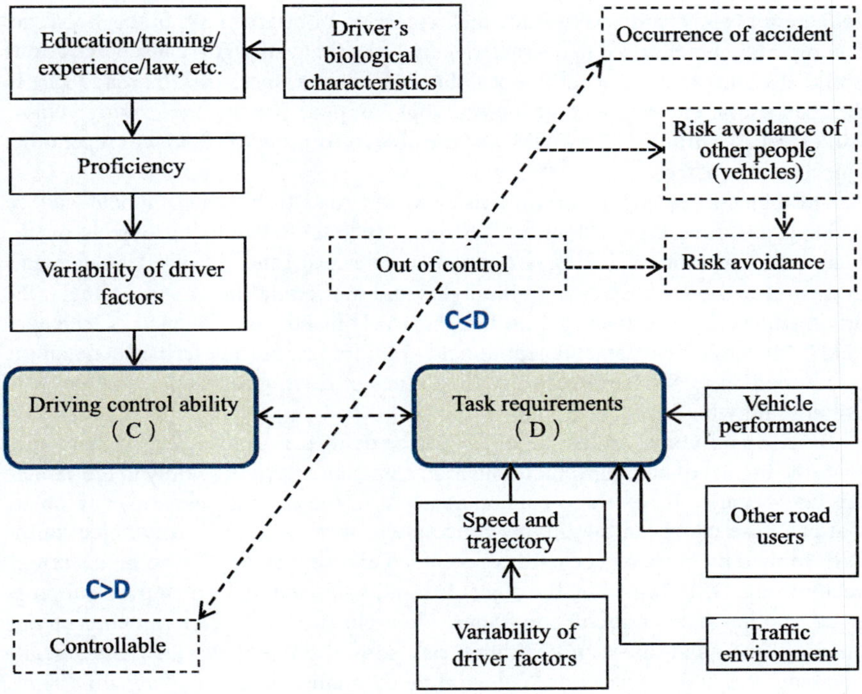

Fig. 3.14 Impact mechanism of various factors on driving risk

great significance to establish and improve the control measures of driving risk. Each component of the transportation system may generate risk, which is objective and universal. In the driving process, vehicle safety is affected by many dynamic factors, so the possibility, degree and category of driving risk are also dynamic and uncertain. The occurrence of traffic accidents is accidental, but through the observation and analysis of various driving risks, it is found that there is a certain causal relationship between these influencing factors and accidents, which indicates the significant regularity and estimability of driving risk.

Similar to the field source in physics, the driving risk is generated by traffic elements, and has essential features similar to the field. In physics, the distribution of a physical quantity in a specific area of space is called a field. As a basic form of material existence, field is possessed with energy, momentum and mass, and can transfer the interaction between objects. In the physical field, the field quantity is usually expressed by space–time function. The farther away from the field source is, the weaker the corresponding field quantity is. Similarly, the impact of traffic factors on driving risk is also related to time and space. The farther the traffic factors are from the driving vehicle, the smaller the driving risk is. In view of the similar feature between the risk caused by traffic factors and the field theory in physics, the idea of field theory can be adopted to study the expression of driving risk. In this book, the driving safety field is used to characterize the risks (field strength) caused by various traffic elements (similar to different field sources) to vehicles on the road [33]. In the road traffic environment, all objects that can pose risks to the driving vehicle will establish a driving safety field, and the objects forming the field will repel other objects close to it.

The objects leading to driving risks in the road traffic environment can be divided into three types: "human" (driver), "vehicle" (all moving objects on the road, including motor vehicles, non-motor vehicles, etc.) and "road" (static elements in the traffic scenario, such as traffic lights and traffic markings). Accordingly, the driving safety fields formed by these three types of objects can be defined as "behavior field", "dynamic field" and "potential field" [33]. Figure 3.15 describes in detail the corresponding relationship between three types of driving risks and three types of driving safety fields.

Based on the above analysis, the DSF can be defined as a physical field that represents the impact of human, vehicle and road elements on driving safety in the vehicle driving scenario. It has three connotations: ① In the driving scenario, any object that can pose danger in the driving process will form a DSF; ② Any object in the DSF formed by other objects will be excluded and kept away; ③ The field strength vector of the DSF formed by the object is jointly affected by its own physical properties, motion state and road conditions. The field strength vector can characterize the magnitude and direction of driving risk posed by the object [33]. Specifically speaking, the DSF consists of "potential field" related to "road", "dynamic field" related to "vehicle", and "behavior field" related to "human".

The potential field determined by "road" is used to describe the impact of static objects on driving risk. Static objects on the road include but are not limited to

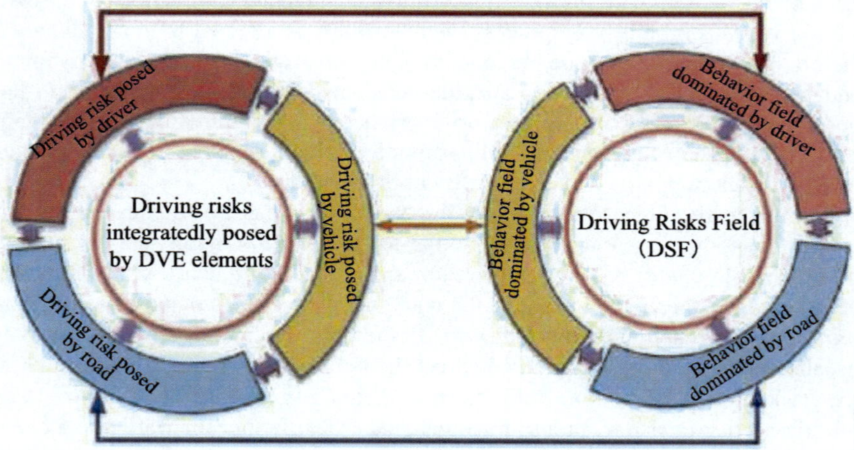

Fig. 3.15 Corresponding relationship between driving risk posed by traffic elements and DSF

vehicles parked on the roadside, isolation belts, deceleration belts, traffic markings, traffic signs, etc. The properties (type, mass) and road conditions of static objects determine the size and direction of potential field strength [41]. The field strength formed by a stationary object a can be expressed by the following formula:

$$\boldsymbol{E}_{\mathrm{R}} = \boldsymbol{E}_{\mathrm{R}}\left(M_a, R_a, \boldsymbol{r}_{aj}\right) = \frac{K \cdot R_a \cdot M_a}{\left|\boldsymbol{r}_{aj}\right|^{k_1}} \cdot \frac{\boldsymbol{r}_{aj}}{\left|\boldsymbol{r}_{aj}\right|} \tag{3.2}$$

where, $\boldsymbol{r}_{aj}$ is the distance vector from the object centroid position to the target point j; M_a is the virtual mass of a static object a; R_a is the influence factor of road conditions at the centroid of object a; $K(>0)$ and $k_1(>1)$ are constants to be determined. The direction of $\boldsymbol{E}_{\mathrm{R}}$ is the decreasing direction of the field strength gradient.

The dynamic field determined by the "vehicle" is used to describe the impact of road moving objects on driving risk. Moving objects on the road include but are not limited to moving vehicles, pedestrians, non-motor vehicles and animals, etc. Object properties (type, mass and speed), motion state and road conditions determine the size and direction of dynamic field strength [41]. The specific formula of dynamic field strength formed by moving objects is as follows:

$$\boldsymbol{E}_{\mathrm{V}} = \boldsymbol{E}_{\mathrm{V}}\left(M_b, R_b, \boldsymbol{r}_{bj}, \boldsymbol{v}_b\right) = \frac{K \cdot R_b \cdot M_b \cdot k_3}{\left(k_3 - |\boldsymbol{v}_b|\cos\theta_b\right) \cdot \left|\boldsymbol{r}_{bj}\right|^{k_1}} \cdot \frac{-\mathrm{grad}\ \boldsymbol{E}_{\mathrm{V}}}{|\mathrm{grad}\ \boldsymbol{E}_{\mathrm{V}}|} \tag{3.3}$$

where, M_b is the virtual mass of the object b; R_b is the influence factor of road conditions at the centroid of object b; $\boldsymbol{r}_{bj}$ is the distance vector from the centroid of object b to the target point j; v_b represents the speed vector of the object b; θ_b is the angle between $\boldsymbol{r}_{bj}$ and $\boldsymbol{v}_b$ (counterclockwise is positive); $K(>0)$, k_3 and $k_1(>1)$ are undetermined constants; the direction of $\boldsymbol{E}_{\mathrm{V}}$ is the decreasing direction of the field strength gradient.

The behavior field determined by "human" is used to describe the impact of driver's behavior characteristics on driving risk. The driver's behavior is mainly affected by psychological factors, cognitive level, driving skill and illegal behavior, and determines the size and direction of the behavior field strength. Generally speaking, the risk level of driving behavior of aggressive driver is higher than that of cautious driver, and the "behavior field" strength made by aggressive driver is higher. The "behavior field" strength of inexperienced driver is usually higher than that of experienced driver [33]. In fact, the driver's influence on driving safety is transmitted to the outside world through the vehicle. Therefore, the driver's behavior field strength can be expressed by the product of the driver's risk factor representing the influence of driver's characteristics and the dynamic field strength formed by the vehicle. The specific formula of the field strength vector $\boldsymbol{E}_{\mathrm{D}}$ of the behavior field formed by the driver of the vehicle c is as follows:

$$\boldsymbol{E}_{\mathrm{D}} = \boldsymbol{E}_{\mathrm{V}} \cdot \mathrm{DR}_c = \frac{-K k_3 R_c M_c \mathrm{DR}_c}{\left(k_3 - |\boldsymbol{v}_c|\cos\theta_c\right) \cdot \left|\boldsymbol{r}_{cj}\right|^{k_1}} \cdot \frac{\mathrm{grad}\ \boldsymbol{E}_{\mathrm{V}}}{|\mathrm{grad}\ \boldsymbol{E}_{\mathrm{V}}|} \tag{3.4}$$

where, DR_c (dimensionless quantity taken in (0, 1]) is the driver risk factor of vehicle c; $\boldsymbol{M}_c$ is the virtual mass of the vehicle c; R_c is the influence factor of road conditions at the centroid position of vehicle c; $\boldsymbol{r}_{cj}$ is the distance vector from the centroid of vehicle c to the target point j; v_v represents the speed vector of vehicle c; θ_v is the angle between $\boldsymbol{r}_{cj}$ and $\boldsymbol{v}_c$ (counterclockwise is positive); $K(>0)$, k_3 and $k_1(>1)$ are undetermined constants; same as the direction of $\boldsymbol{E}_\mathrm{V}$, $\boldsymbol{E}_\mathrm{D}$ is also the decreasing direction of the field strength gradient.

To sum up, the DSF consists of the potential field formed by stationary objects on the road, the kenetic energy field formed by moving objects and the behavior field formed by the driver, that is, the DSF = potential field + dynamic field + behavior field. If $\boldsymbol{E}_\mathrm{S}$ represents the DSF strength, $\boldsymbol{E}_\mathrm{R}$ is potential field strength, $\boldsymbol{E}_\mathrm{V}$ is dynamic field strength and $\boldsymbol{E}_\mathrm{D}$ is behavior field strength, the driving risk based on the DSF can be expressed as the following formula, so as to describe the driving risk degree determined by the possibility and severity of the accident in the objective road scenario:

$$\boldsymbol{E}_\mathrm{S} = \boldsymbol{E}_\mathrm{R} + \boldsymbol{E}_V + \boldsymbol{E}_\mathrm{D} \tag{3.5}$$

The above formula describes the degree of driving risk in the objective road scenario, which is determined by the possibility and severity of the accident.

(3) Evolution Law of Driving Risk

The evolution law of driving risk refers to the mode of action among human, vehicle, road and environment in the road traffic system, and the dynamic change of driving risk by the interaction in a specific driving environment and its feature. The purpose of the research is to analyze the temporal and spatial evolution of driving risk with traffic elements, predict the risk development trend and control the risk within a reasonable range. The core of the research is to explore the relationship, spread derivation and coupling evolution among risk factors based on clarifying risk factors [42]. As the constituent elements of the road traffic system, human, vehicle and road jointly determine the evolution law of driving risk. The harmonious interaction of road traffic elements is the key to ensure the safe operation of the road traffic system. Therefore, the changes of driver behavior, vehicle running state and road traffic environment are closely related to the evolution of road traffic system in three states, i.e., safety, risk and accident. The current research on the evolution law of driving risk mostly focuses on the process of traffic accidents caused by sudden change of risk, that is, the occurrence mechanism of road traffic accidents [43, 44], but it can not reveal the mode and law of risk evolution in the whole driving process. According to the morphological feature of road traffic accident development in the time dimension, some researchers put forward the road traffic accident chain [45] based on the accident chain theory, analyzed the process of traffic or behavior events finally forming road traffic accidents or adjacent road traffic accidents along different risk evolution paths, and studied the coupling relationship among traffic event, driving risk state and vehicle motion state based on Markov chain theory. With the study on the initial probability of different states of traffic events and the transition probability

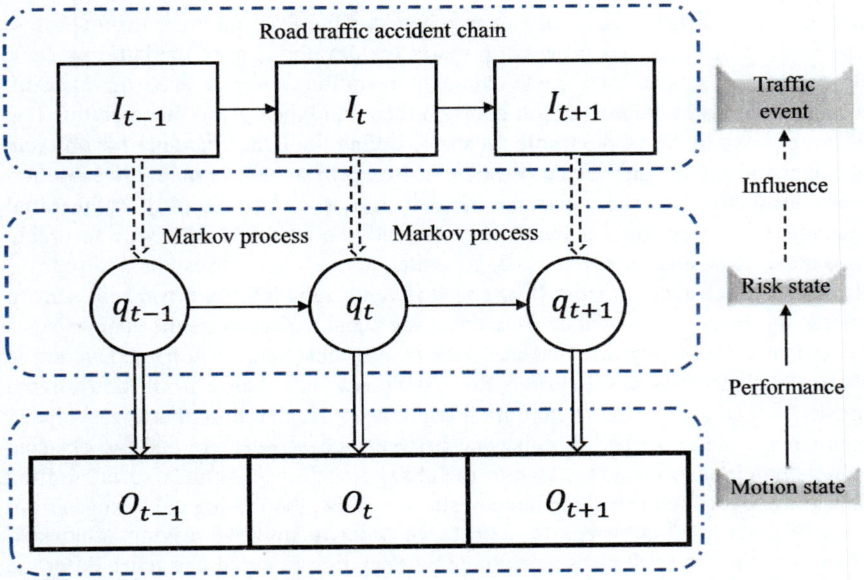

Fig. 3.16 Coupling relationship among traffic event, driving risk state and vehicle motion state [45]

between states, the formalization description of the evolution process of driving risk is made, as shown in Fig. 3.16.

As a dynamic open system, the road traffic system always exchanges material, information and energy with the outside world. In the process of road driving risk evolution, the system safety is not only affected by the changes of drivers and vehicles, but also by the sudden changes of various external factors, so that the original balance of the system is constantly broken and always in the trend of evolving to a new equilibrium. There are two different forms of evolution in the objective world: one is smooth and continuous change, and the other is discontinuously sudden change. The occurrence of road traffic accidents is usually accompanied by the sudden change of driving risk. Therefore, there is a discontinuous change in the evolution process of road driving risk from one form to another. Catastrophe theory is often applied to analyze the evolution law of this discontinuous change process. The key to the application is to construct and evaluate the catastrophe potential function of safety cusp, take the dynamics of traffic system into account, and clarify the causal relationship between the continuous change of any parameter of traffic system and the sudden change of system state in the process of driving risk evolution, on the basis of potential function in state space as well as the characteristic evolution curve and bifurcation set.

The evolution process of driving risk is accompanied by the evolution of the structure, state and behavior of the traffic system. The whole driving process can be decomposed into a continuous process connected by multiple traffic events in series, and there is a certain causal relationship between the front and rear traffic events. The result of the previous process promotes the development of the latter process. By analyzing the state of various elements in traffic system in each time period, we can understand the interaction between them, and then grasp the evolution law of driving risk in the whole traffic scenario. Taking the Lane changing for obstacle avoidance in multi-Lane scenario shown in Fig. 3.17 as an example, the evolution mechanism of driving risk can be described by time-state analysis [44]. At the initial moment, the driver operates the vehicle normally on a straight road with favorable road conditions, and the driving risk generated at this stage corresponds to state A; If there is an obstacle in front (low-speed moving vehicle), the driver will start to release the accelerator pedal or press the brake pedal to decelerate the vehicle, so as to maintain a relatively safe distance from the obstacle vehicle in front. Due to the change of the state of DVE system, the driving risk will change from state a in the previous stage to state B; Then, during continuous deceleration, the driver's speed continues to decrease and the distance between the ego-vehicle and the obstacle vehicle continues to shorten. Due to the change of the ego-vehicle speed and the distance between the vehicle and the obstacle vehicle, the driving risk evolves from state B in the previous stage to state C; In order to improve driving safety and efficiency, the driver chooses lane changing operation to avoid obstacles. After the vehicle reaches the target lane with favorable road conditions, the driver operates the accelerator pedal and make the speed gradually return to the target speed. And The driving risk will evolve from State C to state D.

Driving risk always exists in the road traffic, and its evolution law is affected by traffic elements and their interaction, as shown in Fig. 3.18. During driving, drivers observe the road traffic environment through vision and hearing, and the change of traffic factors will affect their operation intention; The change of driver's handling behavior will further affect the vehicle kinematics, including the vehicle's own speed

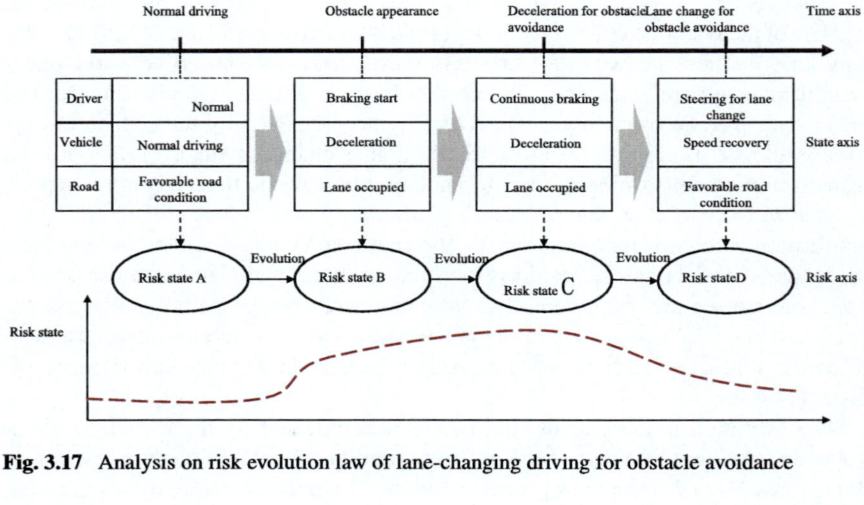

Fig. 3.17 Analysis on risk evolution law of lane-changing driving for obstacle avoidance

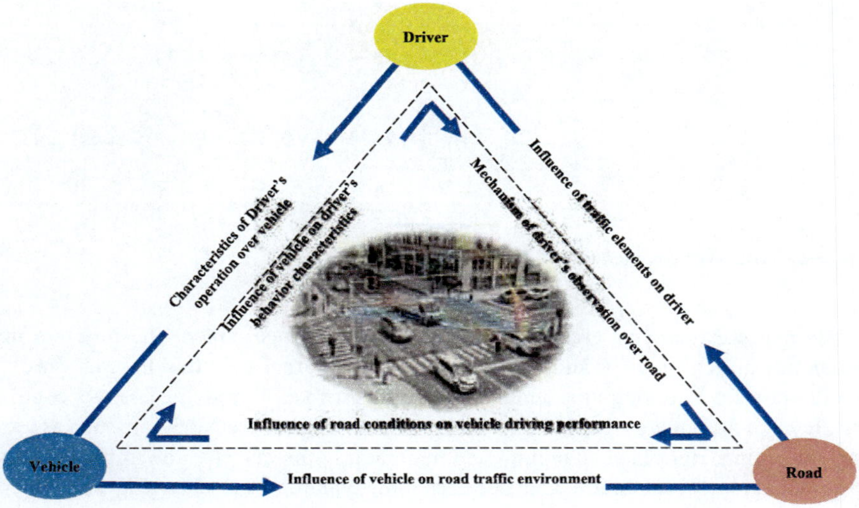

Fig. 3.18 DVE interaction relationship

and traffic flow rate. Meanwhile, the vehicle will also affect the driver's behavior characteristics; The change of vehicle state will affect the traffic environment, and the road conditions will also affect the driving performance of vehicles. Due to the interaction between human, vehicle and road, any change in the state of traffic elements will lead to the change of driving risk.

2. Driver Risk Cognition Mechanism

In the traffic scenario, driver mainly receives the input of external dynamic information through vision, hearing and touch, and processes the information cognitively, to output commands for vehicle control. The behavior characteristics of different drivers are complicated, and there are multiple processes of cognition, decision-making and operation, with diverse features and great differences. The decision-making behavior in the driving process is affected by many external factors, such as road traffic environment and dynamic characteristics of the vehicle. The risk in driving is closely related to the real-time operation of driver and the dynamic change of traffic environment. The driver's cognition of traffic environment coupled with various factors will affect decision-making, and the behavior output also reflects risk cognition ability. Therefore, to explore the relationship between numerous factors such as road conditions and vehicle attributes and the driver risk cognition mechanism, it is necessary to comprehensively explore the driver risk cognition characteristics and internal driving decision-making mechanism.

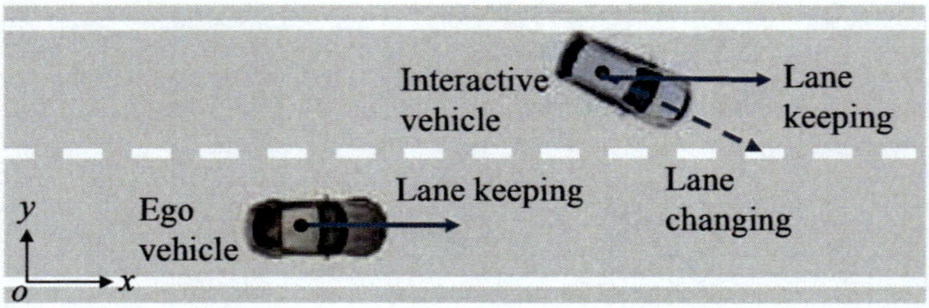

Fig. 3.19 Multi-vehicle interactive lane changing behaviors

Human behavior has great randomness and high uncertainty. The problem of prediction difficulty can be addressed by observing external activities. In other words, the driver's behavior output is understood as transformation from the abstract cognition level to the concrete behavior level. For example, there are behaviors such as left lane changing, right lane changing, and maintaining the current lane in the vehicle lane changing; acceleration and deceleration, dynamic lane changing and other behaviors on the highway for avoiding surrounding vehicles, as shown in Fig. 3.19. Therefore, the driver risk cognition for different scenarios is an abstract cognition process in essence, which is externally reflected in concrete driving behavior. In other words, in the face of complex dynamic scenarios, driver's different behaviors can represent risk cognition level to a certain extent.

At present, the research on driver cognition and behavior law mainly includes driving motivation model based on risk avoidance behavior, information processing model based on driver information processing, and driver behavior decision-making model based on cybernetics [46]. Speicifically, the driving motivation model was initially put forward in the 1960s–1970s, and gradually evolved and put into practical application. Its basic assumption is that driving behavior is a kind of self-control behavior, and driver can choose the bearable degree of risk as well the motivation to determine behavior, mainly including risk homeostasis theory, risk threshold model and risk avoidance model. The driving motivation model generally simulates driver's cognition mechanism by describing driver's dynamic perception of risk and action output [47]. The driving motivation model can reflect the psychological cognition process and better simulate human characteristics, but is difficult to ensure the accuracy.

The information processing model is represented by a series of stages in driver's information processing, including perception, decision-making, choice response and behavior response. The early information processing models and theoretical approaches are not suitable for complicated driving behavior, and are difficult to deal with and analyze the risk cognition process alone, resulting in the outcome inconsistent with the reality to a certain extent [48].

The establishment of driver behavior decision model based on cybernetics has long development history and profound theoretical foundation. It mainly makes systematic and in-depth research on driver decision-making behavior from local and overall aspects, and constructs the theoretical system and method model for behavior analysis from the perspective of "human-centered", which mainly includes the formation process of driving behavior, the output process of operation behavior, the subjective decision-making mechanism and the quantization of relevant parameters. Driver behavior decision-making model and theoretical approach is a new exploration of driving risk cognition theory. Its motivation related to driving purpose and importance will influence driving behavior in the whole driving process, and the disposal of accidents and emergencies encountered in the situation may only be made by driver's instantaneous operation output.

It has become a research hot topic in this field to build the fundamental mechanism and theoretical approach for risk cognition, design the risk cognition model, and realize the effective analysis and accurate prevention and control of traffic accidents. Therefore, the theories and approaches for driver risk cognition are systematically classified, as shown in Table 3.4 in detail.

Table 3.4 Theories and approaches for driver risk cognition

Model type	Typical approach	Advantage	Disadvantage
Driving motivation model	Risk homeostasis theory	① Subjective risk is a relatively stable parameter ② Better control in driver's operation	① Unclear distinction between group and individual differences ② Controversial invalidity ③ Not real driving simulator test ④ Poor accuracy of the model ⑤ Unable to meet the requirement of complex scenarios
	Risk threshold model	① Better simulate the driver's psychology ② Simulate human characteristics	
	Risk avoidance model	Higher accuracy and clear motivation, compared with other motivation models	
Information processing model	Filter theory	① Represented by a series of stages, and capable to realize multi-stage information processing ② Capable to simulate driver risk cognition law in the real scenario	① Not suitable for complex driving behavior ② Unable to simulate unknown scenes ③ Unable to describe the driver's personality
	Attenuation theory		
Driver behavior decision-making model	Hierarchical control theory	① Improve the driver's driving experience ② Improve the acceptance of IV ③ Reduce the possibility of system errors	① Without a dynamic relationship between different levels pointed out ② Not mature research at present ③ Difficulty in personality classification
	General error model system		

(1) Driving Motivation Model

The driving motivation model assumes that driving behaviour is a self-controlled behaviour and that drivers can choose the level of risk they are comfortable with. Risk cognition focuses on temporary or specific situational factors, emphasizes the individual's feeling and understanding of various objective risks in the external world, and also includes the influence of individual's experience obtained from intuitive judgment and subjective feeling on his cognition. The individual driver is an extremely dynamic and complex microsystem. The diversified factors affecting specific motivation include age, gender, driving experience, occupation, personality, etc. Therefore, driver's motivation model for risk cognition can be represented in Fig. 3.20. The more classic models include risk homeostasis theory, risk threshold model and risk avoidance model.

According to the Risk Homeostasis Theory (RHT) proposed by Wilde in 1976 [49], the subjective risk level of drivers during driving is constant. In order to maintain this level, they will eliminate the difference between behavioral risk and established subjective risk by adjusting their behaviors. For example, the distance between the driver and the vehicle in front is determined by assessment on the risk of rear-end collision. In 1982, Wilde [50] further proposed the Risk Compensation Model (RCM), believing that there is a general compensation mechanism for drivers, which can predict the risk, balance the relationship between road situation and subjective risk degree, make the comprehensive judgment of subjective perception and risk acceptability, and finally realize the unity of subjectivity and objectivity.

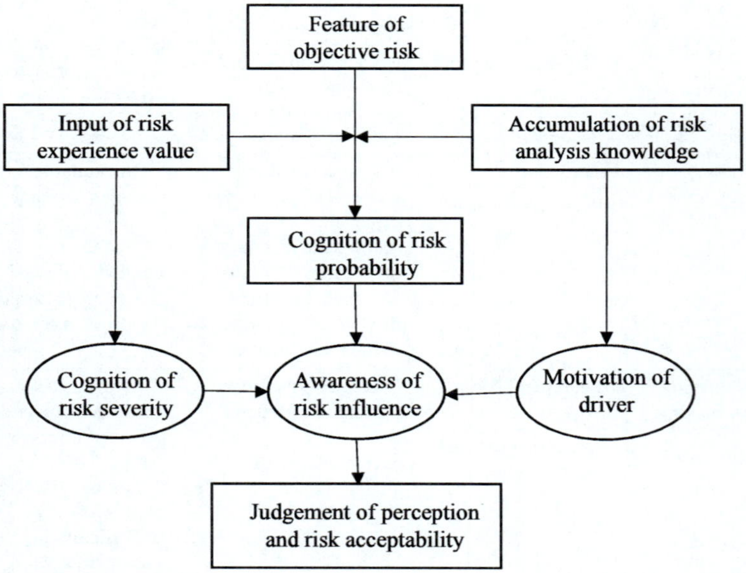

Fig. 3.20 The motivation model of driver risk cognition

In 1976, Naatamn and Summala [51] proposed the Risk Threshold Model (RTM), or Zero-Risk Theory (ZRT). The model assumes that the risk level of the driver in the driving process is not constant, and the driver's risk control is mainly based on his or her own safety margin, such as maintaining a certain space between the ego-vehicle and other vehicles to ensure no collision or scratch. The zero risk theory also believes that the main reason for traffic accidents is that the driver's subjective risk threshold is too high, that is, the safety margin is too wide to accurately reflect the objective risk. The risk threshold model usually ensures driving safety based on safety factors of the operation threshold, in which the safety factor is determined by the time interval and space distance between the vehicle and the hazardous event.

Fuller [52] proposed the Risk Avoidance Model (RAM) in 1984. The model assumes that arriving at the destination and avoiding risks are the two main motivations of drivers. The similarity between the risk avoidance model and the zero risk model is that driving is the reflection of motivation. But the difference is that the risk avoidance model is based on behaviorism learning theory. When people face some dangerous situations during driving, they take some effective measures to avoid accidents, and output different emergency behaviors in different dangerous situations, which can reflect driver's learning ability to a certain extent.

(2) Information Processing Model

Information processing model mainly refers to that in the process of identifying potential hazards, drivers will receive a large amount of external stimulation information, evaluate the degree of driving risk, selectively screen some effective information, and finally implement feasible behavior strategies to deal with potential driving risks. The model attaches importance to the information screening and feedback mechanism in the process of external stimulation. Classical information processing models include the filter model and attenuation model. The driver information processing process is shown in Fig. 3.21.

Broadbent, a British psychologist, proposed the filter model in 1958 [53], which observed that the capacity of the individual nervous system to process information at the same time is very limited, and filters are needed to prevent the central nervous system from being overloaded. Therefore, the information processing model is represented by a series of stages, including perception (reception of external environmental stimulation), decision-making (information filter), selection response and behavior

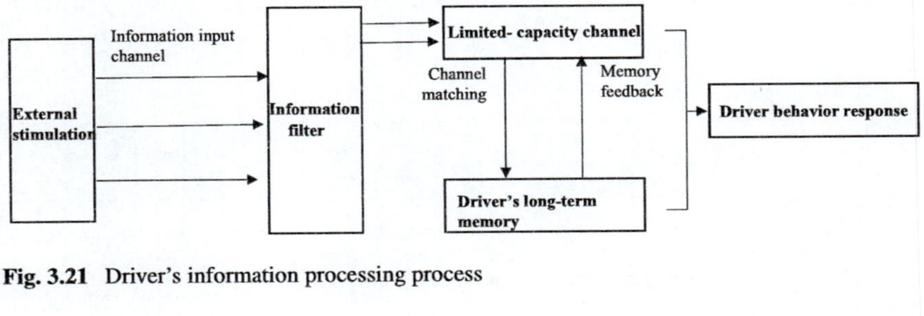

Fig. 3.21 Driver's information processing process

response (memory feedback). In the filter model, many channels are connected to only one, which is why some people call this model the "single channel model".

In 1960, American psychologist Treisman proposed an attenuation model to modify the filter model [53]. According to this model, the actual information processing process usually attenuates rather than cuts off some unfocused information, and important information can still be processed in advance and reflected in consciousness. Although the above theories have made distinctive research on the selection mechanism of attention, they have not yet deeply revealed the internal process and stage of the attention process, and difficult to adapt to the information processing and risk cognition process of a complicated driver system [54].

(3) Driver's Behavior Decision Model

In recent years, scholars at home and abroad have done a lot of research on driver behavior decision-making model and achieved rich results [55]. These studies aim to find the influencing factors, influencing modes and performance characteristics of drivers' behavior decision-making, so that researchers can make a reasonable explanation for the decision output, with driving safety improved and accident rate reduced. The influencing factors of driver's cognition and decision-making behavior mainly include driving skill, risk sensitivity, driving environment, etc. For example, drivers with less driving experience are usually more likely to cause traffic accidents than drivers with more experience [56]. The driver's risk perception and behavior decision-making process is shown in Fig. 3.22. The driver's behavior decision model mainly expresses the level of risk cognition by controlling behavior output (including control command output by hands and feet, such as acceleration, braking and steering), which can usually support IV control.

In 2011, scholars used the Structural Equation Modeling (SEM) to analyze the driver's risk perception, and found that the risk perception ability of different drivers has a significant impact on the process of cognition and decision-making behaviors. For an individual driver, the operation behavior and cognitive characteristics can also reflect the ability of risk perception. After the systematic research on risk cognition, the systematic risk cognition model for drivers has been gradually applied and popularized [57]. The model can effectively respond to the risks in the traffic environment, involving four processes, i.e., risk identification, risk assessment, behavior selection and behavior execution. Based on the systematic analysis, some researchers further refine drivers' behaviors and divide them into skill-based behavior, rule-based behavior and knowledge-based behavior. Skill-based behavior mainly includes

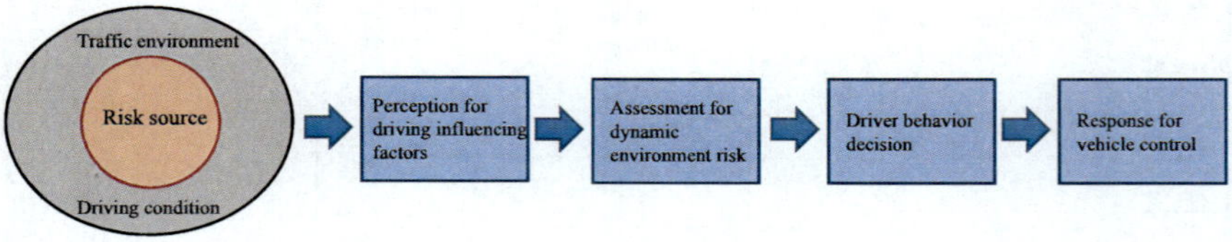

Fig. 3.22 Driver's risk perception and behavioral decision process

learning procedures of various schemata. Rule-based behavior refers to the behavior of rule automation. Knowledge-based behavior refers to the ability to solve problems consciously and explore unknown situations.

Some researchers also describe driver risk cognition and behavior decision-making process in terms of vehicle and driver respectively. Within it, at the vehicle level, the driving events experienced by the driver refer to the operation output during driving, including acceleration, deceleration, lane changing, etc., which are represented by different driving modes. At the driver level, driving behavior mainly emphasizes the process that driver's internal characteristics (such as driving skill and driving style) affect the output of behavior decision-making.

Generally speaking, the current research outcomes on driver's cognition and behavior law can only reflect the relationship between driver's overall driving behavior and cognitive process, and pay less attention to the differences between driver's cognition and behavioral decision-making. The generation of driving risk is a multi-dimensional and diverse process, including perception, cognition, decision-making and other dimensions. Specifically, the insufficient perception will affect the judgment on speed, location, road and traffic signs, resulting in inaccurate input information of risk sources; The lack of cognition leads to the deviation of understanding for risk and its change law; The lack of decision-making makes it difficult to formulate correct strategies in the process of risk avoidance behavior selection and path generation; Insufficient operation leads to poor action and risky behavior. Perception, cognition, decision-making and human risk behavior are indirectly related, whereas operation and human risk behavior are direct. However, abnormalities at any stage will induce driving risk.

Affected by their personality and traffic scenario characteristics, drivers with different styles have different subjective cognition, evaluation and corresponding driving reactions to the actual and potential hazards in the same traffic scenario. Therefore, not all drivers have the same level of hazard perception when encountering the same driving hazard. For the same traffic hazard, the information perceived by drivers with high hazard perception levels and those with low levels is different. Drivers with high levels can avoid danger by identifying the potential hazard state of the scenario and making themselves in a safe driving state.

(4) Driver's Risk Perception Process

The driver is the information processor, decision maker, regulator and controller of road traffic system. With the deepening of driving behavior research, driver risk cognition has become the focus of research. There are many versions of translation for "risk perception" by domestic scholars, such as "risk perception", "risk cognition", "hazard perception" and "hazard awareness". Descriptions of its connotation are also diverse. Among them, hazard awareness refers to the psychological process in which driver's brain identifies the perceived external information after processing. The most accurate translation is risk cognition, which can be defined as the driver's subjective understanding and judgment of the potential objective risk in the traffic situation, and timely feedback for behavioral decision-making. Risk cognition ability affects driver's cognition, judgment and decision-making ability, and thus affects driving safety. Different scenarios have an uncertain impact on driver risk cognition ability. Assessment of driver's risk perception ability is to investigate the ability to identify road traffic situations, that is, driver's ability to identify, predict and evaluate dangerous situations in different traffic scenarios. There are great differences in drivers' skills, preferences and driving styles. Meanwhile, there are complex interactions between drivers in the transportation system. Inappropriate risk cognition leads to driver's inability to recognize or ignore hazardous situations or take dangerous driving behaviors. Therefore, the research on the relationship between driver's risk awareness and driving behavior and the analysis of driver risk cognition process will not only be conducive to deeply analyzing the causes of traffic accidents, but also reducing the occurrence of accidents from the root. In addition, the research outcomes are used to formulate necessary intervention or improvement measures for the driving assistance system, to set up a safe and efficient automated driving system [58].

As shown in Fig. 3.23, driver risk cognition process starts from the driver's perspective in the complex DVE coupling system, and involves stages of conducting behavior reasoning based on individual differences such as driver's cognitive characteristics (observation, judgment, etc.), outputting driver's real-time operation behavior, illustrating the dynamic interference of road traffic environment in combination with driver's visual characteristics, exploring the action mechanism of DVE coupling factors on driver risk cognition, and then refining the risk cognition mechanism. It can be found that the risk mainly arises from the real-time operation of driver and the dynamic interference of traffic environment. The subjective and objective risks interact to form the psychological projection of driver risk cognition, and finally output the comprehensive risk degree. When the risk is higher than the set safety threshold, it will lead to a potential traffic accident. Therefore, risk cognition is defined as the psychological projection area of various elements in the dynamic traffic environment acting on driver risk cognition from the driver's perspective.

Based on Fig. 3.23, a general mathematical framework for comprehensively describing micro-dynamic traffic can be proposed. In the framework, the dynamics of micro traffic environment is generated by multiple interactive driver decision-making behaviors operating in a complex and partially observable environment. The framework is a partially observable stochastic game (POSG) with discrete time and multiple agents. In view of all traffic scenarios, driver's risk perception is constructed into the physical state $x_i^{(t)}$ and internal state $b_i^{(t)}$. The physical state describes the position, direction and speed of the vehicle operated by the driver, whereas the internal state includes the driver's navigation target, behavior characteristics and "psychological model" based on the surrounding environment. As shown in Fig. 3.24, each layer of the graphic model corresponds to a specific driver, and the time increases from left to right. Lines indicate the direction of information flow. Within it, $x_{id}^{(t)}$ is the physical state of driver id, $b_{id}^{(t)}$ is the internal state, $u_{id}^{(t)}$ is the control action, $z_{id}^{(t)}$ is the observation value, $x_{id}^{(t+1)}\ F_{id}(x_{id}^{(t)}, u_{id}^{(t)})$ is the state transfer function, $z_{id}^{(t)}\ G_{id}(x_1^{(t)}, \ldots, x_n^{(t)})$ is the observation function, $b_{id}^{(t+1)}\ H_{id}(b_{id}^{(t)}, z_{id}^{(t)})$ is the internal state update function, and $u_{id}^{(t)}\ \pi_{id}(b_{id}^{(t)})$ is the strategy function.

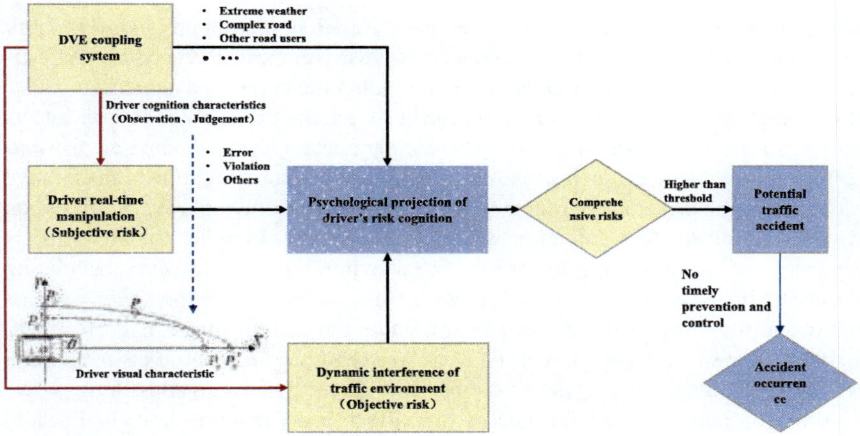

Fig. 3.23 Driver risk cognition process

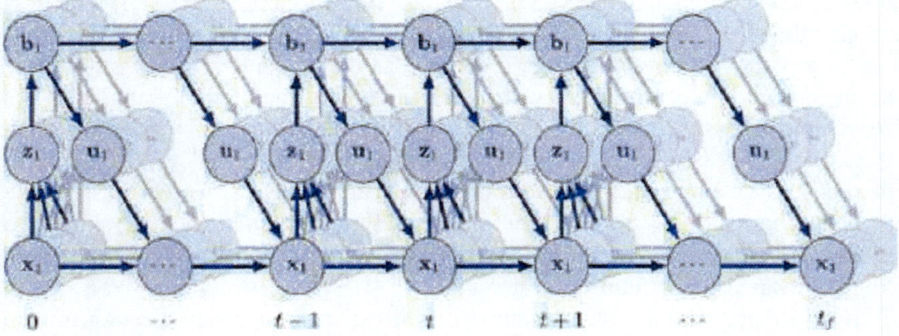

Fig. 3.24 Multi-agent POSG framework

$z_{id}^{(t)}\ G_i(x_1^{(t)}, \ldots, x_n^{(t)})$ is supposed to represent a potentially uncertain process through which the driver id observes the surrounding environment in time t. G_{id} is defined as an observation point function, and $z_{id}^{(t)}$ is the observation value. Generally speaking, the driver id only partially observes the environment (i.e. waste or loss of $z_{id}^{(t)}$). When the driver id processes each newly-observed information, its internal state will evolve over time. Therefore, $b_{id}^{(t+1)}\ H_{id}(b_{id}^{(t)}, z_{id}^{(t)})$ can be used to describe this process, where H_{id} is the internal state update function. At every moment, the driver id selects a control action $u_{id}^{(t)}$ based on $u_{id}^{(t)}\ \pi_{id}(b_{id}^{(t)})$, in which π_{id} is called strategy function, and its parameters reflect the fact that the decision comes from the driver's internal state. According to $x_{id}^{(t+1)}\ F_{id}(x_{id}^{(t)}, u_{id}^{(t)})$, the state evolves with time, and F_{id} is the discrete-time stochastic state transition function.

The POSG model is essentially similar to the specific driver risk cognition modeling tasks that may be involved in driver decision-making and control, including state estimation, intention estimation, feature estimation, motion prediction and risk cognition. State estimation is the process that the driver extracts the consistent estimation of the physical environment state, including the current physical state ${x^{(t)}}_{1:n}$ of surrounding vehicles. The estimation can be deterministic, or can reflect the uncertainty through the probabilistic distribution of parameters (such as Gaussian distribution) or non-parameters (such as particle distribution). State estimation is an effective way to infer the driver's environment, that is, all other modeling tasks are based on the inferred information. Intention estimation is a process in which the driver infers the actions of surrounding vehicles in the near future. It usually involves calculating the probabilistic distribution of a limited set of potential future behavior patterns that the surrounding vehicles may perform under the current situation, correspondingly to different navigation targets (such as lane changing and overtaking). Intention estimation is mainly about the reason why the driver tries to do something, while feature estimation is about the factors that affect how the driver does it. Broadly speaking, features include skills, preferences and styles, as well as attributes such as driving fatigue and distraction. In the POSG framework, feature estimation can be interpreted as the process of inferring the "parameters" of the driver's strategy function according to the observed driving behavior. Motion prediction is a process in which the driver judges the future physical state $x_{1:n}^{(t+1:t_f)}$ of surrounding vehicles. It is the key task of driver behavior modeling, as the state inference of other vehicles in surrounding environment will directly influence the motion planning of the ego-vehicle. Effective prediction process usually requires state estimation, intention estimation and feature estimation. Risk assessment involves quantifying the risks inherent in a given scenario. It is particularly vital in ADAS, which must decide whether, when and how to intervene. Risk assessment is fundamentally linked to motion prediction and other core tasks, for risk is a measure of the "unsafe" degree of one or more drivers' future movements.

Therefore, through the POSG framework, driver risk cognition mechanism can be constructed to a certain extent. And the whole process of driver risk cognition can be mathematically described based on the clear definition of relevant state space, action space, observation space and internal state space, as well as the state transfer function, observation function, strategy function and internal state transfer function related to its model.

3. Driver's Risk Response Decision Mechanism

Driver's improper response to the results of risk cognition is a critical factor threatening DSV. This section mainly refines the features of driver's operation behavior, reveals driver's risk response mechanism, clarifies driver's driving behavior decision-making motivation, and lays theoretical foundation for the study of driving accident causation mechanism.

(1) Driver's Operation Behavior Features

Driver's driving behavior refers to the operating actions such as direction control, driving control and braking control through neuromuscular response after perception, synthesis, judgment, inference and decision-making on information. It has the characteristics of strong randomness, adaptability, discreteness and time variability. Specifically speaking, the driver's operation behavior mainly includes lateral driving behavior dominated by steering manipulation and longitudinal driving behavior dominated by braking and driving manipulation. It is a stress behavior taken to deal with the external environment and the state of the vehicle itself.

Recognition and understanding on the law of driver's operation behavior is a considerable part of realizing the concept of "learning from human, simulating human, surpassing human and serving human" of AV, and is also an important prerequisite for realizing the human-like operation of AV. Therefore, in order to fully analyze the law of driver's operation behavior, the following part will analyze on the evolution law of typical operation characteristic parameters in the process of lane changing and car following based on the ND data of the real vehicle, and illustrate the operation behavior mechanism followed by the driver during driving.

The scenario and quality of ND data set are critical to objectively analyze driving behaviors of drivers. The HighD data set released by the Institute for Automotive Engineering of RWTH Aachen University, Germany, discloses a more comprehensive natural vehicle trajectory data set. Therefore, focusing on the analysis of drivers' operation behavior features in the scenario of high-speed car following and lane changing, more than 3600 complete lane changing trajectories are extracted from the public data set. Through the time response analysis of the lateral displacement, lateral speed and lateral acceleration in the process of vehicle lane changing in this scenario, it is found that the maximum lateral acceleration, the maximum lateral speed has the following relationship with lane-changing time and lateral displacement during lane changing:

$$a_{y\_\max} = \frac{k_1 d}{t_f^2} \tag{3.6}$$

$$v_{y\_\max} = \frac{k_2 d}{t_f} \tag{3.7}$$

where, $a_{y\_\max}$ and $v_{y\_\max}$ are the maximum lateral acceleration and maximum lateral speed during lane changing respectively. These two physical quantities reflect the safety (comfort) and lane-changing efficiency during lane changing; k_1 and k_2 are regulating parameters, and can be calibrated according to the measured data through numerical optimization algorithm. Furthermore, in order to realize the comprehensive measurement of safety, comfort and lane changing efficiency, the above formula can be reformulated as:

$$\frac{a_{y\max} d}{v_{y\max}^2} = \frac{k_1}{k_2^2} = k \tag{3.8}$$

On the premise that the maximum lateral displacement in the lane changing process is already known, the smaller the maximum value of lateral acceleration is, the greater the maximum value of lateral speed is and the smaller the value of the index is, indicating that the lane changing process has higher safety, better comfort and higher efficiency. The better the comprehensive performance of the lane changing process is, the higher the coordination and balance capability of the three performances is. To sum up, the index can directly reflect the multi-performance balance mechanism followed by the driver's lateral handling behavior in a physical sense, as shown in Fig. 3.25.

Therefore, from the perspective of lateral kinematics, the quantitative evaluation index of lateral driving behavior based on driver's coordination mechanism of multi-performance balance such as safety, comfort and efficiency can be put forward by observation and extraction of the driving behavior data. And this index is defined as the Lateral Quantification Balance Index (LQB Index), for the quantitative evaluation of driver's lateral driving behavior [34]. As a dimensionless quantitative evaluation index, it can comprehensively evaluate the geometric performance, motion performance and driver operation conversion efficiency of the candidate lane-changing trajectory.

$$I_{\text{lat}} = \frac{a_{y\,\max} d}{v_{y\,\max}^2} \tag{3.9}$$

Similarly, by analyzing the vehicle longitudinal motion information in the data set, it is found that the driver's longitudinal operation behavior follows the same operation mechanism of time headway and Inverse TTC, which is specifically shown as follows:

$$a_x = k_1 \tanh(k_2 \times RP + k_3) + k_4 \tag{3.10}$$

$$\text{RP} = \frac{k_5}{\text{THW}} + \frac{k_6}{\text{TTC}} \tag{3.11}$$

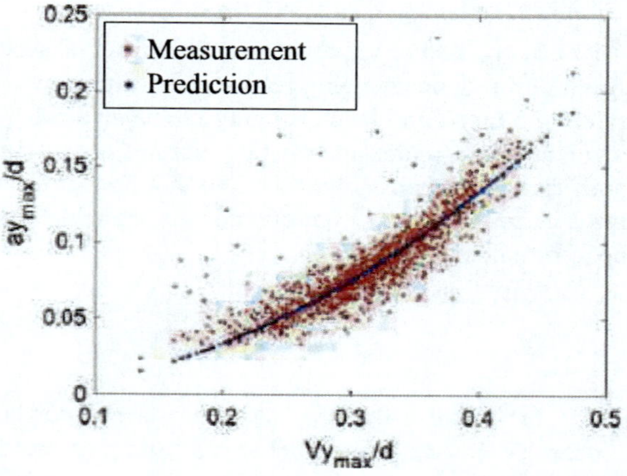

Fig. 3.25 Relationship between maximum lateral speed and maximum lateral acceleration of vehicle under lateral operation of driver

where, THW and TTC are the time headway and time to collision respectively; and $k_i (i = 1, 2, 3, 4, 5, 6)$ is the parameter to be calibrated. By analyzing the above data set, it is found that the braking intensity is always larger than the acceleration intensity during car following. There are two reasons for this phenomenon. First, the wind resistance becomes greater with the increase of vehicle speed during acceleration, resulting in the weakening of vehicle longitudinal acceleration; Secondly, driver's response to risk is usually more timely and rapid than that to driving efficiency. Therefore, the intensity and amplitude of driver's stepping on the brake pedal will be greater than that of driver's stepping on the accelerator pedal.

The above mainly analyzes the multi-performance balance mechanism followed by the driver's lateral and longitudinal operation behaviors in the high-speed straight highway scenario. Driver's operation behavior features in other complex scenarios shall be further studied and explored.

(2) Driver's Risk Response Mechanism

Driver's risk response refers to the corresponding countermeasures taken by drivers according to the scale of various risks after evaluating the generation and evolution of driving risks, which shows significant differences. Specifically speaking, driver's risk response in terms of behavior mainly includes straight ahead, lane changing, U-turn, parking, etc.; In terms of operation, it mainly includes acceleration, deceleration, steering, etc., which are the measures taken by drivers in the face of driving risks.

Exploration of driver's risk response mechanism is the theoretical basis for the implementation of driving assistance and safety warning, and it is also an important prerequisite for improving the humanization of driving assistance system or automated driving system. Straight driving is a common driving condition, and acceleration and deceleration are common driving operations. This section will disclose driver's longitudinal risk response mechanism in detail, and explore driver's response mechanism under the action of longitudinal risk, by establishing the mapping relationship between driving risk and driver's response on the basis of ND data.

Based on driver's ND data and driving environment information under various driving scenarios, the mapping model of driving risk quantification and driver's response is established to explore the action law of driver's response and driving risk and reveal its response mechanism [59]. Specifically speaking, a charge-like risk field model is established in terms of risk representation. Based on the analysis of the impact of distance, vehicle motion state and traffic rules on the risk distribution features, a vector representation model of multiple risk field strength is established, i.e.,

$$a^*_{R,ji} = \frac{E_{ji}}{V_i} = \frac{V_j \cdot D^*_{ji} \cdot D_{\mathrm{mot},ji} \cdot D_{1m,ji}}{V_i} \tag{3.12}$$

where, E_{ji} is the risk field intensity vector generated by the traffic unit j at i; V_i and V_j are the risk quantities of traffic units i and j; D^*_{ji} is the risk distribution intensity vector based on distance; $D_{\mathrm{mot},ji}$ is the characteristic coefficient of risk distribution based on motion state; and $D_{1m,ji}$ is the risk filtering coefficient formed by traffic rules, which can be expressed respectively as follows,

$$V = T \cdot m \cdot \left[1 + k_1 \left(\frac{\upsilon}{\upsilon_{\text{lim it}}}\right)^2\right] \tag{3.13}$$

$$D^*_{ij} = \varepsilon_{ij} \frac{r_{ij}}{\left|r_{ij}\right|^2} \tag{3.14}$$

$$D_{\text{mot},ij} = \exp[|\upsilon_{\text{ret}}| \cdot \cos(\theta)] \tag{3.15}$$

$$D_{1m,ij} = \begin{cases} T_{1m} \cdot \cos\left(\left|r_p\right| \cdot \frac{\pi}{3w}\right) r_p > \frac{w}{2} \\ 1\ r_p \le \frac{w}{2} \end{cases} \tag{3.16}$$

where, $\upsilon_{\text{lim it}}$ is the ultimate driving speed, which is influenced by traffic rules and vehicle performance; k_1 is the speed correction coefficient and m is the mass of the traffic unit; T is the coefficient of traffic unit type, which is affected by the properties of load and vehicle shape; r_{ij} is the distance vector pointing to j from i; ε_{ij} is the constant coefficient to be determined; $\upsilon_{\text{ret}}=\upsilon_i - \upsilon_j$ is the relative speed vector of traffic units i and j; $\theta < 180°$ is the angle between relative speed and distance vector r_{ij}; T_{1m} is the type of lane line; r_p is the distance vector from the vehicle's position to the lane centerline; and w is the lane width. Therefore, the risk action intensity of traffic unit i in multiple interaction scenarios is:

$$a^*_{R,i} = \sum_{j=1}^{N_u} a^*_{R,ji}$$

where, N_u refers to the total number of other traffic units that have risk interaction with traffic unit i in the environment. The research shows that there is a certain lag effect between the driver's operating behavior and the driving comprehensive risk. In other words, driver's risk response at a certain time is the joint action result of the current time and the past period of time. Taking into account the past risk effect of the vehicle, the intensity of comprehensive risk impact on the traffic unit can be modified as follows:

$$a_R = a^*_R \otimes H(n) \tag{3.17}$$

where, $H(n)$ is the half Hanning window function with the length of n frame. Through the setting of n, the past risk action can be integrated into consideration. The relationship between the finally obtained vehicle longitudinal acceleration vector and the normalized comprehensive risk action intensity is shown in Fig. 3.26. The risk point cloud is divided into three parts, i.e., safety area, turning area and risk area. With the increase of risk intensity, driver's driving behavior gradually changes from acceleration to braking. The point with zero acceleration on the mean curve is defined as the equilibrium point, and the abscissa corresponds to the equilibrium value of risk intensity. The equilibrium value describes the driver's response strategy to risk, represents the risk level that the driver can accept and tends to maintain, and is the embodiment of driver's inherent driving state. The actual driving process is affected by quite a number of factors. On the whole, the actual response of driver to risks fluctuates around the average response strategy, as shown by the confidence interval results in the figure [60]. Driver's risk response mechanism can be summarized as follows: the driver tends to take acceleration to achieve efficient driving in the face of low risk, whereas tends to take braking behavior to ensure driving safety in the face of high risk.

The above analyzes the mechanism followed by driver's longitudinal risk response in the car-following scenario. Driver's risk response mechanism in other complex scenarios needs to be further studied and explored.

(3) Driver's Decision Mechanism

Driver's decision-making refers to the driver's decision on the next action of vehicle after pondering, judging and optimizing the factors affecting the realization of the goal, based on the objective possibility, the acquired information and driving experience, as well as corresponding tools, skills and approaches. It is the product of the combination of driver's thinking process and driving behavior.

The in-depth analysis on the internal motivation of driver's decision-making as well as the study and summary of human decision-making wisdom will be helpful for

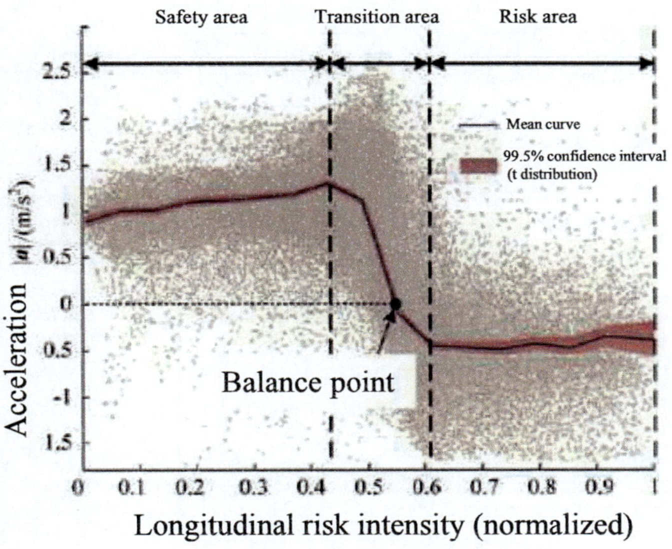

Fig. 3.26 Mapping relationship between vehicle response and comprehensive risk intensity under driver's operation

guiding the independent decision-making of IV. Based on a large number of ND data, this section carries out research on the distribution of characteristic parameters that can objectively reflect driver's decision, and explores the fundamental compliance for driver's decision-making mechanism.

Firstly, it is to collect ND data such as free driving, car following and lane changing without ADAS, in order to cover most driving scenarios; The test route is shown in Fig. 3.27, including highways and urban roads; The collected data includes GPS and vehicle CAN bus data, with a total of 32.5 h of effective experimental data. The statistical distribution of driver's lateral and longitudinal operating behavior is shown in Fig. 3.28, including the distribution law of vehicle THW, lateral position and acceleration.

The test results show that 90% of THW is concentrated around 0.6–3.6 s, 90% of the distance between the vehicle centerline and the lane centerline is maintained within −0.77 to 0.69 m (negative sign indicates left), and 90% of the acceleration is distributed at −0.594 to 0.485 m/s^2, which reflects a certain ultimate phenomenon. These data show that the driver is used to maintaining a certain THW in the longitudinal direction, the acceleration is tried to be stable at a = 0 m/s^2, and the distance between the vehicle and the lane centerline in the lateral direction is basically stable. The extreme value of characteristic parameters not only reflects the driver's behavior features, but also reflects the driver's decision-making objectives. Through the above analysis, the fundamental reason for driver's decision output is its decision-making

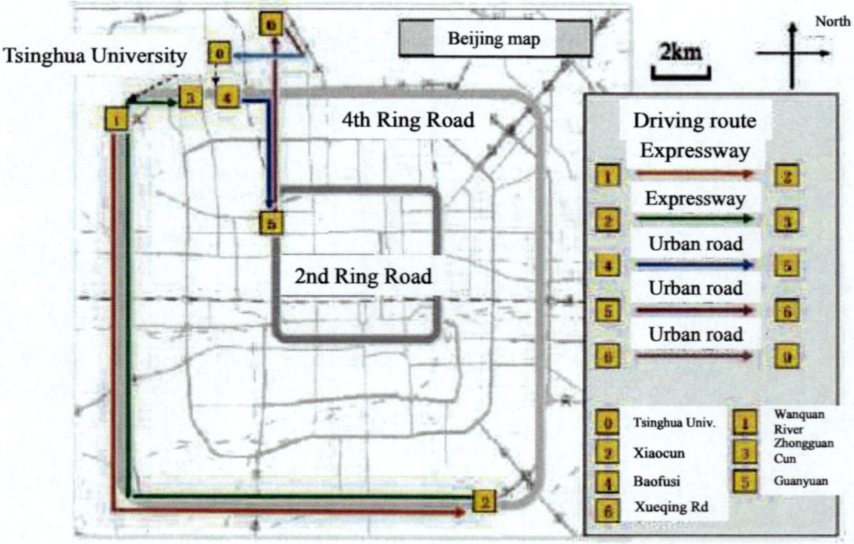

Fig. 3.27 Test route

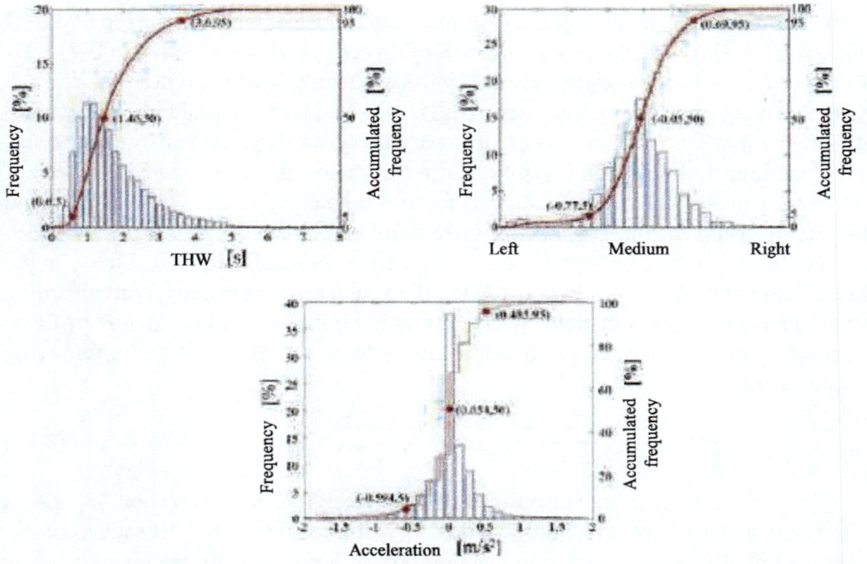

Fig. 3.28 Driver's longitudinal and lateral driving features

mechanism of "pursuing benefit and avoiding harm", and its basic strategy and principle is to avoid collision and violation (harm) while obtaining high mobility and safety (benefit).

This decision-making mechanism is very similar to the extreme value phenomenon in nature. Taking water droplets for instance, when gravity is ignored, the water droplets will remain spherical, that is, the surface area of the ball in all cubes of the same volume is the smallest; If gravity is considered, the gravity center of the spherical water drops under the action of gravity, but the surface tension keeps it spherical as much as possible, so it finally appears as an ellipsoid (regardless of the molecular force between the liquid and the ground). In addition, for the rope with fixed ends and free sag, its final state is the lowest gravity center of the system. No matter how the shape of the iron chain is changed, the center of gravity will always be higher than the real situation. In physics, the principle of least action is often adopted to describe the extreme value process. As a variational principle, the principle of least action is applied to the action of mechanical system, in order to obtain motion equation of the system. In view of the similarity between driver's decision-making mechanism of "pursuing benefit and avoiding harm" and the extreme value phenomenon in nature, the principle of least action in physics is used as reference for explaining the driver's decision-making operation mechanism.

In the process of driving, the vehicle will be affected by the DVE closed-loop system in three aspects: ① driven by driver's expected goal; ② impact of dynamic traffic environment; ③ constraint of traffic rules. Since the traffic flow always moves in a certain direction like the free falling motion of an object [61], the motion of the vehicle is analogous to the process of a small ball rolling down from the top of the U-shaped groove, that is, the physical model of driver's driving, as shown in Fig. 3.29. In the process of rolling down, the ball will be affected by three types of forces: ① virtual driving force driven by driving target; ② virtual risk external force affected by dynamic traffic environment; ③ virtual constraint resistance generated by traffic rules. The functions of the three kinds of forces constitute the virtual mechanical system of the driving process. The virtual driving force is defined as G_i, $G_i = m_i g \sin\theta_i$. And the inclination θ_i of the U-shaped groove is related to the driver's expected speed or mobility, including two parts, i.e., $\theta_{i,x}$ in the longitudinal direction and $\theta_{i,y}$ in the lateral direction. When $\theta_{i,y} \neq 0$, the ball will make lateral movement, corresponding to the lane changing of vehicle. According to the transformation principle of Euler angular coordinate system, the relationship between the three meets the following requirements:

$$\theta_i = \arccos(\cos\theta_{i,x} \cdot \cos\theta_{i,y}) \tag{3.18}$$

When θ_i is slightly increased during lane changing, the corresponding G_i is slightly increased as well. In addition, drivers will face adverse effects such as obstacles, bad road conditions and bad weather during driving. Here, the above adverse effects are described as some small protrusions in the U-shaped groove, as shown in the part marked with blue circle in Fig. 3.29.

In addition, due to the constraints of traffic rules (such as speed limit v_{limit}) in the actual traffic, drivers often tend to avoid violations, and thus virtual constraint resistance R_i is defined for expressing the constraint effect of traffic rules on drivers. For example, when the vehicle stops, that is, $v_i = 0$, the virtual constraint resistance $R_i = 0$, the driver has the highest expectation for mobility; If the vehicle accelerates to the road speed limit, that is, $v_i = v_{\text{limit}}$, the virtual constraint resistance $R_i = G_i$, where G_i is the virtual gravity generated by driver's requirements for mobility during

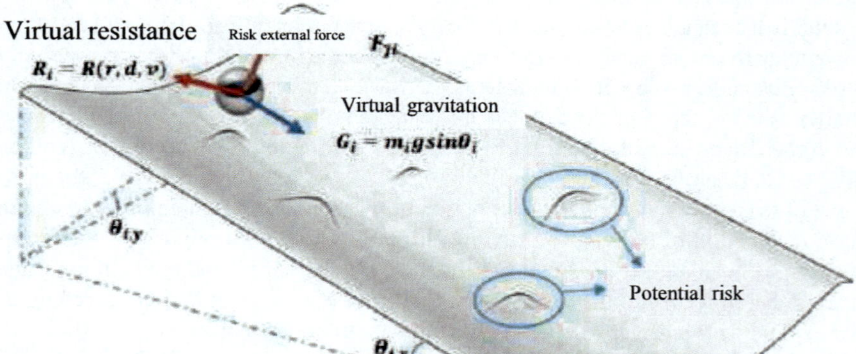

Fig. 3.29 Physical model and stress analysis of driving process

driving. At this time, the virtual gravity representing driver's mobility requirements is balanced with the virtual constraint resistance; If the vehicle continues to accelerate due to some random disturbance, that is, $v_i > v_{\text{limit}}$, the virtual constraint resistance $R_i > G_i$, and thus the driver shall slow down at this time to avoid speeding violation.

Therefore, the physical model and its virtual mechanical system based on the above process are defined according to the principle of least action, including:

$$S_i = \int_{t_0}^{t_f} L_i \mathrm{d}t \tag{3.19}$$

$$L_i = T_i - U_i \tag{3.20}$$

where, S_i is the action of "pursuing benefit and avoiding harm" of driver in the driving process; t_0 is the time of starting position; t_f is the time to arrive at the destination; L_i is the system Lagrangian during driving; T_i and U_i are respectively the kenetic energy and potential energy of the vehicle driven by driver. And U_i is the potential energy formed by the combined forces of R_i, G_i and F_{ji} in the physical model shown in Fig. 3.29.

According to the above analysis, driver's decision-making mechanism of "pursuing benefit and avoiding harm" is that driver always hopes to obtain high mobility and safety and avoid collision and violation during driving. According to Formula (3.19), the smaller $t_f - t_0$ is, the higher the mobility is. In other words, the shorter the travel time is, the higher the efficiency is. From the system Lagrangian expression shown in Formula (3.20), it can be found that kenetic energy T_i has two meanings: ① traffic efficiency. The larger the kenetic energy is, the faster the speed is. And the less the time is required to travel the same distance, the higher the traffic efficiency is, and vice versa; ② driving safety. When the vehicle is out of control due to mechanical fault, human error or environmental factor during driving, the faster the vehicle speed is, the higher the driving risk is. Therefore, on the one hand, increasing kenetic energy (speed) can improve traffic efficiency, but on the other hand, it also shows potential safety hazards.

Potential energy U_i represents the impact of traffic environment on vehicles, and G_i represents the driver's pursuit for mobility in the current road environment, that is, the driver's pursuit for efficiency. During driving, the potential energy $U_{i,G}$ generated by G_i has been decreasing, so as to improve the kenetic energy $U_{i,G}$ of the vehicle, overcome the potential energy $U_{i,R}$ of virtual resistance R_i and the potential energy $U_{i,F}$ generated by risk external force F_{ji}. The potential energy $U_{i,R}$ generated by virtual resistance R_i directly reflects whether the driver is speeding or breaking the rules; The potential energy generated by the risk external force F_{ji} directly affects the driving safety. For example, when following the vehicle, the closer the rear vehicle is to the front vehicle at the same speed, then the larger the external force is, the more the potential energy is, and the higher the risk is. During driving, if $U_{i,R}$ increases, it indicates that the possibility of violation increases, and thus T_i shall to reduced for violation avoidance; If $U_{i,F}$ increases, it means that the driving risk increases, and thus T_i shall be reduced for avoiding rear collision or driving the vehicle to a position with low $U_{i,F}$. The vehicle conducts braking or lane changing, so that the Lagrange quantity of the system becomes smaller and the action is cut down.

Safety and efficiency are the basic actions of the driver in the driving process. The closer it is to the least action of the theory, the higher the safety and efficiency is. Due to the mutual restriction and coupling between safety and efficiency, driver's driving strategy is often expected to ensure safety and improve efficiency as much as possible. The ideal goal of decision-making is to achieve the best balance between safety and efficiency and optimize their comprehensive performance. Therefore, based on the principle of least action, driver's decision-making mechanism of "pursuing benefit and avoiding harm" can be described as that driver expects to reach the destination with the lowest risk (the highest safety) and the shortest time (the highest efficiency) [62].

3.1.3 Mechanism of Driving Accident

Restrained by the road geometry, adhesion conditions, surrounding vehicles and road facilities in the complex driving environment, the vehicle may have dynamic instability due to the lack of adhesion on the one hand, and make collision due to the improper or untimely handling of driving conflict on the other hand. Therefore, the causation mechanism of driving accident focuses on the dynamic instability mechanism and collision conflict mechanism. This section focuses on the forms, analysis approaches and safety boundaries of the two causes, illustrates the generation mechanism of dynamic instability and kinematic collision, further analyzes the coupling mechanism and internal relationship between instability and collision risk, explores the key factors and causes of accidents, examines the occurrence mechanism of typical road traffic accidents, and anticipates the research trend of driving accident generation mechanism, as shown in Fig. 3.30.

1. Driving Instability Mechanism

Driving stability refers to the capability of the vehicle to recover its original driving state and direction without losing control, side-slip, rollover and other dangers after being disturbed by the outside environment change during driving. Vehicle stability mainly depends on road conditions, vehicle structural parameters, driver's operation, etc. Vehicle dynamics instability is the result of the joint action of DVE and other factors, and is also one of the manifestations of closed-loop system instability.

(1) Dynamic Instability Form

The instability of conventional vehicles occurs mainly during side-slip and tail-flick due to insufficient adhesion, with the following specific performance: when the adhesion of the rear axle is insufficient, the vehicle will drift out and further lose stability; when the adhesion of the front axle is insufficient, the vehicle will slip and lose stability. For heavy-duty vehicles with high center of mass, rollover instability

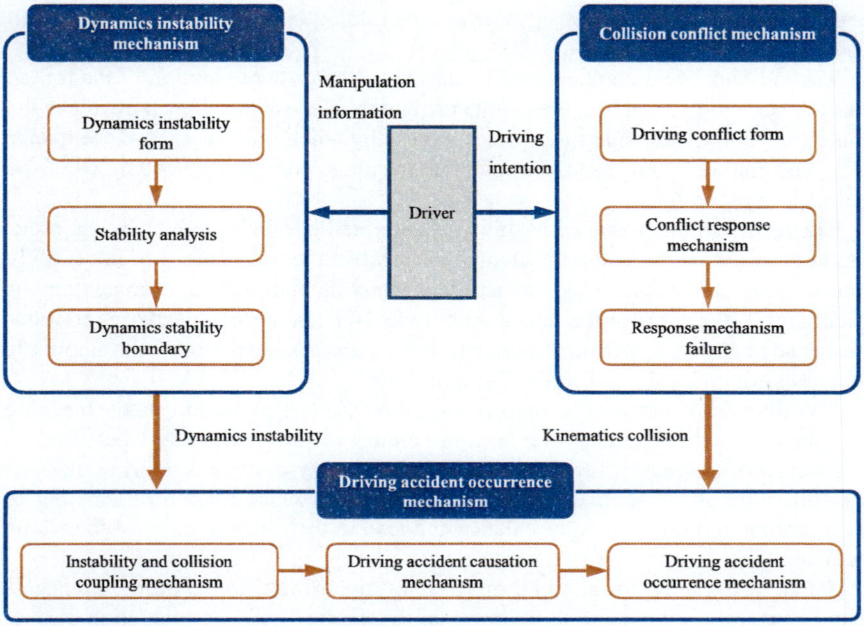

Fig. 3.30 Mechanism of driving accident

often occurs due to uneven distribution of vertical load on the left and right of tires, increase of roll angle and lateral acceleration during steering or other operations. Rollover instability of heavy-duty vehicles is usually evaluated by indicators such as vertical load transfer rate and lateral acceleration.

Compared with ordinary vehicles, semi-trailer/multi-trailer automobile trains include more forms of instability, such as folding of tractor and trailer, tail flick of trailer, lateral shimmy of trailer and rollover instability of the whole vehicle; Especially under high-speed conditions, multi-trailer trains are more prone to lateral swing, which greatly increases the lateral amplitude of trailers, resulting in the instability of the whole vehicle [63]. Therefore, taking the instability form of semi-trailer train as an example, the following analyzes the typical and common vehicle instability forms, including the folding, tail flick and lateral shimmy-snake instability of automobile train, so as to discuss the vehicle dynamics instability more comprehensively.

The folding of automobile train is caused by the unstable movement of tractor. During driving on slippery curves or emergent braking, the tractor turns around the saddle hinge in addition to bearing part of the load of semi-trailer, and the semi-trailer simultaneously brakes with multiple tires and maintains the original forward motion state. These interactions may saturate the force between the tire and road surface. Meanwhile, if the articulation angle between the tractor and the trailer exceeds the corresponding limit value during steering, the folding instability of the vehicle will occur.

The tail flick of automobile train is caused by the unstable motion of the trailer. The reason is that the lateral force of the tire and the force at the hinge point make the trailer move from one side to the other, especially when the rear axle of the tractor is braked, the trailer moves to one side with the hinge point as the center, that is, the tail flick of the trailer.

The lateral shimmy-snake instability of automobile train is induced by the oscillation of trailer on the plane. It often happens when the rear trailer can not quickly return to the stable state after line shifting, which is manifested in the continuous oscillation and change of the lateral amplitude. If the amplitude decreases continuously and finally returns to the stable state, it is a stable snake motion. If the amplitude increases continuously, it is snake instability.

The rollover of automobile train is caused by the lateral motion state exceeding its limit threshold. The automobile train belongs to the heavy-duty truck, and its rollover form and reason are similar to that of the heavy-duty truck. During rollover, the roll angle of the vehicle reaches the limit value, which makes the moment of force around its roll axis out of balance and lead to the roll instability of the whole vehicle.

Based on the above instability forms of semi-trailer/multi-trailer trains, it is necessary to study the vehicle instability mechanism and its dynamics generation mechanism under specific conditions in the future, so as to lay a basis for vehicle stability analysis and active safety control research.

(2) Stability Analysis

The instability of vehicle dynamics is closely related to driver's operation input (front wheel angle δ_f, vehicle speed v_x), road adhesion conditions (road adhesion coefficient μ), vehicle structural parameters (distance from the center of mass to the front and rear axles L_f/L_r, moment of inertia I_z around the axle z), loading conditions (vertical load F_z), etc. The specific is shown in Fig. 3.31, where ρ represents other unlisted parameters, and other parameters involved in the figure will be explained later. Especially under extreme conditions, the vehicle lateral-longitudinal-vertical dynamics coupling is strong, and the vehicle operation and stability change significantly. Therefore, to study and clarify the instability mechanism of vehicle dynamics under extreme conditions and then accurately judge the stable area of vehicle dynamics is an important topic of the research on the DSV under all conditions and the causation mechanism of accidents.

With the stable system motion, the system state can always converge to the stable equilibrium point. For the linear system, the condition of system stability is that the zero point of the system characteristic polynomial is located in the left half plane of the complex plane. The early research on the instability mechanism of vehicle dynamics is mainly based on the linear analysis. By constructing the linear vehicle motion model, the lateral instability mechanism of vehicle is analyzed by Hurwitz criterion [64]. With the development of nonlinear dynamics theory and its application in complex systems, the research on the nonlinear mechanism of vehicle dynamics

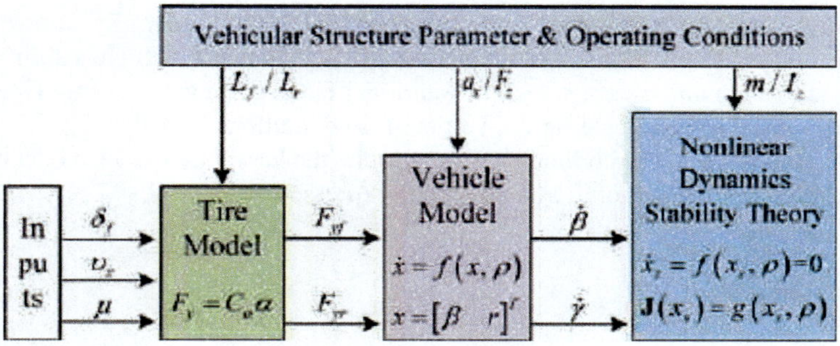

Fig. 3.31 Research idea of vehicle dynamic stability

instability has made great progress. The adopted analysis methods include energy method, phase plane method, bifurcation method, etc. [41]. The energy method is to express the energy of the system as a function of dynamics stability indicators such as yaw rate and centroid sideslip angle. The larger the energy is, the easier it is to break away from the attraction of the equilibrium point and lose stability. The phase plane method is a graphical way to solve ordinary differential equations. The dynamic process of the system is drawn into a phase plane in the form of motion trajectory, and then the stable area of the vehicle driving state is determined according to the geometric properties of the phase plane.

The stable area is closely related to the road adhesion, driver's operation and vehicle structural parameters. Bifurcation theory can better describe the dynamics behavior of nonlinear system before and after bifurcation, i.e., linear stability behavior before bifurcation and instability behavior after bifurcation. The relevant research mainly involves two means, i.e., open-loop analysis without considering the driver and closed-loop analysis considering the driver. The open-loop research uses the Routh-Hurwitz stability standard and local bifurcation theory to derive the system stability conditions at constant speed. The results show that the vehicle speed has an important impact on the position of the bifurcation point of system: the larger the vehicle speed is, the smaller the front wheel angle at the bifurcation point is, which is also one of the reasons for the instability of high-speed driving.

With the large-scale application, research and test of active safety system represented by electronic stability control, the phase plane has been proved to be one of the most effective approaches for studying the instability boundary of vehicle dynamics. The commonly-used phase planes corresponding to vehicle dynamics include [65]: centroid sideslip angle-centroid sideslip angle velocity phase plane, centroid sideslip angle-yaw angle velocity phase plane, and front and rear wheel sideslip angle phase plane. In order to describe the maneuverability and stability of the vehicle at the same time, the phase plane based on centroid sideslip angle and yaw rate is more widely applied in the study of vehicle dynamic instability. In the following, the dynamic equations of vehicle centroid sideslip angle and yaw rate will be established first.

Secondly, according to the nonlinear characteristics of the system, Lyapunov indirect method is used to reveal the internal relationship between vehicle stability and tire sideslip state under pure sideslip conditions; Finally, the mechanism of vehicle dynamics instability under different vertical load conditions is analyzed.

Regardless of the longitudinal DOF of vehicle, the lateral dynamics model of the 2-DOF vehicle with monorail system can be expressed as:

$$\begin{cases} \dot{\beta} = \frac{F_{yf}+F_{yr}}{mv_x} - r \\ \dot{r} = \frac{F_{yf}L_f - F_{yr}L_r}{I_z} \end{cases} \tag{3.21}$$

where, β and r respectively represent the vehicle centroid sideslip angle and yaw rate, as shown in Fig. 3.32; v_x represents the longitudinal speed of the vehicle; L_f and L_r represent the distance from the centroid to the front and rear axles; F_{yf} and F_{yr} represent the lateral force of front axle tire; m is the mass of the vehicle; and I_z is the moment of inertia around the axis z. The tire lateral force can be modeled as:

$$\begin{cases} F_{yf} = C_{yf}\alpha_f = C_{yf}(\beta + \frac{rL_f}{v_x} - \delta_f) \\ F_{yr} = C_{yr}\alpha_r = C_{yr}(\beta - \frac{rL_r}{v_x}) \end{cases} \tag{3.22}$$

where, C_{yf} and C_{yr} represent the cornering stiffness of front and rear axle tires; and δ_f indicates the front wheel angle.

Therefore, the vehicle dynamics equation (3.5) can be rewritten as:

$$\dot{x} = \begin{bmatrix} \dot{\beta} \\ \dot{r} \end{bmatrix} = \begin{bmatrix} \frac{(C_{yf}+C_{yr})\beta v_x + (C_{yf}L_f - C_{yr}L_r)r - C_{yf}v_x\delta_f}{mv_x^2} - r \\ \frac{(C_{yf}L_f - C_{yr}L_r)v_x\beta + (C_{yr}L_r^2 + C_{yf}L_f^2)r + C_{yf}L_f\delta_f}{I_z v_x} \end{bmatrix} \tag{3.23}$$

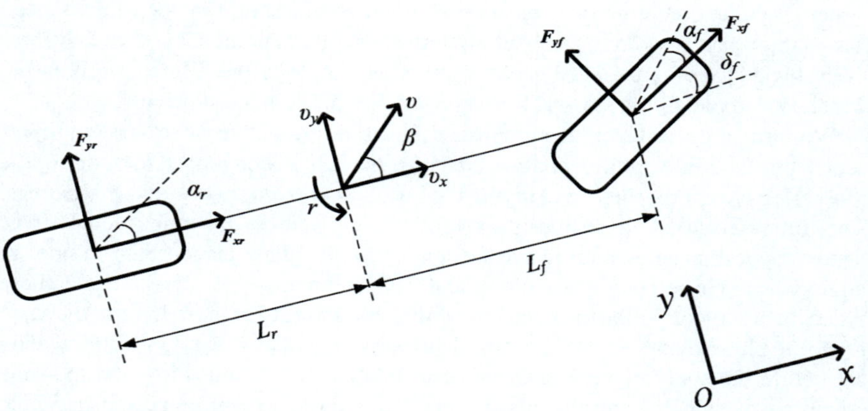

Fig. 3.32 Monorail vehicle dynamics model

The Jacobian matrix at the equilibrium point of the system (distinguished by *) can be obtained by calculating the partial derivative of the above state equation, i.e.,

$$J_{\mathrm{F}}(\beta^*, r^*)=\begin{bmatrix} \frac{C^*_{\mathrm{yf}}+C^*_{\mathrm{yr}}}{mv_x} & \frac{C^*_{\mathrm{yf}}L_{\mathrm{f}}-C^*_{\mathrm{yr}}L_{\mathrm{r}}}{mv_x^2}-1 \\ \frac{C^*_{\mathrm{yf}}L_{\mathrm{f}}-C^*_{\mathrm{yr}}L_{\mathrm{r}}}{I_z} & \frac{C^*_{\mathrm{yf}}L_{\mathrm{f}}^2+C^*_{\mathrm{yr}}L_{\mathrm{r}}^2}{I_z v_x} \end{bmatrix} \tag{3.24}$$

According to the Lyapunov indirect method, the condition for the stability of the system equilibrium point is that the two characteristic values of the Jacobian matrix have negative real parts, that is, the following conditions are satisfied:

$$\begin{cases} \mathrm{Tr}(J_{\mathrm{F}})=\frac{I_z(C^*_{\mathrm{yf}}+C^*_{\mathrm{yr}})+m(C^*_{\mathrm{yf}}L_{\mathrm{r}}^2+C^*_{\mathrm{yr}}L_{\mathrm{f}}^2)}{I_z m v_x}<0 \\ \mathrm{Det}(J_{\mathrm{F}})=\frac{C^*_{\mathrm{yf}}C^*_{\mathrm{yr}}L^2+mv_x^2(C^*_{\mathrm{yf}}L_{\mathrm{f}}-C^*_{\mathrm{yr}}L_{\mathrm{r}})}{I_z m v_x^2}>0 \end{cases} \tag{3.25}$$

where, $\mathrm{Tr}(J_{\mathrm{F}})$ and $\mathrm{Det}(J_{\mathrm{F}})$ represent the trace and determinant of Jacobian matrix J_{F} respectively, $L = L_{\mathrm{f}} + L_{\mathrm{r}}$.

According to the Magic Formula tire model, the variation relationship of tire lateral force with tire sideslip angle and tire vertical load can be obtained, as shown in Fig. 3.33. Due to the difference between the vertical loads of the front and rear axles (generally, the front axle load is larger than the rear axle load), the vehicle dynamics instability mechanism is analyzed in the following tire cornering state of the front and rear axles:

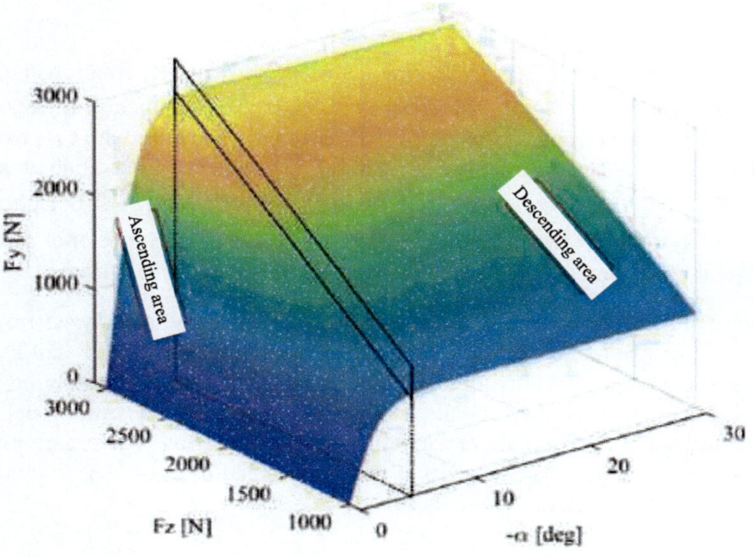

Fig. 3.33 Tire lateral force under pure cornering condition

i. When the tire cornering states of front and rear axles are in the rising zone, i.e., $C^*_{\mathrm{yf}} < 0$, $C^*_{\mathrm{yr}} < 0$, the system characteristic value is obviously less than 0, and the positive and negative conditions of $\Delta = C^*_{\mathrm{yf}} L_{\mathrm{f}} - C^*_{\mathrm{yr}} L_{\mathrm{r}}$ in the system shall be further analyzed: when $\Delta \geq 0$, the system is stable, and the vehicle has understeer or neutral steering (if and only if $\Delta = 0$) characteristics; When $\Delta < 0$, the vehicle has excessive steering characteristics. At this time, the condition for system stability is that the longitudinal speed is less than its critical speed, that is:

$$\upsilon_x < \sqrt{\frac{C^*_{\mathrm{yf}} C^*_{\mathrm{yr}} L^2}{-m\Delta}} \tag{3.26}$$

ii. The front axle cornering state is in the rising area, and the rear axle cornering state is in the descending area, i.e., $C^*_{\mathrm{yf}} < 0$, $C^*_{\mathrm{yr}} > 0$. At this time, $\mathrm{Det}(J_{\mathrm{F}}) < 0$ indicates that the system has two characteristic values with different symbols. If the equilibrium condition is not satisfied, the stability bifurcation point (saddle point) will appear.

iii. The front axle cornering state is in the descending area and the rear axle cornering state is in the rising area, i.e., $C^*_{\mathrm{yf}} > 0$, $C^*_{\mathrm{yr}} < 0$. Since the slope of the falling zone is small and meets the requirements, then $\mathrm{Tr}(J_{\mathrm{F}}) < 0$, the conditions for system stability at this time are:

$$\upsilon_x > \sqrt{\frac{C^*_{\mathrm{yf}} \left| C^*_{\mathrm{yr}} \right| L^2}{m\Delta}} \tag{3.27}$$

In this interval, $\left| C^*_{\mathrm{yf}} \right|$ is in the linear area and is relatively small; while $\left| C^*_{\mathrm{yr}} \right|$ is in the nonlinear region and is relatively large; The tail flick of the rear axle can be avoided through acceleration, so the stability conditions are easy to meet.

iv. The cornering states of the front and rear tires are in the descending area, that is, $C^*_{\mathrm{yf}} > 0$, $C^*_{\mathrm{yr}} > 0$ and $\left| C^*_{\mathrm{yr}} \right| \approx \left| C^*_{\mathrm{yf}} \right|$. At this time, $\mathrm{Tr}(J_{\mathrm{F}}) > 0$, $\mathrm{Det}(J_{\mathrm{F}}) > 0$, the system has two positive characteristic values, which does not meet the equilibrium condition. This area is the instability area of the system.

The nonlinear mechanism of vehicle dynamics instability is analyzed from the perspective of tire cornering. When the vehicle structural parameters and loading conditions are determined, the descending area of rear axle wheel cornering is the main reason for its instability, and the ground adhesion conditions, front wheel angle and vehicle speed are the main factors affecting tire cornering. Therefore, in-depth study of tire cornering characteristics is the premise and basis of vehicle dynamics instability mechanism and stability boundary research.

(3) Dynamics Stability Boundary

The dynamics stability boundary of vehicle is closely related to its longitudinal speed, front wheel angle and road surface adhesion. In particular, the cornering of tire is mainly influenced by vehicle steering. Therefore, taking the longitudinal speed and front wheel angle as the input, by analyzing the influence law of different inputs under the known road adhesion coefficient on the vehicle dynamics stability boundary, the determination approach of vehicle dynamics stability boundary is established to clarify the vehicle dynamics stability area, so as to provide theoretical reference for the research of dynamics stability boundary of different vehicle types.

According to the motion form, the vehicle dynamics stability boundary can be divided into longitudinal dynamics instability and lateral dynamics instability. The longitudinal dynamics instability is mainly manifested in longitudinal slip, and the lateral dynamics instability is mainly in sideslip and yaw. For heavy vehicles with high center of mass, there is also rollover instability. For the sideslip and yaw stability area, the stability boundary designed by most studies only allows the vehicle to move in the state stability area or boundary, and there are few studies on the limit state stability boundary of vehicles outside the state stability area. The traditional state stability area is defined by the open-loop control of the actuator. For the IV with the closed-loop feedback control, the stability boundary obtained by the open-loop control is difficult to adapt to the state stability boundary of the IV under extreme conditions. Reference [66] designed an expanded vehicle dynamics stability boundary based on the envelope idea, but which is effective only when the front wheel angle is less than the maximum stable front wheel angle. Further, Ref. [67] can restore the limit motion state outside the traditional stability boundary to the stable state through reverse direction operation, which proves that there is a controllable vehicle state outside the stable region of vehicle dynamics state under front wheel angle closed-loop control, and puts forward a new idea for the research of IV motion stability expansion based on front wheel angle closed-loop feedback control.

Based on the above ideas, taking the initial longitudinal speed, road adhesion coefficient and the limit value of front wheel angle as the adjustable parameters of the stability boundary model can improve the universality and reusability of the model. Firstly, according to the nonlinear dynamics theory and the characteristics of the equilibrium point, the saddle point in the fourth quadrant of the phase plane composed of the centroid cornering angle and yaw angular speed has the following characteristics: (1) the cornering angular speed of the mass center at this point is zero; (2) the yaw angular acceleration at this point is zero. At this time, the analytical expression of the saddle point is:

$$\omega_{r\_saddle} = -\frac{\mu g}{v_x} \tag{3.28}$$

$$\beta_{\_\mathrm{saddle}} = \frac{L_f \mu g}{v_x^2} + \tan\left(\alpha_{r\_st} + \delta_{f\_\max}\right) \tag{3.29}$$

where, $\omega_{\_saddle}$ and $\beta_{\_saddle}$ are the yaw angular speed and centroid cornering angle at the saddle point of the fourth quadrant respectively; α_{r_st} is the cornering angle when the lateral force of the rear wheel is saturated; δ_{f_max} is the limit value of front wheel angle, and other parameters are defined in the previous section. When the change rate of yaw angle is zero, the set of yaw angular speed rate and centroid cornering angle is approximately a straight line on the phase plane. By numerical analysis and tests, the analytical solution of the linear boundary slope of the stable boundary model is:

$$k_b = \frac{1}{2}\left(\frac{\omega_{s1} - \omega_{s2}}{\beta_{s1} - \beta_{s2}} + \frac{L_r}{v_x}\right) \tag{3.30}$$

where,

$$\omega_{s1} = \frac{v_x}{L}\left[\tan\left(\alpha_{f\_st} + \delta_{f\_\max}\right) - \tan\left(\alpha_{r\_st}\right)\right] \tag{3.31}$$

$$\omega_{s2} = \frac{\mu g}{v_x} \tag{3.32}$$

$$\beta_{s1} = \frac{L_r}{L}\left[\tan\left(\alpha_{f\_st} + \delta_{f\_\max}\right) - \tan\left(\alpha_{r\_st}\right)\right] + \tan\left(\alpha_{r\_st}\right) \tag{3.33}$$

$$\beta_{s2} = \frac{L_r \mu g}{v_x^2} - \tan\left(\alpha_{r\_st}\right) \tag{3.34}$$

where, α_{f_st} is the cornering angle when the lateral force of front wheel is saturated. The boundary of the controllable area of vehicle dynamics state in the third quadrant passes through the intersection of the fourth quadrant boundary and the longitudinal axis, and is parallel to the lateral axis. According to the symmetrical properties of the boundaries of the controllable areas in the first and second quadrants and the third and fourth quadrants about the origin, the boundary of the vehicle dynamics stability area can be obtained, as shown in the black dotted line in Fig. 3.34. Within it, the state stability area of the black solid envelope is determined through the stability analysis of the open-loop dynamics system, and the intervention of the closed-loop feedback control of the actuator may expand this area.

During the actual driving of vehicle, the uncertainty of the model and the interference of the external environment can easily lead to the too large controllable area of the defined state stability. Therefore, for the sake of safety, the stability boundary is moved to the origin, and the translation factor ε_1 and ε_2 are introduced into the controllable area boundary model. The reference boundary model is established as follows:

$$\frac{\beta_{\text{boundary}} - \beta_{\text{saddle}} + \varepsilon_1}{\omega_{\text{boundary}} - \omega_{\text{saddle}} + \varepsilon_2} = k_{\text{b}} \tag{3.35}$$

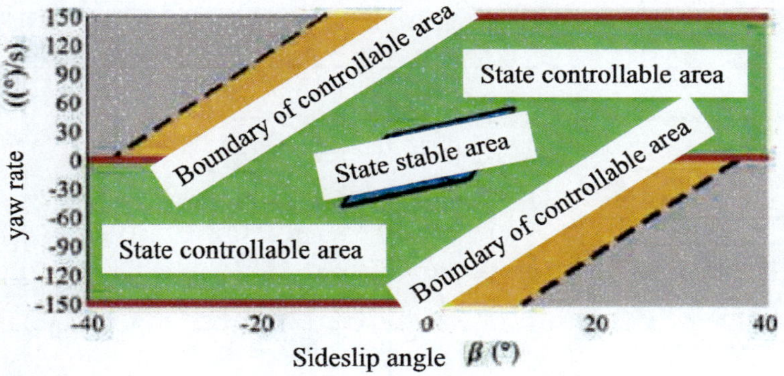

Fig. 3.34 Vehicle dynamics stability area boundary [68]

where, β_{boundary} and ω_{boundary} are the vehicle states on the boundary. The controllable area after translation is shown by the red solid line in Fig. 3.34, in which the translation factor can be comprehensively determined by integrating vehicle test data and control system design objectives.

2. Collision Conflict Mechanism

In the actual traffic, the vehicle runs in the environment containing static and dynamic obstacles (i.e. other traffic participants), so the vehicle and obstacles have a large number of potential conflicts [69], that is, the risk of collision accidents. In most cases, the driving conflict will not be transformed into a collision accident under driver's effective conflict response driving behavior. On the contrary, when the response mechanism fails, the driving conflict may lead to a collision accident.

(1) Form of Driving Conflict

Driving conflict is common in road traffic and has many forms. For example, there may be potential conflicts between vehicles and road boundaries, and there may also be conflicts among different traffic participants in scenarios such as lane changing, ramp merging and crossing intersections. There are many ways to classify driving conflicts. Based on the objects involved in the conflict, conflicts can be divided into "vehicle-static obstacle conflict" and "vehicle-traffic participant conflict", in which the latter includes "leader–follower conflict" and "interactive conflict" as shown in Fig. 3.35.

In the road driving environment, ordinary static obstacles mainly include static obstacles, such as road boundary, road scattering and parked vehicles. In addition, traffic markings representing traffic restrictions (such as double yellow lines and one-way signs) can also be regarded as a class of static obstacles. Violation of traffic restrictions may not lead to collision accidents, but will constitute a traffic rule violation. Therefore, crossing such markings should also be avoided in the design

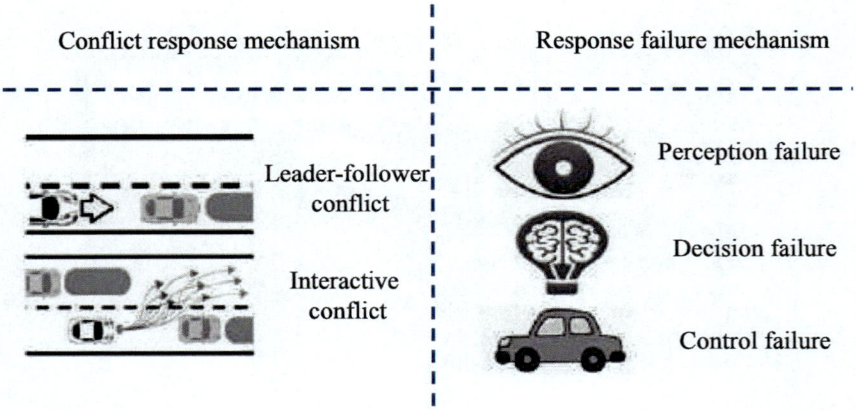

Fig. 3.35 Driving conflict form and collision generation process

of IV. The road environment is segmented based on the position of the above static obstacles, and the free space of the vehicle can be designated.

Compared with static obstacles, the movement of traffic participants has higher time-varying and uncertainty, so the analysis and modeling of their driving conflict mechanism is more challenging. A group of "vehicle-traffic participant conflicts" includes at least one ego-vehicle and another traffic participant. In the leader–follower conflict, the behavior of the leader is not affected by the other party, while the follower responds based on the behavior of the leader. For example, in the following driving scenario, the driving behavior of the front vehicle (leader vehicle) is generally not affected by the rear vehicle, while the rear vehicle (follower vehicle) needs to adjust the speed to keep the distance from the front vehicle. In interactive conflict, behaviors of both parties are mutually affected by each other. For example, in the lane changing driving scenario, the behaviors of both lane changing vehicles and lateral/rear vehicles are affected by the conflict, and form the combination of rushing-yielding behaviors. In the current academic study, researchers mainly define the "interaction" in driving conflict based on whether the behaviors of both parties (or multiple parties) are affected by the conflict at the same time [70].

In the face of driving conflict, the driver can generally avoid the conflict from evolving into a collision accident through reasonable response. However, in some cases, such as perceptual blind spots, aggressive driving, control failure and other situations, the conflict response mechanism may fail, resulting in collision accidents, as shown in Fig. 3.35. The study of conflict response mechanism and response failure-based collision mechanism will be instructive to the analysis of driving conflict.

(2) Conflict Response Mechanism

When facing the driving conflict, drivers can avoid the conflict from turning into collision accident through effective driving responses. In the face of static obstacles, drivers usually avoid obstacles by keeping a certain distance between the vehicle and obstacles. In contrast, the movement of traffic participants has the characteristics of time-varying and uncertainty, and all parties take their own passage according to their driving goal in the process of conflict. Therefore, the analysis of vehicle-traffic participant conflict is more complex and difficult. Vehicle-static obstacle conflict can be regarded as a specially simplified situation in which traffic participants remain stationary.

In terms of leader–follower conflict, longitudinal speed control in car-following scenario is a classical problem. Based on the study and imitation of the driver's behavior, researchers proposed a variety of car-following models and developed ADAS functions such as ACC (Active Cruise Control) and AEB (Autonomous Emergency Braking). At present, the design of car-following model is mainly based on two kinds of indicators: safety distance (SD) and time to collision (TTC). Among them, SD is the distance threshold is calculated based on such information as the front and rear vehicle speed, vehicle distance, driver's response time and vehicle control performance, and then the subsequent longitudinal control method is designed according to this threshold. Common SD algorithms include MAZDA model [71], GM model [72], IDM model [73], etc. Take Mazda model as an example:

$$d_{\mathrm{br}} = 0.5\left[\frac{v^2}{a_1} - \frac{(v - v_{\mathrm{rel}})^2}{a_2}\right] + vt_1 + v_{\mathrm{rel}}t_2 + d_0 \tag{3.36}$$

where, a_1 refers to the maximum deceleration of the ego-vehicle; a_2 indicates the maximum deceleration of the front vehicle; t_1 indicates the driver's response delay time; t_2 represents the brake delay time; v_{rel} represents the relative speed of two vehicles; and d_0 Indicates the minimum stopping distance.

When the vehicle distance is less than the safety distance d_{br}, braking shall be adopted to make the vehicle distance return to above d_{br}.

TTC model is expressed as:

$$\mathrm{TTC} = \frac{D}{v_{\mathrm{rel}}} \tag{3.37}$$

where, D represents the relative distance.

When designing response logic, longitudinal control logic is designed based on different TTC thresholds, such as "partial braking when TTC is less than 1.6 s, and full braking when TTC is less than 0.6 s".

In addition, some scholars further considered the vehicle multi-directional risk, extracted the driver's personality parameters based on the actual following process of a large number of drivers, and finally proposed the human-like longitudinal control algorithm based on the driver's risk response mechanism. In the section "Driver's Risk Response and Decision-making Mechanism" of this chapter, an acceleration response model of driver's response risk is introduced. Based on this model, the researcher designed the corresponding longitudinal car-following model RRDM (Risk-Response-Driver-Model) [74].

In algorithm application, firstly, the personalized identification of RRDM model parameters is conducted based on the driver's actual driving data, and then the human-like car-following behavior closest to the driver's behavior (i.e. vehicle longitudinal acceleration) is generated in the vehicle autonomous driving stage. In order to illustrate the anthropomorphism of this method, the researchers compared the driving behavior based on RRDM with the human driving behavior in this scenario and the classical driver models such as GM and IDM.

Figure 3.36a, b show the identification results of driving characteristic parameters of a driver and the simulation results of car-following process. It can be seen from the headway curve and longitudinal acceleration curve that the longitudinal behavior of RRDM model is closer to the real driving behavior of the driver than GM and IDM model. Figure 3.36c further statistically analyzes the average headway of 24 drivers in the following process. It can also be seen that the RRDM model better simulates the THW characteristics of each driver in the following process.

In interactive conflict, both parties are mutually affected. The driver's response is mainly influenced by three aspects: the rules in traffic regulations or social norms, the information transmission between the conflicting parties based on behavior (such

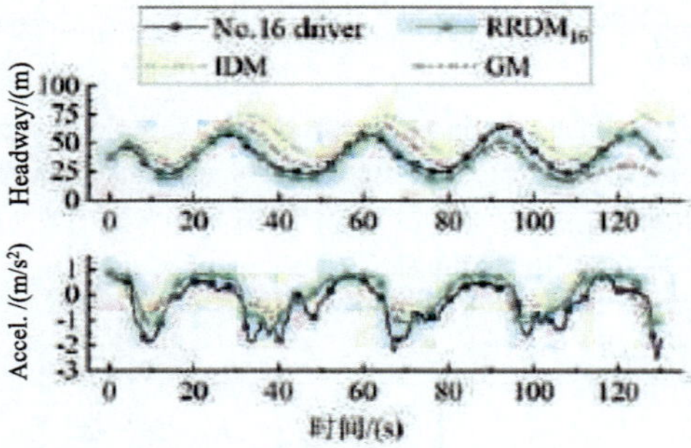

(a) Headway and Acceleration of A Driver (No.16) in Car-following Scenario

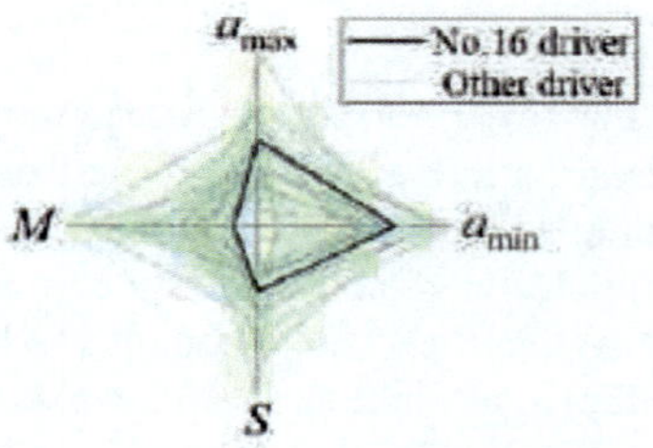

(b) Driving Characteristic Parameters of The Driver

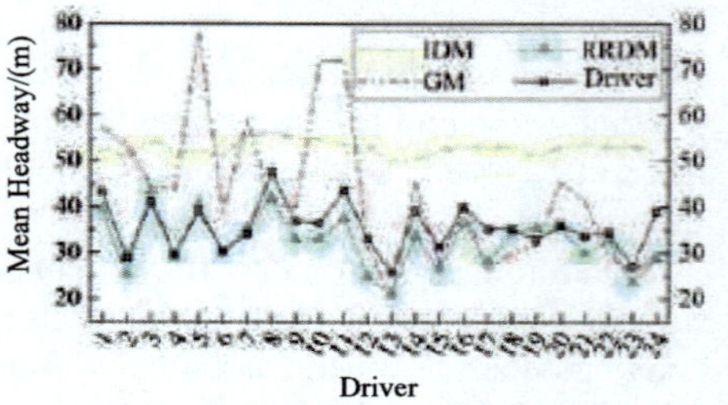

(c) Mean Headway under Control of Driver, RRDM, GM and IDM Figure 3-36 RRDM Model Simulation Results

Fig. 3.36 RRDM model simulation results

as gesture, light signal, acceleration and deceleration and steering), and their own driving objectives and behavioral characteristics. Therefore, there are three main modeling approaches for the driving response process in interactive conflict: rule-based model, probability reasoning model and game-based model. The principles and characteristics of the approaches are shown in Table 3.5.

As shown in Fig. 3.37, taking the left turn scenario at an unprotected intersection (without guidance such as traffic lights or stop signs) for instance, the output results and expressions of the three modeling approaches based on rule, probabilistic reasoning and game can have the forms summarized in Table 3.6 (the behaviors taken by left turning vehicles and straight driving vehicles respectively).

Table 3.5 Principles and characteristics of modeling approaches for conflict response process

Approach	Basic principle	Characteristic
Based on rule	Design rules for conflict scenarios and assume that all traffic participants make decisions based on rules. For example, turning gives way to going straight, and vehicle gives way to pedestrian	(1) The model design is simple and logical (2) On the one hand, it is difficult to design rules in complex scenes. On the other hand, traffic participants do not necessarily act based on rules in real traffic environment
Based on l probability reasoning	Build a probabilistic graph model and understand the intention semantics behind the driver's behavior by means of conditional probability modeling and parameter-learning	(1) The model has strong logic and is suitable for complex scenarios and uncertainty of interaction (2) It is difficult to explain the internal motivation of decision-making of traffic participants, and difficult to infer and deal with abnormal situations deviating from normal conditions
Based on game	Design the objective function and game model of the conflict participants, and assume that all parties will make decisions based on the Nash Equilibrium of the game	(1) The game-based approach is helpful to analyze and explain the internal motivation of behavior (2) Traffic participants are not fully rational and have inconsistent information with each other. The prediction based on game theory is difficult to correspond with the real traffic data

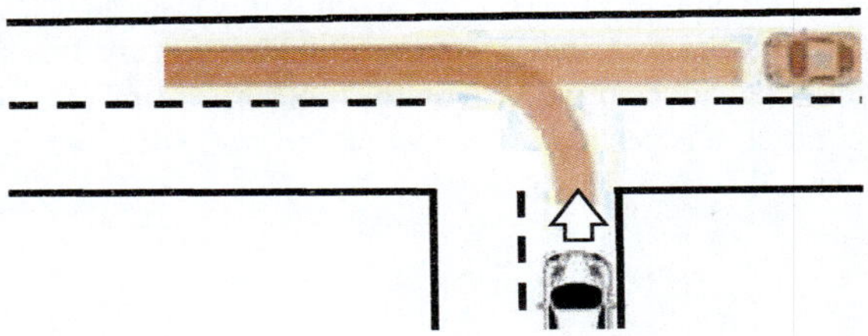

Fig. 3.37 Left turn scenario at unprotected intersection

Table 3.6 Model output results and expressions of conflict response process

Approach	Model output (example)	Expression
Based on rule	Left-turn vehicle gives way to going straight vehicle	$B_{\text{left}} \in B(B_{\text{left}} \vert \text{yield})$ $B_{\text{str}} \in B(B_{\text{str}} \vert \text{rush})$
Based on probabilistic reasoning	Based on the historical trajectory of the two vehicles, the confidence of the corresponding rushing-yielding intention of the two vehicles is deduced, and then the corresponding driving behavior is generated	$B_{\text{left}} \in B(P_{\text{left-grab}} \vert history)$ $B_{\text{str}} \in B(P_{\text{str-grab}} \vert history)$
Based on game	The combination of two vehicles' rushing-yielding constitutes the Nash equilibrium of the game	$\frac{\delta \text{Reward}_{\text{left}}(B_{\text{left}}, B_{\text{str}})}{\delta B_{\text{left}}} = 0$ $\frac{\delta \text{Reward}_{\text{str}}(B_{\text{left}}, B_{\text{str}})}{\delta B_{\text{str}}} = 0$

In the section "Identification of Traffic Participants' Intentions", several classical modeling approaches will be further analyzed. The behavioral interaction between vehicles and traffic participants widely exists in the real road traffic environment. Effectively identifying and estimating the mutual impact of each other's behavior significantly influences the safety and efficiency of IV in dealing with conflicts. In a survey report of 2018 [75] regarding the test on Waymo of Google in Phoenix, USA, an important factor for residents' dissatisfaction is its excessive hesitation in unprotected left-turn scenario. If IV is difficult to estimate the impact of self-driving behavior on other traffic participants, it is likely to make the judgment that "all behavior choices are unsafe" in high-density traffic scenarios and thus difficult to make effective driving decisions, that is, the problem of "Freezing Robot" with over conservative and inefficient driving [76].

3. Driving Accident Occurrence Mechanism

The occurrence of traffic accident is due to the failure of the DVE closed-loop system, which is mainly manifested in the dynamics instability and kinematics collision of vehicles. Capturing the causes and occurrence mechanism of driving accidents based on the coupling mechanism of instability and collision risk as well as its comprehensive evaluation approach is of great significance for the prevention and precaution of driving accidents.

(1) Instability and Collision Risk Assessment

Traffic accidents with heavy casualties often occur in extreme conditions, where vehicles face not only the risk of instability, but also the risk of collision with other vehicles or road facilities. Instability and collision risk are closely related to driver's operation and road adhesion. Based on the mechanisms of vehicle instability and collision risk conflict, it is particularly necessary to study the comprehensive evaluation of vehicle instability and collision risk.

In view of the limitation that it is difficult to obtain the vehicle instability boundary and collision risk boundary accurately, this section proposes a vehicle motion risk assessment approach based on safety margin. In terms of the motion risks of two different time-scales, the comprehensive risk assessment based on double-layer fuzzy mechanism is proposed [77]. As shown in Fig. 3.38, firstly, it is to establish the vehicle dynamics instability index and its change rate, input it into the first layer of fuzzy, and obtain the vehicle instability risk level with the formulated fuzzy rules; Similarly, it is to establish the vehicle kinematics collision risk index and its change rate, and obtain the vehicle kinematics collision risk level with the corresponding fuzzy rules. Through the first layer of fuzzy mechanism, the two risks are transformed into the same scale for evaluation. And through the second layer, the comprehensive risk fuzzy rules are formulated based on the coupling of the two risks and the causation of accidents, and the comprehensive driving risks are acquired by means of fuzzy reasoning and anti fuzziness.

In order to achieve the above purpose, it is necessary to establish the vehicle dynamics instability evaluation index primarily. According to the distance between the centroid sideslip angle and yaw rate and instability boundary, the vehicle dynamics instability index is established, and the change rate of vehicle instability is obtained in accordance with the change rate of vehicle motion state; Similarly, the ratio of the risk field strength of the current moment to the critical risk field strength when the vehicle is about to collide is defined for describing the safety margin of vehicle collision risk.

And then the fuzzy set of instability and collision risk is defined as {low, medium low, medium, medium high and high}, and the fuzzy set of its change rate as {negative large, negative small, zero, positive small and positive large}. Combined with the fuzzy rules defined in Table 3.7, the vehicle dynamics instability risk level (IRL) and kinematics collision risk level (CRL) can be obtained. The output variables of risk level include safety (SA), low risk (LR), medium risk (MR), medium high risk (MHR) and high risk (HR).

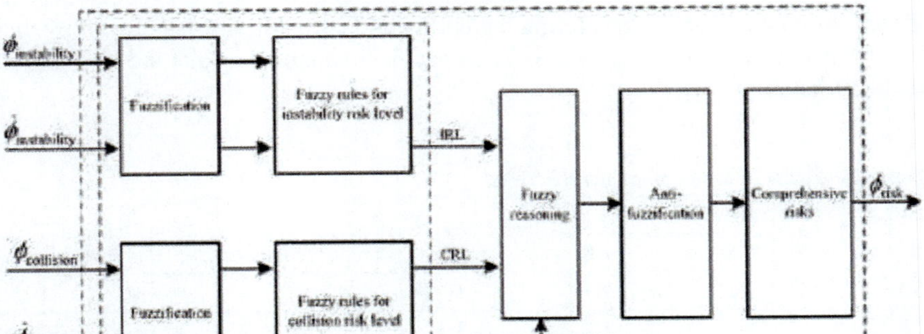

Fig. 3.38 Research framework of comprehensive risk assessment

Table 3.7 Fuzzy rules of CRL

$\dot{\phi}_{\text{collision}}$ ($\dot{\phi}_{\text{instability}}$)	$\phi_{\text{collision}}$ ($\phi_{\text{instability}}$)				
	L	LM	M	MH	H
NL	SA	SA	LR	MR	MHR
NS	SA	LR	MR	MHR	MHR
ZE	SA	LR	MR	MHR	HR
PS	LR	MR	MHR	HR	HR
PL	MR	MHR	HR	HR	HR

After obtaining the risk level of vehicle instability and collision, it is necessary to formulate fuzzy rules for comprehensive risk assessment. Although both risks are related to the steering, driving, braking and other operation inputs of the ego-vehicle, the instability risk and collision risk caused by them are on two different time scales: the instability risk of vehicle dynamics is at the millisecond level, while the collision risk of vehicle is at the second level. Therefore, during the design of comprehensive risk fuzzy rules, the safety margin and time scale should be considered as the whole part. The overall idea of fuzzy rule design is as follows: when a single risk is high, the comprehensive risk of vehicle is medium high; when the risk of vehicle instability is high, the comprehensive risk of vehicle shall not be lower than the risk level of vehicle instability; and when instability and collision risks are medium high, the comprehensive risk is high. The comprehensive risk fuzzy set is input with IRL = {SA, LR, Mr, MHR, HR} and CRL = {SA, LR, Mr, MHR, HR}. The comprehensive risk fuzzy rules formulated by means of accident causation analysis are shown in Table 3.8.

The output fuzzy sets of comprehensive risk assessment Z = {H1, H2, H3, H4, H5}, H1/H2/H3/H4/H5 represent SA(H1), LR (H2), MR(H3), MHR (H4) and HR(H5). With the input of the system instability risk level and collision risk level, as well as the comprehensive risk fuzzy rules and inference, the comprehensive risk level of the vehicle is finally obtained by the centroid anti-fuzzification, that is, the coupling evaluation of driving instability and collision risk is realized by the fuzzy rules of comprehensive risks. The specific evaluation results are still under exploration.

Table 3.8 Fuzzy rules of comprehensive risks

CRL	IRL				
	SA	LR	MR	MHR	HR
SA	H1	H2	H2	H3	H4
LR	H2	H2	H3	H4	H4
MR	H2	H3	H3	H4	H4
MHR	H3	H3	H4	H5	H5
HR	H4	H4	H4	H5	H5

(2) Driving Accident Causation Analysis

The causation of driving accident is a general term for the factors or combination of factors that lead to driving accidents, including road factors, driver's factors, vehicle factors and environmental factors. Specifically, it can be divided into two categories: one is the driver's factors, which are called internal factors; and the other is environmental factors, called external factors. Driver's factors mainly includes two aspects: first, the driver is unable to normally perceive and identify the risk and other subjective errors, such as failing to yield as required, speeding and other violations of traffic rules; second, the driver cannot correctly handle internal and external interference with limited ability, such as failing to pay attention to behaviors of other traffic participants and low driving proficiency. Environmental factors mainly include two aspects: one is the dangerous driving of other vehicles, such as the risking behavior of suddenly rushing into the lane from the opposite lane; and the other is the influence of road factors, such as bad weather conditions and complex road geometry.

The driving accident is the result of the comprehensive DVE actions. The above driver's factors and external environmental factors act on the vehicle together, so that the vehicle may have dynamics instability due to insufficient adhesion on the one hand, and kinetics collision due to improper handling of interference or conflict on the other hand. In terms of the causation of dynamics instability, the driver operates the vehicle through steering, braking, driving, etc. The vehicle responds to the driver's manipulation through its own movement and interacts with the surrounding driving environment. Driving behavior is essentially a behavior decision made by the driver based on perception of the vehicle and environment. Accordingly, due to the complexity and abruptness of the environment, the driver's behavior decision is not always optimal. Reference [78] pointed out that fatigue driving or drunk driving and the decline of driver's perception ability are the main reasons for rear-end collision accidents, and the poor lighting in driving environment is the main reason for driving accidents at night. Under extreme conditions, the vehicle's own problems may also lead to driving instability accidents, such as abnormal tire pressure and brake failure. With the increase of temperature in summer, the pressure inside the tire increases and the tire body becomes thinner. In addition, the heat generated by friction with the ground makes the vehicle prone to tire burst. According to statistics, 46% of traffic accidents on highway are related to tires, and tire burst accidents account for 70% of tire-related accidents, which has become the main cause of driving accidents on highway. In addition, road factors are also very important causes of accidents. Non-ideal weather conditions and complex road geometry may also lead to traffic accidents. It is found that when driving on ice and snow roads, vehicles are more prone to sideslip accidents, and the sideslip threshold is greatly affected by the road adhesion coefficient; Under the condition of relatively high road adhesion (above 0.5), the road curvature and lateral slope are the main factors affecting sideslip threshold; On high-adhesion roads (above 0.75), the driving speed is the main factor affecting the rollover threshold [79]. Driving dynamics instability is mostly the result of the coupling and combined actions of the above-mentioned DVE factors. Therefore, the analysis on the causation of vehicle dynamics instability under the interaction of

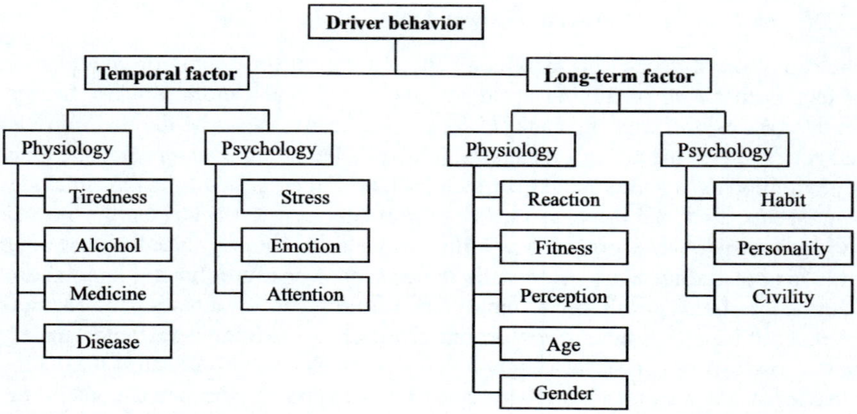

Fig. 3.39 Factors affecting driver's ability to respond to conflict and countermeasures [81]

human, vehicle and road environment has become a significant direction of future research.

The causation of collision accidents usually include three types of factors: driver's factors, such as inattention, lack of skills and being too reckless; vehicle factors, such as mechanical failure; and environmental factors, such as backlight, blind area of vision and extremely low road adhesion. In view of the above factors, researchers in the field of transportation carried out further statistics and cluster analysis on various factors based on traffic accident data. Some scholars [80] analyzed the traffic accident data at home and abroad and summarized the factors affecting the driver's ability to respond to conflict and countermeasures, which are summarized in Fig. 3.39 [81].

In addition, although with relatively low proportion, vehicle and environmental factors can also be the inducement of collision accidents. For instance, some studies [82] further divided vehicle factors into three categories: machinery and braking-tire system-driving equipment (such as lighting and rearview mirror), and categorized road and environmental factors into geometric features-intersection and road segment setting-pavement conditions-traffic facilities-natural environment-artificial environment, etc. During a collision accident, the above influencing factors will further lead to the failure of perception, decision-making and control, and risky behavior of vehicle towards collision. For example, the "poor attention" in driver factors may lead to the failure to observe the conflicting vehicle (perception failure), then the failure to take risk avoidance in time (decision failure), and the result of the collision due to the failure to slow down (high-risk behavior).

In addition to the analysis of accident factors, the occurrence process of accidents, i.e., the theory describing the causal chain, mutual influence mode and influence degree between factors, is also an important topic for accident causation.

(3) Driving Accident Occurrence Mechanism

The occurrence mechanism of driving accident refers to the internal action mode of human, vehicle, road and other elements in the closed-loop system in the process of traffic accidents due to specific causation, as well as the rules, the development and change laws of mutual connection and action between elements under the environmental conditions of the accident. The explanation of the causation of dynamics instability and kinematics collision is helpful for understanding the accident occurrence mechanism in accordance with external factors. To fully comprehend the accident mechanism, it is necessary to understand and grasp the evolution and occurrence process of vehicle instability and collision accident in accordance with both internal and external factors.

According to the different causes and action laws, the occurrence mechanism of driving accidents can be divided into three categories, i.e., information processing error, sudden change of environmental information and loss of system (element) function. The driving process of driver involves multi-input and multi-output information processing. If the driver makes mistakes in any section such as perception, decision-making and control, it may lead to driving accidents. The occurrence mechanism of such accidents can be summarized as the occurrence mechanism of information errors. In addition, sudden changes in such driving environment as road conditions, traffic flow conditions and weather conditions can lead to driving accidents, which can be summarized as the occurrence mechanism of sudden changes in environmental information, including the sudden entry of vehicles into the ponding road, the sudden braking and parking of vehicles in front, the sudden appearance of pedestrians in front, etc. In addition, the abnormal function or insufficient function of a subsystem or element in the closed-loop system will also lead to accidents. Such occurrence mechanism can be summarized into the category of system (element) function loss.

The driving behaviors of drivers varies greatly due to low road adhesion and decreasing controllability under the condition of ice and snow, which is a typical scenario with high incidence of driving accidents. The following will take the driving accidents on urban icy and snowy roads for example to clarify the accident mechanism by analyzing its occurrence process. Research shows that speeding is a critical causation of driving accidents on urban icy and snowy roads. Speeding vehicles generally go through the reaction stage, steering or braking stage and accident formation stage from "hazard perception" to "accident occurrence", as shown in Fig. 3.40. Within the same perception time and braking reaction time, high-speed vehicles are often prone to accidents with insufficient braking distance due to the rapid reduction of vehicle spacing at high driving speed and the low friction coefficient on icy and snowy roads. The accident mechanism can be explained in combination with the state axis. When driver perceives the hazard at the beginning, the vehicle is running normally with low road adhesion performance; In the reaction stage, the driver realizes that measures should be taken, and the vehicle is still running normally; At the stage of taking measures, the driver keeps the vehicle away from hazard by stepping on the brake pedal or turning the steering wheel. At this time, the vehicle is in the state of

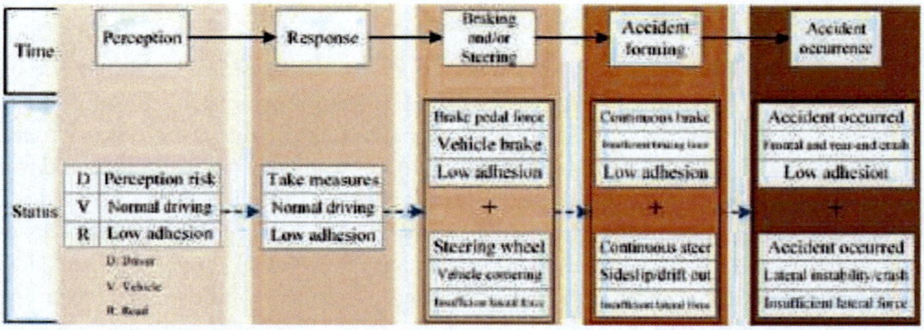

Fig. 3.40 Occurrence mechanism of overspeed driving accidents on icy and snowy roads

braking or steering. With the increase of braking strength and steering strength, the vehicle has the risk of driving collision and instability due to insufficient braking performance and lateral force, which eventually lead to the occurrence of collision and instability accidents.

Furthermore, scholars have carried out research on the mechanism of accidents with data analysis. For example, some researchers have analyzed the causation of accidents based on the data of 28 traffic accidents at intersections in Norway from 2005 to 2007 [83]. As shown in Fig. 3.41, with regard to 14 traffic accidents caused by the conflict between left-turn vehicles and opposite vehicles, the occurrence frequency and correlation of various relevant factors are counted and analyzed. It can be found that there is a clear causal relationship among the three factors, i.e., accident inducement, response failure and high-risk behavior. The most common accident causation process can be described as: the driver's inattention-failure to notice the conflicting vehicle-misjudgment toward the driving situation—too early entrance to the intersection-occurrence of driving conflict-formation of traffic accident (i.e. the process marked by the thick and solid line in the figure).

With the analysis on influencing factors and process of accidents, researchers further modeled the accident mechanism. In the field of road traffic, methods based on accident causal chain theory are often used to study the causal logic and risk transformation path of accidents, such as fault tree analysis (FTA) [84], Bayesian network model [85], neural network model [86] and Petri net model [87].

As shown in Fig. 3.42, an accident evolution process is described based on event tree model [84]. Within it, *T*, *I* and *X* events correspond to accident occurrence, intermediate state (which can be classified as perception failure, decision failure and control failure) and accident inducement respectively, and the arrow indicates the probability of state transition between events *P*. Based on the analysis and training of a large number of accidents (including high-risk scenarios of adjacent accidents), the event tree can be established and its parameters can be identified, which can not only effectively describe the accident process, but also help to quantitatively identify the potential risk before the accident.

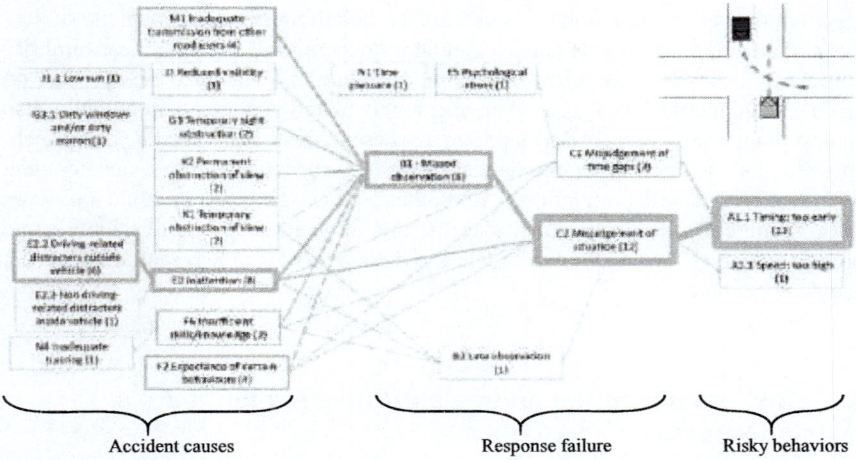

Fig. 3.41 Analysis on influencing factors of collision accidents at intersections [83]

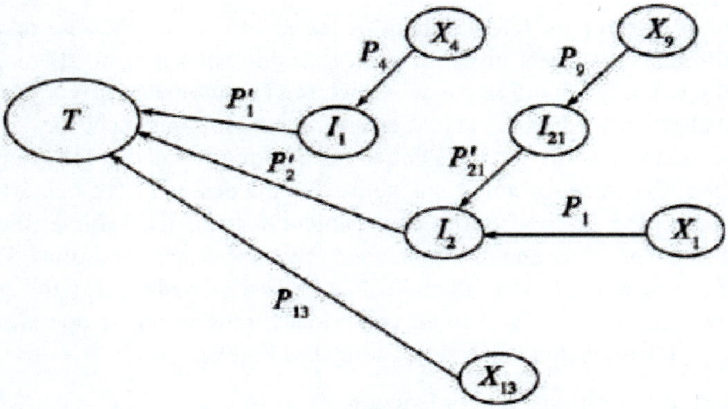

Fig. 3.42 Accident evolution process modeling based on Markov model [84]

In addition to obtaining the parameters of state transition probability between factors through data training, some researchers also use traffic accident data to train the structure of the network model. For example, the structural learning of Bayesian network for traffic accident causation analysis can be carried out by means of correlation analysis and K2 algorithm [85], which is used to make the quantitative analysis on the correlation between environment-road-vehicle factors and accident severity.

In the current research of driving accident mechanism, the influencing factors are still divided into four categories according to the traditional classification [88]. It will be difficult to explore specific cause of the accident on this basis. Therefore, it is urgent to find a new causation classification method to further coordinate the subjective and objective, inevitable and accidental influencing factors, which can simplify the research on the mechanism of driving accidents. Meanwhile, the research on driving accident mechanism based on accident data and mathematical model is also an important trend of future development: at present, the research on accident mechanism mainly focuses on mathematical modeling, but with the rapid development of data mining, accident monitoring and other technologies, the dynamic and static information collection and statistics at the time of accident are more scientific and precise, which also provides more tools and data support for the research on driving accident mechanism.

3.2 Risk Identification and Quantitative Evaluation Approaches

3.2.1 Design of Operational Safety Domain

The design of operational safety domain is the key to ensure the safe operation of IV with different intelligent levels in specific traffic environment. Its main task is to define the working conditions under which the IV automated driving system can work normally. Only when the actual operational conditions meet the set working conditions, can the auto drive system be activated normally and operate safely. Generally speaking, the working conditions defined in the design of the operation safety domain include the external traffic environment state of the vehicle, the state of vehicle, the state of driver and passenger, and other necessary conditions. Therefore, the design of operational safety domain shall be firstly conducted in the automated driving system, that is, to clarify all the conditions for the system to operate normally and safely perform the dynamic driving task (DDT) [89].

1. Definition of Operational Safety Domain

In the *Taxonomy of Driving Automation for Vehicles,* the operational design domain (ODD) is defined as the external environmental conditions applicable to its functional operation determined during the design of automated driving system. The typical external driving environmental conditions include road, traffic, weather and light [33, 90]. The definition of operational safety domain (OSD) is different from that of ODD. As a sub-interval of ODD, OSD refers to the environmental conditions that ensure the safe operation (acceptable risk range) of automated driving system within the scope of ODD. Figure 3.43 shows the diagram of driving operation between the two places. In the process of driving, since the driving environment before and after the highway does not meet the external driving environment conditions specified in the OSD, the vehicle shall to be driven manually. On the highway, the driving environment meets the requirements of OSD, and thus the vehicle can choose automated driving mode.

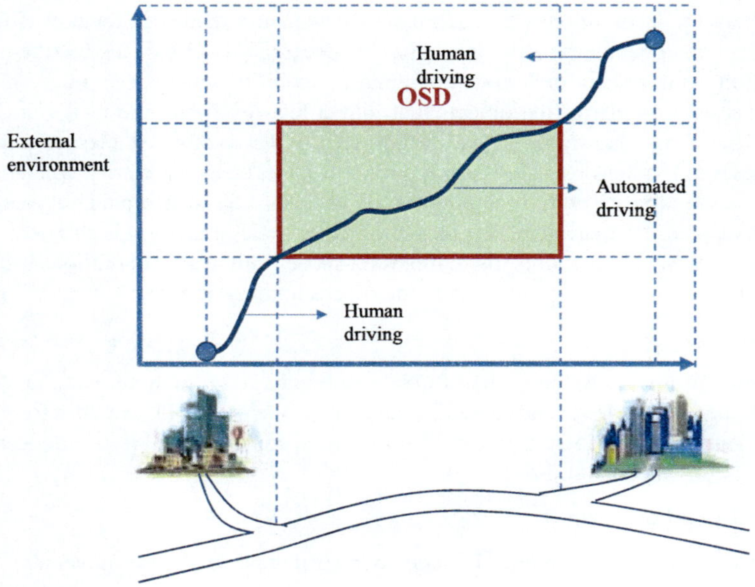

Fig. 3.43 Example of OSD

2. Operational Safety Condition and Scope

Setting operational safety conditions is the first step in the development of automated driving system. By defining the conditions, the functions and limitations of IV automated driving system are specified, and are transmitted to the system or users. Within the given conditions, IV can safely start and run the automated driving function. Beyond this condition, there will be risks in running.

(1) Definition of Operational Safety Condition

In the *Taxonomy of Driving Automation for Vehicles*, the operational design condition (ODC) is defined as the general term of various conditions applicable to its functional operation determined during the design of the automated driving system, including the designed operation range, vehicle state, state of driver and passenger and other necessary conditions [89]. Operational safety condition (OSC) refers to the necessary conditions for the operation of automated driving system at a theoretically low risk when the design operation conditions are met. During the design of OSC, all conditions for the normal startup and safe operation of automated driving system are clearly defined, including the external traffic environment (OSD), state of the vehicle, state of driver and passenger and other necessary conditions.

OSC can be divided into five parts: place (where), time (when), object (what), interaction (who) and constraint (how). First of all, it is necessary to specify the operational locations of AV, which usually include road types (such as highway and rural road), pavement conditions (such as concrete pavement), road edge features (such as shoulder), road geometry, etc.; Secondly, it is necessary to clarify the feature of the operation environment, including temperature, visibility, wind speed, etc.; Thirdly, it is necessary to clarify the objects that may affect driving in the traffic environment, including road traffic signs, traffic participants (motor vehicles, non-motor vehicles, etc.), other obstacles, etc.; Fourthly, it is necessary to clarify which facilities to communicate with, including motor vehicles, roadside sensing equipment, remote dispatching management information system, etc.; Finally, it is also necessary to clarify relevant constraints, including road speed limit, traffic conditions (normal and crowded), traffic regulations and other operational conditions.

(2) Scope of OSC

As shown in Fig. 3.44, the scope of OSC can be divided into three parts, i.e., road environment OSD, driver and passenger, and the ego-vehicle. OSD includes five types of elements: physical infrastructure, environmental constraints, objects, connectivity and operational constraints.

1) Physical infrastructure

Infrastructure elements mainly include four attributes: road type, pavement type, road edge and road geometry.

① Road type: expressway, highway, secondary trunk road, rural road, etc.
② Pavement type: asphalt pavement, concrete pavement, and block pavement, gravel pavement, soil pavement, deceleration belt, grassland, off-road pavement, etc.
③ Road edge: lane line, shoulder, concrete guardrail, fence, handrail, curb, etc.

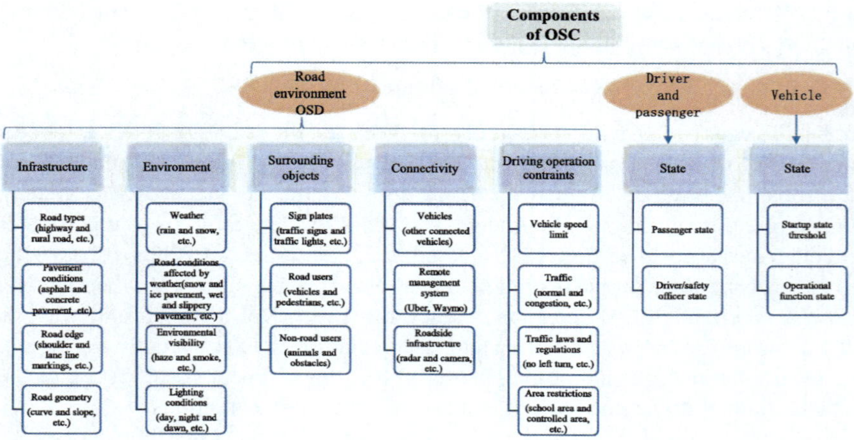

Fig. 3.44 Composition of OSC

④ Road geometry: refers to the shape of road, including straight road, curve, ramp, corner, lane width, etc.

2) Environmental constraints

Environmental constraints mainly include four attributes: weather, road conditions affected by weather, particulate matter and illumination.

① Weather: weather conditions have a great impact on the sensor. The design of safety zone mainly considers the weather conditions such as wind, rain, snow, sleet and temperature.
② Road conditions affected by weather: ponding, inundation, icing, snow and other conditions on the road due to rain and snow will reduce the road adhesion of tires, and then affect the motion control capability of intelligent driving system.
③ Particulate matter: fog, haze, smoke and dust in the atmosphere will affect visibility, as well as the sensing capacity of the sensor.
④ Illumination: too strong or too weak illumination will affect the perception capability of the camera. The actual driving time (day, dawn, dusk, night), the lighting of surrounding road facilities (street light, building light) and vehicle light will affect the lighting conditions.

3) Objects

The objects that may appear around the vehicle in the driving scenario include three categories: signs, traffic participants, and obstacles/objects of non-traffic participants.

① Signs: various traffic signs, traffic lights, zebra crossings, railway crossings, call-for-help signs, construction signs, signals and gestures sent by other traffic participants, etc.
② Traffic participants: various vehicles (cars, light trucks, buses, engineering vehicles, emergency vehicles, and special vehicles, etc.), pedestrians (adults, children), cyclists of two-wheeled vehicles (bicycles, battery cars, motorcycles), etc.
③ Obstacles/objects of non-traffic participants: animals, engineering equipment, road obstacles, etc.

4) Connectivity

In vehicle safety design, connectivity can be regarded as V2X (vehicle to everything), that is, the communication between vehicle and anything, including vehicle to vehicle (V2V), vehicle to infrastructure (V2I), vehicle to person (V2P) and vehicle to network (V2N), as shown in Fig. 3.45.

5) Operational Constraints

Operational constraints mainly include speed limit and traffic conditions.

① Speed limit: includes the allowable maximum speed and minimum speed.
② Traffic conditions: includes traffic flow (trough, peak, normal) and changes of traffic conditions (accidents, emergency vehicle passage, maintenance, road closure, special events, etc.).

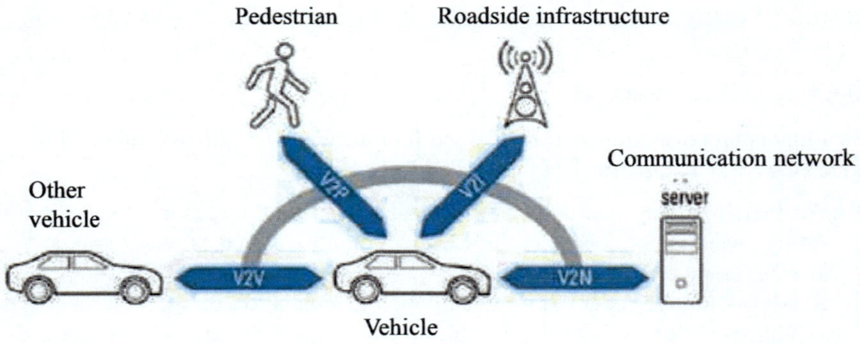

Fig. 3.45 Components of connectivity

3. Minimum Risk Condition

The design process of OSC defines all the conditions for the normal startup and safe operation of automated driving system, mainly including OSD of the system, vehicle state, driver's and passenger's state and other necessary conditions. According to the mechanism of driving risk generation, different operational conditions will lead to driving risks with different degrees. When the actual operational conditions of IV do not meet the designed OSC, the system will face too much risk to start normally or complete DDT safely. The minimal element set of OSC to be clarified for the minimal operational safety requirements is defined as the minimal risk condition (MRC) [89]. When the system detects that the operational conditions are about to exceed the OSC boundary during operation, it will activate the minimal risk manoeuvre (MRM) or adjust the operation model to the degraded mode so as to ensure the safe operation of system in accordance with MRC, or activate the manual driving takeover request to transfer driving control. Generally speaking, different stakeholders (enterprises, users, testing agencies, regulators, etc.) may choose different element sets to design OSC in accordance with MRC. When formulating element sets, the following design principles should be adopted according to the design elements contained in OSC [90]:

1) Design principle for operational environment elements in accordance with MRC. In view of the fact that the operational environment information of automated driving system comes from the on-board sensing system or V2X, when designing the minimal element set of the operational environment, it is necessary to consider the types of information that IV can obtain in real time, as well as the random measurement errors of the on-board sensing system or V2X, so as to ensure that the system can accurately judge the OSD boundary.
2) Design principle for state of driver and passenger in accordance with MRC. In view of the fact that when the systems of L3 and below request the timely manual takeover of the driver and passenger, it is necessary to ensure that the driver and passenger abide by the relevant traffic rules for the normal start of system, when designing the minimal element set of driver and passenger state.

The states required by MRC include that the driver and passenger have fastened their seatbelts, and have no preventive measures; For level L3 and below system, it is also necessary to take into account the conditions of driver and passenger requested by timely takeover, including no fatigue driving, no alcohol driving, sitting in the driving position correctly, etc.

3) Design principle of the vehicle state in accordance with MRC. The design of minimal element of the vehicle state must consider the threshold value of the vehicle state and operational function state for system start. Among them, the vehicle state threshold for system start includes vehicle speed threshold, gear state, parking brake state, etc. Whether the vehicle meets the conditions for activating the automated driving function is determined by whether the vehicle reaches the set state threshold. The operational function state is related to the operational safety state of automatic system. Usually, the system is equipped with software and hardware modules with self-test function to transmit monitoring information to the drivers and passengers or to the system.

During the design of OSC, all necessary conditions for the normal startup and safe operation of the system have been determined. However, different enterprises or research institutions differ from each other in detailed selection of the minimal element set of OSC. At present, no unified standard has been formed in the industry. In 2020, the *White Paper on Operational Design of Automated driving System* issued by the ICV Sub Technical Committee of the National Technical Committee of Auto Standardization elaborated the minimal element set defined by enterprises in accordance with the minimal safety requirements, and allowed different stakeholders (users, testing agencies or regulators, etc.) to refer to the conditions of the minimal element set, and add new layers for longitudinal expansion or add new elements for lateral expansion [89]. Next, with reference to the white paper, the minimal element set of OSC that meets the MRC is demonstrated.

(1) Minimal Element Set of OSD in Accordance with MRC

In line with the design principle of operational environment elements, the minimal element set of OSD should also be designed from five dimensions, according to the OSD constituent elements shown in Fig. 3.44. With reference to the minimal element set defined in the minimal safety requirements of enterprises in Ref. [89], the design idea for minimal elements should be illustrated in terms of five levels. Road types in physical infrastructure and weather in environmental constraints are taken for examples, as shown in Table 3.9.

(2) Minimal Element Set of Vehicle State in Accordance with MRC

In line with the design principle of vehicle state elements, the design idea of minimum elements is illustrated from five levels by taking the starting state threshold and operational function state as examples, with reference to the minimal element set defined by enterprises for minimal safety requirements in Ref. [89], as shown in Table 3.10.

Table 3.9 Minimal element set of OSD [89]

1st level	2nd level	3rd level	4th level
State of drivers and passengers	State of drivers or safety officers	Fatigue state	No fatigue
			General fatigue
			Very tired
			Severe fatigue
		Attention state	Concentration
			Transient distraction
			Periodic distraction
			Sustained distraction
		Driving posture state	Normal driving state
			Abnormal driving posture
			Proper driving position
			Improper driving position
		Physical/mental state	Normal physiological indicators
			No drunk driving behavior
			No road rage behavior
			No toxic driving behavior
		Seatbelt state	Fastened
			No fastened
	Passenger state	Posture	No intervention in driving
			With intervention in driving
		Child restraint system state	Correct use
			Incorrect use
		Seatbelt state	Fastened
			No fastened

3.2.2 *Dynamic Cognition of Driving Environment*

According to the DSD and its OSC, there are driving risks induced by many factors such as the ego-vehicle and its surrounding environment in actual driving. Therefore, it is necessary to dynamically recognize the risk influencing factors in the driving environment, which is also the premise and basis for subsequent risk assessment and risk avoidance. Therefore, it is feasible to start from the two aspects of vehicle

Table 3.10 Minimal element set of vehicle state [89]

1st level	2nd level	3rd level	4th level	5st level
Vehicle state	Starting state threshold	Control function	Drive	No input in accelerator pedal
			Braking	No input in brake pedal
			Steering	No input in steering wheel
			Gear	Forward gear: D gear
	Operational function state	Perception function	Perception range	Requirement fulfilled
		Positioning function	Positioning accuracy	Requirement fulfilled
		V2X function	Communication quality	Requirement fulfilled
		High-precision map function	Data format	Requirement fulfilled
		Decision planning function	FuSa	Requirement fulfilled
		Control function	Steering	≥Minimal steering radius
			Braking	≤Maximal braking strength
			Drive	≤Maximum acceleration
			Thermal management and monitoring	Requirement fulfilled
			Tire pressure monitoring	Normal tire pressure
		HMI interaction	Interactive mode	Normal interaction
		Vehicle body	Sensor cleaning	Requirement fulfilled
			Signal light	Normal function
			Vehicle door	Door closed
			Wipers	Requirement fulfilled
		FuSa	Passive safety system	Requirement fulfilled
			Active safety system	Requirement fulfilled
			Cybersecurity system	Requirement fulfilled

collision and instability. First, it is to analyze the intention identification and motion prediction of traffic participants in consideration of the collision risk between vehicles and traffic participants, and then analyze the estimation of key vehicle–road state parameters from the perspective of instability risk.

(1) Identification of Traffic Participants' Intention

In the section "Driving Conflict and collision Mechanism", two main types of driving conflict are analyzed, i.e., vehicle-static obstacle conflict and vehicle-traffic participant conflict. Among them, the motions of traffic participants are time-varying and uncertain, and may interact with the ego-vehicle. Traffic participants mainly include vehicles and VRUs (vulnerable road users, referring to vulnerable traffic participants such as pedestrians and cyclists). In the road traffic environment, the position and speed of traffic participants change dynamically. Therefore, ICV is required not only to estimate the driving risk of their current motion states, but also their future motion trends.

The intention identification of traffic participants includes three aspects, i.e., destination, behavior and interaction, which constitute a hierarchical and progressive relationship (Fig. 3.46). Destination identification is to estimate the long-term traffic objectives of traffic participants, such as the driving direction of vehicle at the intersection, or the destination of a pedestrian. Behavior recognition is to identify the short-term behaviors of traffic participants, such as lane changing behavior of vehicles in multi-lane roads or crossing behavior of pedestrians on the basis of destination identification. Interactive intention identification is important when multiple traffic participants are affected by conflicts, which is often reflected by the choice of rushing-yielding behaviors. For example, in dense traffic, a merging vehicles can only change to a new lane if the straight-going vehicles give way. Although there might be clear rules for the right of way, vehicles with lower prior right shall still demonstrate their own intentions and pass the conflict scenario once others' yielding intentions are identified. Destination, behavior and interaction have a hierarchical relationship. For example, when a vehicle leaves the ramp, its driving intention can be expressed as: leaving the main road-merging into the ramp-entering the ramp before its conflicting vehicle.

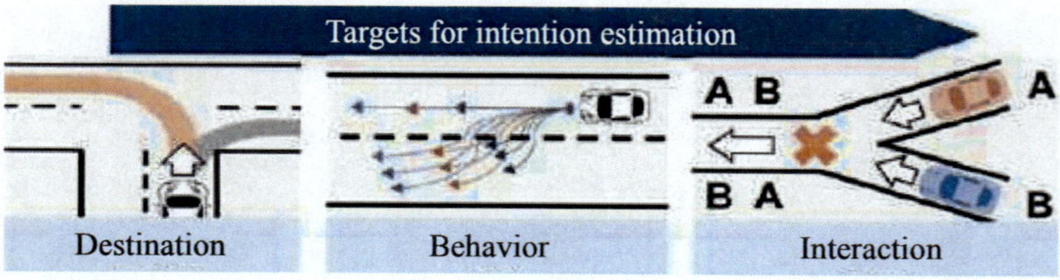

Fig. 3.46 Typical objectives of traffic participants' intention identification

At present, scholars in related fields have proposed a variety of approaches for intention identification. In terms of destination, it is feasible to analyze the left turn—right turn-straight intention of vehicles at the intersection [91], or infer the potential destination of pedestrians [92]. In terms of behavior, relevant research covers the identification of vehicle acceleration-braking-constant speed behaviors in the longitudinal direction [93], or the identification of vehicle lane-change intention in the lateral direction [94], etc. Currently, the research on interactive intention identification is relatively preliminary, such as modeling the competitive driving behaviors of vehicles based on game model [95], or identifying the characteristics of drivers in conflict by game approach, followed by behavior prediction [96].

In terms of intention, the widely employed means include Hidden Markov Model (HMM) [97], deep learning [98], game theory and so on.

HMM model is established based on the probabilitical relationship among various factors, and the connection relationship between nodes represents its conditional probability. For example, HMM as shown in Fig. 3.47 can be established for identifying pedestrians's crossing intention [99].

In the figure, the implicit variables represented by the box have discrete semantics (in a time segment, from left to right are destination G, crossing intention C, longitudinal risk LonC and location in the crossing area or not Pos). The explicit variables in the circle can be observed directly (from left to right are pedestrian orientation BO, movement behavior B, pedestrian-vehicle distance LonD and pedestrian-road distance LatD respectively). Each arrow represents the conditional probability relationship between variables. There are many ways to infer hidden variables in HMM, such as the approximate inference method ADF (Assumed Density Filtering):

$$\text{To forecast: } \overline{P}(C_t, C_{t-1}, E_t, E_{t-1}) = P(C_t|C_{t-1}, E_t) \times \overline{P}(E_t|E_{t-1})$$

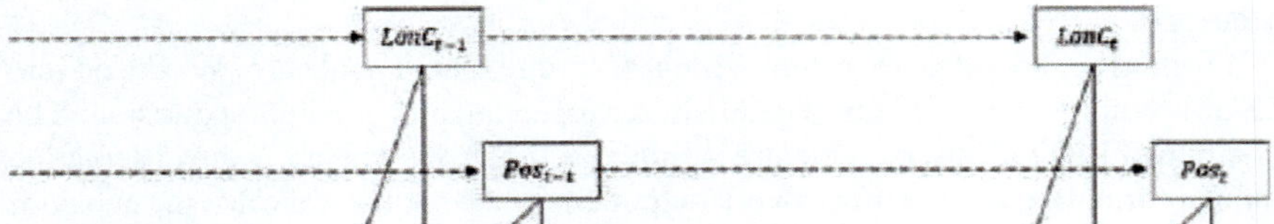

Fig. 3.47 Pedestrian's crossing intention modeling based on HMM [100]

$$\times \widehat{P}(C_{t-1}, E_{t-1})$$

$$\overline{P}(C_t, E_t) = \sum_{C_{t-1}} \sum_{E_{t-1}} \overline{P}(C_t, C_{t-1}, E_t, E_{t-1}) \tag{3.38}$$

To update: $\widehat{P}(C_t, E_t) \propto P(O_t|E_t) \times P(B_t|C_t) \times \overline{P}(C_t, E_t)$

$$\widehat{P}(C_t) = \sum_{E_t} \widehat{P}(C_t, E_t) \tag{3.39}$$

where, C represents pedestrian's passing intention; E represents hidden variables other than intention (destination, longitudinal risk and location in a safe area or not); and O represents all observable variables.

In addition, the inference task of HMM can also be solved by precise inference methods such as forward–backward algorithm. The conditional probability parameters in HMM can be manually labeled based on experience or trained with ND data based on expectation-maximum (EM) algorithm. The advantage of HMM model lies in its strong interpretability, and the method based on model + data training also have good identification effect. However, due to the network scale, HMM modeling process usually includes artificial simplification and assumptions for a series of environmental information.

In terms of intention identification based on deep learning, long short-term memory (LSTM) is a common neural network technique. As a recurrent neural network, LSTM shows certain advantages in learning the nonlinear feature of sequences.

In addition, Forget Gate, external Input Gate and Output Gate are introduced into LSTM to solve the problem of gradient disappearance or gradient explosion. The diagram of LSTM network structure is shown in Fig. 3.48, where x is the observation input information and h is the network output. Since the output of the driving behavior model is classified, the loss function of the model adopts Cross-entropy loss function:

$$L(a, y) = -[y * \ln a + (1 - y)\ln(1 - a)] \tag{3.40}$$

where, a represents the predicted value of the model; and y represents the tag value.

It is worth noting that LSTM also has shortcomings, including the lack of interpretability, demand of a large amount of labeled data for network training, and difficulty in multi-scenario generalization.

In terms of interactive intention identification, game theory is a common modeling method. As shown in Fig. 3.49, in the scenario of left turn without signal light, the reward matrix of the ego vehicle-other vehicle behavior can be given based on the

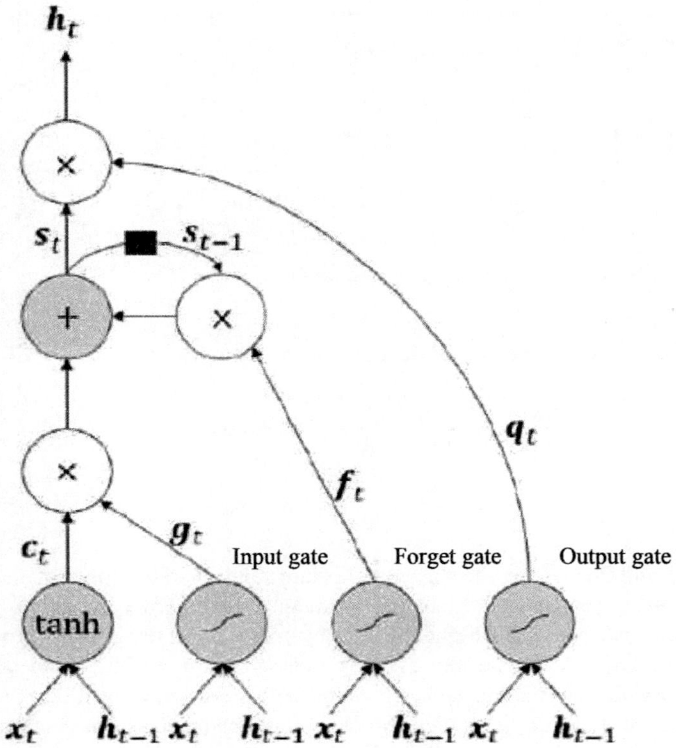

Fig. 3.48 LSTM network structure

right of way, distance to the intersection, speed and other conditions. In the mixed-strategy game analysis, it is assumed that the probability of vehicles A and B adopting the yielding strategy is *a* and *b* respectively, that is, the probability of rushing is (1 − a) and (1 − b) respectively. And it is assumed that both parties (rational individuals) intend to maximize their own benefits and understand each other's benefits (public knowledge). Firstly, the best response is defined as the best strategy taken by one party when the other party takes a certain strategy. In game theory, Nash Equilibrium is the strategy combination with the best response to both sides. Under the Nash equilibrium, neither side can obtain more benefits by unilaterally changing the strategy, so the combination constitutes the consistent prediction of both rational sides.

From the reward matrix of the example, the expected income function of two vehicles can be written:

$$E_A(a,b) = ab + (-3)a(1-b) + 3b(1-a) + (-5)(1-a)(1-b) \tag{3.41}$$

$$E_B(a,b) = ab + 2a(1-b) + 0 \times b(1-a) + (-1)(1-a)(1-b) \tag{3.42}$$

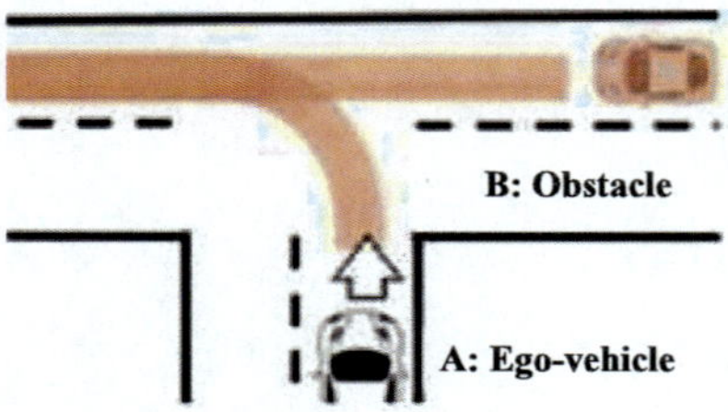

Reward Matrix (A,B)

A \ B	Yield	Rush
Yield	(1,1)	(1,5)
Rush	(4,1)	(-10,-1)

Fig. 3.49 Analysis of reward matrix based on game theory

Then the optimal responses of the corresponding parties are:

$$\frac{\delta E_A(a,b)}{\delta a} = 0 = \frac{\delta}{\delta a}(-4ab + 2a + 8b - 5) \tag{3.43}$$

$$\frac{\delta E_B(a,b)}{\delta b} = 0 = \frac{\delta}{\delta a}(-2ab + 3a + b - 1) \tag{3.44}$$

A = 0.5 and B = 0.5 are the Nash equilibrium solution of the above game process, representing the probability of yielding behavior of both parties. At this equilibrium point, neither side can obtain higher benefits by unilaterally changing its own decision (behavior probability). In the above game process, it is assumed that both sides make only one-step decision at the same time, and both sides know each other's benefits and actions. For more complex game process modeling (such as dynamic game [101] and incomplete information game [102]), it is necessary to further refer to relevant research work. Game theory is based on the assumption of rational decision-makers, but drivers are not fully-rational [103]. Therefore, modeling the irrational features of drivers and increasing the consistency between models and ND data are key research directions in related field.

(2) Trajectory Prediction of Traffic Participants

Trajectory prediction is used to output the spatial distribution of traffic participant's future motion state, which is a direct expression of their motion prediction. The results are generally expressed as the position (and its distribution) of the vehicle in each time step within a certain time domain. After intention identification, the motion trajectory corresponding to behavior intention can be further predicted. In addition, some prediction algorithms can also predict the trajectory without intention identification. At present, there are mainly four trajectory prediction approaches, respectively based on physical model [104], behavioral cognition [105], interaction [100], and deep learning [106], as shown in Table 3.11.

In terms of physical models, common models include constant speed model (CV) and constant acceleration model (CA) for longitudinal direction, as well as constant turn rate and acceleration model (CTRA) for lateral direction, etc. The above physical model is often applied in a series of filter-based trajectory prediction methods such

Table 3.11 Main approaches for target trajectory prediction

Approach	Characteristic	Deficiency
Based on physical model	Give a simple motion model, for short-term prediction	It is difficult to be used for long-term motion prediction and complex behavior process
Based on behavioral intention	Based on the identification of behavior intention, it can generate a more accurate and explanatory motion prediction	It depends on accurate behavior modeling and effective intention identification
Based on interaction	In conflict including multi-participants, the interaction model is established to predict the motion during interaction process	The construction of interaction model is relatively complex and the accuracy is relatively insufficient
Based on deep learning	Trajectory prediction can be achieved without intention identification	With high data dependence and insufficient interpretability, it is difficult to generalized application

as Kalman filter, which can also model and describe the uncertainty of prediction. The prediction based on physical model only considers the current motion state of the vehicle and its motion model. It is often used in short-term motion prediction (such as less than 1 s). For scenarios with longer time domain and more complex vehicle behavior, it is difficult to achieve high-precision of prediction.

The method based on behavioral intention takes the above intention identification as the conditional input. It is assumed that the motion of the target corresponds to a series of behavioral strategies for its driving purpose, and then trajectory prediction is carried out based on this assumption. Due to the prior information, the behavior-based trajectory prediction is more reliable than the physical model-based method in the long-term domain. Behavior-based approach usually include two steps: behavior intention identification and trajectory prediction. And trajectory prediction ways include prototype trajectory clustering, Gaussian process regression (GPR), etc.

As shown in Fig. 3.50, the prototype trajectory clustering [107] is to generate multiple cluster trajectories offline in advance on the road network, in which each cluster trajectory corresponds to a behavioral intention (such as turning left, and going straight). The trajectory prediction process is transformed into the online process matching the actual trajectory and the offline trajectory.

GPR is a classical trajectory prediction based on machine learning. In order to realize trajectory prediction based on behavior intention, firstly, GPR model is established:

$$f \sim GP(u, K)$$

$$u = \{m(t_i), t_i = 1 : T\}$$

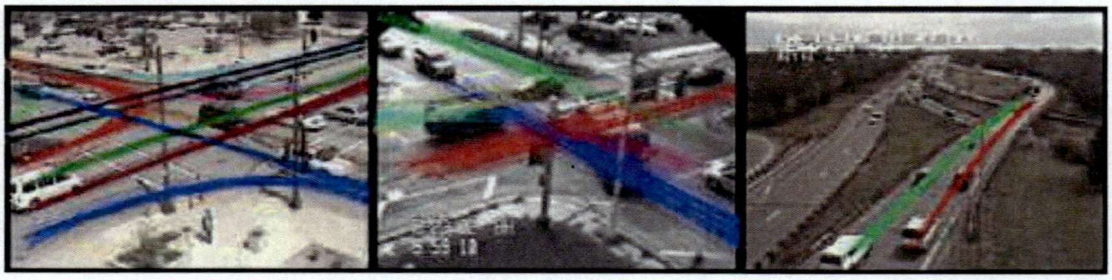

Fig. 3.50 Trajectory cluster corresponding to intention [107]

$$K = \{\kappa(t_i, t_j), i, j = 1 : N\}$$

where, u is the mean vector (representing the predicted position at each time); and K represents the covariance matrix, which is used to describe the distribution of predicted positions.

The maximum likelihood estimate (MLE) algorithm can be adopted to train the model, based on the data set corresponding to each behavior (for example, classify the trajectory in highway into left line merging, straight line and right line merging).

In recent years, deep learning-based approaches have also been widely used in trajectory prediction, including LSTM [108], graph neural network (GNN) [27], generative adversarial network (GAN) [109], etc. Taking LSTM as an example, compared with behavior intention identification based on LSTM, the network changes from "multiple input and single output" (input position according to timing, and output intention) to "multiple input and multiple output" (input the past timing position, and output future timing position). The model structure is shown in Fig. 3.51.

The final output of the trajectory prediction model is the sequential result representing the position at each time step. The mean-square error (MSE) can be selected as the loss function of the model:

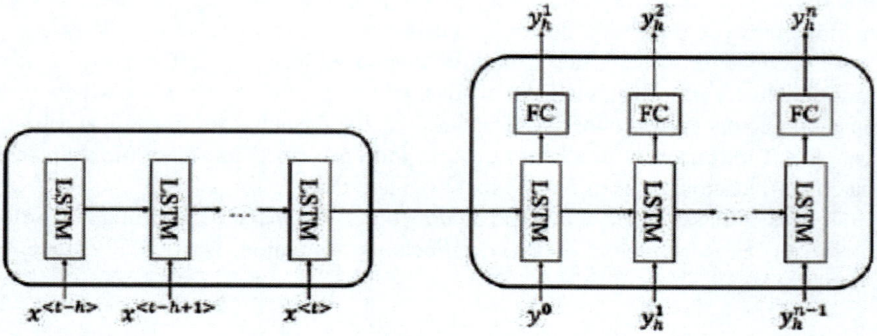

Fig. 3.51 Diagram of LSTM structure with "multiple input to multiple output"

$$\text{Loss} = \text{MSE} = \frac{\sum_{i=1}^{n}(y_i - y_{pi})^2}{n} \tag{3.45}$$

where, y_i represents the true value of the sample (i.e. the position coordinates of the trajectory point); y_{pi} represents the predicted value of the output; and n represents the number of samples.

At present, the frontier research on trajectory prediction further considers the interaction among traffic participants and the multi-modal expression of predicted results. Multi-modal means that the vehicle may have different behavior modes such as acceleration, deceleration and lane changing. As shown in Fig. 3.52, the researchers of Uber [110] adopt the GNN to input the information of vehicle-lane, lane-lane, lane-vehicle and vehicle-vehicle into the network, and output a specific number of predicted tracks in the future 3 s and the confidence corresponding to each trajectory. Compared with the previously-introduced interactive behavior modeling, the vehicle-vehicle interaction modeling based on GNN is essentially an implicit modeling way. Specifically speaking, the potential impact of interaction on vehicle motion is characterized by neural network, but the concrete strategies adopted by each driver in the interaction and the intention of rushing-yielding are not modeled and identified.

In practical application, the predictor based on deep learning can achieve high accuracy in short-term trajectory prediction, but also has such deficiencies as data dependence, difficulty to explain data and transfer between different scenarios. In addition, this method is often trained with the third-person perception data collected from the roadside, which is difficult to be used in the vehicle decision-making process

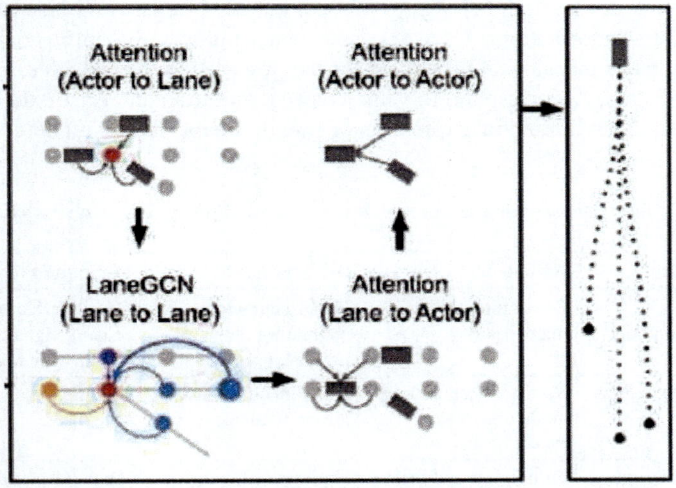

Fig. 3.52 Multi-modal vehicle trajectory prediction based on GNN [110]

to predict others' response to different behaviors of the ego-vehicle. Therefore, trajectory prediction is still a technical difficulty to be further explored by researchers.

(3) Estimation of Vehicle–Road Dynamics Parameters

Accurate and real-time acquisition of key vehicle–road dynamic parameters is critical for the comprehension and identification toward dynamic environment. Vehicle–road dynamics parameters can be divided into state parameters and structural parameters. State parameters mainly include vehicle centroid sideslip angle, yaw rate, etc. And structural parameters mainly include mass and slope, etc. In the actual operation, the yaw rate can be obtained more accurately by gyroscope, while the acquisition of centroid sideslip angle is often obtained by observation due to limitations of measurement cost and installation position. There are three kinds of observation approaches for centroid sideslip angle, respectively based on kinematics and motion geometry, dynamics model, and multiple information fusion. The specific analysis is shown in Table 3.12.

The estimation of road adhesion coefficient influences the risk identification of dynamics instability. Especially under emergent conditions, the longitudinal and lateral coupling effect of vehicles dynamics is strong, which emphyasizes the importance of the real-time estimation of road adhesion coefficient. Based on the vehicle longitudinal and lateral coupling dynamics, the tire dynamics model is first explored, and the estimation method of nonlinear tire-road adhesion coefficient is further proposed [115].

The establishment of tire dynamics model is essential for the observation of dynamics-based tire-road adhesion coefficient. So far, a variety of tire dynamics models have been established [116]. The two-dimensional LuGre tire mechanical model shown in Fig. 3.53 belongs to a brush model. It regards the tire-road interaction as the deformation of bristles, and then simulates the spring damping to describe the tire mechanical properties for the observation of road adhesion coefficient. v_x, F_x and z_x correspond to vehicle speed, adhesion and bristle deformation respectively. The differential expression of LuGre tire mechanics model is:

Table 3.12 Observation approaches for vehicle centroid sideslip angle and its characteristics [111–114]

Classification	Approach	Advantages	Limitations
Based on kinematics model	Direct integration, fuzzy logic	High accuracy in estimating linear and nonlinear regions	Greatly affected by the sensing signal, and with results easy to drift
Based on dynamics model	Kalman filter, sliding mode observation, robust observation	Clear mechanism and strong robustness	Complex observer structure
Based on multiple information fusion	Pseudo integral fusion, hybrid filtering	High precision and strong robustness	Weak adaptability

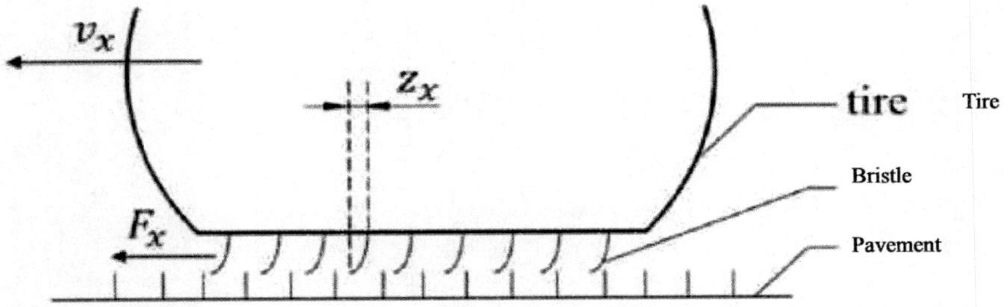

Fig. 3.53 LuGre tire mechanics model

$$dF_i = (\sigma_{0,i} z + \sigma_{1,i}\frac{dz_i}{dt} + \sigma_{2,i} v_{r,i})dF_N \tag{3.46}$$

where, $\sigma_{0,i}$, $\sigma_{1,i}$ and $\sigma_{2,i}$ represent the tire stiffness, damping coefficient and viscous damping coefficient respectively; and $v_{r,i}$ represents the relative speed of the tire tread in the longitudinal direction ($i = x$) and lateral direction ($i = y$).

LuGre tire dynamics model can be expressed as:

$$\begin{bmatrix} \dot{\bar{z}}_x \\ \dot{\bar{z}}_y \\ \dot{\theta}_x \\ \dot{\theta}_y \end{bmatrix} = \begin{bmatrix} v_{r,x} - \gamma_x(v_r, \theta)\bar{z}_x - \frac{k\omega r}{L}\bar{z}_x \\ v_{r,y} - \gamma_y(v_r, \theta)\bar{z}_y - \frac{k\omega r}{L}\bar{z}_y \\ 0 \\ 0 \end{bmatrix} \tag{3.47}$$

$$\begin{bmatrix} F_x/F_N \\ F_y/F_N \end{bmatrix} = \begin{bmatrix} [\sigma_{0,x} - \sigma_{1,x}\gamma_x(v_r, \theta) - \sigma_{1,x}\frac{k\omega r}{L}]\bar{z}_x + (\sigma_{1,x} + \sigma_{2,x})v_{r,x} \\ [\sigma_{0,y} - \sigma_{1,y}\gamma_y(v_r, \theta) - \sigma_{1,y}\frac{k\omega r}{L}]\bar{z}_y + (\sigma_{1,y} + \sigma_{2,y})v_{r,y} \end{bmatrix} \tag{3.48}$$

where, $\bar{z}_x$ and $\bar{z}_y$ respectively represent the average deformation of the bristles in the x and y directions; θ_x and θ_y represent the coefficients derived from the Coulomb friction coefficient and static friction coefficient of tread in x and y directions respectively; γ is a function derived from relative speed and coefficient θ; k represents the coefficient related to the distribution of tread deformation; ω, r and L represent the wheel's rotation rate, radius and grounding length respectively.

It can be proved that the necessary condition for the system observability is that the relative speed between the tread and the road surface is not equal to zero. Because the tire-road relative speed is high during rapid acceleration, braking and steering of vehicle, the convergence speed and accuracy of adhesion estimation are faster and higher, as is proved in the simulation and real vehicle experiment.

Due to the strong nonlinear feature of tire mechanics model, unscented Kalman filtering (UKF) algorithm is adopted in this section to realize the second-order accuracy estimation, and the final result is weighted and output by means of the lateral-longitudinal estimation in the fusion estimation. The estimated maximum ground adhesion coefficient is:

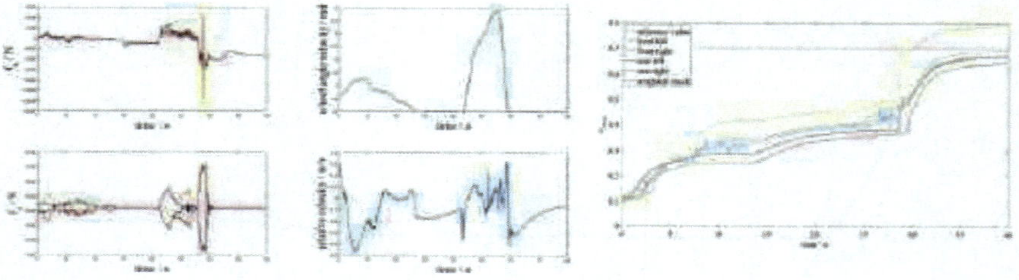

Fig. 3.54 Actual test results of vehicle adhesion estimation

$$\mu_{\max} = \frac{\overline{F}_x g_x(v_{r,x}, \theta_x) + \overline{F}_y g_y(v_{r,y}, \theta_y)}{\overline{F}_x + \overline{F}_y} \tag{3.49}$$

where, $g_i(v_{r,i}, \theta_i) = \theta_i(\mu_c + (\mu_s - \mu_c)e^{-|v_{r,i}/v_s|^{0.5}}$, μ_s, μ_c and v_s represent Coulomb friction coefficient, minimal static friction coefficient and Stribeck relative speed respectively. The above algorithm is verified based on the real-vehicle test of adhesion. Figure 3.54a shows the wheel longitudinal force, vertical force, wheel speed and tire-road relative speed during the test, in which wheel longitudinal force and vertical force respectively include the response of four wheels. Figure 3.54b shows the fusion estimation results of the corresponding adhesion coefficients, including the maximum pavement adhesion coefficient of four wheels, the weighted estimation results and reference values. The test results indicate that the weighted estimation results are basically consistent with the reference values. Especially in the case of sudden change of pavement adhesion coefficient, the accuracy of the proposed weighted estimation is higher. Hence, the advantages of this estimation method are more obvious under extreme conditions such as sudden change of pavement adhesion coefficient.

The test results indicate that the proposed estimation of adhesion coefficient performs well under different conditions. And the method converges faster during braking or accelerating of vehicle, which proves that the algorithm theory is reliable.

The dynamics-based adhesion estimation has excellent real-time performance, but poor robustness of observed results; The method based on visual sensing has relatively accurate results, but has the shortcoming in large amount of calculation. Therefore, the identification combining dynamics and vision is one of the development directions in the future.

3.2.3 *Quantitative Evaluation of Driving Risk*

In the transportation system, risk is an objective quantity for describing the degree of system risk. Generally speaking, risk is defined from two aspects: one is as a measure of the possibility of hazardous events or abnormal events in systems; and the other is as the product of the consequence severity and the probability of an accident. For the sake of simplicity, it is expressed as the second definition in most references. However, since its defined loss or severity is only applicable to accidents, it is difficult to comprehensively express the process of risk generation and evolution by measuring system risk only by accidents. The hazard source in the system is the root of the accident and the risk. Willie Hammer defined hazard sources as potential unsafe factors that may lead to driver's injury or property loss. According to the role in the occurrence and development of accidents, hazard sources can be divided into two categories: the first refers to the energy or dangerous substances that may be accidentally released into the system; and the second refers to various unsafe factors that lead to the failure or destruction of measures for energy restriction.

To comprehensively characterize the impact of hazard sources during driving and quantitatively evaluate the driving risk, firstly, the existing real-time risk quantification methods (time logic, kinematics, collision probability, statistical learning and APF) are summarized and compared. However, in the complex traffic environment with mixed traffic of motor vehicles and vulnerable traffic participants, the current classical risk quantification methods are difficult to accurately identify the potential risks of dynamic and random traffic participants, comprehensively evaluate the driving risk of DVE coupling, and support real-time decision-making in dynamic and complex scenarios. Therefore, through the systematic quantification of single-factor risks in terms of vehicle, road traffic facility and driver behavior respectively, a unified modeling of DSF is proposed to comprehensively evaluate the overall driving risks of DVE coupling from the perspectives of vehicle and traffic. Finally, employing the comprehensive driving risk assessment, the application examples and result analysis under typical scenarios are put forward, as shown in Fig. 3.55.

1. Classical Risk Quantification Methods

The concept of risk first appeared in the western economic field at the end of the nineteenth century and has been widely used in economics, sociology, engineering science, environmental science, disaster science and other fields [117]. Different studies have different definitions of risk. For example, risk is the uncertainty of the possible future results of events, the comprehensive evaluation of the scale and possibility of loss, risk the result of the interaction among risk constituent elements, as well as the random feature of the uncertainty. In the field of FuSa, the International Organization for Standardization standard defines risk as the combination of consequences of the incident (including changes of environment) and the occurrence of relevant incidents [118]. Incident-based risk can be modeled by the combination of expected damage and occurrence of the actual event. Therefore, the definable risk model under the traffic scenario mainly includes two parts, i.e., the occurrence of critical incidents and the expected loss of the actual incident. The driving risk is mainly composed of risk factors, accident losses and other elements. According to the concept of risk in engineering, the concept of driving risk expressed by probabilistic elements and severity elements is proposed, and can be expressed as follows:

$$R_i = P_i \cdot S_i \tag{3.50}$$

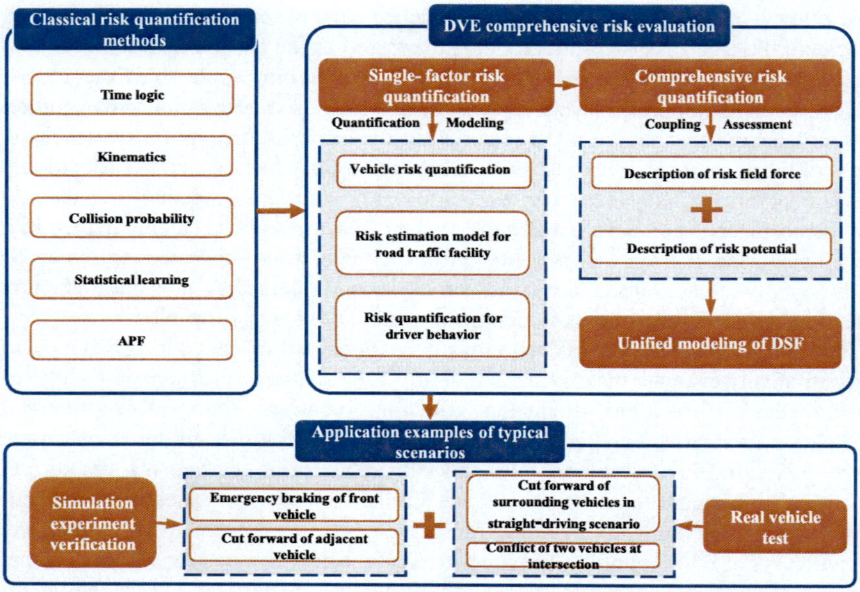

Fig. 3.55 Logic diagram of quantitative evaluation of driving risk

where, P_i is the occurrence of vehicle accident; S_i is the possible consequences of accidents; R_i is the risk.

With the continuous maturity of theories, multiple potential risks during driving are comprehensively considered, and the risk measurement can be expressed as follows:

$$R_i = \sum \sum P_{ij} \cdot S_{ij} \tag{3.51}$$

where, P_{ij} refers to the occurrence of accidents caused by the interaction between vehicle i and vehicle j; S_{ij} refers to possible consequences of accidents after interaction between vehicle i and vehicle j.

The above risks are all defined for accidents, including accidents that have occurred and that will occur. From the perspective of the whole system, risk is the function of influencing factors of system hazard, and can be expressed as [119]:

$$R = f(R_1, R_2, R_3, R_4, R_5) \tag{3.52}$$

where, R_1 is the vehicle factor; R_2 is the driver-related factor; R_3 is the environmental factor; R_4 is the traffic management factor; and R_5 refers to other factors.

It is necessary to quantitatively evaluate the risk in the driving process, and transform the qualitative description into quantitative output, to better assist the vehicle safety and reasonable decision-making. The real-time risk assessment usually adopts ways based on the microscopic perspective of vehicle, which is the quantitative output process of risk with vehicle as the center in accordance with physical properties such as vehicle kinematics and dynamics. The current research on driving risk from the microscopic perspective is conducted in terms of five aspects (Fig. 3.56), i.e., time logic (car-following time headway, collision avoidance time, etc.), kinematics (distance, acceleration, etc.), collision occurrence, statistical learning, and APF [120].

(1) Time Logic-based Approach

The time-based risk assessment algorithm takes time as the basis for driving risk assessment, mainly including: time headway (THW), time to collision (TTC), inverse time-to-collision (TTCi), TTC extension, etc. Among them, TTC and its extended indicators are mainly used for collision risk assessment between vehicles and obstacles, while THW indicator is only applicable to the behavioral characteristics and risk analysis of drivers in the car-following scenario [121].

THW indicator can be defined as:

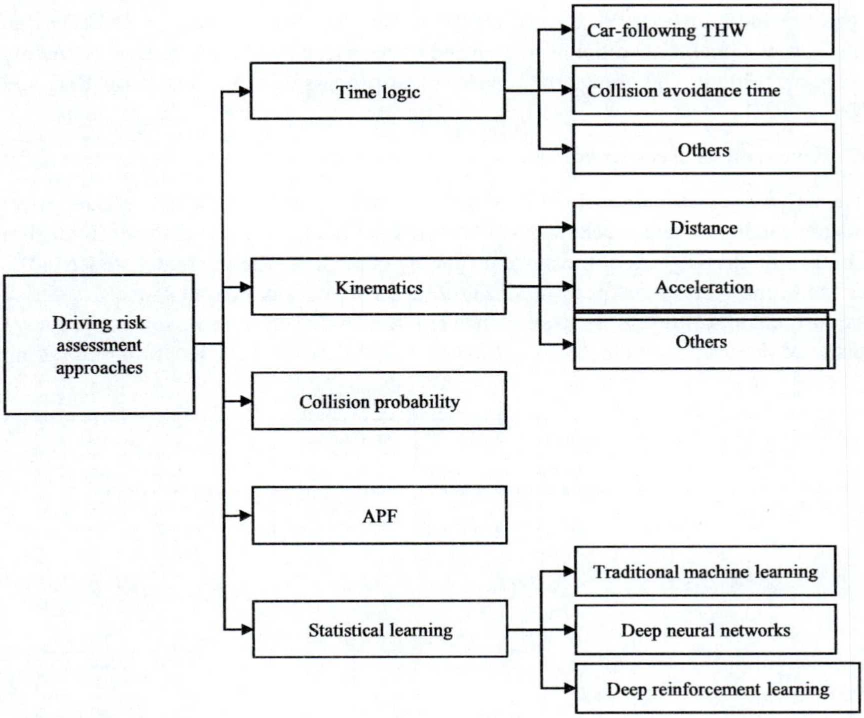

Fig. 3.56 Classification of driving risk assessment approaches

$$THW = t_1 - t_h \tag{3.53}$$

where, t_1 indicates the time when the head of the front vehicle passes through the same observation position; and t_h indicates the time when the head of the ego-vehicle passes through the same observation position.

TTC indicators can be defined as:

$$\text{TTC} = \{t|d_r^t = 0\} = \begin{cases} -\frac{d_r^0}{v_r^0} & (v_r^0 < 0, a_r^0 = 0) \\ -\frac{v_r^0}{a_r^0} - \frac{\sqrt{(v_r^0)^2 - 2d_r^0 a_r^0}}{a_r^0} & (v_r^0 < 0, a_r^0 \neq 0) \\ -\frac{v_r^0}{a_r^0} + \frac{\sqrt{(v_r^0)^2 - 2d_r^0 a_r^0}}{a_r^0} & (v_r^0 \geq 0, a_r^0 < 0) \\ \text{no defination} & \text{others} \end{cases} \tag{3.54}$$

where, d_r^t represents the relative distance; d_r^0, v_r^0 and a_r^0 represent the relative distance, relative speed (the speed of the front vehicle minus the speed of the ego-vehicle) and relative acceleration at the initial time.

In particular, TTC is defined only when the relative speed is negative. TTC can measure the collision risk caused by the change of relative speed, and THW can measure the potential collision risk caused by too close distance in the car-following scenario. Figure 3.57 shows the development process of the research on TTC and THW [122].

(2) Kinematics-based Approach

The risk assessment algorithm based on kinematics mainly includes two categories: distance indicator and acceleration indicator. Specifically, the safe distance algorithm obtains the driving risk at the current time by comparing the relative distance at the current time with its risk threshold. Figure 3.58 shows a typical distance-based risk assessment algorithm. Relative-distance risk threshold refers to the minimal relative distance required for vehicle avoidance of collision with obstacles through braking

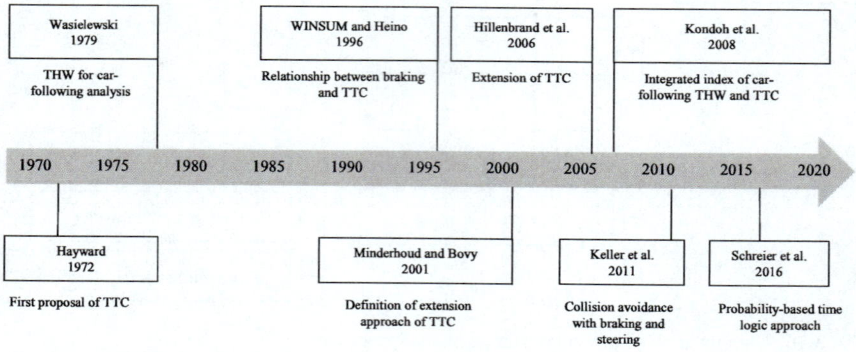

Fig. 3.57 Driving risk assessment algorithm based on time indicators

under current conditions. Take the safe distance algorithm developed by Mazda as an example [123]:

$$d_b = \frac{1}{2}\left(\frac{v_h^2}{2a_h} - \frac{v_l^2}{2a_l}\right) + \tau_1 v_h - \tau_2 v_r + R_{min} \tag{3.55}$$

where, d_b represents the threshold value of starting distance of automatic braking; v_h, v_l and v_r respectively represent the ego-vehicle speed, the front vehicle speed and the relative speed between the ego-vehicle and the front vehicle; a_h and a_l represent the acceleration of the ego-vehicle and the front vehicle respectively; τ_1 is the reaction time of the driver of the ego-vehicle; τ_2 indicates the duration for the front vehicle to keep moving at a constant speed; and R_{min} indicates the safe distance at the stop moment.

The acceleration-based risk assessment algorithm obtains the current driving risk by comparing the acceleration at the current time with the acceleration risk threshold, which is the minimal braking deceleration required by the rear vehicle to brake for collision avoidance with the front vehicle. The following describes the brake thread number (BTN), an acceleration risk assessment indicator:

$$\text{BTN} = \left|\frac{a_s}{a_{h\_max}}\right| \tag{3.56}$$

$$a_s = a_l - \frac{v_r^0 |v_r^0|}{2d_r^0} \tag{3.57}$$

where, a_s is the acceleration risk threshold of rear vehicle; a_{h_max} is the maximum braking deceleration of the rear vehicle; a_l refers to the acceleration of the front

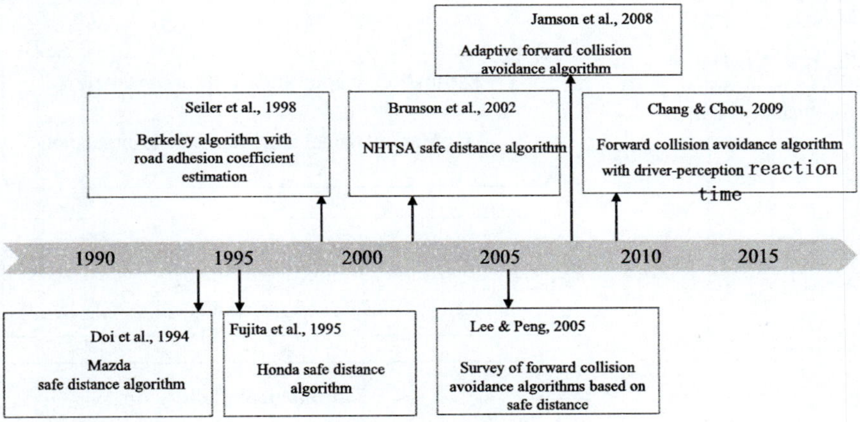

Fig. 3.58 Distance-based driving risk assessment algorithm

vehicle; And v_r^0 and d_r^0 respectively represent the relative speed and distance between the ego-vehicle and the front vehicle at the current time.

The acceleration-based risk assessment algorithm shall estimate the braking capacity of the vehicle in real time according to road adhesion conditions, and the deceleration of the front vehicle cannot be measured by the sensor. It still takes the uniform acceleration model as the obstacle motion hypothesis, which is difficult to deal with the sudden change of the scenario (such as the emergency braking of the front vehicle).

(3) Collision Occurrence-based Approach

The collision occurrence algorithm shall calculate the total probabilities of all trajectories that may lead to collisions, which can be described in three steps (Fig. 3.59):

1) Occurrence density function estimation is to estimate the state distribution of each object with collision possibility, mainly including Markov model and Monte Carlo estimation;
2) Collision detection is to detect whether the ego-vehicle and surrounding objects intersect in space or trajectories;
3) Occurrence summation is to calculate the percentage of trajectories with possible collision, mainly through integral or discrete summation [124]. The collision occurrence usually can be expressed as:

$$P = \iint I_c(x_h, x_o) p(x_h, x_o) dx_h dx_o \tag{3.58}$$

where, $p(x_h, x_o)$ is the joint occurrence density function of the state of the ego-vehicle h and surrounding objects; o is the other surrounding objects; x_h represents the state of the ego-vehicle; x_o represents the state of surrounding

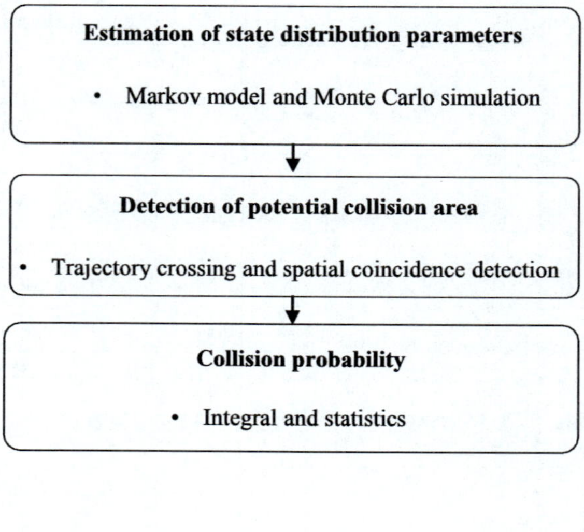

Fig. 3.59 Driving risk assessment algorithm based on collision occurrence

objects; and $I_c(x_h, x_o)$ represents the collision indication function. If there is a spatial intersection between the two, the collision indication function value is 1, otherwise is 0.

(4) Statistical Learning-based Approach

The risk assessment of IV can also be modeled as discrete classification tasks in machine learning. Different from the previous continuous indicators with clear risk quantification formula, machine learning algorithm obtains discrete indicators based on the classification of risk incidents. Then, the features extracted from the learning model can be considered as the implicit key measure for risk assessment.

1) Traditional Machine Learning

Traditional machine learning algorithms include supervised learning and unsupervised learning, both of which have been applied to the risk assessment of IV. For example, in accordance with braking features, unsupervised K-means clustering is used to quantify the risk level of near collision scenarios. Supervised neural networks require well-defined input–output pairs and have been used in anti-collision systems. Even if there is no independent threat assessment module, the machine learning algorithm can still evaluate the risk level based on the learned features.

2) Deep Neural Networks

The latest development of artificial intelligence pushes forward the application of deep neural networks (DNN) in IV. Common DNN applied to AV, such as the end-to-end DNN model, directly maps the original input to the driving action (such as acceleration and steering angle), and finally outputs the action behavior that can avoid collision. DNN can learn many features, within which there may be ones responsible for risk assessment and regarded as key measures [125].

3) Deep Reinforcement Learning

Similarly, deep reinforcement learning for risk assessment is mainly used in robotics research, involving the task of avoiding collision [126, 127]. For example, some researchers have proposed a deep reinforcement learning method for avoiding collision, which allows the robot to complete the expected task while experiencing a slight collision. A series of deep learning methods in view of collision avoidance have also been widely studied in the research of IV. Researchers propose a deep q-network for autonomous braking system, in which the reward function is designed by balancing the collision accident damage and the reward from risk avoidance. In some cases, even if there is no independent threat assessment module in deep reinforcement learning, collision-free driving strategies can be gained by embedding safety factors in rewards.

(5) APF-based Approach

The artificial potential field (APF) model was first applied in the field of robot research and is widely used in mobile robot motion planning, as shown in Fig. 3.60. The main idea of the APF model is to assume that the vehicle runs in the potential field formed by the target point and the obstacle, in which the target point generates a gravitational field to attract the vehicle to approach, and the obstacle generates a repulsive field to prevent collision. In the field of transportation, APF is used to build a risk map to describe the driving risk constraints of surrounding traffic elements, including lane lines, obstacles and traffic signs. Generally speaking, moving along the lowest point of the risk potential map is the safest driving path [128]. For example, the APF model is used to establish the vehicle-centered traffic risk potential map to realize the navigation of AV, with the risk map to fuse such information as the road, the relative motion state with obstacles and the traffic sign, and with the APF risk potential to reflect the risk distribution of various points in the surrounding environment. In the development of a driving assistance system controller based on APF model, a variety of driving assistance systems can be comprehensively controlled by the superposition of potential energy and damping function. In the design of lane keeping system, the boundary constraints of lane line and the collision constraints of surrounding vehicles can be described by APF, which can integrate environmental constraints and vehicle dynamics properties for the safety of driving process [129]. In the follow-up study, the driving risk field model based on APF theory and traffic accident causation mechanism is proposed, to realize the quantification and unification of driving risks related to drivers' characteristics, road conditions and surrounding obstacles [130, 131]. However, it is still difficult to construct APF mathematical model for complex scenarios. The most fundamental APF model can be expressed as:

$$U_{\text{art}}(x) = U_{x_{\text{d}}}(x) + U_{\text{O}}(x) \tag{3.59}$$

$$U_{x_{\text{d}}}(x) = \frac{1}{2}k_p(x - x_{\text{d}})^2 \tag{3.60}$$

$$U_{\text{O}}(x) = \begin{cases} \frac{1}{2}\eta\left(\frac{1}{\rho} - \frac{1}{\rho_0}\right)^2, \rho \le \rho_0 \\ 0, \rho > \rho_0 \end{cases} \tag{3.61}$$

where, $U_{x_{\text{d}}}(x)$ is the attraction generated by the target position during driving; $U_{\text{O}}(x)$ is the repulsive force generated by the surrounding obstacles during driving; x_{d} indicates the target position; x indicates the actual position of the vehicle; η is the ultimate distance affected by the APF; ρ_0 is the ultimate distance affected by the APF; and ρ represents the shortest distance to the obstacle.

To sum up, the research on risk quantification algorithms has been conducted for many years. The vehicle driving risk assessment models in the current literature are comprised of the above five categories, i.e., time logic, kinematics, collision occurrence, statistical learning and APF models, which are widely used in IV with their characteristics. Among them, time logic and kinematics methods are mostly based on vehicle kinematics and dynamics theory, and the expression of vehicle driving safety is mostly based on vehicle state information and relative motion relationship information between two vehicles. These methods have been applied in a certain range because of their simple parameters and physical meanings in line with human

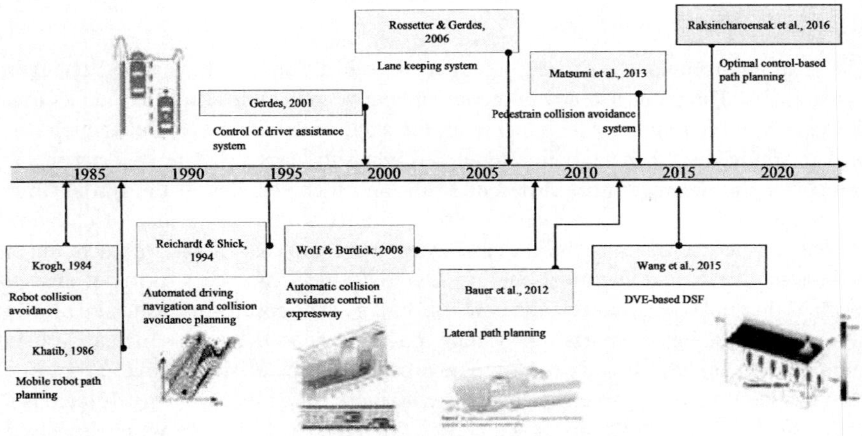

Fig. 3.60 Driving risk assessment algorithm based on APF

subjective feelings, but are difficult to realize the two-dimensional uncertainty risk assessment in the actual traffic scenario with limited practical application. Compared with the quantitative methods based on time or dynamics, the collision occurrence takes into account the probabilities of different trajectories and can deal with the uncertainty to a certain extent. However, it has some difficulties for application in practice, due to more computing resources in calculating the predicted state distribution and collision detection. Statistical learning can quantify risk into discrete classification tasks, but its interpretability still shall be improved. The APF theory is gradually mature and effective in car following, path planning and collision avoidance algorithms, but without a comprehensive consideration of the risk factors in the existing research, including a lack of consideration of the driver's own psychological, physiological and behavioral characteristics, and ignorance of the impact of complex weather, complex road factors and the coupling of various factors in the DVE system on driving risks.

2. DVE Comprehensive Risk Assessment

In the complex DVE system, the traffic elements are diverse, dynamic and random. Generally, the road traffic risk occurs between the traffic participants (such as the ego-vehicle and its surrounding vehicles), or between the traffic participants and the road environment, instead of existing independently. To comprehensively evaluate the risks generated by various risk sources and quantitatively output the safety state of traffic environment, the definition of physical field will be referred to characterize risk, which is defined as the field interaction between the research objects. Then in view of the coupling of DVE, a comprehensively quantitative evaluation model of driving risk is proposed.

(1) Vehicle Risk Quantification

The road environment of driving is engaged with multiple vehicles, and changes dynamically. The inconsistency of movement between vehicles leads to the fact that the state of each vehicle will change with the state of the other vehicles, which can lead to traffic disturbances being generated when the driving state of a particular vehicle is significantly more different from the other vehicles in the traffic flow. From the third-party perspective of traffic administrators, the essence of collision in road traffic accidents is the process when the energy of moving objects (such as kinetic energy and chemical energy) is transformed into other forms of energy (such as thermal energy and elastic potential energy) and results in elastic and plastic deformation. In the transportation system, moving objects (such as vehicles, cyclists and pedestrians) have kinetic energy. The process of accelerating from the start to a certain speed is an energy storage process with increasing kinetic energy. If there is a collision accident with an object at a certain time, the stored energy will be released and transferred, resulting in elastic and plastic deformation. In the collision process, the larger the kinetic energy conversion is, the larger the loss is. Therefore, the risk of the vehicle can be described following the relationship between force acting and energy conversion in collision. For vehicle j in a single movement, there is:

$$E_{j,0} = \frac{1}{2} m_j v_j^2 = \frac{m_j v_j^2}{2 r_{ji}} \cdot r_{ji} \tag{3.62}$$

where, $E_{j,0}$, m_j and v_j respectively represent the kinetic energy, mass and speed of a moving object; and r_{ji} refers to the distance between any point i in the speed direction of the moving object j and itself.

$$E_{j,0} = F_{ji} \cdot r_{ji} \tag{3.63}$$

where, F_{ji} represents the equivalent effect and the risk of moving vehicle j on the point i in the traffic environment, with the unit of Newton.

In consideration of the certain range of mutual influence among traffic objects as well as the constraint of driver and traffic environment, the DSF model is defined as [132]:

$$F_{ji} = \begin{cases} E_{j,0} & r_{ji} \in [0, r_{\min}) \\ E_{j,0} r_0 \left(\frac{1}{r_{ji}^2} - \frac{1}{r_{\max}^2} \right) & r_{ji} \in [r_{\min}, r_{\max}] \\ 0 & r_{ji} \in (r_{\max}, +\infty) \end{cases} \tag{3.64}$$

where, when $r_{ji} \in [0, r_{\min})$, F_{ji} is equal to $E_{j,0}$ in numerical value, with the unit of Newton; r_0 refers to the scope of the driver's attention to the risk, which is related to the driver's car-following distance; $r_{\max}$ is the distance between free flow vehicles, which indicates the maximum impact range of risk, and $r_{\min}$ is the minimal impact range of risk.

The car-following of the driver is subject to the current traffic environment. According to the traffic flow theory, the relationship between the macro traffic flow and the flow velocity in the car-following process can be expressed as:

$$q = \frac{v}{\left(\gamma v^2 + \tau v + l\right)\left[1 - \ln\left(1 - \frac{v}{v_f}\right)\right]} \tag{3.65}$$

where, q is the traffic flow; v is the flow velocity; γ represents the driver's overall aggressiveness; τ refers to the average reaction time; and l is the effective vehicle length.

According to Formula (3.65), the car-following distance of the driver is:

$$r_0 = \left(\gamma v^2 + \tau v + l\right)\left[1 - \ln\left(1 - \frac{v}{v_f}\right)\right] - l \tag{3.66}$$

where, the driver's overall aggressiveness γ is taken as $\gamma \in [-0.03, 0]\ \mathrm{s}^2/\mathrm{m}$. The smaller the value of γ is, the higher the driver's degree of aggressiveness is, that is, the smaller the car-following distance is; the average reaction time τ is taken as $\tau = 1\ \mathrm{s}$.

With the free flow velocity, the maximum value of traffic flow meets the following formula:

$$q_{\max} = 3100 - 54v_{\mathrm{f}} \tag{3.67}$$

where, $q_{\max}$ is the maximum flow under free traffic flow, with the unit of pc/h/ln; and v_{f} is the free flow velocity, with the unit of m/s.

Therefore, the vehicle spacing in the free flow is:

$$r_{\max} = \frac{1800v_{\mathrm{f}}}{1550 - 27v_{\mathrm{f}}} \tag{3.68}$$

where, l is the vehicle length, take $l = 6\mathrm{m}$.

And, $r_{\min}$ and $r_{\max}$ are related to the value of r_0, expressed in the following formula.

$$r_{\min} = r_{\max}\sqrt{\frac{r_0}{r_{\max}^2 + r_0}} \tag{3.69}$$

Since F_{ji} takes a constant value within the range of $r_{ji} \in [0, r_{\min}) \cup (r_{\max}, +\infty)$ and is only related to the change of r_{ji} within $r_{ji} \in [r_{\min}, r_{\max}]$, the subsequent analysis mainly focuses on the interval of $r_{ji} \in [r_{\min}, r_{\max}]$ and carry out calculations directly by the constant value within the range of $r_{ji} \in [0, r_{\min}) \cup (r_{\max}, +\infty)$.

If the vehicle j moves freely at a constant speed in a borderless environment and is considered as a particle, the driving risk made by the vehicle in the environment meets the isotropy on the plane because the vehicle can choose any direction. Therefore:

$$F_{ji,0} = E_{j,0} r_0 \left(\frac{1}{x_{ji}^2 + y_{ji}^2} - \frac{1}{r_{\max}^2} \right) \tag{3.70}$$

where, x_{ji} and y_{ji} represent the distance between the vehicle and any point i in the environment in the longitudinal and lateral directions, respectively. In the real road traffic environment, due to the direction-oriented vehicle movement, its risk to the outside world is not isotropic, i.e. the risk to the outside world in the positive direction of movement direction is greater than that in the negative direction. This phenomenon is similar to the Doppler shift effect of wave. In the Doppler shift effect, the movement of the wave source will increase the frequency received by the observer on one side of the moving direction and decrease the frequency received by the observer in the negative direction:

$$f_s' = \frac{v_{s,0} \pm v_o(t)}{v_{s,0} \mp v_s(t)} f_s \tag{3.71}$$

where, f_s' is the frequency at the observer; f_s is the initial frequency of wave source; $v_{s,0}$ is the wave velocity; $v_o(t)$ is the movement speed of the observer, which is positive when approaching the wave source and negative when away from the wave source; $v_s(t)$ is the movement speed of the wave source, negative when approaching the observer and positive when away from the observer.

Therefore, the above view that the risk to the outside world caused by the vehicle in the positive direction of moving direction is greater than that in the negative direction can be described as that the gradient decline degree of the DSF force is related to the moving direction of the vehicle. Specifically speaking, when the vehicle j approaches the point, i the gradient decline of the DSF force slows down. Therefore, if the longitudinal and lateral gradient adjustment coefficient is set based on Formula (3.71), then:

$$F_{ji,0} = E_{j,0} r_0 \left(\frac{1}{k_{x,0} x_{ji}^2 + k_{y,0} y_{ji}^2} - \frac{1}{r_{\max}^2} \right) \tag{3.72}$$

where, $k_{x,0}$ and $k_{y,0}$ are the longitudinal and lateral gradient adjustment coefficients respectively. Obviously, parameters $k_{x,0}$ and $k_{y,0}$ have a direct impact on the distribution of DSF, and play a vital role in accurately quantifying driving risks.

If the vehicle travels along x in the positive direction, with the Doppler frequency shift principle shown in Formula (3.71), $k_{x,0}$ and $k_{y,0}$ are defined as follows:

$$k_{x,0} = \frac{\left[v_{\max} - v_j \cdot \text{sign}(x_i - x_j) \cdot \text{sign}_0(v_j - v_i)\right]^2}{\left[v_{\max} + v_i \cdot \text{sign}(x_i - x_j) \cdot \text{sign}_0(v_j - v_i)\right]^2} \tag{3.73}$$

$$k_{y,0} = 1 \tag{3.74}$$

$$\mathrm{sign}(x) = \begin{cases} 1 & x \geq 0 \\ -1 & x < 0 \end{cases} \tag{3.75}$$

$$\mathrm{sign}_0(x) = \begin{cases} 1 & x > 0 \\ 0 & x = 0 \\ -1 & x < 0 \end{cases} \tag{3.76}$$

where, x_j and v_j represent the coordinate and speed of the vehicle j in the direction x, respectively; i represents any point in the road environment or other vehicles. When it represents a fixed point in the environment, x_i is the fixed coordinate of the point i in the direction x, and $v_i = 0$. On the contrary, when i represents the vehicle, x_i is the coordinate of the vehicle in the direction x and v_i is the moving speed; and $v_{\max}$ is the spread speed of risk. The risk of moving objects to the outside world is generally related to their attributes.

From the perspective of vehicle, the root cause of traffic disturbance lies in the driver's response to the stimulation of changes in the external traffic environment. Usually, the driver assesses the current safety situation according to the road environmental conditions and the movement state of other vehicles during driving. Generally speaking, drivers are most sensitive to the relative distance and speed of other vehicles. Therefore, based on Formula (3.72) and from the perspective of the driver, the vehicle risk in the traffic environment is described by the DSF as:

$$F_{ji,p} = E_{j,p} r_0 \left(\frac{1}{k_{x,p} x_{ji}^2 + k_{y,p} y_{ji}^2} - \frac{1}{r_{\max}^2} \right) \tag{3.77}$$

$$E_{j,p} = \frac{1}{2} m_j \left(\vec{v}_j - \vec{v}_i \right)^2 \tag{3.78}$$

$$k_{x,p} = k_{x,0} \tag{3.79}$$

$$k_{y,p} = k_{y,0} \tag{3.80}$$

where, $E_{j,p}$ represents the increment of disturbance risk source of vehicle; $\vec{v}_j$ and $\vec{v}_i$ represent the vector speed of the vehicles j and i; $k_{x,p}$ and $k_{y,p}$ are the longitudinal and lateral gradient adjustment coefficients respectively.

In the car-following scenario, if i is the front vehicle and j is the rear vehicle, when $v_j > v_i$, observe the vehicle j from the vehicle i, and will find that the risk caused by the vehicle j to the front is greater than that to the rear, reflecting that the front vehicle i has the risk of being hit by the vehicle j; On the contrary, if i is the rear vehicle and j is the front vehicle, when $v_j < v_i$, observe the vehicle j from the vehicle i, and will be found that the risk caused by the vehicle j to the rear is greater than that to the front [132].

Similarly, the impact of a vehicle j in the traffic environment on the whole traffic flow can be analyzed from the perspective of traffic flow. If the average speed of traffic flow is expressed as $\overline{v}$, Formula (3.73) can be written as:

$$k_{x,p} = \frac{\left[v_{j,\max} - v_j \cdot \operatorname{sign}\left(x_i - x_j\right) \cdot \operatorname{sign}_0\left(v_j - \overline{v}\right)\right]^2}{\left[v_{j,\max} + \overline{v} \cdot \operatorname{sign}\left(x_i - x_j\right) \cdot \operatorname{sign}_0\left(v_j - \overline{v}\right)\right]^2} \tag{3.81}$$

It can be seen from Formula (3.81) that when the vehicle speed $v_j > \overline{v}$, the risk to the front of the vehicle j is greater than its impact to the rear; on the contrary, when the vehicle speed $v_j < \overline{v}$, the risk caused by the vehicle j to the rear is greater than its impact to the front; when the vehicle speed $v_j = \overline{v}$, the vehicle j will not disturb the traffic flow.

By observing vehicle j from any vehicle i in the traffic environment, it can be found that the risks caused by vehicle j are significantly different. To more intuitively analyze the impact of the vehicle on other vehicles in the environment, the traffic flow is analyzed here. When the average speed of the traffic flow $\overline{v} = 25\,\text{m/s}$ and the speed of vehicle j is $v_j = 30\,\text{m/s}$, the DSF generated by the vehicle in the traffic environment is shown in Fig. 3.61, and the field force gradient is shown in Fig. 3.62.

If the speed of vehicle j is $v_j = 20\,\text{m/s}$ and other conditions remain unchanged, the DSF generated by the vehicle in the traffic environment is shown in Fig. 3.63, and the field force gradient is shown in Fig. 3.64. It can be seen from the figure that when $v_j = 30\,\text{m/s}$, since the speed of the vehicle j is larger than the average speed of traffic flow, the disturbance made by the vehicle in the positive direction of its driving is greater than that in the negative direction; when $v_j = 20\,\text{m/s}$, as the speed of the vehicle j is less than the average speed of the traffic flow, the disturbance caused by the vehicle to its rear is greater than the driving direction.

There are certain limitations when observing a vehicle from the perspective of traffic management or vehicle driving alone (such as risk quantification for two relatively stationary vehicles and a single stationary vehicle). Therefore, it is necessary to observe the risk from one angle and consider the influencing factors from another angle. In other words, based on the vehicle driving perspective, the driving risk shall

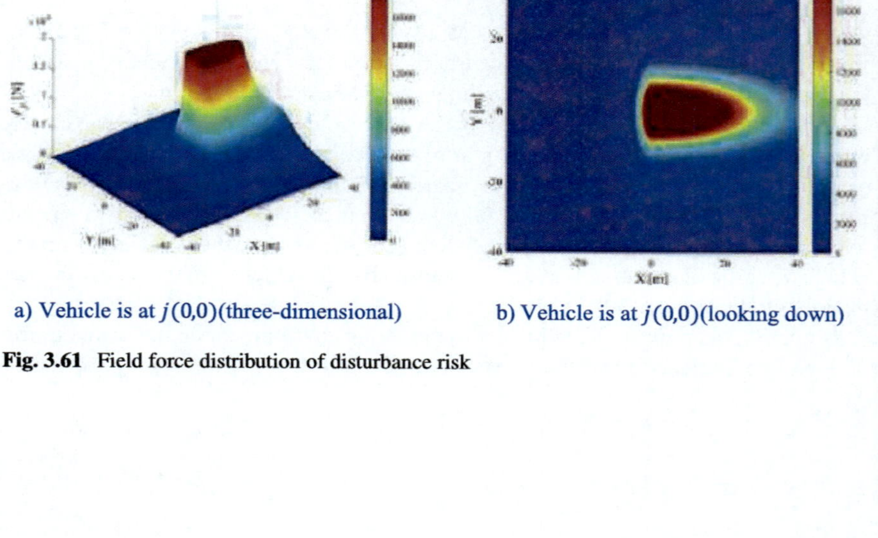

a) Vehicle is at j(0,0)(three-dimensional) b) Vehicle is at j(0,0)(looking down)

Fig. 3.61 Field force distribution of disturbance risk

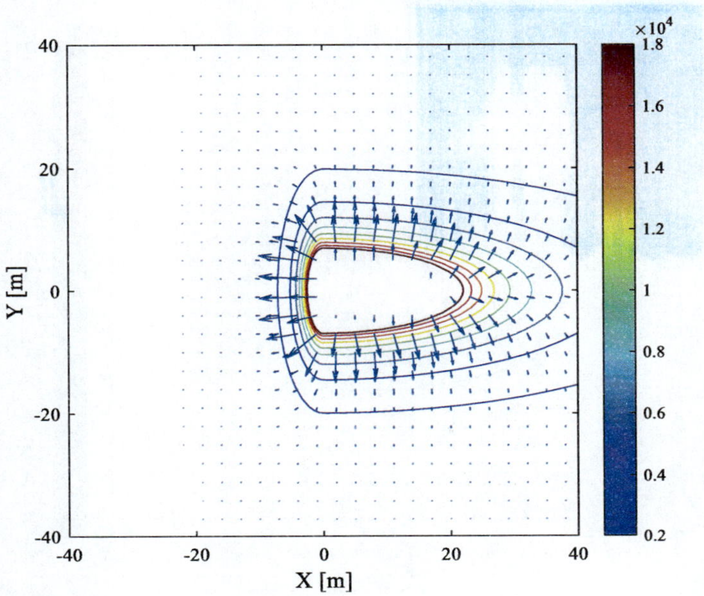

Fig. 3.62 Field force gradient distribution of disturbance risk

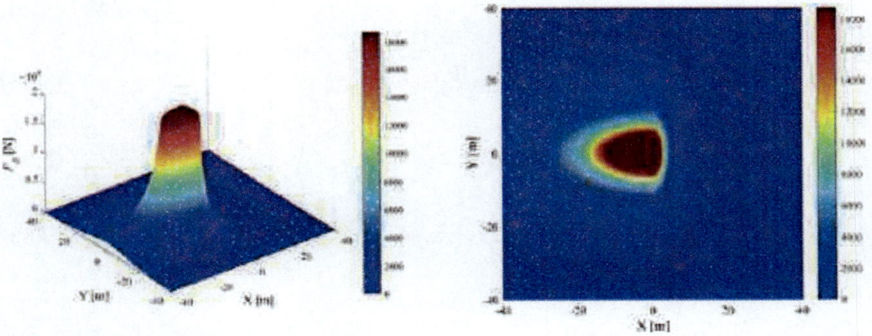

a) Vehicle is at $j(0,0)$(three-dimensional) b) Vehicle is at $j(0,0)$(looking down)

Fig. 3.63 Field force distribution of disturbance risk

be quantitatively described from the perspective of traffic management, so as to form a unified form of comprehensive driving risk combining the vehicle's own attributes and the interaction in the traffic flow [133]:

$$F_{ji} = \left(E_{j,0} + E_{j,p}\right) r_0 \left(\frac{1}{k_{x,0} k_{x,p} x_{ji}^2 + k_{y,0} k_{y,p} y_{ji}^2} - \frac{1}{r_{\max}^2} \right) \quad (3.82)$$

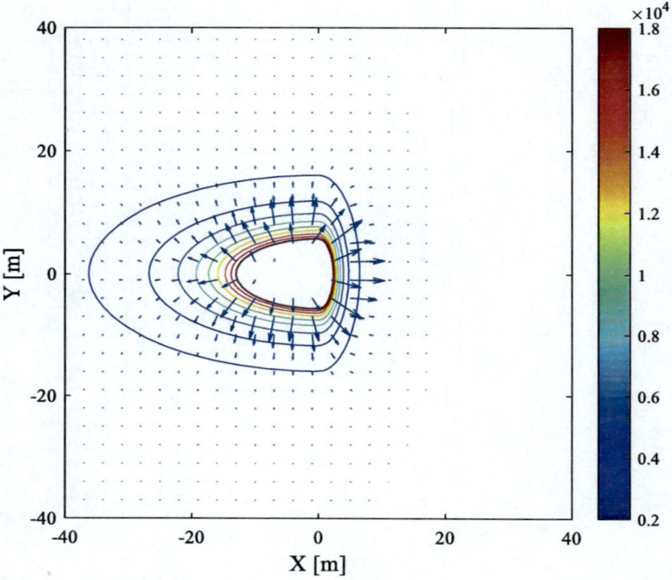

Fig. 3.64 Field force gradient distribution of disturbance risk

If:

$$E_j = E_{j,0} + E_{j,p} \tag{3.83}$$

$$k_x = k_{x,0} k_{x,p} \tag{3.84}$$

$$k_y = k_{y,0} k_{y,p} \tag{3.85}$$

Then Formula (3.82) can be written as:

$$F_{ji} = E_j r_0 \left(\frac{1}{k_x x_{ji}^2 + k_y y_{ji}^2} - \frac{1}{r_{\max}^2} \right) \tag{3.86}$$

where, E_j is the risk source of the vehicle j during driving; k_x and k_y are the change trends of the risk in the lateral/longitudinal directions generated by the vehicle j during driving. It should be emphasized that when i represents a specific target (such as vehicle i, cyclist i, pedestrian i, and stationary obstacle i), F_{ji} represents the DSF force corresponding to the risk of vehicle j to i.

(2) Risk Constraint Model of Road Traffic Facilities

Serving for driving safety improvement, road traffic facilities is divided into four categories in this study. The first category is the facilities that isolate vehicles with possibility of trajectory crossing in space, such as overpasses, pedestrian overpasses and pedestrian underpasses; The second category is the facilities that prevent the intersection of vehicle trajectories by forcing one party to stop, such as traffic lights, and road traffic signs that prohibit or restrict vehicle behavior (such as stop and yielding signs); The third category is facilities that reduce the speed of vehicle and driving risk by warning or increasing the right of way of the other vehicle, such as crosswalk lines and other directional markings, and road traffic warning signs. The fourth category is the facilities that control the driving risk by separating the traffic flow in the same direction or in different directions and guiding the moving direction of vehicles, such as road traffic markings [132].

1) Traffic Lights

The national standard *Specifications for Road Traffic Signal Setting and Installation* (GB 14886-2016) stipulates that the yellow light duration of the signal light should be 3–5 s. If the vehicle is far enough from the intersection when the yellow light is on, it has at least 3 s to slow down to the stop line. Then, vehicles with enough time for slowing down will be subject to the equivalent constraint resistance F_{sj} from the red light of the traffic signal:

$$F_{sj} = -R_{sj} r_0^2 \left(\frac{1}{k_{sx} x_{sj}^2} - \frac{1}{r_{\max}^2} \right) \tag{3.87}$$

$$R_{sj} = m_j a_{jb,\max} \tag{3.88}$$

where, $a_{jb,\max}$ represents the maximum deceleration of the vehicle during braking; $\overline{v}_j$ is the average speed of the vehicle during driving; x_{sj} indicates the longitudinal distance between the vehicle and the traffic lights; k_{sx} is the longitudinal gradient adjustment coefficient, which is related to the motion state of the vehicle; and the negative sign indicates that its direction is opposite to the direction of speed v_j.

Therefore, the equivalent constraint resistance field generated by the red light of the traffic signal is shown in Fig. 3.65. It can be seen from the Figure that if the driving speed is constant, the closer the vehicle is to the traffic lights, the larger the equivalent constraint force is.

2) Crosswalk Line

Pedestrian crossing facilities, including pedestrian overpasses, underpasses, crosswalk lines, etc. The first two directly isolate the trajectories of pedestrians and vehicles in space, and obviously reduce the driving risk and enhance the safety of traffic environment. For the crosswalk line, the reasonable setting of the mark can reduce the sensed risk as well as the actual risk of pedestrians when crossing the road. Therefore, the crosswalk line is regarded as a facility to restrict the driving speed of

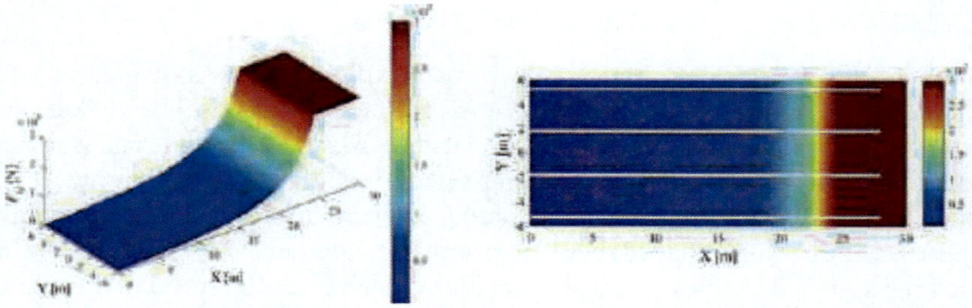

a) Red light is at $x = 30\text{m}$(three-dimensional) b) Red light is at $x = 30\text{m}$ (looking down)

Fig. 3.65 Equivalent constraint resistance field generated by red light

vehicles and reduce the driving risk. Then, for moving vehicles, the closer they are to the crosswalk line before passing through, the greater the traffic constraint resistance they face. Therefore, it is necessary to reduce the speed for driving safety. According to China's traffic regulations, the speed of vehicle passing on the crosswalk line shall not exceed 30 km/h. Therefore, the equivalent constraint resistance F_{cj} generated by the crosswalk x_{cj} before passing through of the vehicle j is defined as:

$$F_{cj} = -R_{cj} r_0^2 \left(\frac{1}{k_{cx} x_{cj}^2} - \frac{1}{r_{\max}^2} \right) \tag{3.89}$$

$$R_{cj} = G_j \frac{v_j}{v_c} \tag{3.90}$$

$$G_j = m_j a_{ja} \frac{v_{j,der}}{v_c} \tag{3.91}$$

where, m_j is the mass of the vehicle j; a_{ja} is the maximum acceleration expected by the driver during starting; v_j is the speed of the vehicle j during driving; $v_c = 30\,\text{km/h}$ refers to the maximum speed limit of vehicle on the crosswalk line; x_{cj} is the longitudinal and lateral distance between the vehicle and the crosswalk; k_{cx} is the longitudinal gradient adjustment coefficient, which is related to the driver's behavior. $\overline{E}_{cj}$ indicates the increment of violation risk when the vehicle speed is higher than the road speed limit. When the vehicle speed is higher than the crosswalk speed limit, $\overline{E}_{cj} < 0$; The negative sign indicates that its direction is opposite to the direction of speed $\overline{v}_j$; On the contrary, when the speed of the vehicle is less than or equal to the speed limit of the crosswalk, then $\overline{E}_{cj} = 0$.

3) Road Traffic Speed Limit Sign

According to the relevant provisions of the national standard *Road Traffic Signs and Markings Part 5: Speed Limits* (GB 5768.5-2017) and the *Law of the People's Republic of China on Road Traffic Safety*, the road traffic speed limits in China mainly include posted speed limits and legal speed limits. Road traffic speed limit signs include maximum speed limit and minimum speed limit. In general, the speed limit of the road is related to the functional type and geometric line type of the road, and is mainly expressed as the maximum speed limit; But on the highway, in addition to the maximum speed limit, there is also the minimum speed limit. Setting up road traffic speed limit signs with national standards and regulations to make the driving speed of vehicles more consistent with the average speed of traffic flow in the road traffic environment is a simple and effective means for driving safety.

The road speed limit sign can restrict the driving speed of the vehicle, and the risk to the outside world in a short time is less than the two violations of running the red light and not slowing down in front of the crosswalk. When drivers find their overspeed, they can quickly reduce violation risk by dynamically adjusting speed. Compared with traffic lights and crosswalk lines, road speed limit signs are quite different. Since the road speed limit rule is effective everywhere on the road, it is only related to speed. If v_l is used to represent the speed limit value of the lane, where $v_{l,m}$ represents the minimum speed limit of the lane and $v_{l,h}$ represents the maximum speed limit, the equivalent constraint resistance F_{lj} of the vehicle j generated by the road traffic speed limit sign is:

$$F_{lj} = -R_{lj} = -G_j \frac{v_j}{v_l} sign\left(v_j - v_l\right) \tag{3.92}$$

$$G_j = m_j a_{ja} \frac{v_{j,der}}{v_l} \tag{3.93}$$

where, $v_{j,der}$ is the expected speed of the driver of vehicle j; and v_l is the speed limit of the road.

4) Road Traffic Marking

Road traffic markings include white dotted line/solid line, double white dotted line/ solid line, yellow dotted line/solid line, double yellow dotted line/solid line, etc., which act to restrain the driving risk of vehicle with different indications. Taking the common lane as an example, the lane width has an important impact on driving safety. In road traffic engineering, the appropriate lane width is usually designed according to the road width to improve the traffic efficiency as much as possible on the premise of ensuring the road traffic safety. The lane lines influence the driver's behaviour in a lateral direction and thus influence the driving process of the vehicle. The lane lines do not directly affect driving risk of vehicles, and crossing it will not directly lead to traffic accidents. Road traffic markings are generally regarded as a virtual constraint force (such as lane keeping) on the lateral direction of vehicles during driving. The constraint force of road traffic markings on vehicles is defined as:

$$F_{mj} = k_m E_{j,0} r_0 \left(\frac{1}{k_{my} y_{mj}^2} - \frac{1}{r_{\max,m}^2} \right) \tag{3.94}$$

where, k_m is a constant coefficient, which can be calibrated by the righting lateral acceleration of the vehicle at different speeds; k_{my} is the lateral gradient adjustment coefficient, which is related to the driver behavior; $r_{\text{max,m}}$ indicates the influence range of road traffic markings, taken as $r_{\text{max,m}} = 0.5l_w$, which means that vehicles are not affected by traffic markings when driving on the lane centerline.

(3) Risk Quantification of Driver Behavior

1) Driver's Normal Driving Behavior

In the normal driving process, the interaction between the driver and the environment and vehicle mainly reflects the driver's perception of the environment and dynamics feedback of manipulating the vehicle. Within it, the perception is closely related to its visual properties. The visual recognition ability of driver's naked eyes is greatly affected in driving. With the increase of vehicle speed, the field angle becomes narrower. Many researchers regard driver's field of vision as an ellipse. The dynamics feedback is related to physiological constraints. In the process of driving, the constraints by road traffic markings and by vehicle mechanical structure in the lateral direction and movement lead to notably different risks in the road traffic environment in the longitudinal and lateral directions [132]. Without considering the driver and environmental factors, the risk posed by the vehicle to the outside world can be regarded as isotropic. According to the analysis of driver's normal driving behavior, the vehicle-related risk distribution in the traffic environment can be represented by the ellipse as shown in Fig. 3.66. A_1A_2 and B_1B_2 are the major axis and minor axis of the ellipse respectively, and $A_1A_2 = 2A_1j = 2jA_2 = 2A_j$, $B_1B_2 = 2B_1j = 2jB_2 = 2B_j$. And the ellipse shown in Fig. 3.66 is a contour line of the risk field generated by the vehicle in the environment.

The driver always abides by the rules and ensures safe driving as much as possible, and usually maintains a certain THW in driving. Meanwhile, the traffic rules stipulate that the vehicle is not allowed to change lanes continuously. Combined with the geometric dimension of the vehicle, the lengths of the elliptical semi-major axis and semi-minor axis in Fig. 3.66 are set as follows:

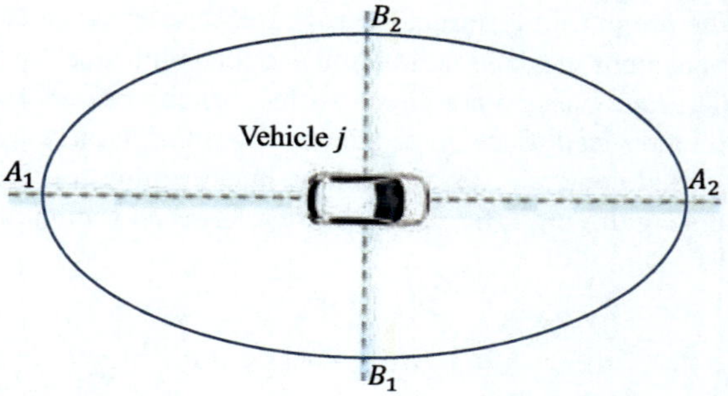

Fig. 3.66 Elliptical constraint of road traffic markings on driving risk

$$A_j = r_{\max} + l_1 \tag{3.95}$$

$$B_j = l_w + l_2 + l_{cj} \tag{3.96}$$

where, A_j is the semi-major axis of ellipse; l_1 is half the length of the vehicle; B_j is the semi-minor axis of ellipse; l_w is twice the lane width (usually taken as $l_w = 3.5$ m); l_2 is half the width of the vehicle; l_{cj} is the distance between the vehicle and the centerline of the lane. It is worth noting that the long axis of the ellipse is a function related to the vehicle speed. The lower the vehicle speed is, the shorter the long axis is. Therefore, in order to keep the length of the long axis more than the short axis, $r_0 \geq l_w$ is specified.

Due to the lateral restraint of lane line, there are obvious differences in the longitudinal and lateral risk distribution during vehicle driving. The essence of the ellipse shown in Fig. 3.66 is that after considering the safe time distance and traffic flow speed in the longitudinal direction and the influence of lane constraints in the lateral direction, the isotropic circular distribution is compressed into an elliptical risk distribution with dynamic change of long minor axis, as shown in Fig. 3.67. When compressed horizontally, the orange circular contour line is compressed into a blue elliptical contour line. Although $B_1' B_2'$ is shortened to $B_1 B_2$, the two field force contours represent the same risk value. Therefore, the risks generated by vehicles are distributed according to the blue contour:

$$k_{x,d} x_{ji}^2 + k_{y,d} y_{ji}^2 = r_{ji'}^2 \tag{3.97}$$

where, $k_{x,d}$ and $k_{y,d}$ are the gradient adjustment coefficients in the longitudinal and lateral directions respectively. According to the contour variation shown in Fig. 3.67, it can be gained as:

$$k_{x,d} = 1 \tag{3.98}$$

$$k_{y,d} = \frac{A_j^2}{B_j^2} \tag{3.99}$$

where, A_j and B_j are the semi-major axis length and semi-minor axis length of ellipse respectively.

According to the nature of ellipse:

$$x_{ji} = \frac{r_{ji'}}{\sqrt{k_{x,d}}} \cos t \tag{3.100}$$

$$y_{ji} = \frac{r_{ji'}}{\sqrt{k_{y,d}}} \sin t \tag{3.101}$$

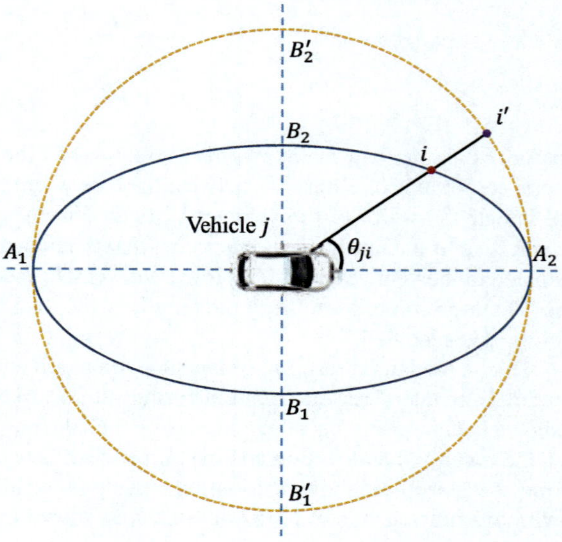

Fig. 3.67 Ellipse constraint principle

$$\frac{y_{ji}}{x_{ji}} = \tan\theta_{ji} = \sqrt{\frac{k_{x,d}}{k_{y,d}}}\tan t \tag{3.102}$$

where, t is a parameter and $t \in [0, 2\pi]$.

Then the distance r_{ji} from the center of the vehicle j to any point i on the ellipse is:

$$r_{ji} = \frac{r_{ji'}\cos t}{\sqrt{k_{x,d}}\cos\theta_{ji}} \tag{3.103}$$

where, θ_{ji} is the included angle between the line connecting the vehicle j and the point i and the speed direction of vehicle j.

According to Formula (3.102):

$$\frac{k_{y,d}}{k_{x,d}}\tan^2\theta_{ji} = \tan^2 t \tag{3.104}$$

Namely:

$$\frac{k_{y,d}}{k_{x,d}}\sin^2\theta_{ji} + \cos^2\theta_{ji} = \frac{\cos^2\theta_{ji}}{\cos^2 t} \tag{3.105}$$

Combining Formulas (3.103) and (3.105), it can be obtained:

$$r_{ji} = \frac{r_{ji'}}{\sqrt{k_{x,d}\cos^2\theta_{ji} + k_{y,d}\sin^2\theta_{ji}}} \tag{3.106}$$

Combining Formulas (3.70), (3.97) and (3.106), it can be obtained:

$$F_{ji,0} = E_{j,0}r_0\left[\frac{1}{r_{ji}^2\left(k_{x,d}\cos^2\theta_{ji} + k_{y,d}\sin^2\theta_{ji}\right)} - \frac{1}{r_{\max}^2}\right] \tag{3.107}$$

Write it as rectangular coordinate system to get:

$$F_{ji,0} = \eta E_{j,0}r_0\left(\frac{1}{k_{x,d}x_{ji}^2 + k_{y,d}y_{ji}^2} - \frac{1}{r_{\max}^2}\right) \tag{3.108}$$

Combining Formulas (3.86) and (3.108), the DSF force in view of the impact of traffic management, vehicle driving and lane line on driving risk can be expressed as:

$$F_{ji} = \left(E_{j,0} + E_{j,p}\right)r_0\left(\frac{1}{k_{x,0}k_{x,p}k_{x,d}x_{ji}^2 + k_{y,0}k_{y,p}k_{y,d}y_{ji}^2} - \frac{1}{r_{\max}^2}\right) \tag{3.109}$$

Therefore, if the driver strictly abides by the traffic rules, the DSF generated by the vehicle in the road traffic environment is shown in Fig. 3.68, the gradient distribution is shown in Fig. 3.69, with the arrow indicating the gradient change direction of the DSF. It can be found from the Figure that under the constraints of road traffic markings, the vertical and horizontal risk distribution is obviously different.

2) Driver Violations

Drivers' violations include speeding, running red light, and changing lanes against regulations, etc. The reasons for these behaviors include driver's misperception of the environment, improper operation of the vehicle, and aggressive driving. As the impact

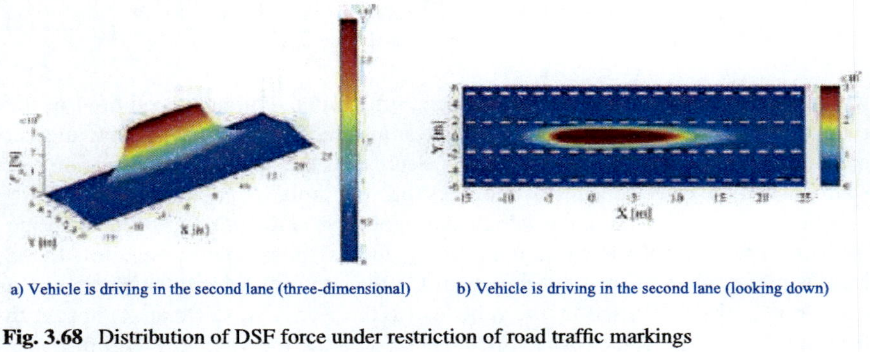

a) Vehicle is driving in the second lane (three-dimensional) b) Vehicle is driving in the second lane (looking down)

Fig. 3.68 Distribution of DSF force under restriction of road traffic markings

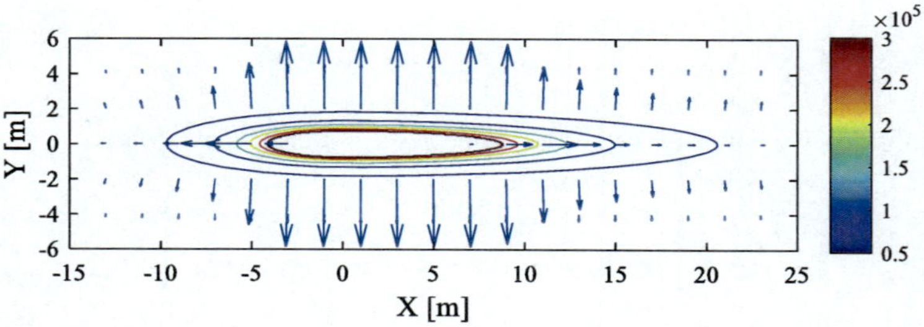

Fig. 3.69 Gradient change of DSF force under restriction of road traffic markings

of road traffic facilities on driving safety is reflected in the constraints on driving risks, driver's violations indicate that the driver intentionally or unintentionally breaks the constraints for some reason. Therefore, it is necessary to analyze the motion state of the vehicle to judge whether the driver's behavior has a tendency of rule violation. Here, four kinds of road traffic facilities can be divided in terms of function: ① longitudinal speed limit, including traffic signal lights, crosswalk lines and road traffic speed limit signs; ② lateral position restrictions, mainly referring to road traffic markings.

i. Longitudinal Speed Limit

The risk caused by the vehicle disturbance in the traffic environment is related to the speed of vehicle and the average speed of traffic flow. When the vehicle speed is over the maximum limit or lower than the minimum limit, the risk for the road traffic environment due to violation can be expressed by the following formula:

$$F_{jl} = E_{j,l} r_0 \left(\frac{1}{k_{xl} x_{ji}^2 + k_{yl} y_{ji}^2} - \frac{1}{r_{\max}^2} \right) \tag{3.110}$$

where,

$$E_{j,l} = \begin{cases} \frac{1}{2} m_j \left(v_j - v_l\right)^2 & v_j \in \left[0, v_{l,m}\right] \cup \left[v_{l,h}, v_{j,max}\right] \\ 0 & v_j \in \left(v_{l,m}, v_{l,h}\right) \end{cases} \tag{3.111}$$

where, $v_{l,m}$ and $v_{l,h}$ are the minimum speed limit and maximum speed limit of the road respectively; k_{xl} and k_{yl} are gradient adjustment coefficients in longitudinal and lateral directions respectively; and E_{jl} represents the increment of violation risk source when the drive speed is over or less than the limit.

Due to driver's violation, the vehicle does not drive at the limited speed, resulting in risks. When the vehicle speed is $v_j > v_{l,h}$, the risk posed by the vehicle j to the front is greater than its impact to the rear; On the contrary, when the vehicle speed is $v_j < v_{l,m}$, the risk posed by the vehicle j to the rear is greater than its impact to the front; When the vehicle speed is $v_j \in \left(v_{l,m}, v_{l,h}\right)$, the vehicle j will not pose violation risk to its road traffic environment. Therefore, for vertical violations, the main risks reflect the comprehensive risks of the vehicle to the outside world, and the DSF force F_{ji} corresponding to the driving risks is:

$$F_{ji} = \left(E_{j,0} + E_{j,p} + E_{j,l}\right) r_0 \left(\frac{1}{k_{x,0}k_{x,p}x_{ji}^2 + k_{y,0}k_{y,p}y_{ji}^2} - \frac{1}{r_{\max}^2} \right) \tag{3.112}$$

ii. Lateral Position Limit

The function of road traffic markings is to restrict and reduce the impact of vehicles on the traffic environment by restricting the movement of vehicles in the lateral direction. When the vehicle runs stably in the lane formed by lane line 2 and lane line 3, as shown in Fig. 3.70, the elliptical restraint effect of the two lane lines on the vehicle follows Formula (3.95) and Formula (3.96) respectively in the longitudinal and lateral directions, and the distribution of DSF to the outside is shown in Fig. 3.71. When the vehicle runs on a road marking line, e.g., vehicle j on lane line 3 as shown in Formula 3.70 for a long time, it is believed that lane line 3 does not restrict the vehicle j, and lane line 2 and lane line 4 generate the elliptical restraint of road traffic markings on the vehicle. Vertically, the semi-major axis of the ellipse still follows Formula (3.96); and horizontally, the semi-minor axis of the ellipse follows the following formula:

$$B_j = l_w + l_2 + l_{jc} \tag{3.113}$$

where, l_{jc} refers to the distance from the vehicle j to the centerline.

Therefore, assuming the mass of the vehicle j is $m_j = 1500\,\text{kg}$ and the speed is $v_j = 20\,\text{m/s}$, the safe THW is taken as $THW_j = 1.5\,\text{s}$. When the driver rides the vehicle j illegally on lane line 3, the DSF generated by the vehicle in the road traffic environment is shown in Fig. 3.72. Compared with Fig. 3.68, it can be found that illegal line riding poses greater risk to the traffic environment than normal driving.

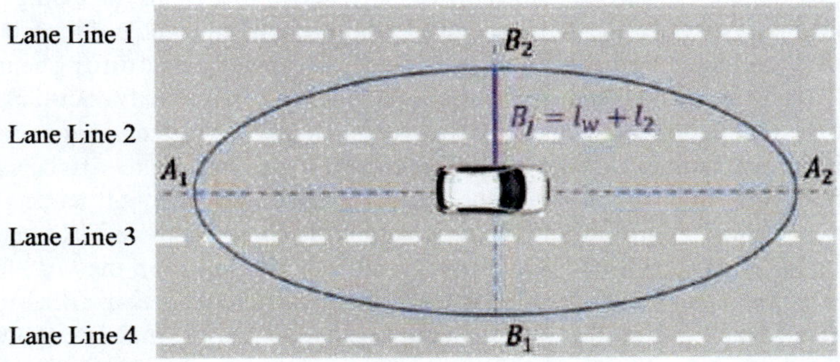

Fig. 3.70 Vehicle j normal driving in the lane

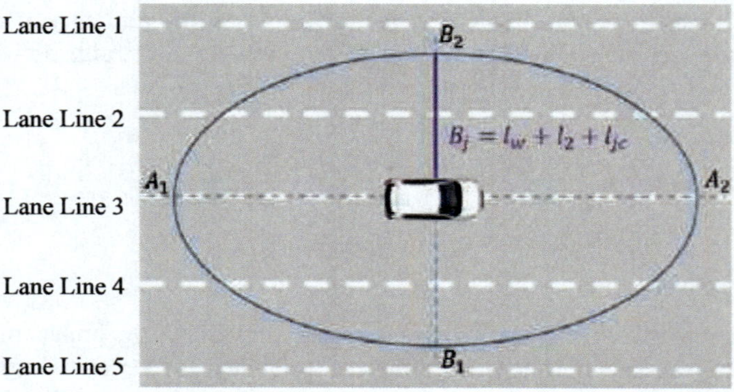

Fig. 3.71 Vehicle j driving on lane line 3

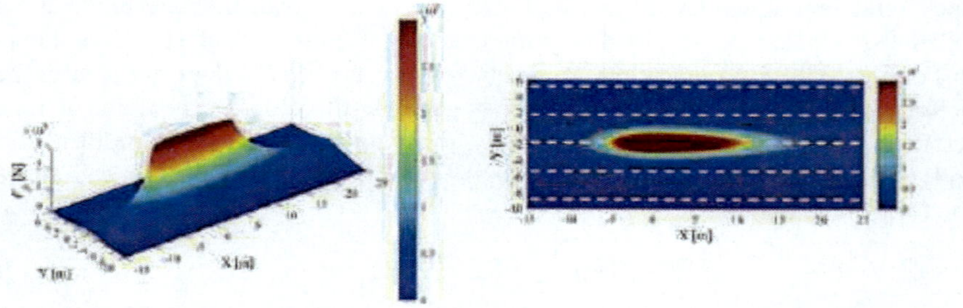

a) Vehicle j drives at j (0, −1.75) of lane line 3 (looking down) b) Vehicle j drives at j (0, −1.75) of lane line 3 (three-dimensional)

Fig. 3.72 Distribution of DSF force generated by illegal ride on line of vehicle

(4) Comprehensive Risk Quantification for DVE

In the complex DVE environment, driving risks constantly appear in the road traffic. The generation of driving risk is related to such factors as vehicle kinematics factors including vehicle's speed and traffic flow velocity, traffic rules factors including traffic signal lights, crosswalk lines, road traffic speed limit signs and road traffic markings, human factors including driver's normal driving behavior and illegal driving behavior, as well as environmental factors including road boundary, pavement conditions, lane curvature, road slope and environmental visibility. The impact of various factors on driving risk is mainly reflected in the impact of the risk source itself and the gradient change trend of the risk in the vertical or horizontal direction starting from the risk source. To uniformly characterise the DVE comprehensive risk, a unified model of DSF is established. The driving risk is quantified from the vehicle and traffic perspectives through field force description and potential energy description. Vehicle perspective refers to taking the vehicle as the center, and mainly considering the relative attributes (including relative position, speed, etc.) of surrounding traffic participants and the vehicle in the micro traffic system. And traffic perspective refers to quantifying the absolute attributes (location, speed, etc.) of traffic participants from the perspective of macro traffic flow.

1) DSF Force Description

Based on the analysis of risk sources leading to generation and change, vertical gradient and horizontal gradient of driving risks posed by various main factors, the comprehensive driving risk in the complex DVE environment can be characterized by field force [33]. It takes the vehicle as the core, and forms the overall framework of the risk posed by the vehicle to the outside world, as shown in the following formula:

$$F_{ji} = E_j r_0 \left(\frac{1}{k_x x_{ji}^2 + k_y y_{ji}^2} - \frac{1}{r_{\max}^2} \right) \tag{3.114}$$

where,

$$E_j = \sum E_{j,\text{fac}} \tag{3.115}$$

$$k_x = \prod k_{x,\text{fac}} \tag{3.116}$$

$$k_y = \prod k_{y,\text{fac}} \tag{3.117}$$

In Formula (3.115), $E_{j,\text{fac}}$ includes the kinetic energy of the vehicle $E_{j,0}$ and the relative kinetic energy $E_{j,p}$, $E_{j,\varphi a}$,$E_{x,c}$, $E_{j,s}$ and $E_{j,l}$ among traffic flow, road adhesion coefficient, road curvature, slope and road speed limit rules. When any of these factors is ignored, the corresponding relative kinetic energy value is 0. In Formulas (3.116) and (3.117), any parameter corresponds to the impact of a specific influencing factor on the risk of the vehicle in the longitudinal or lateral direction. When any of these factors is ignored, the corresponding parameter value is 1.

2) Description of Potential Energy of Driving Risk

When a point in the road environment is affected by multiple vehicles, it will be acted by multiple forces. Since the force is a vector, it is unreasonable to use the resultant force of multiple forces to describe the coupling action of multiple vehicles at this point. For example, when the point is subjected to the force in opposite direction, the vector superposition will result in the offset of the force. Therefore, it has certain limitations in directly using force to describe the impact of multiple vehicles on a certain point in the road environment. Accordingly, from the perspective of traffic, potential energy is further used to describe the interaction of DVE elements in the road traffic environment, to realize the quantitative description of driving risk.

In the DSF, the potential energy obtained by the vehicle at point $P(P \in [r_{\min}, r_{\max}])$ under the influence of the field source is equal to the work done at $r_{ji} \geq r_{\max}$ by moving the unit-mass object m_e from a certain point P to the place where potential energy is 0, with unit of joule (J). According to Formula (3.106), the distance r_{ji} between r_j, corresponding to the medium potential line in DSF generated by the field source, and the vehicle j and any point i is expressed in the following formula:

$$r_j = \sqrt{k_x x_{ji}^2 + k_y y_{ji}^2} = r_{ji}\sqrt{k_x \cos^2\theta_{ji} + k_y \sin^2\theta_{ji}} \tag{3.118}$$

Therefore, the potential energy U_{jp} at the point P is:

$$U_{jp} = \int_{(p)}^{r_{\max}} E_j r_0 \left(\frac{1}{r_j^2} - \frac{1}{r_{\max}^2}\right) dr_j = E_j r_0 \cdot \left(\frac{1}{p} + \frac{p - 2r_{\max}}{r_{\max}^2}\right) \tag{3.119}$$

Therefore, the potential energy U_{ji} of the vehicle i subjected to the field source j is:

$$U_{ji} = E_j r_0 \left(\frac{1}{r_{ji}\sqrt{k_x \cos^2\theta_{ji} + k_y \sin^2\theta_{ji}}} + \frac{r_{ji}\sqrt{k_x \cos^2\theta_{ji} + k_y \sin^2\theta_{ji}} - 2r_{\max}}{r_{\max}^2}\right) \tag{3.120}$$

If Formula (3.120) is written in the form of rectangular coordinate system, there is:

$$U_{ji} = E_j r_0 \left(\frac{1}{\sqrt{k_x x_{ji}^2 + k_x y_{ji}^2}} + \frac{\sqrt{k_x x_{ji}^2 + k_x y_{ji}^2} - 2r_{\max}}{r_{\max}^2}\right) \tag{3.121}$$

Then the gradient change of potential energy is:

$$\nabla U_{ji} = E_j r_0 \left[\frac{-r_{\max}^2 + \left(k_x x_{ji}^2 + k_x y_{ji}^2\right)}{r_{\max}^2 \left(k_x x_{ji}^2 + k_x y_{ji}^2\right)^{\frac{3}{2}}}\right] \left(k_x x_{ji} \cdot \vec{i} + k_y y_{ji} \cdot \vec{j}\right) \tag{3.122}$$

The generation mechanism of driving risk lies in the comprehensive interaction of various factors affecting driving safety. In the integrated DVE environment, the unified model of DSF reflecting the interaction relationship among DVE elements can effectively reveal the generation mechanism of driving risk and quantify the comprehensive driving risk of DVE in the traffic environment.

3. Examples of Application in Typical Scenarios

To verify the effectiveness, accuracy and real-time of the proposed quantitative evaluation of driving risk, firstly, the simulation test is carried out to quantitatively evaluate the risk situation of the traffic system composed of the target vehicle and its adjacent vehicles from the perspective of the driver, to support the decision-making of the target vehicle and verify the effectiveness and accuracy of the driving risk evaluation from the functional point of view. Through the real vehicle test platform, the test of the driving risk assessment is conducted to verify the effectiveness of the algorithm.

(1) Verification of Simulation Test

In the process of simulation test verification, the driving scenario is quickly constructed and generated by MATLAB Driving Scenario Designer. The AV module equipped with radar and camera for detection of surrounding objects is adopted to replace the local network environment, record the position data and vehicle dynamics of all vehicles for algorithm verification.

In the straight-ahead scenario of highway, due to the fast driving speed, once the motion state of other surrounding vehicles changes, an accident will be likely to happen if the vehicle does not respond correctly in a short time. Therefore, in the process of driving, it is necessary to obtain the motion information of surrounding vehicles in real time and make corresponding decisions after accurately evaluating the current driving risk, to effectively ensure driving safety.

The distribution of other vehicles that have a direct impact on the driving safety of the vehicle i, shown in white, in Fig. 3.73. Covering the lane where the vehicle is located and the left and right adjacent lanes, the road environment can be divided into 8 areas as shown below. The change of the movement state of any vehicles other than vehicle i in the area may affect the driving safety of vehicle i. According to the position relationship between vehicles shown in Fig. 3.73, the simulation test is designed for the straight driving scenario of highway, involving the free driving of left front vehicle, front vehicle and right front vehicle; the left and right adjacent vehicles and the left and right rear vehicles can choose to follow and change lanes during driving; the rear vehicle keeps following the vehicle i. In consideration of the driving randomness of all vehicles, it is to evaluate the safety situation of the vehicle i during driving, and facilitate decision-making of the vehicle.

According to the interaction relationship between the ego-vehicle and other vehicles in Fig. 3.73, it can be found that the risks encountered by vehicles during driving mainly come from several types of conflict scenarios as shown in Fig. 3.74: ① Scenario 1, the front vehicle takes emergency braking; ② Scenario 2, the rear

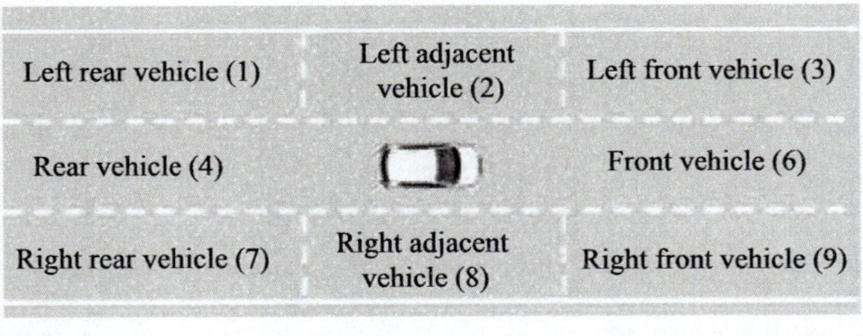

Fig. 3.73 Relationship between any vehicle and other vehicles on straight road

vehicle loses control and hits the front vehicle; ③ Scenario 3, the adjacent vehicle cuts forward; ④ Scenario 4, the adjacent vehicle in the rear urgently requires lane changing. In these scenarios, if the ego-vehicle fails to make a corresponding safety decision without discovering the upcoming risk in time, the accident is bound to occur. Based on these typical conflict scenarios, the proposed unified model of DSF is applied and compared with the current cutting-edge driver models, to functionally verify the accuracy and effectiveness of driving risk assessment approaches proposed in this study.

Among the four conflict scenarios, scenario 1 and scenario 3 are more common. Therefore, these two types of scenarios are mainly designed and compared. Firstly, the white ego-vehicle is controlled according to the DSF-based decision model (DSFDM), and its motion changes are observed in two scenarios; Then the most advanced longitudinal control model (LCM) which can better reflect driver's behavior is applied to simulate under the same condition and observe the motion changes of the ego-vehicle based on the LCM model shown as follows [134]:

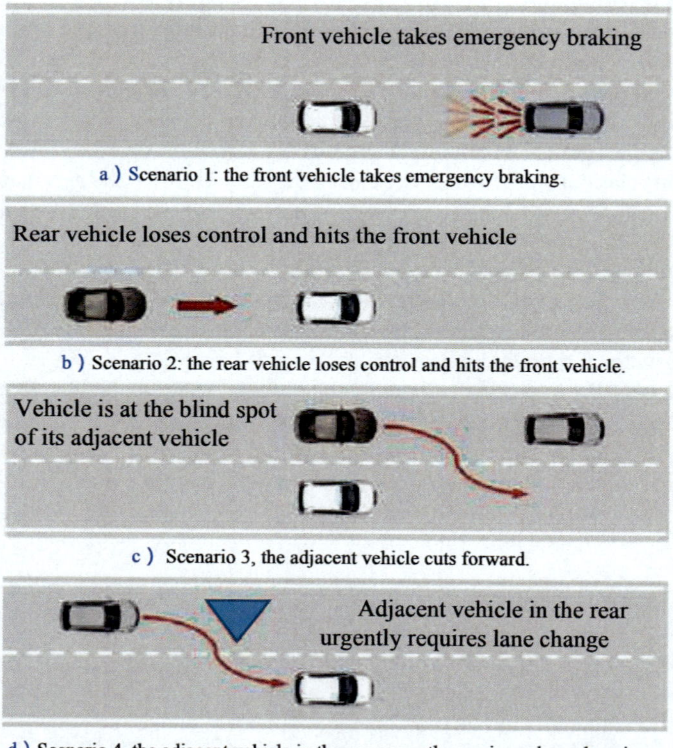

a) Scenario 1: the front vehicle takes emergency braking.

b) Scenario 2: the rear vehicle loses control and hits the front vehicle.

c) Scenario 3, the adjacent vehicle cuts forward.

d) Scenario 4, the adjacent vehicle in the rear urgently requires a lane changing.

Fig. 3.74 Classic conflict scenario

$$\ddot{x}_r(t+\tau_r) = A_r\left[1 - \frac{\dot{x}_r(t)}{v_r} - e^{1-\frac{d_{rf}}{d_{rf}^*}}\right] \tag{3.123}$$

where, r and f represent the rear vehicle and the front vehicle respectively; τ_r is the reaction time of the driver in the rear vehicle; A_r is the maximum acceleration expected by the driver behind when starting; v_r is the expected speed of the driver behind; d_{rf} is the THW between the front and rear vehicles; d_{rf}^* is the desired distance of driver, which is related to the kinematics and dynamics parameters of vehicle, as follows:

$$d_{rf}^* = \frac{\dot{x}_r^2(t)}{2|b_r|} - \frac{\dot{x}_f^2(t)}{2|b_{fe}|} + \dot{x}_r\tau_r + l_f \tag{3.124}$$

where, b_r is the maximum braking deceleration adopted by the driver of the rear vehicle during the emergency braking of the front vehicle, taken in the test as $b_r = 5\,\mathrm{m/s^2}$; b_{fe} is the estimation of emergency braking deceleration of the front vehicle made by the driver of the rear vehicle, taken as $b_{fe} = 4\,\mathrm{m/s^2}$; and l_f refers to the effective length of the front vehicle, taken as $l_f = 5\,\mathrm{m}$.

1) Test 1: Emergency Braking of Front Vehicle

In Scenario 1, both the ego-vehicle and the front vehicle drive at a uniform speed, and the distance between the two vehicles is 100 m. The front vehicle brakes for 3 s with the braking deceleration of $3\,\mathrm{m/s^2}$, $6\,\mathrm{m/s^2}$ and $7.8\,\mathrm{m/s^2}$ respectively. The performance of DSFDM model and LCM model in this scenario is observed.

First, it is necessary to set the braking deceleration of the front vehicle. The test results when the front vehicle decelerates with $3\,\mathrm{m/s^2}$ are shown in Fig. 3.75, where Fig. 3.75a, b and c display displacement–time curve, speed-time curve and THW-time curve respectively. It can be seen from the figure that both DSFDM model and LCM model perform braking during emergency braking of the front vehicle. As shown in Fig. 3.70b, DSFDM model responds quickly according to the speed of the front vehicle, reduces the speed to 21.3 m/s within 8 s, and performs stable following. In contrast, the speed reduction process of the rear vehicle made by the LCM model is smoother, and stable following is achieved within 60 s. It can also be seen from Fig. 3.75a that the distance between the DSFDM model and the vehicle in front in the subsequent driving process is more consistent with the initial time, while the LCM model gradually shortens the following distance. This phenomenon can also be seen from Fig. 3.75c, where DSFDM model still maintains the distance at the initial time when the speed decreases and thus the THW increases from 3.38 to 4.4 s, while LCM model slowly reduces the THW to 1.44 s. It reflects the advantage of LCM model in car-following, which behavior is more in line with the actual driver's behavior. Although DSFDM model and LCM model have different performance in car-following, they can ensure their safe driving of themselves in this scenario.

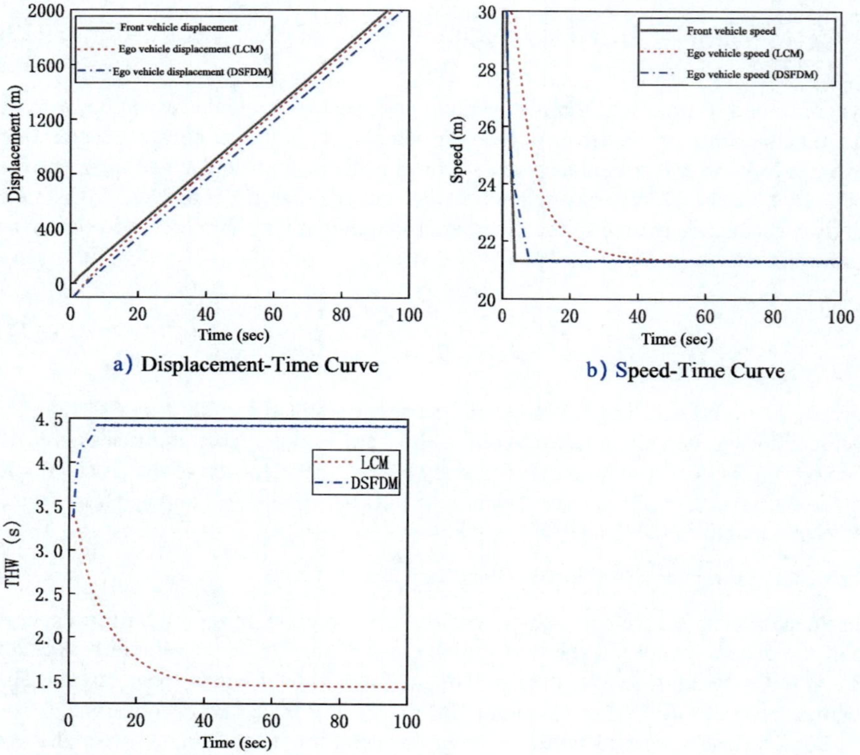

a) Displacement-Time Curve

b) Speed-Time Curve

c) THW-Time Curve

Fig. 3.75 Comparison between DSFDM model and LCM model in scenario 1 (deceleration of front vehicle is 3 m/s^2)

When the front vehicle takes extreme emergency braking at the deceleration of 7.8 m/s^2, the test results are shown in Fig. 3.76, where Fig. 3.76a and b are displacement–time curve and speed-time curve respectively. It can be seen from the figure that both DSFDM model and LCM model perform emergency braking. As shown in Fig. 3.76a, DSFDM model responds quickly according to the speed of the front vehicle, reduces the speed to 7 m/s within 10 s, and then performs stable following after adjustment for about 5 s. Although the LCM model also makes emergency braking, it can be seen from Fig. 3.76b that its deceleration process is abnormal, and the speed even decreases to a negative value. And it can be seen from Fig. 3.76a that in fact, the LCM model has collided with the front vehicle at about 8 s.

In scenario 1, the front vehicle is braked at two different decelerations for 3 s. The simulation results show that when the front vehicle decelerates, the LCM model shows better following performance than the DSFDM model. However, with

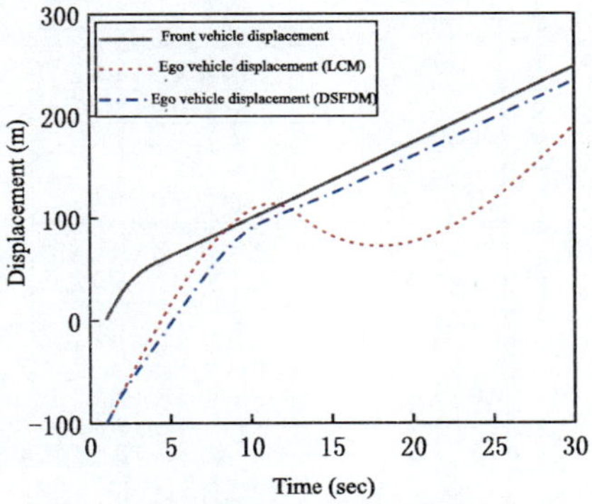

a) Displacement-Time Curve

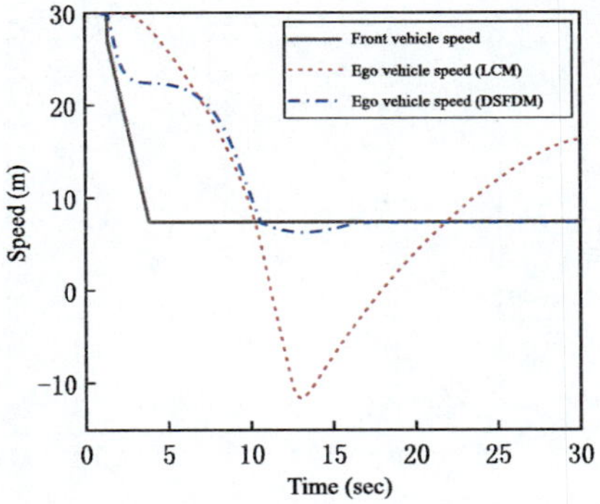

b) Speed-Time Curve

Fig. 3.76 Comparison between DSFDM model and LCM model in scenario 1 (deceleration of front vehicle is 7.8 m/s^2)

the increase of the front vehicle braking deceleration, the DSFDM model gradually shows its excellent performance in ensuring driving safety, which reflects the accuracy of the theory proposed in this study in terms of collision avoidance performance.

2) Test 2: Forward Cut of Adjacent Vehicle

In scenario 3, the adjacent vehicle and the ego-vehicle are set to drive side by side at a uniform speed of 30 m/s, when the adjacent vehicle slowly changes lanes and cuts into the lane of the ego-vehicle. Then the performance of DSFDM model and LCM model in this scenario is observed. The results of test 2 are shown in Fig. 3.77. At the initial time, the relative distance between the adjacent vehicle and the ego-vehicle is quite small. As the adjacent vehicle cuts into the LCM model, there is no response. Therefore, the displacement curve of the vehicle coincides with that of the adjacent vehicle. In this process, the collision may have occurred. However, under the same condition, DSFDM responds quickly, gradually expands the relative distance with adjacent vehicle, and finally follows the vehicle stably, which is displayed in Fig. 3.77b. Firstly, the DSFDM model decelerates to 25.4 m/s within 4 s and then gradually increases the speed to the expected 30 m/s, whereas the LCM model does not take any action, resulting in the coincidence of the speed curve of an adjacent vehicle and the ego-vehicle. According to the THW distribution of the two vehicles in the two processes, as shown in Fig. 3.77c, the DSFDM model can ensure that the THW value of the front and rear vehicles is in a safe range. In contrast, the LCM model does not take any action, so the THW does not change and a collision occurs in the scenario.

The simulation experiments of two conflict scenarios show the collision avoidance effectiveness of DSFDM model in all directions. It functionally verifies the

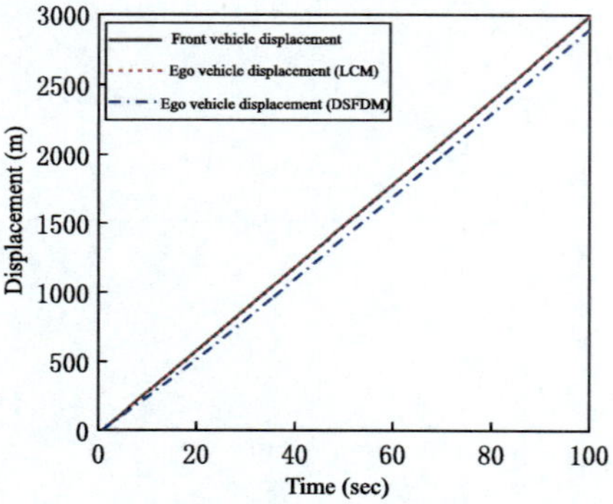

a) Displacement-Time Curve

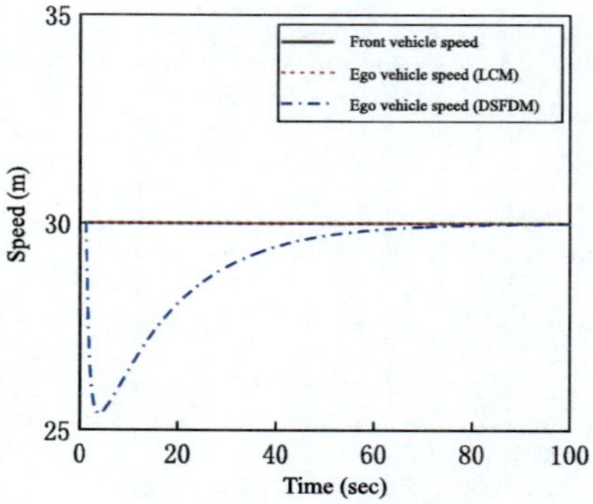

b) Speed-Time Curve

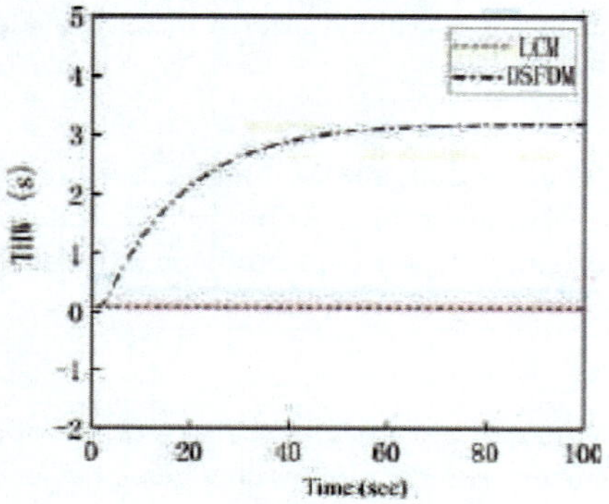

c) THW-Time Curve

Fig. 3.77 Comparison between DSFDM model and LCM model in scenario 3

effectiveness and accuracy of the proposed DVE comprehensive driving risk assessment, and lays a theoretical basis for the development of a unified intelligent driving algorithm.

(2) Real Vehicle Test Verification

To further verify the application effect of the proposed quantitative evaluation of driving risk on real vehicles, Qichen test platform is taken as the test vehicle and Ruicheng test platform is taken as the conflict vehicle in this test (Fig. 3.78a). The rear compartment equipment of two test vehicle platforms is shown in Fig. 3.78b. The test is conducted in a two-way and two-lane road segment with flat pavement and low pedestrian and vehicle traffic flow. Meanwhile, the section is not blocked by tall buildings, which is conducive to Wi-Fi signal transmission and V2V-based real vehicle test. In addition, the road segments cover straightway and intersections, which is fully applicable for the real vehicle test [132].

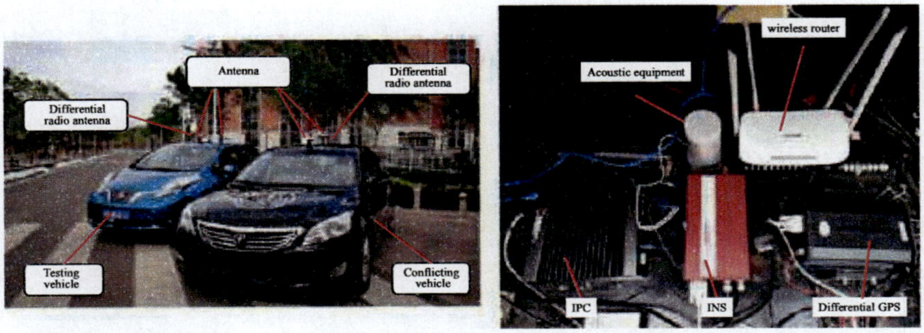

a) Qichen Vehicle Platform and Ruichi Vehicle Platform b) Rear Compartment Equipment of Vehicle Platform

Fig. 3.78 Real vehicle verification platform

Firstly, the surrounding environment is divided into 8 areas centered on the vehicle, as shown in Fig. 3.79: front, left front, right front, left, right, left rear, right rear and rear. According to the driver's visual characteristics, the included angle of the front and rear areas is set as 20°, the included angle of the left and right areas as 30°, and the included angle of other areas is 65°.

The vehicle at the speed of 15 km/h is approaching the conflict vehicle running at a constant speed of 10 km/h. At the initial time, the two vehicles are 20 m apart. The driver operates according to the warning tone in this process. The test process is shown in Fig. 3.80.

To evaluate the safety degree of driving process, D_{SE} is defined as the indicator representing the safety degree. D_{SE} is a quantitative index based on the unified model of DSF. The greater its value is, the higher the degree of danger is. A detailed illustration of this indicator is made in relevant papers [132, 135]. Figure 3.81 shows the speed-time curve, D_{SE}-time curve and brake oil pressure–time curve of the vehicle and conflicting vehicle during the test. In the figure, the situation of $D_{SE} > 0.5$ occurs three times, and generates three warnings at 2.8 s, 7 s and 27 s respectively. When

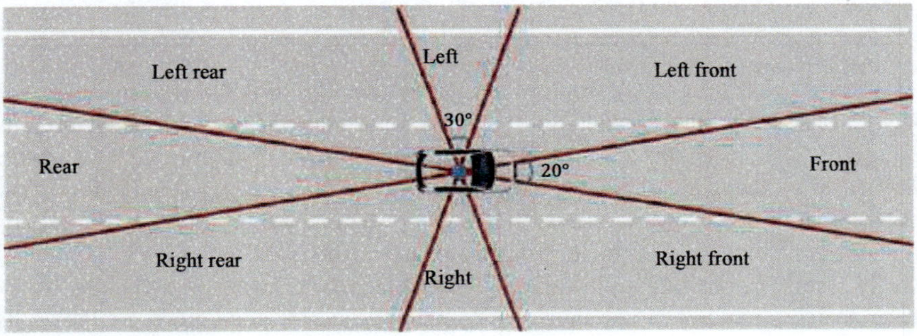

Fig. 3.79 Division of areas around vehicle

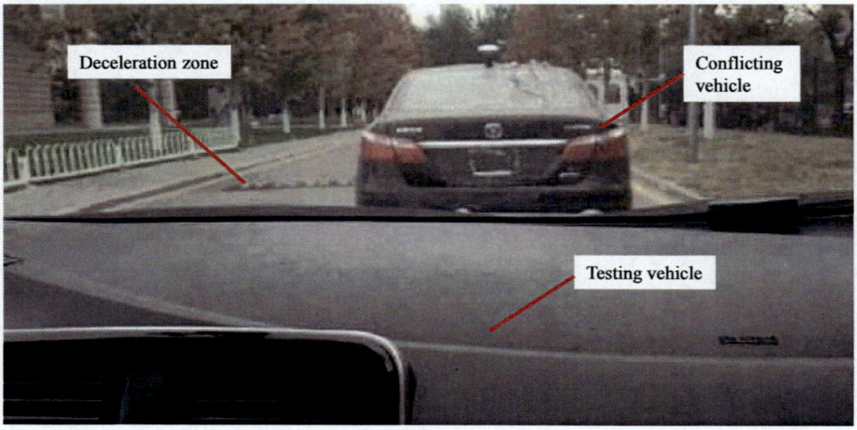

Fig. 3.80 Vehicle test process

$t = 2.8$ s, the driver steps on the brake pedal shortly after hearing the warning tone; When $t = 7$ s, the driver first releases the accelerator pedal after hearing the warning tone, and the vehicle speed decreases slowly due to driving resistance. When $t = 11$ s, the driver recognizes the obvious risk, and takes braking to continuously decrease D_{SE}; Later when $t = 27$ s, the situation occurs similarly to when $t = 7$ s. The reason is that with low vehicle speed and lowly-set threshold of D_{SE}, the driver has enough confidence to ensure driving safety.

The real vehicle test shows that in the V2V environment, the proposed driving risk assessment approach can accurately identify the scale and direction of the risk. When applied to the driving risk early warning system, it can provide the driver with warning information, the risk at the vehicle location, and control suggestions for driving safety assurance.

3.2.4 Analysis of Driving Risk Evolution

To ensure the safety of driving process, the capacity for fully comprehending key factors and predicting collision risks has been essential for various preventive measures. It is necessary for IV to understand the behavior intention and evolution law of other traffic participants, and accurately evaluate driving risk in the traffic environment. However, in a mixed traffic environment, it is difficult to accurately model the behaviour of dynamic and stochastic traffic participants, to comprehensively assess the risk of coupled driver, vehicle and environment, and to realise real-time decision-making in dynamic and complex scenarios. Therefore, as shown in Fig. 3.82, the physical model of risk evolution is established for the driving process, which is divided into stable and unstable state. The influence of dynamic factors

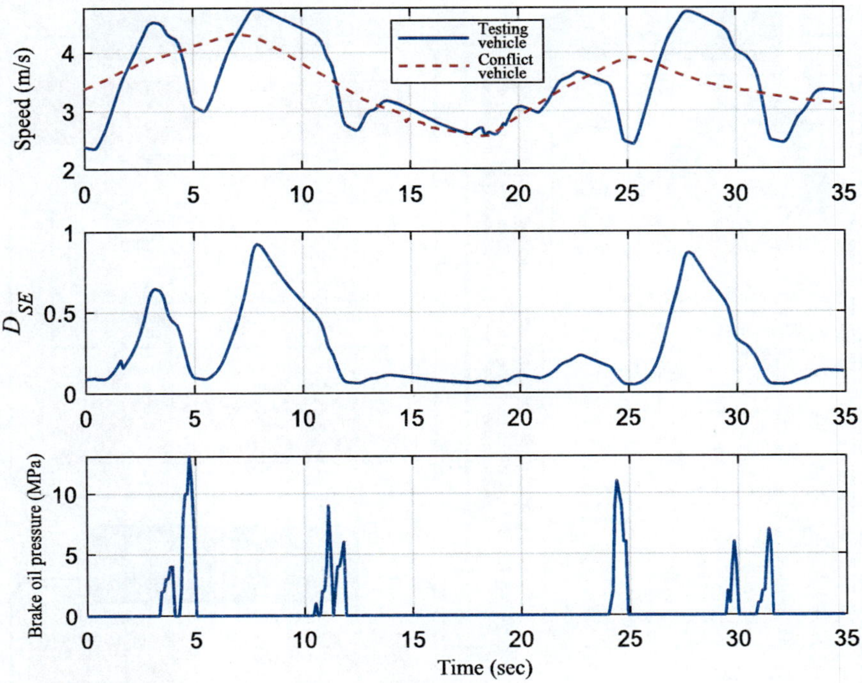

Fig. 3.81 Results of scenario 1 test

such as driver's attributes, road environmental conditions and vehicle attributes on the future evolution direction and path of risk in the long-term and short-term complex environment is analyzed. To describe the dynamic evolution trend of driving risk, it is necessary to analyze the collision risk within the prediction range and the collision risk outside the prediction range in accordance with such variables as collision probability, TTC and collision relative velocity, quantify the uncertain cognition risk, and realize the spatio-temporal evolution analysis for multi-vehicle conflict situation. Finally, the impact of uncertainty on the results of IV situation assessment is analyzed, and the application examples and result analysis of typical traffic scenarios involving pedestrian behavior mutation and communication/sensing failure are output.

1. Risk Evolution State Analysis

Comprehension on driving environment situations can be defined as the process of perceiving, recognizing, understanding and predicting the vehicle that moves and changes with the passage of time and space in a complex dynamic traffic. Risk evolution state analysis mainly extracts and analyzes the dynamic state in different stages during driving. In the traffic environment, there are a large number of traffic participants and complicated information. Considering that there are great differences in the causes and consequences of risks from various risk sources, the risk evolution

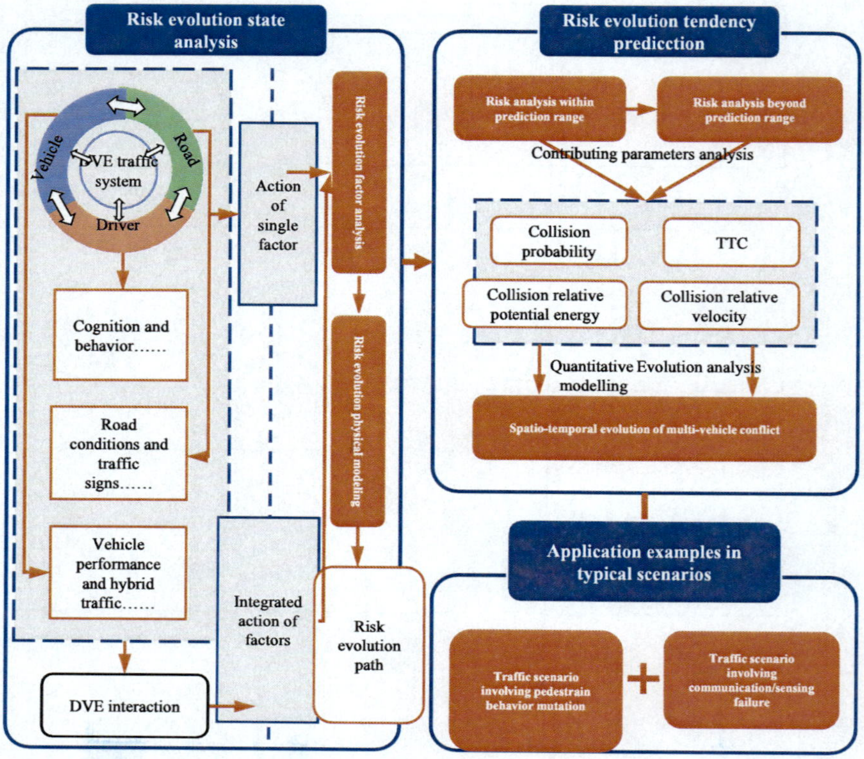

Fig. 3.82 Logic diagram of driving risk evolution analysis

state analysis usually involves the comprehensive application of multidisciplinary methods.

(1) Analysis of Risk Evolution Factors

Road traffic system is a complex and dynamic system composed of drivers, vehicles, traffic participants, road environment (including static and dynamic elements, e.g., traffic facilities, pedestrians and cyclists). There are a lot of uncertainty problems in the DVE traffic environment, which affect not only the situation assessment, but also the decision-making and control of IV. In view of the time-varying, complex and coupling features of driving risk, it is necessary to quantify the driving risk posed by various factors such as the vehicle, road traffic facilities, driver behavior and road environment, and determine the internal factors affecting the evolution of driving risk.

In the road traffic system, due to the dynamic changes of DVE, driving risk has become a potential variable. For the whole road traffic system, the occurrence of accidents is always related to adverse changes of drivers, traffic participants and a certain factor or the interaction of multiple factors. And factors suddenly appearing in a short time due to sudden change of some factors in the traffic environment may lead to the emergence of potential risks, and other subsequent factors will also be mutated without prevention and control, resulting in the accident after step-by-step interaction of factors. The evolution process of traffic collision accident is shown in Fig. 3.83.

To explore the possibility of risk posed by various factors and provide quantitative results, many scholars have paid great attention to such problems. Through the study of a large number of traffic accidents, Researchers Treat from the US and Sabey from the UK have made statistical analysis and acquired their outcome. The accident data is used to quantify the role of three important traffic factors: driver, vehicle and road in the system, and the obtained results are displayed in Table 3.13.

The two researchers concluded that the trend is consistent, and the proportion obtained is representative: that is, in the DVE system, the driver factor is the root of

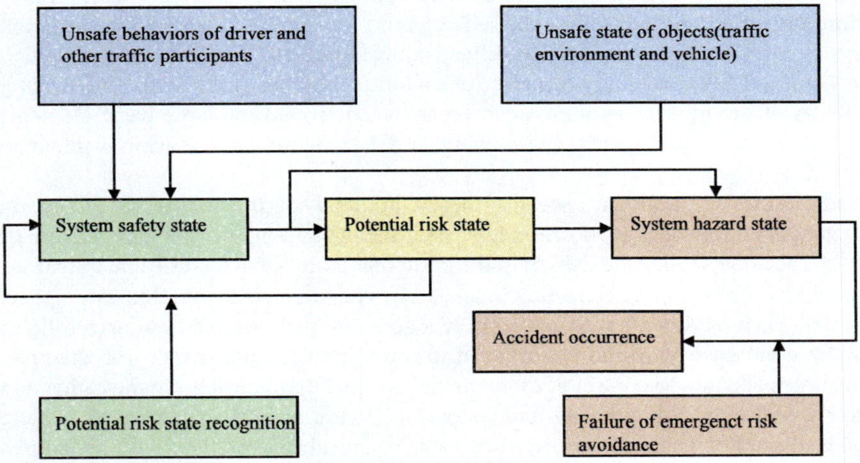

Fig. 3.83 Evolution process of traffic collision accident

Table 3.13 Study on proportion of driving risk influencing factors (British and American countries)

Reason	Sabey conclusion (%)	Treat conclusion (%)
Single road	2	3
Single human	65	57
Single vehicle	2	2
Road and human	24	37
Vehicle and human	4	6
Road and vehicle	1	1
Driver, vehicle, road	1	3

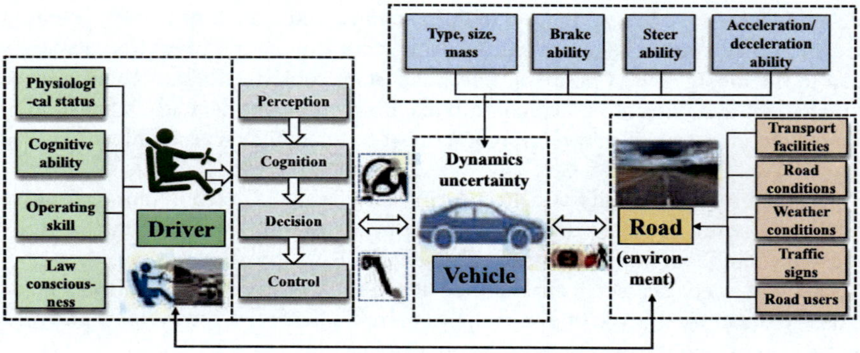

Fig. 3.84 Main influencing factors of driving risk

the instability of risk state, and other factors account for a relatively small proportion. Therefore, when studying the self-organization evolution law of transportation system, it is necessary to consider driver, vehicle, road and environment as a unified system, and take the study of driver behavior as the core. That is, the influencing factors of driving risk evolution can be analyzed around the three basic elements of driver-traffic participants-environment, and the structural relationship is shown in Fig. 3.84 [136].

Based on the accident causality theory, the analysis process of risk evolution factors is established, as shown in Fig. 3.85. It can be found that the direct cause of traffic accidents is that the risks of traffic participants are out of control, and the causes include: ① Driver's risk behaviors (mistakes or violations) lead to traffic participants at risk, such as driver's fatigue and distraction, and pedestrian's running red light; ② External environmental factors lead to traffic participants at risk, for example, the brake and out-of-control of motor vehicles under extreme abnormal weather and road conditions; ③ Internal system factors lead to traffic participants (motor vehicle and non-motor vehicle) at risk, such as mechanical or electrical system failures and other internal reasons. The causal relationship of driver risk behavior, external environmental factors, internal system factors with risk state of traffic participants will be discussed below respectively.

1) Driver factor

Driver's dynamic decision-making on the driving environment leads to the increase of driving risk. Meanwhile, the existing data can also indicate that the driver factor is the main internal cause of accidents. According to the results of investigation on 1500–2000 drivers in different countries as shown in Table 3.3 in the previous chapter, although the statistics of driver factors in different countries account for different proportions of traffic accidents, drivers' cognitive levels, driving skills, as well as psychological and physiological states will affect the stability of driving process to a certain extent, and their abnormal behaviors have become the main cause for risk's evolution into accident. When the driver's psychological and physiological

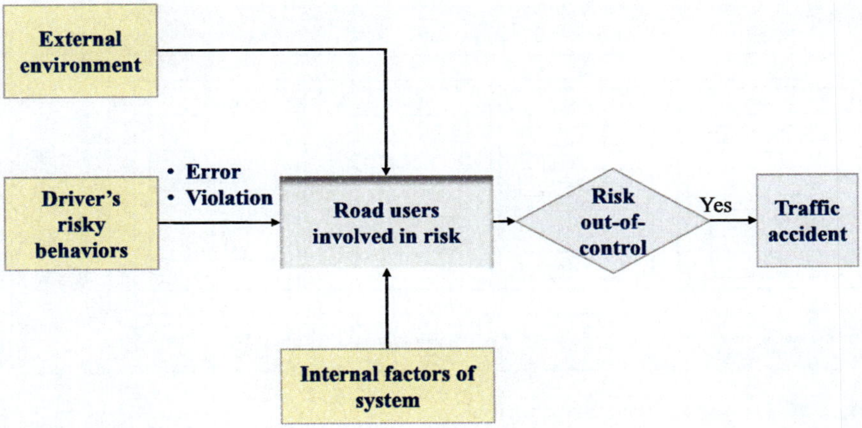

Fig. 3.85 Analysis process of risk evolution factors

state is abnormal, it will affect the normal perception, cognition, decision-making and operation. For example, driver's insufficient perception of the traffic environment, cognitive and decision-making errors due to fatigue driving or distraction, mental tension and sudden physical diseases will lead to operation errors. Driver's cognitive ability is related to intelligence, educational background and experience. The lack of cognitive ability is easy to lead to decision-making mistakes. Driver's control skills are related to driving skills. Unqualified skills are easy to lead to control errors. Weak awareness of traffic laws and regulations is prone to illegal decisions, such as speeding, retrograde, and running red lights. Perception, cognition, decision-making and control are the four inevitable stages of risk behaviors (mistakes or violations). To analyze the role of drivers in risk evolution, it is necessary to analyze the problems that drivers shall solve and the abilities they shall have in the four stages, and then analyze the causal relationship between the four stages and driving risk.

Perception is the process of obtaining information from the surrounding traffic environment. Perception ability can be summarized into position perception ability, speed perception ability, road perception ability and traffic sign perception ability, as shown in Fig. 3.86. Position perception ability represents driver's ability to perceive the position of the vehicle and surrounding objects; Speed perception ability represents driver's estimation of the speed of the vehicle and surrounding moving objects; Road perception ability represents driver's estimation of pavement material, roughness, curvature, road adhesion coefficient and slope through vehicle vibration and such operations as braking, acceleration and steering; Traffic sign perception ability represents driver's sensitivity to traffic signs and signal lights. For cyclists or pedestrians, it mainly includes position perception, speed perception and traffic sign perception.

Cognition is the process of processing information based on perception results, including understanding, selective attention, memory, reaction time, etc., as shown

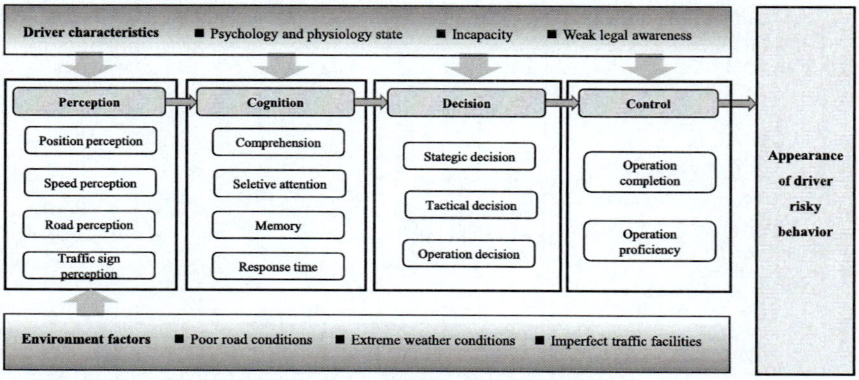

Fig. 3.86 Formation of driver's risk behavior

in Fig. 3.86. Specifically, understanding ability is the comprehension of the essence and development law according to own experience or knowledge base; Selective attention refers to the conscious screening toward important information; Memory includes instantaneous memory and short-term memory. Instantaneous memory is the simple storage of information, and short-term memory is the further processing of information; Reaction time reflects the ability to respond to external incentives [136].

Decision-making is the process of making behavioral strategies based on cognitive results. Driver's decision-making process can be described as a top-down hierarchical framework (Fig. 3.86), which mainly includes three aspects: strategic decision, tactical decision and operational decision. The strategic decision is to optimize driving route; Tactical decision is to choose driving behavior (such as following, changing lanes, and overtaking); Operational decision is to generate driving path. Similarly, the decision-making stage of cyclists and pedestrians includes two processes: behavior selection and path generation. For example, in the scenario of pedestrian's crossing the road, the decision-making process of pedestrian can be described as follows: first, select the behavior category, that is, decide whether to pass or not; secondly, choose path mode, that is, decide how to pass, with a slow walk or run.

The task of the operation stage is to execute the strategy generated in the decision-making phase. The driver's ability to control the vehicle can be reflected by driving action completion and driving action familiarity, as shown in Fig. 3.86. Integrity and accuracy are the indicators to evaluate the completion of driving movements, while efficacy, smoothness and coordination are the indicators to evaluate the familiarity of driving movements. Integrity is considered for examining whether the driver performs the whole driving actions completely; Accuracy is for whether the driving actions achieve the expected effect; Efficacy is for whether the driver completes corresponding actions within the specified time; Smoothness is for whether the driver operates stably and smoothly; And coordination is mainly for whether various basic actions are coordinated, e.g., the cooperative operation of accelerator pedal, gear and clutch.

In conclusion, the insufficient perception will affect judgment of speed, location, road and traffic signs, and lead to inaccurate input information of risk sources; Insufficient cognition leads to the deviation of understanding on risk and its change law; Insufficient decision-making leads to the failure to formulate correct strategies in the process of risk avoidance behavior selection and path generation; Insufficient operation leads to poor action and risky behavior. Among them, perception, cognition and decision-making act indirectly with risky behavior, while operation acts directly with risky behavior. However, abnormalities at any stage will lead to risky behavior. Driver's characteristics (abnormal psychological and physiological state, lack of ability, and weak awareness of laws and regulations) will affect the four stages of perception, cognition, decision-making and operation, and are the subjective factors for risky behavior; The environment mainly affects perception stage, and is the objective factor for risky behavior. For example, poor road conditions and extreme weather conditions will lead to insufficient perception, as shown in Fig. 3.89. Risky behavior is the man-made performance resulting in risk state of traffic participants.

2) Vehicle

The differences in driving speed, acceleration and deceleration of different types of vehicles and their mutual influence constitute the impact of vehicle type distribution on traffic. The specific impact can be divided according to the size of the vehicle body (wheelbase, and the number of axles). And the physical properties of the vehicle include the vehicle size and individual space demand, operation performance (maximum speed, acceleration, deceleration performance, etc.), as well as dynamics differences of vehicles (endurance capacity, acceleration and deceleration performance, fuel consumption, etc.). When the dynamics and kinematics properties of the vehicle change, it is easy to result in the instability of the DVE system. That is, vehicle basic attributes (mass, size, and type, etc.) and motion state (position, speed, acceleration, etc.) are the influencing factors for driving risk. In addition, the mechanical system (braking, steering, and tires), electrical system (lighting, signal device, and control system) and sensors of motor vehicles are also critical influencing factors for driving risk.

3) Road Environment

Broadly speaking, road environment refers to various external and internal conditions that affect drivers and change their driving actions. It is mainly divided into road conditions and environmental conditions. Road conditions mainly include road adhesion coefficient, road alignment, road anti-sliding performance, road slope, number of lanes, etc. And environmental conditions mainly include surrounding environmental conditions (static obstacles), weather conditions (such as day and night, cloudy and sunny, snow, rain, fog, wind, temperature and air pressure) and various signal signs and traffic signs. Environmental factors are the objective inducements of traffic accidents, mainly including road traffic signs, road design, road static obstacles and weather. The environmental factors that influence driving risk include: ① imperfect road traffic signs, e.g., insufficient installation of sharp turns, continuous curves, road narrowing, steep slopes, pedestrian haunting and stopping and other traffic signs; ② unreasonable road design, e.g., non-conforming alignment, sight distance, lane width and curvature radius of the road, low road adhesion coefficient, excessive slope, and poor line of sight; ③ collision with static obstacles on the road, including crossing obstacles (lane line, deceleration belt, etc.) and non-crossing obstacles (protective fence, green belt, cone barrel, etc.); ④ deterioration of road conditions and poor sight under extreme weather conditions [136].

(2) Physical Model for Risk Evolution

The change of the transportation system state can usually be simulated by the physical model, and the evolution of different states can be characterized according to the changes of parameters in the physical model. In accordance with the goal of driving, i.e., safe passage through a complete distance, the driving process is divided into the stable and unstable state. With the stable state, the vehicle will have no traffic accident. While with unstable state, it will be involved in accident eventually. The driver will face a series of adverse factors during driving. With correct operation in face of risk, the driving process will not be greatly affected, and no traffic accident occurs at this time. In the process of driving, the vehicle will be affected by the DVE closed-loop system in three aspects: ① driven by driver's driving goal; ② affected by dynamic traffic environment; ③ constrained by traffic rules. According to this idea, the driving process is described vividly. In this section, the model shown in Fig. 3.87 will be used to describe the risk evolution process: when the small ball slides down from the above of the U-shaped groove, it smoothly slides to the bottom under the joint action of equivalent attraction G_i, road resistance R_i and external force F_{ij}, which is regarded as a steady-state process. As shown in Fig. 3.87, the vehicle state is always controlled in this process.

On the contrary, when the driver has no time to make correct operation or even wrong operation in the face of adverse factors, the driving process will be greatly affected, and traffic accidents will occur. Therefore, when the small ball encounters a large external force F_{ij} in the process of sliding, the balance between the original equivalent attraction Gi and the road resistance Ri is broken, resulting in the following phenomenon due to unbalanced state of the small ball during sliding: ① The small

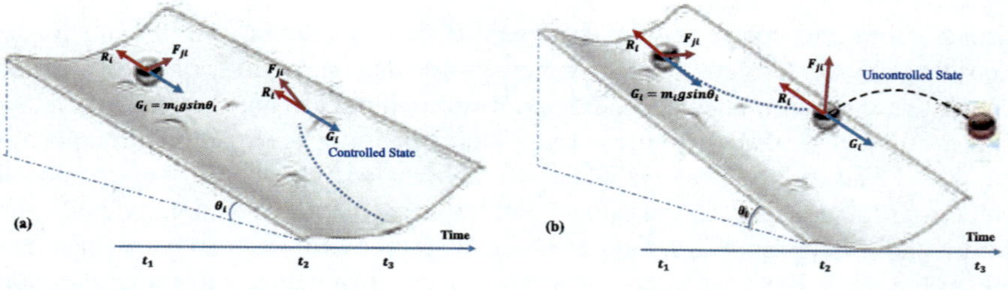

Fig. 3.87 Different states in driving

ball flies out of control under the action of the force pointing out of the groove; ② In the face of the too large obstacle, the small ball can not reach the bottom of U-shaped groove due to insufficient force, which is regarded as an unsteady process. As shown in Fig. 3.87b, the vehicle state is uncontrolled in this process [132].

In Fig. 3.87, the angle θ_i between the U-groove and the plane XOY reflects the driver's overall pursuit of efficiency; $\theta_{i,x}$ and $\theta_{i,y}$ correspond to the driver's pursuit for efficiency in the longitudinal and lateral directions respectively. According to the transformation principle of Euler angular coordinate system, the relationship among the three is derived by the following formula:

$$\theta_i = \arccos\left(\cos\theta_{i,x} \cdot \cos\theta_{i,y}\right) \tag{3.125}$$

First, the relationship of $\theta_{i,x}$ and $\theta_{i,y}$ with θ_i is analyzed. As shown in Fig. 3.87, according to the principle of Euler angle, the U-shaped groove first rotates the axis Y with $\theta_{i,x}$ to change the coordinate system OXYZ into OX′YZ′; then rotates the axis X′ with $\theta_{i,y}$ to change the coordinate system into OX′Y′Z″. Therefore, the rotation matrix is:

$$Y\left(\theta_{i,x}\right) = \begin{bmatrix} \cos\theta_{i,x} & 0 & \sin\theta_{i,x} \\ 0 & 1 & 0 \\ -\sin\theta_{i,x} & 0 & \cos\theta_{i,x} \end{bmatrix} \tag{3.126}$$

$$X'\left(\theta_{i,y}\right) = \begin{bmatrix} 1 & 0 & 0 \\ 0 & \cos\theta_{i,y} & \sin\theta_{i,y} \\ 0 & -\sin\theta_{i,y} & \cos\theta_{i,y} \end{bmatrix} \tag{3.127}$$

For vector $\mathbf{z}(0,0,1)^{\mathrm{T}}$, $\mathbf{z}''$ is obtained after rotation of $Y\left(\theta_{i,x}\right)$ and $X'\left(\theta_{i,y}\right)$:

$$\mathbf{z}'' = Y\left(\theta_{i,x}\right) \cdot X'\left(\theta_{i,y}\right) \cdot \mathbf{z} = \begin{bmatrix} \sin\theta_{i,x}\cos\theta_{i,y} \\ \sin\theta_{i,y} \\ \cos\theta_{i,x}\cos\theta_{i,y} \end{bmatrix} \tag{3.128}$$

The included angle θ_i between $\mathbf{z}$ and $\mathbf{z}''$ meets the following requirement:

$$\cos\theta_i = \frac{\mathbf{z} \cdot \mathbf{z}''}{|\mathbf{z}| \cdot |\mathbf{z}''|} = \frac{\cos\left(\theta_{i,x} + \theta_{i,y}\right) + \cos\left(\theta_{i,x} - \theta_{i,y}\right)}{2} \tag{3.129}$$

For vectors $\mathbf{y}(0, 1, 0)^{\mathrm{T}}$, $\mathbf{y}'$ is obtained after rotation of $Y\left(\theta_{i,x}\right)$ and $X'\left(\theta_{i,y}\right)$:

$$\mathbf{y}' = Y\left(\theta_{i,x}\right) \cdot X'\left(\theta_{i,y}\right) \cdot \mathbf{y} = \begin{bmatrix} -\sin\theta_{i,x}\sin\theta_{i,y} \\ \cos\theta_{i,y} \\ -\cos\theta_{i,x}\sin\theta_{i,y} \end{bmatrix} \tag{3.130}$$

The included angle α_i between $\mathbf{y}$ and $\mathbf{y}'$ meets the following requirement:

$$\cos\alpha_i = \frac{\mathbf{y} \cdot \mathbf{y}'}{|\mathbf{y}| \cdot |\mathbf{y}'|} = \cos\theta_{i,y} \tag{3.131}$$

I.e.,

$$\alpha_i = \theta_{i,y} \tag{3.132}$$

Obviously, when the vehicle changes lanes, θ_i increases slightly, together with the corresponding G_i, which is in line with driver's behavior of accelerating lane changing in the real driving process. In addition, driver will face "adverse effects" such as obstacles in front, bad road conditions and bad weather. Here, the above "adverse effects" are described as some small protrusions in the U-shaped groove, as shown in the part marked with a blue circle in Fig. 3.87. It is worth noting that these small protrusions represent not only some real obstacles, but also some potential "adverse effects", which can be regarded as potential risks during driving. It can be defined that the impact of this potential risk on the vehicle is a virtual external force F_{ji}. Different types of risks will exert different external forces on the driving process, such as the constraint force of lane line and road boundary, and the repulsive force of stationary objects and moving objects. Therefore, the mathematical models of risk external forces F_{ji} will also be diverse. The physical model can vividly reflect the real driving process and reveal the mechanism of accidents. For example, when the driver's requirement for vehicle mobility is not high and θ_i is small, the driver has enough reaction time to make correct operation in the face of risk and keep the vehicle moving smoothly; When the driver desires high speed and θ_i is large, the driver does not have enough reaction time in the face of risk and will be easily involved in an accident.

2. Prediction for Risk Evolution Trend

It is essential for vehicle safety control decisions to accurately measure the risk and its dynamic change trend in complex road environment. The existing risk assessment models are usually based on vehicle kinematics and dynamics theory, and apply vehicle state information (speed, acceleration, yaw speed, etc.) and relative motion information (relative speed, relative distance, etc.) to quantify and characterize the degree of risk. However, these methods and models are difficult to reflect the comprehensive risk level of vehicles in traffic system, describe the dynamic evolution trend of driving risk, and lay a continuous judgment basis on risk dynamic evolution in time and space for driving decision and vehicle control. Therefore, it is urgent to study the generation mechanism of driving risk and establish a driving safety model that can describe the generation, diffusion and evolution of driving risk, to lay the foundation for intelligent decision-making and control of vehicles. Interacting with the driver is one of the major challenges faced by advanced driving assistance systems and AV [133, 137]. Although the driver intention modeling is very complex, human drivers can often accurately predict the intention of surrounding vehicles. The possible reasons are that human beings have the following abilities: ① accurate estimation of the internal attributes of driver's decision-making behavior; ② inference on other drivers' risk avoidance behaviors from the perspective of others. In accordance with these characteristics of drivers, risk evolution modeling in the uncertain environment will be better conducted.

Model-based scenario prediction usually assumes that all intelligent agents take risk minimization behaviors (encoded in the model), and cannot predict dangerous operations that cannot be explained by the model (for example, risky lane changing in highway). On the contrary, the dynamics-based prediction is to estimate the trajectory of the vehicle motion model without inferring the traffic scenario (that is, ignoring the interaction between vehicles). In the past ten years, the approaches for IV behavior inference and motion prediction are usually based on three categories: physical model, behavior model and interaction model. These three types are deterministic, and ignore the uncertainty of the environment. Therefore, to analyze the impact of environmental uncertainty on situation assessment, the state distribution of various elements in the traffic environment is described through the probability distribution function based on the stochastic environment model, which facilitates the model-based situation assessment analysis and decision-making for IV.

(1) Risk Analysis within Scope of Trajectory Prediction

Within the scope of trajectory prediction, the collision probability is calculated and evaluated based on the stochastic environment model and the trajectory prediction results. As shown in Fig. 3.88, truck A turns left at the T-junction and vehicle B passes through the intersection. Vehicle A drives towards the intersection and keeps approaching vehicle B. When A and B are close, B is limited in acquiring the information of A in the case of a non-internet connection. In view of the dynamic impact of A on B, it is necessary to predict the trajectory change of A to avoid the possible risk behavior (left turn or right turn) of B and the increase of collision probability between two vehicles. Therefore, the collision probability analysis based on trajectory prediction is introduced firstly, and then the collision probability based on behavior and trajectory planning is analyzed.

The probability of vehicle collision in traffic scenario can be expressed as $P(C_{V_i,V_j})$. Within the prediction range, the probability of vehicle collision at a

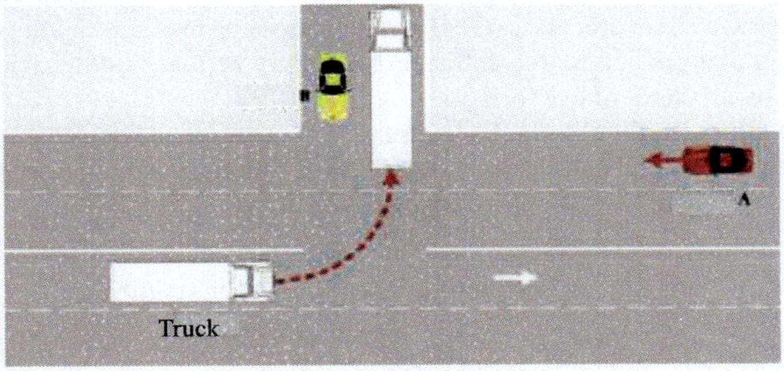

Fig. 3.88 Situation assessment based on environment change

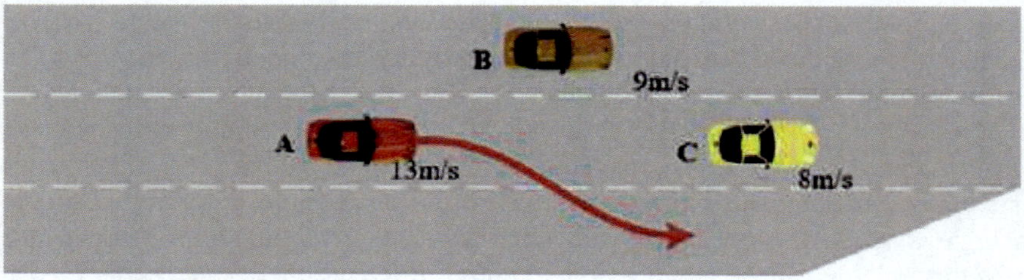

Fig. 3.89 Risk assessment scenario without considering risk beyond prediction range

specific time point can be calculated as follows:

$$P[C_{V_i,V_j}(t)] = \iint C\big[b_{V_i}(t), b_{V_j}(t)\big]p[b_{V_i}(t), b_{V_j}(t)]db_{V_i}db_{V_j} \tag{3.133}$$

where, V_i is vehicle i; t refers to time; $b_{V_i}(t)$ indicates the predicted position of the vehicle V_i in the time t; $p[b_{V_i}(t), b_{V_j}(t)]$ represents the probability of the vehicle i and vehicle j at the position $b_{V_i}(t)$ and $b_{V_j}(t)$ at the time t; t_0 indicates the starting prediction time; T_p represents the range of prediction time, $t \in [t_0, t_0 + T_p]$; $C[b_{V_i}(t), b_{V_j}(t)]$ represents the collision factor in accordance with the physical size of the vehicle. The calculation formula of $C[b_{V_i}(t), b_{V_j}(t)]$ is as follows:

$$C[b_{V_i}(t), b_{V_j}(t)] = \begin{cases} 1, & O[b_{V_i}(t)] \cap O[b_{V_j}(t)] \neq \varnothing \\ 0, & \text{others} \end{cases} \tag{3.134}$$

where, $O[b_{V_i}(t)]$ represents the plane space occupied by the vehicle i. The formula shows that if the plane space occupied by two vehicles V_i and V_j intersects, the two vehicles collide at time t, and the collision factor $C[b_{V_i}(t), b_{V_j}(t)]$ is 1; If there is no intersection in the plane space occupied by V_i and V_j, that is, the intersection is empty, the two vehicles will not collide at time t, and the collision factor $C[b_{V_i}(t), b_{V_j}(t)]$ is 0 [138].

Driving behavior is an abstract representation of vehicle motion, and can be expressed as Gaussian processes (GPs) regarding existing research. Within the scope of planning and prediction, the collision probability of planning behavior can be estimated based on the above formula.

Meanwhile, according to the above assumptions, the planned trajectory is a determined path, so the collision between the planned trajectory and other surrounding vehicles at a certain time point is estimated as follows:

$$P[C_{V_i,V_0}(t)] = \int C\big[b_{V_i}(t), b_0(t)\big]p_i[b_{V_i}(t)]db_{V_i} \tag{3.135}$$

where, $t \in [t_0, t_0 + T_p]$. T_p represents the prediction time range.

Based on the prediction of driving trajectory, this chapter carries out the collision risk assessment in terms of collision probability, TTC, vehicle mass and relative driving speed. At a certain prediction time, the risk function is expressed as follows:

$$\mathrm{Risk}_{\mathrm{in}}[v_i(t), v_j(t)] = \mathrm{P}[C_{v_i,v_j}(t)]\mathrm{cost}_{\mathrm{coll}}(t) \tag{3.136}$$

where, $\mathrm{Risk}_{\mathrm{in}}[v_i(t), v_j(t)]$ represents the collision risk at the prediction time t; $\mathrm{cost}_{\mathrm{coll}}(t)$ represents the risk factor function relative to the collision. $\mathrm{cost}_{\mathrm{coll}}(t)$ can be expressed as follows:

$$cost_{coll}(t) = \frac{1}{2}\frac{m_i m_j}{m_i + m_j}\|v_r(t)\|^2 \cdot \frac{1}{t_p} \tag{3.137}$$

where, m_i represents the mass of the object i; m_j represents the mass of the object j; $v_r(t)$ indicates the relative speed between vehicles; t indicates the evaluation time. $t_p = t - t_0$, $E = \frac{1}{2}\frac{m_i m_j}{m_i+m_j}\|v_r(t)\|^2$ is called internal energy.

Therefore, within the prediction range, the collision risk between vehicles $\mathrm{Risk}_{in}[v_i(t_0 : t_0 + T_p), v_j(t_0 : t_0 + T_p)]$ can be expressed as follows [138]:

$$\mathrm{Risk}[v_i(t_0 : t_0 + T_p), v_j(t_0 : t_0 + T_p)] = \int_{t_0}^{t_0+T_{\max}} \frac{1}{2}\frac{m_i m_j}{m_i + m_j}\|v_r(t)\|^2 \mathrm{P}[C_{v_i,v_j}(t)]\frac{1}{t_p}dt \tag{3.138}$$

$$T_{\max} : \mathrm{P}[C_{v_i,v_j}(T_{\max})] = \max_{t\in}\mathrm{P}[C_{v_i,v_j}(t)] \tag{3.139}$$

where, $T_{\max}$ represents the time when the maximum value is reached in the collision risk prediction sequence.

In the complex traffic environment, the risk of vehicle v_i in the scenario s_i is expressed as follows:

$$\mathrm{RA}_{\mathrm{in}}(v_i, s_i) = \max_j\{\mathrm{Risk}[v_i(t_0 : t_0 + T_p), v_j(t_0 : t_0 + T_p)]\} \tag{3.140}$$

where, s_i represents the scenario i, v_i and v_j represent the vehicles in the scenario. The above formula shows that in the scenario s_i, the risk of the vehicle v_i is expressed as the maximum value of the risk of collision with other vehicles.

(2) Risk Analysis beyond Scope of Trajectory Prediction

In addition to estimating the collision risk among vehicles within the prediction range, it is also necessary to estimate the collision risk outside the prediction range. In the scenario shown in Fig. 3.89, vehicle A follows vehicle C in the middle lane and slow vehicle B runs in the left lane. Vehicle A can change lanes to the right within forecast and planning constraints. Without consideration of risk assessment beyond the prediction range, vehicle A will be close to the end of the right lane. At this time, vehicle A has to reduce its speed to avoid the collision. Therefore, the risk assessment beyond the prediction range plays an important role in the environmental cognition and decision-making of IV. Based on the driving trajectory prediction as well as the collision probability, TTC and collision energy in the future, this section briefly introduces a risk assessment model beyond the prediction time range.

In this part, the risk beyond the prediction time range can be calculated as follows [138]:

$$\mathrm{RA}_{\infty}(v_i, s_i) = \int P(t) I\left(v_i, s_i, m_t^i\right) \frac{1}{2} \frac{m_i m_j}{m_i + m_j} \|v_r(t)\|^2 dp \tag{3.141}$$

where, the risk $I\left(v_i, s_i, m_t^i\right)$ beyond the prediction range $\mathrm{RA}_{\infty}(v_i, s_i)$ is expressed as follows:

$$I\left(v_i, s_i, m_t^i\right) = \begin{cases} \frac{\Delta v}{\Delta d} \frac{\Delta v}{\Delta d} > 0; \\ 0 \frac{\Delta v}{\Delta d} \leq 0; \end{cases} \tag{3.142}$$

where, Δv and Δd are the predicted relative speed and relative distance between the vehicle i and the vehicle j. According to the formula, when the relative speed between the vehicle i and the vehicle j is 0 or negative, $\mathrm{I}\left(v_i, s_i, m_t^i\right)$ is 0; When the relative speed between the vehicle i and the vehicle j is positive, $\mathrm{I}\left(v_i, s_i, m_t^i\right)$ is the reciprocal of TTC.

(3) Temporal and Spatial Evolution of Multi-vehicle Conflict Situation

The comprehensive collision risk assessment of IV can be divided into collision risk assessment within the prediction time range and beyond the prediction time range. The comprehensive collision risk assessment $\mathrm{RA}(v_i, s_i)$ in the scenario s_i of vehicle v_i can be expressed as follows:

$$\mathrm{RA}(v_i, s_i) = \mathrm{RA}_{in}(v_i, s_i) + \mathrm{RA}_{\infty}(v_i, s_i) \tag{3.143}$$

The environmental uncertainty is represented by Gaussian distribution $\mathcal{N}$. Therefore, the driving state of vehicle in the traffic environment at a certain time can be expressed as:

$$X(t) \sim \mathcal{N}(\mu(t), \Sigma(t)) \tag{3.144}$$

where, $\mu(t)$ represents the mean value of vehicle state, and $\Sigma(t)$ is the covariance. $X(t)$ can be obtained employing the perception tracking module.

Based on the historical data of perception and tracking results, the uncertainty of traffic environment state can be predicted. In the prediction time T_p, the prediction result can be expressed as:

$$\left\{X(t+1), X(t+2), \ldots, X\left(t+T_p\right)\right\} \tag{3.145}$$

Therefore, the multi-vehicle conflicting situation can be obtained by calculating RA(vi, si). Speicifically, the trajectory prediction combines the prediction with physical model and behavior cognition model, which can ensure the prediction accuracy in the short-term domain and the rationality of the prediction trend in the long-term domain. When the onboard environment sensor or communication fails, the prediction result can be used as the perception result. However, there will be a large difference between the prediction result and the latest perception or communication result. At this time, the collision probability will be updated again.

The multi-vehicle conflicting situation can assess the risk posed by unforeseen/unexpected obstacles. Assuming that the obstacle has a typical motion mode (e.g., pedestrian moves across the road), the probability of unexpected obstacles in the traffic scenario can be expressed as the following Homogeneous Poisson process:

$$P_{\mathrm{u}}[N_k(t_1) - N_k(t_0) \geq 1] = 1 - e^{-\lambda_k \tau} \tag{3.146}$$

where, $(t_0, t_1]$ represents a time interval; $\tau = t_1 - t_0$. $N_k(t_1) - N_k(t_0) \geq 1$ indicates that the number of unexpected obstacles in a time interval is greater than or equal to 1. λ_k is a parameter representing the number of unexpected obstacles in unit time, and k represents different scenario. Different scenario k has different parameter λ_k, and its value can be obtained by observing whether there are unexpected obstacles at a certain time and place.

3. Application Examples in Typical Scenarios

The uncertainty of dynamic traffic environment is widespread, including various uncertainties in perceived information, risk assessment and driving behavior. Therefore, the uncertainty of environmental elements is expressed based on the uncertain environmental prediction, and then the collision risk is estimated based on the probabilistic distribution. Finally, the accuracy of driving risk evolution is verified by analyzing the impact of uncertainty on collision risk assessment in different scenarios (such as pedestrian behavior mutation and communication sensing failure).

(1) Traffic Scenario Involving Sudden Change of Pedestrian Behavior

Unexpected obstacles are prevalent in traffic scenarios. It is of great significance to analyze the impact of the uncertainty of the obstacles on the rational decision-making of intelligent driving. Therefore, it is necessary to first describe the typical scenarios with unexpected obstacles, and then analyze the uncertainty situation assessment results of the above scenarios. In traffic scenarios, especially in urban conditions, it is very common for unexpected obstacles (such as pedestrians) to appear in imperceptible areas. As shown in Fig. 3.90, the three vehicles are vehicle A, vehicle B and the white truck respectively. When the white truck stops on the right side of the road, it is difficult for vehicle A to perceive moving objects such as pedestrians in an imperceptible area. However, vehicle A is likely to collide with unexpected pedestrians later. The higher the speed of vehicle A is, the higher the risk is. Meanwhile, as vehicle A advances, the imperceptible area of vehicle A changes.

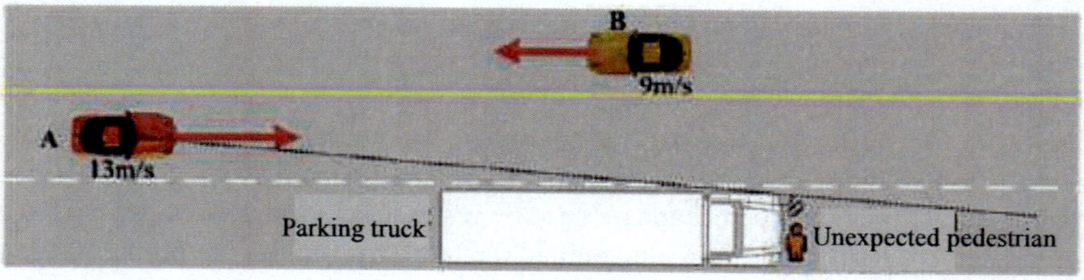

Fig. 3.90 Traffic scenario involving unexpected obstacle

In the scenario shown in Fig. 3.90, unexpected pedestrian appears in front of the white truck, but has not crossed the street yet. However, vehicle A must take into account the collision risk of pedestrian crossing in the imperceptible area in front of the truck, that is, the scenario involving accidental obstacle. In this part, the probability of pedestrian crossing the road in each cycle can be expressed as P_u. Therefore, the collision risk at each prediction time can be described as [138, 139]:

$$\text{Risk}_\text{u}[v_i(t), v_j(t)] = P_\text{u} \cdot \text{Risk}[v_i(t), v_j(t)] \tag{3.147}$$

where, $\text{Risk}_\text{u}[v_i(t), v_j(t)]$ is to consider the collision risk of unexpected obstacles at the prediction time t; P_u is the occurrence of unexpected pedestrian; $\text{Risk}[v_i(t), v_j(t)]$ refers to the risk of collision during pedestrian's crossing.

As vehicle A approaches the stationary white truck, the observation range of vehicle A will also change. Its imperceptible area will become smaller and smaller, and the lateral displacement of unexpected pedestrian will become larger. In this case, the lateral displacement can be expressed as the following formula:

$$l_s = l_o + w \cdot \cot\theta \tag{3.148}$$

where, l_s represents the lateral distance between vehicle A and the starting crossing point; l_o is the lateral distance between vehicle A and large truck; w is the physical width of pedestrian; θ is the angle of the imperceptible area as shown in Fig. 3.90.

It is assumed that vehicle A remains constant speed in the middle of the lane. To facilitate the comparison of results, the same components in the situation evaluation are defined as 1. In this scenario, the results of situation awareness are shown in Fig. 3.91.

As shown in Fig. 3.91, the horizontal axis represents the driving speeds with different levels, and the ordinate represents the risk assessment results. Figure 3.91 compares and analyzes the risk assessment results of vehicle passing through the unexpected obstacle with speeds of 5 m/s, 10 m/s, 15 m/s and 20 m/s respectively. The changes in risk assessment are represented by different colors. It can be seen from the figure that the risk assessment result at the driving speed of 20 m/s is larger than that at 5 m/s. Within the same level of speed, the darker the color is, the higher the risk is, e.g., blue represents low risk and red represents high risk. In other words,

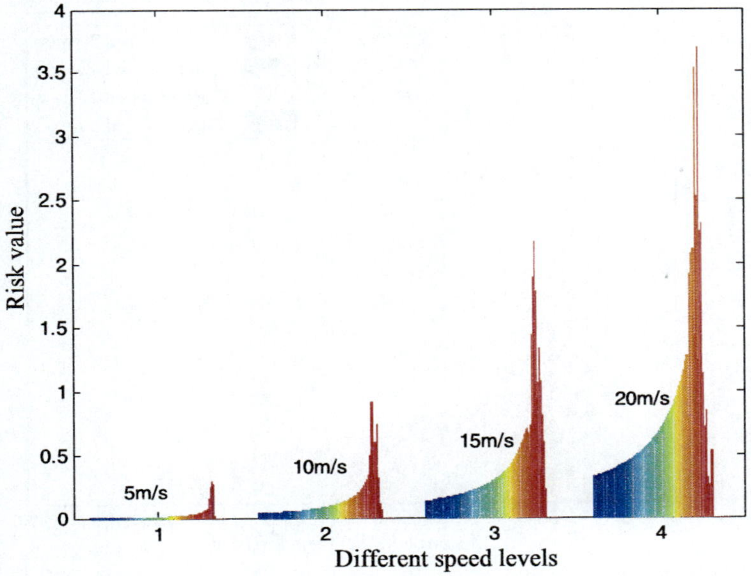

Fig. 3.91 Cognition comparison among situation of different vehicle speeds in scenario with sudden change of pedestrians behavior

the higher the speed is, the higher the risk of a sudden change in pedestrians behavior is, which is consistent with actual driving experience. When a driver passes a parked bus, the higher the speed is, the more unsafe the driver will feel, as unexpected pedestrians may appear nearby.

To analyze the impact of different distances from the truck parked on the roadside on the risk, the results in Fig. 3.92 are as follows: the horizontal axis is the longitudinal distance between the vehicle and the left front end of the truck; different colors in the figure represent the risk assessment results of the vehicle at different speeds under unexpected obstacle scenario. It can be found that the risk will gradually increase with the vehicle approaching from a farther place, but it will decrease when reaching the maximum due to the reduction of the undetected area. Meanwhile, the risk also decreases with the reduction of vehicle speed, and the closer the maximum risk point is to the truck, which is also in line with the daily driving experience of most drivers.

(2) Traffic Scenario Involving Communication/Sensing Failure

It is assumed that during communication, relevant status and uncertainty information will be broadcast in intelligent and connected vehicles. If the communication is lost or the sensor fails in a certain period, the predicted result will be calculated as the latest initial state. When communication or perception is restored, the estimated environmental parameters and related uncertainty information are calculated as new

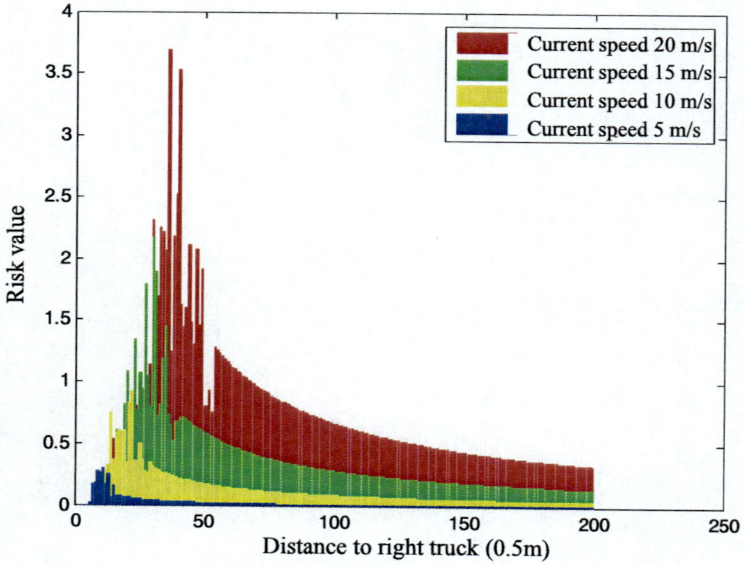

Fig. 3.92 Cognition on situation and analysis on distance between vehicle and large truck in scenario with sudden change of pedestrian behavior

initial information. Therefore, it is necessary to first describe the scenario of communication loss or sensor failure, and then analyze the uncertain situation assessment results under this scenario.

As shown in Figs. 3.93 and 3.94, the situation assessment is carried out in view of the uncertainty due to sensor failure or communication loss. In the two figures, two vehicles are driving in the right lane, where Δt_{loss} indicates the time of sensor failure or communication loss. Before sensor failure or communication loss, the behavior probabilistic distribution estimated based on a dynamic Bayesian network is taken as the initial behavior probabilistic distribution, which is expressed as follows [138, 139]:

$$P_m^0 = (P_1^0, \ldots, P_n^0, \ldots, P_N^0) \tag{3.149}$$

where, P_m^0 is the behavior probabilistic distribution at the initial time $t = 0$; P_n^0 indicates the probability of behavior n; and N indicates the size of the behavior.

Based on the first-order Markov theory, the probabilistic transition of behavior during sensor failure or communication loss is expressed as follows:

$$P_m^k = M^{\mathrm{T}} \tag{3.150}$$

where, P_m^k is the probabilistic distribution of sensor failure or communication loss at the k-th type of risk; M is the probabilistic transition matrix. Therefore, the risk of

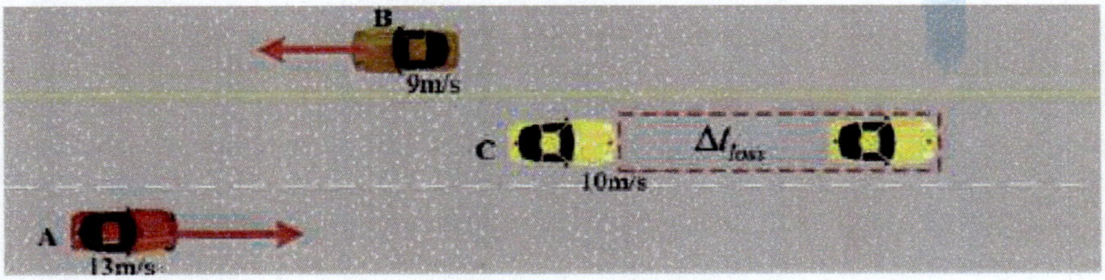

Fig. 3.93 Scenario of sensor failure or communication loss under lane keeping

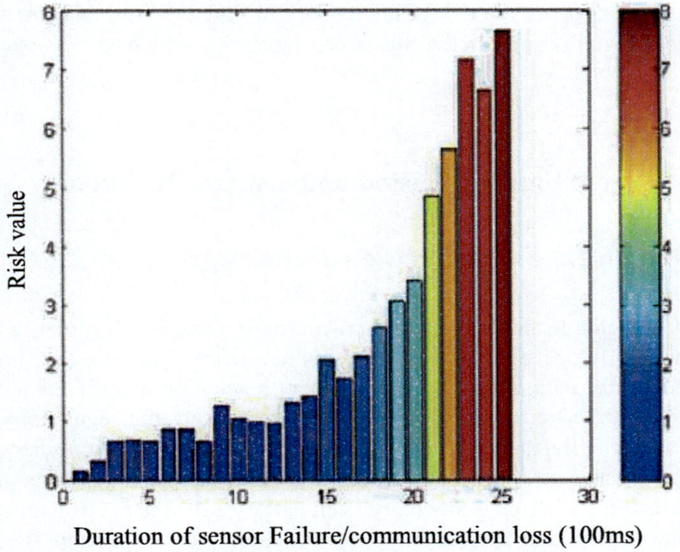

Fig. 3.94 Risk analysis on scenario of sensor failure or communication loss under lane keeping

sensor failure or communication loss can be expressed as follows:

$$\mathrm{RA}_k(v_i, s_i) = \sum_{n=1}^{N} P_n^k \mathrm{RA}_k\big(v_i, s_i | m_j = n\big) \tag{3.151}$$

where, $\mathrm{RA}_k(v_i, s_i)$ is the k-th type of risk under sensor failure or communication loss; N represents the size of behavior sequence of vehicle j, and $\mathrm{RA}_k\big(v_i, s_i | m_j = n\big)$ is the risk of behavior $m_j = n$ of vehicle j.

The following is an example to illustrate the above idea. In the lane keeping scenario shown in Fig. 3.93, vehicle C is remaining the current lane, in which the driving speed of vehicle A is 13 m/s and that of vehicle C is 10 m/s. During sensor failure or communication loss, vehicle A cannot obtain information related to vehicle C. Therefore, vehicle A can only obtain relevant historical information and use the predicted results as perception information for iterative calculation. In the lane keeping scenario, the risk assessment results considering sensor failure or communication loss are shown in Fig. 3.94, which indicates the relationship between the time of sensor failure or communication loss and the risk. The horizontal axis represents the time length of communication loss or sensor failure, with different risk values represented by different colors. The color from blue to red indicates that the greater the risk is. The results show that the longer the time of sensor failure or communication loss is, the higher the risk is. In the scenario, vehicle C may also change lanes in case of sensor failure or communication loss. That is, the risk assessment takes into account the risk of uncertainty in sensor failure or communication loss. If the uncertain risk from the side lane is ignored, it may lead to a serious traffic accident.

3.3 Intelligent Safety Decision and Control Technology

3.3.1 Intelligent Safety Decision Technology

Intelligent decision technology plays a crucial role in icv, which serves as the brain of the vehicle. Before making decisions, it is necessary to acquire the surrounding environment information from the environment perception module, including road topology, real-time traffic situation, traffic participant, traffic regulations, state of ego-vehicle, etc. Driving risk is also an important basis for making decisions. Based on the above input information, the decision-making system generates the target driving behavior and driving trajectory, while afterwards, the control system drives the vehicle to track and execute the decision results, so as to ensure the safe and efficient drive to the destination.

The decision-making system takes the environmental perception information, risk assessment results and driving objectives (including destination, personalized demand of passengers, etc.) as the input, and the trajectory planning results (target path and speed in a period of time) as the output. Separated by decision-maker and the controlled object, the decision-making technology can be divided into single-vehicle decision-making, he and multi-vehicle cooperative decision-making. Among them, for single-vehicle decision-making, the ego-vehicle is the decision-maker and controlled object, while taking the safety and efficiency Search instead for efficienicy of driving as the goal; multi-vehicle cooperative decision-making, is based on vehicle-vehicle and vehicle–road network communication. Decision makers include central decision-making unit and on-board decision-making unit, aiming at the safe and efficient passage of all vehicles under control. Figure 3.95 shows the content structure of intelligent safety decision-making technology.

In intelligent safety decision-making system, single-vehicle technology is the basis of decision-making function, and multi-vehicle technology is used to jointly

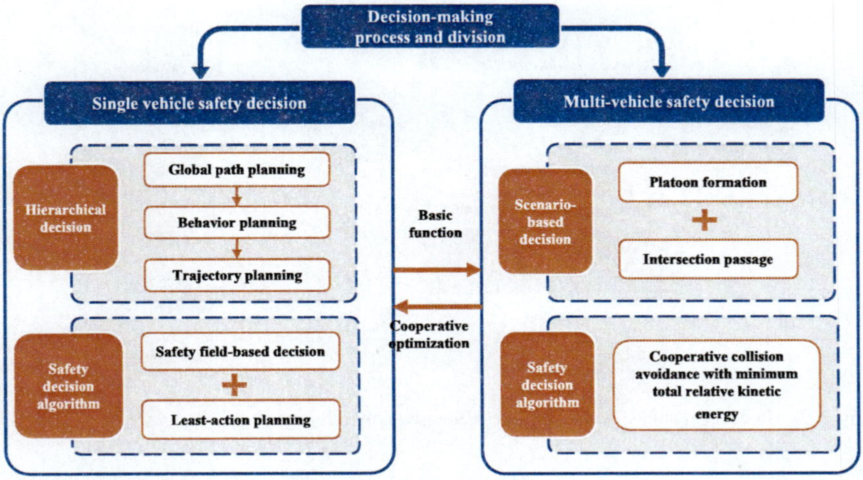

Fig. 3.95 Content structure of intelligent safety decision technology

optimize system performance. Figure 3.96 introduces the task division of typical intelligent decision-making process, which includes V2V communication, central computing and on-board computing unit. Driving behavior and trajectory can be generated by the two decision techniques, and shall be further divided into specific decision-making tasks. For example, based on the global performance optimization of the transportation system, the cooperative decision system sends the expected behavior instructions to each vehicle (such as the target time of passing through the intersection, and driving lane changing). And each vehicle judges the safety of instructions by the single vehicle decision technology and plans the specific driving trajectory accordingly. The vehicle actuators (steering, acceleration, braking, etc.) is generally controlled by on-board computing unit, which will be introduced in the intelligent safety control technology in the next section.

1. Intelligent Safety Decision of Single Vehicle

After years of development, single-vehicle decision making technology has been applied in simple traffic scenarios and driving tasks successfully. For example, L2-level ADAS system integrating adaptive cruise and lane keeping functions on highway has been widely commercialized in the automotive industry. However, the interaction between vehicle is complex, and traffic elements are diverse. For example, the uncertain behavior and interactive game of other traffic participants, the non-ideal conditions of weather and road, and the personalized driving demand of the vehicle driver (or passengers) pose challenges to the commercial application of higher-level intelligent driving system.

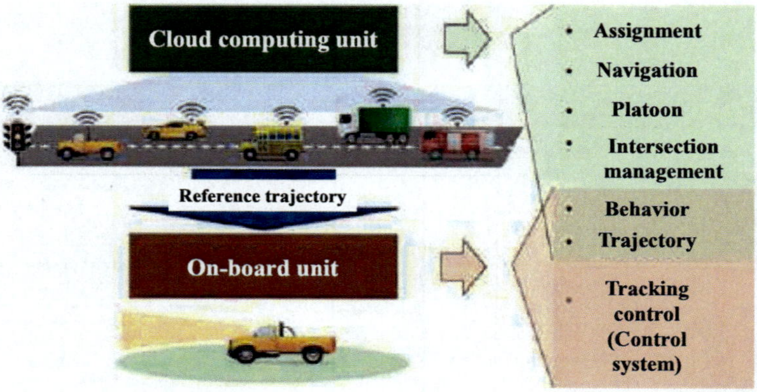

Fig. 3.96 Intelligent safety decision process and division of work

In order to decouple the decision-making process and simplify algorithm design, single-vehicle decision making technology often performs in a hierarchical way. The decision-making process can be divided into three main steps, i.e., global path planning, behavior planning and trajectory planning. A variety of safety evaluation indicators, such as distance to obstacles, TTC, THW, and risk field, are critical in decision-making process for driving safety. In recent years, during the development of higher-level intelligent driving technology, researchers not only explore and innovate in risk assessment, but also put forward a variety of new approaches in decision-making technology, such as machine learning and the application of human-like driving model.

(1) Vehicle Intelligent Hierarchical Decision

The single-vehicle decision-making algorithms are usually implemented in a hierarchical manner, from macro to local, and from long-term to short-term, mainly including global path planning, behavior planning and trajectory planning, as shown in Fig. 3.97. Global path is an ideal driving path generated by icv based on HD map, and connects the starting position and target position on the map. The global path planning takes the macro traffic information as the input, and updates the results according to the changes of traffic conditions along driving. In the actual driving procedure, ICV shall deal with the changing factors such as pedestrians, vehicles and traffic lights in the micro traffic environment, which requires adjustment of driving behavior and trajectory in real time, that is, behavior planning and trajectory planning. The behavior planning layer generates semantic driving behaviors, such as car-following, changing lanes to overtake, slowing down and yielding; The trajectory planning layer generates the target trajectory corresponding to the driving behaviors. Trajectory planning can be divided into two parts, i.e., local path planning and speed planning. The complete planning result coupled by the target path and its corresponding speed is called trajectory.

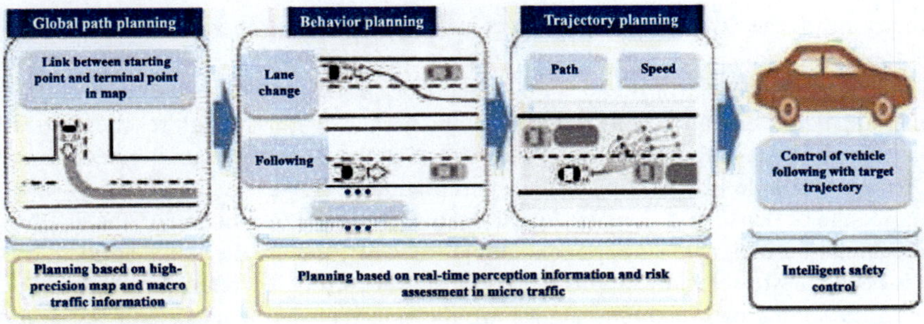

Fig. 3.97 Hierarchical decision process

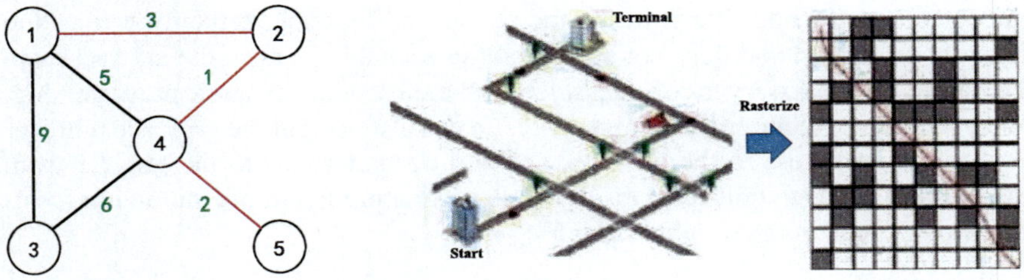

Fig. 3.98 Scene recognition results of Floyd (left) and Dijkstra (right)

1) Global Path Planning

The global path planning takes the map information as the input to plan the desired path connecting the starting point and the target point. Global path planning algorithms mainly include Floyd and Dijkstra. Figure 3.98 shows their typical planning results.

Floyd algorithm: also known as point insertion, mainly uses nodes to represent different locations and their relative positions in the global environment according to the characteristics of High-Definition (HD) map, and abstracts the road information into the node connection path represented by line segments, so as to convert the complex map information into basic map structures. Then, based on the idea of dynamic programming, it will find the shortest path from the beginning to the end through the recursion of adjacency matrix in the weighted and directed graph. Due to the feature dimension reduction brought by the abstracting process, the path planning of Floyd method is characterized by small amount of calculation and high efficiency. Its disadvantage lies in the inaccurate node position that is prone to cumulative error. For global path planning, Floyd is usually sufficient in accuracy. Therefore, owing to small amount of calculation, it is widely used in global path planning of large-scale map.

Dijkstra algorithm: most rasterize the global environment, but some only mark the key nodes in the global environment. During rasterizing, Dijkstra will make binary fill in the grid according to perceived results. The grid with obstacles is set as the unreachable area, and the grid without obstacles is set as the reachable area. Then, based on the idea of Breadth-First-Search, Dijkstra gradually expands its global planning path from the starting point to the end point, and finally acquires the shortest path from the starting point i to the end point j in the established undirected graph. Dijkstra is simple in operation and easy to maintain. However, in the case of large map range or dense grid division for the sake of accuracy improvement, the amount of calculation will increase exponentially [140]. Therefore, in the actual application, it is necessary to select the appropriate grid size according to the map size.

In the global path-planning algorithm, the safety of the planned path is mainly ensured in consideration of the unreachable area. For Floyd algorithm, all nodes and line segments are reachable areas in the graph structure obtained by scenario abstraction. Therefore, there is no need to consider safety after completion of map information extraction. For Dijkstra, the scenario abstraction step only divides the scenario into grids, and does not remove all unreachable areas in the abstract map. Therefore, it is necessary to consider safety in the subsequent dynamic programming. When the safety requirement is high, Dijkstra can also weight the grid, add a higher cost to the grid close to the obstacle, and add a smaller cost to the grid far from the obstacle, so as to transform the dynamic programming into the minimum (cost) trajectory search in the weighted graph.

2) Behavior Decision

In the face of various risk conditions in the real traffic environment, the vehicle cannot always run according to the expected global path. Behavior decision-making and trajectory planning shall work for driving decisions with safety and efficiency based on the real-time state of micro traffic environment. The output of behavior decision-making is semantic driving behavior, such as “left lane changing”, “accelerate for overtaking”, and “deceleration to avoid collision”. At present, the widely-used methods mainly include finite state machine, game theory and Bayesian network.

Finite state machine: is a directed graph composed of state nodes and corresponding state transition functions. The four basic elements of the state machine are the current state of the target, the state transition conditions, the actions executed in the state transition and the new state transferred to after the action. Figure 3.99a shows a basic state machine model: the current state is state 1. If condition 1 is met, it will be transferred to state 2 with operation 1. Similarly, in state 2, if condition 2 is met, it will be transferred to state 1 with operation 2. For some complicated applications, a state machine can be further designed within a state to form a hierarchical finite-state-machine (HFSM) with composite structure. This kind of HFSM is used in driving behavior decision. Its outer layer is responsible for driving scenario identification and its inner layer is for driving behavior decision-making. The scenario identification system continuously subdivides and determines the scenarios where the vehicle is located according to the rules (such as “Lane-obstacle in front”), and then plans the behaviors to be taken (such as “deceleration” or “lane changing”). In spite of being simple and easy, the finite state machine model is insufficient in

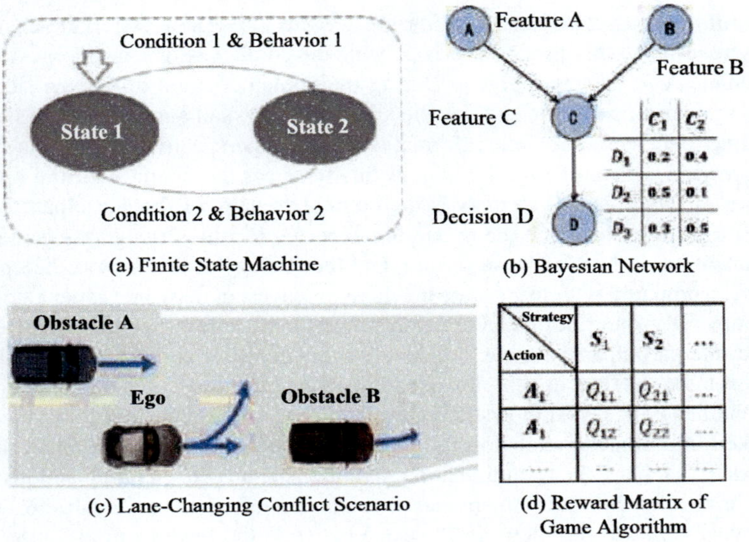

Fig. 3.99 Reward matrix of different algorithms

consideration of environmental dynamics and uncertainty, and thus not competent for complex driving scenarios such as urban roads.

Bayesian network: can generate behavior decisions based on knowledge inference. This kind of decision model simulates the behavior decision-making process of drivers by establishing the mapping relationship between "scenario features" and "driving actions". Its mapping is mainly stored and trained through driving knowledge base. Finally, the driving action in the current scenario is inferred from the knowledge base through the query mechanism [141]. Driving knowledge can be stored in the knowledge base through neural network or in the form of function through data fitting. Specifically, the Bayesian network constructs a directed and acyclic graph as shown in Fig. 3.99b, and uses the conditional probability table to describe the probabilistic relationship between the defined features and decisions. The corresponding parameters in the conditional probability table are obtained through driving knowledge base training. Bayesian network conforms to human decision logic and has strong interpretability. However, the training process relies heavily on the existing driving data, and the parameters of the conditional probability table are obtained through training, which shows "black box" problem and poor transparency. Therefore, Bayesian network is effective in decision-making for general scenarios, but not sufficient for extreme driving scenarios with less driving data.

Game theory: mainly considers the impact of the game process between the ego-vehicle and other surrounding traffic participants on vehicle decision-making. For the game theory, participant, strategy and income are the most basic elements. These elements constitute the game process with action and information, and finally achieve the equilibrium, that is, Nash equilibrium. Figure 3.99c is a basic lane-changing game process. At this time, there is a behavior conflict between the ego-vehicle and vehicle A in the environment. During the implementation of the vehicle lane-changing decision task, the ego-vehicle and vehicle A shall simulate driver's action according to the environmental information from the perspective of game, as shown in the reward matrix of Fig. 3.99d, and finally select the action with the optimal income. At this time, the strategy is the optimal strategy of the participant. When all traffic participants reach the maximum expected income through the game, i.e., the realization of the Nash equilibrium, and the strategy at this time is the optimal strategy. Compared with finite state machine or Bayesian network, game algorithm can more fully describe the interaction process and results between vehicle and other traffic participants around, effectively avoid behavior conflict, and have better decision-making effect in complex interaction scenarios.

In the behavior decision system, the environment and other traffic participants are taken into consideration for collision avoidance and safety assurance. In the finite state machine, the state transition conditions and corresponding actions shall be set in view of the environment and the behaviors of other traffic participants, so as to avoid conflict with them. In Bayesian network, the features used for decision inference generally include environment and other vehicle information, such as TTC, THW, lateral and longitudinal distance and relative speed. In the game theory, for the optimal game that reaches Nash equilibrium, all traffic participants reach the optimal decision in this scenario. Among them, "optimal" is mainly evaluated by economic indicator, safety indicator, comfort indicator, efficiency indicator, etc. According to the emphasis on safety of vehicles, the evaluation mode can be adjusted for driving safety assurance. Compared with the way of considering the unreachable area in the global path planning, the behavior decision system requires higher real-time performance of the scenario, and the decision process shall consider multiple performance indicators such as efficiency, safety and comfort. Therefore, the safety is mainly taken into account through indirect ways (cost/income function, safety properties, etc.).

3) Trajectory Planning

After the target driving behavior is determined by behavior decision layer, the decision-making system shall further plan the vehicle trajectory corresponding to the target behavior. Trajectory planning is mainly divided into two parts: local path planning and vehicle speed planning. In the planning process, local path planning is usually carried out first, and then the corresponding vehicle speed planning is completed based on the path. In order to further improve the decision-making effect, the horizontal-vertical multi-directional risk evaluation based on risk field can also realize the cooperative planning of path and speed.

i. Local Path Planning

Local path planning algorithms usually involve two scenarios: structured road and unstructured road. Structured road includes highway scenarios with clear structure and simple geometric feature such as road marking lines. The planning process is mainly based on the road centerline or reference route. Unstructured road mainly includes less-structured scenarios such as parking lots, which does not have a clear reference route and requires a feasible path in view of objectives and obstacles. The path planning algorithm of IV originated from the research field of mobile robot. Common local path planning includes algorithms based on roadmap, fixed trajectory, grid, sampling, etc.

Roadmap algorithm: is to describe the obstacle as an unreachable area composed of polygons, take the straight-line distance as a weight, and construct a smooth local path from the starting point to the end point based on geometric diagram theory. Common methods mainly include visual diagram and Voronoi diagram, etc. The basic principle of Voronoi diagram is to model the known environment, establish a local planning space map, and divide the planning space into multiple Voronoi areas according to the location of obstacle. Each area has only one obstacle (if it is intersected with the boundary of Voronoi area, it will be fused). The larger the Voronoi area is, the larger the safety boundary of the obstacle is [142]. In this way, the local path planning algorithm shall only search the boundary line of Voronoi area to find the collision-free path from the beginning of the path to the end of the path. Voronoi diagram has the advantages of simple modeling and small amount of data, which can greatly reduce the amount of spatial calculation in complex scenarios and improve the real-time performance of the algorithm. Its disadvantage is that it can only carry out path planning in two-dimensional plane space, and can not deal with the high-dimensional path planning space.

Fixed trajectory algorithm: is to generate intermediate points and form smooth local trajectory by interpolation after determining the starting point and end point of local path. It mainly includes spline curve, Bezier curve, helix, etc. Spline curve is a curve with continuous curvature, which is spliced by high-order polynomials. Common spline curves mainly include cubic spline curve, quartic spline curve and quintic spline curve. The higher the order of the polynomial constituting the spline curve is, the larger the amount of calculation and the smoother the planned path is. Bezier curve is a non-interpolating curve, which forms a path with smooth curvature through several control points. In addition to the starting point and end point, the curvature of Bezier curve can also be controlled through intermediate control points. The trajectory generation of the fixed trajectory is a relatively random process, in which it is difficult to consider the path safety. Therefore, it is mainly used for local path optimization and less used for path planning.

Grid algorithm: is to process the road map with grid or raster and conduct path planning by means of map search according to the obtained grid or rasterized map. It mainly includes A* algorithm and D* algorithm, etc. As a classical algorithm in the field of robot path planning, A* algorithm has the core of heuristic evaluation function $f(n)$, mainly including two parts, i.e., the actual cost of node and starting point $g(n)$, and the predicted cost of node and end point $h(n)$. The smaller the evaluation function $f(n)$ is, the smaller the comprehensive cost of node and the higher the planning priority is [143]. With optimality and completeness, A* algorithm can control the importance of the algorithm to safety and efficiency according to the setting of heuristic evaluation function $f(n)$. It is widely used in the field of path planning. Its defect is that with the increase of nodes in the planning space, its calculation time increases rapidly. Therefore, a variety of optimized A* algorithms such as sparse A* algorithm has come into being with corresponding improvements.

Sampling algorithm: is to connect the starting point and ending point of the required path through node sampling in the planning space and the expansion of nodes in the search space. It mainly includes state grid and RRT, etc.

RRT algorithm: is to take the starting point of the path as its root node, and then generate a random tree to expand the planning space through the random increase of leaf nodes in the space. In order to improve the planning efficiency, the algorithm can also set the expansion range of the random tree in the direction of the target node. When the target node is included in the random tree, the algorithm ends. With fast search speed and no requirement for space modelling, RRT algorithm is suitable for path planning in high-dimensional space. However, it has no optimality. The generated path is only a feasible path, not an optimal or sub-optimal path [144]. Therefore, the algorithm is more available for obstacle avoidance and other functions in emergent situations. When the planning time is sufficient, its path is random and non-optimal, which limits its application.

Optimal planning based on Frenet coordinate system: the above algorithms are all based on path planning in ground coordinate system. At present, the path planning is often modeled as an optimization in Frenet coordinate system. In the Frenet coordinate system shown in Fig. 3.100 (left), the coordinate of any point (s, d) is represented by its distance d to the projection point of the reference path (also known as the lateral distance) and the distance s from the reference point to the original point along the reference path (also known as the longitudinal distance). In current engineering applications, designers often model the path planning as the optimization in Frenet coordinate system: under the constraints of vehicle model and obstacles, path planning facilitates the optimization of such objectives as path smoothness, less vibration, and closeness to the reference path. Figure 3.100 (right) shows a path planning method adopted in a structured road with reference path. Firstly, the Frenet coordinate system is established based on the reference path (i.e. lane centerline), and n reference points are selected longitudinally and equidistantly along the reference path from the origin. The lateral distances of the actual path points relative to their corresponding reference points are $d(i)$ respectively. According to the goal of path planning, the cost function and constraint of the example can be obtained, and a series of path points can be generated to minimize the cost:

$$\min_{d(i)} \sum_{i=0}^{n-1} \{k_1 \cdot d(i)^2 + k_2 \cdot [d(i) - \frac{l(i) + r(i)}{2}]^2 + k_3 \cdot [d(i+1) - d(i)]^2 + \cdots\} \tag{3.152a}$$

Then,

$$\min\{|d(i) - l(i)|, |d(i) - r(i)|\} > C_1 \tag{3.152b}$$

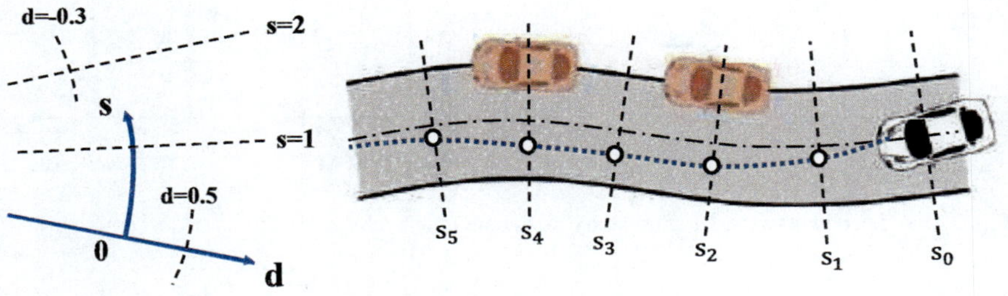

Fig. 3.100 Frenet coordinate system

$$d(i) < C_2 \tag{3.152c}$$

$$[d(i+1) - d(i)] - [d(i) - d(i-1)] < C_3 \tag{3.152d}$$

In the cost function (3.152a) of the example, $l(i)$ and $r(i)$ are the driving ranges on the left/right sides corresponding to the i-th reference point. The parameter $k_1 \sim k_3$ corresponding terms respectively consider the targets of "close to the center line", "passing through the middle of the left and right driving boundaries" and "small lateral distance difference between adjacent path points". The three constraints (b), (c) and (d) correspond to "the distance to the left and right boundaries is more than the safety value", "the deviation from the center line is less than the safety value" and "the lateral distance difference between the adjacent three points is less than the safety value (to avoid excessive bending)".

In local path planning, the safety of vehicle mainly depends on the realization of real-time obstacle avoidance. The roadmap algorithm describes the surrounding areas of obstacles as unreachable paths, which maximizes the generation of non-collision local paths, but reduces the driving area of vehicle. The fixed trajectory algorithm ignores the collision factor, and thus is mainly used for trajectory optimization rather than generation. By gridding or rasterizing the road environment, the grid algorithm represents the obstacle boundary in the raster, and then determines the local path with the cost function. While ensuring the safety boundary, it also obtains the optimal local planning path. In the process of node expansion, the sampling algorithm removes the area where the obstacle is located, and can quickly acquire a safe and feasible path, but its non-optimal characteristics restrict its application. The algorithm based on Frenet coordinate system ensures driving safety through the constraint of cost function, and the constraint mode can be determined freely according to safety requirement.

ii. Speed Planning

Vehicle speed planning refers to the planning of the vehicle speed curve on the target path made by the decision-making system, which is based on behavior decision and local path planning, and aims to ensure safety and optimize efficiency and comfort.

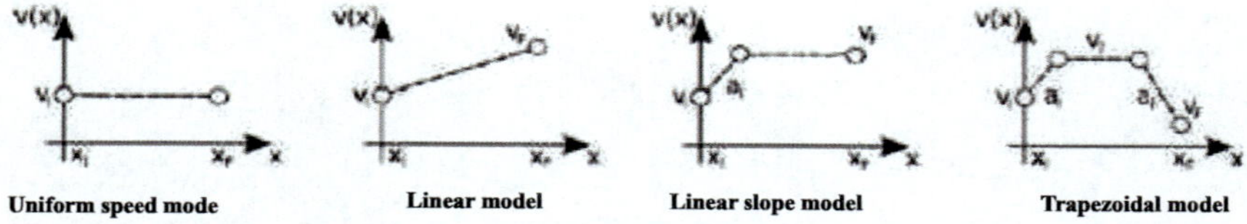

Fig. 3.101 Speed curve model adopted by Boss unmanned vehicle [145]

Speed model: is the speed planning based on specific speed curve model. Figure 3.101 shows various speed curve models adopted by Boss unmanned vehicle [145]. The model takes the position distance on the path as the abscissa and the speed corresponding to the position as the ordinate. In terms of different curves, it can be divided into uniform speed model, linear model, linear slope model and trapezoidal model.

In the planning process, the method first selects the curve model based on the target driving behavior, and then determines the speed and position of each key point on the curve model. The considered information includes: target speed, feasible speed range, legal speed limit, and maximum feasible speed determined by the curvature of the road centerline, etc.

The speed model is convenient to design, and oriented to relatively simple traffic scenarios and driving tasks, such as car following and intersection parking. It can simply calculate the key points on the speed curve based on the speed position information such as the speed of the vehicle in front and the distance from the stop line. However, in contrast, in the face of more complex scenarios and tasks, there is a lack of universal mapping relationship between such environmental information as vehicle speed and such risk assessment information as TTC and key points of the curve, resulting in limited planning capability.

Optimal planning based on S-T coordinate system: in order to overcome the defects of traditional model, some studies transform speed planning to optimization in time–space coordinate system, instead of using specific speed model. Similar to the operation that local path planning can be transformed to optimization in Frenet coordinate system, the speed planning can be transformed to optimization in $s-t$ graph (ST-Graph). Where, s represents the distance of running along the path and t represents the time. In addition, in order to express the speed constraint on the path (for example, the speed of overtaking cannot be too low and the speed of passing pedestrians cannot be too high), it is necessary to supplement the $v-s$ graph to show the speed constraint. Referring to the Apollo case [146], the speed planning for the scenario of adjacent vehicle cut-in on highway is shown in Fig. 3.102. The left side of the figure shows that the ego-vehicle runs straight along the center of the lane, and the obstacle vehicle on its left side is ready to cut in; The upper right shows the maximum speed limit of the vehicle at each position of the path and the lower right shows the speed planning in ST-Graph.

In the lower right of the figure, the gray quadrilateral area represents the space that obstacle vehicles may occupy in ST-Graph, which is also the risk representation

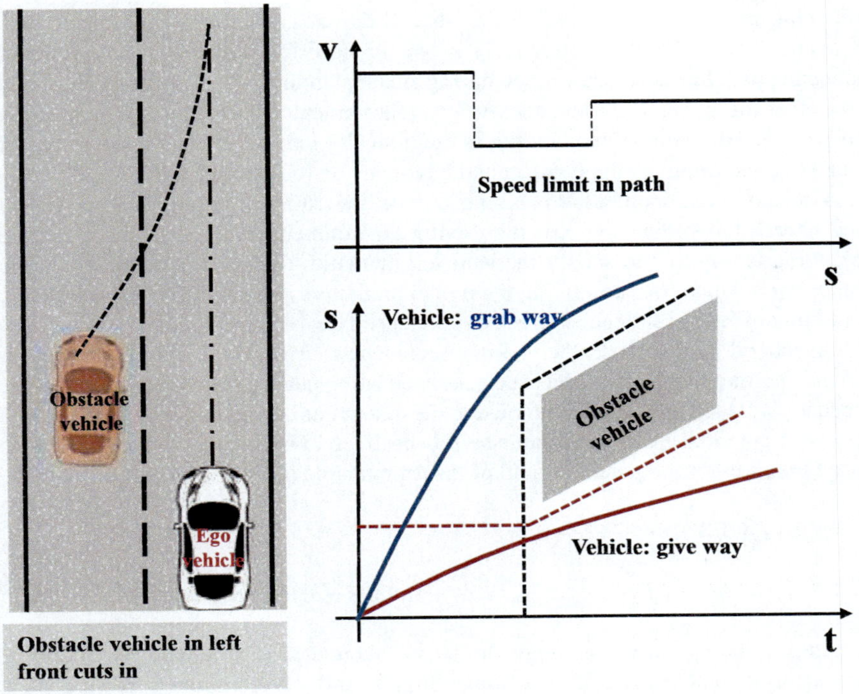

Fig. 3.102 Speed planning based on ST-graph

based on speed planning. Collision avoidance requires the planned speed curve shall not enter the gray area. In order to reasonably express the gray area, ensure driving safety and avoid over-conservative decision, it is necessary to predict the movement of obstacle vehicles as accurately as possible. Meanwhile, in view of the uncertainty of vehicle behavior, the collision risk can also be described by potential field and other means, and the speed curve with the best comprehensive safety and efficiency is finally selected.

Speed planning is constrained by behavior decision (rushing or yielding) in the process. For example, in the case of giving way, the curve shall be below the red dotted line. After that, by inputting the above speed position constraints and position time constraints into the optimizer and considering the optimization objectives such as "speed as stable as possible" and "away from obstacles", the optimal speed curve can be planned in ST-Graph (red curve indicates yielding; blue curve indicates rushing).

iii. Local Path-Speed Cooperative Planning

Path-speed hierarchical motion planning is conducive to reducing the computational complexity of the planning process, and ensuring the safety, comfort and driving efficiency of the results in the simple scenarios. However, the above hierarchical planning has such defects as follows: due to the lack of information during path planning, it is difficult to effectively avoid the collision with dynamic obstacles; since the path has been determined during speed planning, the planning results may not meet the dynamics/kinematics safety requirements. Therefore, the realization of path-speed cooperative planning is theoretically helpful to ensure the safety of planning and optimize the comfort and efficiency of vehicle driving.

For path-speed cooperative planning, the risk assessment is required to fully reflect the lateral-longitudinal risks of the driving environment on the one hand, and the planning is required to satisfy the vehicle kinematics/dynamics constraints on the other hand. Under this condition, the model predictive control (MPC) based on risk field model [147] can realize effective path-speed cooperative planning.

In general, ICV shall deal with obstacles (such as surrounding vehicles and pedestrians), surmountable obstacles (such as speed bumps) and road boundaries in traffic scenarios. Based on the risk field model, the mathematical models of the above three types of potential fields are established respectively. Firstly, hyperbolic function is used to represent the potential field of insurmountable obstacles, namely:

$$\begin{aligned} U_{NC_i}(X,Y) &= \frac{a_i}{s_i\left(\frac{dX}{X_{si}},\frac{dY}{Y_{si}}\right)^{b_i}} \\ X_{si} = X_0 + uT_0 + \frac{\Delta u_{ai}^2}{2a_n}, &\ Y_{si} = Y_0 + (u\sin\theta + u_{oi}\sin\theta)T_0 + \frac{\Delta v_{ai}^2}{2a_n} \end{aligned} \tag{3.153}$$

where, a_i and b_i are respectively the shape parameter and density parameter of the insurmountable potential field function; dX and dY respectively represent the longitudinal distance and lateral distance between the ego-vehicle and the obstacle vehicle; s_i represents the SD distance [147], X_0 and Y_0 respectively represent the minimum longitudinal distance and minimum lateral distance to be maintained between the ego-vehicle and the obstacle; X_{si} and Y_{si} respectively represent the longitudinal and lateral safety distance to be maintained between the ego-vehicle and the obstacle; T_0 represents the safety time interval; u and u_{oi} respectively represent the vehicle speed and obstacle speed; θ is the angle between vehicle and the obstacle; u_{ai} and v_{ai} respectively represent the longitudinal relative speed and lateral relative speed between the ego-vehicle and the obstacle.

Secondly, the exponential function is used to express the potential field of surmountable obstacle:

$$U_{Cj}(X,Y) = a_j e^{-b_j s_j} = a_j e^{-b_j s_j\left(\frac{dX}{X_{si}},\frac{dY}{Y_{si}}\right)} \tag{3.154}$$

where, a_j and b_j respectively represent the shape parameter and density parameter of the potential field; s_j represents the normalized distance between the ego-vehicle and obstacle vehicle.

Then, the quadratic function is used to express the road boundary:

$$U_{Rq}(X,Y) = \begin{cases} a_q[S_{Rq}(X,Y) - D_a]^2 & [S_{Rq}(X,Y) < D_a] \\ 0 & [S_{Rq}(X,Y) > D_a] \end{cases} \tag{3.155}$$

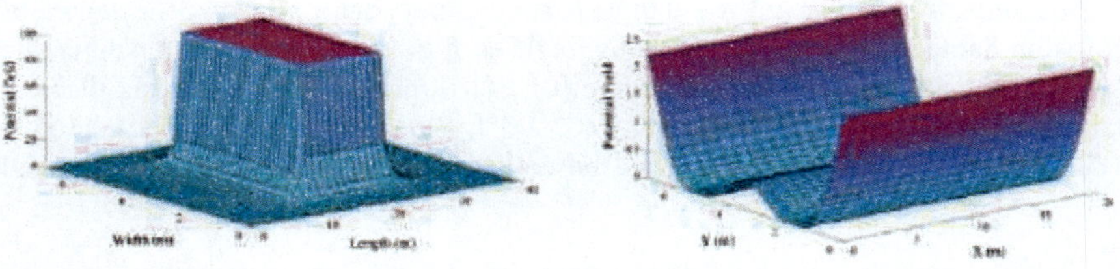

a) Insurmountable obstacle b) Lane line and road boundary

Fig. 3.103 Risk field modeling

where, S_{Rq} represents the SD value of the vehicle from the road boundary; D_a indicates the distance that the vehicle is allowed to approach the road boundary; Subscripts $q = r$ and l represent the left and right boundaries of the road respectively; a_q represents the density parameter of the road potential field. Figure 3.103 shows the risk field modeling.

In addition, in the face of inevitable collision, the planner shall also alleviate the damage caused by the collision. Figure 3.104 shows the hazard scenario faced by a smart car (silver). Vehicle 1 (Orange) is changing lanes to the right when vehicle 2 is driving in front of vehicle 1 at a lower speed. At this time, there are two selection strategies for the ego-vehicle, namely strategy 1 trajectory and strategy 2 trajectory in the figure. If the vehicle chooses the route of strategy 1, it will collide with vehicle 2 at a certain angle; If it chooses the route of strategy 2, it will collide with vehicle 1 in the rear in spite of its deceleration with the maximum braking intensity.

The potential crash severity index (PCSI) is used as the main indicator to measure the degree of collision damage. First, the item related to speed in PCSI index is defined as [143]:

$$PCSI(\Delta V) = \frac{\Delta V}{D} \tag{3.156}$$

where, ΔV and D respectively represent the relative speed and vehicle spacing between the ego-vehicle and obstacle vehicle.

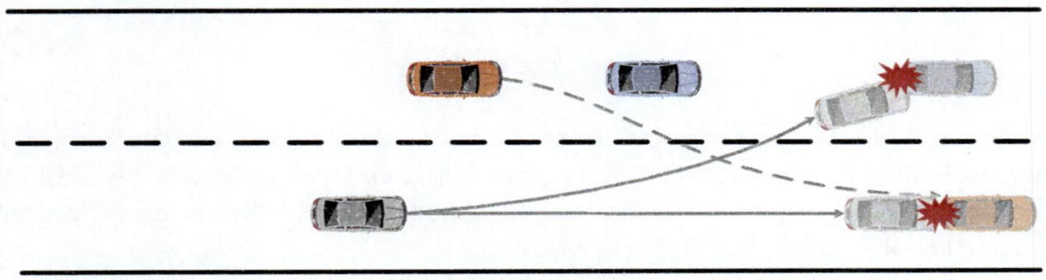

Fig. 3.104 Hazard scenario with collision damage

Secondly, the relative collision angle is another key factor affecting the degree of collision damage. The relative heading angle θ of the two vehicles is taken as the relative collision angle with value range [0°, 180°], and is divided into six discrete ranges, namely [0°–15°), [15°–45°], [45°–90°], [90°–135°], [135°–165°], [165°–180°]. Therefore, the item related to the collision angle in PCSI index is defined as:

$$\mathrm{PCSI}(\theta) = \begin{cases} 1, \theta \in \{0°, 180°\} \\ 4, \theta \in (0°, 15°] \cap [165°, 180°) \\ 2, \theta \in (15°-45°] \cap [135°-165°) \\ 3, \theta \in (45°, 135°) \end{cases} \tag{3.157}$$

In addition, the mass ratio between the two vehicles will also be affected. Therefore, the items related to PCSI index and quality ratio are defined as:

$$\mathrm{PCSI}(W) = \frac{W_0}{W} \tag{3.158}$$

where, W_0 represents the mass of the obstacle vehicle; and W Indicates the mass of the ego-vehicle.

To sum up, the calculation formula of PCSI index is:

$$\mathrm{PCSI} = k_{\Delta v}\mathrm{PCSI}(\Delta V) + k_\theta \mathrm{PCSI}(\theta) + k_w \mathrm{PCSI}(W) \tag{3.159}$$

where, $k_{\Delta v}$, k_θ and k_w respectively represent the weight coefficients corresponding to different factors.

Therefore, the established potential field function can be expressed as:

$$U = \sum_i U_{NC_i} + \sum_j U_{C_j} + \sum_q U_{R_q} \tag{3.160}$$

MPC is adopted to realize the trajectory planning of ICV. First, it is to define the system state as $\boldsymbol{x}$, the system control quantity as $\boldsymbol{u}$ and the system output quantity as $\boldsymbol{y}$, then the system state space model can be expressed as:

$$\begin{aligned} \dot{\boldsymbol{x}} &= \boldsymbol{A}\boldsymbol{x} + \boldsymbol{B}\boldsymbol{u} \\ \boldsymbol{y} &= \boldsymbol{C}\boldsymbol{x} + \boldsymbol{D}\boldsymbol{u} \end{aligned} \tag{3.161}$$

where, $\boldsymbol{A}$, $\boldsymbol{B}$, $\boldsymbol{C}$ and $\boldsymbol{D}$ represent vehicle state matrix, input matrix, output matrix and feedforward matrix respectively. The matrix form is related to the adopted vehicle dynamics model. Common vehicle dynamics models will be further introduced in the section of "Vehicle Intelligent Control".

Finally, the cost function of MPC is defined, including potential field U, collision severity PCSI, expected output $\boldsymbol{y}_{\mathrm{des}}$, penalty term of control quantity change $\boldsymbol{Q}$, relaxation variable $\boldsymbol{R}$ and weight matrix $\boldsymbol{S}$,$\boldsymbol{P}$ of corresponding items. The cost function is defined as follows:

$$\begin{aligned}
\min J &= \min_{u_c,\varepsilon} \sum_{k=1}^{N_p} (U^{t+k,t} + PCSI^{t+k,t} + \|y^{t+k,t} - y_{\text{des}}^{t+k,t}\|_Q^2 + \|\varepsilon_k\|_P^2) \\
&+ \sum_{k=1}^{N_c} (\|u_c^{t+k-1,t}\|_R^2 + \|u_c^{t+k-1,t} - u_c^{t+k-2,t}\|_S^2) \\
st&: \\
x^{t+k,t} &= Ax^{t+k-1,t} + Bu_c^{t+k-1,t} \\
y^{t+k,t} &= Cx^{t+k,t} + Du_c^{t+k,t} \\
\mathbf{u}_{\text{c_min}} &< u_c^{t+k-1,t} < u_{c_\max}
\end{aligned} \tag{3.162}$$

where, N_p represents the prediction horizon; N_c represents the control horizon; in superscript $t + k - 1, t$ indicates the predicted value of the forward k steps; ε_k represents the relaxation variable of the k-th step. In addition, the value of system control quantity u shall meet the response constraints.

So far, the IV decision planning has been transformed to the optimal control in Formula (3.162).

The following test scenarios are designed to verify the effectiveness of the proposed safety decision algorithm. As shown in Fig. 3.105, the ego-vehicle (silver) runs on Road 1 at the initial speed of 100 km/h, and obstacle vehicle 1 and obstacle vehicle 2 run on Road 2 at the speed of 60 km/h and 45 km/h respectively. The initial positions of the ego-vehicle, obstacle vehicle 1 and obstacle vehicle 2 are [2, 1.75] m, [22, 5.25] m and [30, 5.25] m respectively. Obstacle vehicle 1 changes lanes from road 2 to road 1, and obstacle vehicle 2 keeps going straight.

In this scenario, the ego-vehicle cannot avoid collision with obstacle vehicle 1 and obstacle vehicle 2. The simulation environment is built based on Matlab/Simulink, and the decision algorithm without considering the degree of collision damage is designed to illustrate the effectiveness of the proposed safety decision algorithm. Figure 3.106 shows the driving trajectory using the decision algorithm without considering the degree of collision damage, that is, the blue driving route in the figure. The ego-vehicle selects emergency braking at $t = 1$ s, and finally collides with obstacle vehicle 2 at a certain angle at 2.05 s. During collision, the positions of the ego-vehicle and the obstacle vehicle 2 are [51.84, 3.25] m and [55.63, 5.25] m respectively. However, the two vehicles collide obliquely and laterally, and the speed

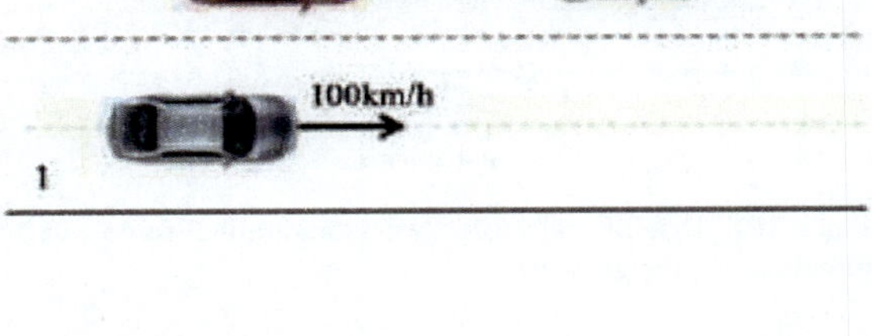

Fig. 3.105 Collision risk scenario

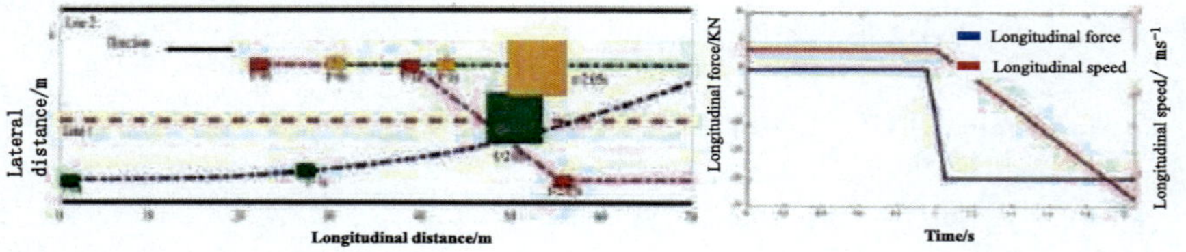

Fig. 3.106 Driving trajectory (left), longitudinal force and driving speed (right) without considering collision damage degree

difference is large at this time, which may lead to vehicle rollover and secondary collision, with severity much higher than that of frontal collision.

Figure 3.107 shows the driving trajectory of the vehicle using the decision algorithm in consideration of the degree of collision damage. It can be seen from the figure that the ego-vehicle selects the driving route of rear-end collision with obstacle vehicle1, which reduces the degree of collision damage, and the collision time is $t = 2$ s. The ego-vehicle and the obstacle vehicle 1 are at the positions of [51.00, 1.75] m and [55.33, 1.75] m during the collision, when the relative speed is low and the corresponding damage risk is low. The experimental results show that when the collision is inevitable, the decision algorithm considering the degree of collision injury is helpful to optimize the form of collision accident, reduce accident damage and improve vehicle safety.

(1) Novel Vehicle Safety Decision Methods

With the rapid development of artificial intelligence in recent years, many novel methods have emerged in the field of vehicle safety decision-making. These methods adopt decision frameworks completely different from the existing rule-based methods mentioned above. These methods made relatively better decision-making performance in some scenarios, and achieved better efficiency and comfort while ensuring safety.

There are mainly two kinds of novel decision methods. One is based on machine learning from the perspective of algorithm optimization; The other kind starts from the decision cognitive mechanism and generates human-like driving strategies by learning and simulating the driver's decision-making mechanism, such as the decision model based on the least action mechanism of driver behavior.

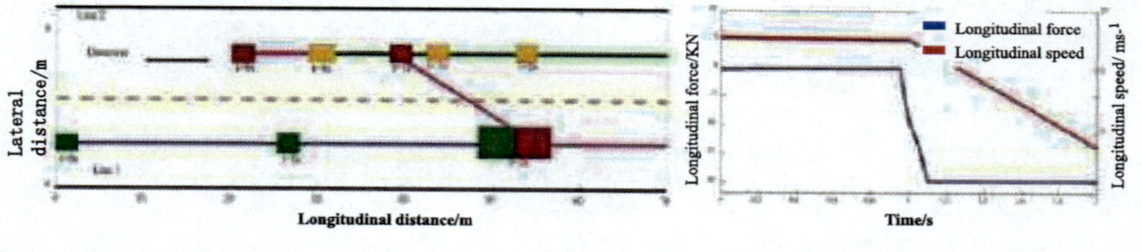

Fig. 3.107 Driving trajectory (left), longitudinal force and driving speed (right) in consideration of collision damage degree

1) Decision Technology Based on Machine Learning

Since the publication of the milestone article "Deep Learning" in *Nature*, 2015, machine learning and deep learning have entered a stage of rapid development. At the end of last century, although multilayer perception (MLP), long short-term memory (LSTM) and many well-known machine learning algorithms have been proposed, these algorithms did not perform well at that time due to insufficient computing power and imperfect hardware platform. However, with the explosive improvement of the performance of GPU, CPU and other computing units in recent decades, large-scale data mining is no longer an impossible task, and the effectiveness of machine learning methods have gradually showed their potential in various fields of artificial intelligence and began to occupy the mainstream, such as natural language processing, intelligent medicine and computer vision.

Automated driving is also a key field of artificial intelligence application. IOVs are considered to be capable to greatly improve traffic safety, reduce accident occurrence, speed up urban transportation efficiency and better serve mankind. In all aspects of automated driving, decision-making is generally regarded to be a more complex and key part. The main challenges faced by current intelligent safety decision include:

i. Requirement diversity of drivers: different drivers have different habits and needs. Some drivers pay more attention to safety, while others pay more attention to efficiency or fuel consumption. Therefore, different requirements of different drivers shall be met.
ii. Complicated and endless traffic elements: the driving environment faced by automated driving includes quite a number of elements such as VRUs (pedestrians and cyclists), road infrastructure and other vehicles. The types are endless and change dynamically at the same time. The inference and understanding of the relationship between each object and the vehicle are very essential for safety decision.
iii. Incompatibility among the decision-making systems: due to difference in interfaces and protocols, how to realize decision exchange among the AV with different architectures is also one of the difficulties.

Facing the above three challenges, some decision-making methods based on machine learning have solved part of solutions. Although the specific algorithms used are different (the algorithms will be introduced later), the way of considering safety in machine learning decision is unified, that is, safety is modeled in the objective function, and then different algorithms are applied to minimize the objective function. In this way, the decision output will naturally ensure driving safety. Typical quantitative methods for driving safety are as follows:

i. Distance-based safety quantification: e.g., the distance from the front vehicle, the distance from the rear vehicle, the distance from the lane line, the lateral deviation from the scheduled route, and time headway.
ii. Speed-based safety quantification: e.g., the speed difference with the relative speed of the vehicle, gap between the current speed and the reference speed (such as road speed limit).

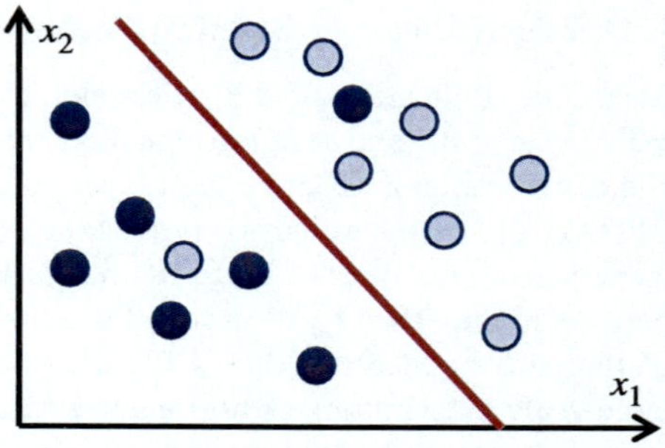

Fig. 3.108 SVM decision

iii. Collision probability-based safety quantification: time to collision (TTC, which has been described in detail in the previous article), collision penalty term (in case of collision, the constant value is subtracted from the objective function).
iv. External constraint-based safety quantification: traffic rule reward (subtract the constant value from the objective function when a certain rule is broken), traffic rule genetic strategy, safety potential field, etc.

In accordance with the safety indicators defined above, some decision-making methods perform optimization iterations in the defined time domain and space domain based on machine learning algorithm to find the optimal solution. In terms of specific classification, the algorithms based on machine learning for safety performance can be divided into naive learning and deep learning, which are introduced accordingly in this study.

A) Naive Learning

Naive machine learning methods include support vector machine (SVM, as shown in Fig. 3.108), naive Bayesian neural network, PCA dimensionality reduction, decision tree, random forest and other classical machine learning methods, with the following characteristics:

i. The parameters of the model are obtained automatically through data training without manual adjustment.
ii. In order to ensure the performance of the model, a certain amount of data is required as input. The closer the distribution of data is to the assumed distribution of the model, the stronger the performance of the model is.
iii. The essence of the model is to fit the input data set in probability.

Compared with deep learning, naive learning has no back propagation process, and thus all parameters can be obtained at one time after determination of the data input.

Support vector machine (SVM): SVM is a commonly-used classification method, and is also used for decision-making in specific scenarios in many studies. For example, in terms of lane changing, the output category of decision-making can be divided into two categories: "lane changing" and "no lane changing". Such multiple classifications can be solved by SVM.

Bayesian network: Bayesian network is another commonly-used method. For example, the idea of Bayesian network with normal distribution is that each decision output corresponds to a multivariate normal distribution, fits the mean and variance parameters of the multivariate normal distribution through the input data, and finally makes the decision corresponding to the distribution by judging which distribution is the closest to the environment.

Random forest and decision tree: the idea of random forest and decision tree is similar to SVM, and the decision is regarded as a classification with multiple inputs. The decision tree shall be constructed with definition of the fuzzy division of input variables, as shown in Table 3.14. The final decision is whether to change lanes. The input variables are front vehicle distance, lateral vehicle speed and lateral vehicle position, marked with three levels of high, medium, low or near, middle and far respectively. Then, the information entropy gain is calculated at each layer, and the feature with the largest gain is used as the classification feature to realize multi-tree classification until the decision output of leaf nodes is exactly the same. Random forest goes further on the basis of decision tree. Multiple trees are constructed in the same problem. Each tree uses part of the training data, and thus is independent with different structure. When making a decision, the current information is input into each tree, and the final result lies in the decision input of most trees. For example, there are 1100 pieces of data in the current training data set. A decision tree is constructed for every 100 pieces, and a total of 11 trees are constructed. During decision-making, if the results of 7 trees are lane-change and the results of 4 are no lane-change, the overall forest decision is lane-change.

Table 3.14 Example of decision tree data input

Front vehicle distance	Lateral vehicle speed	Lateral vehicle position	Lane changing
Near	Low	Near	Change
Near	High	Far	Change
Near	High	Near	No change
Middle	Low	Far	Change
Middle	Low	Near	No change
Middle	High	Near	No change
Far	Low	Far	Change
Far	Low	Near	Change
Far	High	Far	No change
Far	High	Near	No change

B) Deep Learning

At present, there are few practical applications of deep learning for decision-making, but there have been some attempts in academic research, such as convolutional neural network and recurrent neural network.

Convolutional neural network (CNN): A typical example is that in the car-following scenario, CNN can be used to take the continuous image in front of the vehicle as the input and directly output the ideal acceleration of the ego-vehicle [148]. The function of CNN is to simulate the driver's attention mechanism, obtain the focus in the space, and establish the mapping between the comprehensive state of the front vehicle and the ideal operation of the ego-vehicle. CNN is also adopted as a separate data preprocessing module to extract spatial features in complex traffic environment, so as to determine the areas of concern for the ego-vehicle decision-making and simulate the cognitive process of drivers.

Recurrent neural network (RNN): Time sequence model is also a popular research field, and is suitable for solving decision-making, which is a continuous event in time domain as well. For example, RNN and long-short term memory model (LSTM) can extract valuable information for decision-making in time sequence in the form of forgetting gate, output gate and input gate. Time sequence network can not only extract the time features of the past, but also predict the future state, so as to provide a valuable reference for decision-making. In addition, CNN and LSTM can also be combined into ConvLSTM network, which takes the long-time continuous video as the input, and uses sequential networks to extract the temporal correlation of spatial states and obtain the spatio-temporal inference results.

Reinforcement learning: In addition, as a branch of deep learning, reinforcement learning has been widely concerned in the field of intelligent decision. The core framework of reinforcement learning is self-learning based on the defined reward function in the predefined state quantity and operation set. It is necessary to ensure that the training environment is quite similar to the actual environment. The weight of each item in the reward function can be trained by inverse reinforcement learning for the balance among safety, timeliness and energy consumption. Reinforcement learning has made some progress in single scenarios such as car following, lane changing and intersection, but has the major deficiency in its low algorithm generalization [149]. How to improve its scenario generalization is also one of the hot topics in current frontier research.

2) Decision Technology Based on the Principle of Least Action

In the previous article, the principle of least action has been briefly introduced. And in this section, how to apply it to intelligent safety decision technology will be described in detail. There is an extreme value phenomenon in the driver's decision-making process [150], which is not a simple single quantity. Therefore, it is inaccurate to use only a single quantification such as THW and passage time. According to the driver's cognition, the extreme value should be the quantity with comprehensive consideration of all risks during driving. Based on this action quantity and the driving characteristics of "pursuing benefit and avoiding harm" analyzed above, the driver's intelligent decision process can be simulated through the principle of least action [151].

Assuming that the comprehensive quantity is defined as S, and the decision of the vehicle is d, the principle of least action can be expressed indirectly by the following formula, and the ideal decision behavior shall make the action quantity change to 0.

$$\delta S(d) = 0 \tag{3.163}$$

In the process of driving, the risk related to intelligent safety can be divided into two categories: dynamic risk posed by dynamic behavior, and static risk posed by relative position. Combined with the concepts in Physics (mainly Mechanics), the former can be quantified by kinetic energy, and the latter can be represented by potential energy. The potential energy comes from the existence of force. Therefore, in the least action decision system, it is necessary to define the virtual force F, and then integrate in the virtual displacement direction Δx to obtain the potential energy change, as shown in Formula (3.164), in which there may be multiple virtual forces. Therefore, it is necessary to add the vectors of virtual forces and then integrate the resultant force.

$$U = \int_0^{\Delta x} \sum F_i d\Delta x \tag{3.164}$$

Potential energy shall include many factors to ensure the effectiveness and safety of decision. By means of experimental verification and analysis, it shall at least include the following factors:

i. Target driving force: traffic flow always moves in a certain direction like an object in free fall. The driver's target driving force makes the vehicle move from the starting position to the ending position. When there is no lane changing, driver's target driving force only exists in the longitudinal direction; When the driver chooses to change lanes, driver's target driving force will have a component in the lateral direction due to the lateral movement. Therefore, the target driving force always ensures that the result of the algorithm can make the vehicle get closer to the target instead of being involved into local optimization.
ii. Traffic environment constraint resistance: The previous chapter systematically models various typical risk constraints, including physical constraints (pedestrians, bicycles, fixed traffic facilities, etc.) and virtual constraints (signal lights, traffic rules, additional risks posed by characteristics of the object, etc.). Based on the risk value of DSF, the constraint resistance made by the traffic environment can also be decomposed in the lateral and longitudinal directions in order to quantify the potential energy.

iii. Risk external force posed by the static state of other vehicles: The static state here is similar to other static obstacles (trees, sign barrels, etc.) in the traffic environment.

The definition of kenetic energy is relatively simple and direct. It can be defined directly with the kenetic energy of vehicle motion. Here, the kenetic energy brought by its own rotation angular speed during steering is ignored in modeling. If the vehicle motion is regarded as translational, the kenetic energy is expressed as follows:

$$T = \frac{1}{2} m_i v_i^2 \tag{3.165}$$

where, i is the number of the vehicle; m_i is the mass of the vehicle i; and v_i is the speed of the vehicle i.

In order to obtain the ideal decision output at the next moment, the Lagrange quantity L can be defined as the difference between kenetic energy and potential energy with reference to the Lagrange equation in mechanics, as shown in the following formula:

$$L = T - U \tag{3.166}$$

If the current time is t_o and the decision ending time is t_f, the actual action S of the vehicle driving process can be defined as:

$$S = \int_{t_o}^{t_f} L dt \tag{3.167}$$

If S^* is used to represent the least action corresponding to the actual action during vehicle driving in the virtual mechanical system, there are:

$$S^* = \int_{t_o}^{t_f} L^* dt \tag{3.168}$$

where, L^* is the Lagrangian quantity that strictly follows the virtual mechanical system.

In order to evaluate the safety degree of driver's driving process, D_{SE} is defined as the index. When the traffic efficiency is certain, the greater its value is, the higher the risk degree is. Its expression is as follows:

$$D_{SE} = \frac{\exp\left[\frac{\eta(S-S^*)}{S_{max}-S^*}\right] - 1}{\exp(\eta) - 1} \tag{3.169}$$

where, η is the calibration coefficient; S^* is the least action of the driving process; S is the actual action of the driving process; S_{max} is the allowable maximum action. When the actual action exceeds this value, the accident will occur.

The principle of least action is not based on on a specific scenario comprehension and risk assessment algorithm, but is applicable to any risk assessment model with the physical meaning and interpretability. The following is a case of decision-making based on the least action. Here, the elliptical spring model is established for the risk assessment around the vehicle, as shown in Fig. 3.109.

As shown in the above figure, A_1A_2 and B_1B_2 are the major axis and minor axis of the ellipse respectively, and the ego-vehicle j is the center of the ellipse. At any point P on the ellipse, there is a virtual spring between the ellipse and the ego-vehicle, which will generate a virtual compression force when another traffic participant approaches. The virtual spring stiffness at any point on the same ellipse is the same. In the actual decision process, the major axis and minor axis of the ellipse are respectively set as follows, where a is the major axis, b is the minor axis, t_{set} is the constant duration, here set as 1.5 s, and v_j is the speed of the ego-vehicle.

$$a = t_{\text{set}} v_j \tag{3.170}$$

$$b = l_w \tag{3.171}$$

Under this definition, the virtual force is calculated as follows, where d_{ji} represents the distance between the ego-vehicle and another vehicle entering the ellipse, as shown in Fig. 3.110:

$$F_{ji} = k_1 \Delta x_{ji} \tag{3.172}$$

$$\Delta x_{ji} = a - d_{ji} \tag{3.173}$$

$$k_1 = \frac{m_j v_j^2}{2ad_{ji}} \tag{3.174}$$

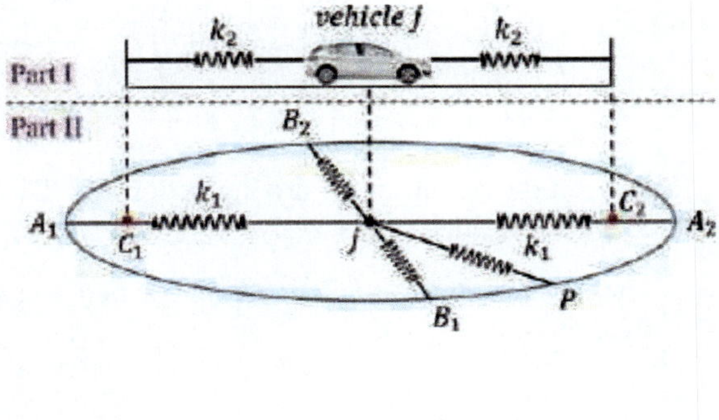

Fig. 3.109 Elliptical spring model

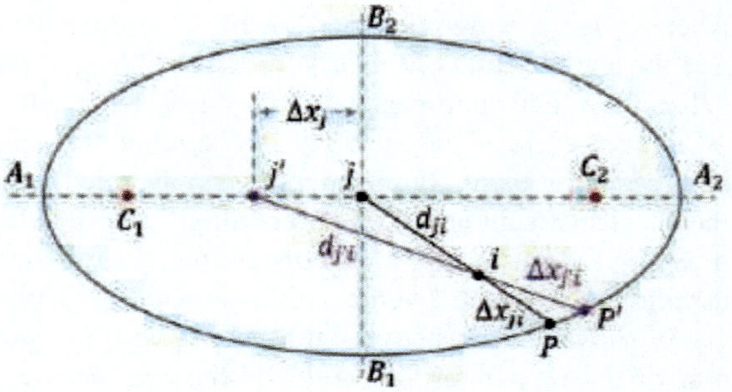

Fig. 3.110 Calculation of elliptical spring model

Therefore, the potential energy made by other traffic participants can be calculated as follows:

$$U_{ji} = \int_{0}^{\Delta x_{ji}} F_{ji} d\Delta x_{ji} \tag{3.175}$$

Then, the driving force is modeled and the attraction is defined as follows, where k is a constant, set as 0.2 here; v_{der} is the expected speed of the driver; v_{limit} is the speed limit of the road; and g is the acceleration of gravity.

$$G_i = m_i g_h \tag{3.176}$$

$$g_h = k \frac{v_{der}}{v_{limit}} g \tag{3.177}$$

If the vehicle motion is still regarded as translational, the expression of kenetic energy is as follows:

$$T = \frac{1}{2} m_j v_j^2 \tag{3.178}$$

Therefore, the driving action of the overall process can be expressed as follows:

$$S = \int_{t_o}^{t_f} L dt = \int_{t_o}^{t_f} \left[\frac{1}{2} m_j v_j^2 - \int_{0}^{\Delta x_{ji}} (F_{ji} + G_i) d\Delta x_{ji} \right] dt \tag{3.179}$$

Based on the definition of driving action, six T-junction scenarios as shown in Fig. 3.111 are verified respectively:

i. The ego-vehicle turns left, and a vehicle in the opposite direction is approaching from the left;

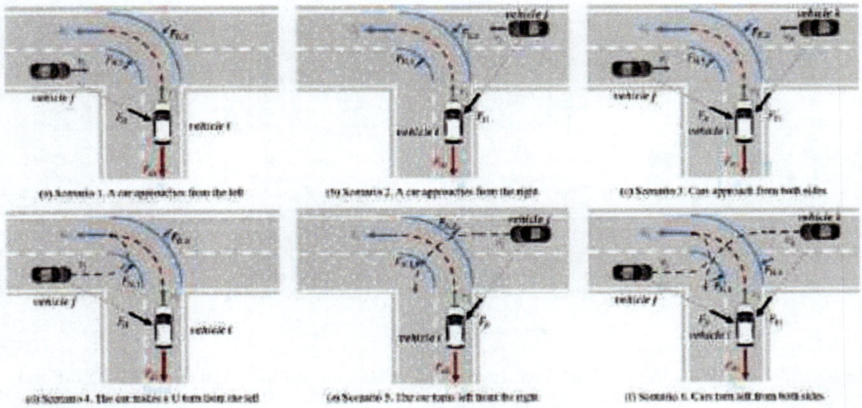

Fig. 3.111 Verification scenario

ii. The ego-vehicle turns left, and a vehicle in the same direction is approaching from the right;
iii. The ego-vehicle turns left, a vehicle in the opposite direction is approaching from the left, and a vehicle in the same direction is approaching from the right;
iv. The ego-vehicle turns left, and a vehicle on the left in the opposite direction is trying to turn around.
v. The ego-vehicle turns left, and a vehicle on the right in the same direction is trying to turn left;
vi. The ego-vehicle turns left, a vehicle on the left in the opposite direction is trying to turn around, and a vehicle on the right in the same direction is trying to turn left.

In this case, the least action is compared with a classic rule-based TTI (time to intersection) decision framework. The decision results are shown in Fig. 3.112, where gray indicates the probability of waiting for the other party to pass, red indicates the probability of collision, and green indicates the probability of passing through the intersection before other vehicles. It can be concluded by the analysis that the decision made with the least action is much better than that with TTI. The former not only realizes zero collision and improves the safety of intelligent decision, but also greatly enhances the traffic efficiency.

In addition, the verification results also reflect the personalization of the least action. It is indicated by the analysis that the least action is more detailed in the operation. In the face of potential collision, it will decelerate in advance, so that even if the other vehicle continues to pass, it can brake in time to prevent collision. When the other vehicle is more conservative, the decision generated by the least action will appear more radical and forces the opposite vehicle to yield. It is proved that the direct passage probability based on the least action is the highest. The result of the least action can adjust driving style of the vehicle according to the behavior of the

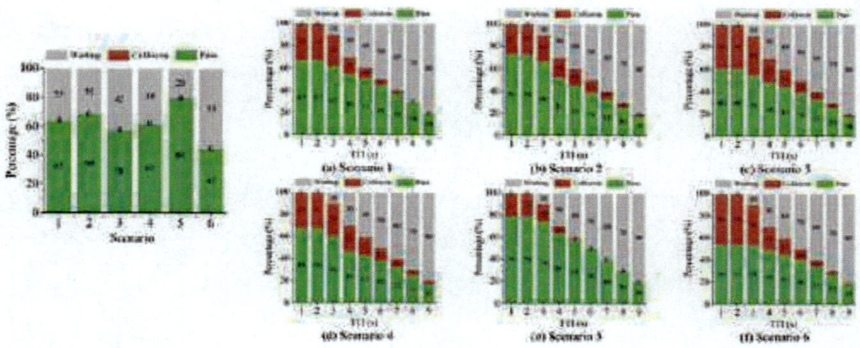

Fig. 3.112 Simulation results based on least action (left), and results based on TTI with different thresholds in six scenarios (six graphs in the right)

other vehicle, which is consistent with driver's cognitive law of "pursuing benefit and avoiding harm" and the concept of decision design. In conclusion, the decision based on least action has a wide application prospect in intelligent decision safety.

2. Multi-vehicle Safety Decision-Making

The intelligent decision-making technology for single vehicle is expected to achieve the goal of energy-saving, efficient and comfortable driving on the premise of ensuring safety. However, vehicles are not isolated individuals, and a traffic system is composed of all vehicles and facilities in the area. With the development of communication technology, researchers further take the vehicle–road-cloud integrated system as the research object, and optimize system safety and efficiency by means of multi-vehicle cooperative decision-making and control.

Multi-vehicle cooperative decision-making is a new theoretical research field in recent years. It is mainly based on the communication between vehicle-vehicle or vehicle-infrastructure, and coordinates the vehicle behavior and signal timing in a certain area. The research for multi-vehicle cooperative safety decision algorithm mainly focuses on two types of traffic scenarios: platoon formation decision-making and intersection passing decision-making. Two types of scenarios and corresponding decisions are shown in Fig. 3.113.

(1) Platoon Formation Decision

The decision-making content of vehicle platoon formation often includes three types [152]: ① Formation generation and maintenance. Each ICV is involved in a formation according to its initial position; ② Formation maintenance during driving; ③ Formation change management. The platoon transformation and recovery can be quickly completed in the case of obstacles avoiding or vehicles entering and leaving the platoon. The above decision tasks can be accomplished with three common methods based on leader following, virtual structure and behavior [153], as shown in Table 3.15.

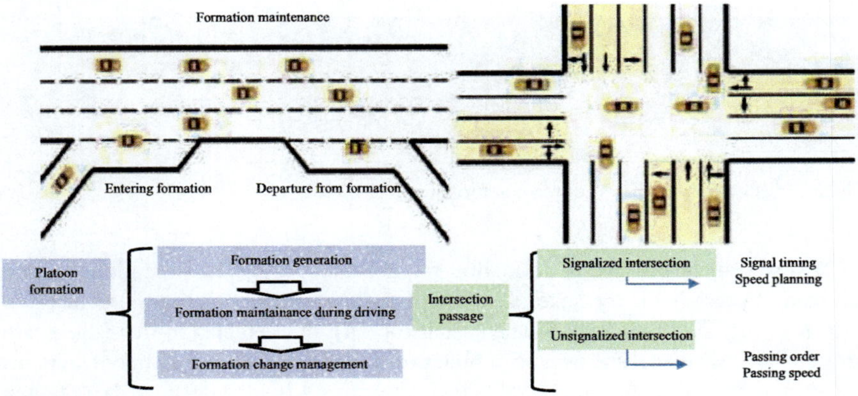

Fig. 3.113 Two typical scenarios based on multi-vehicle cooperative decision

Table 3.15 Vehicle platoon formation

Method	Basic principle	Characteristic
Leader–follower	There is one leading vehicle in the formation, and the rest following vehicles track the leader's position and speed based on the geometry of the platoon	(1) The algorithm is simple in design, and is helpful to reduce tracking errors (2) It relies too much on leading vehicle and is lack of feedback channels from followers to leaders
Virtual structure	Based on the rigid structure representing each agent, the expected motion of the rigid structure is designed for controlling the vehicle trajectory in the formation	(1) It is easy to plan the overall cooperative behavior and is applicable for dealing with multi-task scenarios (2) It is in the centralized form. Single node failure may lead to system failure, and the computing and communication burden of the central controller is heavy
Behavior-based method	The behavior coordinator presets a variety of basic behaviors such as obstacle avoidance, formation maintenance and target search, and outputs the final behavior according to the actual scenario	It is suitable for complex scenarios, but is complicated in system model and is difficult to design

A formation generation process of platoon based on virtual structure is introduced below. The process generally includes three steps: generating the target virtual platoon, arranging the target position of each vehicle, and planning the trajectory of each vehicle reaching the target position, as shown in Fig. 3.114.

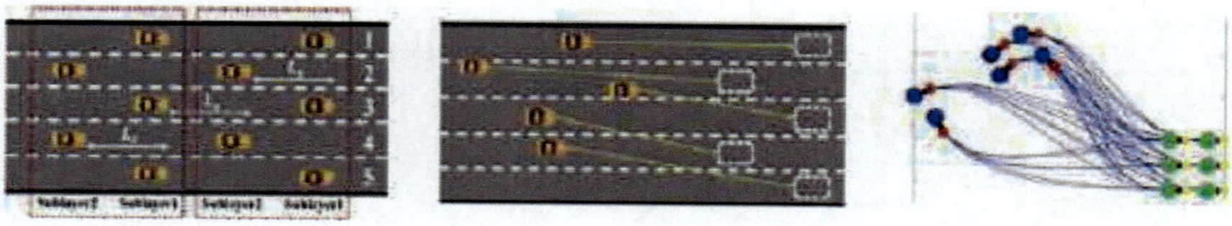

Fig. 3.114 Generation process of platoon formation [154, 155]

Taking vehicle passage in formation for instance, staggered formation is often adopted at present for the sake of safety in terms of virtual platoon structure. As shown in Fig. 3.114a, in the 5-lane scenario [154], the vehicles in the 5 lanes are arranged in a W-shape, and every 5 vehicles form a basic unit (the number of vehicles n in the unit is equal to the number of lanes). The target virtual platoon can be obtained by calculating the longitudinal distance L_s between adjacent sub-layers based on the information such as vehicle speed and road conditions. As shown in Fig. 3.114b, the purpose of vehicle location arrangement is to make the vehicles in each unit form a target formation based on arbitrary location distribution, and achieve the objectives of shortest running distance in total and minimum lane-change times [154]. Assuming that the cost of vehicle i reaching the target position j is c_{ij}, the solution to the optimal $i-j$ matching can be transformed into 0–1 linear programming:

$$\min_{A} \sum_{i=1}^{n} \sum_{j=1}^{n} (c_{ij} \times a_{ij}) \tag{3.180}$$

Enables,

$$\sum_{i}^{n} a_{ij} = \sum_{j}^{n} a_{ij} = 1 \tag{3.181}$$

$$A = [aij], \text{ and } aij = \begin{cases} 1 \text{ , when vehicle } i \text{ is assigned in the position of } j \\ 0 \text{ , others} \end{cases} \tag{3.182}$$

In accordance with the assigned position of each vehicle, as shown in Fig. 3.114c, the trajectory planning can be realized based on the single-vehicle trajectory planning [155]. In order to avoid collision in the formation generation process, a common way is to solve the conflict-free path through the iterative process: first, plan the trajectory of each vehicle independently; In the face of conflict, re-plan the trajectory of the conflicting vehicle applying the conflict point as the constraint; For the newly-generated conflict points, repeat the above process until the conflict-free trajectories of all vehicles are generated.

(2) Intersection Passing Decision

Compared with platoon formation in road segment, the conflict of traffic flow at the intersection is likely to lead to congestion and even collision. Two scenarios are often analyzed in terms of cooperative decision-making: signal-controlled intersection passage, which optimizes the signal light timing, as well as the speed curve for all vehicles; non-signal controlled intersection passage, which plans for the sequence and speed of each vehicle to pass through the intersection.

The decision-making for signal-controlled intersection passing mainly includes two parts: the optimal timing of signal lights and the optimal trajectory of each vehicle passing through the intersection. Considering that there are many vehicles in the traffic flow, complex nonlinear optimization for huge state space and decision space is required. Therefore, the existing research often decouples the optimization model, and adopts a centralized-distributed decoupling approach as shown in Fig. 3.115 [156]. The centralized planning is to solve the signal light timing and the passing time of each vehicle, in accordance with the vehicle position and speed input, on the premise of no collision and the goal of improving the overall traffic efficiency. Based on the received timing information, each vehicle plans its own optimal speed curve for improving energy efficiency.

With the continuous development and popularization of ICV technology, the high penetration rate of ICV will help to free the traffic system from the dependence on signal lights and further improve the traffic efficiency. In the existing research on non-signal controlled intersection, the decision-making task can also be undertaken

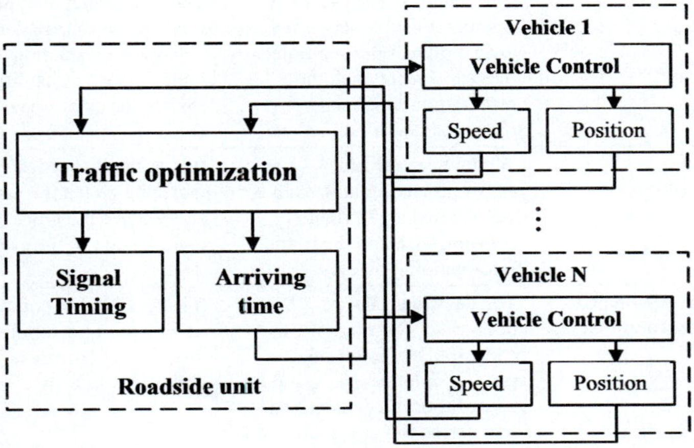

Fig. 3.115 Cooperative decision for signal-controlled intersection based on model decoupling [156]

by roadside facilities (centralized) or vehicles (distributed) [157]. Its characteristics are shown in Table 3.16.

Next, the passage decision-making process at unsignalized intersection will be introduced by means of the "virtual platoon" based on centralized-distributed coupling decision [158].

In the traffic scenario shown in Fig. 3.116a, the driving targets and positions of each traffic flow are shown. Figure 3.116b shows the traffic flow conflict generated from the scenario, including four types of relationships: path-crossing, merging to the same lane, diverging to different lanes and no path conflict (not drawn). As shown in Fig. 3.116c, "virtual platoon" is the projection of the two-dimensional traffic flow at the intersection on the one-dimensional virtual platoon. The platoon order expresses the order in which vehicles pass through the intersection (when the vehicles pass in order, there will be no collision, and the vehicles without conflict can be in the same order and pass at the same time). At this time, the centralized decision mainly plans the vehicle sequencing on the virtual platoon, and the distributed decision of each vehicle carries out specific trajectory planning accordingly. Xu et al. [155] adopted the depth-first spanning tree for central planning, and established the linear feedback control for designing the distributed controller for vehicle platoon.

(3) Multi-vehicle Cooperative Safety Decision Algorithm

Compared with single vehicle decision technology, the application of vehicle–vehicle and vehicle–road communication technology enables vehicles to have the stronger environmental perception capability and coordinated driving behaviors, which is

Table 3.16 Methods and characteristics of decision at unsignalized intersection

Method	Basic principle	Characteristic
Centralized decision-making	The central system globally optimizes the passing order, arrival time or driving trajectory of each vehicle, including reservation-based and optimization-based methods	The centralized method is relatively simple in design, but is large in calculation and low in fault-tolerant capability in the case of the existence of many vehicles
Distributed decision-making	Without centralized decision-making unit, each vehicle makes distributed decision based on the traffic information it obtains	It is difficult for each vehicle to grasp the global information and achieve the global optimization of the system
Centralized-distributed coupling decision-making	The decision tasks are decoupled. For example, the center only generates the passing order decision, and the task of trajectory planning is transferred to the vehicle end for distributed implementation	It is beneficial to combine the advantages of centralized and distributed algorithms, but the design is relatively complex

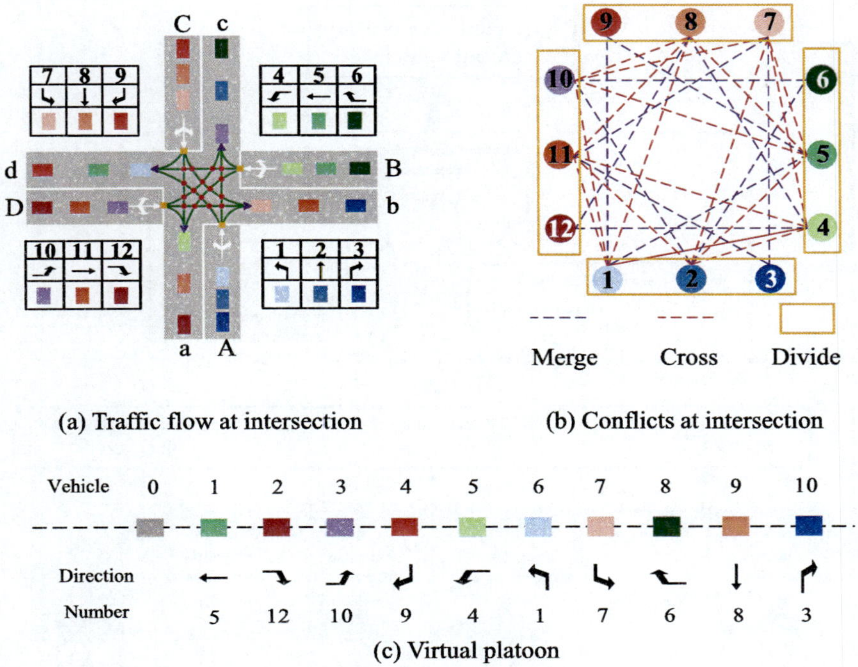

Fig. 3.116 Passage decision of unsignalized intersection based on virtual platoon [155]

conducive to the driving safety improvement of multiple vehicles. This section introduces a multi-vehicle cooperative algorithm for collision avoidance [159], mainly based on the scenario of the platoon-driving. When the leading vehicle of the platoon adopts emergency braking due to special circumstances, it is necessary to plan the longitudinal driving behaviors of other vehicles behind the platoon to avoid (or reduce) collision accidents.

As shown in the principle block diagram of multi-vehicle cooperative collision avoidance algorithm in Fig. 3.117, the first vehicle in the platoon takes emergency braking due to an emergency. At this time, it is necessary to plan the braking acceleration for each vehicle behind to minimize the risk of collision according to their motion state and performance constraints. In the cooperative collision avoidance algorithm, the first thing to be designed is the index function for evaluating collision risk of each vehicle.

In some traditional cooperative collision avoidance algorithms, TTC or DTC (distance to collision) between adjacent vehicles is often taken as the risk evaluation index. For example, each vehicle in the platoon calculates the target speed based on its TTC (or DTC) with the vehicle in front. This kind of algorithm has safety problems. For example, in a three-vehicle scenario, when the first vehicle makes emergency braking, we assume the second vehicle collides with it due to weaker

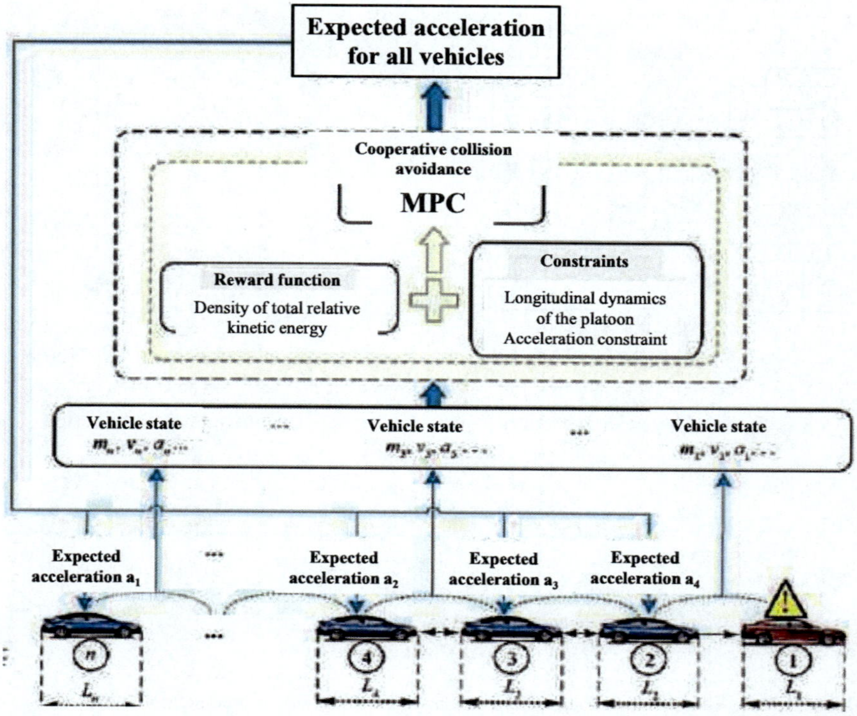

Fig. 3.117 Principle of multi-vehicle cooperative collision avoidance algorithm [159]

braking capacity. For the third vehicle, however, before the rear-end collision, it may judge itself relative safe based on TTC, and doesn't take the maximum braking. After the collision, the speed of the second vehicle decreases sharply, which may result in a subsequent rear-end collision between the third vehicle and the second vehicle. Therefore a new safety evaluation index should be introduced to systematically evaluate the collision risk of the whole platoon. Based on this idea, the index of "total relative kenetic energy" can be established [160]:

$$E(t) = \frac{1}{2}\sum_{i=2}^{N} m_i(v_{i-1}(t) - v_i(t))^2,\ v_{i-1}(t) \leq v_i(t) \tag{3.183}$$

where, the vehicle $i - 1$ is the vehicle in front of the platoon; the vehicle i is the rear vehicle; m and v represent mass and speed respectively; "Total relative kenetic energy" is expressed as the sum of the relative kenetic energy of the front and rear vehicles in each group of the platoon (the speed of the current vehicle is greater than that of the rear vehicle, and the speed of the front vehicle is equal to that of the rear vehicle in calculation). Different from the traditional way, it can measure the overall collision risk from the energy perspective and reduce the collision damage.

In addition to mass and relative speed, the distance between vehicles also affects the accident risk. Accordingly, the relative kenetic energy density can be obtained by dividing the relative kenetic energy of front and rear vehicles by the distance between vehicles. Since the higher the relative kenetic energy density is, the greater the collision risk is, the goal of cooperative collision avoidance is to minimize the total relative kenetic energy density of the platoon:

$$\min_{a_{des}} \frac{1}{2} \sum_{i=1}^{N-1} \frac{m_{i+1}}{x_i(t) - x_{i+1}(t) - L_i} \cdot [v_i(t) - v_{i+1}(t)]^2 \tag{3.184}$$

Of which:

$$\begin{cases} a_{i,\min} \le a_{i,\text{des}}(t) \le a_{i,\max} \\ a_{1,\min} \le a_{1,\text{des}}(t) \le C_1 < 0 \,, i = 2, 3, \ldots, N-1 \\ C_N \le a_{N,\text{des}}(t) \le a_{N,\max} \end{cases}$$

In the objective function, N is the total number of vehicles, m_i is the mass of the i-th vehicle, L is the length of the i-th vehicle, $x_i(t)$ and $v_i(t)$ are the front position and speed of the i-th vehicle at time t, $v_{i-1}(t) \le v_i(t)$, $i = 1, 2, \ldots, n-1$. When $v_{i-1}(t) > v_i(t)$, $v_{i-1}(t) = v_i(t)$. And the target minimum acceleration of each vehicle a_{des} corresponding to the collision risk shall be solved.

In the constraint function, $a_{i,\min}$ and $a_{i,\max}$ are the minimum and maximum accelerations of the i-th vehicle respectively, C_1 is the minimum deceleration of the emergency braking of the first vehicle, and C_N is the maximum braking deceleration that the N-th vehicle can take to avoid collision with vehicles further behind.

With regard to the longitudinal acceleration planning, the multi-vehicle cooperative collision avoidance algorithm for minimizing total relative kinetic energy density (TKED algorithm for short) is designed and solved, based on model predictive control (MPC). Figure 3.118 shows the application effect of TKED algorithm in 10 fleet scenarios (each vehicle has different mass and braking capacity), i.e., the realization of collision avoidance of all vehicles.

The TKED algorithm is compared with the traditional driver-reaction based brake control (DRBC) algorithm and the cooperative collision avoidance algorithm with the minimum total relative kinetic energy (TKE algorithm for short). In the simulation experiment of 1000 groups with random initial conditions, the collision avoidance rate of TKED algorithm is 93.7%, and its performance is better than traditional algorithms in collision avoidance rate and distance control. Table 3.17 shows statistics on the distance between vehicles in the platoon after they stop, and its variance based on the three algorithms. It can be seen that the distance between vehicles in the platoon based on the TKED algorithm is more uniform.

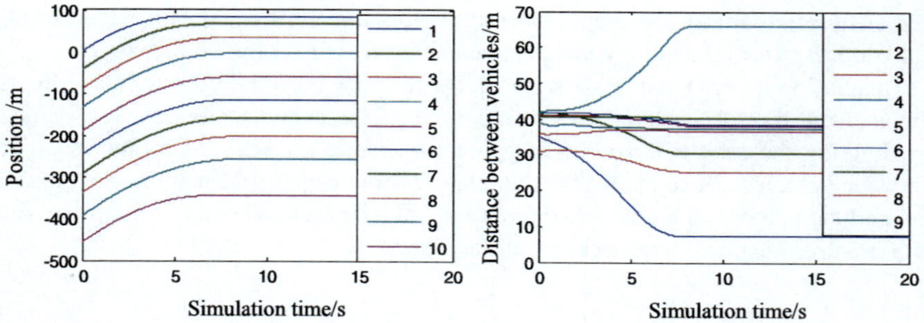

Fig. 3.118 Simulation effect of obstacle avoidance of platoon based on TKED algorithm

Table 3.17 Statistics of parking distance and its variance based on three algorithms

Algorithm	Mean value of distance between vehicles (m)	Mean value of variance (m)	Minimum times of variance (times)	Maximum times of variance (times)
DRBC	18.2	657.7	11	978
TKE	36.1	159.5	225	18
TKED	36.1	152.2	764	4

3.3.2 *Intelligent Safety Control Technology*

Intelligent vehicle control technology is one of the key technologies for intelligent vehicle driving. It mainly includes single vehicle intelligent safety control and multi-vehicle intelligent safety control. Oriented in trajectory tracking control, the single vehicle intelligent control is to control vehicle's steering, driving and braking systems and make the vehicle drive in accordance with the reference trajectory output by the decision system. Multi-vehicle intelligent control technology takes multiple vehicles in the domain as the cooperative control object, regards multiple vehicles as a multi-agent dynamic system with a certain communication topology, and aims at the safe and efficient operation of the multi-vehicle system. As shown in Fig. 3.119, this section will introduce single vehicle intelligent safety control and multi-vehicle intelligent safety control respectively, including the three main parts of control object model, safety control index and control algorithm.

In the process of intelligent vehicle control, the single vehicle safety control technology provides the basic function, that is, the target driving trajectory is realized by controlling the actuators such as power and steering of the vehicle. The multi-vehicle cooperative control system takes each vehicle as the control object, and generally takes the target driving trajectory of the vehicle as the output. Compared

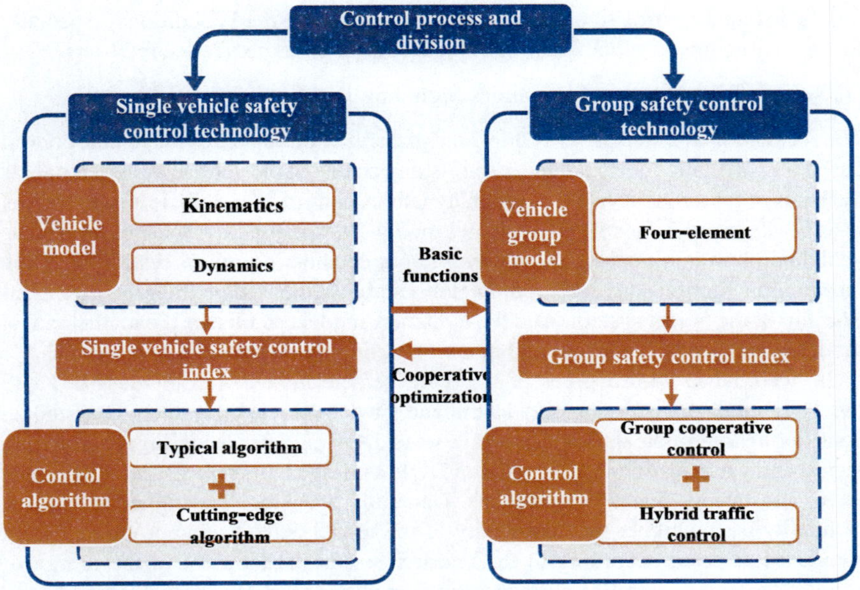

Fig. 3.119 Structure of intelligent safety control technology

with the multi-vehicle cooperative safety decision in the previous section, the multi-vehicle cooperative control technology and decision-making technology are similar in function. In the relevant research at home and abroad, the relevant research of "cooperative/coordinated control" does not distinguish between decision and control. In this study, multi-vehicle cooperative control technology has three characteristics: taking the collaborative system with system dynamics modeling as research object; taking the control performance of system stability, robustness and cohesion as part of indicators; taking the classical control, optimal control and other control algorithms as solution methods. Multi-vehicle cooperative decision technology may not have the above characteristics.

1. Single Vehicle Intelligent Safety Control Technology

IV safety control aims to control the action of the chassis actuator in accordance with the instructions of the decision system and the real-time driving state of the vehicle, so that the IV can track the target trajectory accurately and safely. As a key IV technology, the performance of IV safety control directly affects the driving safety. Therefore, the research in this field has both theoretical and engineering application significance. In order to realize the intelligent safety control of single vehicle, it is necessary to firstly establish the vehicle system model that can accurately describe the vehicle motion state, then apply reasonable and feasible performance indicators to evaluate the control performance for representing the safety of the control system, and finally design a control algorithm for all weather and all road conditions, especially for restricting the state boundary of vehicle control under extreme conditions.

(1) Vehicle Kinematics and Dynamics Modeling

The realization of IV decision system and other functions depends on vehicle control based on kinematics or dynamics. Establishing a reasonable vehicle system model is the basis of research on vehicle system dynamics control. The vehicle system model includes kinematics model and dynamics model. The kinematics model describes the vehicle motion in a mathematical way without considering the forces affecting the motion. The motion equation is completely established based on the geometric relationship of the object system. And the dynamics model establishes the mathematical model by analyzing the forces and moments acting on the vehicle.

In accordance with degrees of freedom, vehicle models can be classified into two DOF model considering only lateral and yaw of vehicle body, three DOF model considering longitudinal, lateral and yaw, seven DOF model considering longitudinal, lateral and yaw plus four-wheel rotation DOF, as well as higher eight DOF, fourteen DOF, and twenty-seven DOF models. Generally speaking, the higher the degree of fidelity is, the higher the complexity of the model is. Hence, appropriate model should be selected according to the research needs. In addition, in the design of vehicle dynamics controller, corresponding simplified models are usually established for different control requirements in view of the difficulty of solution and real-time control. A simple kinematics model can be usually applied under low-speed conditions, but is not applicable in medium and high-speed or extreme conditions, and thus the dynamics model should be selected. The following will briefly introduce vehicle kinematics modeling and vehicle dynamics modeling.

1) Vehicle Kinematics Modeling

i. Vehicle Longitudinal Kinematics Modeling

The vehicle longitudinal kinematics modeling usually regards the vehicle as a particle, uses the double integrator model to describe the vehicle motion, and ignores the factors such as power system. The specific expression is as follows

$$\begin{cases} \dot{x} = v \\ \dot{v} = u \end{cases} \tag{3.185}$$

where, x is vehicle position, v is vehicle speed, u is vehicle acceleration and system control quantity.

In addition, in order to facilitate the theoretical analysis, some studies on platoon also use the single integrator model, i.e.

$$\dot{x} = v \tag{3.186}$$

The control quantity of the model is equivalent to the vehicle speed. In practical application, different vehicle longitudinal kinematics models shall be selected according to specific scenarios and requirements.

ii. Vehicle Lateral Kinematics Modeling

As one of the most classical models in vehicle lateral kinematics, the bicycle model is widely applied in IV planning and control, and is generally applicable for medium and low speed conditions. Its assumptions include:

① The vehicle moves in a plane;
② The steering angles of the left and right front wheels are approximately equal;
③ The lateral force on the tire is ignored;
④ The transfer of front and rear axle load is ignored;
⑤ The vehicle body and suspension system are rigid.

The diagram of the bicycle kinematics model for the vehicle is shown in Fig. 3.120. Under the inertial coordinate system *OXY*, *C* is the vehicle centroid, *A* and *B* the axes of the front and rear axles of the vehicle respectively, *P* is the instantaneous rotation center of the vehicle, *R* is the turning radius, β is the side deflection angle of the centroid, ψ is the heading angle relative to the *X* axis, δ_f is the steering angle of the front wheel, and v is the speed of the centroid.

According to the trigonometric function:

$$\frac{\sin(\delta_f - \beta)}{l_f} = \frac{\sin(\frac{\pi}{2} - \delta_f)}{R} \tag{3.187}$$

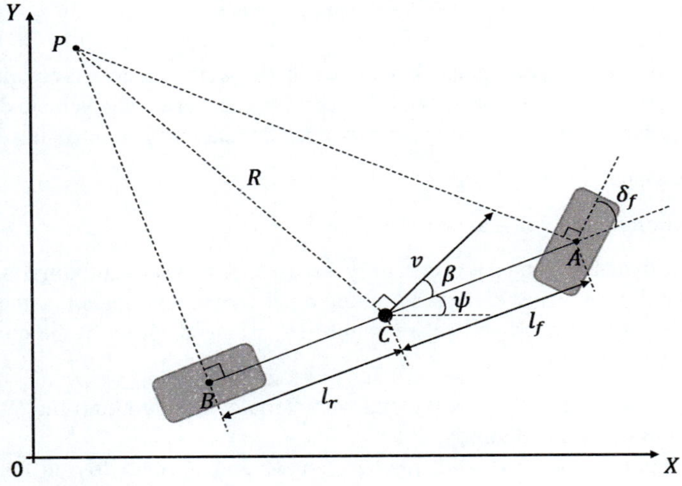

Fig. 3.120 Vehicle lateral kinematics model

$$\frac{\sin(\beta - \delta_r)}{l_r} = \frac{\sin(\frac{\pi}{2} + \delta_r)}{R} \tag{3.188}$$

Simultaneous Formulas (3.187) and (3.188) can obtain:

$$\left(\tan\delta_f - \tan\delta_r\right)\cos\beta = \frac{l_f + l_r}{R} \tag{3.189}$$

Under the low-speed condition, the turning radius of the vehicle's driving path changes slowly. Assuming that the direction change rate of the vehicle is equal to the angular velocity of the vehicle, the angular velocity can be expressed as:

$$\dot{\psi} = \frac{v}{R} \tag{3.190}$$

Simultaneous Formulas (3.189) and (3.190) can obtain:

$$\dot{\psi} = \frac{v\cos\beta}{l_f + l_r}(\tan\delta_f - \tan\delta_r) \tag{3.191}$$

To take the position of vehicle centroid x and y, yaw angle ψ and speed v as the state quantity, take acceleration a and front wheel steering angle δ_f as the control quantity, and take the vehicle centroid as the coordinate origin, the vehicle lateral kinematics model can be obtained:

$$\begin{cases} \dot{x} = v\cos(\psi + \beta) \\ \dot{y} = v\sin(\psi + \beta) \\ \dot{v} = a \\ \dot{\psi} = \frac{v\cos\beta}{l_f + l_r}(\tan\delta_f - \tan\delta_r) \end{cases} \tag{3.192}$$

Generally speaking, the vehicle kinematics model is applicable to medium and low speed scenarios. For high-speed and other complex scenarios, the vehicle dynamics model shall be established. The vehicle dynamics model will be introduced below.

2) Vehicle Dynamics Modeling

i. Vehicle Longitudinal Dynamics Modeling

The vehicle dynamics model is also mainly discussed from the longitudinal and lateral aspects. Firstly, in terms of vehicle longitudinal dynamics modeling, the vehicle longitudinal dynamics properties are simplified in consideration of the complexity and accuracy as follows:

① Only the longitudinal motion of the vehicle is considered, and the lateral and vertical motion are ignored;
② It is applicable to favorable road pavement and normal driving conditions, ignoring tire slip;

③ The vehicle is regarded as a rigid body, the left and right are completely symmetrical, and the load transfer of the front and rear axles is ignored;
④ The input and output characteristics of vehicle power system are approximately represented by the first-order inertia segment.

Based on the above simplification, the nonlinear longitudinal dynamics model for vehicle nodes is established:

$$\begin{cases} \dot{x} = v \\ \frac{\eta_T}{r} T = m\dot{v} + C_A v^2 + mgf \\ \tau \dot{T} + T = T_{\text{des}} \end{cases} \tag{3.193}$$

x and v represent the displacement and speed of the vehicle respectively, m is the mass of the vehicle, C_A is the lumped air resistance coefficient, g is the gravitational acceleration constant, f is the rolling resistance coefficient, T is the actual driving force or braking force moment, T_{des} is the expected driving force or braking force moment, τ is the time delay constant of the longitudinal power system, r is the wheel radius, and η is the mechanical efficiency of the transmission system.

The nonlinear model can accurately describe the longitudinal dynamics of vehicles. In some studies, in order to facilitate the theoretical analysis, the feedback linearization technology is used to linearize the above model into a third-order state equation. The derivation process of feedback linearization is given below for deriving the output:

$$\ddot{x} = a \tag{3.194}$$

$$\begin{aligned} \dddot{x} &= \frac{1}{m}\left(\eta_T \frac{\dot{T}}{r} - 2C_A v\dot{v}\right) = \frac{1}{m}\left(\eta_T \frac{T_{des} - T}{\tau r} - 2C_A v\dot{v}\right) \\ &= -\frac{1}{\tau}a + \frac{1}{\tau}\frac{1}{m}[\eta_T \frac{T_{des}}{r} - C_A v(2\tau\dot{v} + v) - mgf] \end{aligned} \tag{3.195}$$

In Formula (3.194), a refers to the acceleration of vehicle nodes. The selection of feedback linearization strategy is as follows:

$$T_{des} = \frac{1}{\eta_T}[C_A v(2\tau\dot{v} + v) + mgf + mu]r \tag{3.196}$$

where, u is the control input of vehicle nodes after feedback linearization. Therefore, Formula (3.195) is transformed into:

$$\tau\dot{a} + a = u \tag{3.197}$$

Then the third-order state space model of nodes in the vehicle platoon is obtained:

$$\frac{d}{dt}\begin{bmatrix} x \\ v \\ a \end{bmatrix} = \begin{bmatrix} 0 & 1 & 0 \\ 0 & 0 & 1 \\ 0 & 0 & -\frac{1}{\tau} \end{bmatrix}\begin{bmatrix} x \\ v \\ a \end{bmatrix} + \begin{bmatrix} 0 \\ 0 \\ \frac{1}{\tau} \end{bmatrix} u \tag{3.198}$$

ii. Vehicle Lateral Dynamics Modeling

Bicycle model is widely applied in the process of vehicle lateral dynamics modeling, and the following assumptions shall be made in the modeling process:

① The nonlinear effects such as suspension motion, road inclination and aerodynamics are ignored;
② The longitudinal and lateral coupling relationship of tire force is ignored.

As shown in Fig. 3.121, the bicycle model takes the vehicle centroid as the reference point, the center line of the front and rear wheels as the x-axis, and the line perpendicular to the center line of the front and rear wheels as the y-axis for stress analysis. Considering the influence of tire lateral force, the lateral dynamics relationship of the vehicle is:

$$ma_y = F_{yf} + F_{yr} \tag{3.199}$$

where, F_{yf} refers to the cornering force of the front wheel; F_{yr} refers to the cornering force of the rear wheel; m is the mass of the whole vehicle; a_y is the lateral acceleration, expressed as follows:

$$a_y = \ddot{y} + v_x\dot{\psi} \tag{3.200}$$

where, v_x is the velocity component of the vehicle centroid on the connecting line between the front and rear wheels, then Formula (3.196) can be written as:

$$m(\ddot{y} + v_x\dot{\psi}) = F_{yf} + F_{yr} \tag{3.201}$$

Furthermore, the vehicle yaw dynamics formula is obtained through analysis as:

$$I_z\ddot{\psi} = l_f F_{yf} - l_r F_{yr} \tag{3.202}$$

where, I_z is the moment of inertia. In addition, according to the stress analysis, F_{yf} and F_{cr} can be expressed as follows:

$$\begin{cases} F_{yf} = F_{cf}\cdot\cos\delta_f + F_{lf}\cdot\sin\delta_f \approx F_{cf} \\ F_{yr} = F_{cr} \end{cases} \tag{3.203}$$

According to the tire mechanics, the tire stress is shown in Fig. 3.122. For calculation of the cornering forceF_{yr} and F_{yf}, the tire cornering angle should be given. With δ_{f} as the steering angle of front wheel, θ_{Vf} as the speed deflection angle of front

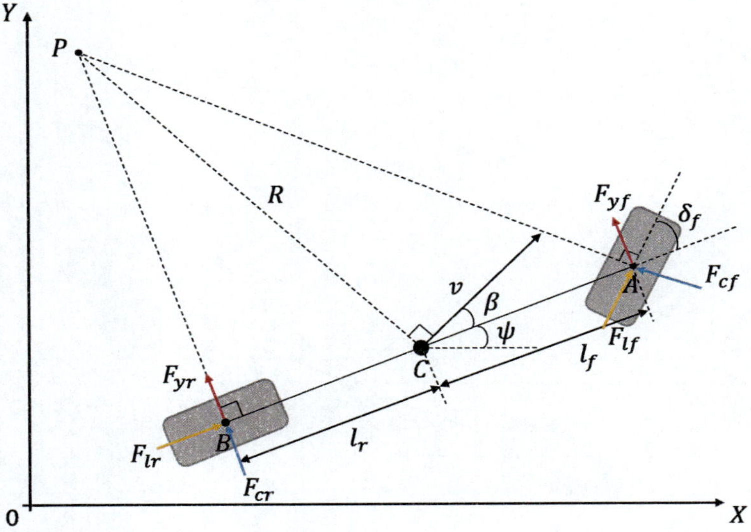

Fig. 3.121 Vehicle lateral dynamics model

wheel, θ_{Vr} as the speed deflection angle of rear wheel and α_{f} as the cornering angle of front wheel, the cornering angles of the front and rear wheels are respectively:

$$\begin{cases} \alpha_{\mathrm{f}} = \delta_{\mathrm{f}} - \theta_{\mathrm{Vf}} \\ \alpha_{\mathrm{r}} = -\theta_{\mathrm{Vr}} \end{cases} \quad (3.204)$$

Considering that there are two wheels at the front and rear of the actual vehicle, it is obtained that:

$$\begin{cases} F_{\mathrm{yf}} = 2C_{\mathrm{af}}\alpha_{\mathrm{f}} = 2C_{\mathrm{af}}(\delta_{\mathrm{f}} - \theta_{\mathrm{Vf}}) \\ F_{\mathrm{yr}} = 2C_{\mathrm{ar}}\alpha_{\mathrm{r}} = -2C_{\mathrm{ar}}\theta_{\mathrm{Vf}} \end{cases} \quad (3.205)$$

where, C_{af} and C_{ar} are the cornering stiffness of the front and rear wheels respectively. Then, according to the geometric relationship of stress analysis, it is obtained that:

$$\begin{cases} \theta_{\mathrm{Vf}} = \frac{v_{\mathrm{fy}}}{v_{\mathrm{fx}}} = \frac{\dot{y}+l_{\mathrm{f}}\dot{\psi}}{v_x} \\ \theta_{\mathrm{Vr}} = \frac{v_{\mathrm{ly}}}{v_{\mathrm{lx}}} = \frac{\dot{y}-l_{\mathrm{r}}\dot{\psi}}{v_x} \end{cases} \quad (3.206)$$

Combining Formulas (3.205) and (3.206), the vehicle lateral dynamics modeling is obtained:

$$\begin{cases} m(\ddot{y} + v_x\dot{\psi}) = 2C_{\mathrm{af}}\left(\delta_{\mathrm{f}} - \frac{\dot{y}+l_{\mathrm{f}}\dot{\psi}}{v_x}\right) + 2C_{\mathrm{ar}}\left(-\frac{\dot{y}-l_{\mathrm{r}}\dot{\psi}}{v_x}\right) \\ I_z\ddot{\psi} = 2l_{\mathrm{f}}C_{\mathrm{af}}\left(\delta_{\mathrm{f}} - \frac{\dot{y}+l_{\mathrm{f}}\dot{\psi}}{v_x}\right) - 2l_{\mathrm{r}}C_{\mathrm{ar}}\left(-\frac{\dot{y}-l_{\mathrm{r}}\dot{\psi}}{v_x}\right) \end{cases} \quad (3.207)$$

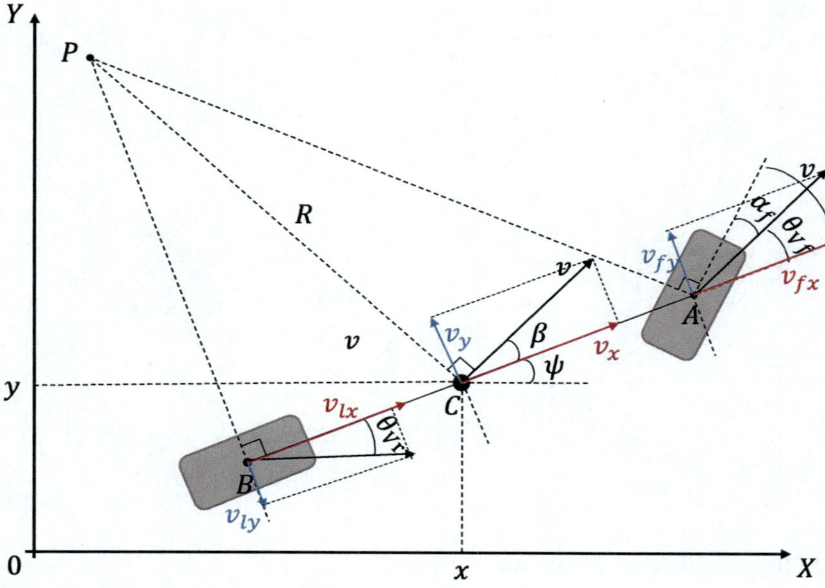

Fig. 3.122 Tire stress of vehicle lateral dynamics model

With the state vector as $X = [y\dot{y}\psi\dot{\psi}]^{\mathrm{T}}$ and the control quantity as the steering angle of the front wheel δ_{f}, the standard state space form is obtained as follows:

$$\frac{\mathrm{d}}{\mathrm{d}t}\begin{bmatrix} y \\ \dot{y} \\ \psi \\ \dot{\psi} \end{bmatrix} = \begin{bmatrix} 0 & 1 & 0 & 0 \\ 0 & -\frac{2C_{\mathrm{af}}+2C_{\mathrm{ar}}}{mv_x} & 0 & -v_x-\frac{2C_{\mathrm{af}}l_{\mathrm{f}}-2C_{\mathrm{ar}}l_{\mathrm{r}}}{mv_x} \\ 0 & 0 & 0 & 1 \\ 0 & -\frac{2C_{\mathrm{af}}l_{\mathrm{f}}-2C_{\mathrm{ar}}l_{\mathrm{r}}}{I_z v_x} & 0 & -\frac{2C_{\mathrm{af}}l_{\mathrm{f}}^2+2C_{\mathrm{ar}}l_{\mathrm{r}}^2}{I_z v_x} \end{bmatrix}\begin{bmatrix} y \\ \dot{y} \\ \psi \\ \dot{\psi} \end{bmatrix} + \begin{bmatrix} 0 \\ \frac{2C_{\mathrm{af}}}{m} \\ 0 \\ \frac{2l_{\mathrm{f}}C_{\mathrm{af}}}{I_z} \end{bmatrix}\delta_{\mathrm{f}} \quad (3.208)$$

Formula (3.208) is the state space expression of vehicle lateral dynamics model based on bicycle model, which is widely adopted in the research of vehicle safety control.

(2) Vehicle Intelligent Control

1) Control Objectives and Indicators

For general control systems, stability is the most basic control objective. In the field of vehicle intelligent control, stability is also a necessary condition for safety assurance, especially under complex road environment, such as high speed, low road adhesion coefficient and other conditions. Stability means that if the control system is disturbed and deviates from the original equilibrium state, the system can return to the original equilibrium state with sufficient accuracy after the disappearance of disturbance. System instability means that the system will deviate from the equilibrium state to a great extent or make continuous vibration, which indicates a great safety risk in vehicle intelligent control or even a subsequent accident. At present, Lyapunov stability analysis is one of the most commonly-used methods for analyzing the stability of control systems. It mainly includes Lyapunov's first method (indirect method) and Lyapunov's second method (direct method).

For vehicle intelligent control, researchers also put forward some more specific stability indicators, which mainly include two aspects: yaw stability and roll stability. At present, typical vehicle stability control modules usually only aim at one aspect of yaw stability or roll stability, which can not meet the requirements for driving safety. Under all weather and all road conditions, it is necessary to comprehensively take into account the vehicle stability index to fully ensure the vehicle driving safety. Common vehicle stability indicators are shown in Table 3.18.

On the basis of ensuring system stability, analyzing the relevant parameters in the time-domain response process of general control systems is also a common means to measure the control performance. Specifically speaking, the control system analyzes the transient and steady-state performance of the system according to the output time-domain response curve under certain inputs (such as step input, ramp input and pulse input). An example of a typical system state response curve under step input is shown in Fig. 3.123. The basic performance parameters under this response include

Table 3.18 Vehicle stability indicators

Stability indicators	Type	Meaning
Yaw stability	Centroid sideslip angular velocity—centroid sideslip angular phase plane	By selecting different initial states, multiple phase trajectories are drawn, and the vehicle control stability is determined according to the equilibrium saddle point in the phase plane
	Front axle sideslip angle—rear axle sideslip angular phase plane	
	Yaw rate-centroid sideslip phase plane	
	Yaw rate-lateral velocity phase plane	
Roll stability	Lateral acceleration	The lateral acceleration at the beginning of rollover is used to estimate the anti-rollover capability
	Lateral load transfer ratio (LTR)	Ratio of the difference between the vertical loads of the left and right wheels to the sum of the vertical loads of the left and right wheels
	Rollover protection reserve energy	Energy required for the vehicle to reach the rollover critical state from the current state
	Time to rollover (TTR)	Time when the vehicle is about to roll over

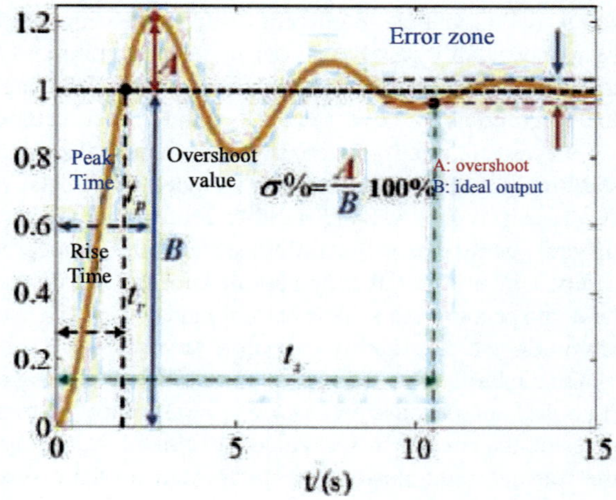

Fig. 3.123 System state response curve under step input

t_r (rise time), t_s (response time), ess (steady-state errors) and σ% (overshoot), the meaning of each indicator is as follows.

Rise time t_r: the time required for the actual output of the system to rise from 10 to 90% of the normal output.

Response time t_s: the time required for the actual output value of the system to stabilize within 5% or 2% of the normal output value.

Overshoot $\sigma\%$: The ratio of the difference between the maximum value and the normal value of the actual output of the system to the normal value.

Steady-state error *ess*: the ratio of the absolute value of the difference between the output value and the normal value when the system reaches the steady state to the normal value.

These indicators in the time-domain response process have intuitive physical significance in the field of vehicle intelligent control. Rise time and response time reflect the sensitivity of vehicle tracking target state; Overshoot reflects the maximum offset in the tracking process, and excessive offset may lead to potential safety hazards; Steady-state error directly reflects the control accuracy and is closely related to safety. Generally speaking, in order to improve the effective tracking of the system state to the target state and avoid conflicts in the response process, the system response process shall involve small response time, steady-state error and overshoot.

In addition to the performance indicators listed above for time-domain response analysis, there are other unique time-domain performance indicators for vehicle intelligent control. The typical representative is tracking accuracy, which is one of the main indicators for evaluating the performance of trajectory tracking controller and speed controller. Root mean square error (RMSE) is usually taken as its quantitative form. Taking lateral tracking control as an example, the control accuracy can be expressed by the following formula:

$$\text{RMSE} = \sqrt{\frac{1}{M}\sum_{i=1}^{M}(y_{r,i} - y_i)^2} \tag{3.209}$$

where, M is the number of path points; y_r is the desired lateral displacement in the target trajectory; and y is the actual lateral displacement of the vehicle. The lower the indicator is, the higher the control accuracy is.

2) Typical Control Methods

According to different operating conditions and actual requirements, researchers have carried out a lot of relevant research on IV safety control. Taking trajectory tracking control as an example, the proposed control methods mainly include PID control, optimal control [161], MPC, and adaptive control [162]. The above methods will be briefly introduced below.

i. PID Control

PID control, namely proportional–integral–derivative control, is one of the most basic controllers in feedback control. Firstly, the vehicle state is measured by the sensor, and then the error is obtained by making a difference with the input reference trajectory; After the controller performs proportional, integral and derivative operations on the error and sums them, it outputs control signals for controlling the vehicle movement. The control principle is shown in Fig. 3.124.

When designing PID trajectory tracking controller, preview mechanism should be adopted for better control. Taking the trajectory tracking in lateral control for instance, the vehicle position and the reference trajectory obtained from the decision are shown in Fig. 3.125. The error e_t is defined as the included angle between the heading angle and OB at time t. Within it, point A is the closest point between the vehicle and the decision trajectory, point B is the preview point, i.e., the point on the target trajectory S with a given preview distance L from point A, $\tilde{AB} = L$, and point O

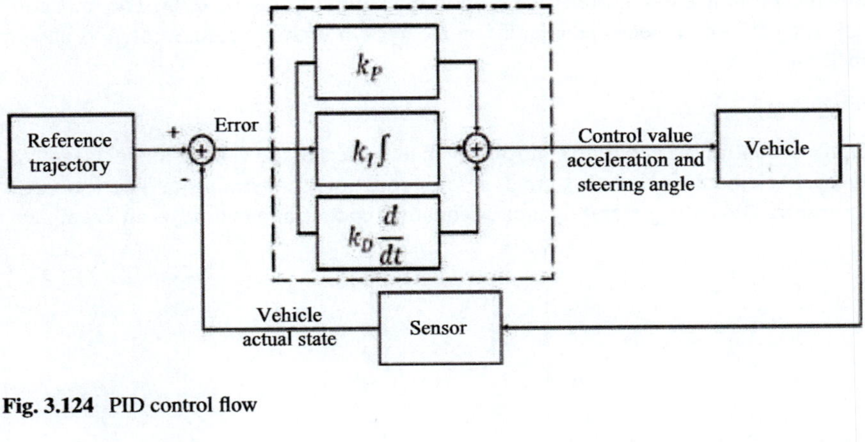

Fig. 3.124 PID control flow

Fig. 3.125 Preview mechanism

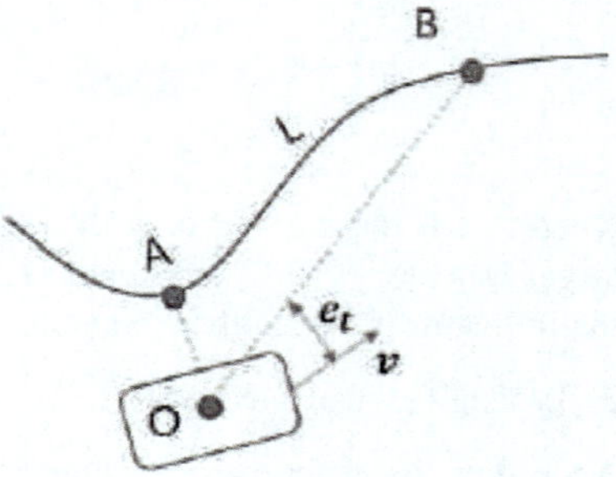

is the vehicle centroid. Error integral $e_\Sigma = \Sigma_{k=0}^{t} e_k$, error derivative $\Delta e = e_t - e_{t-1}$, then the current given lateral control quantity is:

$$u = k_P \cdot e_t + k_I \cdot e_\Sigma + k_D \cdot \Delta e \tag{3.210}$$

The above three terms are proportional term P, integral term I and derivative term D. By adjusting the coefficient k_P, k_I and k_D, the control performance can be adjusted. The proportional term is proportional to the error, indicating the severity of the tracker's response to the error. It is the most basic term in PID control. The larger k_P is, the larger the controller output will be at the same error, but too large value will easily lead to system instability, such as a large snake forward of the vehicle. The integral term is the integral of the cumulative errors of the system. If there is a constant system deviation between the system and the trajectory, the integral term will increase linearly with time. At this time, it can be eliminated by the feedback control of the integral term. The larger k_I is, the faster the steady-state error elimination of the tracker is. The steady-state error refers to the error between the actual output and the expected output when the system output reaches stability under step input. The derivative term is equivalent to the damping term of the control system. Its increase can make the output of the controller more stable and eliminate the instability due to the excessive proportional term or integral term. The impact of adjusting PID parameters separately on the system within a certain range is shown in Table 3.19.

ii. Optimal Control

Optimal control is a common control way in the case that the vehicle model is known. Its basic elements include system dynamics equations, constraints and performance indicators. Take the general dynamics equation in continuous time as an example:

$$\dot{x} = f(x, u, t) \tag{3.211}$$

Table 3.19 Impact of increasing A PID parameter alone on system indicators

Increase coefficient	Rise time	Overshoot	Response time	Steady-state error	Stability
k_P	Decrease	Increase	Small increase	Decrease	Decrease
k_I	Decrease	Increase	Increase	Substantial decrease	Decrease
k_D	Little decrease	Decrease	Decrease	No significant impact	Increase when k_D is smaller

where, x and u represent system state quantity and control quantity respectively. Constraint condition refers to the constraint on system state and control quantity, and its specific form includes initial constraint, terminal constraint, process constraint, etc. Constraints are generally expressed as

$$x \in \mathcal{X}, u \in \mathcal{U}, C(x, u) \leq 0 \tag{3.212}$$

where, $\mathcal{X}$ and $\mathcal{U}$ represent the state constraint set and control constraint set of the system; $C(x, u)$ represents the constrained objective function in the control process. In vehicle intelligent control, these constraints are directly related to safety. For example, the following distance of the constrained vehicle is not less than the safety distance, which can effectively ensure driving safety.

Control performance indicator is a quantitative evaluation of system state properties and control input properties in the control process, and generally can be described as

$$J = \Phi\left[x\left(t_f\right), t_f\right] + \int_{t_0}^{t_f} L(x, u, t)dt \tag{3.213}$$

where, t_0 and t_f are the start time and end time of the control respectively; Φ is the system terminal performance indicator; L is the system process performance indicator. In vehicle intelligent control, specific indicators such as fuel economy, comfort and safety can be reflected in the performance indicator function as the optimization objective.

By synthesizing the above elements, the general expression of optimal control can be obtained as

$$\begin{gathered} \min_u J = \Phi\left[x\left(t_f\right), t_f\right] + \int_{t_0}^{t_f} L(x, u, t)dt \\ \text{s.t.} \dot{x} = f(x, u, t), x \in \mathcal{X}, u \in \mathcal{U} \end{gathered} \tag{3.214}$$

The optimal control input can be obtained by solving the optimization. It can be found that the optimal control explicitly represents the performance indicator and constraints of the control system, so that the control input with the optimal performance indicator can be obtained on the premise of satisfying the constraints. In the field of vehicle intelligent safety, the safety indicator can be reflected in both the objective function, such as punishing the following distance in the following control, and the constraints, such as the following distance with no less than a certain limit, so as to ensure driving safety.

In solving optimal control, there are three basic methods, including variational method, minimum principle and dynamic programming, which have their own adaptive conditions. However, these methods can not conclude the existence and uniqueness of the solution, and thus further analysis and discussion are required for practical problems. However, there is a special kind of optimal control problem which can be easily solved. The optimization can be transformed into a derivative equation, even with an analytical solution. It is common in the regulation control design of the system. This kind of problem is linear quadratic regulator (LQR). The following is an introduction to the infinite time-domain LQR of continuous linear time-invariant systems.

In view of linear time-invariant systems:

$$\dot{x}(t) = Ax(t) + Bu(t) \tag{3.215}$$

Given the initial time t_0 and initial state $x(t_0)$, the performance indicator is designed in quadratic form:

$$J = \frac{1}{2}\int_{t_0}^{\infty} \left[x^{\mathrm{T}}(t)Qx(t) + u^{\mathrm{T}}(t)Ru(t)\right]dt \tag{3.216}$$

where, Q and R represent the weight matrix of state quantity and control quantity, which are positive semi-definite and positive definite constant matrix respectively. When (A, B) is a stabilizable pair, there is a unique optimal solution in the form of $u^*(t) = -Kx(t)$, where the feedback gain is:

$$K = -R^{-1}B^{\mathrm{T}}P \tag{3.217}$$

where, P is the non-negative definite solution of the following algebraic Riccati equation:

$$PA + A^{\mathrm{T}}P - PBR^{-1}B^{\mathrm{T}}P + Q = 0 \tag{3.218}$$

The following illustrates the application of LQR with an example of adaptive cruise control (ACC) in the adaptive cruise system, as shown in Fig. 3.126.

The distance between the ego-vehicle and the front vehicle is recorded as d. The basic control objective of ACC system is to maintain a certain desired distance

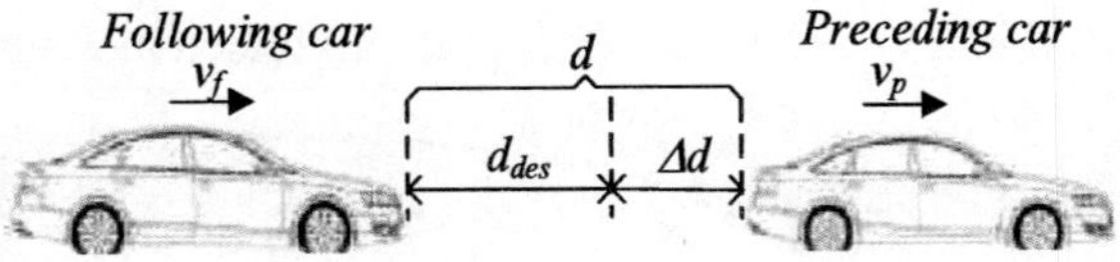

Fig. 3.126 Model predictive control

d_{des} between the ego-vehicle and the front vehicle. For this system, the state space equation as follows can be constructed:

$$\dot{x}(t) = Ax(t) + Bu(t) + Ha_{\text{p}} \tag{3.219}$$

where, $x = \begin{bmatrix} \Delta d \ \Delta v \ a_f \end{bmatrix}^{\text{T}}$ includes Δd representing the difference between the following distance and the expected distance, Δv representing the relative speed between the ego-vehicle and the front vehicle, and a_{f} representing the ego-vehicle acceleration. The control input refers to the expected acceleration of the ego-vehicle; a_{p} is the acceleration of the front vehicle, which can be regarded as external input or noise in the system. The expression of coefficient matrix is:

$$A_{\text{c}} = \begin{bmatrix} 0 & 1 & -\tau_h \\ 0 & 0 & -1 \\ 0 & 0 & -\frac{1}{T_{\text{L}}} \end{bmatrix}, B_{\text{c}} = \begin{bmatrix} 0 \\ 0 \\ \frac{K_{\text{L}}}{T_{\text{L}}} \end{bmatrix}, H_{\text{c}} = \begin{bmatrix} 0 \\ 1 \\ 0 \end{bmatrix} \tag{3.220}$$

where, τ_h represents the expected time interval; K_{L} and T_{L} are parameters of the vehicle first-order longitudinal dynamics model. For the adaptive cruise control system, the diagonal weight coefficient matrix can be designed in the performance indicator as follows:

$$Q = \begin{bmatrix} w_d & & \\ & w_v & \\ & & w_a \end{bmatrix}, R = w_u \tag{3.221}$$

where, w_d, w_v, w_a and w_u correspond to weight coefficients of Δd, Δv, a_p and u respectively, with all in positive numbers. With specific parameters, the feedback gain of the control quantity can be obtained by solving the Riccati matrix algebraic equation. The specific solution can be directly conducted by the commercial solution software, such as the care function of MATLAB.

Although the LQR is simple to solve, it should be noted that it cannot directly deal with the problem with constraints. Therefore, in the vehicle intelligent control, the safety requirements can only be reflected in the performance indicators for optimization, instead of being directly ensured by binding constraints.

iii. Model Predictive Control

As a common approach in the field of vehicle intelligent control, MPC can deal with constrained optimization problems, and adopt the concept of receding horizon

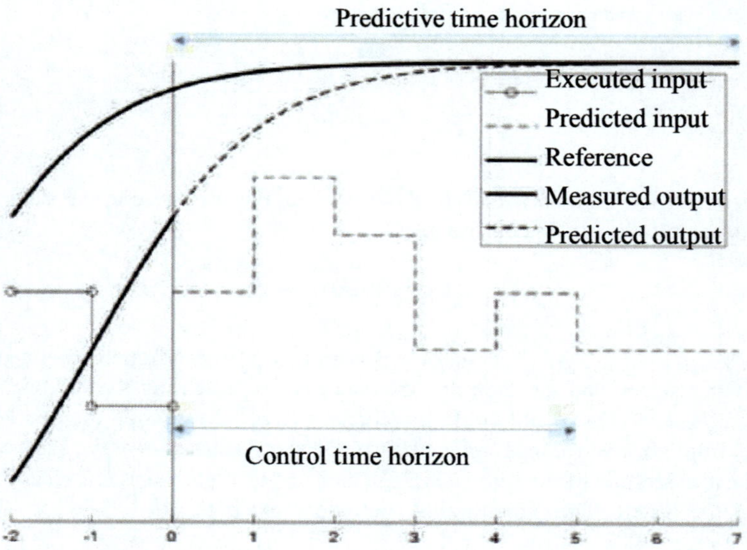

Fig. 3.127 Model predictive control

optimization to update the control sequence in real time according to system state output. Therefore, it is widely applied in ensuring vehicle driving safety.

Based on the predictive model, it can predict the output of the system under a given control sequence in the future time domain. For the optimization objective, MPC solves an optimal control problem at the current time on the basis of satisfying the constraints. The first step or the first several steps in the optimal control sequence are actually executed as control quantities. After execution, the optimal control problem is re-solved according to the latest state information obtained at the new time, and the above steps are repeated, so as to achieve an effect similar to feedback optimization. The diagram is shown in Fig. 3.127.

Next, the basic problem description and solution idea of MPC are explained by taking the linear time-invariant system as an example.

At the current moment, the discrete state space equation is used:

$$x_{k+1|t} = Ax_{k|t} + Bu_{k|t} \tag{3.222}$$

where, the subscript $k|t$ indicates the quantity of the next k-th step from the time t. If the predictive time domain is recorded as N, the predictive equation of the system from the current state $x_{0|t}$ to the next N-th step can be obtained:

$$\begin{bmatrix} x_{1|t} \\ x_{2|t} \\ \cdots \\ x_{N|t} \end{bmatrix} = \begin{bmatrix} B & 0 & \ldots & 0 \\ AB & B & \ldots & 0 \\ \cdots & \cdots & \cdots & \cdots \\ A^{N-1}B & A^{N-2}B & \ldots & B \end{bmatrix} \begin{bmatrix} u_{0|t} \\ u_{1|t} \\ \cdots \\ u_{N-1|t} \end{bmatrix} + \begin{bmatrix} A \\ A^2 \\ \cdots \\ A^N \end{bmatrix} x_{0|t} \tag{3.223}$$

The system performance indicator in quadratic form is used here:

$$J = x_{N|t}^{\mathrm{T}} P x_{N|t} + \sum_{k=0}^{N-1} \left(x_{k|t}^{\mathrm{T}} Q x_{k|t} + u_{k|t}^{\mathrm{T}} R u_{k|t} \right) \tag{3.224}$$

In addition, the system state quantity and control quantity are required to meet the following general constraints:

$$x_{k|t} \in \mathcal{X}, u_{k|t} \in \mathcal{U}, \forall k \in \{0, 1, \ldots, N-1\} \tag{3.225}$$

Then the optimal control can be constructed as:

$$\begin{aligned} &\min_{u_{0|t},\ldots,u_{N-1|t}} \quad J = x_{N|t}^{T} P x_{N|t} + \sum_{k=0}^{N-1} \left(x_{k|t}^{T} Q x_{k|t} + u_{k|t}^{T} R u_{k|t} \right) \\ &\text{s.t.} \begin{bmatrix} x_{1|t} \\ x_{2|t} \\ \cdots \\ x_{N|t} \end{bmatrix} = \begin{bmatrix} B & 0 & \ldots & 0 \\ AB & B & \ldots & 0 \\ \cdots & \cdots & \cdots & \cdots \\ A^{N-1}B & A^{N-2}B & \ldots & B \end{bmatrix} \begin{bmatrix} u_{0|t} \\ u_{1|t} \\ \cdots \\ u_{N-1|t} \end{bmatrix} + \begin{bmatrix} A \\ A^2 \\ \cdots \\ A^N \end{bmatrix} x_{0|t} \\ &x_{k|t} \in \mathcal{X}, u_{k|t} \in \mathcal{U}, \forall k \in \{0, 1, \ldots, N-1\} \end{aligned} \tag{3.226}$$

The optimal control sequence $u_{0|t}^{*}, \ldots, u_{N-1|t}^{*}$ is obtained by the above solution. At the current time t, the first control input $u_{0|t}^{*}$ is applied to the system. Then, when the time is up to $t + 1$, repeat the above steps at the current time to complete the dynamic update of the system.

(3) Frontier Approach for Vehicle Safety Control

1) Vehicle Safety Control under Constraint Information

With the development of 5G and related communication technologies, V2X technology further facilitate the widening application scope of single vehicle intelligent safety technology. However, some inherent uncertainties of existing communication technologies (such as packet loss or delay) and communication channel properties (such as bandwidth limitation) may reduce the control accuracy and affect driving safety of ICV. Therefore, the research on the design of ICV controller in non-ideal communication environment is of practical significance in the field of vehicle intelligent safety.

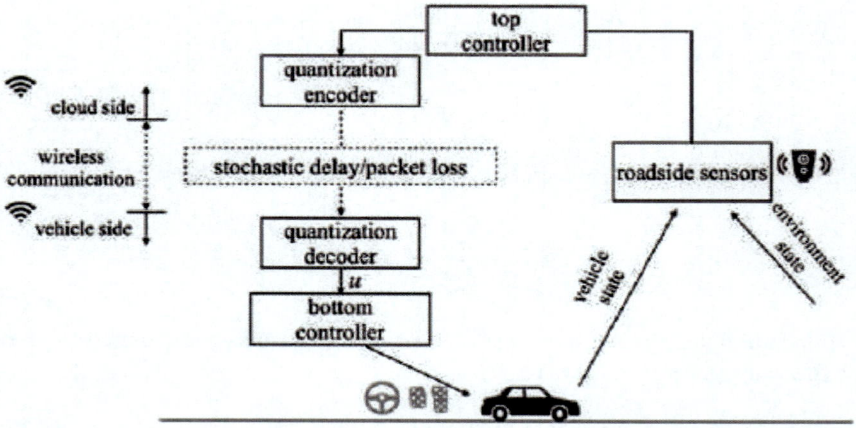

Fig. 3.128 Typical ICV control scenario

For ICV control system, it is necessary to focus on the design of quantitative controller under random communication packet loss or delay, model the system with unreliable communication conditions as a jump system, propose ICV control method suitable for non-ideal communication environment and the controller solution based on linear matrix inequality to ensure the stability of the system. A typical control system framework is shown in Fig. 3.128. In such applications, the status of the vehicle and the status of the environment are obtained by the roadside sensor and sent to the network controller on the roadside or in the cloud. The corresponding longitudinal and lateral control commands are calculated and sent to the vehicle for execution through the network wireless communication. In this process, the non-ideal properties such as quantization, packet loss or time delay of communication channel may affect the control performance of vehicles.

The controller design is illustrated by an example below. In order to study the controller design under the influence of non-ideal communication in the above ICV control scenario, such time-varying and non-ideal communication factors as packet loss and time delay are modeled as Markov jump process based on the network control system theory, and the controller is solved for discrete Markov jump linear systems. Taking the discrete jump linear feedback control system as the object, the system equation can be expressed as:

$$\boldsymbol{x}(k+1) = \boldsymbol{A}_{\theta(k)}\boldsymbol{x}(k) + \boldsymbol{B}_{\theta(k)}\boldsymbol{u}(k) \tag{3.227}$$

$$\boldsymbol{y}(k) = \boldsymbol{C}_{\theta(k)}\boldsymbol{x}(k)$$

where, for all $\theta(k) \in \{1, \ldots, N\}$, there are $\boldsymbol{A}_{\theta(k)} \in \mathbb{R}^{n\times n}$, $\boldsymbol{B}_{\theta(k)} \in \mathbb{R}^{n\times n_u}$ and $\boldsymbol{C}_{\theta(k)} \in \mathbb{R}^{n_y\times n}$. When $\theta(k) = i$, the system is in the i-th modal. $\theta(k)$ is a random process of a range in the set$\{1, .., N\}$. If the jump of the system over time follows a Markov process, the above system is called a Discrete Markov Jump Linear System. The state transition matrix between different models is:

$$\boldsymbol{P} = \left[p_{ij}\right]_{N \times N} \tag{3.228}$$

where, p_{ij} is the probability of transmission from modal i to j. For all $i,j \in \{1, \ldots, N\}$, meets $p_{ij} \geq 0$ and $\sum_{j=1}^{N} p_{ij} = 1$.

In the network control system, the network bandwidth limits the amount of data transmitted in each cycle of communication, so it is necessary to quantify the transmission information to realize the effective transmission of information. The logarithmic quantization can make full use of the limited bandwidth to quantify the relative error in the control feedback, and can also use the logarithmic discrete control quantity to stabilize the closed-loop system and derive the quantization conditions. Therefore, it is also applicable here.

To sum up, a linear feedback controller for logarithmic quantization associated with system modal is adopted as follows:

$$u = \begin{cases} -\beta_i {\rho_i}^l, \ {\rho_i}^{l+1} < \boldsymbol{H}_i x \leq {\rho_i}^l \\ 0, \ \boldsymbol{H}_i x = 0 \\ \beta_i {\rho_i}^l, -{\rho_i}^l \leq \boldsymbol{H}_i x < -{\rho_i}^{l+1} \end{cases} \tag{3.229}$$

where, i is the system modal; $\boldsymbol{H}_i$ is the mode-related control gain; β_i is the correction quantity of control gain; ρ_i is the logarithmic quantization density at the current time. The above quantization feedback involves two processes: logarithmic encoding process and logarithmic decoding process. The logarithmic coding process mainly occurs at the output side of the controller, and determines the quantized coding value l according to the control quantity. The decoding process mainly occurs at the receiving end of the controlled object and reconstruct ${\rho_i}^l$ according to the received quantized feedback information and calculate the actual control quantity by controlling gain correction.

Taking the vehicle lateral control for example, the discrete system is obtained by using the state feedback controller and the zero-order holder to discretize the sampling time T, without considering the wind disturbance in the environment:

$$\boldsymbol{x}(k+1) = \boldsymbol{A}_{\text{lat}} \boldsymbol{x}(k) + \boldsymbol{B}_{\text{lat}} \delta(k) + \boldsymbol{P}_{\text{r}} \rho(k) \tag{3.230}$$

$$\delta(k) = -\boldsymbol{K}_{\text{lat}} \boldsymbol{x}(k) \tag{3.231}$$

$$\boldsymbol{x}(k) = [\beta(k), r(k), \psi_{\text{L}}(k), y_{\text{L}}(k)]^{\text{T}} \tag{3.232}$$

where, $\boldsymbol{A}_{\text{lat}}$ is the state matrix of path tracking discrete system; $\boldsymbol{B}_{\text{lat}}$ is the control matrix; $\boldsymbol{P}_{\text{r}}$ is the input coefficient matrix of control target; $\boldsymbol{K}_{\text{lat}}$ is the feedback controller gain; $\boldsymbol{x}(k)$ is the system state quantity, as shown in Formula (3.165), which includes $\beta(k)$, $r(k)$, $\psi_{\mathrm{L}}(k)$, and $y_{\mathrm{L}}(k)$, respectively representing the centroid sideslip angle, yaw rate, heading angle error and lateral error at the preview distance l_{s}; $\delta(k)$ is the steering angle of the front wheel; $\rho(k)$ is the road curvature. The matrix parameters in the above discrete system are calculated continuously as:

$$\boldsymbol{A}_{\mathrm{lat0}} = \begin{bmatrix} a_{11} & a_{12} & 0 & 0 \\ a_{21} & a_{22} & 0 & 0 \\ 0 & 1 & 0 & 0 \\ v & l_{\mathrm{s}} & v & 0 \end{bmatrix}, \boldsymbol{B}_{\mathrm{lat0}} = \begin{bmatrix} b_1 \\ b_2 \\ 0 \\ 0 \end{bmatrix}$$

$$\boldsymbol{P}_{r0} = \begin{bmatrix} 0 \\ 0 \\ -v \\ -vl_{\mathrm{s}} \end{bmatrix} \tag{3.233}$$

$$a_{11} = -2\frac{c_{\mathrm{f}} + c_{\mathrm{r}}}{mv}, a_{12} = -1 + 2\frac{c_{\mathrm{r}}l_{\mathrm{r}} - c_{\mathrm{f}}l_{\mathrm{f}}}{mv^2}$$

$$a_{21} = 2\frac{c_{\mathrm{r}}l_{\mathrm{r}} - c_{\mathrm{f}}l_{\mathrm{f}}}{I_z}, a_{22} = -2\frac{c_{\mathrm{f}}l_{\mathrm{f}}^2 + c_{\mathrm{r}}l_{\mathrm{r}}^2}{vI_z}$$

$$b_1 = 2\frac{c_{\mathrm{f}}}{mv}, b_2 = 2\frac{c_{\mathrm{f}}l_{\mathrm{f}}}{I_z} \tag{3.234}$$

where, c_{f} and c_{r} are the sideslip stiffness of the front and rear wheels respectively; l_{f} and l_{r} are the distances from the centroid to the front and rear axles respectively; m is the mass of the whole vehicle; I_z is the moment of inertia of the vehicle around the vertical axis; v is the longitudinal speed; l_{s} is the preview distance. Specific parameter settings are shown in Table 3.20.

According to the system equation, the linear feedback system architecture involving random packet loss process is shown in Fig. 3.129, and its system equation can be expressed as:

$$\boldsymbol{x}(k+1) = \boldsymbol{A}\boldsymbol{x}(k) + \beta\boldsymbol{B}\boldsymbol{u}(k) \tag{3.235}$$

where, $\beta \in \{0,1\}$ is the packet loss state coefficient. If $\beta = 0$, packet loss occurs, the control information is not received by the controlled object, and the system degenerates into an open-loop system; Otherwise, there is no packet loss. The Bernoulli process commonly involved in packet loss research can be regarded as a simplified two-state Markov process. Given the packet loss rate is p_d, its state transition matrix

Table 3.20 Parameters of vehicle lateral tracking control system

l_{f} (m)	l_{r} (m)	m (kg)	I_z (kg m^2)	v (m/s)	c_{f} (N/rad)	c_{r} (N/rad)
1.14	1.4	1500	2420	10	44,000	47,000

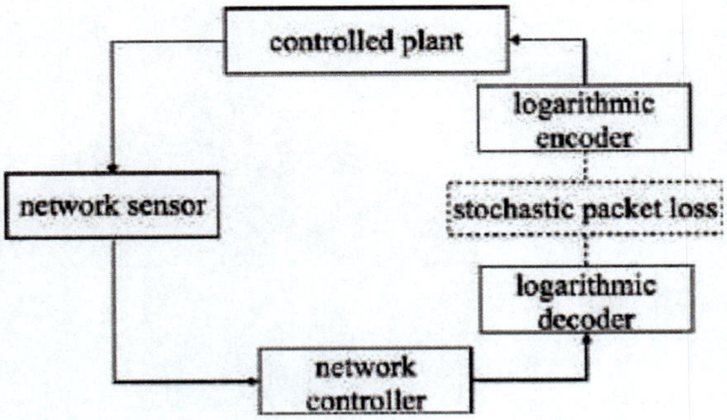

Fig. 3.129 Logarithmic quantization feedback control system with random packet loss

can be expressed as:

$$P = \begin{bmatrix} 1 - p_d & p_d \\ 1 - p_d & p_d \end{bmatrix} \quad (3.236)$$

Therefore, under the premise of given quantitative density and packet loss rate, the method of solving linear matrix inequality is used to stabilize the control parameters without packet loss, and then stabilize the linear logarithmic quantization feedback control system with packet loss. It should be noted that although the above system is a two-state jump system, the other state is a packet loss without feedback control. Therefore, it is only necessary to calculate the feedback control parameters under the no-packet loss state.

With this method, the parameters of corresponding trajectory tracking control quantization feedback controller can be solved. In the simulation test, the controlled vehicle traces a sinusoidal trajectory at the constant speed, and the trajectory function is as follows:

$$y_{\text{ref}} = 4 \cdot \sin\left(x_{\text{ref}} \frac{\pi}{50}\right)(x_{\text{ref}} \in [0, 100]) \quad (3.237)$$

The control performance is mainly reflected by the lateral distance error and heading angle error of the reference trajectory and the actual trajectory. In this method, the normalized tracking error index (TEI) is set, and its calculation formula is:

$$\text{TEI} = \frac{1}{T_s} \int_0^{T_s} (\alpha_y |y_r(t)| + \alpha_a |\psi_L(t)|) dt \quad (3.238)$$

where, T_s is the total simulation time, α_y and α_a are the weight coefficients of lateral distance error and heading angle error, set as $\alpha_y = 1, \alpha_a = 0.5$.

Figure 3.130 shows the difference of TEI between ordinary controller and the controller proposed in this method under different quantization densities and packet

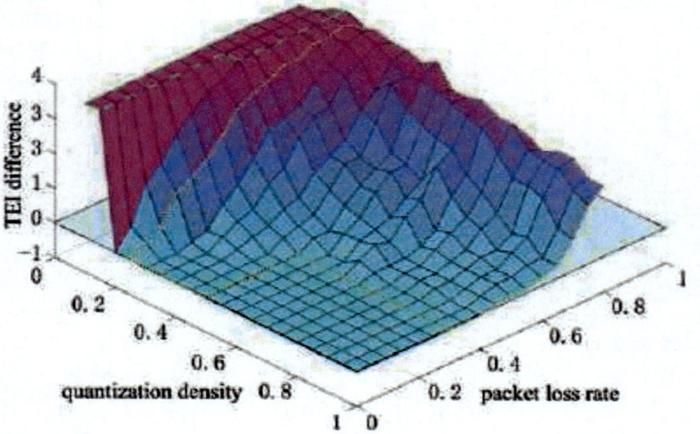

Fig. 3.130 TEI difference of two controllers under different quantization densities and packet loss rates

loss rates. It can be seen from the figure that both the ordinary controller and the controller proposed here can stabilize the system well under a small packet loss rate and a fine quantization density, as shown in the lower plane of the figure. With the increase of packet loss rate and the decrease of quantization density, both perform worse in the average tracking error, but the ordinary controller fluctuates greatly at a lower packet loss rate. As shown in the figure, the control performance of the proposed controller is much better than that of the ordinary controller, especially within a certain range of packet loss rate and quantization density.

2) Safe Control for Cyber Attack

The IOV environment may make the control system of ICV vulnerable to the intrusion of adversarial information, and then affect the safety of vehicle control. On the one hand, the classical control method is not capable to ensure the safety of vehicle control. On the other hand, there are still some problems in the existing research on the control of adversarial information (such as the treatment of constraints, the optimization of control performance, and the integration with vehicle control), which shall be further studied and improved.

To solve the above problems, it is necessary to establish an ICV control approach in the face of the bounded adversarial information, so that the vehicle controller can ensure the safety of vehicle control, as well as the performance of vehicle control. The typical research focuses on the solution of vehicle control strategy in finite time domain and safety verification. According to the boundary of the difference reachable set, the real safety constraint set is contracted, and the non-adversarial information constraint set is obtained for eliminating the uncertainty of the system. Based on the non-adversarial information constraint set and vehicle dynamics model, the existence

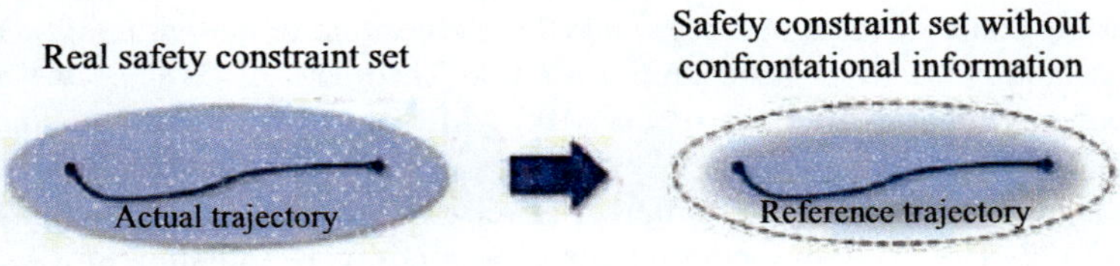

Fig. 3.131 Conversion and contraction of safety constraint set

based on the satisfiability model theory is constructed to deal with the existence of vehicle reference trajectory control strategy under the non-adversarial information constraint. Furthermore, the rolling optimization solution of vehicle control strategy is carried out for improving the applicability of this scenario.

It is a common processing way to convert the reference trajectory corresponding to the actual environment to the reference trajectory without adversarial information, which is conducive to eliminating the uncertainty of bounded adversarial information. Meanwhile, the safety constraint set for the actual trajectory of the vehicle state should also be transformed accordingly. Obviously, when the adversarial information is uncertain, the safety constraint set of the reference trajectory shall be contracted compared with that of the actual trajectory, as shown in Fig. 3.131.

In a typical car-following scenario, the real safety constraint set of the ego-vehicle is determined by the kinematics of the front vehicle. Before calculating the cyber-security constraint set without adversarial information, the motion trajectory of the front vehicle within the prediction time of a single planning cycle (the predicted total step size is 6, and the total time is 0.6 s). Based on the known movement trajectory of the front vehicle in a single planning cycle, the longitudinal model of the vehicle is shown in Formula (3.134), and the real safety constraint set of the ego-vehicle over time can be set, as shown in Formula (3.172)–(3.174).

$$\left|\left[s_p(k) - s(k)\right] - d_{des}\right| \le d_{safe} \tag{3.239}$$

$$\left|v_p(k) - v(k)\right| \le \Delta v_{safe} \tag{3.240}$$

$$|a(k)| \le a_{max} \tag{3.241}$$

where, k is the prediction time, $k = 1, 2, \ldots, 6$; s_p and v_p respectively represent the position and speed of the front vehicle, s, v and a are the state quantities of the ego-vehicle, representing the position, speed and acceleration respectively; d_{des} is the expected follow-up distance of the vehicle, taken as the constant value of 5 m here; d_{safe} is the safety threshold of the deviation between the actual follow-up distance and the expected follow-up distance, which is taken as the constant value of 1 m here; Δv_{safe} is the safety threshold value of the deviation between the speed of the ego-vehicle and the speed of the front vehicle, taken as the constant value of 0.5 m/s here; a_{max} is the limit value of the maximum acceleration of the ego-vehicle, taken as the constant value of 5 m/s^2 here. The algorithm proposed in this study is the Reachability Analysis-Satisfiability Modulo Theories (RA-SMT). One of the corresponding algorithms is the robust MPC (rMPC), and the other is the normal MPC (nMPC).

The simulation results under different adversarial information types are shown in Fig. 3.132. The diagram includes three sub-graphs, corresponding to RA-SMT, rMPC and nMPC algorithms respectively; Each control algorithm also includes upper, middle, and lower parts, which respectively represent different front vehicle kinematics. The upper part of the sub-graph indicates that the front vehicle moves at a constant speed, the middle part indicates that the front vehicle moves at a sinusoidal speed, and the lower part indicates that the front vehicle moves at a measured speed; Each simulation scenario contains ten different adversarial information, with input corresponding to the simulation results (a pair of curves in the figure represent the following position error and speed error).

It can be seen from the figure that RA-SMT algorithm and rMPC algorithm can ensure that the state of the vehicle is always controlled within the safety boundary under the input of different front vehicle kinematics and different bounded adversarial information, while nMPC algorithm cannot ensure the control safety of the vehicle

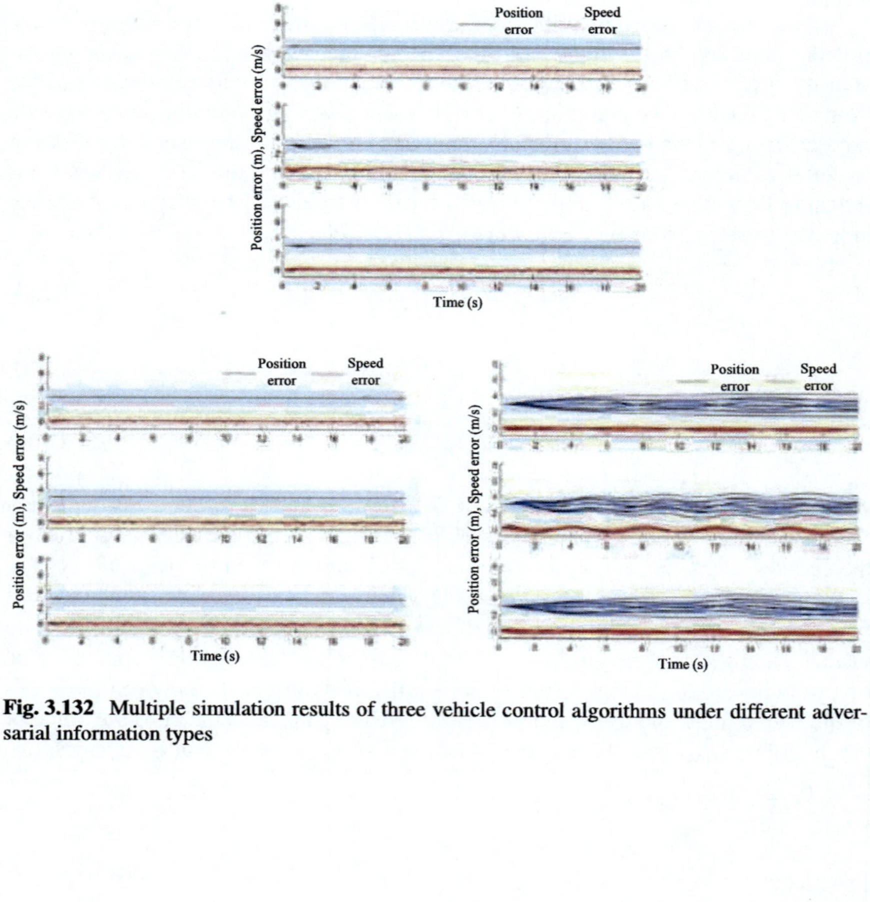

Fig. 3.132 Multiple simulation results of three vehicle control algorithms under different adversarial information types

under the input of some bounded adversarial information, where the vehicle breaks through the state safety boundary. On the other hand, in terms of tracking error, the position error and speed error of the ego-vehicle and the front vehicle controlled by RA-SMT algorithm and rMPC algorithm fluctuate within a small range, with the curve clusters relatively coincident; The tracking error controlled by nMPC algorithm is large, which is reflected in the large fluctuation of position error curve cluster and velocity error curve cluster, as well as the low coincidence degree. It can be found that that RA-SMT algorithm and rMPC algorithm can make better tracking performance under different types of bounded adversarial information input, and significantly superior to nMPC algorithm.

3) Safety Sandbox and Control Arbitration

The control quantity (vehicle speed, steering angle of front wheel, etc.) of ICV is calculated by the controller that is usually developed according to different scenarios, such as platoon following and speed guidance at intersection. It is designed merely based on the feasibility of driving in specific scenarios. However, it cannot ensure the DSV in complex coupled scenarios where multiple control objectives and instructions may conflict with each other. Therefore, it is necessary to design a vehicle safety arbitration mechanism to judge whether the vehicle control input can ensure DSV in real time, and to timely plan a new control input when DSV cannot be guaranteed.

Safety sandbox is a typical technology applied to ensure system safety in the computer field. It provides an execution environment that restricts program behaviors according to safety policies, that is, a virtual system program. It allows various user programs to run in the sandbox environment, and the changes resulting from running can be deleted later. It creates a sandbox-like independent working environment, where the running programs cannot have a permanent impact on the hard disk. As an independent virtual environment, it can be used to test untrusted applications or Internet behaviors.

The framework of ICV safety assurance techniques based on safety sandbox consists of five parts: safety monitor, safety sandbox, sensor interface, network information interface and vehicle control interface, as shown in Fig. 3.133. Among them, the sensor interface and network interface are responsible for transmitting the information collected by the sensor and the information from the network to the safety monitor and the user program in the safety sandbox. The user program includes the program stored locally and the control program from the cloud. It calculates the control input according to the driving purpose and the transmitted environmental state information, and sends the control input to the safety monitor, which judges whether the control input calculated by the user program can ensure the vehicle safety according to the current status of vehicle and environment. If yes, it sends an enabling command to the safety sandbox, and the results calculated by the user program are sent to the vehicle for control. If not, the safety monitor recalculates the control input that can ensure the vehicle safety and sends it to the vehicle for control, as shown in Fig. 3.134.

Figure 3.135 shows an example for judging the current vehicle safety status, where the vehicle drives forward at a constant speed. The longitudinal and lateral

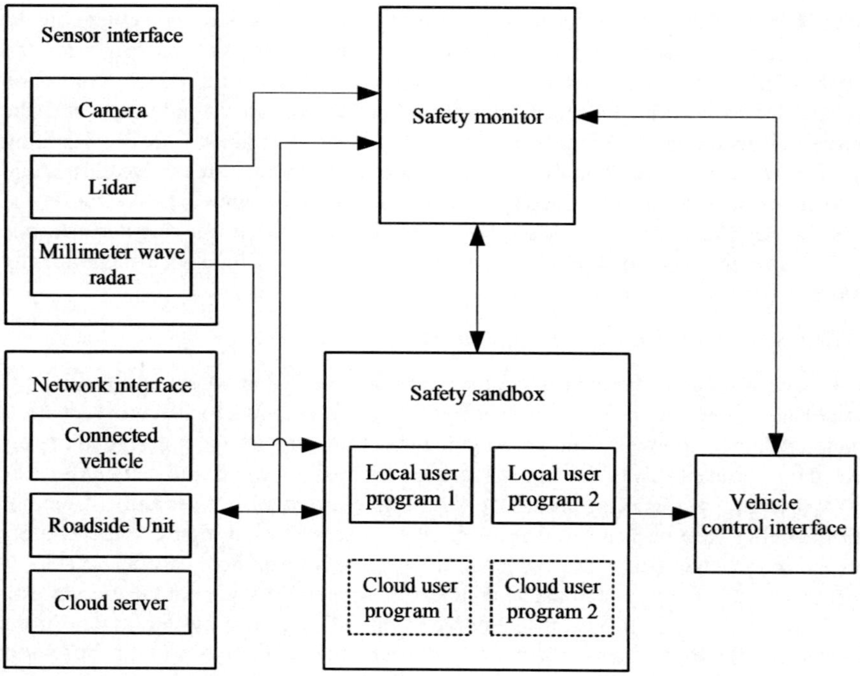

Fig. 3.133 Framework of ICV safety assurance techniques based on safety sandbox

coordinates of the figure respectively represent the distance between the ego-vehicle and the obstacle in front of the vehicle, and the included angle between the vehicle-obstacle line and the front of the vehicle. By checking whether the vehicle will collide with obstacle when it continues to drive forward, when it turns and keeps running at a constant speed, and when it decelerates according to the maximum braking deceleration, the safe state area of the vehicle at the current speed can be divided. The two figures on the left and right respectively show the division results of the safe state area when the vehicle speed is 1 and 2 m/s.

2. Multi-vehicle Intelligent Safety Control

Multi-vehicle intelligent control technology takes multiple vehicles into consideration as the cooperative control objects, and constitutes a multi-agent dynamic system with a certain information topology. In the current research on multi-vehicle cooperative control, some scholars also collectively refer to "cooperative/coordinated control" as "cooperative decision and control", with certain similarities in research issues (such as improvement of the energy efficiency of vehicle platoons) and control objects (such as the acceleration of vehicles in platoons). However, multi-vehicle cooperative control often has following features. It is aimed at the cooperative system

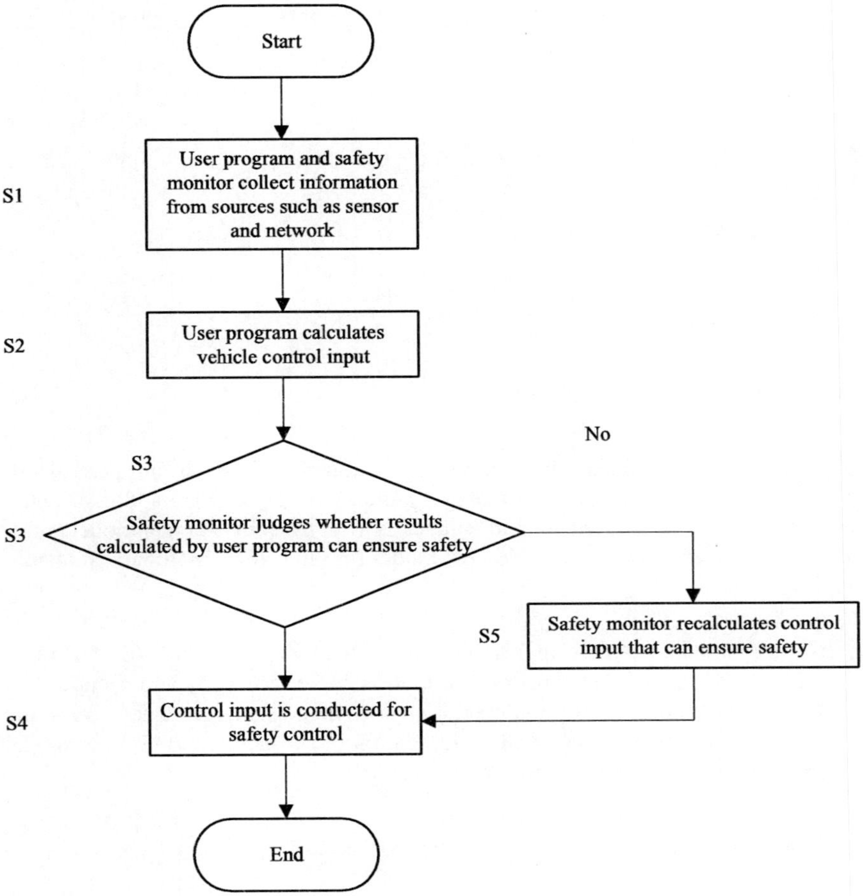

Fig. 3.134 Flow diagram for safety control based on safety sandbox

with system dynamics modeling, and further realizes the safe and efficient operation of the system through optimization control and other methods on the basis of ensuring the stability, robustness, cohesion and other basic performance of the closed-loop control system. The above process is more conducive to analyzing the impact of disturbance and various non-ideal factors on system safety and traffic efficiency. Multi-vehicle cooperative safety control technology mainly includes three aspects: system dynamics modeling, multi-vehicle cooperative safety control performance indicators, and multi-vehicle cooperative safety control methods.

(1) System Dynamics Modeling

The multi-vehicle cooperative control system can be regarded as a dynamic system which is controlled through information communication and then established through

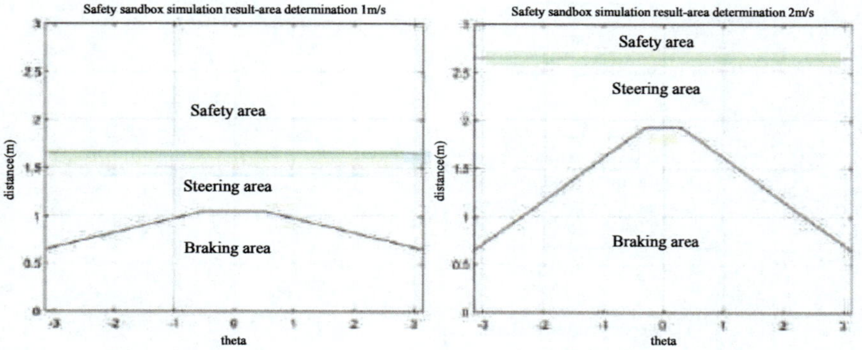

Fig. 3.135 Instance of safety arbitration

coupling of multiple vehicles. Its dynamics modeling often includes four parts [163]: node dynamics, communication topology, multi-vehicle geometry and distributed controller, and can be described mathematically with the four-element model architecture. The diagram of the four-element model for the vehicle platoon is shown in Fig. 3.136.

1) Node dynamics model is applied to describe the dynamics behavior of a single vehicle, and its modeling is similar to that of single vehicle dynamics model. In view of the complexity of nonlinear dynamics model in large-scale control, the existing research commonly adopts single integral model, second-order model and other ways to linearize the vehicle model.

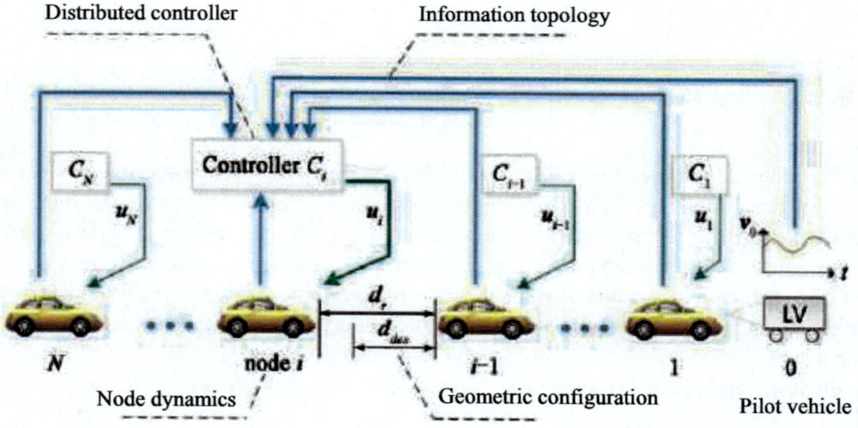

Fig. 3.136 Four-element system dynamics model under platoon scenario [163]

2) Communication topology structure is used to describe the topological relationship of information transmission between vehicle nodes, and represents the information adopted by each vehicle, which has a significant impact on the stability, stability margin and other features of the platoon. The communication topology shall be designed in accordance with the requirements of scenario application. Taking the platoon scenario with leading and follow-up vehicles as an example, the information transfer from leading vehicle to follow-up vehicles should include at least one directed spanning tree with the leading as the root node, that is, any follow-up vehicle can directly or indirectly obtain the information of the leading vehicle. Communication topology is often expressed by adjacency matrix and traction matrix. Taking the predecessor-leader following (PLF) structure for instance, each vehicle establishes communication with the leading vehicle and its front vehicle. In the PLF structure, nodes can obtain node information, so the neighborhood set of nodes is:

$$(\mathbb{N}_i)_{\text{PLF}} = \begin{cases} \emptyset, & i = 1 \\ \{i-1\}, & i = 2, 3, \ldots, N \end{cases} \tag{3.242}$$

Accordingly, its adjacency matrix can be expressed as:

$$(\mathcal{A}_N)_{\text{PLF}} = \begin{bmatrix} 0 & & & \\ 1 & 0 & & \\ & \ddots & \ddots & \\ & & 1 & 0 \end{bmatrix} \tag{3.243}$$

In addition, since each vehicle can obtain the leading vehicle information, the set of leading node information can be as follows:

$$(\mathbb{P}_i)_{\text{PLF}} = \{0\}, \quad i = 1, 2, \ldots, N \tag{3.244}$$

Accordingly, the traction matrix can be written as:

$$(\mathcal{P})_{\text{PLF}} = \begin{bmatrix} 1 & & & \\ & 1 & & \\ & & \ddots & \\ & & & 1 \end{bmatrix} \tag{3.245}$$

3) Multi-vehicle geometry is used to describe the position relationship between two adjacent vehicles in a group. Taking the platoon control problem as an example, the platoon geometry can be expressed as follows: the following vehicle and the leading vehicle have the same speed, and the desired distance between adjacent vehicles is maintained. The expected distance between vehicles is designed from the three perspectives: the constant distance, i.e., the distance between vehicles

is a constant larger than 0; constant time distance, i.e. constant TTC with vehicle distance larger than 0; and nonlinear distance, i.e., the expected vehicle distance is expressed as a nonlinear function of the ego-vehicle speed. Different approaches proposed in references are designed in different forms based on the goal of ensuring platoon stability and improving traffic efficiency.

In the actual control process, it is difficult to keep the vehicle speed and distance stably at the expected value. Therefore, effective controller design and reasonable geometric configuration design are the premise to avoid collision and ensure the safety of multi-vehicle cooperative operation.

4) Distributed controller is used to describe the feedback control law with the vehicle information in the neighborhood. Similar to the multi-vehicle cooperative decision technology, the multi-vehicle cooperative control includes centralized control and distributed control. Considering that the centralized control is difficult to solve large-scale control problems with insufficient robustness, distributed controller is usually applied in the existing research. By designing the feedback control based on communication topology for each vehicle, the overall coordination goal of the system can be achieved.

(2) Multi-vehicle Cooperative Safety Control Performance Indicators

Multi-vehicle cooperative safety control theoretically belongs to the cooperative control of multi-agent systems. In the field of multi-agent cooperative control, the control objectives include consensus control, rendezvous control, flocking control and formation control. Among them, the most basic one is consensus control, that is, the state convergence among individuals is achieved through individual information interaction and distributed control. Platoon control is regarded as the consensus control. Its control objectives can be expressed as:

$$\lim_{t \to +\infty} = x_j(t) - x_i(t) = 0 \quad (\forall i, j \in \{1, 2, \ldots, N\}) \tag{3.246}$$

where, $x_i(t)$ represents the state of individual i; and N represents the number of individuals in the system.

The flocking control of IV focuses on the implementation of the control research according to the specific multi-vehicle geometry aggregation and collision avoidance. Taking platoon control as an example, the above formula can be changed in accordance with the expected vehicle distance d to:

$$\lim_{t \to +\infty} = x_j(t) - x_i(t) = d \quad (\forall i, j \in \{1, 2, \ldots, N\}) \tag{3.247}$$

Compared with the multi-vehicle cooperative safety decision technology, which usually assumes that each vehicle can run according to the expected planning results, the multi-vehicle cooperative safety control shall consider the deviation between its actual movement. In addition, unlike single vehicle control, multi-vehicle cooperative control takes multi-vehicle system as the control object, and further considers the transmission of disturbance in the system. For example, in the platoon control scenario mentioned above, on the one hand, it is necessary to consider that the vehicle will not fully follow the expected geometric configuration. And on the other hand, it is also necessary to avoid small disturbances being transmitted and amplified in the group, so as to avoid the subsequent collision accidents.

Therefore, on the basis of the safety control performance indicators proposed in the single vehicle control section, new operational safety indicators for the multi-vehicle cooperative system shall be introduced. The commonly used closed-loop control system performance indicators and their meanings are shown in Table 3.21.

Similar to multi-vehicle cooperative decision-making, multi-vehicle cooperative control also aims to improve the system traffic efficiency and reduce energy consumption. These objectives are generally reflected in the design of the objective function of the optimal control. However, in the design process of system control, the aforementioned performance indicators related to the operational safety of the multi-vehicle cooperative system shall be satisfied first.

Taking the platoon stability indicator as an example, the current schemes for improving the platoon stability mainly include the following three categories: (1) slack platoon configuration, that is, design a large follow-up distance in the platoon geometry, or adopt a non-linear vehicle distance model in design; (2) heterogeneous controller architecture, that is, to set the control gain of different controllers for different vehicle nodes; (3) topology for information flow expansion, that is, to enable the vehicle obtain the information of multiple vehicles in the platoon.

(3) Multi-vehicle Cooperative Safety Control

Linear feedback control is the most common and mature means for multi-vehicle cooperative control. Linear feedback control is characterized by simple form, easiness in implementation, and easiness in analyzing and optimizing the dynamics

Table 3.21 Operational safety performance indicators of closed-loop control system

Indicator	Meaning	Impact on operational safety of multi-vehicle cooperative system
Internal stability	It is the basic indicator of the system, ensuring the stable operation to the target state. For example, all characteristic roots of linear systems shall have negative real parts	To enable multi-vehicles to reach the target operation state, internal stability must be achieved first
Stability margin	On the basis of internal stability, the attenuation velocity of disturbance is described. Linear systems are defined by the absolute value of the real part of the minimum eigenvalue (the most unstable eigenvalue)	The system with high stability margin has faster attenuation of state disturbance and stronger stability
Platoon stability	Platoon stability requires that the amplitude of any frequency disturbance will not be amplified by vehicle transmission process in the platoon	Internal stability cannot guarantee the platoon stability, so the disturbance transmission process shall be further considered for collision avoidance

performance of the system. Based on the above four-element model, the consensus control of platoon system is taken for example:

$$u_i(t) = -\sum_{j\in\mathbb{I}_i}\Big[k_{ij,p}[p_i(t) - p_j(t) - d_{i,j}] + k_{ij,v}[v_i(t) - v_j(t)] \\ +k_{ij,a}[a_i(t) - a_j(t)]\Big] \tag{3.248}$$

On the left side of the equation, vehicle i is the controlled vehicle, $u_i(t)$ is the control input of the controlled vehicle at time t. On the right side of the equation, p, v and a correspond to the longitudinal position, speed and acceleration respectively, and $d_{i,j}$ is the expected distance between vehicles i and j. Vehicle j is the vehicle from which vehicle i can obtain information based on the communication topology, and $j \in \mathbb{I}_i$ is the neighborhood information that vehicle i can use. $k_{ij,\#}$, $(\# = p, v, a)$ is the gain of the controller. It can be found that each vehicle in the controller can generate control information through linear negative feedback based on its available neighborhood vehicle information.

In multi-vehicle cooperative control, how to deal with constraints is also one of the key issues. System constraints come from the requirements of control enforceability and safety, while linear feedback control is difficult to fuse the constraints. Therefore, the research based on distributed model predictive control (DMPC) is becoming the focus of multi-vehicle cooperative control.

Taking the platoon control for instance, the basic form of DMPC controller can be expressed as:

$$\min_{U_i} J_i[\boldsymbol{y}_i^p(:|t), \boldsymbol{u}_i^p(:|t), \boldsymbol{y}_i^a(:|t), \boldsymbol{y}_{-i}^a(:|t)] \\ = \sum_{k=0}^{N_p-1} l_i[\boldsymbol{y}_i^p(k|t), \boldsymbol{u}_i^p(k|t), \boldsymbol{y}_i^a(k|t), \boldsymbol{y}_{-i}^a(k|t)] \tag{3.249}$$

To make:

$$\begin{aligned} \dot{\boldsymbol{x}}_i^p(k+1|t) &= \phi_i[\boldsymbol{x}_i^p(k|t)] + \boldsymbol{\psi}_i \cdot \boldsymbol{u}_i^p(k|t) \\ \boldsymbol{y}_i^p(k|t) &= \boldsymbol{\gamma}\boldsymbol{x}_i^p(k|t) \end{aligned} \tag{3.250}$$

Of which:

$$k = 0, \ldots, N_p - 1$$

$$\boldsymbol{u}_i^p(k|t) \in \mathcal{U}$$

$$\boldsymbol{y}_i^p(N_p|t) = \frac{1}{|\mathbb{I}_i|}\sum_{j\in\mathbb{I}_i}[\boldsymbol{y}_j^a(N_p|t) - \widetilde{\boldsymbol{d}}_{j,i}]$$

$T_i^P(N_p|t) = h_i[v_i^P(N_p|t)]$. DMPC can minimize the total loss $\sum_{k=0}^{N_p-1} l_i$ of the system by solving the optimal control U_i of each vehicle. Each constraint corresponds to the motion constraint, the amplitude constraint of control input, the terminal position constraint and the terminal state constraint of each vehicle based on the kinematics model, as well as the guarantee of the asymptotic stability of the system.

Compared with linear feedback control, the modeling of DMPC is more complex and the solution is more difficult. Interested readers can further refer to Ref. [164] for details.

During the development of ICV, there will be a long-term mixed traffic condition with co-existence of ICV and human-driven vehicles (HDV). Understanding and evaluating the performance of mixed traffic system is of great significance to the promotion of ICV.

Taking the mixed traffic condition of platoon driving as an example, it is necessary to model the driver first. At present, the commonly-applied models are the optimal vehicle model (OVM) and intelligent driver model (IDM), which are close to the car-following behavior of real drivers and can approximate the simulation of manual driving. Such models can be expressed as:

$$\begin{aligned} \dot{d}_i(t) &= v_i(t) - v_{i-1}(t) \\ \dot{v}_i(t) &= F[d_i(t-\tau), \dot{d}_i(t-\tau), v_i(t-\tau)] \end{aligned} \quad (3.251)$$

where, d_i represents the headway between the vehicle i and the front vehicle $i-1$, v_i is the speed of the vehicle i. $\dot{v}_i$ represents the acceleration of the vehicle i, in m/s^2. τ indicates the reaction time of the driver, in s, which is usually a constant. In order to analyze the performance of ICV technology in mixed traffic, a large number of studies make analysis on Vissim and SUMO traffic simulation platforms, and adopt such indicators as traffic flow, traffic flow stability and traffic flow uniformity to study the impact of ICV on traffic performance under different penetration rates [165]. In terms of theoretical analysis, Head-to-tail Platoon Stability is a classical analysis indicator for describing whether the disturbance generated by the head vehicle will be amplified when it is propagated to the tail vehicle in a traffic flow composed of multiple vehicles [166]. Some scholars have proposed the possibility of making a single ICV to stabilize traffic flow, and proved that the linear closed-loop system can achieve stability under specific parameter conditions. Such studies reflect the potential of ICV to stabilize traffic flow under low penetration rate. For example, researchers at the University of Illinois at Urbana Champaign (UIUC) proposed the FolloweStopper control strategy [167], and conducted tests on a long ring road of 260 m involving 21 HDVs and 1 ICV to verify the capability of the ICV in suppressing traffic disturbance and stabilizing traffic flow, as shown in Fig. 3.137.

The following is a specific case to analyze the application effect of multi-vehicle cooperative control technology in mixed traffic [168]. As shown in Fig. 3.138, a mixed platoon composed of HDVs and ICVs will pass through the intersection controlled by traffic lights. At this time, the multi-vehicle cooperative controller shall improve

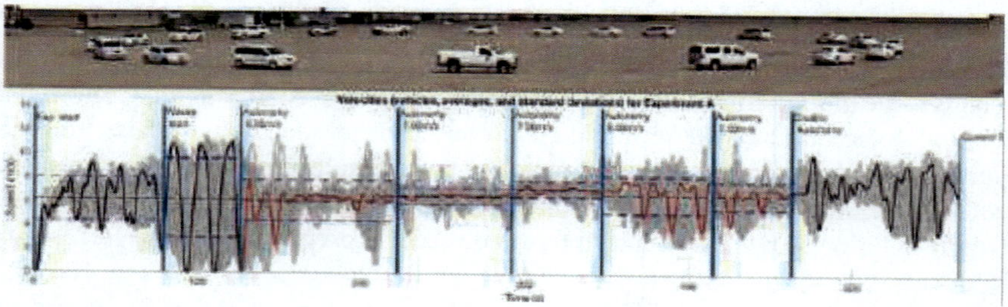

Fig. 3.137 Real vehicle experiment on ring road conducted by UIUC researchers [167]

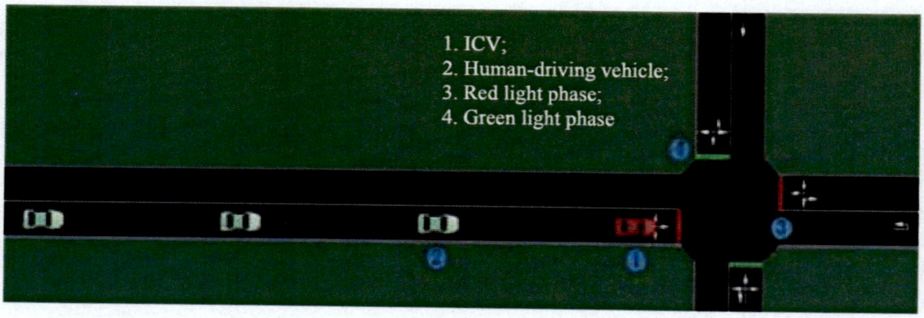

Fig. 3.138 Hybrid platoon with HDVs-ICVs passing through intersection controlled by traffic lights

the traffic efficiency of the platoon and reduce fuel consumption under the constant traffic light timing.

When the platoon passes through the intersection, it is only necessary to select the traffic lights with constant time interval for optimization under the stable traffic flow, instead of allocating traffic light phase. ICV can obtain the constant traffic light timing, and then carry out trajectory planning according to the constant timing. The cooperative optimization framework is mainly adopted. And the optimization is divided into two basic steps: using the constant traffic timing information to optimize the time of vehicles arriving at the intersection; taking the intersection passage time as the terminal constraint to optimize the vehicle trajectory.

In the general process of controller design, firstly, necessary traffic information is extracted from the traffic environment, including signal light timing, status of HDV and ICV, etc. Then the platoon length is optimized by timing information, vehicle and environmental constraints to determine the time when ICV arrives at the parking line. Next, according to the final state time and final state speed, the optimal control model is constructed and solved by Gaussian pseudospectral for optimizing the fuel economy of the platoon. Finally, the obtained trajectory information is sent to ICV for speed control.

Taking the ICV penetration rate of 0% as the benchmark evaluation index, the optimization results of the hybrid platoon algorithm under different traffic flows are calculated. The calculation results are shown in Fig. 3.139, where the red solid line represents the benchmark performance corresponding to 0% of ICV market penetration, the part below the red solid line is the traffic scenario that can reduce and improve traffic efficiency or fuel consumption, and the part above the red solid line is the traffic scenario that has not been improved. It can be found that with the increase of ICV penetration rate, the optimization effect of the algorithm often does not increase with positive correlation. The possible reason is that in the single lane scenario, the control of ICV will always bring about the speed fluctuation of HDV. Under low penetration rate, the bullwhip effect caused by the accumulation of multiple HDVs gradually amplifies the speed fluctuation, and the decline of traffic efficiency and fuel economy of HDV is almost inevitable. Under the medium and high penetration rate, the hybrid platoon algorithm can limit the speed fluctuation within the expected range by restricting the maximum platoon length and the cooperative control of ICV, so as to improve the performance of the algorithm for the single ICV.

At present, the research on hybrid traffic flow is still being pushed forward. The relevant research still focuses on how to make the driver model more close to the actual driving behavior of human beings, and how to design control methods to make the hybrid traffic system with better control performance under various ICV penetration rates.

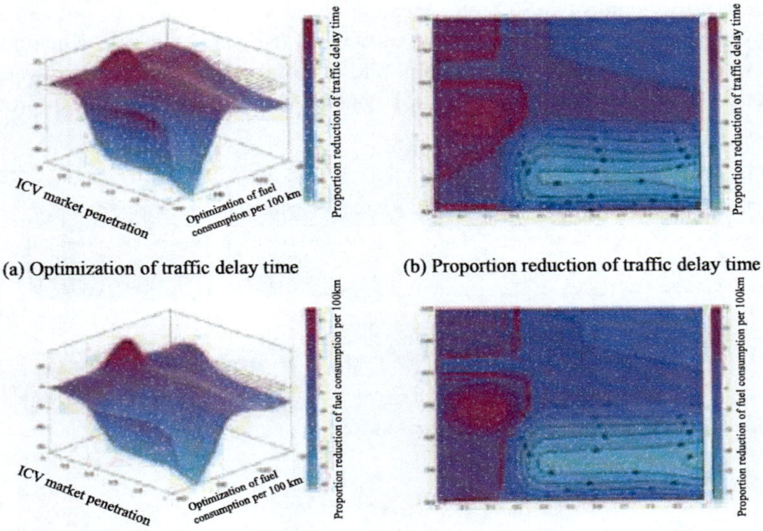

(a) Optimization of traffic delay time (b) Proportion reduction of traffic delay time

(c) Optimization of fuel consumption per 100 km (d) Proportion reduction of fuel consumption per 100km

Fig. 3.139 Optimization of traffic delay time and fuel consumption per 100 km under different ICV penetration rates [168]

3.4 Intelligent Driving Safety System

The previous chapters respectively introduce the theory of intelligent driving safety, risk identification, and intelligent safety decision and control technology. This section will summarize the safety systems in the process of IV operation from the perspective of application, including active safety control systems, assistance driving safety systems, autonomous driving safety systems and intelligent driving cooperative safety systems. It mainly introduces the background, structure and typical application scenarios of each safety system, and points out the development direction of intelligent driving safety system. Four types of systems in line with the safety system level will be introduced below. The development and evolution of vehicle intelligent safety system is shown in Fig. 3.140.

3.4.1 *Active Safety Control System*

Compared with passive safety systems such as seatbelts and airbags, active safety control system refers to the safety control measures taken to prevent vehicle instability or collision accidents. It mainly includes anti-lock braking system (ABS), electronic stability control (ESC), etc. Meanwhile, in view of the drift out caused by the front axle side slip and the spin caused by the rear axle side slip under extreme working conditions, major automobile manufacturers compete to develop and produce a variety of electronic control systems that can significantly improve

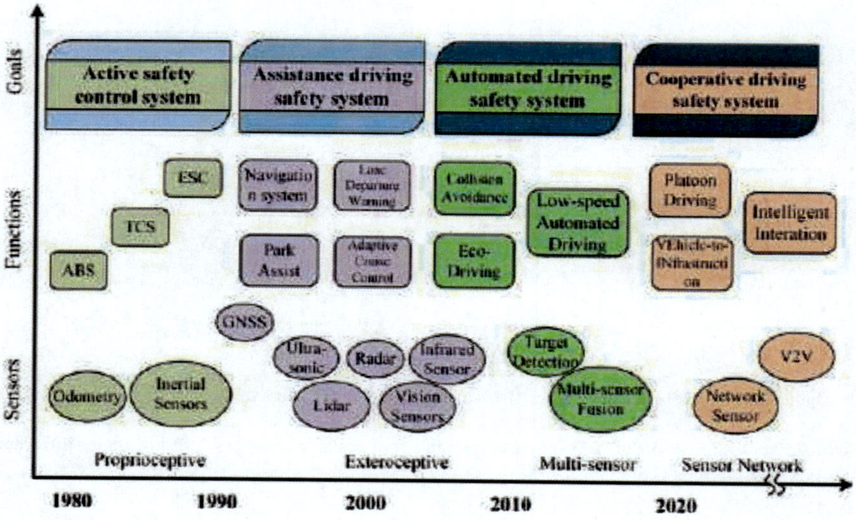

Fig. 3.140 Development and evolution of vehicle intelligent safety system [169]

the operational stability, such as four wheel steering system (4WS), vehicle stability control (VSC) system, and vehicle dynamics control (VDC) system.

1. System Composition

The functions of the current active safety control systems are designed according to effective working range, specifically divided by friction (ellipse), as shown in Fig. 3.141. The effective range of 4WD for improving the stability of curve motion is the center of the attached circle, and the tire lateral force and longitudinal force are within the linear variation range; The effective working range of TCS is the limit area near the large driving force, such as accelerating start; The effective working area of ABS is the limit area near the large braking force, such as emergency braking; The effective working area of ESC is in the limit area of large sideslip force, such as emergency steering. Other electronic control systems, such as driving force distribution control, braking force distribution control and sideslip stiffness distribution control [170], mainly work in the tire nonlinear area with large ground reaction force. In particular, the working principle of the typical electronic stability system ABS is to transmit the wheel rotation signal to the electronic control unit (ECU) through the wheel speed sensor. The ECU compares the wheel speed signal with the vehicle speed signal, monitors the wheel drift rate, and sends a command to the ABS actuator to adjust the brake pipeline pressure, so as to ensure that the wheels are not locked up. At present, ABS is a mature vehicle active safety control and has become a standard configuration for medium and premium vehicles. The structure, principle and design of ABS are introduced in Ref. [171].

With the vigorous development of computer, sensor and actuator technologies, the electronic control system represented by ESC has become another milestone active safety system after ABS. Research conducted by scientific research institutions in the United States, Europe, Japan and other countries and regions shows that ESC can effectively avoid traffic accidents, especially single vehicle rollover accidents. Based

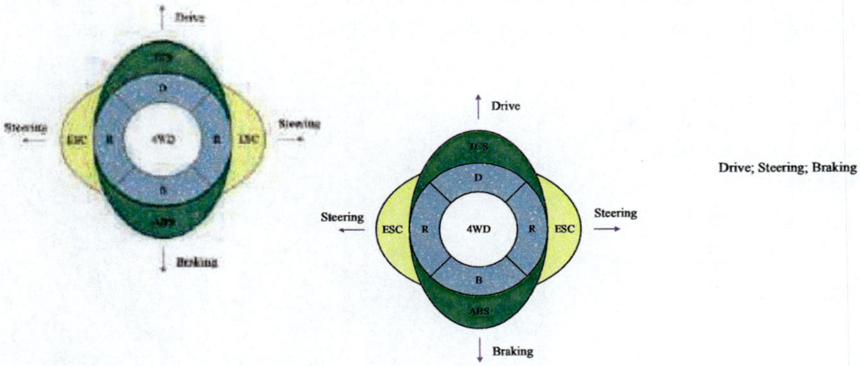

Fig. 3.141 Effective working range of electronic control system for improving operational stability. D-drive force distribution control; R-sideslip stiffness distribution control; B-braking force distribution control

on the above remarkable effects, the specification and definition of ESC were added to the revision of the passenger car braking standard ECE R13H in 2009, where the features of ESC were clarified: ① By applying the braking moment to each wheel separately, the yaw moment of the correcting vehicle motion is generated, so as to improve the directional stability of the vehicle; ② The closed-loop algorithm is applied to control the vehicle oversteer or understeer; ③ It can measure the yaw rate of the vehicle, and estimate the sideslip angle of the centroid and its change rate with time; ④ It can monitor the driver's steering wheel input; ⑤ It can judge whether the engine torque shall be adjusted, and adjust it when necessary to assist the driver to maintain control over the vehicle; ⑥ It is available within the full speed range of the vehicle (except when the vehicle speed is lower than 15 km/h and reversing). Since ESC has become a standard system for medium and high-class vehicles with more and more mature technology, it will not be discussed in detail here.

2. System Application Scenarios

The function of ABS is to prevent the wheel from being locked by the brake during braking, improve the directional stability and steering control capability, and shorten the braking distance of the vehicle. Vehicles in the past may have the following problems during driving: slipping or even turning around on slippery and muddy asphalt roads in rainy or snowy days; losing direction control in the case of emergency braking, with the left wheel and right wheel separately on roads with different adhesion coefficients, e.g., one wheel on snow road and the other on asphalt road; sliding to the roadside or into the opposite lane, in the case of emergency braking during high-speed driving on a curve; losing steering capability to avoid obstacles and other dangerous situations in the case of emergency braking on straight road [171]. The emergence of the above-mentioned dangerous situations has led to the development of ABS. The specific control effect is shown in Fig. 3.142. Vehicle equipped with ABS can be prevented from driving out of the lane due to saturation of lateral force.

ESC mainly controls the longitudinal and lateral stability to ensure the stable driving of the vehicle. For example, when the vehicle turns left and oversteers on a

Fig. 3.142 Typical application scenario of ABS: preventing vehicle from sideslip

Fig. 3.143 Typical application scenario of ESC

slippery road (turning too fast), it will be involved in the state of right tail flick. After its sensor senses the sliding of the tire, ESC quickly issues a command to brake the right front wheel to generate an opposite torque and keep the vehicle in the original lane. The specific control effect is shown in Fig. 3.143.

3.4.2 Assistance Driving Safety System

Driver's misoperation is one of the major causes of traffic accidents. The application of electronic assistance devices for reducing or correcting drivers' wrong driving behavior can effectively ensure driving safety. Therefore, researchers have developed various assistance driving safety systems, which relies on the on-board sensor system to sense the environment, predicts potential risks, and takes measures to warn or intervene driver's operation for ensuring driving safety. It can be mainly divided into two types, i.e., early warning system and control system.

1. System Structure

The assistance driving safety system is mainly composed of sensors, computing units and actuators, as shown in Fig. 3.144. The sensor senses the surrounding environment and vehicle status; The calculation unit judges driver's status or predicts the potential collision risk of the vehicle, such as possible collision time and collision severity, and then determines appropriate intervention measures to make warning or forced control, etc.; The actuator executes the instructions of the computing unit, sends warning messages to the driver or directly controls the vehicle.

Sensors can be mainly divided into three categories. The first category is used to detect the surrounding environment information of vehicles, such as the speed of vehicles in front and behind and the position of lane line. Common sensors include

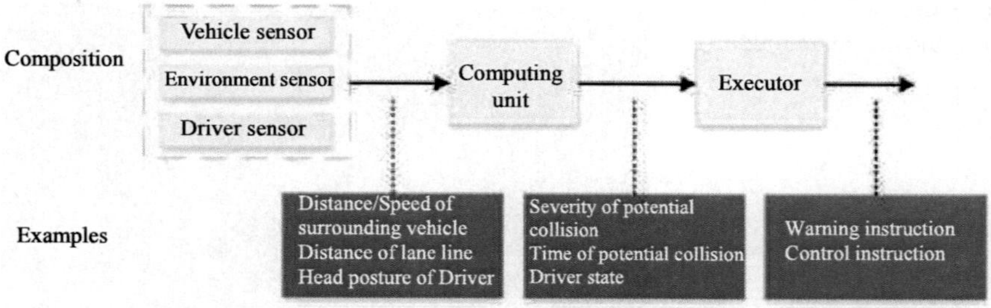

Fig. 3.144 Structure of assistance driving safety system

millimeter wave radar, laser radar, camera, rain sensor, etc.; The second is used to measure the ego-vehicle status, such as driving speed and acceleration. Common sensors include global positioning system and acceleration sensor, etc.; And the third is used to measure the driver's state, such as driver's head movement feature and EEG signals. Common sensors include in on-board cameras and EEG signal collectors, etc.

The sensors applied in the assistance driving safety systems with different functions or the safety systems with the same assistance function based on different technical schemes are different. On the one hand, various sensors have different detection ranges and detection accuracy, so the sensor configuration selected for various specific assistance driving safety systems are different. Table 3.22 shows the common sensor configuration schemes in some assistance driving safety systems. On the other hand, there may be different sensor schemes for systems with the same function. For example, driver's fatigue state can be judged and warned based on the driver's physiological signals (EEG, ECG, etc.), the driver's physiological response features (eye movement, head movement, etc.), the driver's operational behavior (steering wheel operation, pedal operation, etc.) or the vehicle's trajectory (trajectory change, trajectory deviation, etc.) [172]. Different sensors shall be correspondingly applied to collect the required information in these four technical solutions.

2. System Application Scenarios

Assistance driving safety systems are widely adopted. Common warning systems include FCW, LDW, DFW, TVS, blind spot detection (BSD), tire pressure monitoring system (TPMS), etc. Common control systems include ACC, APS, lane keeping system (LKS), AEB, etc. These systems basically realize the multi-directional and full-range safety protection for vehicles. For multi-directional protection, ACC and FCW are available in the forward direction, BSD and LDW are available in the lateral direction, APS and TVs are available in the backward direction; For full range protection, APS and BSD are available in short distance, LDW and FCW are available in medium distance, and ACC is available in long distance, as shown in Fig. 3.145.

Table 3.22 Common sensor schemes in some assistance driving safety systems

System	Sensor scheme
Lane Departure Warning (LDW)	Monocular camera, binocular camera
Auto/Assistance Parking System (APS)	Ultrasonic radar, monocular camera
Forward Collision Warning (FCW)	Millimeter wave radar, laser radar, monocular camera
Adaptive Cruise Control (ACC)	Millimeter wave radar, laser radar, monocular camera
Driver Fatigue Warning (DFW)	Interior camera
Lane Changing Assistance (LCA)	Millimeter wave radar, laser radar, monocular camera
Top View System (TVS)	Camera array

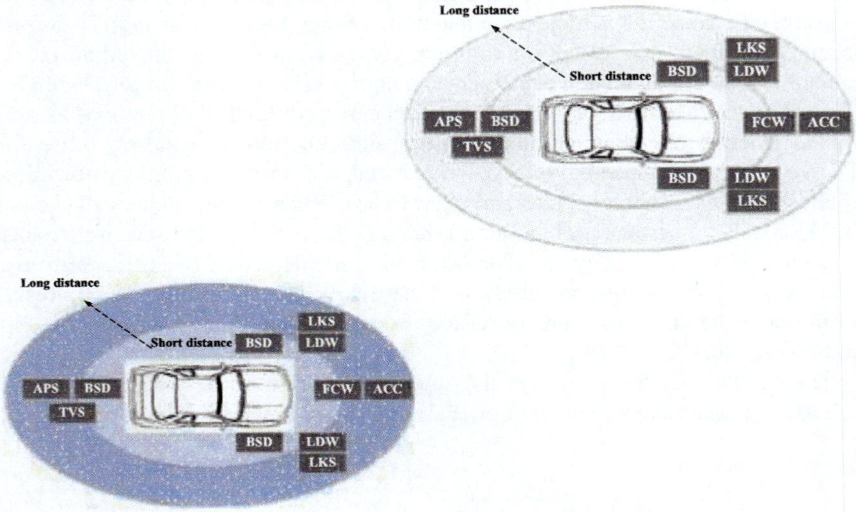

Fig. 3.145 Multi-directional and full-range protection for vehicles by assistance driving safety system

3.4.3 Autonomous Driving Safety System

Due to the development of artificial intelligence, chips and other software and hardware technologies, researchers anticipate to realize autonomous driving and completely liberate human from driving tasks, so as to further avoid the hazardous scenarios induced by driver misoperation. Autonomous driving safety system refers to a new generation of automobile system that is added with advanced sensors (radar, camera, etc.), controllers, actuators and other devices on the basis of traditional HDVs. With the capability of intelligent environmental perception, it can automatically analyze the safety and dangerous status of vehicle driving, arrive at the destination according to human's will, and finally realize the operation instead of human.

The main advantages of autonomous driving safety system include that it has a wider environmental perception capability than the driver; it can avoid the risks caused by distraction, fatigue and other dangerous human driving situations; In addition, its decision-making and execution cycle is shorter than the driver's response time, as a higher response speed is vital to deal with emergencies.

1. System Structure

The system structure of the autonomous driving safety system is similar to that of the assistance driving safety system. It is mainly composed of three units, i.e., sensor, computing unit and actuator, but the structural complexity of each unit is significantly enhanced.

The sensing system of the autonomous driving safety system shall be capable to accurately sense the vehicle state and surrounding environment information in an all-round way. It is usually a complex combination scheme, including radar, camera, V2X communication equipment and high-precision positioning system. The sensing system normally provides only the necessary information required for the system function. For example, in the AEB system, the sensor shall only sense the relative position and speed of the ego-vehicle and its front vehicle. In consideration of advantageous technology, cost and other issues, different companies will choose different sensing system configuration schemes. For example, the sensing scheme of Google Waymo is mainly based on laser radar, supplemented by millimeter wave radar, global position system (GPS), and inertial measurement unit (IMU); Tesla, on the other hand, relies more on vision, assisted by millimeter wave radar and ultrasound, etc. (Fig. 3.146).

The central decision system of the autonomous driving safety system shall be able to accurately comprehend the current traffic situation, assess vehicle operational

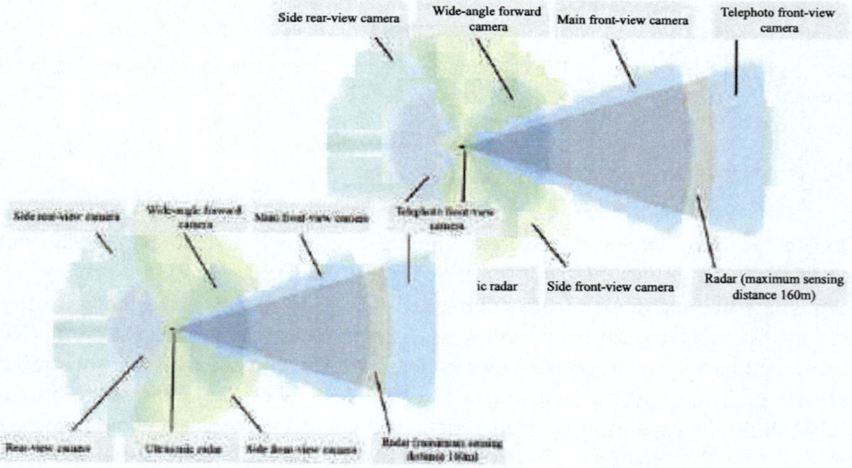

Fig. 3.146 Tesla autoleading sensor scheme [173]

risks, and make reasonable decisions and instructions according to the information collected by the sensing system. In terms of hardware architecture, the decision unit has a large demand for computing power. The optional chip solutions include GPU, FPGA, DSP and ASIC, e.g., NVIDIA's GPU-based computing solution DRIVE PX2, TI's DSP-based solution TDA2x SoC, Altera's FPGA-based solution CycloneV SoC, and Mobileye's ASIC-based solution EyeQ®5 SoC. Two exampled solutions are shown in Figs. 3.147 and 3.148.

The executive systems in the autonomous driving safety include steering power motor, brake master cylinder motor, drive motor, etc. In addition, some active safety control systems, such as electronic stability system and electric power steering system, will also play a role in the implementation. Due to different chassis structures of fuel vehicles and electric vehicles, the feasible execution modes provided for autonomous driving system are also different. For example, in terms of speed control, the former is realized by controlling the throttle opening of the engine, while the latter is realized by controlling the speed and torque of the motor.

In terms of system functions, the functions of the autonomous driving safety system usually include perception, decision, planning and control units. Different

Fig. 3.147 NVIDIA's GPU-based computing solution DRIVE PX2

Fig. 3.148 Altera's FPGA-based solution cycloneV SoC

teams carry out specific design for driving tasks according to their own understanding, and decompose complicated driving tasks into a combination of relatively simple tasks. In the desert challenge of DARPA in the United States, due to the relatively simple road structure, the logic of autonomous driving system proposed by each team is relatively simple, basically following the route of "perception-path planning-trajectory tracking", such as Stanley [174] developed by Stanford University, and Alice [175] developed by California Institute of technology (Fig. 3.149). In the subsequent urban challenge, because the competition scenario was more similar to the urban traffic scenario and the driving task was more complicated, some teams responded to various complex situations by adding decision units, that is, adopting the route of "perception-behavior decision-path planning-trajectory tracking", such as Boss [176] (Fig. 3.150) developed by Carnegie Mellon University and Odin [177] developed by Virginia Tech.

2. System Application Scenarios

The complexity of traffic environment and the importance of traffic safety put forward high requirements for autonomous driving safety system. At present, the automotive industry has basically reached a consensus on the gradual implementation of the application of autonomous driving safety system: from low-speed enclosed areas, to highways with clear road structure, and then to complex urban traffic environment.

(1) In the enclosed area, the speed requirement is low, the environment is simple, and even the driving route is fixed, which is conducive to full sensing information with high precision, such as high-precision positioning, high-precision map, and V2X communication information. Therefore, the driving safety of AV can be easily guaranteed. In this environment, the autonomous driving safety system can be used in logistics, sanitation, distribution, personnel connection, mobile advertising, etc., so as to save a lot of labor costs and reduce the traffic risk caused by fatigue of drivers after repeated monotonous work.
(2) In the highway area, the lane lines, signs and other road structural features are clear and easy to be accurately perceived; Moreover, the driving laws of surrounding vehicles are simple, and there are few highly interactive and complex traffic scenarios. Therefore, it is easier to realize autonomous driving.

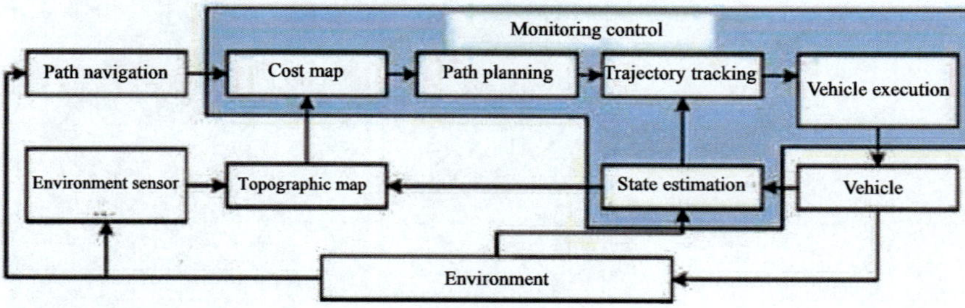

Fig. 3.149 Structure of decision unit of Alice [175]

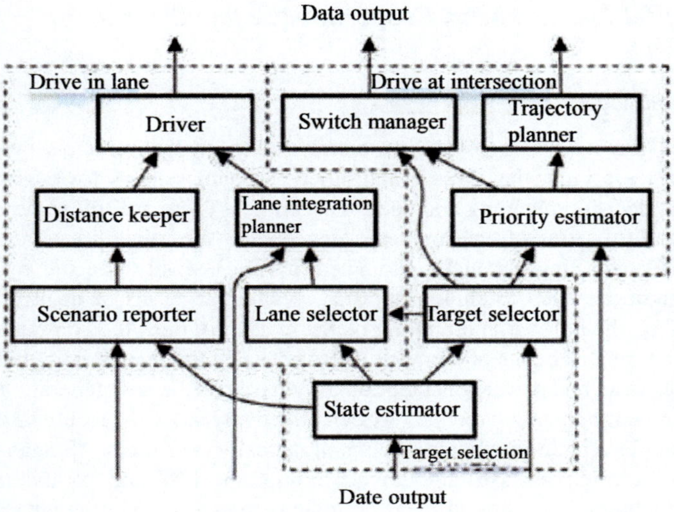

Fig. 3.150 Structure of behavior decision unit of Boss [176]

The challenge of applying autonomous driving system in highway scenario lies in the high driving speed. In this scenario, the autonomous driving safety system can be used in trunk line logistics and other work to reduce the traffic risk posed by driver fatigue and distraction. The rapid execution capability of the autonomous driving safety system is also applicable to emergency traffic scenarios such as emergency collision avoidance.

(3) In the urban traffic scenario, the traffic flow density is high; the road structure is complex, including roundabouts and intersections, etc.; there are many road signs, including traffic lights, no stop signs, tidal Lane signs, etc.; there are many traffic participants, including surrounding vehicles, pedestrians, non-motor vehicles, etc. These factors pose challenges to the application of autonomous driving safety system. Therefore, this is a long-term application goal. In the urban scenario, the autonomous driving safety system can facilitate not only the travel of private vehicles, but also such serves as unmanned taxis. It can not only liberate the productivity, but also reduce the traffic risks posed by human errors such as distracted operation.

3.4.4 Intelligent Driving Cooperative Safety System

1. System Structure

Automated driving of single vehicle mainly relies on the vehicle's own vision, millimeter wave radar, laser radar and other on-board sensors for environmental perception, decision planning and control execution. There are still some problems to be solved. In terms of environmental perception, the reliability of various on-board sensors for single vehicle automated driving is easily affected by environmental conditions such as shelter and bad weather. In terms of decision-making and planning, the on-board computing platform can not meet the computing power consumption requirements of complex algorithms due to the contradiction between computing power and power consumption. ICV based on the new generation of information and communication technology can effectively solve the technical problems faced by single vehicle automated driving, organically connect the "human-vehicle–road-cloud" traffic participation elements through the IOV, and expand and facilitate the capability upgrading of single vehicle automated driving in environmental perception, decision planning and control execution.

The intelligent driving cooperative safety system is mainly composed of four modules, i.e., the vehicle information collection system, network information interaction system, on-board controller, on-board actuator and HMI system, with each undertaking different tasks [178]. For the research, development and application of the system based on ICV, it is necessary to rely on conventional vehicle–road collaboration and IOV information service technology, and explore the deep integration and reconstruction of DVE system, so as to realize the fusion perception, decision and control of complex systems facing various scenarios. Figure 3.151 shows a typical intelligent connected vehicle–road-cloud cooperative control system architecture [179, 180], which is composed of cloud control platform, roadside infrastructure, vehicle and other traffic participants, communication network and related resource platforms [180, 181]. With the cloud control platform as the core, the ICV, roadside infrastructure and resource platform are connected to the cloud control platform through the communication network. For the integrated optimization of ICV and transportation, the physical layer, information layer and application layer of human-vehicle–road-cloud system are connected by using the new generation of information and communication technology to realize fusion perception, decision and control, so as to comprehensively enhance vehicle driving, traffic safety and efficiency.

(1) Cloud Control Platform

The cloud control platform consists of basic platform and cooperative application. It uniformly regulates and manages the basic platform and cooperative application according to the optimization requirements of vehicle and traffic operation. The basis platform provides communication chains, real-time data of all traffic elements

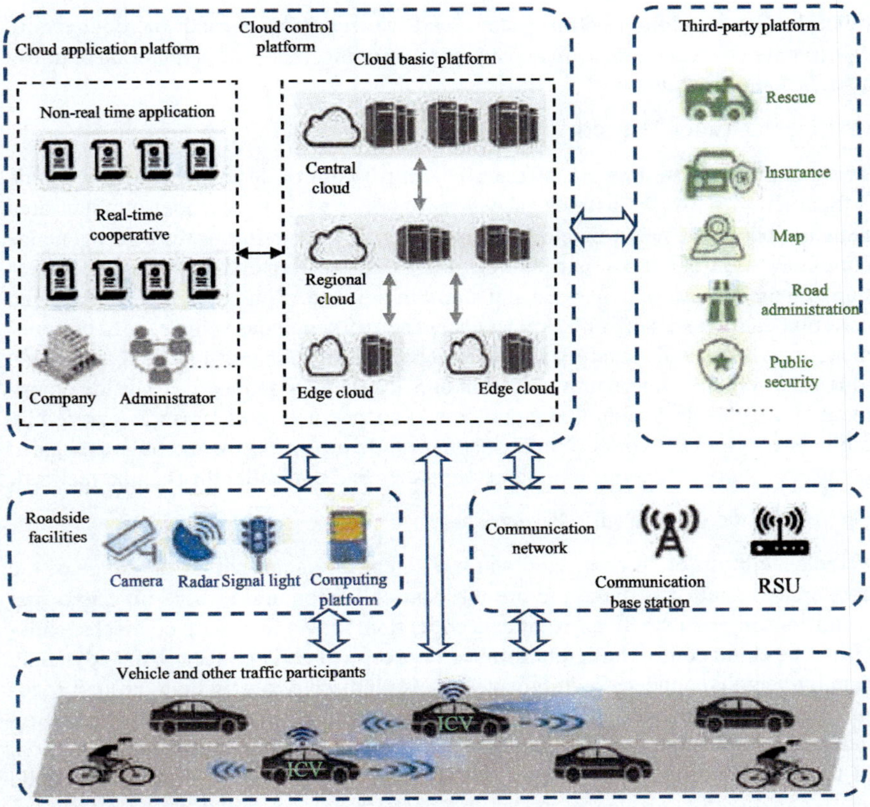

Fig. 3.151 Intelligent and connected vehicle–road-cloud cooperative control system [180, 181]

and real-time operation environment for cooperative application. In accordance with cooperative applications and scenarios, the cloud control platform uniformly optimizes the usage of system computing resources, realizes the high concurrency and on-demand real-time operation of cooperative applications, and ensures the safety and expected performance of cooperative applications for vehicle and traffic operation optimization.

(2) Roadside Infrastructure

Roadside infrastructure is a collection of equipment arranged near the road with such functions as DVE interconnection, fusion perception and local assistance positioning, mainly including road side unit (RSU), roadside computing unit (RCU), roadside sensing equipment (such as camera, millimeter wave radar and laser radar), traffic signal facilities (such as traffic lights), etc. Roadside sensors arranged through optimized design serve the cloud control system's fusion perception on mixed traffic. The roadside communication equipment enhances the communication coverage and reliability of the cloud control system, and realizes vehicle–road interconnection, environmental awareness, local assistance positioning, real-time acquisition of traffic signals and other functions.

(3) Communication Network

The communication network of the cooperative control system includes roadside infrastructure network, wireless access network, core network, metropolitan area network, backbone network, etc. The cloud control system integrates heterogeneous communication networks, and realizes the extensive interconnection communication among human, vehicle, road and cloud in the system based on the standardized communication mechanism. Wireless access network includes vehicle–road communication network and vehicle-vehicle direct communication network (V2I and V2V), to realize vehicles' intercommunication and access to edge cloud; Roadside equipment and all levels of cloud on cloud control infrastructure platform are carried by multi-level wired networks. With 5G, software-defined network and other advanced communication technologies, high performance and controllability can be realized.

(4) Vehicle and Other Traffic Participants

Vehicles with different connected and intelligent levels are within the range of the cooperative control system, and are the data collection and service objects of the cloud control platform. In terms of data collection, for vehicles with connected capability, the cloud control basic platform can not only directly collect vehicle dynamic data by vehicle connected equipment, but also indirectly obtain data through roadside intelligent perception; For vehicles without connected capability, the platform indirectly obtains the vehicle dynamic data through roadside intelligent perception. In terms of cloud control services, vehicles with L3 or above driving automatic levels can directly receive the cooperative decision and control data output from the cloud control platform, and then respond correspondingly by their on-board intelligent computing platforms or controllers; Vehicles with L2 and below automatic levels indirectly receive the cooperative decision data output from the cloud control platform, then receive the decision by on-board HMI platform, and finally realize self-control or control by drivers.

Other traffic participants include pedestrians, cyclists and other vulnerable traffic participants. The cloud control basic platform can collect the location and speed information of other traffic participants by the roadside intelligent systems or the positioning devices of other traffic participants registered on it, and offer safety warning services to these traffic participants through cloud fusion perception.

(5) Resource Platform

The resource platform is a professional platform that provides other data required for cooperative application, involving high-precision maps, ground-based enhanced positioning, meteorology, traffic management, public security, etc. Among them, high-precision dynamic map is the main carrier for the cloud control system in providing dynamic basic data. With map engine, it can timely provide updated dynamic state data for the cloud control basic platform based on basic data; With the global satellite signal receiver, the ground-based enhanced positioning platform provides high-precision positioning services at levels of 1–2 m, decimeter and centimeter for the cloud control basic platform through the ground reference station network; The meteorology platform collects meteorological information such as visibility, rainfall, wind direction, thunderstorm, fog (mass fog) through the meteorological station equipment arranged along the road, so as to provide real-time weather conditions for the cloud control basic platform; The traffic management and public security platform can provide real-time information such as road administration, maintenance, service area and emergencies for the cloud control basic platform.

2. System Application Scenarios

Based on single vehicle intelligent driving, the introduction of networked technology has realized more intelligent connected driving applications for driving safety improvement in terms of environmental perception, decision planning, control execution and failure management. The cooperative safety of intelligent driving includes three parts, i.e., cooperative perception, cooperative decision and cooperative control. Typical application scenarios of intelligent driving cooperative safety system involve variously structured road environments, such as intersections, highways, tunnels, parking places, mines and ports, as well as various general driving environments. The cooperative safety applications of intelligent driving for ICVs with different intelligent levels under typical application scenarios of the system is shown in Table 3.23.

It is shown in Table 3.23 that the intelligent driving cooperative safety system is applicable to various scenarios. For ICVs with different intelligent levels, driving safety can be enhanced from three aspects of perception, decision and control in different road environment. In order to further clarify the application effect of the intelligent driving cooperative safety system, two common scenarios, i.e. unsignalized intersection and straight car-following driving, are selected to illustrate the performance of the system in driving safety improvement.

(1) Cooperative Safety Passage Control of Intelligent Driving at Unsignalized Intersection

In Fig. 3.152, the scenario of intelligent driving cooperative safety passage at an unsignalized intersection is shown. The IV entering the intersection perceive traffic environment information through the network (vehicle-vehicle, vehicle–road), and can predict the trajectory of other vehicles at the intersection and assess the collision risk (Fig. 3.153). The collision conflict at an unsignalized intersection can be resolved through cooperative decision of multiple vehicles. At intersection without signal control, the traffic sequence shall be judged by the driver. If all traffic participants follow the rules, orderly and safe passage can be ensured. If a driver does not comply with the rules, it is easy to result in accident or congestion during the passage. The development of intelligent cooperative safety technology provides a new way for dealing with this problem.

Table 3.23 Cooperative safety application scenarios of intelligent driving

<table>
<tr><th rowspan="2">Automatic level</th><th colspan="3">Cooperative section</th></tr>
<tr><th>Networked perception</th><th>Networked decision</th><th>Networked control</th></tr>
<tr><td>Networked manual driving L0</td><td rowspan="3">(1) Over the horizon/blind spot perception
(2) Forward collision warning
(3) Emergency braking warning
(4) Pedestrian crossing warning
(5) Lane-changing hazard warning
(6) High precision map and positioning
(7) Dangerous driving behavior recognition
……</td><td colspan="2">(1) Left turn decision guidance
(2) Overtaking decision guidance
(3) Obstacle avoidance planning guidance
(4) Mergence and divergence guidance
……</td></tr>
<tr><td>Networked assistance driving L1–L2</td><td colspan="2">(1) Over-the-horizon AEB
(2) Over-the-horizon optimization ACC
……</td></tr>
<tr><td>Networked automated driving ≥L3</td><td>(1) Left turn decision assistance
(2) Overtaking decision assistance
(3) Obstacle avoidance planning assistance
(4) Mergence and divergence coordination
……</td><td>(1) Remote takeover
(2) Parallel driving
(3) Cooperative passage at intersections
(4) Networked formation driving
……</td></tr>
</table>

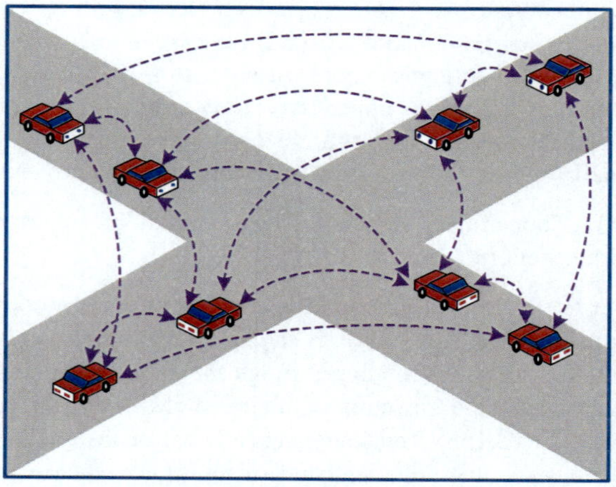

Fig. 3.152 Vehicle-vehicle cooperative safety passage at unsignalized intersection

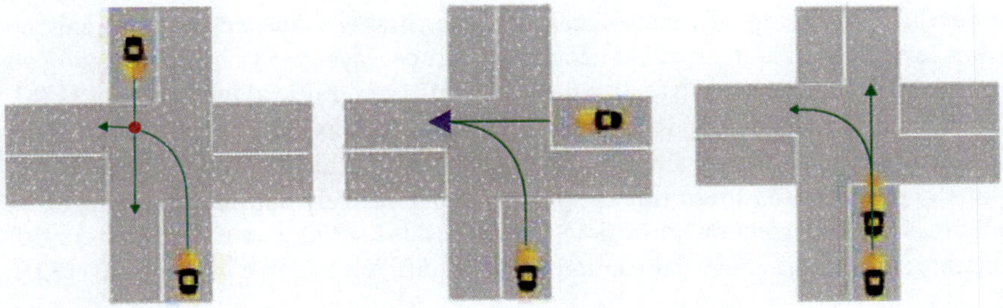

Fig. 3.153 Traffic flow conflict at unsignalized intersections

In the vehicle-vehicle/vehicle–road information interaction, each vehicle can pass through the intersection according to the preset traffic rules or instructive coordination. Currently, vehicles are running in a distributed or centralized way in accordance with a conflict-free scheme, instead of traffic light signals. Regarding cooperation mode, the cooperative control at unsignalized intersections can be conducted in terms of reservation, hierarchical optimization and centralized optimization, etc. [182].

The reserved cooperation control at intersections was proposed by the team of Professor Stone of the University of Texas [183]. The vehicle agent heading for the intersection applies to the traffic management agent on the roadside for certain time and space resources in passing through the intersection, while the traffic management agent confirms or rejects the application according to the priority of agents and the conflict of the trajectories at the intersection. If the application is rejected, the vehicle will re-plan its own motion trajectory, as shown in Fig. 3.154, based on its intersection management strategy and trajectory planning.

The hierarchical optimization control optimizes the passing sequence and trajectories of vehicles respectively through two-level planning. The upper level in view

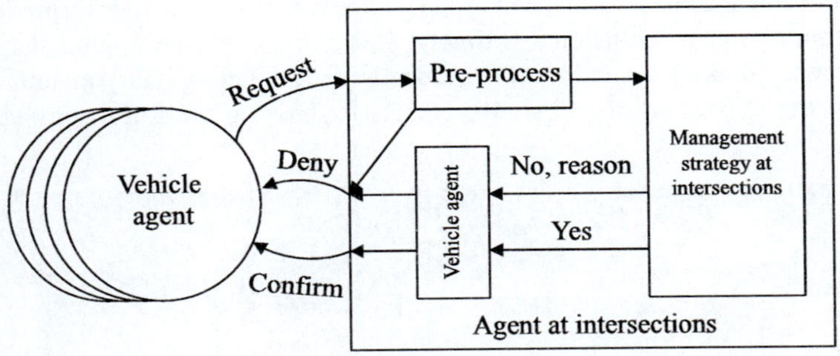

Vehicle agent; Request; Deny; Confirm; Pre-process; Post-process; No, reason; Yes; Management strategy at intersections; Agent at intersections

Fig. 3.154 Reserved cooperation control process at intersections [183]

of conflict relationship of vehicles optimizes the passing sequence and time, and the lower level optimizes the speed trajectory. The upper layer works by constructing an optimization model, so as to realize the highly efficient traffic at intersections [178].

The centralized optimization control adopts a unified optimization model to simultaneously optimize the passing sequence and motion trajectory of vehicles. It can globally optimize the speed trajectory by comprehensively considering the conflict relationship of intersection traffic flow. However, due to a large state space and control variable space, it has huge calculation load and difficult to solve in real time [182].

(2) Intelligent Cooperative Control for Collision Avoidance during Vehicle-following

Car-following driving is the most common traffic phenomenon on the road, and is also the simplest driving behavior. In the process, the driving behavior of the rear vehicle is affected by that of the front. It is necessary to continuously adjust the speed according to the speed, relative distance and other parameters of the two vehicles, in order to ensure the safe following distance. Intelligent cooperative control for collision avoidance in car-following driving refers to that in the driving scenario, the front and rear vehicles in the same direction and the same lane can transmit their own status and environmental information through the network (vehicle-vehicle, vehicle–road), based on which the motion trajectory of other vehicles can be predicted and the driving risk can be quantitatively evaluated. The rear-end collision can be successfully avoided by means of cooperative collision avoidance decision [184] (Fig. 3.155).

With the introduction of vehicle-vehicle/vehicle–road communication, the comprehensive information of multiple vehicles within the communication range can be shared in real time and accurately, and the vehicles within a certain range can know the hazard information. The concept of multi-vehicle cooperation can also be applied to collision avoidance control, creating space for the further development of active collision avoidance control. Research on cooperative collision avoidance based on V2V has been gradually carried out [159]. Gumaste et al. [185] designed a vehicle collision avoidance system based on V2V and GPS information, and proposed the concept of vehicle collision zone, which can realize collision avoidance by predicting the location of vehicle collision and timely adjusting the vehicle motion state. For the emergency braking scenario of the head vehicle, Choi et al. [186] quantitatively evaluated the mitigation of collision accidents by braking the follow-up vehicles

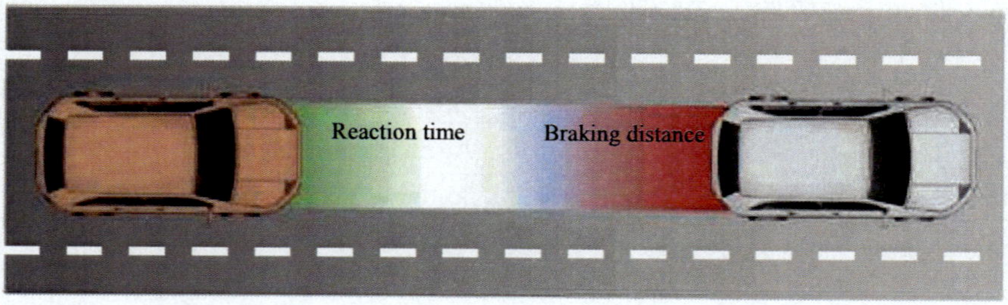

Fig. 3.155 Intelligent cooperative control for collision avoidance during car-following

based on the speed and position information of the head vehicle and the adjacent front vehicle, compared with the traditional way of merely braking the follow-up vehicles with maximum deceleration braking mode. Tatchikou et al. [187] designed a cooperative collision avoidance system based on the broadcast V2V mode. When the head vehicle brakes in an emergency, the rear vehicles can receive the braking information in time by V2V mode and brake before their front vehicles turning on the brake light, so as to effectively avoid the occurrence of serial collisions. Wang Pangwei et al. [188] proposed a multi-vehicle cooperative collision avoidance way by sliding mode control based on V2V, which gives priority to the safety of rear vehicle braking. When the braking distance of rear vehicle is insufficient, the active cooperative collision avoidance is realized by reasonably distributing the acceleration of front and rear vehicles.

3.5 Summary

To improve DSV and ensure the scientificity, timeliness and accuracy of driving risk assessment, this chapter mainly introduces the theories, methods, technical application and system development of DSV during driving.

(1) The mechanism of driving risk generation lies in the comprehensive interaction between various factors in the DVE system affecting driving safety. By proposing a unified form of comprehensive driving risk that simultaneously considers the vehicle's attributes and the interaction in the traffic flow, and establishing a unified driving safety field model that reflects the interaction among DVE, it can effectively analyse the impact of risk sources and their change trends, and organically combine various influencing factors to achieve an accurate assessment of driving risk.

(2) The interaction among DVE in a complex traffic environment can be analyzed by a physical model and virtual mechanical system of driving process. Meanwhile, the hypothesis that the driver's decision making mechanism follows the principle of least action is verified based on the phenomenon that many physical features and biological behaviour in nature have extreme values and follow the principle of least action. However, a large number of ND data extract that the vehicle motion feature parameters that reflect the driver's decision-making behavior have extreme values. The internal motivation is that the driver follows the basic decision mechanism of "pursuing benefit and avoiding harm", which follows the principle of least action.

(3) The proposal and application of intelligent decision and control technology for a single vehicle and multiple vehicles can improve the rationality and safety of single vehicle decision-making, as well as the ability of vehicle-road-cloud integrated cooperative control technology. Within it, the hierarchical decision plays an important role in reference path planning, behavior planning and trajectory planning. For cooperative decision and control technology, the application of

the four-element model of networked system can improve driving safety and efficiency in platoon scenarios and intersection scenarios.

(4) The safety system in intelligent driving mainly includes active safety control systems, assistance driving safety systems, autonomous driving safety systems and intelligent driving cooperative safety systems. The theoretical methods and technologies in this chapter can be further applied to the intelligent driving safety system, to effectively ensure DSV.

References

1. XIE S, CHEN S, ZHENG N, et al. Modeling Methodology of Driver-Vehicle-Environment System Dynamics in Mixed Driving Situation [C]. 2020 IEEE Intelligent Vehicles Symposium (IV). IEEE, 2020:1984–1991.
2. LU N, CHENG N, ZHANG N, et al. Connected vehicles: Solutions and challenges [J]. IEEE internet of things journal, 2014, 1(4):289–299.
3. EL HAJJAJI A, BENTALBA S. Fuzzy path tracking control for automatic steering of vehicles [J]. Robotics and Autonomous systems, 2003, 43(4):203–213.
4. HORNG W B, CHEN C Y, CHANG Y, et al. Driver fatigue detection based on eye tracking and dynamic template matching [C]. IEEE International Conference on Networking, Sensing and Control, 2004. IEEE, 2004, 1:7–12.
5. CLANTON J M, BEVLY D M, HODEL A S. A low-cost solution for an integrated multisensor lane departure warning system [J]. IEEE Transactions on Intelligent Transportation Systems, 2009, 10(1):47–59.
6. KATSAROS K, KERNCHEN R, DIANATI M, et al. Performance study of a Green Light Optimized Speed Advisory (GLOSA) application using an integrated cooperative ITS simulation platform [C]. 2011 7th International Wireless Communications and Mobile Computing Conference. IEEE, 2011:918–923.
7. LEE J, PARK B. Development and evaluation of a cooperative vehicle intersection control algorithm under the connected vehicles environment [J]. IEEE Transactions on Intelligent Transportation Systems, 2012, 13(1):81–90.
8. Adaptive Integrated Driver-vehicle InterfacE (AIDE) [EB/OL]. [2021-01-19]. http://www.aide-eu.org/objectives.html.
9. CARSTEN O. From driver models to modelling the driver: what do we really need to know about the driver? Modelling driver behaviour in automotive environments [M]. London: Springer, 2007:105–120.
10. JACOBSON B. Vehicle Dynamics Compendium for Course MMF062; edition 2016 [R]. Chalmers University of Technology, 2016.
11. SALVATORE S. The estimation of vehicular speed as a function of visual stimulation [J]. Human factors, 1968, 10(1):27–31.
12. HORSWILL M S, PLOOY A M. Auditory feedback influences perceived driving speeds [J]. Perception, 2008, 37(7):1037–1043.
13. WANG J, WU J, LI Y. The driving safety field based on driver-vehicle-road interactions [J]. IEEE Transactions on Intelligent Transportation Systems, 2015, 16(4):2203–2214.
14. NI D. A unified perspective on traffic flow theory, part I: the field theory [J]. ICCTP 2011: Towards Sustainable Transportation Systems. 2011:4227–4243.
15. HELLIER E, NAWEED A, WALKER G, et al. The influence of auditory feedback on speed choice, violations and comfort in a driving simulation game [J]. Transportation research part F: traffic psychology and behaviour, 2011, 14(6):591–599.

16. TANAKA Y, KANEYUKI H, TSUJIY T, et al. Mechanical and perceptual analyses of human foot movements in pedal operation [C]. 2009 IEEE International Conference on Systems, Man and Cybernetics. IEEE, 2009:1674–1679.
17. STEVENS S S. Psychophysics: Introduction to its perceptual, neural and social prospects [M]. Routledge, 2017.
18. NEWBERRY A C, GRIFFIN M J, DOWSON M. Driver perception of steering feel [J]. Proceedings of the Institution of Mechanical Engineers, Part D: Journal of Automobile Engineering, 2007, 221(4):405–415.
19. BELLER J, HEESEN M, VOLLRATH M. Improving the driver-automation interaction: An approach using automation uncertainty [J]. Human factors, 2013, 55(6):1130–1141.
20. VAN DEN BEUKEL A P, VAN DER VOORT M C, EGER A O. Supporting the changing driver's task: Exploration of interface designs for supervision and intervention in automated driving [J]. Transportation research part F: traffic psychology and behaviour, 2016, 43:279–301.
21. NGUYEN A T, SENTOUH C, POPIEUL J C, et al. Shared lateral control with on-line adaptation of the automation degree for driver steering assist system: A weighting design approach [C]. 2015 54th IEEE Conference on Decision and Control (CDC). IEEE, 2015:857–862.
22. LI R, LI S, GAO H, et al. Effects of human adaptation and trust on shared control for driver-automation cooperative driving [R]. SAE Technical Paper, 2017.
23. JUGADE S C, VICTORINO A C, CHERFAOUI V B. Shared Driving Control between Human and Autonomous Driving System via Conflict resolution using Non-Cooperative Game Theory [C]. 2019 IEEE Intelligent Transportation Systems Conference (ITSC). IEEE, 2019:2141–2147.
24. GUO K, GUAN H. Modelling of driver/vehicle directional control system [J]. Vehicle system dynamics, 1993, 22(3–4):141–184.
25. MACADAM C C. Understanding and modeling the human driver [J]. Vehicle system dynamics, 2003, 40(1–3):101–134.
26. SADIGH D, SASTRY S, SESHIA S A, et al. Planning for autonomous cars that leverage effects on human actions. Robotics: Science and Systems [J]. Ann Arbor, MI, USA, 2016, 2.
27. LI X, YING X, CHUAH M C. Grip: Graph-based interaction-aware trajectory prediction [C]. 2019 IEEE Intelligent Transportation Systems Conference (ITSC). IEEE, 2019:3960–3966.
28. LI J, YANG F, TOMIZUKA M, et al. Evolvegraph: Multi-agent trajectory prediction with dynamic relational reasoning [J]. arXiv preprint arXiv:2003.13924, 2020.
29. ODHAMS A M C, COLE D J. Models of driver speed choice in curves [C]. Proceedings of the 7th International Symposium on Advanced Vehicle Control. Citeseer, 2004.
30. Yu Zhisheng. Vehicle theory [M]. 5th Edition. Beijing: China Machine Press, 2009 [in Chinese].
31. WANMING Z. New advance in vehicle-track coupling dynamics [J]. China Railway Science, 2002, 23(2):1–14.
32. YANG S, CHEN L, LI S. Dynamics of vehicle-road coupled system [M]. London: Springer, 2015.
33. Wang Jianqiang, Wu Jian, Li Yang. Concept, principle and modeling of driving risk field based on DVE cooperation [J]. China Journal of Highway and Transport, 2016, 29 (01): 105–114 [in Chinese].
34. Liu Qiaobin, Liu Ke, Wang Tao, Gao Ming, Yang Lu, Xu Qing, Wang Jianqiang, Li Keqiang. Human-like trajectory planning for IVs based on lateral quantitative balance index [P]. Beijing: CN113771884A, 2021-12-10 [in Chinese].
35. LAM L T. Distractions and the risk of car crash injury: The effect of drivers'age [J]. Journal of safety research, 2002, 33(3):411–419.
36. DAHLEN E R, MARTIN R C, RAGAN K, et al. Driving anger, sensation seeking, impulsiveness, and boredom proneness in the prediction of unsafe driving [J]. Accident analysis and prevention, 2005, 37(2):341–348.

37. CHARLTION S G, BAAS P H. Road User Interactions: Patterns of Road Use and Perceptions of Driving Risk [C]. Institution of Professional Engineers New Zealand (IPENZ) Transportation Group. Technical Conference Papers 2002.
38. KNEE C R, NEIGHBORS C, VFETOR N A. Self-Determination Theory as a Framework for Understanding Road Rage 1 [J]. Journal of Applied Social Psychology, 2001, 31(5):889–904.
39. GREENE K, KRCMAR M, WALTERS L H, et al. Targeting adolescent risk-taking behaviors: the contributions of egocentrism and sensation-seeking [J]. Journal of adolescence, 2000, 23(4):439–461.
40. CHARLTON S G, STARKEY N J, PERRONE J A, et al. What's the risk? A comparison of actual and perceived driving risk [J]. Transportation Research Part F: Traffic Psychology and Behaviour, 2014, 25:50–64.
41. Li Xiaoyu, Xu Nan, Guo Konghui, et al. An adaptive SMC controller for EVs with four IWMs handling and stability enhancement based on a stability index [J]. Vehicle System Dynamics, 2020: 1–24.
42. Zhao Xianli. Research on evolution mechanism of airport runway safety risk [D]. Wuhan: Wuhan University of Technology, 2017 [in Chinese].
43. Xiong Xiaoxia. Research on evolution model and blocking strategy of road traffic accident chain based on Markov Chain Theory [D]. Zhenjiang: Jiangsu University, 2018 [in Chinese].
44. Huang Fei. Research on evolution mechanism of traffic accidents on urban expressway with ice and snow [D]. Changchun: Jilin University, 2017 [in Chinese].
45. Li Xiaoyu. Research on instability mechanism and handling stability control of distributed-drive electric vehicles under combined working conditions [D]. Jilin University, 2020 [in Chinese].
46. Wang Wuhong, Guo Hongwei, Guo Weiwei. Traffic behavior analysis and safety assessment [M]. Beijing: Beijing University of Technology Press, 2013 [in Chinese].
47. OU YK, LIU YC, SHIH FY. Risk prediction model for drivers' in-vehicle activities-Application of task analysis and back-propagation neural network [J]. Transportation Research Part F: Traffic Psychology and Behaviour, 2013, 18:83–93.
48. MCCARTT A T, SHABANOVA V I, LEAF W A. Driving experience, crashes and traffic citations of teenage beginning drivers [J]. Accident Analysis and Prevention, 2003, 35(3):311–320.
49. WILDE G. Does Risk Homeostasis Theory Have Implications for Road Safety? [J]. Education and Debate, 2002, 32(4):1149–1152.
50. WILDE, G. The theory of risk homeostasis: implications for safety and health. Risk Analysis. 1982, 2:209–225.
51. SUMMLA H. Risk Control is not Risk Adjustment: the Zero-risk Theory of Driver Behaviour and Its Implications [J]. Ergonomics, 1988, 31(4):491–506.
52. Fuller R. Towards a general theory of driver behaviour. Accident Analysis and Prevention. 2005, 37:461–472.
53. Anderson. Cognitive psychology and its implications [M]. 7th Edition. Translated by Qin Yulin, Cheng Yao, et al. Beijing: People's Posts and Telecommunications Press, 2012 [in Chinese].
54. VAN DER MOLEN H H, B?TTICHER A M. A hierarchical risk model for traffic participants [J]. Ergonomics. 1988, 31:537–555.
55. Ren Futian, Liu Xiaoming. Analysis of road traffic system safety--Road traffic safety [M]. Beijing: People's Communications Press, 2001 [in Chinese].
56. POLLATSEK A, NARAYANAAN V, PRADHAN A, et al. Using Eye Movements to Evaluate a PC-Based Risk Awareness and Perception Training Program on a Driving Simulator [J]. Human Factors: The Journal of the Human Factors and Ergonomics Society, 2006, 48(3):447–464.
57. MICHON J A. 1985. A Critical View of Driver Behavior Models: What Do We Know, What Should We Do? [M]. EVANS L, SCHWING R C. Human Behavior and Traffic Safety. Boston, SPRINGER, 1985.

58. MICHON J A. A Critical View of Driver Behavior Models: What Do We Know, What Should We Do? [G]. EVANS L, SCHWING R C. Human Behavior and Traffic Safety. Boston: Springer, 1985: 485–524.
59. He Ren, Zhao Xiaocong, Wang Jianqiang. Modeling of driver risk responsiveness under DVE interaction [J]. China Journal of Highway and Transport, 2020, 33 (09): 236–250 [in Chinese].
60. ZHAO X, HE R, WANG J. How do drivers respond to driving risk during car-following? Risk-response driver model and its application in human-like longitudinal control [J]. Accident Anal. Prev., 2020, 148: 105783.
61. NI D. A unified perspective on traffic flow theory, part I: The field theory [J]. Appl. Math. Sci, 2013, 7(39): 1929–1946.
62. Zheng Xunjia, Huang Heye. Driver's driving decision-making mechanism follows the principle of least action [J]. China Journal of Highway and Transport, 2020, 33 (04): 155–168 [in Chinese].
63. Zhang Yihua. Analysis on lateral instability mechanism and research on in-loop control strategy of double-trailer train [D]. Jilin University, 2017 [in Chinese].
64. Yang Xiujian, Li Yaoping, Xiong Jian. Analysis on lateral stability and instability mechanism of semi-trailer train [J]. Automotive Engineering, 2011, 033 (006): 486–492 [in Chinese].
65. Xiong Lu, Qu Tong, Feng Yuan, Deng Luhua. Criteria of vehicle driving stability under extreme working conditions [J]. Journal of Mechanical Engineering, 2015, 51 (10): 103–111 [in Chinese].
66. Bobier C G. A phase portrait approach to vehicle stabilization and envelope control [D]. Stanford University, 2012.
67. Goh J Y, Goel T, Christian Gerdes J. Toward automated vehicle control beyond the stability limits: drifting along a general path [J]. Journal of Dynamic Systems, Measurement, and Control, 2020, 142(2): 021004.
68. Zheng Sheng. Research on lateral motion control of automated driving vehicle under extreme conditions [D]. Tsinghua University, 2022 [in Chinese].
69. HYDE N C. The development of a method for traffic safety evaluation: The Swedish Traffic Conflicts Technique [J]. Bulletin Lund Institute of Technology, Department, 1987 (70).
70. MARKKULA G, MADIGAN R, NATHANAEL D, et al. Defining interactions: A conceptual framework for understanding interactive behaviour in human and automated road traffic [J]. Theoretical Issues in Ergonomics Science, 2020, 21(6):728–752.
71. Hu Yuanzhi, Lv Zhangjie. Longitudinal collision avoidance algorithm and simulation verification of AEB system based on PreScan [J]. Journal of Automotive Safety and Energy, 2017, 8 (02): 136 [in Chinese].
72. CHAKROBORTY P, KIKUCHI S. Evaluation of the General Motors based car-following models and a proposed fuzzy inference model [J]. Transportation Research Part C: Emerging Technologies, 1999, 7(4):209–235.
73. TREIBER M, HENNECKE A, HELBING D. Congested traffic states in empirical observations and microscopic simulations [J]. Physical review E, 2000, 62(2):1805.
74. Zhao Xiaocong. Research on the flexible switching of human-machine driving control rights under the DVE interaction [D]. Jiangsu University, 2020 [in Chinese].
75. AMIR E. Waymo's Big Ambitions Slowed by Tech Trouble. [EB/OL]. [2018-08-28]. https://www.theinformation.com/articles/waymos-big-ambitions-slowed-by-tech-trouble.
76. TRAUTMAN P, KRAUSE A. Unfreezing the robot: Navigation in dense, interacting crowds [C]. 2010 IEEE/RSJ International Conference on Intelligent Robots and Systems, October 18–22, 2010, Taipei, Taiwan. IEEE, 2010.
77. Wang Jianqiang, Yang Lu, Cui Mingyang, Huang Heye, Lin Xuewu, Xu Qing. Comprehensive risk assessment ways and devices for vehicle instability and collision under extreme working conditions [P]. Beijing: CN113370980B, 2021-11-02 [in Chinese].
78. Li Yibing, Sun Yueting, Xu Chengliang. Analysis of development trend of vehicle safety technology based on traffic accident data [J]. Journal of Automotive Safety and Energy, 2016, 7 (03): 241–253 [in Chinese].

79. Li Fangyuan. Study on causation mechanism and risk behaviors of major and extraordinary traffic accidents [D]. Chang'an University, 2014 [in Chinese].
80. Ren Futian, Liu Xiaoming, Rong Jian, et al. Traffic engineering [M]. Beijing: People's Communications Press, 2008 [in Chinese].
81. Sun Yixuan. Research on road traffic accident analysis based on data mining [D]. Beijing Jiaotong University, 2014 [in Chinese].
82. Pei Yulong. Road traffic safety [M]. Beijing: People's Communications Press, 2004 [in Chinese].
83. AUST M L, FAGERLIND H, SAGBERG F. Fatal intersection crashes in Norway: Patterns in contributing factors and data collection challenges [J]. Accident Analysis and Prevention, 2012, 45:782–791.
84. WANG W, JIANG X, XIA S, et al. Incident tree model and incident tree analysis method for quantified risk assessment: an in-depth accident study in traffic operation [J]. Safety Science, 2010, 48(10):1248–1262.
85. Xu Hongguo, Zhang Huiyong, Zong Fang. Bayesian network modeling for traffic accident causation analysis [J]. Journal of Jilin University: Engineering Edition, 2011 (S1): 89–94 [in Chinese].
86. DELEN D, SHARDA R, BESSONOV M. Identifying significant predictors of injury severity in traffic accidents using a series of artificial neural networks [J]. Accident Analysis and Prevention, 2006, 38(3), 434–444.
87. Mou Haibo, Yu Jianning, Liu Linzhong. Modeling and analysis of causes of traffic accidents based on fuzzy Petri nets [J]. Chinese Journal of Safety Science (12): 93 [in Chinese].
88. Li Shuqing, Peng Weilang, Xiao Liying, et al. Research status and trend analysis of road traffic accident occurrence mechanism [J]. Journal of Safety and Environment, 2014, 14 (03): 14–19 [in Chinese].
89. ICV Sub Technical Committee of National Technical Committee of Auto Standardization. White paper on design operating conditions of automated driving system [R/OL]. (2020-09) http: ‖ catarc.org.cn/2009151518412898.pdf [in Chinese].
90. China National Standardization Administration. 192,314-T-339, Classification of vehicle driving automation [S]. Beijing: China Standards Press, January 8, 2020 [in Chinese].
91. Phillips D J, Wheeler T A, Kochenderfer M J. Generalizable intention prediction of human drivers at intersections [C]// 2017 IEEE intelligent vehicles symposium (IV). IEEE, 2017: 1665–1670.
92. Rehder E, Wirth F, Lauer M, et al. Pedestrian prediction by planning using deep neural networks [C]// 2018 IEEE International Conference on Robotics and Automation (ICRA). IEEE, 2018: 5903–5908.
93. Agamennoni G, Nieto J I, Nebot E M. Estimation of multivehicle dynamics by considering contextual information [J]. IEEE Transactions on robotics, 2012, 28(4): 855–870.
94. Xin L, Wang P, Chan C Y, et al. Intention-aware long horizon trajectory prediction of surrounding vehicles using dual lstm networks [C]// 2018 21st International Conference on Intelligent Transportation Systems (ITSC). IEEE, 2018: 1441–1446.
95. Liniger A, Lygeros J. A noncooperative game approach to autonomous racing [J]. IEEE Transactions on Control Systems Technology, 2019, 28(3): 884–897.
96. Schwarting W, Pierson A, Alonso-Mora J, et al. Social behavior for autonomous vehicles [J]. Proceedings of the National Academy of Sciences, 2019, 116(50): 24972–24978.
97. KOOIJ J F P, SCHNEIDER N, FLOHR F, et al. Context-Based Pedestrian Path Prediction [C]. European Conference on Computer Vision. Springer, Cham, 2014:618–633.
98. HUANG H, WANG J, FEI C, et al. A probabilistic risk assessment framework considering lane-changing behavior interaction [J]. Science China Information Sciences, 2020, 63(9):1–15.
99. WU H, WANG L, ZHENG S, et al. Crossing-Road Pedestrian Trajectory Prediction Based on Intention and Behavior Identification [C]. 2020 IEEE 23rd International Conference on Intelligent Transportation Systems (ITSC). IEEE, 2020:1–6.

100. Mohamed A, Qian K, Elhoseiny M, et al. Social-stgcnn: A social spatio-temporal graph convolutional neural network for human trajectory prediction [C]// Proceedings of the IEEE/CVF Conference on Computer Vision and Pattern Recognition. 2020: 14424–14432.
101. Bahram M, Lawitzky A, Friedrichs J, et al. A game-theoretic approach to replanning-aware interactive scene prediction and planning [J]. IEEE Transactions on Vehicular Technology, 2015, 65(6): 3981–3992.
102. Schwarting W, Pierson A, Karaman S, et al. Stochastic Dynamic Games in Belief Space [J]. IEEE Transactions on Robotics, 2021.
103. A. I. Goldman et al., "Theory of mind," The Oxford handbook of philosophy of cognitive science, vol. 1, 2012.
104. SCHLECHTRIEMENJ, WEDELA, BREUELG, et al. A probabilistic long term prediction approach for highway scenarios [J]. in IEEE Conference on Intelligent Transportation Systems, 2014:732–738.
105. LI Y, LU X Y, WANG J, et al. Pedestrian Trajectory Prediction Combining Probabilistic Reasoning and Sequence Learning [J]. IEEE Transactions on Intelligent Vehicles, 2020, 5(3):461–474.
106. HU Y, ZHAN W, TOMIZUKA M. Scenario-transferable semantic graph reasoning for interaction-aware probabilistic prediction [J]. arXiv preprint arXiv:2004.03053, 2020.
107. ATEV S, MILLER G, PAPANIKOLOPOULOS N P. Clustering of vehicle trajectories [J]. IEEE transactions on intelligent transportation systems, 2010, 11(3):647–657.
108. RIDEL D, DEO N, WOLF D, et al. Scene compliant trajectory forecast with agent-centric spatio-temporal grids [J]. IEEE Robotics and Automation Letters, 2020, 5(2):2816–2823.
109. LI J, MA H, ZHAN W, et al. Coordination and trajectory prediction for vehicle interactions via bayesian generative modeling [C]. 2019 IEEE Intelligent Vehicles Symposium (IV). IEEE, 2019:2496–2503.
110. Liang M, Yang B, Hu R, et al. Learning lane graph representations for motion forecasting [C]// European Conference on Computer Vision. Springer, Cham, 2020: 541–556.
111. Gao Bolin, Xie Shugang, Gong Jinfeng. Vehicle sideslip angle estimation based on kinematics-dynamics fusion [J]. Journal of Automobile Safety and Energy, 2015, 000 (001): 72–78 [in Chinese].
112. Li X, Xu N, Li Q, et al. A fusion methodology for sideslip angle estimation on the basis of kinematics-based and model-based approaches [J]. Proceedings of the Institution of Mechanical Engineers, Part D: Journal of Automobile Engineering, 2020, 234(7): 1930–1943.
113. Li B, Du H, Li W, et al. Non-linear tyre model-based non-singular terminal sliding mode observer for vehicle speed and side-slip angle estimation [J]. Proceedings of the Institution of Mechanical Engineers, Part D: Journal of automobile engineering, 2019, 233(1): 38–54.
114. Li Xiaoyu, Xu Nan, Guo Konghui. Estimation of centroid sideslip angle based on fusion of kinematics and motion geometry [J]. Journal of Mechanical Engineering, 2020, 56 (02): 121–129 [in Chinese].
115. Lin X, Wang J, Xu Q, et al. Real-Time Estimation of Tire-Road Friction Coefficient Based on Unscented Kalman Filtering [C]// 2020 IEEE 5th International Conference on Intelligent Transportation Engineering (ICITE). IEEE, 2020: 376–382.
116. Lee H, Taheri S. Intelligent tires? A review of tire characterization Reference [J]. IEEE Intelligent Transportation Systems Magazine, 2017, 9(2): 114–135.
117. He Yong. Status and countermeasures of road traffic safety in China [J]. Journal of Highway and Transportation Research and Development, 2003, 20 (1): 119–122 [in Chinese].
118. FRANKJ G, CAROLS, MOSLEHA. QRAS-the quantitative risk assessment system [J]. Reliability Engineering and System Safety, 2006, 91:292–304.
119. Yin Jingbo. Quantitative risk assessment in maritime safety management [M]. Shanghai: Shanghai Jiaotong University Press, 2015 [in Chinese].
120. LI Y, LI K, ZHENG Y, et al. Threat Assessment Techniques in Intelligent Vehicles: A Comparative Survey [J]. IEEE Intelligent Transportation Systems Magazine, 2021, 13(4): 71–91.

121. ARCHIBALD J K, HILL J C, JEPSEN N A, et al. A Satisficing Approach to Aircraft Conflict Resolution [J]. IEEE Transactions on Systems, Man, and Cybernetics, Part C (Applications and Reviews), 2008, 38(4):510–521.
122. ALLEN C, EWING M, KESHMIRI S, et al. Multichannel sense-and-avoid radar for small UAV [C]. 2013 IEEE/AIAA 32nd Digital Avionics Systems Conference (DASC), 2013:6E2-1–6E2-10.
123. LEE K, PENG H. Evaluation of automotive forward collision warning and collision avoidance algorithms [J]. Vehicle System Dynamics, 2005, 43(10):735–751.
124. ALTHOFF M, STURSBERG O, BUSS M. Model-Based Probabilistic Collision Detection in Autonomous Driving [J]. IEEE Transactions on Intelligent Transportation Systems, 2009, 10(2):299–310.
125. THORSSON J, STEINERT O, Neural Networks for Collision Avoidance [M]. Gothenburg: Chalmers Univ. Technol., 2016.
126. LEVINE S, FINN C, DARRELL T, et al. End-to-End Training of Deep Visuomotor Policies [J]. 2016, 1–40.
127. SALLAB A E, ABDOU M, PEROT E, et al. Deep Reinforcement Learning framework for Autonomous Driving [J]. Electronic Imaging, 2017, 2017(19):70–76.
128. GERDES J C, ROSSETTER E J. A Unified Approach to Driver Assistance Systems Based on Artificial Potential Fields [J]. Journal of Dynamic Systems, Measurement, and Control, 2001, 123(3):431.
129. ROSSETTER E J, GERDES J C. Lyapunov Based Performance Guarantees for the Potential Field Lane-keeping Assistance System [J]. Journal of Dynamic Systems, Measurement, and Control, 2006, 128(3):510.
130. ZHENG X, WANG J, WANG J. A Novel Road Traffic Risk Modeling Approach Based on the Traffic Safety Field Concept [J]. CICTP 2018:263–274.
131. HUANG H, ZHENG X, YANG Y, et al. An integrated architecture for intelligence evaluation of automated vehicles [J]. Accident Analysis and Prevention, 2020, 145:105681.
132. Zheng Xunjia. Driving risk generation mechanism and its quantitative evaluation [D]. Beijing: Tsinghua University, 2020 [in Chinese].
133. HUANG H, WANG J, FEI C, et al. A probabilistic risk assessment framework considering lane-changing behavior interaction [J]. SCIENCE CHINA Information Sciences, 2020, 63(9):190203.
134. Ni D, Leonard J D, Jia C, et al. Vehicle longitudinal control and traffic stream modeling [J]. Transportation Science, 2015, 50(3): 1016–1031.
135. WANG J, WU J, ZHENG X, et al. Driving safety field theory modeling and its application in pre-collision warning system [J]. Transportation Research Part C: Emerging Technologies, 2016, 72:306–324.
136. Li Yang. Intelligent vehicle decision-making based on pedestrian behavior prediction [D]. Beijing: Tsinghua University, 2020 [in Chinese].
137. XIE G, GAO H, HUANG B, et al. A Driving Behavior Awareness Model based on a Dynamic Bayesian Network and Distributed Genetic Algorithm [J]. International Journal of Computational Intelligence Systems, 2018, 11(1):469.
138. Xie Guotao. Research on dynamic environment cognition of intelligent vehicle under uncertainty [D]. Anhui: Hefei University of Technology, 2018 [in Chinese].
139. XIE G, ZHANG X, GAO H, et al. Situational Assessments Based on Uncertainty-Risk Awareness in Complex Traffic Scenarios [J]. Sustainability, 2017, 9(9):1582.
140. Wang Ziqiang, Hu Xiaoguang, Li Xiaoxiao, et al. Overview of global path planning algorithms for mobile robots [J]. Computer Science, 2021, 48 (10): 11 [in Chinese].
141. Huang Shan. Decision control for driving behaviors of multi-ICVs based on game theory [D]. Yanshan University [in Chinese].
142. Ye Mingfei. Path planning of mobile robot based on Voronoi diagram and uncertainty potential field [D]. University of Electronic Science and Technology of China, 2021. https://doi.org/10.27005/d.cnki.gdzku.2021.001074 [in Chinese].

143. Liu Xiang, Ye Xiaoming, Wang Quanbin, Li Weiguang, Gao Hanlin. Overview of research on local path planning algorithms for unmanned surface vehicles [J]. Chinese Journal Ship Research, 2021, 16 (S1): 1–10. https://doi.org/10.19693/j.issn.1673-3185.02538 [in Chinese].
144. Wang Ziqiang, Hu Xiaoguang, Li Xiaoxiao, Du Zhuoqun. Overview of global path planning algorithms for mobile robots [J]. Computer Science, 2021, 48 (10): 19–29 [in Chinese].
145. Urmson C, Anhalt J, Bagnell D, et al. Autonomous driving in urban environments: Boss and the Urban Challenge [J]. Journal of Field Robotics, 2008, 25(8):425–466.
146. Baidu Apollo Developer Center [EB/OL]. https://apollo.auto/devcenter/document_list_cn.html [in Chinese].
147. Wang H, Huang Y, Khajepour A, et al. Crash mitigation in motion planning for autonomous vehicles [J]. IEEE transactions on intelligent transportation systems, 2019, 20(9): 3313–3323.
148. Chen S, Jian Z, Huang Y, et al. Autonomous driving: cognitive construction and situation understanding [J]. Science China Information Sciences, 2019, 62(8): 1–27.
149. Huang Y, Chen Y. Autonomous driving with deep learning: A survey of state-of-art technologies [J]. arXiv preprint arXiv:2006.06091, 2020.
150. Wang Jianqiang, Zheng Xunjia, Huang Heye. Driver's driving decision-making mechanism follows the principle of least action [J]. China Journal of Highway and Transport, 2020, 33 (04): 155–168. https://doi.org/10.19721/j.cnki.1001-7372.2020.04.016 [in Chinese].
151. Zheng X, Huang H, Wang J, et al. Behavioral decision-making model of the intelligent vehicle based on driving risk assessment [J]. Computer-Aided Civil and Infrastructure Engineering, 2021, 36(7): 820–837.
152. LIU Y, BUCKNALL R. Path planning algorithm for unmanned surface vehicle formations in a practical maritime environment [J]. Ocean engineering, 2015, 97:126–144.
153. LUO Y, YANG G, XU M, et al. Cooperative lane-change maneuver for multiple automated vehicles on a highway [J]. Automotive Innovation, 2019, 2(3):157–168.
154. XU Q, CAI M, LI K, et al. Coordinated formation control for intelligent and connected vehicles in multiple traffic scenarios [J]. IET Intelligent Transport Systems, 2021, 15(1):159–173.
155. CAI M, XU Q, LI K, et al. Multi-lane formation assignment and control for connected vehicles [C]. 2019 IEEE Intelligent Vehicles Symposium (IV). IEEE, 2019:1968–1973.
156. XU B, BAN X J, BIAN Y, et al. Cooperative method of traffic signal optimization and speed control of connected vehicles at isolated intersections [J]. IEEE Transactions on Intelligent Transportation Systems, 2018, 20(4):1390–1403.
157. BIAN Y, LI S E, REN W, et al. Cooperation of multiple connected vehicles at unsignalized intersections: Distributed observation, optimization, and control [J]. IEEE Transactions on Industrial Electronics, 2019, 67(12):10744–10754.
158. XU B, LI S E, BIAN Y, et al. Distributed conflict-free cooperation for multiple connected vehicles at unsignalized intersections [J]. Transportation Research Part C: Emerging Technologies, 2018, 93:322–334.
159. Xu Cheng. Research on multi-vehicle cooperative collision avoidance algorithm under partial vehicles' networking [D]. Beijing: Tsinghua University, 2015 [in Chinese].
160. Lu X Y, Wang J, Li S E, et al. Multiple-Vehicle Longitudinal Collision Mitigation by Coordinated Brake Control [J]. Mathematical Problems in Engineering, 2014, 2014.
161. MERABTI H, BELARBI K, BOUCHEMAL B. Nonlinear predictive control of a mobile robot:a solution using meta-heuristic [J]. Journal of the Chinese Institute of Engineers, 2016, 39(3):282–290.
162. FELIPE N, WANDERLEY C, RICARDO C, et al. An adaptive dynamic controller for autonomous mobile robot trajectory tracking [J]. Control Engineering Practice, 2008, 16(11):1354–1363.
163. Zheng Yang. Vehicle platoon dynamics modeling and distributed control based on four-element architecture [D]. Beijing: Tsinghua University, 2015 [in Chinese].
164. Camponogara E, Jia D, Krogh B H, et al. Distributed model predictive control [J]. IEEE control systems magazine, 2002, 22(1):44–52.
165. KERNER B S. Failure of classical traffic flow theories: stochastic highway capacity and automatic driving [J]. Physica A: Statistical Mechanics and its Applications, 2016, 450:700–747.

166. TALEBPOUR A, MAHMASSANI H S. Influence of connected and autonomous vehicles on traffic flow stability and throughput [J]. Transportation Research Part C: Emerging Technologies, 2016, 71:143–163.
167. STERN R E, CUI S, DELLE MONACHE M L, et al. Dissipation of stop-and-go waves via control of autonomous vehicles: Field experiments [J]. Transportation Research Part C: Emerging Technologies, 2018, 89:205–221.
168. Chen C, Wang J, Xu Q, et al. Mixed platoon control of automated and human-driven vehicles at a signalized intersection: dynamical analysis and optimal control [J]. Transportation Research Part C: Emerging Technologies, 2021, 127: 103138.
169. Klaus Bengler, Klaus Dietmayer, Berthold F?rber, et al. Three Decades of Driver Assistance Systems: Review and Future Perspectives [J]. IEEE Intelligent Transportation Systems Magazine, 2014, 6(4): 6–22.
170. Yu Zhisheng. Vehicle Theory. 6th edition [M]. China Machine Press, 2019 [in Chinese].
171. Zhou Zhili, Xu Liyou. Principle and structure of vehicle ABS. 2nd edition [M]. China Machine Press, 2011 [in Chinese].
172. Cheng Bo, Zhang Guangyuan, Feng Ruijia, et al. Status and development of driver fatigue monitoring technology [C]. 2007 China International Conference on Automotive Safety Technology and the 10th Annual Conference on Automotive Safety Technology of China Society of Automotive Engineers. 2007 [in Chinese].
173. Tesla official website [EB/OL]. [2019-05-25]. https://www.tesla.com/autoleading?redirect=no.
174. THRUN S, MONTEMERLO M, PALATUCCI M. Stanley: The Robot That Won the DARPA Grand Challenge [J]. Journal of Field Robotics, 2009, 23(9):661–692.
175. CREMEAN L B, FOOTE T B, GILLULA J H, et al. Alice: An Information-Rich Autonomous Vehicle for High-Speed Desert Navigation [J]. Journal of Field Robotics, 2006, 23(9):777–810.
176. URMSON C, ANHALT J, BAGNELL D, et al. Autonomous driving in urban environments: Boss and the urban challenge [J]. Journal of Field Robotics, 2008, 25(8):425–466.
177. BACHA A, BAUMAN C, FARUQUE R, et al. Odin: Team victortangos entry in the darpa urban challenge [J]. Journal of field Robotics, 2008, 25(8):467–492.
178. AHMANE M, ABBAS-TURKI A, PERRONNET F, et al. Modeling and controlling an isolated urban intersection based on cooperative vehicles [J]. Transportation Research Part C: Emerging Technologies, 2013, 28:44–62.
179. Li Keqiang, Li Jiawen, Chang Xueyang, et al. Principle and typical application of cloud control system for ICVs [J]. Journal of Automotive Safety and Energy, 2020, 11 (03): 261–275 [in Chinese].
180. Li Keqiang, Chang Xueyang, Li Jiawen, et al. Cloud control system of ICVs and its implementation [J]. Automotive Engineering, 2020, 42 (12): 1595–1605 [in Chinese].
181. China ICV Industry Innovation Alliance. White paper on vehicle-road-loud integrated control system [R/OL]. (2020-09) [2020-09-28]. http://www.caicv.org.cn/index.php/newsInfo?id=279 [in Chinese].
182. LEE J, PARK B B, MALAKORN K, et al. Sustainability assessments of cooperative vehicle intersection control at an urban corridor [J]. Transportation Research Part C: Emerging Technologies, 2013, 32:193–206.
183. DRESNER K, STONE P. A multiagent approach to autonomous intersection management [J]. Journal of artificial intelligence research, 2008, 31:591–656.
184. Lu Guangquan, Wang Yunpeng, Tian Daxin. Vehicle-vehicle cooperative safety control technology [M]. Beijing: Science Press, 2014 [in Chinese].
185. GUMASTE A, SINGHAI R, SAHOO A. Intellicarts: Intelligent car transportation system [C]. Proc. IEEE LANMAN. 2007.
186. CHOI W, SWAROOP D. Assessing the safety benefits due to coordination amongst vehicles during an emergency braking maneuver [C]. Proceedings of the 2001 American Control Conference, IEEE, 2001, 3:2099–2104.

187. TATCHIKOU R, BISWAS S, DION F. Cooperative vehicle collision avoidance using inter-vehicle packet forwarding [C]. GLOBECOM'05. IEEE Global Telecommunications Conference, 2005. IEEE, 2005, 5(5):2766.
188. Wang Pangwei, Yu Guizhen, Wang Yunpeng, et al. Vehicle-vehicle cooperative active collision avoidance algorithm based on sliding mode control [J]. Journal of Beijing University of Aeronautics and Astronautics, 2014, 40 (2): 268–273 [in Chinese].

Chapter 4
Intelligent Protection

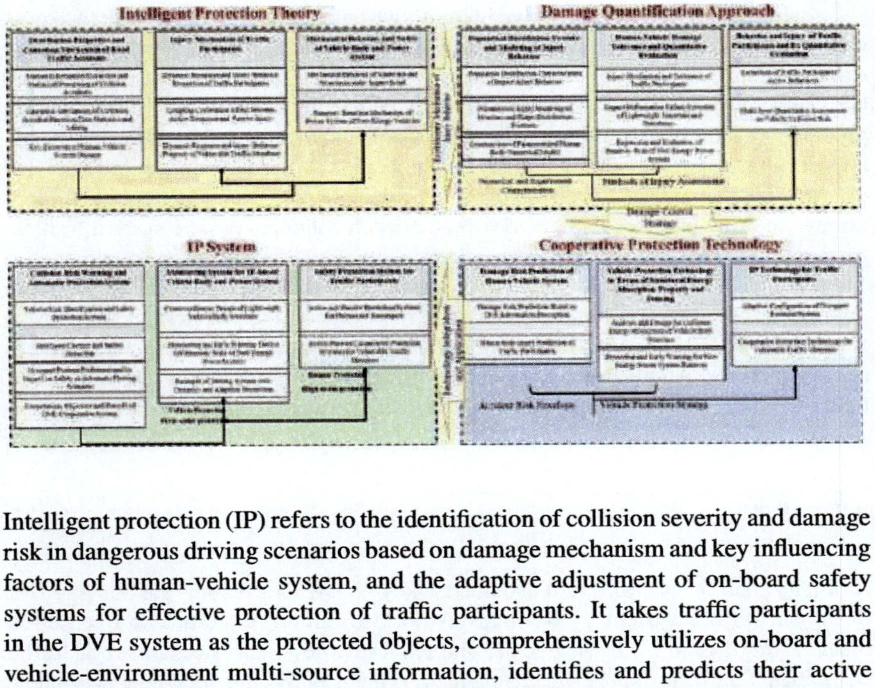

Intelligent protection (IP) refers to the identification of collision severity and damage risk in dangerous driving scenarios based on damage mechanism and key influencing factors of human-vehicle system, and the adaptive adjustment of on-board safety systems for effective protection of traffic participants. It takes traffic participants in the DVE system as the protected objects, comprehensively utilizes on-board and vehicle-environment multi-source information, identifies and predicts their active behaviors and injury features by studying dynamic cognitive responses and biomechanical mechanisms of traffic participants, provides protection-oriented safety decisions and implementation schemes, and ensures adaptive configuration adjustment of vehicle body structure and on-board safety system represented by occupant restraint, so as to minimize the injury risk and severity of traffic participants in accidents.

J. Wang et al., *The Intelligent Safety of Automobile*, Key Technologies on New Energy Vehicles, https://doi.org/10.1007/978-981-99-6399-7_4

The main purpose of protection is to inhibit the damage generation and development of human-vehicle system under dangerous working conditions. The relevant research involves four aspects, i.e., basic theory, damage quantification approach, damage suppression technology, and adaptive protection system. Based on such mechanisms as distribution and evolution of road traffic accidents, behavior mechanism and impact damage of traffic participants, and impact mechanical response of vehicle body and power system, IP systematically develops the behavior feature identification and damage tolerance quantification approach in accordance with population distribution, so as to predict damage risk of human-vehicle system. Facilitated by multi-source sensing information and advanced materials, intelligent safety protection systems such as collision risk warning and adaptive restraint are established. With the development of new energy vehicles, the runaway reaction mechanism and protection technology of vehicle power systems such as batteries are also an essential part of IP.

Structure of this chapter is shown in the above figure. Firstly, based on IP theory as well as the distribution and causation mechanism of road traffic accidents, this paper explores injury mechanism of traffic participants and failure mechanism of vehicle body and power systems; Secondly, the damage behavior and its crowd distribution property are identified by numerical and experimental means, and the damage quantitative evaluation of human-vehicle damage tolerance is made to confirm the behavior of traffic participants and evaluate the risk of vehicle collision; Cooperative protection technology includes damage risk prediction for human-vehicle system, vehicle protection, IP technology for traffic participants, etc. Implementation approaches and features of key technologies for IP improvement is illustrated; Finally, the technology is applied to IP system, i.e., the carrier of integrated application of IP theories, approaches and techniques, including collision risk warning and automatic protection system, etc. And the structural composition, working principle and application field of vehicle and human protection system are summarized and analyzed in turn.

4.1 IP Theory

IP theory mainly identifies and establishes key elements affecting the damage risk of human-vehicle system by the analysis on distribution properties and causation mechanism of road traffic accidents, and explores the injury mechanism of traffic participants and the failure mechanism of vehicle body and new energy vehicle power system based on such theories as impact dynamics, injury biomechanics and electrochemistry, so as to provide a theoretical basis for IP.

4.1.1 Distribution Properties and Causation Mechanism of Road Traffic Accidents

With the attention of different countries to road traffic accidents and the development of accident information extraction technology, different types of database information are gradually developed and enriched. By the analysis of road traffic accident database, the accident overview and its main distribution properties can be obtained. Through database study with statistical learning, data mining and other analyzing approaches, the occurrence and evolution process of accidents can be deduced, and the causation mechanism of collision accidents can be identified in terms of human, vehicle and road. The direct consequence of the accident is the damage of human-vehicle system, including impact injury, contusion, run-over injury and burn of traffic participants (human), vehicle body deformation and damage, power system failure of new energy vehicles, etc.; Further study is conducted to identify key factors affecting the damage, clarify protection requirements, and promote development of accident information extraction technology and the improvement and perfection of the database (Fig. 4.1).

1. Feature Information Extraction and Statistical Processing of Collision Accidents

Statistical analysis of road traffic accident data is the most intuitive and effective way to obtain the general situation and distribution properties of accidents. It not only helps to focus on the key research objects of IP, but also promotes the development of information extraction technology.

(1) Overview of Road Traffic Accidents

Road traffic accident is a major public health problem all over the world. It is emphasized in *Global Plan for the Decade of Action for Road Safety 2011–2020* [1] that improving road safety is a worldwide priority. Road traffic accident is a process in which human or vehicle is damaged due to the coupling imbalance of dynamic and static factors including human, vehicle, road and environment, etc. The *Global Status Report on Road Safety 2018* released by the World Health Organization (WHO) shows that about 1.35 million people die in road traffic accidents worldwide every year (Fig. 4.2), of which more than half are VRU, such as pedestrians and two-wheeler cyclists. The establishment of road traffic accident database provides an important data reference for accurately identifying the feature of vehicle collision accidents and key factors affecting human-vehicle damage behavior.

(2) Accident Information Extraction Technology and Development

In terms of requirements for IP application, the existing databases can be divided into two categories, i.e., one is the road traffic accident database, which takes the collected actual accident information as an important data basis, and can be used to analyze the accident properties and development trend and then formulate corresponding protection measures; the other is Naturalistic Driving (ND) database, which can

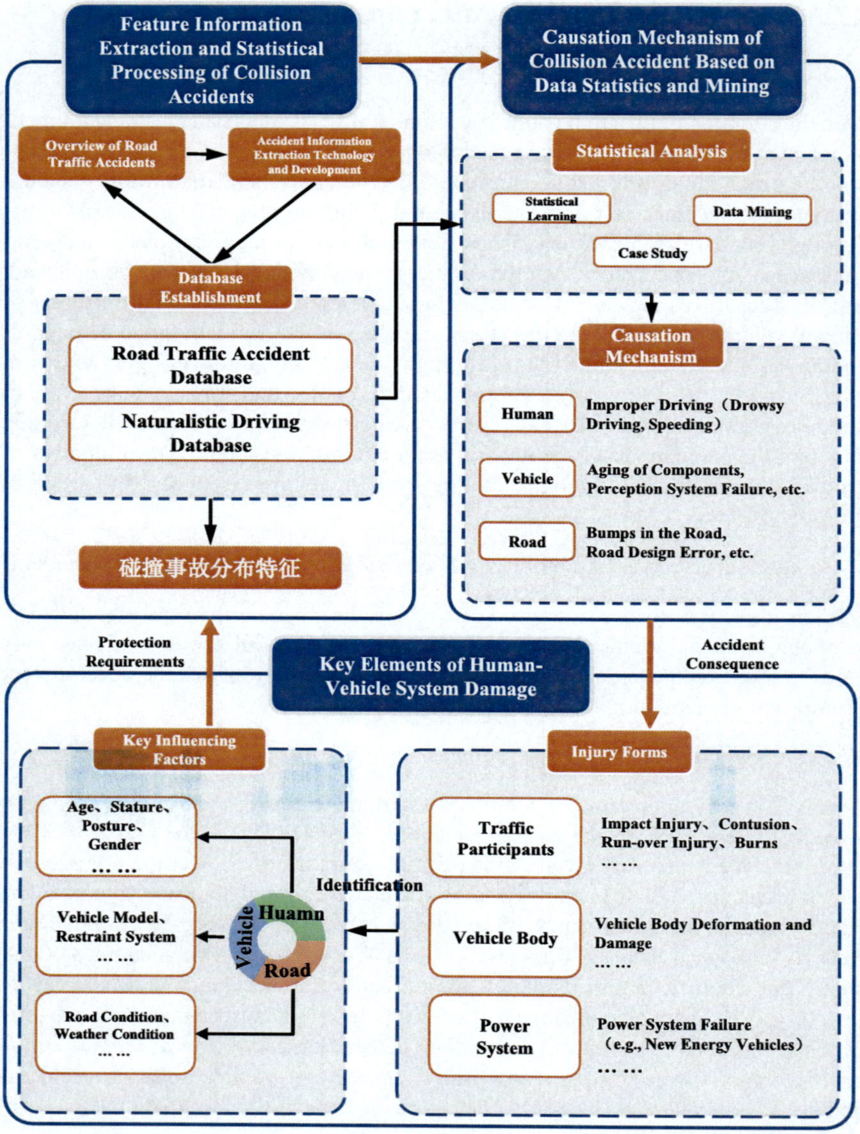

Fig. 4.1 Research framework of "database establishment—data analysis—damage factor identification"

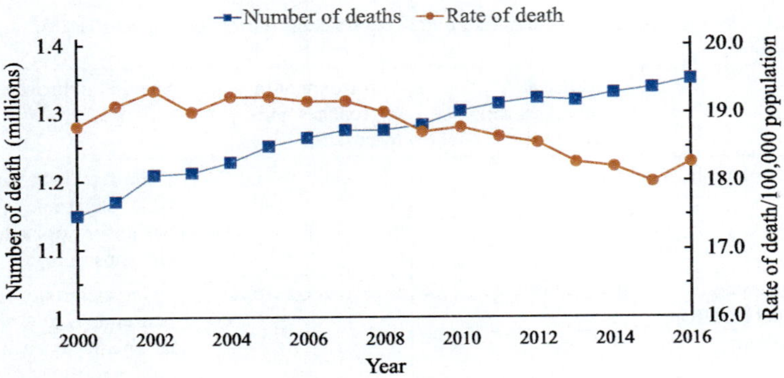

Fig. 4.2 Death toll and mortality rate per 100,000 people in global road traffic within 2000–2016 [2]

be used to study human driving behavior, DVE interaction behavior and automatic driving test.

With the emphasis on road traffic safety and increasing demand for information of accident cases, various countries and regions have successively carried out statistical inclusion of road traffic accident data and established corresponding databases. The traffic accident file contains such DVE information such as accident type, occurrence time, weather condition, and casualty. It is the original material for analyzing traffic accidents. Traditional traffic accident data collection generally relies on police, and accident files are established by means of on-site measurement data and manual sketching. The new accident data collection applies modern technologies, such as video recording, three-dimensional laser scanner and UAV system (Table 4.1), which improves collection speed and accuracy [3] and further promotes the development of road traffic database. For example, using a three-dimensional laser scanner to quickly scan and measure the accident scenario can obtain the point cloud data of the accident scenario element, restore the three-dimensional scenario of the accident, acquire on-site vehicle data, braking traces and overall layout, and then return to the situation at the time of the accident. In traditional accident measurement, it takes a long time with limited ways of obtaining accident panorama. However, the UAV aerial photography technology can quickly and accurately gain road parameters and scenario features of the accident, which is helpful for subsequent detailed analysis on the whole process of the accident [4].

From the perspective of the vehicle, event data recorders (EDR) also serve as an effective data source for extraction of accident information. According to the relevant regulations of NHTSA, most vehicles in North America are equipped with EDR to collect, store and retrieve collision event data after 2012. EDR can recover key collision data parameters from the accident vehicle, including pre-collision data, which is helpful to better master the key factors leading to the collision. It can be used not only to establish collision accident database, but also to collect ND data. In addition, the crash outcome data evaluation system (CODES) can associate the

Table 4.1 Comparison of new and traditional approaches for collecting road traffic accident information

Object	Traditional approaches and techniques	Improvement of approaches and techniques	Specific improvement
Acquisition table	Paper	Electronic format	Compared with paper table, electronic table has lower error rate and HR consuming
Vehicle bar code, personnel certificate information	Manual filling	Chip/barcode reader	Using electronic instruments to automatically scan and obtain certificate information has high efficiency and low error rate
Accident location information	Semantic description	GPS positioning	Obtaining GPS coordinates based on GIS map is more accurate and convenient than semantics
On-site measurement	Total station	3D laser scanner	It is easy to carry and operate, but not easy to cause secondary accident
Accident scenario map	Tape measurement and manual drawing	Laser measuring instruments, professional drawing software	It is with more precise Laser measuring instruments, more standardized professional drawing software, and less HR consuming
Pre-crash information	Semantic description	EDR	Recovering key collision data parameters from the accident vehicle can be of help for collision accident investigation

crash accident record with the personnel injury information collected by emergency medical service on site.

From the perspective of traffic participants, the casualty information is of great significance to the in-depth investigation, research and reconstruction of the accident. The degree of casualties can be described macroscopically with the injury part and its abbreviated injury scale (AIS), max abbreviated injury scale (MAIS) and other information. In some existing databases, more detailed personnel and casualty information, medical treatment and other data have been recorded, such as the length of hospitalization and medical cost of the injured.

(3) Representative Road Accident Databases

There are many kinds of road accident databases at home and abroad. Some representative road accident databases are listed in terms of data volume, information collection richness, recognition in the field, influence and other aspects, as follows.

Built by BASt in 1988, the International Road Traffic and Accident Database (IRTAD) was initially applied to compare international accident data, and then managed and expanded by the International Transport Forum (ITF). With the widest area coverage, IRTAD has collected the latest verified collision accident data from more than 30 countries and regions such as Japan, South Korea, Germany and France [5]. It integrates the annual report according to the data of participating countries every year to display accident trend, and compare road traffic accident mortality, accident distribution and other information in various regions, so as to put forward targeted strategic objectives and legislation.

German In-Depth Accident Study (GIDAS) is the largest accident data collection project in Germany. Different from IRTAD's way of obtaining data worldwide, GIDAS collects about 2000 traffic accidents involving personal injury in the urban areas of Hanover and Dresden every year, and widely gathers data covering all aspects of the accident, such as accident kinematics, injury type, accident cause, accident type and rescue process. The project team drew an accident sketch containing information such as vehicle driving direction, collision location and termination location for subsequent accident reconstruction. GIDAS database has a rich collection of information on single accident cases. Each accident contains about 3,600 parameters (including vehicle model and speed involved, as well as height, weight, age, gender, injured location, injury type of the injured, etc.) and about 150 pictures for research and analysis [6], which can effectively assist to identify significant impact of such key factors as vehicle speed, vehicle model and age on the injury risk of human-vehicle system. GIDAS is a widely recognized database in the field and supportive for many related research.

China In-Depth Accident Study (CIDAS) is a project jointly started by China Automotive Technology and Research Center (CATARC) and several automobile companies in 2011. Accident data have been collected in Changchun, Weihai, Ningbo, Beijing, Foshan, Qianxi'nan and other places, with more than 3000 pieces of information included in each accident case [7]. The database can be used for accident analysis and vehicle structure design optimization, and has been preliminarily reflected in traffic law legislation, vehicle safety technology development and new vehicle evaluation procedure test in China. The National Automobile Accident In-Depth Investigation System (NAIS) is also a relatively mature accident database in China. It is jointly established by the Defective Product Administrative Center of the State Administration for Market Regulation in conjunction with a number of vehicle accident research institutions and automobile judicial appraisal centers, serving for vehicle safety analysis and recall, accident research, automobile product safety improvement, etc. NAIS accident collection mainly include types of road traffic accidents, vehicle fire accidents and other accidents suspected as potential safety hazards [8].

In addition, the Crash Investigation Sampling System (CISS) [9] and the National Automotive Sampling System (NASS) [10] in the charge of NHTSA [11] of the US, the Pedestrian Crash Data Study (PCDS) specially for collection of vehicle–pedestrian collision accidents [12] and the World Accident Data Base (WADB) specially for racing accidents [13], etc. are also important road traffic accident databases and widely used in the evaluation of the overall situation of traffic safety, so as to identify existing and potential traffic safety problems and lay a foundation for the cost and return analysis of traffic safety plan.

(4) ND Database

ND data usually refers to the database that collects and records driver behavior, vehicle operation feature and traffic environment by means of multiple on-board sensors, and includes such information as vehicle speed, steering wheel angle, brake pedal position, and headlamp usage. It can be applied to study driver behavior, interaction between the vehicle and traffic groups outside the vehicle, etc. In the process of ND data collection, it can collect the driver's real driving behavior in the natural state when the driver operates the vehicle habitually, with high reliability and effectiveness. The study of ND is of great significance to the studies of the direct inducement of traffic accidents, the feature of HMI, as well as the next generation intelligent transportation system.

By integration of ND data, driving data and driving behavior can be effectively connected, and the driving risk identification model can be constructed for studying active safety control technology [14] or testing relevant theoretical approaches. The ND databases established earlier and widely used include European Field Operational Test (EuroFOT) and TRB's Second Strategic Highway Research Program Naturalistic Driving Study (SHRP2 NDS), etc. Under the leadership of Ford and 28 partners including European automobile manufacturers, the EuroFOT project tested 1000 vehicles at eight sites in one year and collected more than 100 TB of data (such as driving behavior in various real traffic scenarios, and efficiency of various vehicle safety systems) to evaluate the impact of IV system or active safety system on transportation safety and fuel efficiency [15]. Established by the Transportation Research Board (TRB), the SHRP2 NDS project aims to develop countermeasures for road safety improvement by comprehension of driving behavior. The study involved about 3,100 drivers in six states [16]. In addition, the relatively new ND database also has interaction dataset that records the interaction behavior of various traffic participants in traffic scenarios [17].

(5) Distribution Feature and Case Analysis of Collision Accidents

By statistical analysis, distribution features of accidents can be summarized and extracted, for exploring the correlation between their occurrence frequency and such factors as traffic participants and accident occurrence time, and carrying out targeted prevention accordingly. The accident participants mainly include 4 wheeled vehicles, motorized 2–3 wheelers, cyclists, pedestrians, etc., which have different proportion of casualties in different countries and regions, as shown in Fig. 4.3a. In Europe, due to high rate of car ownership and utilization per capita, car drivers account for 48% of all road traffic deaths, and motorized 2–3 wheelers riders account for 11%. In Southeast Asia, where population density is high, the death rate of car drivers is relatively small, only 16%, while death rate of occupants accounts for as high as 43% due to large ownership rate of motorized 2–3 wheelers. Africa has the highest proportion of pedestrian deaths, up to 40%. According to the data of China National Bureau of statistics [18], 247,646 road traffic accidents occurred in 2019, resulting in 62,763 deaths, among which 90.7% are motor vehicle (including cars, motorcycles and tractors) drivers and passengers and 2.3% are pedestrians, as shown in Fig. 4.3b, c.

Based on statistical analysis of accident distribution properties, the analysis on a single accident case is helpful to find out the cause of the accident and damage characteristics, and offer experience for safety protection. Some studies have used numerical models to reconstruct ten vehicle-VRU collision accidents in the vulnerable road users accident database. By comparing the accident clinical medical injury information and the injury information predicted by simulation, the results show that mechanical indexes such as head linear acceleration, rotation acceleration and shear stress can better predict head landing injury [19].

2. Causation Mechanism of Collision Accident Based on Data Statistics and Mining

Road traffic accident refers to the behavior that a series of events appear in time sequence within a certain space and lead to destruction of stable state of the DVE system through complex interaction, resulting in casualties, vehicle damage and corresponding economic losses. Its essence can be understood as the highly nonlinear change of system state. The data analysis approaches for road traffic accidents can be divided into two types. One is based on traditional mathematical statistical analysis. By such means as database usage statistics, analysis of variance (ANOVA), nonparametric test, and regression on the database, accident distribution properties are obtained, and influence of pre-collision parameters on accident risk is deduced; The other is based on data mining. In view of the contingency, multi-factor and fuzziness of road traffic accidents, it carries out multi-angle and multi-level analysis and research to identify the factors affecting road traffic accidents, reveal potential laws and features of the correlation among historical data of various accidents, and more comprehensively facilitate the study on safety protection strategy for DVE system in road accidents.

(1) Impact of Traffic Accident Distribution and Pre-collision Parameters on Damage Risk

The main causes of accidents are man-made, vehicles, road conditions and environmental factors. For human driving vehicles, main causes of road traffic accidents include driver's perception error, judgment error, slow response, or abnormality of vehicles and road environment. According to the classification of traffic participants, it can be categorized into single vehicle accidents (rolling, falling, hitting fixed

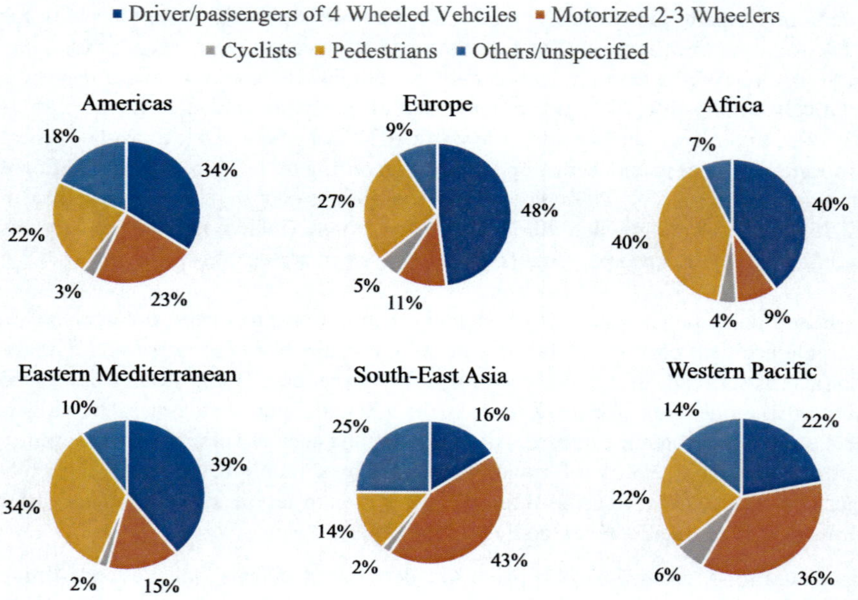

(a) Proportion of Death Toll of Various Participants in Road Traffic Accidents in All Regions of the World (according to the World Health Organization)

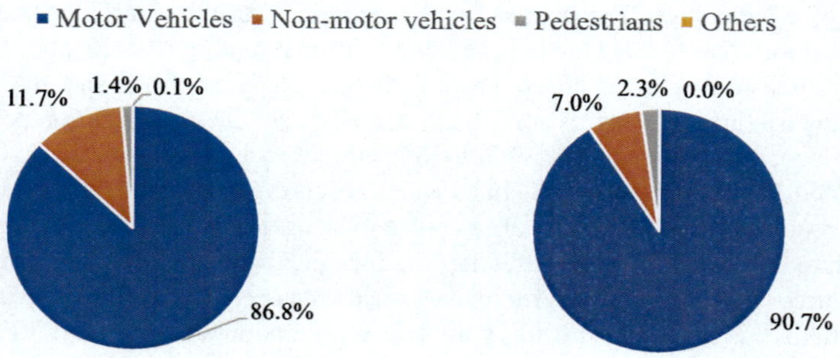

(b) Distribution of Road Traffic Accidents in China in 2019

(c) Distribution of Participants' Death Toll of Road Traffic Accidents in China in 2019

Fig. 4.3 Distribution of participants in road traffic accidents in the world [2] and China [18]

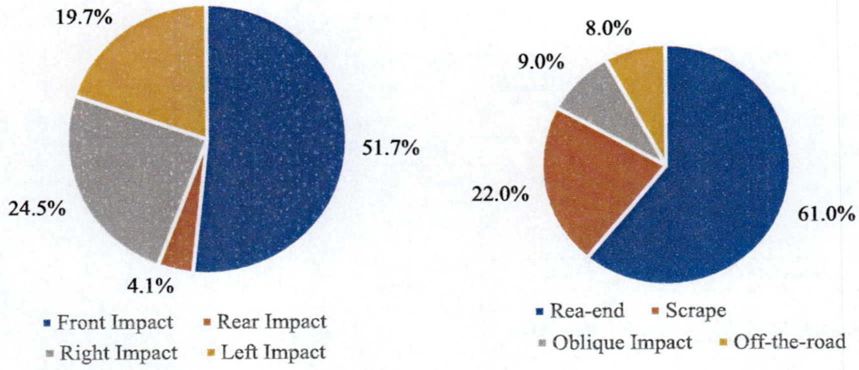

Fig. 4.4 Example of collision accident type distribution

objects, etc.), vehicle accidents, vehicle–pedestrian accidents (collision, rolling), and vehicle-2/3 wheelers accidents. Among them, the collision accidents between vehicles can be divided into frontal collision, rear-end collision, side collision, oblique collision, etc. For example, by analyzing 2538 accidents in the Crash Injury Research (CIREN) database, the distribution of collision scenarios is obtained, as shown in Fig. 4.4a, and more than half of the accidents are front collisions [20]. Another statistical analysis on 113 traffic accidents of AVs from 2014 to 2018 indicates that more than 60% of the accidents are rear-end collisions, as shown in Fig. 4.4b [21].

More than 90% of the existing collision accidents are induced by driver-related factors. The driver's road familiarity, risk-taking behavior tendency and abnormal driving behavior (such as improper driving, fatigue and distraction) all have an impact on the accident risk [22]. For example, the distraction due to high-frequency usage of handheld electronic devices such as mobile phones is a considerable factor affecting driving safety in recent years [23]. Driver's age is also one of the major factors related to collision risk. Compared with drivers aged 20–24, drivers aged 85 and over have almost doubled the occurrence of collision fatalities at intersections [24].

(2) Data Mining Approaches for Accident Causation Analysis

By discovering the hidden influencing factors related to accident occurrence and damage risk, data mining can be used for accident causation analysis. Apriori Algorithm, Bayesian Network, Clustering, Neural Network and Decision Tree are typical data mining algorithms applied in the analysis and research of accident database (Fig. 4.5), which have been widely applied in accident database and case analysis.

As one of the classical algorithms for association rule learning in the field of data mining, Apriori Algorithm is often used in road traffic accident analysis, such as exploring the relationship between road environmental factors (road surface, weather, light conditions), human factors (such as drunk driving) and accident mortality [26].

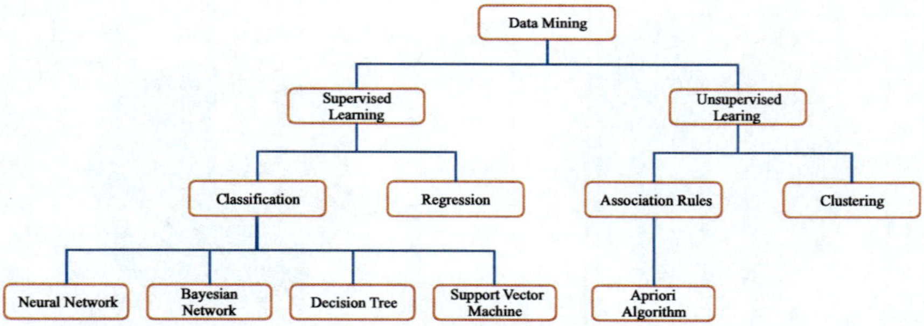

Fig. 4.5 Major algorithms for data mining [25]

A study based on 11,383 rear end crashes in New Mexico from 2010 to 2011 used Bayesian Network to derive the relationship between injury severity and feature. The original data information includes driver behavior, demographic feature, vehicle factors, geometric and environmental feature, etc. The analysis results indicate that existence of trucks among the vehicles in collision, weak light conditions, windy weather and the number of vehicles involved have a significant impact on the severity of driver injury in the rear end collision accident [27]. Another study on autonomous driving accidents classifies accident modes of AV that occurred in California from 2014 to 2019. The research based on Bayesian Clustering analysis show that accidents involving turning, multi-vehicle collision, dim street lighting, side slip and rear end collision usually lead to more serious occupant injury [28].

It should be pointed out that there are quite a number of data mining algorithms applied to the analysis and research of accident causes. The algorithm itself only has the difference of being appropriate and inappropriate, instead of superior or not. And selection of Algorithm is also restricted by the amount of data to a large extent. In addition, complexity of the model and similarity of samples will affect the number of samples required by the model. Generally speaking, algorithm selection mainly depends on trial and error, that is, trying different algorithms to determine the most appropriate one for the problem studied and the database used.

3. Key Elements of Human-Vehicle System Damage

Road traffic accidents can lead to vehicle damage or personal injury. There are usually many factors affecting the damage of human-vehicle system. By identification of key ones, specific object for safety research can be pointed out, and protection can be carried out with more emphasis in the process of road traffic accidents.

(1) Basic Forms of Human-Vehicle System Damage

Human-vehicle system damage mainly includes traffic participant's injury, vehicle body deformation and damage, power system failure and other forms. Among the injuries of traffic participants, taking vehicle occupants for example, head and chest injuries are common and more serious forms. The specific injury forms include head and face impact on the windshield glass or its frame, resulting in skull fracture, etc.; when the compartment is deformed under impact load during the collision, occupant's chest is impacted, or squeezed with steering wheel and other components, resulting in injury; The driver suffers whiplash due to corresponding severe speed change brought by emergency braking or sudden rear end collision with the rear high-speed vehicle, resulting in neck injury.

The injuries of pedestrians, cyclists and other vulnerable road users in human-vehicle collision accidents generally include two categories: one is primary injury (the first injury), which is induced by direct impact. The common types include craniocerebral injury due to head collision with bonnet and windshield, and long bone fracture and knee ligament tear due to lower limb collision with bumper; The second type is secondary injury (the second injury), which refers to the injury due to collision of head and other body parts with the ground or other objects after the impact.

Traffic accident is likely to lead to damage of the paint due to scratching, and damage of the bumper, hood, windshield and other parts due to collision, as well as severe deformation of vehicle body due to extrusion, all of which can be recorded as vehicle body damage; The extrusion deformation of vehicle body may also lead to loss of control in the power battery system of new energy vehicles, subsequently resulting in vehicle fire or even explosion.

(2) Key Influencing Factors of Human-Vehicle System Damage

There are a large number of related studies on the factors affecting damage of human-vehicle system, which analyze correlation with the injury mainly in view of three dimensions: working condition (accident time, road condition, speed, collision angle, etc.), vehicle (restraint system, vehicle model, etc.), and traffic participant (age, body shape, gender, posture, etc.).

In terms of working conditions, vehicle speed is the most important factor affecting injury severity of traffic participants [21, 29]. The higher the collision speed is, the higher the injury risk and severity level is [30, 31]. In terms of vehicles, interior restraint system is an important factor affecting occupant injury, and different constraints have a certain impact on dynamic response of occupants in the process of collision [32]. In addition, vehicle types significantly affect pedestrian injury: the main injury sources of vehicles to pedestrians are windshield and bumper, and the main injury sources of pickup trucks, trucks and SUVs are the hood and the front edge of the hood [33]. Vehicle front shape parameters and front end height have an impact on severity of pedestrian head injury and landing injury [34, 35].

In terms of traffic participants, the age and body stature of traffic participants are important factors affecting injury severity in traffic accidents. The older the age is, the higher the risk of occupant injury is [29, 30]. According to accident data of Sweden in 2004–2008, the elderly over 75 years old account for the vast majority of all pedestrian fatalities [31]. Posture is also an important factor affecting injury.

For example, taller drivers have a relatively high mortality in single vehicle rollover accidents [36], and obese male drivers have a higher risk of injury in upper part of body (including head, chest, spine, etc.) than regular-sized drivers [37]. Based on statistical analysis of accident data, it is necessary to further study how the key factors affect damage of traffic participants and vehicles from the perspective of impact dynamics and injury biomechanics, so as to lay a theoretical basis for protection and loss reduction of human-vehicle system.

4.1.2 Injury Mechanism of Traffic Participants

Research on injury mechanism of traffic participants can be carried out from different dimensions. In terms of time scale, the damage mechanisms involved in different stages of collision accidents include: human body's active response in the pre-collision (also called as pre-crash), injury mechanism of various parts of human body during collision (in-crash), and injury forms and features after collision (post-crash). In terms of geometric scale, damage behavior properties from macro to micro include the whole-body level injury of traffic participants, injury form of key parts such as head, neck and chest, and fracture mechanism of bone and ligament at the tissue. In addition, the drivers and passengers inside the vehicle and those outside the vehicle have different dynamic response features in the accident scenario: driver's pre-crash active response will lead to the changes of posture and joint stiffness, as well change of interaction between passengers and restraint system, which will affect damage evolution; The contact position and impact loading conditions of people outside the vehicle (mainly refer to vulnerable road users such as pedestrians and cyclists) with the vehicle during collision are affected by their initial postures in pre-collision stage, and their main injury forms are also different from those of the driver (Fig. 4.6).

1. Dynamic Response and Injury Behavior Properties of Traffic Participants

As the most core and complex element in DVE system, traffic participants show typically active–passive coupling behavior properties in dangerous scenarios. At the cognitive level, human brain perceives complicated and changeable dynamic traffic scenarios and makes active decision-making in consideration of safety. This active response will exert a significant impact on passive injury behavior during collision; At the vehicle level, there is interaction between occupants and restraint system or vehicle body structure; In terms of driving environment and road, both accident conditions and traffic participants behavior features significantly affect injury risk. Hence, the research on damage mechanism for IP requirements is mainly carried out in the aspects of meso-macro mechanical properties, dynamic response coupling correlation, and damage behavior features of vulnerable road users.

(1) Basic Concepts on Dynamic Response of Traffic Participants

Taking the driver for instance, active response on the time axis includes behavioral processes such as hazard signal perception and recognition, stress decision-making

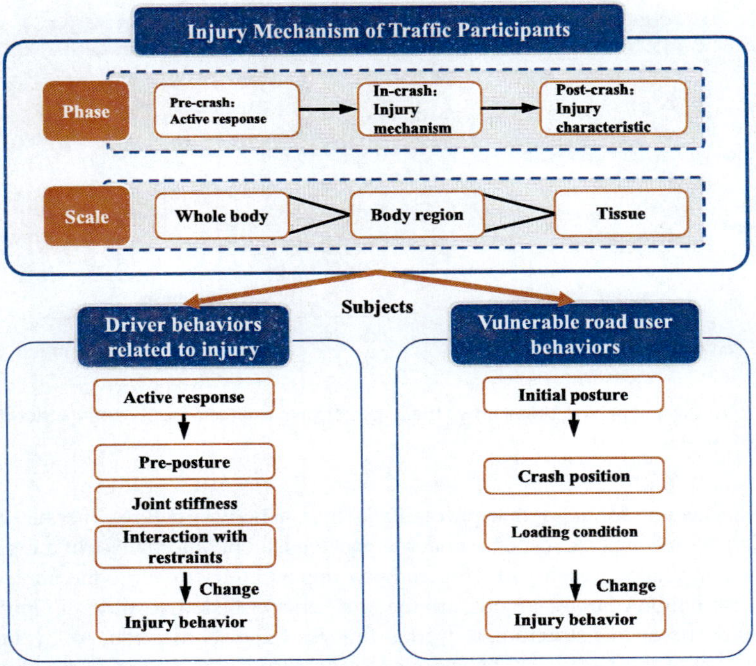

Fig. 4.6 Injury mechanism of traffic participants

and collision avoidance action control (as shown in Fig. 4.7). Among them, rapid environmental perception and understanding with vision as the main attention signal is a unique cognitive attribute in human driving [38]. The brain quickly identifies hazard source based on highly uneven visual attention information and aprior experience, makes judgment from the perspective of safety after understanding and reasonably deducting the scenario, and controls body to make dynamic response (such as steering and braking operation or posture change of pedestrian position by muscle contraction control), so as to change its own dynamic properties and interactively change such boundary conditions as vehicle driving, and significantly affect subsequent accident risk and damage degree.

During passive damage, human and vehicle experience a multi-stage collision process: in primary collision, the passenger compartment makes deceleration, and part of collision energy is absorbed by structural deformation of the front body. The energy absorption property of body collision deformation aimed at occupant protection is called crashworthiness, and vehicle deceleration is determined by vehicle collision property; In secondary collision, occupant is restrained in occupant compartment, and impact force is transmitted to occupant through restraint system and decelerates the occupant. The restraint system can control movement of occupants and

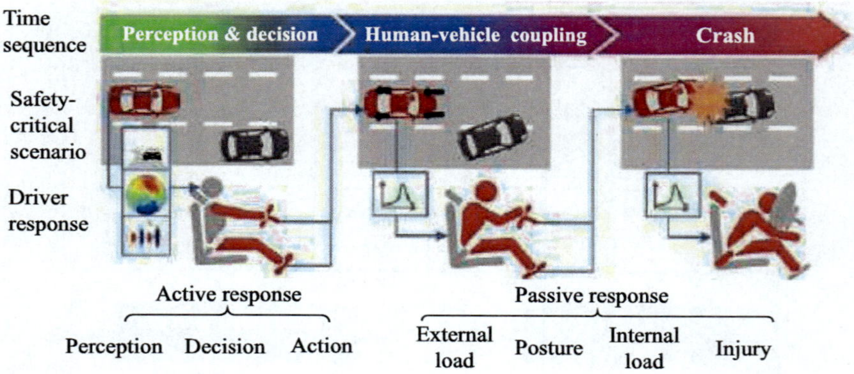

Fig. 4.7 Active and passive behaviors of traffic participants and their coupling properties (taking the driver as an example)

realize protection for them; In triple collision, human body structure generates corresponding biomechanical response under impact load, including changes of movement posture and shape of each part. Human body impact injury refers to the unrecoverable deformation of body structure in the process of collision, resulting in injury of anatomical tissue and structure, as well as failure of normal function. Analyzing the connotation of biomechanical behavior of traffic participants and understanding the injury mechanism is one of the essential parts in vehicle collision safety design.

(2) Impact Damage Mechanism of Key Parts in Human Body

Biomechanics is a biophysics that applies principles and approaches of mechanics to quantitative study of mechanical problems in organisms; As its branch, injury biomechanics uses mechanical principles to study and describe physiological response of loading that may lead to injury. Identification of key injury elements based on analysis of road traffic accident data can facilitate the identification of key body parts (such as head, chest and lower limbs) of traffic participants that are frequently injured in the accident. By analyzing the impact load borne by human body in traffic collision accidents and studying its biomechanical response properties, the correlation between various mechanical and kinematic parameters and human injury forms can be revealed.

1) Mechanism and Evaluation Criteria of Head Injury

As a fatal form of injury, head injuries caused by traffic accidents account for 31.8% of all clinical ones [39]. The head is the most complicated part in body structure, and its collision injury forms are also complex and diverse, generally including facial injury, skull fracture and brain tissue injury. Among them, brain tissue injury refers to the organic injury of brain tissue due to impact force on the head. The main forms of injury include concussion, diffuse axonal injury, subdural hematoma and so on. Taking concussion as an example, its injury mechanism is that relative movement between skull and brain is made in the head under the action of long-time angular acceleration. From the mechanical point of view, since the main damage mechanism is inertia, angular acceleration is used as the damage evaluation criterion. The main mechanism of diffuse axonal injury is also inertial factor, and its severity depends on magnitude and duration of acceleration, initial angular acceleration and direction of motion [40].

2) Mechanism and Evaluation Criteria of Chest Injury

Chest injury is the second leading cause of serious injury and death of occupants in traffic accidents [41]. Collision between chest and vehicle interior trim (including steering wheel, interior trim panel, etc.) and improper use of seatbelts and airbags can result in chest injury. Bone injury (rib fracture and sternum fracture) and thoracic tissue injury (trachea, lung, artery and heart injury) are major forms of injury. The probability of rib fracture in front or side collision is related to age of passengers, which increases significantly with the increase of age. The injury form of thoracic tissue is more complex than that of bone. It is easy to induce lung contusion under high-speed impact, and destroy capillary layer of alveoli when the lung tissue bears pressure or pressure wave. Combined with its damage mechanism, chest injury is mainly evaluated by relevant mechanical indexes such as acceleration, chest compression and viscous deformation [42].

3) Mechanism and Evaluation Criteria of Lower Limb Injury

Data statistics show that lower limbs have gradually become a body part with high injury rate in vehicle collision accidents [43]. 36% of all front-row passengers injured by AIS 2+ were lower limb injuries [44]. Lower limb fracture is one of the common forms of injury, including tibial, patellar, femoral and pelvic fractures, etc. In the process of collision, the axial bending load borne by the front passenger's tibia in contact with instrument panel and pedal will lead to fracture under the combined action of loads in multiple directions. Therefore, the injury evaluation criteria mainly include force or bending moment. In addition, the passenger contacts the front seat or instrument panel, and the axial force is transmitted to the femur through the patella, so that the femur is subjected to bending load generated by axial compression and eccentric loading, resulting in femoral shaft fracture and hip dislocation. The pelvis mainly bears axial force from the femur and the vertical force of vehicle seat, which leads to the fracture of acetabulum and pubic branch under the combined action of multi-directional loads.

(3) Mechanical Behavior and Constitutive Model of Ligament Injury

Biomechanical behavior of human body tissue injury can be studied from different dimensions such as material and structure. In terms of material, ligament tissue is an anisotropic multistage composite. In order to explore its failure forms in different geometric scales, researchers have proposed mechanical constitutive models in different scales that reflect properties of materials or structures. Early mechanical experiments of biological tissue materials followed the experimental approach of solid mechanical materials: the biological tissue was cut into standard shape specimens and loaded with material experimental machine, and the "force–displacement" curve was recorded and the "stress–strain" relationship curve was calculated. Based on material mechanics experiment, mechanical properties of biological tissue at the meso level (Table 4.2) can be obtained to elaborately describe its mechanical response.

In terms of structure, bone ligaments connect adjacent bone structures in the joint area, which can constrain the movement between joint bones and maintain joint stability. The ligament is mainly composed of collagen (forming collagen fiber bundle), elastin (forming elastic fiber network), fibroblasts, water and matrix (composed of glycoprotein, etc.), and its macro mechanical properties are determined by the microstructure comprised of fiber bundle and matrix. Elastic fibers provide the ligament with elasticity and non-fiber direction load, and collagen fibers offer the ligament tensile stiffness and strength. Different ligaments are different in composition proportion, leading to different mechanical behaviors. In active joints of human body, the ligament structure cooperates with the adjacent joint surface to determine passive motion trajectory of the joint. In collision accident, excessive movement of the joint under impact load will result in varying degrees of ligament damages.

In normal use, the ligament only bears load in a specific direction. From the perspective of micro evolution, fibrous structure is formed by gradual upward elongation and thickening of collagen in the fiber mode, and wavy curling occurs from the fiber level. The multi-level fiber structure leads to transverse isotropy of ligament mechanics, with fiber length direction as its main bearing direction. In view of the relationship between mechanical properties of the ligament and its microstructure changes, the "load–displacement" curve of the ligament under external load along the fiber direction can be divided into three stages as a whole (Fig. 4.8) [45]:

Table 4.2 Summary of mechanical properties of knee ligament materials in current studies [47]

Specimen object	Tensile tangent modulus (MPa)	Ultimate tensile stress (MPa)	Maximum tensile strain (%)
Medial collateral ligament (MCL)	332.2 ± 58.3	38.6 ± 4.8	17.1 ± 1.5
Anterior bundle of posterior cruciate ligament (PCL)	248.0 ± 119.0	35.9 ± 15.2	18.0 ± 5.3
Anterior cruciate ligament (ACL), medial collateral ligament (MCL), and lateral collateral ligament (LCL)	345.0 ± 22.4	36.4 ± 2.5	15.0 ± 0.8
Patellar ligament	643.1 ± 53.0	68.5 ± 6.0	13.5 ± 0.7

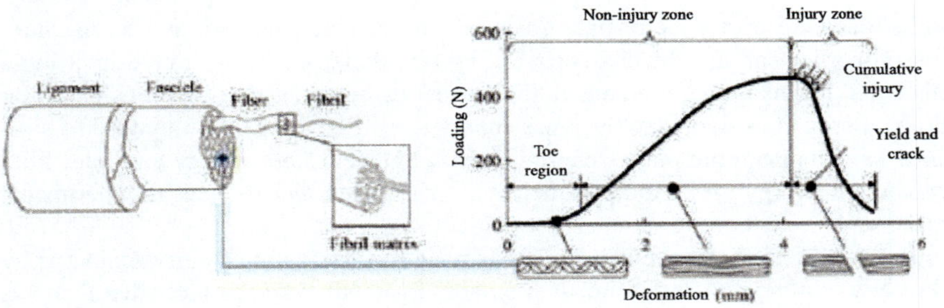

Fig. 4.8 Mechanical properties of human ligaments (macro structure and micro fiber) [45]

(1) Since the initial wave shape of fiber bundle is gradually straightened, the force rises slowly with deformation, and the corresponding bearing capacity is low, which is defined as zero-force region, also known as toe region in some studies.
(2) After fiber bundle is straightened, the ligament structure shows obvious bearing capacity, the load level rises rapidly and enters loading area of approximate linear elasticity. Both the zero-force region and the approximate linear elastic loading region correspond to the non-damage deformation region.
(3) When ligament reaches the bearing limit, the fiber tissue begins to appear damage and gradually accumulates, i.e., it enters the damage and deformation area. With fibers breaking in turn, the load decreases rapidly until the overall failure of the structure [46], which corresponds to the range of ligament damage and deformation.

In the process of establishing mechanical constitutive model of ligament, due to its transverse anisotropy and weak bearing capacity in the non-fiber direction [48], transverse connection between fibers is generally ignored and only stress change in the fiber direction is characterized. Based on this mechanical property, one-dimensional element can be used for modeling. In order to show the zero-force region caused by wavy structure, the axial force can be expressed in segments, and the stiffness coefficients K1 and K2 of the zero-force region and the elastic region are defined respectively.

$$\begin{cases} f = 0 & (\varepsilon < 0) \\ f = k_1(L - L_0)^2 & (0 \le \varepsilon \le 2\varepsilon_l) \\ f = k_2[L - (1 + \varepsilon_l)L_0] & (\varepsilon > 2\varepsilon_l) \end{cases} \tag{4.1}$$

where, f represents the bearing capacity of a single ligament; L and L_0 represent the current length and relaxation length of ligament respectively; and $2\varepsilon_l$ represents the strain threshold from the zero-force zone to the elastic zone, which is determined by the constitutive equation of ligament mechanics.

One dimensional element is mainly used to describe mechanical properties of ligament tissue in the main bearing direction. At the tissue level, distribution of strain and stress field during deformation can be further characterized, and its mechanical properties can also be characterized by two-dimensional or three-dimensional elements. For example, the constitutive model of hyperelastic properties based on strain energy considers that the stress energy (recorded as W) is composed of such three parts as isotropic matrix energy matrix, ligament fiber energy and inter-fiber interaction energy [49]. And parameters are calibrated by mechanical experiments based on ligament tissue samples.

In addition to the above-mentioned damage mechanical behavior dominated by elasticity commonly used in impact damage, ligament tissue as biomaterial shows certain viscoelastic properties, i.e., deformation of ligament lags behind change of stress during loading. The viscoelastic properties of ligament have a significant influence in the strain rate range of 0.0001–1/s. Beyond the range, nonlinear elastic behavior can be used to represent material properties of ligament. First proposed by Fung et al. [50], Quasi-linear Viscoelastic (QLV) model can accurately describe properties of soft tissue during large deformation. Its Cauchy stress formula is as follows:

$$T(\varepsilon) = A\left(e^{B\varepsilon} - 1\right) \tag{4.2}$$

where, ε indicates ligament strain; A and B represent the coefficients of the instantaneous elastic response function.

One of the impacts of ligament viscoelasticity on actual injury is that mechanical behavior has strain rate effect. Taking lateral collateral ligament of knee joint as an example, in the range of 0.01–100/s strain rate, with the increase of strain rate, the elastic modulus rises from 288 to 905 Mpa, the ultimate stress rises from 39.9 to 77.3 Mpa, and the rising range is mainly concentrated in the range of 0.01–1/s, beyond which there is a certain strain rate saturation effect. In the numerical model, viscoelasticity of ligament can be quantitatively characterized by QLV model, and bearing capacity of ligament under different strain rates can be calculated.

(4) Mechanical Behavior and Constitutive Model of Bone Injury

As a natural biological properties, bone is mainly divided into cortical bone and cancellous biological composite material with special mechanical and bone in structure. Macro, micro and nano structure morphological parameters and mechanical property parameters are comprehensive indicators to measure and evaluate biomechanical properties of bone [51]. The overall strength of bone is affected by complex non-independent factors on different scales, which also shows different mechanical behaviors [52].

Cortical bone accounts for 80% of the mass of human bones and is the main load-bearing component. The load mode borne in actual service conditions is usually complex [53]. Taking the stress of pedestrian's Lower Limbs in human-vehicle collision accident as an example, the contact process between lower limbs and vehicle front bumper can be analyzed. The front bumper of vehicle first contacts with the lower leg area, which makes the lower limbs bear macro impact load. Under the combined action of human gravity and ground reaction force, the bearing form of lower limb cortical bone can be regarded as a three-point bending loading mode under precompression. Cortical bone materials possess different mechanical properties at different levels. The outer-ring bone plate first contacts the front-end of vehicle and transmits load to the bone unit inside the cortical bone. Its micro hierarchical structure primarily generates deformation and fracture, which is manifested in crack initiation, propagation and finally fracture failure.

In view of mechanical properties of cortical bone, it is necessary to construct corresponding multi-scale mechanical model. The most common simplified constitutive model of cortical bone is elastic–plastic material model [54]. Strain ε is composed of Elastic strain component ε_e and plastic strain component ε_p. The linear stress–strain relationship in the elastic stage is expressed as follows:

$$\sigma = D\varepsilon \tag{4.3}$$

where, σ and ε represent six-dimensional stress vector and strain vector respectively; D represents 6×6 matrix composed of material parameters, called elastic matrix. For isotropic materials, D matrix is calculated as follows:

$$D = \frac{E}{1+v}\begin{bmatrix} K+\frac{4G}{3} & K-\frac{2G}{3} & K-\frac{2G}{3} & 0 & 0 & 0 \\ K-\frac{2G}{3} & K+\frac{4G}{3} & K-\frac{2G}{3} & 0 & 0 & 0 \\ K-\frac{2G}{3} & K-\frac{2G}{3} & K+\frac{4G}{3} & 0 & 0 & 0 \\ 0 & 0 & 0 & G & 0 & 0 \\ 0 & 0 & 0 & 0 & G & 0 \\ 0 & 0 & 0 & 0 & 0 & G \end{bmatrix} \tag{4.4}$$

where, E is the elastic modulus, K is the bulk elastic modulus and G is the shear modulus. Parameters of biomaterials are related to individual age, gender and other factors.

In consideration of the effect of strain rate, the stress of cortical bone increases with the increase of strain rate. The constitutive of elastic materials can not effectively characterize real mechanical behavior of bones, so the part with increasing elastic modulus shall be paid attention to. By paralleling viscosity coefficient η with the elastic model, Kelvin elastic hysteretic medium is constituted [55]. For Kelvin model, the total stress σ is comprised of two parts, i.e., stress corresponding to elastic deformation σ_e and stress corresponding to viscous resistance σ_v; and the total strain ε includes strain of elastic elements ε_e and strain of viscous element ε_v. It can be expressed by the following formula:

$$\sigma = E\varepsilon + \eta\frac{d\varepsilon}{dt} \tag{4.5}$$

where, G is the shear modulus.

2. Coupling Correlation Effect Between Active Response and Passive Injury

During the normal operation of DVE system, driver's operation behavior has the highest control priority for vehicle, and is also one of the main ways to realize collision avoidance control. Driver's active response includes scenario cognition, decision-making and operation behavior. Its mechanism is a significant theoretical basis for accurate prediction of damage under collision conditions.

(1) Driver's Active Cognition and Behavior Response in Hazardous Scenarios

From the perspective of cognitive science, the attention mechanism and decision-making process related to tasks and apriori knowledge involve complex neuronal excitation conduction and signal communication in the brain based on vision (Fig. 4.9). From the perspective of group behavior reflected by ND data, dangerous driving operation is positively correlated with traffic accident rate [56]. From the perspective of individual behavior, driver's physiological feature, vehicle dynamics and traffic scenario signals are important data to identify and predict driver's active behavior. Based on such sign as physiological signals, body posture or vehicle dynamics, the driving intention can be predicted several seconds in advance to avoid dangerous operation behavior. For example, feature of event related potentials (ERP) is extracted from electroencephalography (EEG), so as to establish the driving decision behavior pattern recognizer of regularized linear discriminant classifier, which can accurately predict driver's braking intention in hazardous scenarios [57].

In recent years, with integration development of cognitive science and deep learning, together with the help of environmental perception capability of intelligent transportation system, some deep neural network models with human perception properties provide a possible numerical simulation tool for parametric reconstruction and analysis of human behavior in specific traffic scenarios. For example, by making use of driver's gaze behavior data collected in ND, a deep neural network model with visual perception properties can be built [58]. Using large-scale real accident videos as incentive to establish a danger gaze database can train the model to learn experienced drivers' ability of perceiving potential hazards in traffic scenarios, so as to better predict driving risks [59].

(2) Coupling Effect of Active Response on Collision Risk

The essence of active response is that human body carries out stress state. After receiving stimulation signal of the input nerve, the central nerve controls the skeletal muscle to make contractile force through the output nerve and generate corresponding motion posture and internal load. For instance, vehicle occupants will take a series of reactions when perceiving the danger before collision and being impacted during collision (Fig. 4.10), including driver's hand supporting steering wheel, trunk expansion and contraction led by resisting inertial movement of neck and trunk, and physical tension behaviors such as pedaling on the ground (Bracing).

The degree of active response is related to individual difference and level of consciousness. Generally speaking, it will change the body dynamics and kinematics, which results in change of stiffness and posture and affects degree of human injury in collision (Table 4.3). Taking lower limb as an example, the muscle contraction

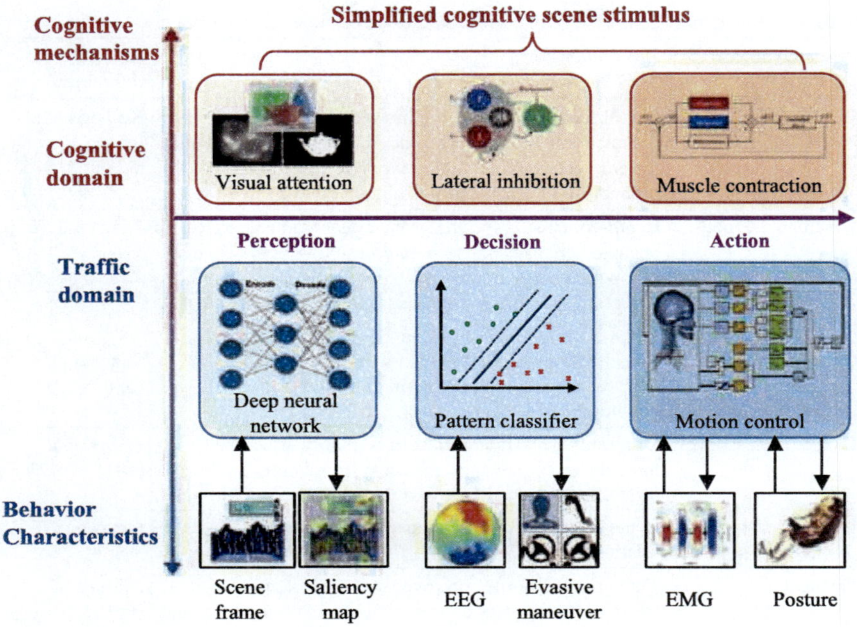

Fig. 4.9 Driver's active response mechanism and feature signals at different stages of hazardous scenarios

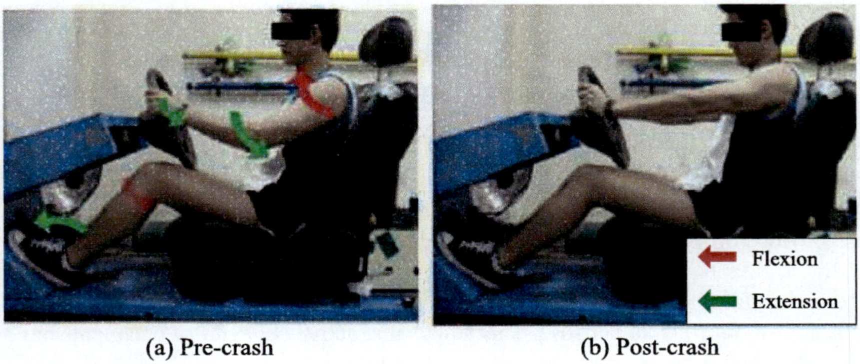

Fig. 4.10 Driver's stress behavior [60]

behavior leads to increase of its stiffness and internal load, and transmits momentum through structures such as knee joint. In this process, body motion state and binding force distribution change accordingly, which can reduce overall submarining risk. However, it may also lead to the increasing risk of lower limb fracture.

Table 4.3 Key conclusions of several current studies on occupant active behavior

Research object	Key conclusion	Related references
Damage form	In real traffic accidents, the limbs and head of occupants are relatively more likely to be injured. Compared with the corpse test under similar working conditions, the probability of chest injury is higher	Kallieris et al. [61]
Kinematic response	In pre-collision stage, the ground pushing behavior of occupant changes body posture, resulting in the change of kinetic energy transmission path in the collision process, which has a significant impact on occupant injury risk	King et al. [62]
Chest stiffness	The effect of muscle tension on occupant contact stiffness is restrained to low-intensity collision conditions	Shaw et al. [63]
Consciousness level	Different levels of consciousness mainly affect the moment when passengers receive stimulation, but have little effect on amplitude of muscle contraction response	Huber et al. [64]
Muscle contraction property	In the actual braking, the muscle activation level is relatively low (13–44%)	Östh et al. [65]
Gender	There are still some controversies on the significance of the impact of gender on stress response, but in the actual accident data, degree of female occupant displacement is higher in general	Ejima et al. [66]
Active safety system	The pretension of active seatbelt makes muscle tension response time advance by about 20–30 ms	Östh et al. [65]

Considering the safety of volunteer experiment, it is impossible to study the coupling effect of active response on injury by means of direct living control experiment. Therefore, relevant biomechanical studies have been applied to reconstruct real traffic accident. Based on accident investigation information, the studies set the cadaveric pulley test according to the same collision boundary, and compare the injury forms obtained in the test with real traffic accident. And it is found that there are obvious differences in the location and form of injury between the two: in real traffic accidents, the head and limbs of occupants have a higher risk of injury; in the test without active response, the proportion of fracture of chest bones (such as ribs and sternum) is higher [61]. This shows that active behavior of occupants has a non-negligible impact on the risk and form of injury.

3. Dynamic Response and Injury Behavior Property of Vulnerable Road Users

Vulnerable road users mainly include those who bear high risk of casualties in traffic accidents, represented by pedestrians and two-wheeler cyclists. Compared with occupants inside the vehicle effectively protected by on-board safety protection device, people outside the vehicle directly collide with vehicle in traffic accident and bear a higher risk of casualties. According to data released by the China National Bureau of statistics, vulnerable road users accounted for 26% of the deaths caused by traffic accidents in 2019 [18]. In collision accidents, the injury mechanism of vulnerable road users is more complicated due to higher difficulty in determining their movement state.

(1) Dynamic Response of Pedestrians, Cyclists and Two-Wheeled Motor Cyclists in Human-Vehicle Collision

Collision between the side of pedestrian and the front of car is the most typical condition of human-vehicle collision accident. Pedestrians are usually in a standing, gait or stress avoidance posture. Their lower limbs first contact the front bumper of vehicle. During the collision, their bodies rotate around the front structure of vehicle body, and then followed in turn by collisions of the thigh and pelvis with the front edge of hood, the chest with the hood, and the head with the front windshield or the rear of the hood (Fig. 4.11). Due to the difference of initial collision speed, pedestrian collision kinematics and corresponding injury forms are also different [67].

Compared with pedestrians, cyclists have different injury mechanism when colliding with the front of vehicles, mainly reflected in the impact of lower limb posture on injury risk. In the process of pedaling, knee joint shows periodic bending and extension behavior, and the maximum bending angle will reach about 90°; When knee is bent to the maximum extent, its height may exceed the leading edge of bonnet. In the process of collision, when the rider's knee is higher than the front edge of bonnet (foot up, forward leaning posture), the front edge of bonnet will contact with the lower leg and collide, causing transverse shear to knee joint and fracture of anterior cruciate ligament [68]. In addition, in the posture of stepping on pedal forward and down, lower limbs of the impacted side will be affected by the impact force of the bicycle frame relative to the opposite direction of vehicle. Due to the reaction force of bicycle frame, the bending moment of tibia will be reduced. Therefore, the risk of tibial injury of cyclists is affected by bending angle of knee joint and collision position with vehicle.

Compared with cyclists, the initial movement of two-wheeled motor cyclists is faster, and initial posture and kinematics is also different. In addition, the mass distribution of power battery and drive motor of two-wheeled motorcycle is also different from that of bicycle (Fig. 4.11). Some research results show that head injury risk arising from the collision between two-wheeled motorcyclists and windshield glass is slightly lower, but the risk is likely to be higher in the secondary collision with the ground [69].

(2) Injury Mechanism and Tolerance Limit

Investigation of road traffic accidents show that among AIS 2+ injuries suffered by pedestrians, the injury frequency of lower limbs is the highest, accounting for 32.6% of all injury parts, followed by head (31.4%), hip and abdomen (11.7%) and chest (10.3%) [70]. Lower limb injury often leads to lifelong disability, and head injury is easy to evolve into serious injury and even life-threatening. The following takes pedestrians as an example to introduce injury mechanism and tolerance limit of commonly-injured body parts such as lower limbs and head.

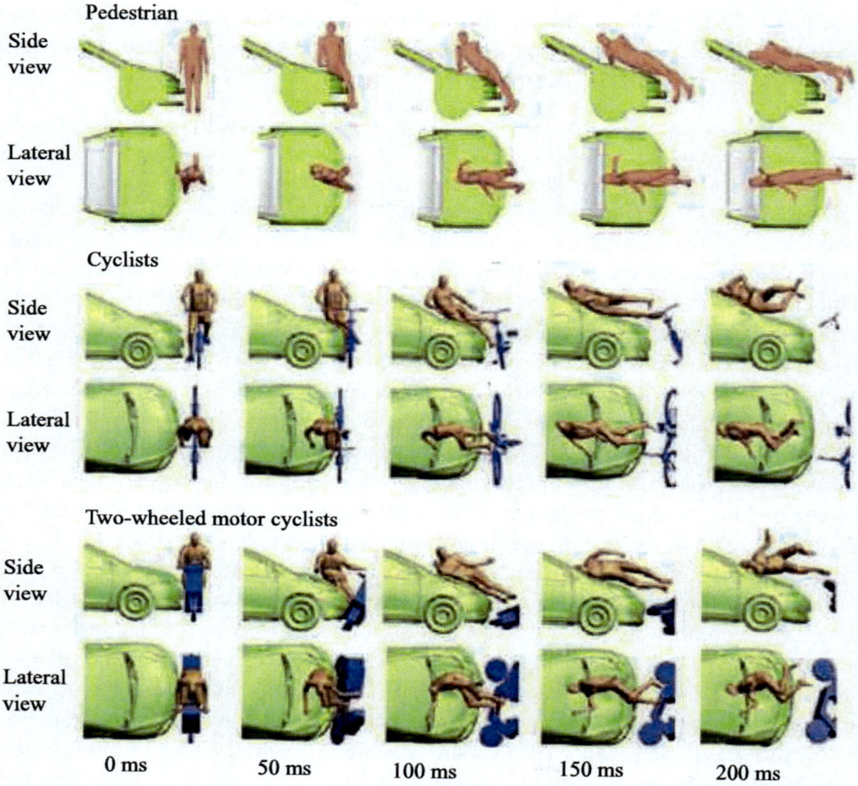

Fig. 4.11 Kinematics of pedestrians, cyclists and two-wheeled motor cyclists during collision with vehicles (*Note* In the illustrated case, collision speed of cars is 40 km/h, and initial speeds of pedestrians, cyclists and two-wheeled motorcycles are 0, 20 and 40 km/h respectively)

Lower limb injury is mainly made by direct contact impact of front bumper system and front edge of hood on the lower limb, or load transmitted to other parts of lower limb through knee joint. The most common injured positions are lower limb long bone, knee joint, hip, foot, etc., among which long bone (femur, tibia, fibula, etc.) fracture and knee ligament injury are the most frequent forms [71]. When pedestrian's lower limbs collide with the front of vehicle, the external load on long bones of lower limbs is mainly lateral bending, and the backbone as the main stress area is fractured under bending or compression loading.

A typical three-point bending experiment can be carried out for studying load distribution of pedestrian's lower limbs, selecting the section bending moment as evaluation index of fracture risk. Early researchers conducted pendulum impact test on the tibia and calibrated tibial fracture tolerance under medium and low-speed loading conditions [72]. After that, more studies adopted high-speed impact test and quasi-static test to study the damage tolerance of long bone fracture with different strain rates. The pedestrian injury tolerance limit is mainly determined by the cadaver experiment under typical loading conditions and the injury severity distribution obtained from accident investigation and statistics. Common knee joint injuries are mainly induced by two deformation modes, i.e., shear and bending. The shear deformation of knee joint occurs in the case of primary contact between the bumper and lower leg, and the impact between knee joint and bumper leads to relative displacement of thigh bone and lower leg bone. With the continuation of collision and the movement of lower limbs, knee joint turns to the bending deformation mode.

The form of pedestrian injury is significantly related to collision speed. Under low-speed impact, main injury of knee joint is ligament tear; Under high-speed impact, risk of long bone fracture is increased, possibly associated with ligament tear. Under the condition of lateral impact on pedestrian's lower limbs, there is a correlation between knee ligament tear and lower limb long bone fracture. According to PCDS database, only 15% of the conditions with fracture have ligament injury at the same time [73]. It should be noted that human-ground collision is also likely to occur during human-vehicle collision. Due to large stiffness of ground, the peak acceleration of head collision happens in extremely short time, but with high peak value. In order to evaluate the risk of pedestrian's head injury, some studies suggest to continue using the previous evaluation approach, but reduce the time window of acceleration integration [74]. Currently, there is no recognized conclusion.

(3) Impact Dynamics Model and Approximate Analytical Solution for Lower Limb

Experiment can be made to obtain real pedestrian injury data, but is difficult to characterize and predict pedestrian injury mechanism from the perspective of kinematics and dynamics. In vehicle conceptual design, a simplified mechanical model can also be used to preliminarily analyze key dynamic parameters, in order to study the correlation between design parameters and collision damage. For example, aiming at the geometric structure of pedestrian lower limbs and the relative motion relationship of various parts, it is abstracted as a multi-rigid body structure, and the dynamic model of the collision between lower limbs of pedestrian and front end of vehicle is established. The shear force, bending moment and forward inclination angle of lower limbs are solved in analytical form. Based on the analytical solution, the change mechanism of lower limb injury index can be analyzed, and the correlation between structural parameters of vehicle front end and index of lower limb injury (such as force) can be characterized [75].

Figure 4.12 shows the stress analysis diagram of pedestrian's lower limbs in the collision process (taking the collision process of pedestrian's lower limb impactor as an example, see Sect. 5.2 of Chap. 5 for details). Firstly, it is assumed that lower limb impactor receives three concentrated forces from the front end of vehicle, which are respectively from the front edge of hood $F_u(t)$, the front bumper $F_m(t)$ and the front auxiliary bumper $F_l(t)$ (Fig. 4.12c). It is defined that the distance between $F_u(t)$ action point and thigh centroid is d_u (the action point above thigh centroid is positive), the distance between $F_m(t)$ and $F_l(t)$ action point and calf centroid is d_m (the action

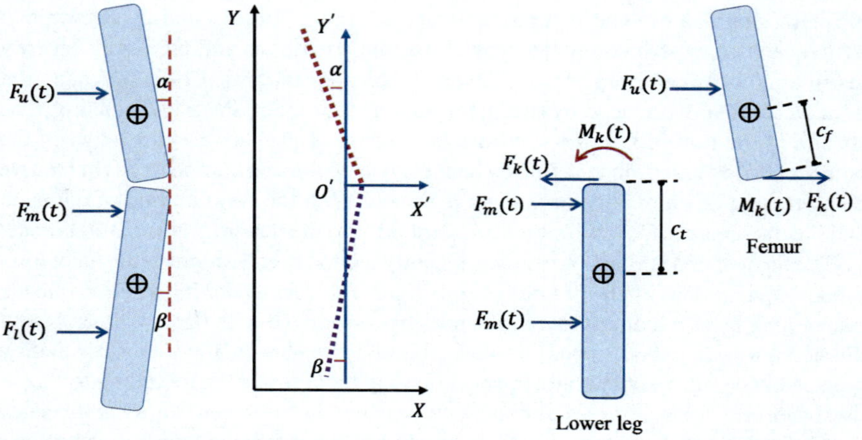

(a) loading condition (b) coordinate system (c) lower limbs

Fig. 4.12 Stress analysis of pedestrian lower limb impactor during collision with vehicle

point above calf centroid is positive) and d_l (the action point below calf centroid is positive), and the distance between thigh and calf centroid and knee joint center is c_f and c_t respectively.

Supposing that pedestrian's lower limb collides with stationary vehicle on the ground at a certain speed, the simplified form of its mechanical balance equation is given, and a new inertial coordinate system ($O'X'Y'$) fixed in the ground coordinate system (OXY) is established. The origin O' of the new coordinate system coincides with knee joint center of lower limb before collision. The included angles between the axes of thigh part and calf part of lower limb and the ordinate axis $O'Y'$ of the coordinate system are α and β respectively, as shown in Fig. 4.12b with positive direction. The section shear force between thigh and calf is $F_k(t)$, and the section bending moment is $M_k(t)$ (Fig. 4.12c). The lower limb motion is simplified to plane motion, ignoring the vertical component of collision force between the front-end structure of vehicle and lower limb and the force between thigh and calf.

Taking the structural design parameters of the front end of vehicle as independent variables, the analysis and derivation are carried out, and the mass parameters of pedestrian lower limbs are substituted. The expression of relevant mechanical quantities can be obtained as follows (taking the collision speed of 40 km/h as an example):

$$\begin{aligned}
F_k &= 1.592[F_u(d_u - 0.064) + F_m(d_m + 0.114) - F_l(d_l - 0.114)] \\
\alpha &= 1.148[-F_u(d_u + 0.161) - F_m(d_m - 0.289) + F_l(d_l + 0.289)] \\
M_k &= 0.491[-F_u(1.035 \cdot d_u - 0.003) + F_m(d_m - 0.005) - F_l(d_l + 0.005)]
\end{aligned} \tag{4.6}$$

The acceleration of the center point of knee joint can be further solved as:

$$x_{acc} = 0.140[-F_u(d_u - 0.372) - F_m(d_m - 0.822) + F_l(d_l + 0.822)] \quad (4.7)$$

The approximate solution obtained by the model quantifies the correlation between structural parameters of vehicle front end and lower limb injury index, which is helpful to understand influence trend of the change of vehicle front end parameters on lower limb injury index. The above approximate analytical solution ignores knee joint deformation and energy absorbed by knee joint bending deformation. It is simplified as a rigid body, which affects the flow direction of energy in the collision process as well as the accurate prediction of the peak acceleration of tibia and the peak bending angle of knee joint. It has certain limitations in practical application. However, the approximate analytical solution intuitively reflects influence trend of vehicle front-end structural parameters on lower limb injury index. In terms of mechanism analysis, it can assist engineers in understanding influence mechanism of vehicle front-end structural parameters on knee injury, assist designers in quickly analyzing influence sensitivity of structural parameters on injury index, and assist in determining improvement schemes for safe vehicle structure design.

4.1.3 Mechanical Behavior and Safety of Vehicle Body and Power System

The threat to vehicle safety arising from external impact load in road traffic accidents is mainly reflected in two aspects: on the one hand, deformation and failure of vehicle body and the transmission of part of impact load to drivers and passengers result in damage risk. And safety protection performance of vehicle body will be affected by such factors as the mechanical behavior of materials and structures under impact load (including common metal materials of vehicle body and lightweight materials for vehicle, as well as their basic mechanical properties), and the dynamic load effect of metal materials under dynamic impact conditions. Taking the circular tube structure commonly used in vehicle body as an example, the mechanical characterization of axial crushing of circular tube is introduced; On the other hand, great deformation occurs in power system of new energy electric vehicle. Due to mechanical, electrical and thermal abuse, thermal runaway and related risks will be caused. Under impact load, power battery will result in thermal runaway because of mechanical abuse (Fig. 4.13).

1. Mechanical Behavior of Materials and Structures Under Impact Load

It is a typical impact dynamics problem that vehicle body is subjected to impact load, and makes deformation and fracture failure in road traffic accidents. The body structure maintains a certain crashworthiness, which can effectively ensure safety of people and goods inside and outside vehicle. Structural crashworthiness is affected by

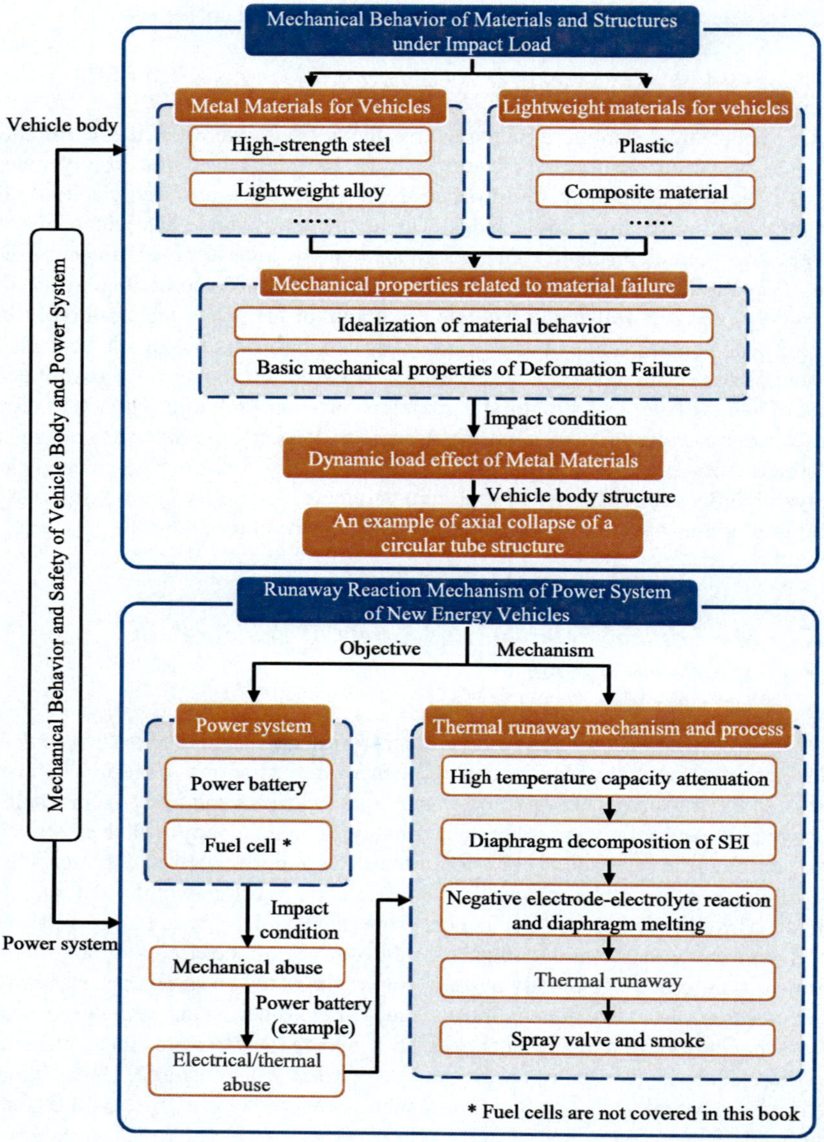

Fig. 4.13 Logic diagram of mechanical behavior and safety of vehicle body and power system

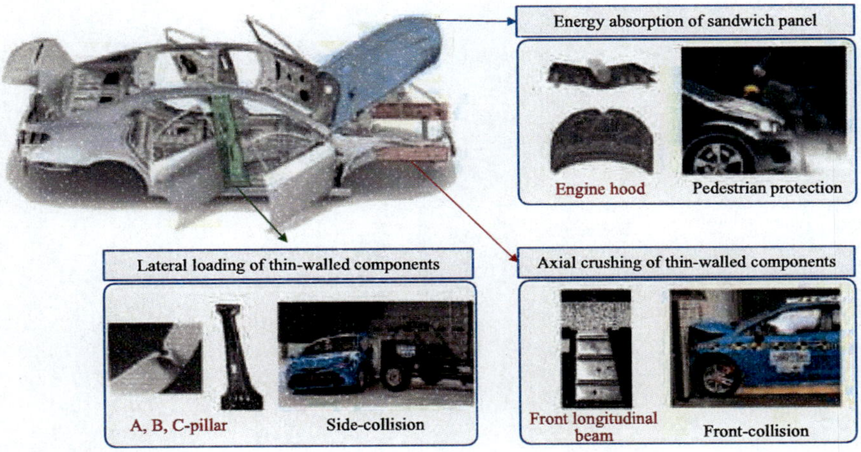

Fig. 4.14 Deformation and energy absorption of thin-walled components in vehicle body structure [76]

the impact dynamics properties of materials and components. Therefore, mastering mechanical properties of body materials is the key to safety design. Most of vehicle bodies are metal thin-walled structures (Fig. 4.14). For example, under the condition of front collision with the highest frequency, the front longitudinal beam will collapse axially, which is essentially the mechanical behavior of typical thin-walled pipe fittings under impact load. The mechanical behavior and constitutive model of common metal materials (low carbon steel, aluminum alloy), advanced lightweight materials (plastics, composites) and structures are gradually built on the basic theory of material mechanics. Taking the axial crushing of circular tubes commonly used in body structure as an example, its mechanical model provides a theoretical basis for establishment of impact deformation failure criterion, energy absorption design and crashworthiness structure design of lightweight body structure.

(1) Mechanical Properties and Ideal Models of Typical Automotive Metal Materials

Basic theories and approaches for analyzing impact response properties of vehicle body structure come from material mechanics and impact dynamics, including basic mechanical properties of materials, idealization of material behavior, large deformation effect, limit analysis and boundary theorem, dynamic load effect, strength theory, energy method, etc. (Fig. 4.15) [76]. Basic mechanical properties of materials, idealization of material behavior and dynamic load effect are mainly introduced as follows.

Taking the uniaxial tensile test of cylindrical specimen as an example, some common mechanical definitions are introduced. Stress σ, refers to the internal force concentration at one point on the section; Strain ε, is the elongation or shortening of unit length of test piece. When vehicle is impacted in traffic accident, its body materials, components and devices usually undergo large plastic deformation. In these

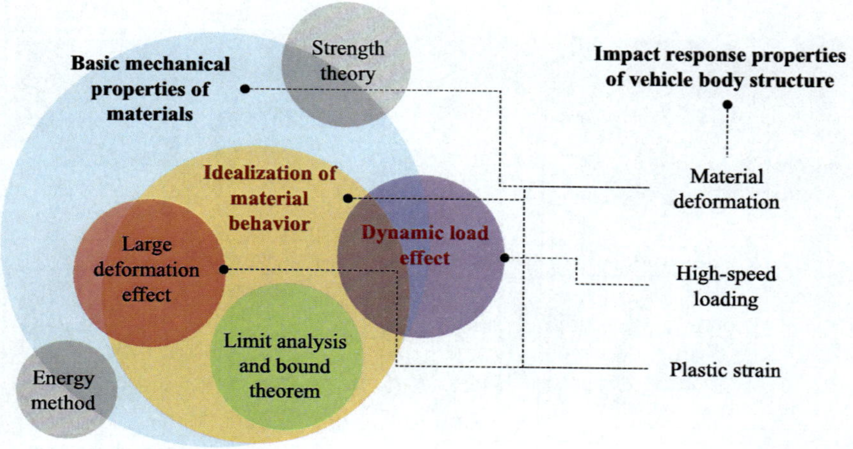

Fig. 4.15 Fundamental theories of material mechanics and impact dynamics involved in vehicle impact loading process

cases, the plastic strain is much larger than elastic strain, so the latter can usually be ignored in the analysis. Before the initial yield, material shows an approximate rigid behavior. Its stress–strain curve is simplified and approximated in general to obtain the perfectly rigid plastic, rigid-linear hardening and rigid-power hardening models (Fig. 4.16). The stress–strain relationships of the three models can be expressed as follows:

$$\begin{cases} \sigma \le Y \ (\varepsilon = 0) \\ \sigma = Y \ (0 < \varepsilon < \varepsilon_{\mathrm{f}}) \end{cases} \tag{4.8}$$

$$\begin{cases} \sigma \le Y \qquad\quad (\varepsilon = 0) \\ \sigma = Y + E_{\mathrm{p}}\varepsilon \ (0 < \varepsilon < \varepsilon_{\mathrm{f}}) \end{cases} \tag{4.9}$$

and

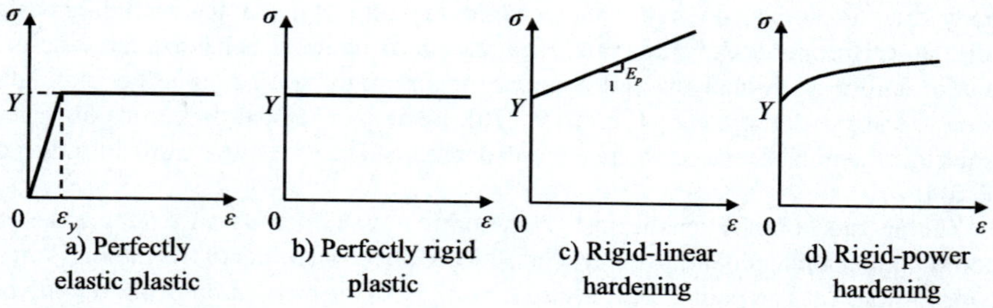

Fig. 4.16 Idealized tensile stress–strain curve of materials

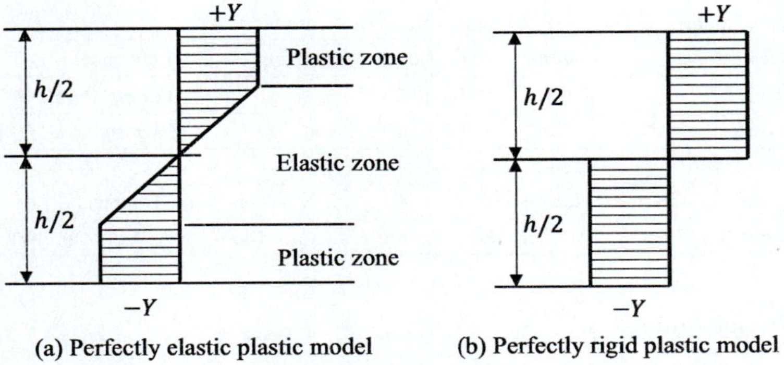

Fig. 4.17 Stress distribution along thickness in the case of bending beam

$$\begin{cases} \sigma \leq Y & (\varepsilon = 0) \\ \sigma = Y + K\varepsilon^q & (0 < \varepsilon < \varepsilon_f) \end{cases} \tag{4.10}$$

The idealization of material behavior greatly simplifies quite a number of mechanical problems, and can grasp its main feature at the same time. For example, taking beam bending as an example, mechanical behavior of rectangular section beam made of ideal elastic–plastic material under the action of bending moment M is considered. Stress distribution along the thickness of beam is shown in Fig. 4.17a, including two plastic zones and one elastic zone. If $M < M_e$ (M_e is the maximum elastic bending moment), The relationship between M and its axis curvature κ is linear. If $M > M_e$, then the relationship of M–κ is nonlinear. If the beam is made of perfectly rigid plastic material, its plastic deformation is concentrated on one or a few sections (Fig. 4.17b), and the section plastic ultimate bending moment $M_p = Ybh^2/4$, with b and h representing the width and height of the rectangular section respectively. If any bending moment less than M_p is applied, the beam will not make plastic deformation.

(2) Dynamic Load Effect of Common Metal Materials

Collision is a typical dynamic impact loading condition. Under the load, dynamic load borne by vehicle body include wave effect, strain rate effect and inertia effect. At the initial stage of structural dynamic response, elastic wave and plasticity affect the energy absorption of materials and structures in a variety of complicated ways under the joint influence of dynamic load itself, structural form and material property.

Strain rate is the time rate of strain change ($\varepsilon = \mathrm{d}\varepsilon/\mathrm{d}t$) Being acted by strong dynamic load, the structure deforms rapidly with a high strain rate. The mechanical properties of many engineering materials are related to strain rate. More available in engineering applications is the phenomenological explicit relationship, that is, the relevant constitutive equation considering influence of strain rate on material yield stress and flow stress. Most widely used in structural collision problems, Cowper-Symonds relationship represents the perfectly rigid plastic material with dynamic yield stress or flow stress related to strain rate. The ratio of dynamic yield stress Y^d

Table 4.4 Parameters of several materials with sensitive strain rate

Material	B(1/s)	q	Bq(1/s)	References
Mild steel	40	5	65	Forrestal et al. [77]
Stainless steel	100	10	160	Forrestal et al. [78]
Titanium (Ti50A)	120	9	195	
Aluminum 6061-T6	1.70 × 106	4	2.72 × 106	Symonds [79]
Aluminum 3003-H14	0.27 × 106	8	0.44 × 106	Bodner et al. [80]

to static yield stress *Y* is:

$$\frac{Y^{\mathrm{d}}}{Y} = 1 + \left(\frac{\dot{\varepsilon}}{B}\right)^{1/q} \ (\dot{\varepsilon} > 0) \tag{4.11}$$

where *B* and *q* are material parameters.

In fact, *B* represents the characteristic strain rate when $Y^{\mathrm{d}} = 2Y$, and the material constant *q* is a measure of material sensitivity. When applying the Cowper-Symonds relationship, some representative values of *q* are shown in Table 4.4, in the case that he strain is $\varepsilon = 0.05$; For strains with large difference from $\varepsilon = 0.05$, the obtained value is likely to be inaccurate.

(3) Mechanical Properties of Lightweight Materials

Lightweight materials represented by plastics and composites have been widely used in vehicle components in recent years, and their mechanical properties also affect collision safety of vehicles. As an ideal lightweight material, plastic has the characteristics of light weight, rust prevention, vibration absorption, large design freedom, fine coloring, excellent processing performance and recyclability. Plastic materials can be divided into brittle and ductile according to their mechanical properties. Figure 4.18 shows the uniaxial tensile stress–strain curves of a brittle plastic and a ductile plastic respectively.

The dynamic response of ductile plastics under loading generally goes through five stages: elastic stage (R1), yield, strain softening, cold drawing stage (R2) and strain strengthening stage (R3). Taking the engineering stress–strain curve of polycarbonate as an example (Fig. 4.19) [81], necking begins to occur at the yield point, and further displacement loading will untie the long molecular chain and reduce the stress to drawing stress σ d, which is the strain softening phenomenon of plastics. In cold drawing stage (R2), winding molecular chains are arranged in order. At this time, material is further deformed and necking begins to propagate along measuring section of specimen, when the increase of deformation does not lead to increase of stress. After molecular chains are completely untied and further stretched, stress will increase linearly, which is the strain strengthening.

As the lightweight material with high specific strength, high specific stiffness and strong corrosion resistance, composites are usually composed of a variety of components with different properties or structures, and can be divided into reinforcement and matrix. The reinforcement material is responsible for bearing most of the load

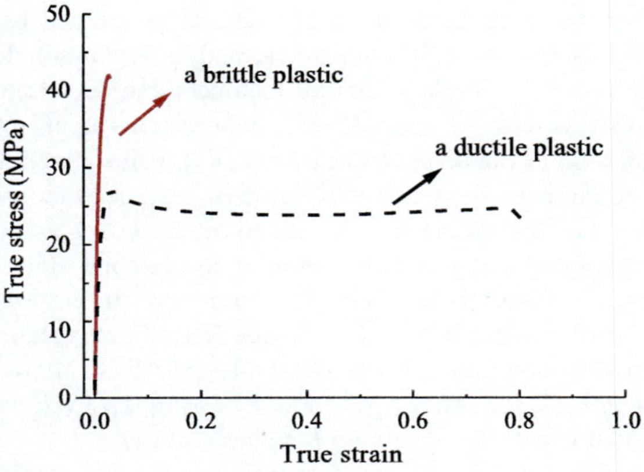

Fig. 4.18 Tensile curve of a brittle plastic and ductile plastic

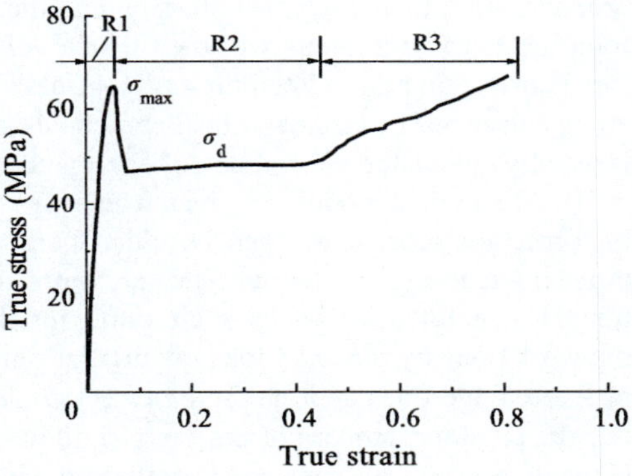

Fig. 4.19 Engineering stress–strain curve of polycarbonate

and providing stiffness and strength. The matrix material is mainly fixed and support in the fiber reinforcement direction and plays the role of protecting the fiber. The composite material can be equivalent to a structure composed of different materials. Its overall mechanical properties are affected by mechanical properties of the composite material, proportion of each material and mode of composition. Different from traditional vehicle materials, composites have stronger design and expansibility. The laying mode, direction and number of layers of constituent materials will affect their overall performance. In addition, compared with traditional vehicle materials that need to be connected by riveting, composite materials can be formed as a whole by means of reasonable design, for improvement of component performance and reduction of cost.

(4) Theoretical Model of Energy Absorption of Structural Parts: Taking Axial Collapse of Circular Tube as An Example

According to description of material properties based on theories of material mechanics and impact dynamics, analytical solution derivation of energy absorption behavior of simple structural members can be further completed [82]. Taking the axial crushing of thin-walled circular tube as an example, the failure mode depends on ratio of diameter to thickness (*D*/*h*), mainly including axisymmetric ring mode, accordion mode or non-axisymmetric diamond mode. For some round pipes with *D*/*h* value, the failure may be mixed mode of ring and diamond, i.e., beginning with ring mode and gradually changing to diamond mode. Based on a large number of crushing test results of circular tubes with various sizes, the failure mode classification of circular tubes can be made [83]. To be specific, when $D/h > 80$, it will fail by diamond mode; When $D/h > 80$ and $L/D < 2$, it will fail by ring mode (L is the length of pipe fitting); When D/h < 50 and $L/D > 2$, mixed mode failure occurs, and Euler instability occurs for long circular tubes.

The typical force–displacement curve of circular tube during axisymmetric crushing is shown in Fig. 4.20. In the initial stage, the axial pressure first reaches a primary peak, then decreases sharply, and then fluctuates. The fluctuation of axial force leads to continuous wrinkle of tube collapse, with each peak value of axial force corresponding to beginning of a wrinkle. The area under curve represents the energy absorbed by circular tube. In practical engineering applications, average force is usually calculated as an index of energy absorption capacity.

The theoretical model is derived from ring model (Fig. 4.21) [84]. Corresponding to formation process of each wrinkle, there will be three circumferential plastic hinges. Assuming that the wrinkles are outward, materials in the area where plastic hinge is generated will undergo circumferential tensile strain. It can be regarded that the work done by the axial load on circular pipe is dissipated by the plastic bending of three hinge lines and the circumferential elongation of materials between them. For the crushing process of perfectly rigid plastic materials, there is no interaction between bending and tension. Therefore, the yield of materials is caused by a single factor of bending or tension. In accordance with the bending and crushing energy

Fig. 4.20 Force–displacement curve corresponding to typical round pipe crushing (the diagram takes aluminum round pipe with D = 97 mm, L = 196 mm and h = 1.0 mm as an example)

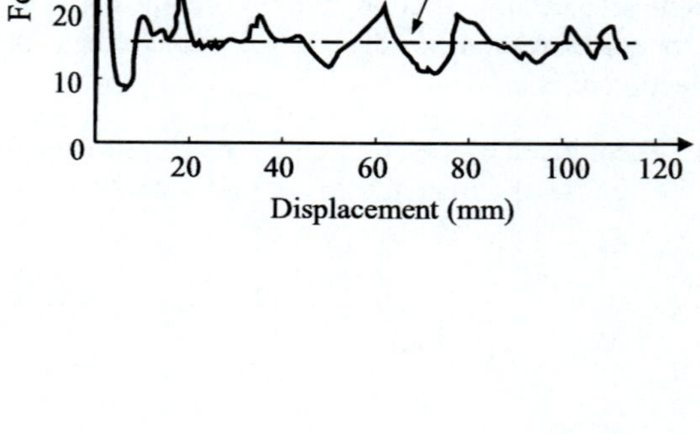

consumption of the material in the crushing process, as well as the yield film force YH per unit length and energy balance properties, average external force of circular tube for completion of the whole crushing and folding process in the ideal state can be approximately expressed as:

$$P_{\mathrm{m}} \approx 6Yh\sqrt{Dh} \tag{4.12}$$

where, P_{m} represents the average external force for completion of the whole folding process; Y is the yield stress; h is the thickness; D is the diameter of the round pipe.

Under dynamic loading, the strain rate effect enhances the material, which is manifested by the increase of yield stress. According to the Cowper-Symonds relationship, the average force expression in consideration of the strain rate effect can be approximately as follows:

$$P_{\mathrm{m}} = 6Yh\sqrt{Dh}\left[1 + \left(\frac{V_0}{2B_c D}\right)^{1/q}\right] \tag{4.13}$$

where, V_0 represents the initial velocity of circular tube during compression; B_{c} is the material parameter and represents the characteristic strain rate when $Y^{\mathrm{d}} = 2Y$. Y^{d} is the dynamic yield stress and Y is the static yield stress; q represents a measure of material sensitivity, which can be obtained by referring to the table [85].

For the non-axisymmetric mode (diamond mode), it is difficult to deduce a more accurate theoretical model. Most models involve triangular elements with respect to smaller bends and assume that the midplane is not extensible.

On the basis of model research [86], Guillow et al. conducted systematic experiments [87], covering a wide range of *D/h* and *L/D* and different crushing modes in a single experimental plan. For the circular tube made of aluminum metal, the variation of its average force with D/h can be plotted in logarithmic coordinates (Fig. 4.22). Regardless of any crushing mode, all points fall on a single curve approximated by a straight line, so that an experimental-based fitting equation is obtained:

$$\frac{P_{\mathrm{m}}}{M_0} = 72.3\left(\frac{D}{h}\right)^{0.3} \tag{4.14}$$

where, M_0 is the plastic ultimate bending moment per unit width.

In addition to the axial collapse of thin-walled circular tubes, vehicle collision safety design also involves deformation and energy absorption of some other structures, such as collapse of square tubes, collapse of multi-cell thin-walled members, and bending failure of thin-walled pipe fittings. (Fig. 4.23). The analysis principle toward their impact dynamic behavior is similar.

2. Runaway Reaction Mechanism of Power System of New Energy Vehicles

New energy vehicles refer to vehicles that use unconventional vehicle fuels as power sources, including pure electric vehicles, hybrid electric vehicles and fuel cell electric

Fig. 4.21 Theoretical analysis model of ring mode failure

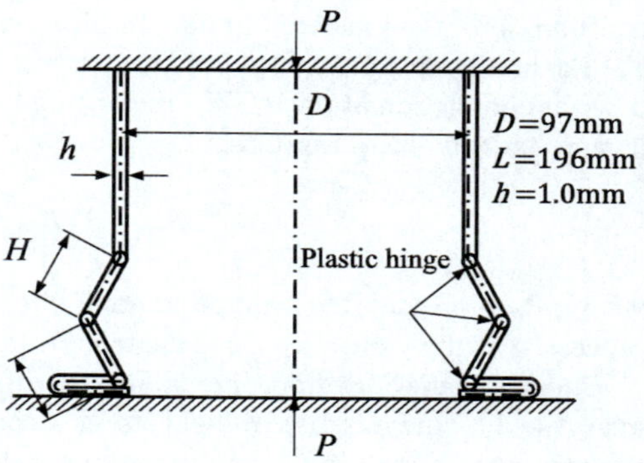

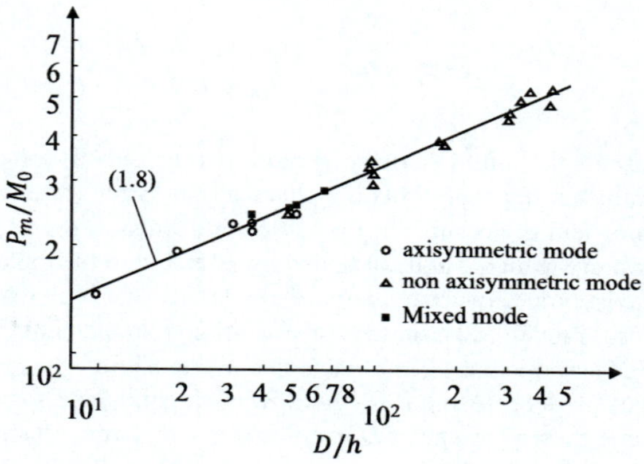

Fig. 4.22 Dimensionless curve of average force of aluminum round pipe varying with D/h

(a) collapse of square tubes (b) multi-cell thin-walled members (c) bending failure of thin-walled components

Fig. 4.23 Deformation and energy absorption forms of common structures

vehicles, in which electric vehicles are the main body. Compared with traditional vehicles, safety design of new energy vehicles should also consider safety problems brought by new power sources, such as fire and explosion accidents induced by thermal runaway of power batteries, in addition to energy absorption and damage of body structure under accident impact loading. The fire accident of lithium-ion battery of electric vehicle (Tesla Model s) in Seattle on October 11, 2013 is taken as an example here. The reason is that after the vehicle drove on the highway and hit the road object, the protective layer of vehicle battery pack was pierced by the road object, resulting in serious deformation and internal short circuit of internal battery, followed by out-of-control heating and final fire. According to incomplete statistics, there were 124 electric vehicle fire accidents reported by the media in China in 2020. Exploring its safety performance in terms of the runaway mechanism of power system is essential for safety protection of new energy vehicles.

(1) Brief Description of Vehicle Power Battery and Fuel Cell

The main power sources of new energy vehicles are power batteries and fuel cells. In spite of both providing electric energy as power, they are different in principle. Vehicle power battery generally refers to the components or equipment used for power storage and power supply in electric vehicles. Lithium-ion battery is the most practical power battery for electric vehicles because of its high specific energy, low self-discharge rate and long service life. Fuel cell is a chemical device that converts chemical energy into electrical energy based on proton membrane exchange technology. It is equivalent to a generator, but cannot store electricity by itself. Taking hydrogen fuel cell as an example, its essence is to generate electricity by means of reverse electrolysis of hydrogen and Oxygen Catalyzed by proton membrane. There is no discharge other than water in this process. Compared with pure electric vehicles, the market share of fuel cell vehicles is not high at present, and safety accidents led by corresponding out-of-control have not been publicly reported. Hence, details will not be covered in this paper.

(2) Thermal Runaway of Power Battery System

The safety accident of lithium-ion power battery characterized by thermal runaway is one of the main safety problems faced by new energy vehicles. Thermal runaway refers to the overheating phenomenon in which the temperature rise rate of the battery changes sharply due to exothermic chain reaction in the power battery. Figure 4.24 compares the reaction and relative heat release of battery component materials at different temperatures, that is, the relative value of heat release of each reaction [88]. In the process of thermal runaway, the lithium-ion power battery will go through the following processes in turn, and the temperature will change from low to high accordingly:

(1) High temperature capacity attenuation: under normal working conditions of lithium-ion battery, high ambient temperature will aggravate side reactions of the battery, resulting in battery capacity loss and capacity attenuation, which constitutes the so-called high-temperature durability or high-temperature self-discharge problem.
(2) Diaphragm decomposition of Solid Electrolyte Interphase (SEI): with increase of temperature, the internal chemical reaction of lithium-ion battery begins to

produce heat. The first exothermic reaction is decomposition of SEI diaphragm, which occurs at 80–120 °C.

(3) Negative electrode–electrolyte reaction and diaphragm melting process: due to decomposition of SEI diaphragm on the negative electrode surface, the negative active material is out of protection, and the negative lithium metal will react with the electrolyte, making the temperature of lithium-ion battery continue to rise. When the temperature reaches the melting point of the diaphragm, the diaphragm melts and shrinks to absorb part of the heat and temporarily reduce the temperature rise rate of the battery. However, the contraction of diaphragm may lead to local short circuit, heat release and acceleration of diaphragm disintegration.
(4) Thermal runaway process: after decomposition of diaphragm, the battery has an internal short circuit and releases a lot of heat. The temperature rises rapidly from 120 to 300 °C or even higher, and the battery quickly reaches thermal runaway. At this time, a variety of chemical reactions occur inside the battery, including positive decomposition, electrolyte solution (electrolyte) decomposition and adhesive reaction.
(5) Spray valve and smoke: the decomposition reaction of electrolyte will produce a large amount of gas, with sharply rising internal pressure of the battery and expanding volume of the battery. If its internal pressure exceeds the safety limit, the battery may explode. Therefore, vehicle battery is currently equipped with safety valve. When the internal pressure of battery exceeds the opening pressure of safety valve, the safety valve will open, so that the battery can enter the injection process and rapidly spew high-pressure gas. If the maximum internal temperature of the battery does not reach the melting point 660 °C of the positive collector aluminum foil, the smoke emitted is in gray white. However, after the heat release of mixed reaction, the internal temperature of the battery may reach more than 800 °C, the aluminum foil will melt and disintegrate, and the positive active material will be brought out of the battery by gas to form black smoke, accompanied by silver aluminum foil fragments.
(6) Electrolyte combustion: the spewed gas mainly composed of electrolyte decomposition products will burn violently and lead to fire if certain condition is satisfied.

Smoke, fire and explosion are major manifestations of thermal runaway of lithium-ion battery. There are three causes of thermal runaway, i.e., mechanical abuse, electrical abuse and thermal abuse [89], with a close internal correlation among their inducements (Fig. 4.25) [88]. The impact on vehicle leads to extrusion and acupuncture of power battery, resulting in mechanical abuse, deformation of the battery and internal short circuit. And overcharge and overdischarge lead to the occurrence of electric abuse. The battery generates high heat energy and overheats, which leads to thermal abuse, followed by thermal runaway chain reaction of lithium-ion battery, and finally the occurrence of thermal runaway. Among them, mechanical abuse refers to the mechanical deformation of the battery or battery pack caused by vehicle collision and subsequent extrusion. Mechanical deformation will have two consequences: one is that the internal diaphragm of the battery is crushed and there is an internal

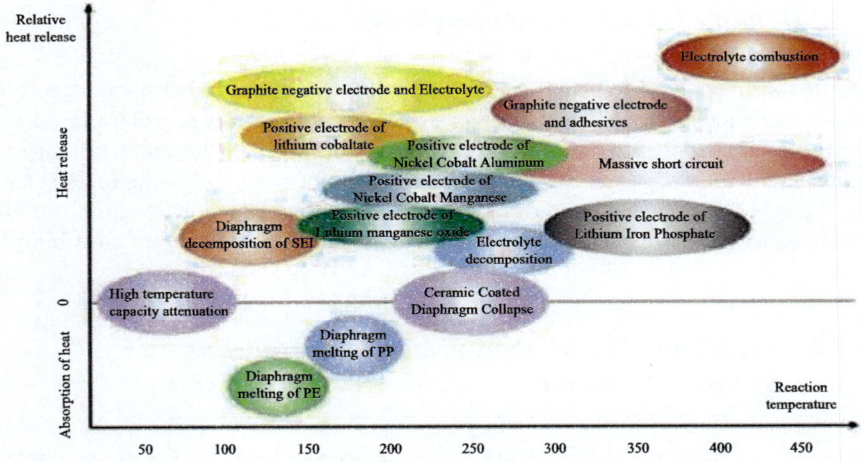

Fig. 4.24 Thermal runaway reaction mechanism of component materials of lithium-ion power battery [88]

short circuit between the positive and negative electrodes; The other is the leakage of flammable electrolyte inside the battery, which leads to secondary harm. In addition to squeeze and deformation of the battery and corresponding internal short circuit, vehicle collision may also result in battery puncture and fire due to internal short circuit.

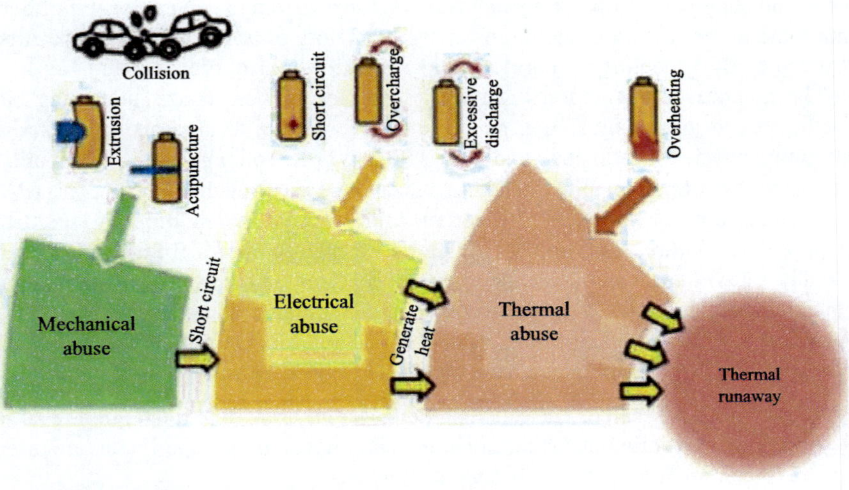

Fig. 4.25 Mode and correlation of thermal runaway inducement [88]

4.2 Damage Quantification Approach

Due to the diversity of collision accident conditions and individual characteristics of traffic participants, the damage behavior of people and vehicles in real traffic accidents presents the property of probability distribution. Based on the understanding of human-vehicle damage mechanism, it is necessary to further establish approaches for the numerical representation of different population, the extraction of natural stress behavior, as well as the quantitative evaluation of human-vehicle injury probability.

4.2.1 Population Distribution Feature and Modeling of Injury Behavior

A number of people bear high injury risk under impact load, such as aging, obese and female population. The specific reasons include the decline of bone-bearing capacity with age, the increase of inertial mass with obesity, and the incomplete matching of body features with current restraint system. Accordingly, probabilistic damage model shall be built for different populations to reflect the distribution feature of impact damage behavior. The specific process is as follows: damage indexes of different populations are mapped by means of dimensional scaling, optimal damage risk distribution function is fitted with experimental data, and damage risk probability and its threshold are established. Probabilistic damage model is also the basis for establishing parametric human body numerical model, including probability of human body micro-geometric structure and material properties from the organization level, and parameterization of posture and body shape from the macro level; Furthermore, the numerical reproduction of active behaviors of traffic participants can be realized by the modeling of muscle and its activation control process (Fig. 4.26).

The essence of population diversity is complex human factors, including but not limited to age, body shape, gender, posture, muscle tension and other stress reactions, physical health state and so on [90]. In order to quantify the distribution characteristics of human injury behavior, researchers use Probabilistic Analysis (PA) to establish the risk function of a certain injury behavior. Traditionally, the data of population injury distribution characteristics mainly comes from experiments. Usually, individuals with different body shape features are taken as the experimental objects for obtaining the damage tolerance of a single individual's specific damage behavior, and then all experimental samples are fitted for gaining the damage risk function. The experimental data have high reliability, but also obvious deficiencies:

(1) Specimens are scarce. The establishment of injury risk function requires a large number of experimental data, and human biological tissue experiments are often restricted by the extremely limited sample size.

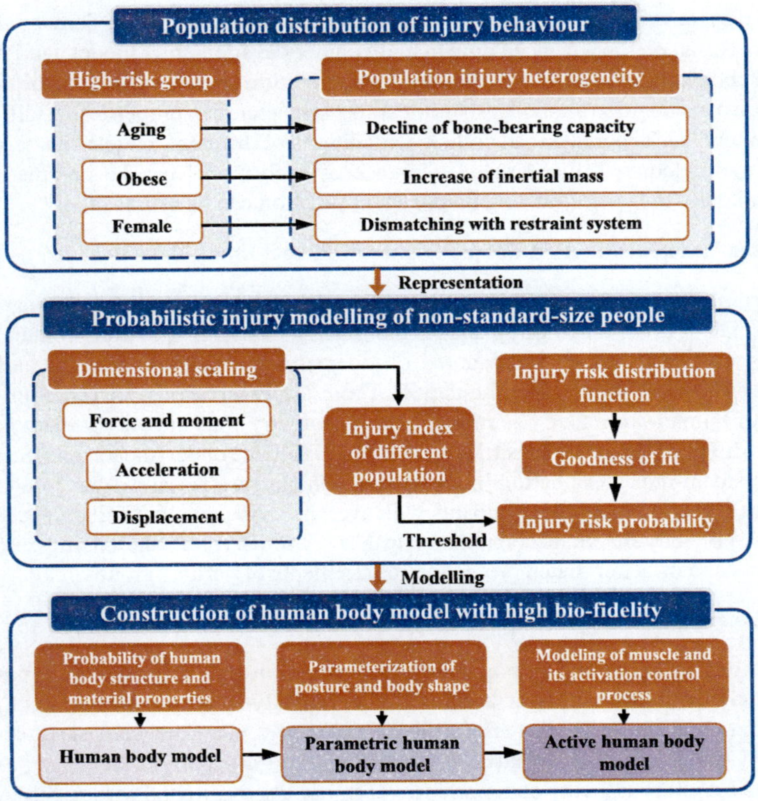

Fig. 4.26 Population distribution feature and modeling of injury behavior

(2) Method is complicated. The geometric irregularity of human tissue and heterogeneity of materials challenge the clamping and loading of the experiment. And it is difficult to ensure the consistency of the experimental setting.
(3) Parameter determination is difficult. The injury in situ often occurs inside the body. The specific form shall be determined by anatomical observation after the experiment. And it is difficult to make sure the damage evolution process during the experiment and the accurate load level corresponding to the time of injury.

In recent years, multi-scale and high-precision human body numerical simulation technologies have been developing consistently. Among them, the dummy model can represent the macro structure and multi-body kinematics, while the finite element human model has the capability to represent the level of micro-tissue damage, which provides a new idea for studying the characterization of population damage distribution. Based on the advanced human body numerical simulation model and PA, the human body numerical model representing distribution characteristics can be generated in consideration of body shape features and material properties of different human bodies. The multi-load collision conditions of different models can be simulated and calculated, the damage tolerance of each model can be obtained, and subsequently the corresponding damage risk function can be established.

1. Population Distribution Characteristics of Impact Injury Behavior

According to statistics of accidents involving drivers and passengers, compared to adult male with average body shape, the elderly, obese people and women have higher rate of serious injury and mortality under similar accident conditions, and thus they are listed as vulnerable population. There are two reasons: first, the physical state of human body (such as material, and geometry) has crowd correlation, and shows different response properties and damage tolerance under impact load; Second, the occupant injury evaluation index used in vehicle passive safety development at present is mainly the male occupants with average shape, without fully considering diversity of body shape and composition in the crowd. Therefore, the existing restraint system can not better match protection requirements of various groups.

(1) Correlation Between Impact Injury Response and Age

With the aging of the world's population, the proportion of elderly drivers and passengers and their injuries in traffic accidents will gradually increase. In various collision accident conditions, the increase of age significantly increases the risk of serious injury (AIS 3+) in all body parts of occupants [91]. For elderly drivers and passengers, chest injury is a very common form of injury. As mentioned earlier, the energy absorption capacity of bone structure tends to decrease with age. The elastic modulus of compact bone and cancellous bone will decrease significantly, especially after the age of 60 (Fig. 4.27). Taking the lower limb bone as an example, its strength reaches the maximum at about 35 years old [92], and then the fracture strain decreases by 5 and 7% respectively for every 10 years [93]. Geometric changes due to the increase of age also affect the risk of human injury. The changes of thoracic vertebra geometry and rib angle of the elderly (such as the common elderly hunchback) will affect the chest load under impact load [94].

(2) Impact Injury Effect in Terms of Obese Shape

The data released by the World Health Organization show that the obesity rate in the world has increased sharply since 1980. The body shape of obese people affects its matching with the existing vehicle occupant restraint system. The data from experiment on a large number of volunteers confirmed that with the increase of body mass index (BMI) of volunteers by 10 kg/m^2, the position of abdominal belt of seatbelt

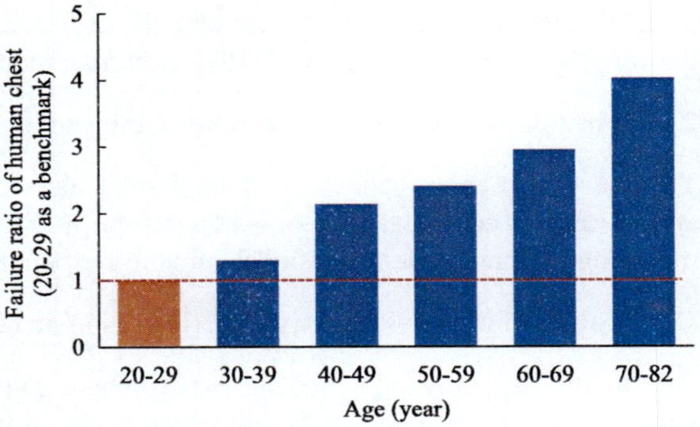

Fig. 4.27 Relative occurrence of chest injury of passengers with different ages under restraint of standard seatbelt [92]

moved forward and upward by 43 and 21 mm respectively relative to the anterior superior iliac spine, and the extension length of the corresponding abdominal belt increased by 130 mm [95]. By comparing the data of different BMI groups in the typical front crash sled test, it is found that obese occupants usually have greater relative displacement than ordinary passengers, and the trunk has an obvious tendency of backward inclination. During similar collisions, the chest compression of obese occupants is 44%, while that of normal weight ones is 29% [96]. The geometric change of obese shape also changes the restraint conditions, making the position of seatbelt far from the position of pelvis [97]. At the same time, it increases the effective mass of the restraint system and alters the biomechanical response of the body in the process of collision.

(3) Impact Injury Effect in Terms of Gender

Accident statistics show that compared with male occupants, the probability of traffic accident and serious traffic accident of female is about 30 and 60% lower respectively [98], but the probability of serious injury of female in the same degree of collision accident is higher than that of male. Mainly taking men with standard shape as the matching object, the existing restraint system can not fully adapt to women's shape features, which leads to an increased risk of injury. According to the traffic accident statistics of NASS-CDC in 1998–2008, the probability of AIS 2+ injury to women's chest and spine is 38 and 67% higher than that of men respectively, in the case of keeping other collision conditions unchanged. At the same time, the probability of MAIS 3+ and MAIS 2+ injury to women is 47 and 71% higher than that of men respectively [98].

(4) Impact Injury Effect in Terms of Occupant Height

The results of sled test indicate that the head and chest injury indexes of mannequins with different heights are more discrete. Similar to the influence of gender factors on injury, individual height of passengers directly affects interaction between passengers and restraint system in the collision stage, resulting in the difference of injury forms. Under the same parameter configuration of seatbelt-airbag restraint system, it can be found that the injury index increases with the decrease of body shape, by comparing the weighted injury criteria (WIC) [99] of human models with different shapes.

2. Probabilistic Injury Modeling of Structure and Shape Distribution Features

The injury mechanical behavior of body tissue under external load is highly nonlinear and presents a certain randomness. Therefore, it is necessary to adopt probability modeling to obtain a damage model suitable for population distribution features.

(1) Scaling Measures for Injury Criteria of Non-standard Sign Mannequin

Due to the difficulty of acquiring experimental samples, injury criteria of non-standard sign mannequin is generally obtained by scaling experimental data of standard sign mannequin indirectly. Scaling is a means based on dimensional analysis. For two similar systems, the proportion coefficient between their responses can be established by basic attribute parameters. While ignoring the change of human tissue density with age, the scaling between different populations is mainly based on the ratio of geometric size to material elastic modulus (recorded as λL, λ_E). With these two basic proportion coefficients, the proportion relationship of some fundamental physical quantities is calculated as follows [100]:

$$\text{Time: } \lambda_{\text{T}} = \lambda L/(\lambda_{\text{E}})^{\wedge l/2} \tag{4.15}$$

$$\text{Acceleration: } \lambda_{\text{A}} = \lambda_{\text{E}}/\lambda L \tag{4.16}$$

$$\text{Force: } \lambda_{\text{F}} = \lambda L^{2*}\lambda_{\text{E}} \tag{4.17}$$

$$\text{Moment: } \lambda_{\text{M}} = \lambda L^{3*}\lambda_{\text{E}} \tag{4.18}$$

In the actual application, the injury thresholds of different percentile shapes can be deduced in terms of dimensions on the basis of proportion coefficients given above. The specific calculation formula is shown in Table 4.5.

(2) Injury Probability Function and Goodness-of-Fit Evaluation of Different Shapes Distribution

The research on material properties of soft tissues such as ligaments shows that the initial wave state and fracture property of fiber bundles vary with the individual, joint and even different ligaments in the same joint, which is finally reflected as the differences in the overall mechanical properties of ligaments [102]. The complexity is described statistically in terms of population distribution, involving different geometric micro and macro-scales. On the micro-scale, some scholars try to describe the distribution characteristics of fiber deformation initiation zone with normal distribution function to establish an analytical model of ligament fracture behavior [103]; On the structural scale, the shape and size of bone near the joint, the thickness of

Table 4.5 Calculation formula of proportional scaling coefficient of injury value and threshold of mannequin model [101]

Injury criteria	Injury threshold of hybrid III 50% quantile dummy	Calculation formula of scaling factor
Head injury criterion (HIC) HIC_{36}	1000	$(\lambda_E)^2/(\lambda L)^{1.5}$
Chest 3 ms acceleration a_{3ms}	60 g	$\lambda_E/\lambda L$
Chest compression C_{comp}	63 mm	λ_L
Neck forward bending moment $M_{in\_flex}(N_{ij})$	310 N m	$\lambda L^3 * \lambda_E$
Neck backward bending moment $M_{int\_ext}(N_{ij})$	135 N m	
Neck tensile force $F_{int\_ten}(N_{ij})$	6.81 kN	$\lambda L^2 * \lambda_E$
Neck compression force $F_{int\_comp}(N_{ij})$	6.16 kN	
Maximum tensile force of neck F_{neck_ten}	4.17 kN	
Maximum compressive force of neck F_{neck_comp}	4 kN*	
Peak thigh force (left, right) F_{fl}, F_{fr}	10 kN	

cortical bone and cartilage, the geometry of ligament and the shape of ligament all show obvious individual differences and distribution characteristics; On the overall scale, the geometric features of human bones or organs change with age, height and other body feature parameters [104]. The geometric distribution characteristics under the influence of population diversity can be illustrated by statistical shape analysis (SSA) [105].

Under the comprehensive action of influencing factors with different geometric scales, the individual differentiation in terms of mechanical behavior of body joint structure are reflected in impact response property and injury tolerance [106, 107]. Due to the fact that existing biomechanical experimental samples can only represent an extremely limited population, it is almost impossible to carry out experimental research on the differences of injury behaviors among people. With the development of numerical characterization technology and computing hardware in recent years, numerical approaches such as PA and grid transformation are expected to become a significant breakthrough for studying and indicating the uncertainty and distribution characteristics of biomaterials. In the injury research of body materials, certain errors inevitably exist in loading, boundary conditions, measuring equipment and test data used for model development and verification, Therefore, random approaches are applied for probability analysis, as it is believed that the input of body numerical model is subject to a specific random distribution, instead of being a certain value [108]. In the research of structural geometry, the human body model with parameterized adjustment of contour geometry can be established by the grid transformation,

Table 4.6 Probability distribution functions and cumulative distribution functions of four distributions

Distribution	Probability density function	Cumulative distribution function
Weibull distribution	$f(x) = \frac{\alpha}{\beta}\left(\frac{x-\gamma}{\beta}\right)^{\alpha-1} e^{\left[-\left(\frac{x-\gamma}{\beta}\right)^{\alpha}\right]}$	$F(x) = 1 - e^{\left[-\left(\frac{x-\gamma}{\beta}\right)^{\alpha}\right]}$
Normal distribution	$f(x) = \frac{e^{-\frac{1}{2}\left(\frac{x-\mu}{\sigma}\right)^2}}{\sigma\sqrt{2\pi}}$	$F(x) = \emptyset\left(\frac{x-\mu}{\sigma}\right)$
Logarithmic distribution	$f(x) = \frac{\mathrm{e}^{\left\{-\frac{1}{2}\left[\frac{\ln(x-\gamma)-\mu}{\sigma}\right]^2\right\}}}{(x-\gamma)\sigma\sqrt{2\pi}}$	$F(x) = \emptyset\left(\frac{\ln(x-\gamma)-\mu}{\sigma}\right)$
Logistic Distribution	$f(x) = \frac{\mathrm{e}^{\left(-\frac{x-\mu}{\sigma}\right)}}{\sigma\left[1+\mathrm{e}^{\left(-\frac{x-\mu}{\sigma}\right)}\right]}$	$F(x) = \frac{1}{1+\mathrm{e}^{\left(-\frac{x-\mu}{\sigma}\right)}}$

Note x represents the independent variable, α, β and γ are the shape functions of Weibull, μ and σ are the mean and standard deviation of the corresponding distribution

including the layer model of body parts (such as chest and lower limb bones, and internal organs) and the overall layer model [109].

The mechanical properties of biological tissue materials (elastic modulus, fracture strain, fracture stress, etc.) vary widely. In the field of biomechanics, the distribution functions commonly used to fit its injury risk include Weibull distribution, Normal distribution, Lognormal distribution and Logistic distribution (Table 4.6). These distribution forms represent the real potential damage distribution characteristics, that is, the functional relationship between injury index as the failure prediction factor or the observation X_i to infer whether injury occurs. Taking the movable joint as an example, it has the risk of injury (such as ligament rupture, and fracture) under the external rotating load, for which the structural failure risk function with angle and torque as the experimental observation value can be established [110]. Based on experimental design, the key ligament structure and mechanical parameters are sampled in the apriori parameter space, and the large-scale simulation matrix is built to obtain the original data with population injury uncertainty. And afterwards, the above four distribution functions are selected for fitting respectively, with the rotation angle and torque as failure predictors and the ligament fracture damage as observed values.

The fitting of injury risk function generally adopts maximum likelihood estimation to determine parameters. After obtaining the function, KS test (Kolmogorov–Smirnov test) is usually applied in statistical analysis to select the best fitting result with the observation data, so as to quantitatively compare the fitting degree between the test observation value and the hypothetical distribution. Assuming that the cumulative probability distribution function $F_n(x)$ obtained from the sample data comes from the same distribution as the specified cumulative distribution function $F(x)$, the observation value X_i (external rotation angle and external rotation torque) are sorted in turn to obtain the cumulative probability distribution function $F_n(x)$.

Where, $I_{[-\infty,x]}(X_i)$ (X_i) represents the indication function. If the observation value X_i is less than the value of the probability density function x, i.e., when $X_i \leq x$, $I_{[-\infty,x]}(X_i)$ $(X_i) = 0$, otherwise it is equal to 1.

The statistic D_n is calculated from the specified cumulative distribution function $F(x)$.

$$D_n = \sup_x |F_n(x) - F(x)| \tag{4.19}$$

where, D_n represents the upper bound of the distance between the observation data $F_n(x)$ and the given $F(x)$.

In Formula (4.19), n represents the sample size of sampling, and the critical value $D_n(\alpha)$ is obtained with reference to the table. When $D_n > D_n(\alpha)$, the original hypothesis is rejected. By selecting the distribution function corresponding to the smallest D_n, the cumulative probability distribution function that best fits the observation data can be chosen. An analysis based on the study of foot and ankle joint injury of vehicle occupants shows that Normal distribution has the best fitting result with external rotation torque as failure prediction factor to predict structural failure; For using external rotation angle as failure prediction factor to predict structural failure, Lognormal distribution has the best fitting effect [110] (Fig. 4.28).

3. Construction of Parameterized Human Body Numerical Model

With the development of numerical simulation technology, numerical approaches represented by finite element can accurately reproduce the mechanical behavior in the real world. In the field of traffic safety, the refined finite element mannequin verified by a large number of biomechanical tests has been widely used to illustrate the impact injury properties [111]. The establishment of a parameterized and controllable numerical body model facilitates the accurate and efficient analysis for the study of injury in different populations and situations.

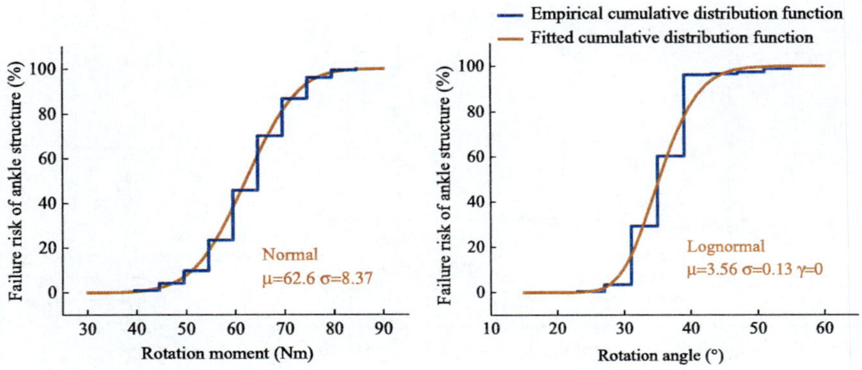

Fig. 4.28 Fitting results of failure risk function (taking ankle structure as an example)

(1) Fine Reconstruction of Soft and Hard Tissue Injury Behavior

The fine reconstruction of soft and hard body tissues (bone, muscle, ligament, cartilage, etc.) is the basis for the establishment of parametric numerical model as well as the study of probabilistic injury modeling. The establishment of refined numerical simulation model generally includes three parts, i.e., geometric reconstruction, material parameter determination and model verification. The following takes the foot and ankle joint as a typical movable joint to illustrate the fine modeling process of human body.

Based on the previous description of mechanical behavior of ligament tissue, it can be known that the mesostructure of ligament is composed of short bundles of fibrous connective tissues, with mechanical properties of transverse anisotropy. In order to truly simulate the contact between bones and the load transfer between bones, the mesostructure of ligament can be regarded as a bunch of fiber bundles parallel to each other [112, 113]. The benchmark model (usually a medium-sized adult model) selects major ligaments in the joint. For each ligament, a bundle of parallel and evenly distributed one-dimensional beam elements are used to represent its micro-fiber structure (beam element number is j, $j = 1, 2, \ldots, n_i$, n_i represents the number of beam elements in the i-th ligament). The width of ligament and the connection between its two ends and bone are determined according to anatomical feature of the bone, and followed by determination of the initial length of each beam element, which is recorded as l_{ij}^0. The finite element expression of ligament mechanical properties in finite elements is shown in Fig. 4.29.

In order to accurately simulate the mechanical response of ligament fiber bundle, the force–deformation curve of beam element is expressed by bilinear function in the simulation model, and the mechanical response features of ligament fiber bundle are simplified into "zero-force zone" and "loading zone" (Fig. 4.29). In a ligament, all beam elements have the same relative zero-force zone length C_i% and loading

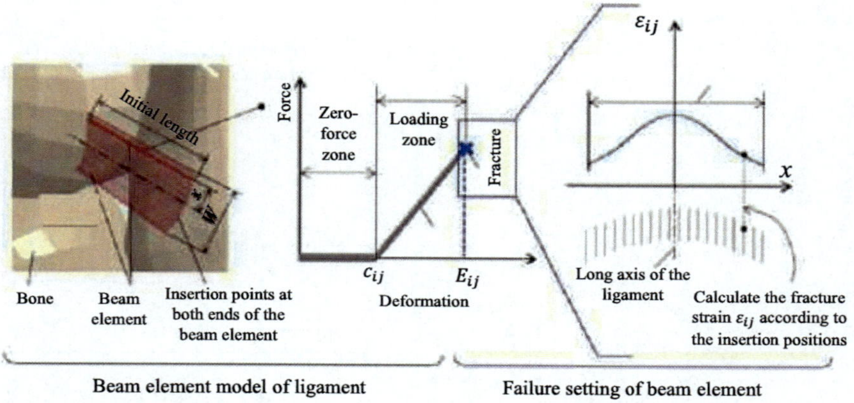

Fig. 4.29 Expression of mechanical properties of movable joint ligament in finite element mannequin

zone stiffness $k = 0.02$ kN/mm. The total stiffness of the ligament is represented by the number of beam elements, $K_i = n_i \cdot k$, and the fracture strain is used to control the fracture of beam element ε_{ij}. The relative zero-force zone $C_i\%$ and the number of beam elements n_i are determined according to the tests under different single loading conditions. The zero-force zone length C_{ij} and the final elongation E_{ij} of a single ligament are calculated by the following formula:

$$C_{ij} = l_{ij}^0 \times (1 + C_i\%) - l_{ij}^0 \tag{4.20}$$

$$E_{ij} = l_{ij} \times (1 + C_i\%) \times \left(1 + \varepsilon_{ij}\right) - l_{ij}^0 \tag{4.21}$$

When the ligament is loaded and enters the state of injury and fracture, the ligament fiber bundle always breaks gradually from the boundary of both sides to the middle due to the effect of stress concentration. In order to reproduce this damage behavior in the model, the fracture strain of beam element in each ligament ε_{ij} is defined as the normal distribution in width, that is, small fracture strain of the edge beam element and large fracture strain of the middle beam element. Based on current experimental results, the variation range of fracture strain is 12–18%. The results obtained from the bone-ligament-bone (BLB) test can verify nonlinear elastic behavior of the model and corresponding instantaneous elastic response coefficient. It is found that the model can also better predict bone motion and ligament fracture time.

The significant effect of age on injury can also be constructed in the numerical model. Taking the lower limb femur for example, a large number of test results have confirmed the influence of age on elastic modulus and ultimate stress, and verified the corresponding femur model [114, 115]. For given ages, the ultimate tensile stress coefficient of bone density is different, but their change trends are similar. That is to say, the ultimate tensile stress coefficient reaches the maximum at about 30 years old, and then decreases gradually with age. On the basis of Formula (4.5), expression relationship between the material mechanical properties and the age of compact bone is as follows:

$$E = 18 - 0.06A \tag{4.22}$$

$$\sigma_\mu = 136 - 0.46A \tag{4.23}$$

where, E represents the elastic modulus of compact bone of femoral skeleton (GPa); A represents age; and σ_{u} is the ultimate stress (MPa).

Researchers believe that the yield stress should be 0.9 of the ultimate stress σ_{y}. The calculation formula is as follows [116, 117]:

$$\sigma_{\mathrm{y}} = 0.9\sigma_{\mathrm{u}} = 0.9 \times (136 - 0.46A) \tag{4.24}$$

(2) Construction Approach of Parameterized Numerical Mannequin

Refined model reconstruction can characterize the mechanical behavior of biological injury in local joints or organs and tissues of human body. In order to further develop the parametric numerical finite-element model of the whole body, which changes with human body features, it is necessary to realize the synchronous customization adjustment for material properties and geometric forms. Applicable to a variety of load conditions, this kind of model fully reflects individual differences in the population, and enriches research means and tools of injury biomechanics. Model parametric adjustment includes three main steps, and the specific process is as follows (Fig. 4.30):

(1) Prediction of component geometric features based on specific anthropometric parameters: for known specific anthropometric parameters, such as height and weight, the corresponding population sign database can be used to convert the parameters into geometric measurement parameters of various body parts, including head circumference, chest circumference, lower limb bone length, etc. The bone geometric information and external geometric contour under given anthropometric parameters (height, BMI, age, gender, etc.) are obtained.
(2) Scaling deformation and posture adjustment: the reference model is scaled according to the external geometric contour and the geometric information of the bone, so as to obtain the mannequin of the target body feature and adjust the posture of mannequin.

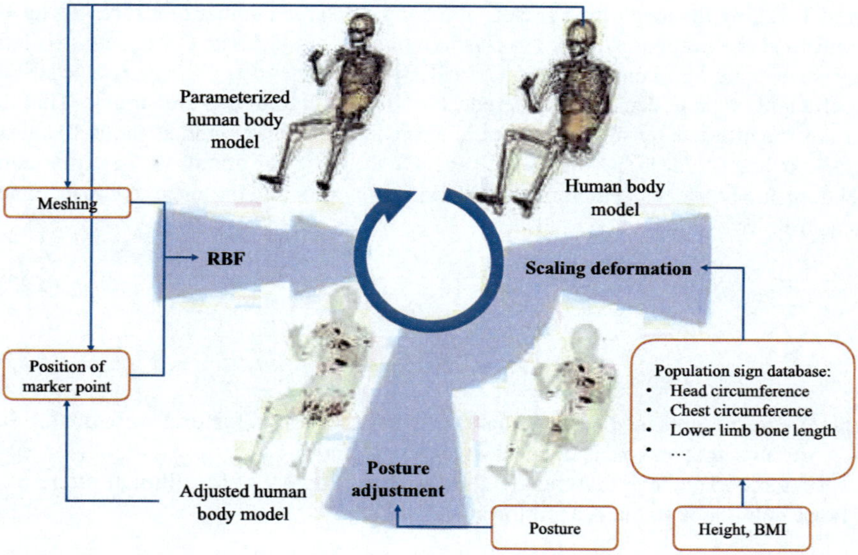

Fig. 4.30 Model parameterization flow diagram in terms of body shape feature and posture

(3) Repair of low-quality grid elements by means of grid transformation: the grid transformation of finite element model generally adopts the radial basis function, that is, the allocation of target position of marker point on the adjusted mannequin (target geometry). Mark points with the same quantity and similar positions are arranged on the mannequin (benchmark model, as the original geometry) with high-quality grid before adjustment as the starting position. Then, the grid sequence and division mode of the benchmark model are transformed to the target position by Radial Basis Function (RBF), and consequently the parameterized mannequin with high-quality grid is obtained. It is a key step to ensure geometric accuracy in the development of parametric mannequin. Its calculation principle is as follows.

Suppose that n marker points are selected in the set S on the original geometry and T on the target geometry. To calculate the weight coefficient λ_i and polynomial $p(x)$ in the interpolation process, the interpolation control points and boundary conditions are included in the formula.

$$\begin{bmatrix} A+\alpha I & B \\ B^T & 0 \end{bmatrix} \begin{pmatrix} \lambda \\ C \end{pmatrix} = \begin{pmatrix} T \\ 0 \end{pmatrix} \begin{pmatrix} T \\ 0 \end{pmatrix} \tag{4.25}$$

where, coordinate information matrix $A_{ij} = \varphi(r_{ij}) = r_{ij}^2 \log r_{ij}$, here i and j refer to the i-th and j-th mark points.

$$B = \begin{bmatrix} 1 & x_1 & y_1 & z_1 \\ \vdots & \vdots & \vdots & \vdots \\ 1 & x_n & y_n & z_n \end{bmatrix} \tag{4.26}$$

where, (x_i, y_i, z_i) is the spatial coordinate of the i-th node in the original geometry.

$$T = \begin{bmatrix} 1 & X_1 & Y_1 & Z_1 \\ \vdots & \vdots & \vdots & \vdots \\ 1 & X_n & Y_n & Z_n \end{bmatrix} \tag{4.27}$$

where, (X_i, Y_i, Z_i) is the spatial coordinate of the i-th node in the target geometry.

As the smoothing factor, α shall be accurately moved to the coordinate position of the corresponding mark point on the target geometry. λ And c is obtained by the solution Formula (4.25). For any point set S^* in the original geometry, the coordinates of the transformed corresponding point set T^* in the target geometry are calculated by the following formula:

$$\begin{pmatrix} T \\ 0 \end{pmatrix} = \begin{bmatrix} A^*+\alpha I & B^* \\ B^{*T} & 0 \end{bmatrix} \begin{pmatrix} \lambda \\ c \end{pmatrix} \tag{4.28}$$

where, A^* and B^* are generated based on the node coordinates of any point set S^* in the original geometry. It should be noted that $A^\times = r_{i,j^2} \log r_{ij}$, but $r_{ij} = \| x_i - \overline{x}_j \|$. x_i is the coordinate of the i-th node of any point set S^* in the original geometry, and $\overline{x}_j$ is the coordinate of the j-th mark point of the original mark point set S. The coordinates of the corresponding point set T^* corresponding to the point set S^* in the target geometry are obtained by Formula (4.28).

(3) Numerical Representation Strategy for Active–Passive Coupling Behavior of Human Body

The active response of vehicle occupants is reflected in the form of muscle contraction. The generated active force of muscle changes body stiffness and internal stress distribution in the process of collision, and affects the specific injury form and severity. It is difficult to investigate the muscle active effect in the study of biomaterial mechanics, which constrains the in-depth analysis of injury mechanism in real traffic collision. Owing to the rapid development of parametric numerical mannequin construction in recent years, the development of active mannequin with active posture control has become one of the most reliable research means to characterize the dynamic response of human in the overall accident scenario.

By means of the control over muscle activation level of numerical mannequin, the traditional mannequin can be used to further develop the active mannequin with the capability to represent active behavior. The muscle activation level can be measured by electromyography (EMG) signals. However, EMG can only collect the surface muscle activation level, instead of the deep muscle activation level. Therefore, it is necessary to determine the muscle activation level of specific body movement with different algorithms [118]. Referring to the establishment approach of muscle activation level, the corresponding behavior control strategies of active mannequin can be divided into open-loop and closed-loop (Table 4.7; Fig. 4.31). The open-loop control mannequin mostly characterizes its constitutive mechanism by Hill-type muscle model, and obtains the preset muscle activation level under specific conditions by using strategies such as optimal muscle contraction energy [119], inverse dynamics [120–122], and deep learning [123, 124]. It is mainly developed and verified based on EMG, motion posture, boundary contact force and other information from volunteer experiments. Most studies focus on the injury properties of occupant models with active muscle contraction, based on the experimental data collected by volunteers in simulated driving scenarios [125]. Taking the lower limb for example, the pushing behavior of occupant's lower limb before collision significantly affects the bone injury threshold or injury risk [126]. In recent years, based on human motion control mechanism and experimental data, researchers have successively proposed closed-loop active controllers [124, 127] with joint angle and muscle fiber length as control objectives to reproduce the active response changes of skeletal muscles in specific scenarios [128]. Taking the mannequin as an effective numerical carrier, the joint modeling of active response and passive injury behavior is carried out to restore the behavior characteristics in real hazardous conditions, to study the body injury impact under different collision strength and occupant posture, or to be used for virtual test on active and passive safety equipment and systems.

Table 4.7 Comparison of main mannequin controllers

Control strategy	Scenario adaptation	Prediction accuracy of muscle activation level	Prediction accuracy of motion posture	Complexity of controller	Adjustment difficulty of control parameters	Notes
Minimum energy consumption	Specific working conditions	Low	Unknown	Open loop	Easy	Too simple and low in prediction accuracy
Inverse dynamics	Specific conditions	High	Relatively high	Open loop	Easy	The algorithm parameters corresponding to the skeletal muscle model need to be established separately
PID	Specific conditions	Low	High	Medium	Medium	The controller parameters under specific working conditions should be calibrated
Optimize PID	Specific conditions	Medium	High	Relatively high	Relatively difficult	Optimize controller parameters
Functional differentiation and closed-loop control	Periodic loading	High	High	High	Difficult	Interdisciplinary field of motor neuroscience and biomechanics

4.2.2 Human-Vehicle Damage Tolerance and Quantitative Evaluation

The human-vehicle damage tolerance and quantitative evaluation includes three parts: traffic participants (human), vehicle body, and power system. The common injury parts of traffic participants in accidents and their corresponding injury degrees can be quantitatively described and evaluated by the injury index, and the injury risk probability can be inferred and established with the research data and empirical formula; The impact failure mode and mechanism of common vehicle lightweight materials can be represented by mechanical parameters, and the failure criterion can be inferred from mechanical properties of materials; A typical mechanism of runaway risk of new energy power system for IP is excessive local extrusion (such as electric core). Equivalent material properties and fracture mode can be obtained by material

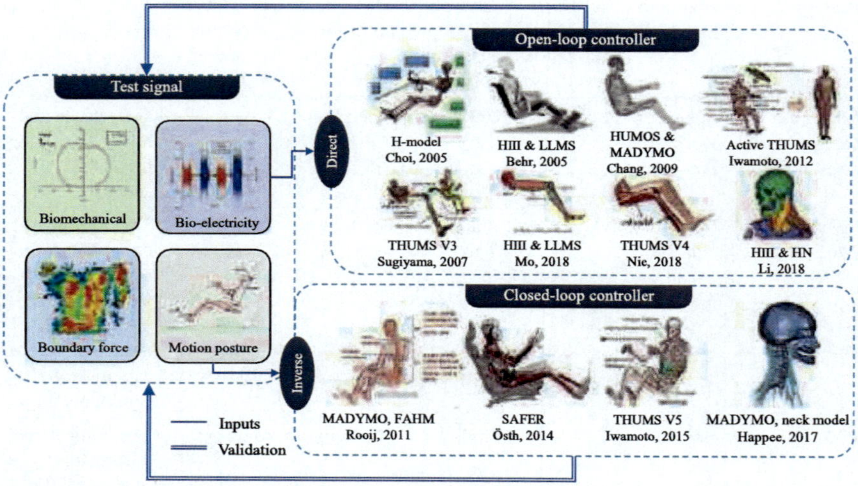

Fig. 4.31 Classification of active mannequin [123, 125–135]

mechanics experiment, then damage tolerance can be established and failure criteria can be inferred and calculated (Fig. 4.32). The construction of quantitative approach for the injury of traffic participants as well as the failure of vehicle body structure and new energy power system, together with the quantitative damage index and its corresponding risk relationship probability, is useful for evaluating injury risk of traffic participants, developing safety research tools, and assessing safety protection performance of vehicles, etc.

1. Injury Mechanism and Tolerance of Traffic Participants

Taking traffic participants as the protection object, injury refers to the deformation of anatomical body structure beyond its failure limit under external load, which results in tissue structure destruction or structural failure. The injury mechanism and failure limit of key body parts are important data references for vehicle safety assessment and traffic participant protection.

(1) Injury Tolerance and Severity

Injury tolerance, also known as injury threshold, refers to the critical value (generally the lowest value) of the injury index when the injury effect is caused by external load, which is used to indicate the bearing capacity of the protected object to external load. It is difficult to directly obtain and evaluate the injury type and severity of traffic participants. Therefore, it is necessary to convert the intuitively measurable physical parameters, such as acceleration, speed, displacement, compression or elongation, force and stress, into injury indexes corresponding to specific body parts (see Fig. 4.33 for common injury indexes).

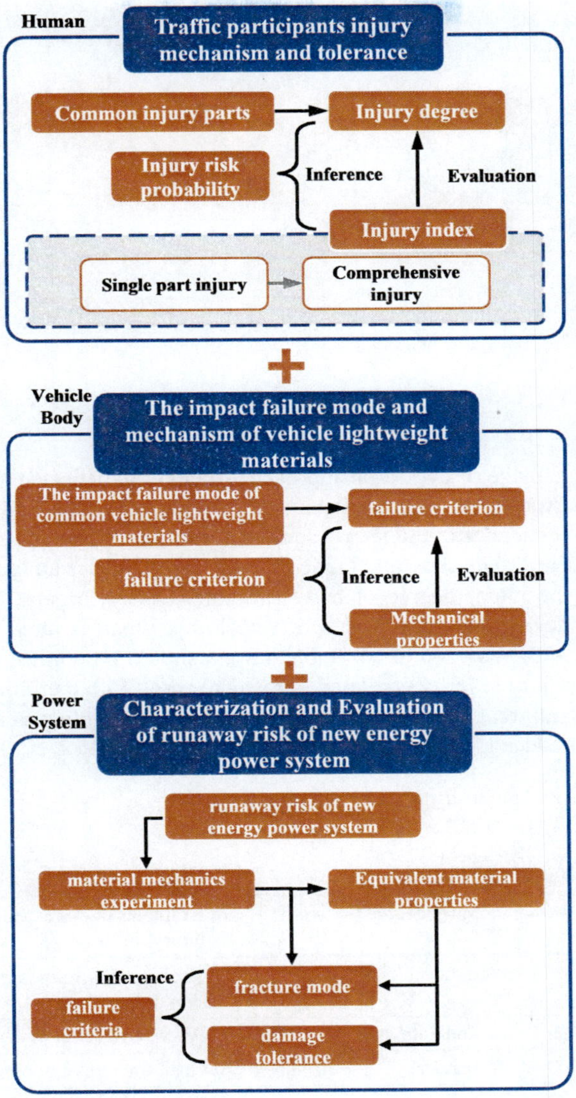

Fig. 4.32 Quantitative evaluation on damage tolerance of human-vehicle system

Brain, neck, thorax, limbs and pelvis are the common injury parts of vulnerable road users such as vehicle occupants and pedestrians in vehicle collision accidents. By analyzing the injury tolerance and establishing the relationship between injury index and injury risk, the injury risk probability of traffic participants in accidents can be intuitively presented and the injury risk of traffic participants in accidents can be quantitatively described. In the field of vehicle safety, the commonly-used

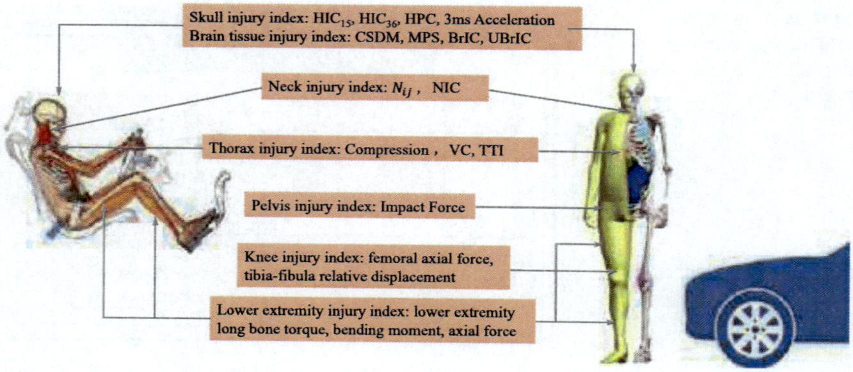

Fig. 4.33 Common injury evaluation indexes of traffic participants

damage risk research approaches include statistical analysis of accident data, corpse experiment, dummy experiment, numerical simulation and so on.

In addition to specific injury forms and indicators, AIS [136] is also commonly used to classify and grade injury severity. AIS is an index to quantitatively evaluate the injury degree of body structure, tissues, organs, etc. According to the injury degree and the threat to life, each injury part is rated as grade 1–6 and marked as AISI (I is 1–6 or 9) (Table 4.8). It should be pointed out that the classification of injury severity is related to medical care. With the improvement of medical quality and treatment system, the AIS level corresponding to specific injury forms may be reduced.

(2) Head Injury

Table 4.8 AIS rating

AIS	Injury degree	Examples of corresponding specific injury forms	Fatality rate (%)
1	Light	Superficial laceration of upper limb skin	0
2	Moderate	Fracture of sternum	1–2
3	Heavier	Open fracture of humerus	8–10
4	Serious, but not life-threatening	Tracheal perforation	5–50
5	Critical, with the possibility of death	Rupture of liver with tissue injury	5–50
6	Extremely heavy, basically unable to rescue	Disconnection of head and neck	100
9	Unknown injury (NFS)	–	–

The injury indexes for skull mainly include HIC, a_{3ms}, peak acceleration, etc., which are calculated by the time course curve of synthetic linear acceleration at the mass center of the head. The injury tolerance of 50% quantile adult men for HIC_{36} and HIC_{15} are 1000 and 700, respectively. Article 208 of the *Federal Motor Vehicle Safety Standards And Regulations* (FMVSS) [137] specifies the HIC threshold corresponding to all shapes of dummies in vehicle test (Table 4.9). In a given condition, the vehicle safety design shall contribute to protecting of the head HIC of corresponding occupant from exceeding this threshold. The injury probability of HIC under corresponding AIS level can be calculated by Formula (4.29). Figure 4.36a shows the injury risk curves of HIC_{36} and AIS 2+, AIS 3+, AIS 4+, and the calculation formula of injury risk probability (p) corresponding to different AIS severity is as follows:

$$P(\text{AIS}X+) = \Phi\left[\frac{\ln(HC) - u_x}{\sigma_X}\right] \tag{4.29}$$

where, Φ Represents the cumulative probability density function of Normal Distribution; $X = 2$–4, $\mu_2 = 6.96352, \sigma_2 = 0.84664$; $\mu_3 = 7.45231, \sigma_3 = 0.73998$; $\mu_4 = 7.65605, \sigma_4 = 0.60580$ [138].

In some regulations and test standards, HIC is also called as Head Performance Criterion (HPC). The Pedestrian Testing Protocol of European NCAP, the European Enhanced Vehicle Safety Committee Working Group 17 (EEVC-WG17) as well as

Table 4.9 HIC thresholds for different body shapes specified in FMVSS 208

HIC type	50% Quantile male	5% Quantile female	Six-year-old children	Three-year-old children	One-year-old baby
HIC_{36} threshold	1000	1000	1000	900	660
HIC_{15} threshold	700	700	700	570	390

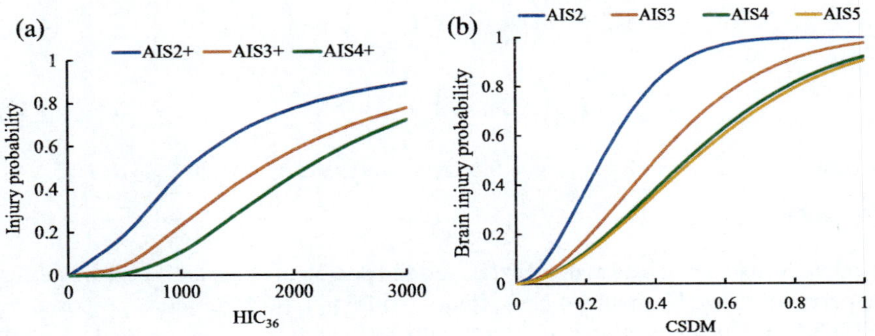

Fig. 4.34 Injury risk relationship curve: **a** head injury criterion HIC skull fracture (applicable to HIC_{15} and HIC_{36}), **b** CSDM brain injury risk curve [138, 139]

China's national standard GB 11551-2014 for the protection of the occupants in the event of a frontal collision for motor vehicle [140] all adopt HPC as the head injury evaluation index. GB 11551-2014 stipulates that the cumulative time of head combined acceleration greater than 80 *g* in collision test shall not exceed 3 ms.

The injury indexes such as HIC and 3 ms acceleration only take into account the head linear load, instead of the head rotation load, which may lead to brain injuries such as concussion, diffuse axonal injury, subdural hematoma, brain contusion and intracerebral hemorrhage. Common brain injury indexes include Cumulative Strain Damage Measure (CSDM), Maximum Principal Strain (MPS), Brain Injury Criteria (BrIC) in terms of rotational angular velocity[139], Universal Brain Injury Criterion (UBrIC) [141], etc. Among them, CSDM is based on the assumption that Diffuse Axonal Injury (DAI) is related to the tensile strain accumulation experienced by brain tissue at a given critical value, and the occurrence of injury can be predicted by monitoring the strain accumulated in the brain. The injury risk probability (*P*) of CSDM and DAI is established by means of Survival Analysis, as shown in Formula (4.30). λ and k are the scale parameters and shape parameters of Weibull Distribution, respectively. The injury risk relationship between CSDM and different AIS levels is calculated as follows [139] (Fig. 4.34b).

$$P(\text{AIS}X{+}) = 1 - e^{-\left(\frac{\text{CSDM}}{\lambda_X}\right)^k} \tag{4.30}$$

where, $X = 2$–5, $k = 1.8$; $\lambda_2 = 0.3$; $\lambda_3 = 0.49$; $\lambda_4 = 0.6$; $\lambda_5 = 0.624$.

BrIC can assess the risk of brain injury due to skull rotation. Formula (4.31) is the calculation of *BrIC*, where ω_x, ω_y and ω_z represent the maximum angular velocity of the head rotating around the three axes of the local coordinate system, ω_{xc}, ω_{yc}, ω_{zc} = s represent the critical value of the maximum angular velocity rotating around the *x*, *y* and *z* axes respectively ($\omega_{xc} = 66.25$ rad/s, $\omega_{yc} = 56.45$ rad/s, $\omega_{zc} = 42.87$ rad/s). The risk of brain injury rises with the increase of *BrIC* value. The probability of injury risk is shown in Formula (4.33).

$$BrIC = \sqrt{\left(\frac{\omega_x}{\omega_{xc}}\right)^2 + \left(\frac{\omega_y}{\omega_{yc}}\right)^2 + \left(\frac{\omega_z}{\omega_{zc}}\right)^2} \tag{4.31}$$

$$P(AISX) = 1 - \mathrm{e}^{-\left(\frac{BrIC}{A_X}\right)^{2.84}} \tag{4.32}$$

where, $X = 1$–5, $A_1 = 0.120$; $A_2 = 0.602$; $A_3 = 0.987$; $A_4 = 1.204$; $A_5 = 1.252$.

(3) Neck Injury

Vehicle collision will lead to rapid extension and flexion of occupant's neck, resulting in cervical injury. Normalized Neck Injury Criterion (N_{ij}) is an index proposed by NHTSA to evaluate neck injury vehicle collision accidents, and is applied to safety regulations or tests such as FMVSS 208. N_{ij} is composed of the axial force and bending moment measured at the occipital condyle of human body, as shown in

Formula (4.33).

$$N_{ij} = \frac{F_z}{F_{\text{int}}} + \frac{M_y}{M_{\text{int}}} \tag{4.33}$$

where, F_z represents the axial force, M_y represents the bending moment of flexion/extension, the subscript "int" refers to the "intercept" of the load and moment intersecting the axis respectively, and j is the four load conditions of the neck, including Tension-Extension (NTE), Tension-Flexion (NTF), Compression-Extension (NCE) and Compression-Flexion (NCF). The critical values of dummies with various body shapes are shown in Tables 4.5 and 4.10. The damage risk probability corresponding to each AIS level is shown in Formula (4.35). FMVSS stipulates that any one of the four loads of N_{ij} shall not be larger than 1.0 [142]. In addition to N_{ij}, studies have also applied the Neck Injury Criterion (NIC) [143] to predict neck whiplash injury, which will not be described here.

$$P(\text{AIS } X+) = \frac{1}{1 + \mathrm{e}^{A_X - B_X \times N_{ij}}} \tag{4.34}$$

where, $X = 2$–5, $A_2 = 2.054$, $B_2 = 1.195$; $A_3 = 3.227$, $B_3 = 1.969$; $A_4 = 2.693$, $B_4 = 1.195$; $A_5 = 3.817$, $B_5 = 1.195$.

(4) Chest Injury

Common chest injury indexes include chest compression rate, Viscous Criterion (VC), Thoracic Trauma Index (TTI), etc. For the chest injury caused by frontal collision, the chest compression rate $C_{\max}$ is the relative ratio of the central deformation of sternum to the chest thickness. It has a high correlation with the forms of chest injury due to compression, such as thoracic fracture or lung injury. Chest acceleration can also be regarded as one of the injury criteria, generally based on the synthetic acceleration measured at the fourth thoracic vertebra (T4) (duration is 3 ms). The amount of chest deformation cannot accurately represent the risk of chest injury during high-speed impact. For example, when the amount of chest compression remains unchanged, the higher the impact speed is, the higher the degree of injury is. Based on this fact, the VC takes into account the influence of loading speed and is obtained by the product of chest deformation speed $V(t)$ and instantaneous

Table 4.10 Critical values of normalized neck injury criterion

Shapes of dummy	Tension (N)	Compression (N)	Flexion (N m)	Extension (N m)
One-year-old baby	1465	1465	43	17
Three-year-old children	2120	2120	68	27
Six-year-old children	2800	2800	93	39
5% Quantile female	3370	3370	155	62
50% Quantile male	4500	4500	310	125

compression $D(t)$, as shown in Formula (4.35). The test data show that when the VC reaches 1 m/s, the probability of serious chest injury (AIS 4+) in adults achieves 25% [144]. According to the corresponding national standard GB 11551-2014, the VC shall not exceed 1 m/s [140].

$$VC = V(\mathrm{t}) \cdot C(\mathrm{t}) = \frac{\mathrm{d}[D(\mathrm{t})]}{\mathrm{d}t} \cdot \frac{D(\mathrm{t})}{D_0} \tag{4.35}$$

where, $D(t)$ is the time-varying chest compression, and D_0 is the initial thickness of the chest.

TTI is the chest injury criterion for side collision, which is defined by the peak acceleration of the lower spine and the maximum acceleration at the ribs, as shown in Formula (4.36). The thoracic injury thresholds of *TTI* in adults and children are 85 and 60 g, respectively.

$$TTI = 1.4 \times AGE + 0.5 \times (RIB_{\mathrm{acc}} + T12_{\mathrm{acc}}) \times \frac{M_{\mathrm{s}}}{M_{\mathrm{std}}} \tag{4.36}$$

where, *AGE* represents the age (year); RIB_{acc} represents the acceleration of the fourth and eighth ribs on the side pushed by the external load (g); $T12_{acc}$ represents the maximum lateral acceleration at the lower spine (g); M_s represents the weight of the subject (kg); M_{std} is the standard body weight (the standard weight of 50% quantile body is 75 kg).

(5) Pelvic Injury

Pelvic fracture and dislocation are also common forms of injury in vehicle collision accidents. Pelvic fracture will seriously affect driver's ability for vehicle control. The main injury mechanism is the eccentric femoral axial load led by the collision between knee and knee pad. The pelvis is a relatively rigid bone structure. Compared with the injury index based on acceleration or compression, the correlation between pelvic fracture and impact force is higher. There are few biomechanical data of direct loading on the front of pelvis in the existing studies. The current tests show that in the frontal collision, the impact force range leading to hip fracture is very large, which is 6.2–25.6 kN. The large difference may be due to the different impact speed and stiffness of impact device used by different testers. Therefore, although the load-based pelvic frontal impact injury standard has been proposed, the injury tolerance has not been evaluated.

The pelvic injury criterion and tolerance capacity in side collision shall also be calibrated by a considerable number of impact tests. In the existing test reports, the minimum load of female pelvic fracture is between 4.6 and 4.9 kN. The tolerance corresponding to injury level AIS 1–3 is 4, 6 and 10 kN respectively (Table 4.11). The pelvic side collision load and the injury risk probability of AIS 2+ and AIS 3+ are shown in Formulas (4.37) and (4.38).

$$P(\mathrm{AIS}\ 2+) = \frac{1}{1 + \mathrm{e}^{6.806 - 0.0089 \cdot \mathrm{age} - 0.0007424 \cdot F}} \tag{4.37}$$

$$P(\text{AIS }3+) = \frac{1}{1 + e^{9.7023-0.04678\cdot\text{ age }-0.0005\cdot F}} \tag{4.38}$$

where F is the pelvic side collision load (N).

(6) Injury of Long Bone in Lower Limb

The injuries of lower limbs in vehicle collision accidents include long bone fracture and knee ligament fracture. The common injury indexes of long bone fracture include bending moment of femur and tibia section, axial compression force, bending load, etc. Injury tolerance is shown in Table 4.12 [145, 146].

Formulas (4.39) and (4.40) show the calculation for the injury risk probability of the femoral axial load and AIS 2+ and AIS 3+ respectively (where F is femoral axial load). Formulas (4.41) and (4.42) are the injury risk probability of axial load at the upper end of tibia and axial load at the lower end of tibia and AIS 2+ and AIS 3+, respectively.

$$P(\text{AIS }2+) = \frac{1}{1 + e^{5.7949-0.5196\cdot F}} \tag{4.39}$$

$$P(\text{AIS }3+) = \frac{1}{1 + e^{4.9795-0.326\cdot F}} \tag{4.40}$$

$$P(\text{AIS }2+) = \frac{1}{1 + e^{(0.5204-0.8189\cdot F+0.068\cdot\text{mass})}} \tag{4.41}$$

$$P(\text{AIS }2+) = \frac{1}{1 + e^{4.572-0.670\cdot F}} \tag{4.42}$$

(7) Knee Injury

Table 4.11 Tolerance for pelvic side collision injury criteria

Injury level	Injury tolerance lateral load (kN)	Corresponding AIS level	Corresponding regulations	Euro NCAP side collision
0 (No injury)	<4.0	0		
1 (No injury)	<4.0	1		<3 kN: Green 4 kN: Yellow
2 (Minor pelvic fracture)	4.0–6.0	2–3	6kN European side collision (96/27/EC)	5 kN: Orange 6 kN: Brown
3 (Large pelvic fracture)	6.0–10.0	4–5	–	>6 kN: Red
4 (Comminuted pelvic fracture)	>10.0	6	–	–

Table 4.12 Injury tolerance of long bone of lower limb [151]

Long bone of lower limb	Femur		Tibia		Fibula	
Gender	Male	Female	Male	Female	Male	Female
Torque (N m)	175	136	89	56	9	10
Bending load (kN)	3.92	2.58	3.36	2.24	0.44	0.30
Average bending moment (N m)	310	180	207	124	27	17
Axial compression (kN)	7.72	7.11	10.36	7.47	0.60	0.48

For vulnerable road users such as pedestrians in collision accidents, the direct impact force of the bumper on the knee joint or the force transmitted by the long bone load through the knee joint will result in knee ligament injury, such as the rupture of anterior cruciate ligament (ACL), medial collateral ligament (MCL), posterior cruciate ligament (PCL), lateral collateral ligament (LCL), etc. [146, 147] Their injury tolerance is shown in Table 4.13. In addition, relevant studies have tested the ultimate tensile stress and maximum tensile strain of knee ligament soft tissue, and the fracture strain is at the level of 12–18% (47) (Table 4.14).

(8) Whole-Body Injury

The above contents are distinguished in terms of different body parts, and the injury tolerance and risk curve of each part are quantified respectively. Traffic participants usually have more than one injury site in the accident. Based on the AIS rating approach, three comprehensive injury severity indicators for the whole body have

Table 4.13 Injury tolerance of knee joint and ligament [146]

Position	Injury index	Injury tolerance		
		5% percentile adult female	50% percentile adult male	95% percentile adult male
Knee joint	Axial load of femur	6510 N	9040 N	10580 N
Knee ligament	Relative translation of tibia-fibula	13 mm	15 mm	16.5 mm

Table 4.14 Ultimate stress and strain of knee ligament [47]

Position	Tensile tangent modulus (MPa)	Ultimate tensile stress (MPa)	Maximum tensile strain (%)
MCL [148]	332.2 ± 58.3	38.6 ± 4.8	17.1 ± 1.5
PCL [149]	248.0 ± 119.0	35.9 ± 15.2	18.0 ± 5.3
ACL, MCL and LCL [150]	345.0 ± 22.4	36.4 ± 2.5	15.0 ± 0.8
Patellar ligament [150]	643.1 ± 53.0	68.5 ± 6.0	13.5 ± 0.7

been evolved: the Maximum Abbreviated Injury Scale (MAIS), Injury Severity Score (ISS) and New Injury Severity Score (NISS). MAIS directly takes the highest level of AIS among all injuries of whole-body. ISS more comprehensively considers the multiple injured parts of the body [151], and sums the AIS ratings of the three most seriously injured parts of the body, as shown in Formula (4.43). With a value between 1 and 75, it stipulates that when AIS is 6 in any part, ISS is 75.

$$
\begin{aligned}
ISS = &\left(MaxAIS\ severity_{\text{bodyregion1}}\right)^2 + \left(MaxAIS\ severity_{\text{bodyregion2}}\right)^2 \\
&+ \left(MaxAIS\ severity_{\text{bodyregion3}}\right)^2
\end{aligned}
\tag{4.43}
$$

NISS squares the AIS levels of the three most serious injuries, regardless of whether these injuries occur at the same injured body part [152, 153]. Studies have shown that NISS performs better than ISS in predicting the viability of the injured [152, 154]. In addition, NISS can better predict the length of hospital care and the organ failure degree of patients with multiple fracture injuries [155, 156].

1. Impact Deformation Failure Criterion of Lightweight Materials and Structures

The performance of vehicle collision safety protection is affected by the body material and structure. In order to evaluate the damage risk of human-vehicle system, it is necessary to quantify the impact deformation failure risk of body materials or structures. The application of lightweight materials is a significant aspect of the clean development of vehicles in the future. On the one hand, vehicle lightweight technology is an important means to achieve energy conservation and emission reduction. For every 100 kg reduction in vehicle mass, 0.6 L fuel can be saved per 100 km; On the other hand, from the perspective of safety, lightweight technology is conducive to reduce its own impact kinetic energy under collision conditions. On the premise of reasonable design, it is expected to reduce the risk of personal injury. The development and application of lightweight materials is one of the key technologies in vehicle design. Since the twenty-first century, lightweight materials such as high-strength steel, aluminum alloy, magnesium alloy, plastic materials and composites have been more and more widely used in vehicles. Figure 4.35a, b show the proportion of vehicle materials used in 1977 and 2007 respectively [157]. The wide application of lightweight materials also puts forward high requirements for component design. Accurate characterization of material mechanical properties is the premise of fine analysis on vehicle collision safety. However, the mechanical properties and yield criteria of lightweight materials are obviously different from those of traditional metal materials. Their basic mechanical properties have been introduced earlier. Here, the failure criteria of vehicle high-strength steel, aluminum alloy, magnesium alloy, plastic and composite materials are mainly illustrated.

(1) Failure Criteria of High-strength Steel and Aluminum Alloy

Widely used in modern vehicle body structure, high-strength steel, aluminum alloy and other metal materials are the major collision energy absorption and strengthening components, and play an important role in vehicle safety protection performance.

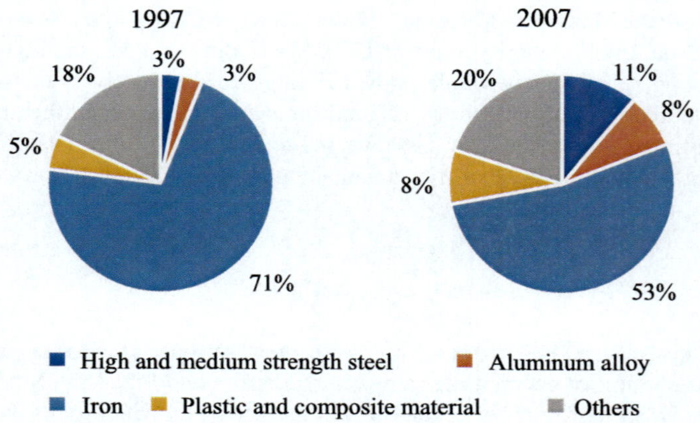

Fig. 4.35 Distribution of proportion of used vehicle materials in 1977 and 2007

After years of experimental research, the corresponding material failure criterion model is also relatively mature, and has been applied in collision safety design and safety evaluation. Related models are usually named after the originator. When describing the mechanical behavior of some new lightweight materials used in vehicles, the existing material models are often used or partially improved as a basis. High-strength steel and aluminum alloy materials have the property of toughness, which failure is often illustrated by ductile fracture criteria in current studies. Considering whether the plastic behavior is affected by material damage, the characterization strategies can be divided into damage coupling and damage decoupling type. Represented by Gurson model, the damage coupling model shows that after a certain degree of plastic deformation, the damage gradually accumulates a certain amount and the material breaks. For example, an index D is defined to measure damage. When the value of D reaches 1 from 0, the material breaks, and the stress during this period shall be converted according to the value of D. The calculation formula is as follows:

$$\sigma = (1 - D) \cdot \overline{\sigma} \tag{4.44}$$

where, $\overline{\sigma}$ represents the equivalent stress without damage.

The damage decoupling model considers that the plastic behavior and fracture behavior are independent from each other, and does not consider the stress conversion, so it is easier to use the model. There are many fracture criteria that can be used in the damage decoupling model. For example, the criteria include such simple forms as maximum shear stress and maximum plastic strain, and such more complex ones as Johnson–Cook model, Xue–Wierzbick model and Cockcroft–Latham model. Commonly applied as fracture criterion at present, Mohr–Coulomb criterion is regarded as an extension of the maximum shear stress model, and its expression is:

$$\left(\sqrt{1+c_1^2}+c_1\right)\sigma_1^p - \left(\sqrt{1+c_1^2}-c_1\right)\sigma_3^p = 2c_2 \quad (4.45)$$

where, c_1 and c_2 represent undetermined parameters; σ_1^p and σ_3^p represent the maximum and minimum principal stresses respectively. It is not difficult to find that when c_1 is equal to 0, the criterion degenerates to the maximum shear strain criterion.

Bai and Wierzbicki et al. [158] modified the criterion with the plastic model related to stress state, and obtained the Modified Mohr–Coulomb (MMC) model. The analytical form of the empirical model is:

$$\overline{\varepsilon}_{\mathrm{f}} = \left\{\frac{A}{c_2}\left[\sqrt{\frac{1+c_1^2}{3}}\cos\frac{\overline{\theta}\pi}{6} + c_1\left(\eta + \frac{1}{3}\sin\frac{\overline{\theta}\pi}{6}\right)\right]\right\}^{-\frac{1}{n}} \quad (4.46)$$

where, A, k, n, c_1 and $c_2$represent the parameters that shall be fitted and estimated according to the experimental results.

Vehicle body parts are often in the form of thin-walled structure, in which the stress state can be approximately regarded as plane stress state. At this time, the relationship between stress triaxiality and Lode angle parameters is as follows:

$$\xi = \cos\left[\frac{\pi}{2}(1-\overline{\theta})\right] = -\frac{27}{2}\eta\left(\eta^2 - \frac{1}{3}\right) \quad (4.47)$$

By combining Formulas (4.46) and (4.47) and eliminating the Lode angle parameter, the fracture criterion curve of equivalent fracture strain varying with stress triaxiality of MMC model under plane stress state can be deduced. The test indicates that MMC model can be used to describe the fracture criteria of various grades of aluminum alloy [159] and high-strength steel [160], and can better predict the fracture behavior of these materials under different stress states.

(2) Failure Criteria of Magnesium Alloy

Die-casting magnesium alloy is one of the lightweight materials widely used in vehicle parts at present. With fine fluidity and high production efficiency in the casting process, magnesium alloy materials are conducive to the mass production of vehicle parts with complex shapes. The constitutive models characterizing the elastic–plastic mechanical behavior of die-casting magnesium alloy are mostly directly based on the models for traditional body structural materials, and also are improved accordingly. The following will take AM60 (Mg–Al–Mn Series) die-casting magnesium alloy as an example to introduce its elastic–plastic mechanical behavior and failure criteria.

Dørum et al. [161] carried out four experiments on AM60 die-casting magnesium alloy, including unidirectional tension, plane strain tension, shear and notch tension. At the same time, Weber distribution is applied for the random distribution and description on the discreteness of its fracture behavior, and the Cockcroft–Latham fracture criterion involving the influence of deformation zone size is further built. When the volume of material deformation zone is V and the plastic work is W, the calculation of fracture probability P is shown in Formula (4.48).

$$P(W) = 1 - \exp\left[-\left(\frac{V}{V_0}\right)\left(\frac{W}{W_{c0}}\right)^m\right] \tag{4.48}$$

where, V_0 and W_{c0} represent reference volume and reference plastic work respectively; m represents Weibull modulus, which shall be determined according to the experimental results.

The calculation formula of plastic work (W) is as follows:

$$W = \int \max(\sigma_1, 0)\mathrm{d}\overline{\varepsilon}_\mathrm{p} \leq W_\mathrm{c} \tag{4.49}$$

where, σ_1 represents the first principal stress at the corresponding position in the material, that is, the maximum principal stress; $\mathrm{W_c}$ represents the key fracture parameter of the criterion, and $\overline{\varepsilon_p}$ is the equivalent plastic strain.

According to Cockcroft Latham fracture criterion, if the first principal stress σ_1 is less than 0 under the stress state dominated by compression, the material will not break, which may be inconsistent with the actual situation. Dørum et al. believe that the criterion has limited application, as it does not take the shear fracture mechanism into account [162].

(3) Failure Criteria of Plastic Materials

There have been a lot of researches on the mechanical properties of plastic materials for vehicle collision, including plastic deformation, mechanical response under dynamic loading, toughness failure and so on. The yield behavior of plastic materials is affected by hydrostatic pressure, temperature and strain rate [163]. The yield criterion of some amorphous plastics and crystalline plastics is different from that of traditional metal materials, and their yield stress increases with the increase of hydrostatic pressure [164]. In order to describe the effect of hydrostatic pressure on yield behavior, the first stress invariant is introduced into the yield criterion. Formulas (4.50) and (4.51) are the modified von Mises yield criterion and the modified Tresca yield criterion proposed earlier [165].

$$\begin{aligned} \tau_\mathrm{oct} &= \tau^0_\mathrm{MMC} - \frac{\mu_\mathrm{MMC}}{3} J_1 \\ \tau_\mathrm{oct} &= \frac{\sqrt{6J_{2\mathrm{D}}}}{3} \end{aligned} \tag{4.50}$$

where, τ_oct represents octahedral shear stress; J_1 represents the first stress invariant; $J_{2\mathrm{D}}$ represents the second invariant of deviatoric stress tensor, $I_{2D} = \frac{1}{2}S_{ij}S_{ij}$, $S_{ij} = \sigma_{ij} - \frac{\sigma_{kk}}{3}\delta_{ij}$, S_{ij} represents the component of deviatoric stress tensor, δ_{ij} represents Kronecker function; τ^0_MMC and μ_MMC are all material parameters.

$$\begin{aligned} \tau_{\mathrm{T}} &= \tau_{\mathrm{MT}}^{0} - \tfrac{\mu_{\mathrm{MT}}}{3} J_1 \\ \tau_{\mathrm{T}} &= \tfrac{1}{2}\sup(\sigma_i - \sigma_j) \end{aligned} \tag{4.51}$$

where, τ_{MT}^{0} and μ_{MT} refer to material parameters; σ_i and σ_j refer to the principal stress.

The experimental data of various plastics indicate that the yield behavior of all plastics with relatively uniform deformation, such as polyvinyl chloride (PVC), epoxy (EP) and high-density polyethylene (HDPE), conforms to the modified von Mises criterion described in Formula (4.50). For materials with uneven deformation due to shear band, such as polyethylene terephthalate (PET), their yield behavior conforms to the modified Tresca yield criterion described by Formula (4.51).

In order to describe the viscoplastic response of thermoplastic under any stress state, the third invariant of deviatoric stress j J_{3D_2} $J_{3D} = \det|S_{ij}|$ is introduced into the expression of yield function. J_{3D} is introduced into the yield function, considering that the stress of plastic flow caused by shear is less than the tensile or compressive stress. The expression of yield function is as follows:

$$f = \frac{3J_{2D}}{\sigma_{st}}\psi(J_{2D}, J_{3D}) + \frac{7(m-1)}{8}J_1 - \frac{7m}{8}\sigma_{st} \tag{4.52}$$

where, $\psi\left(J_{2D,} J_{3D}\right) = 1 - \frac{27}{32}J_{3D}^2/J_{2D}^3$; $m = \sigma_{sc}/\sigma_{st}$, σ_{sc} and σ_{st} represent compressive yield stress and tensile yield stress respectively.

The introduction of the third invariant of deviatoric stress does not affect the yield stress under uniaxial tension, uniaxial compression and pure hydrostatic pressure. For the experimental data of yield point of such materials as polymethylmethacrylate (PMMA), polycarbonate (PC) and polystyrene (PS), the improved yield criterion is obviously more accurate than other criteria.

Although the yield strength of plastic materials is closely related to hydrostatic pressure, the relationship between yield function and hydrostatic pressure is different for different plastic materials. Therefore, for different plastic materials, it is necessary to carry out material experiments under various stress states to determine the yield function as accurately as possible.

(4) Mechanical Properties and Failure Criteria of Composites

With the progress and popularization of composite technology, composites are more and more frequently used in vehicle components, such as engine hood, interior trim panel and suspension spring. Compared with traditional metal materials, the deformation and fracture of composite materials is a mechanical process with countless damage forms of micro cracks with complex geometry, and its mechanical behavior mechanism is very complicated. Continuous damage mechanics can be better adopted to characterize the damage behavior of composites. By considering the micro crack propagation from the macro point of view, the difficulty of characterization due to the complex geometry and uneven spatial distribution of micro cracks is solved. The continuous damage mechanics introduces the damage factors in the direction of fiber and matrix. Taking the model of Formula (4.53) as an example, the model contains three damage factors, two of which represent the damage in the direction of fiber and matrix and the other represents the damage in the shear direction. Each damage factor has the following forms:

$$d = 1 - \exp\left[-\frac{1}{me}\left(\frac{\varepsilon}{\varepsilon_f}\right)^m\right] = 1 - \exp\left[-\frac{1}{me}\left(\frac{E\varepsilon}{X}\right)^m\right] \tag{4.53}$$

where, E is the modulus of elasticity; ε_f represents normal failure strain; X represents the normal failure strength; e is the natural logarithm base; The parameter m determines the shape of the damage evolution curve. The larger m is, the more significant the strain softening phenomenon (i.e. the stress gradually decreases after the initial peak stress) is. While the smaller m is, the more stable the strain softening process is.

If the bimodal Weber distribution is further used to describe the strength distribution of fibers, the damage evolution equation is as shown in Formula (4.54). According to the model theory, unidirectional composites are fiber bundles wrapped around the matrix, which are expressed by one-dimensional damage constitutive equation.

$$d = 1 - \exp\left[-\left(\frac{E\varepsilon}{\sigma_{01}}\right)^{m_1} - \left(\frac{E\varepsilon}{\sigma_{02}}\right)^{m_2}\right] \tag{4.54}$$

where, m_1, m_2 and σ_{01}, σ_{02} represent the shape and scale parameters of the double Weber distribution.

With the continuous improvement of damage mechanics model, it has become an important way to study the damage failure process and failure criterion of composite materials with numerical methods such as finite element. For example, Lavedeze model [166] is widely used to study fiber-reinforced composites. It contains two damage factors representing the matrix and shear direction. In view of the related effects of compression and temperature, it can better predict the collision response of braided fiber-reinforced composites.

3. Expression and Evaluation of Runaway Risk of New Energy Power System

The thermal runaway of new energy power system will result in serious damage to the vehicle body and occupants. If occupants do not leave the spot in time in a collision accident, they are likely to be burned by the fire due to the runaway power battery. Therefore, it is necessary to describe and evaluate the runaway risk of power system dominated by lithium batteries, so as to design a reasonable layout of power batteries and facilitate subsequent runaway early warning and flame retardant and explosion suppression technique.

A typical damage condition of electric vehicle in collision accident is short-circuit failure caused by local extrusion of electric core, which is also the most dangerous loading form. Based on a large number of experimental observations and a detailed analysis of the internal damage accumulation of different types of batteries such as soft pack battery and square hard shell battery under extrusion condition, it is found that the change feature and fracture mode of macro mechanical response are highly related to the feature of short circuit in the battery. There is a multi-layer structure inside the battery, with feature scale across multiple orders of magnitude. The thickness of electrode collector is 10–20 μm, and the length of the battery cell is 100–200 mm. The mechanical behavior of lithium-ion power battery has the following three remarkable characteristics: (1) the mechanical response of battery component materials in tension and compression is significantly different; (2) during the compression, the cell structure becomes denser and the hardening rate increases gradually; (3) the mechanical properties of battery structure vary greatly in different directions. Corresponding research is also carried out at different levels such as battery component materials, monomers and modules.

Based on the research and application requirements at different levels, the battery cell can be divided into Detailed model, Semi-homogenized model and Homogenized equivalent model according to the refinement degree of modeling from high to low [167] (Fig. 4.36). The refined model truly restores geometric features of internal section of the cell, including the positive and negative active coating, the positive and negative collector (copper foil, aluminum foil) and the diaphragm. It takes into account the material properties of different components, and the contact features between the diaphragm and coating as well as the collector and coating. The simulation results of refined model can accurately predict the macro extrusion force displacement response, and more precisely describe the deformation and failure characteristics under different conditions, which is conducive to analyzing the mechanical deformation and failure feature of electric core under complicated conditions and identifying the internal short-circuit triggering mechanism under different loading conditions. The refined model based on the finite element modeling is exemplified here. In view of the grid quality, the number of elements on a single electrode shall be tens of thousands, and the number of elements of a complete cell model shall be more than one million, costing much for calculation. On the premise of ensuring the required simulation accuracy, appropriate element types and such approaches as reduction of the number of elements can be adopted for improvement of calculation efficiency [167–169].

Based on the detailed model, the semi-homogenization model assumes that the thickness of each layer of material increases in proportion, but the total thickness of the core remains unchanged, and the number of core layers is reduced accordingly. On the premise of ensuring the existence of multi-layer structure, the number of elements is decreased and the calculation efficiency is improved. Its calibration approach and process are basically the same as that of the refinement model. Generally by illustrating the fracture behavior of metal collector, the fracture criterion in terms of stress state and equivalent strain is mostly selected and used to accurately predict the load displacement response of cell under different working conditions, so as to judge the local structural failure.

The homogenization model is established according to analysis requirements of structural levels such as power battery module and battery pack. Generally, it ignores the multi-layer structure of the cell, yet regards the whole battery or the coil core

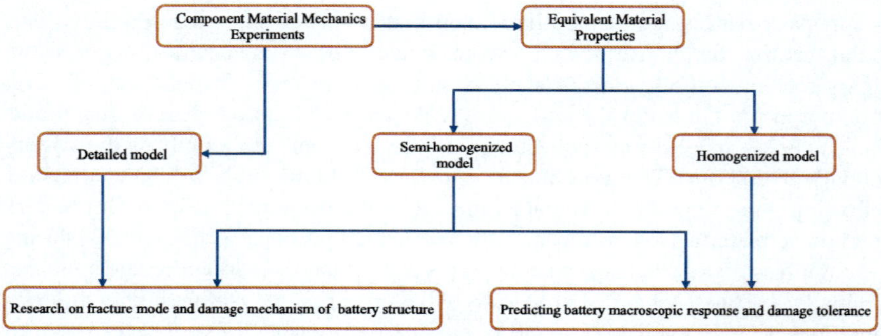

Fig. 4.36 Research and application of refined, semi-homogenized and homogenized cell simulation models [167–169]

inside the battery as a single uniform material. The calibration of material parameters is usually determined by the overall mechanical loading experiment of the whole battery. Compared with the detailed model, it has higher efficiency in computation, but lower accuracy in predicting local fracture behavior of battery. The homogenization model can effectively simulate the load–displacement curve of the cell under various extrusion and puncture conditions, and is mostly used to quickly predict the load–displacement response of the cell under different working conditions. By the introduction of the MPS failure criterion or stress-based MC failure criterion into the homogenization model, a reasonable estimation of the safety threshold of battery deformation can also be obtained.

The quasi-static test and impact test under controllable conditions of the laboratory are essential sources to obtain the deformation and failure mechanical behavior data of power battery modules, monomers and component materials. Figure 4.37 shows the impact test system of battery module based on drop hammer device. By means of the experimental system, perfect and general test procedures can be set up to conduct impact experiments on power battery modules with different configurations, such as adoption of different punches, impact speeds and impact directions, so as to obtain the impact dynamic response of modules under various complicated working conditions, and further establish the detailed model or simplified model of battery modules (Fig. 4.38). At the module level, each battery unit in the detailed model generally adopts the homogenization model (Fig. 4.38), which is used to deeply analyze the internal deformation and damage behavior of battery module under the collision condition, and determine the module collision failure threshold under each condition by combining the test results. The simplified model focuses on significantly improving the calculation efficiency through reasonable simplification, and is used as the simulation module for collision analysis and structural design of power battery system and electric vehicle. In a word, it has the critical demand and application value in the field of intelligent safety protection to carry out experimental test and simulation analysis around different levels of power battery and build a collision safety simulation analysis and test evaluation system.

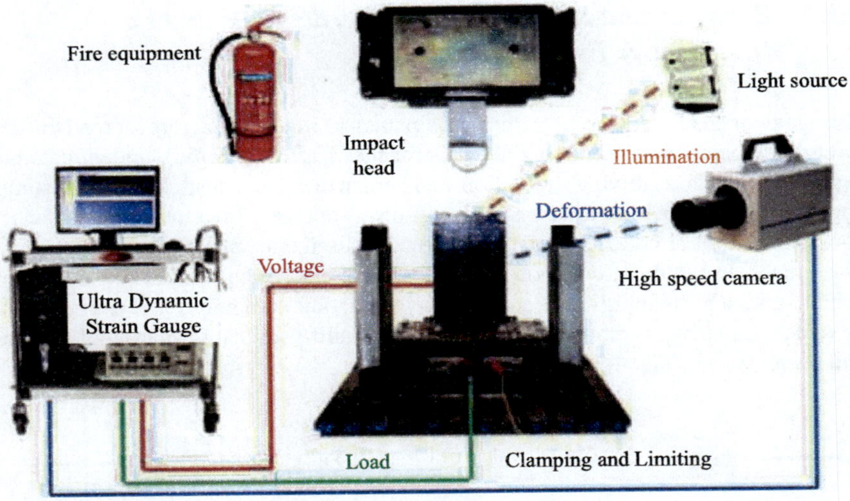

Fig. 4.37 Impact test system for power battery module (modified from literature [167])

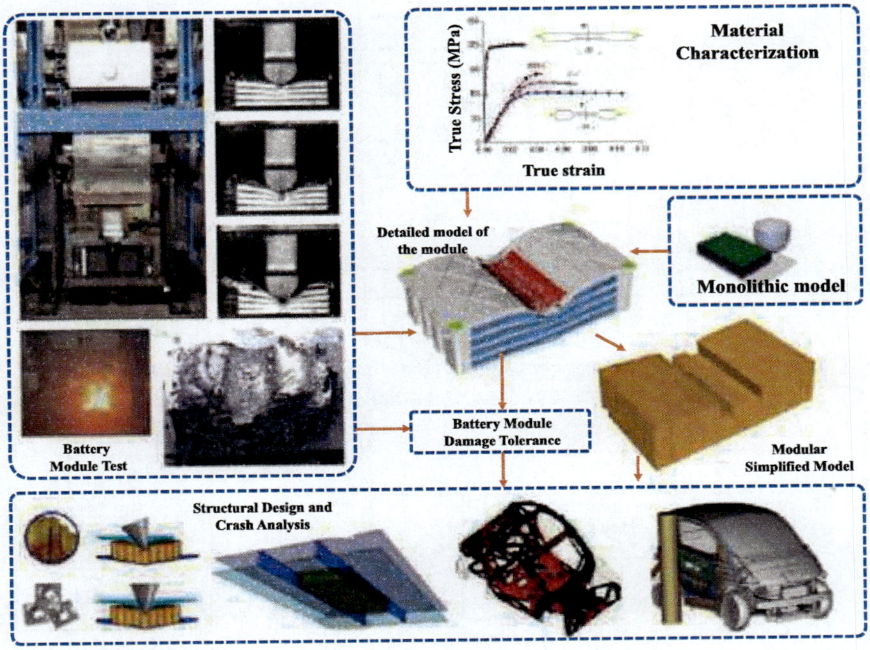

Fig. 4.38 Impact safety research system of power battery module (modified from Refs. [167–169])

4.2.3 Behavior and Injury of Traffic Participants and Its Quantitative Evaluation

The behavior characteristics of traffic participants are important factors for road traffic safety. The behavior research involves approaches based on traffic accident data or experiments (such as driving simulation, and human in the loop test). Different testing approaches have their own advantages and disadvantages. Integration of behavior data from different collection methods provides a reliable source for collision risk assessment. In addition, collision risk assessment also requires the data from the vehicle end, and its multi-level quantitative evaluation is mainly reflected in terms of vehicle kinematics parameters and integrated multi-source information of human and vehicles, etc. (Fig. 4.39).

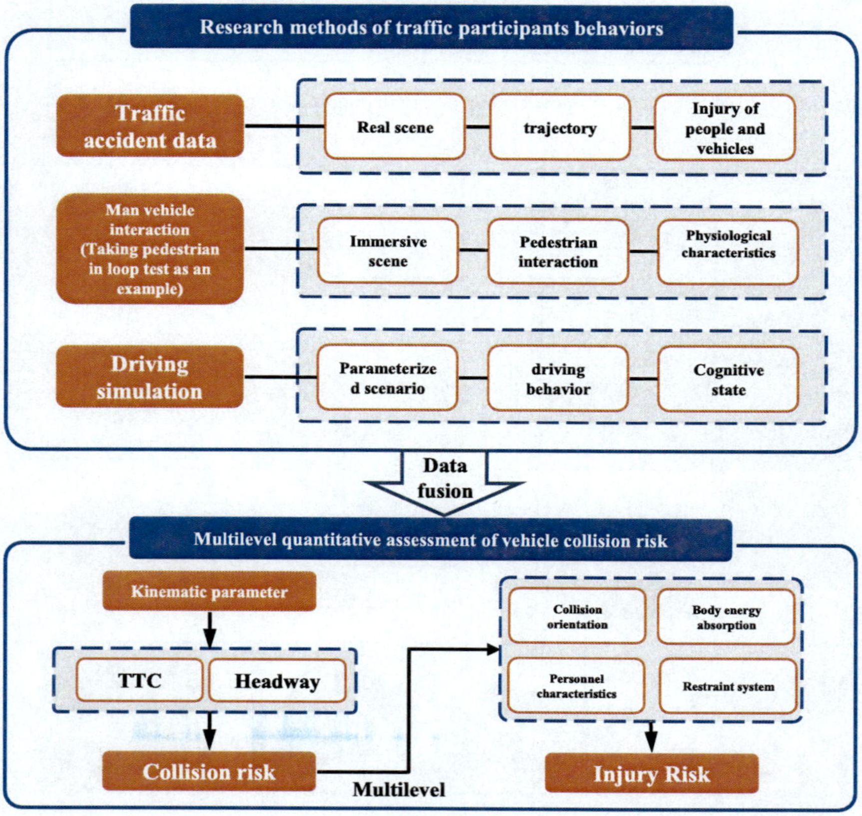

Fig. 4.39 Traffic participants' behaviors and quantitative evaluation

1. Extraction of Traffic Participants' Active Behaviors

Active response/behavior refers to that in the hazard, traffic participants perceives the complicated and changeable dynamic traffic scenario in brain, foresees the possible danger, and adaptively takes reaction or action to a certain extent based on biological instinct or aprior knowledge, so as to avoid the danger subjectively. Some studies also define driver's behavior pattern as hazard perception. As a kind of natural behavior, active behavior includes perceiving danger, making decision, generating reflex response, etc., on the time axis. It not only changes the interaction state between the ego-vehicle and other vehicles, but also affects the passive injury outcome in inevitable accidents. On the premise of ensuring the safety of volunteers, researchers shall quantitatively extract the active behavior data of traffic participants in dangerous traffic scenarios, such as visual attention distribution, kinematics, dynamics and other physiological signals. The behavior characteristics of these traffic participants serve as valuable reference for understanding the behavior mechanism and establishing the mechanism model with clear physical significance.

(1) Active Behaviors of Traffic Participants in Hazardous Scenarios

A series of active responses of traffic participants in the face of hazard will not only affect the accident probability, but also change the state and injury form of human body in the initial stage of collision. Accurately identifying the intention of traffic participants and predicting their behavior provide key information for reducing the accident damage risk of DVE system. Taking the driver for example, the behavior process specifically includes: when the hazard appears in the driver's field of vision, the brain actively and quickly identifies the hazard, deduces the scenario evolution, makes reasonable decisions for collision avoidance, and controls the body to make dynamic response (such as muscle contraction control) and driving operations (such as steering or braking based on muscle contraction), so that the driver's own dynamic property is changed. Then through the interaction between human and vehicle, the boundary conditions such as vehicle driving conditions are changed, which significantly affects the probability, type, damage risk and severity of subsequent traffic accidents. Therefore, in order to establish an accurate behavior recognition and prediction model, it is necessary to extract and quantify human's active behavior in natural state.

Compared with occupants in the vehicle, pedestrians have higher degrees of freedom of movement. In human-vehicle collision accidents, pedestrians often make a reflex avoidance response different from their normal posture if they have noticed the danger before the accident. Analysis based on GIDAS database (2000–2015) shows that pedestrians have different collision postures in vehicle–pedestrian collision accidents, such as walking, running, standing and a small number of lying down states (Fig. 4.42). Pedestrian initial posture has a significant impact on injury risk, but due to the complexity of body posture, it is difficult to accurately and quantitatively

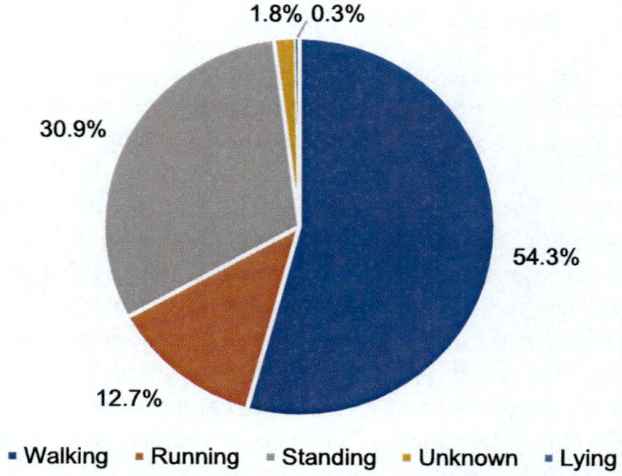

Fig. 4.40 Distribution of pedestrian's postures in collision based on vehicle–pedestrian collision accident data of GIDAS (2000–2015) [34]

characterize it [170], resulting in insufficient research on the impact of pedestrian initial avoidance posture on injury risk.

(2) Experiment and Modeling of Driver's Active Behaviors

The main ways to obtain driver behavior data include natural driving and driving records, simulated driving, road monitoring, accident statistics, etc. Natural driving is the most direct way to collect behavior data, including capturing driver's posture, driving operation, facial expression and other behavior characteristics by vehicle equipped camera and other sensing devices. However, because the emergent collision in the real scenario is an event with small probability, it is difficult to obtain driver behavior data in large-scale dangerous or accident scenarios through natural driving sets [171]. The statistical data of traffic accidents focus on the damage of human and vehicle after the collision. The scenario is reconstructed and analyzed according to the situation of the accident. However, there is still a certain difference between the reconstructed scenario and the real accident, that is, existence of a large amount of noise data. Another data acquisition way is to record by the road monitoring equipment, which can obtain the hazard information and vehicle trajectory in the scenario. However, without the capability of collecting driver's physiological feature signals, it is difficult to facilitate behavior modeling and analysis [172].

Simulated driving is a widely-used experimental approach for the study of driver's behavior. The driving simulation experiment platform reproduces the actual traffic environment in real time by means of visual simulation, auditory simulation and driving control systems (Fig. 4.41), and enables the test object (driver) to have the operation and motion feeling close to reality, so that the driving-related data for experimental research in a relatively stable and safe environment can be collected. The research on simulated driving offers valuable data and approaches for the research on IV safety and design [173].

Although the traditional simulated driving can provide the subjects with a quasi-real driving experience, it can not meet the requirement for model training data

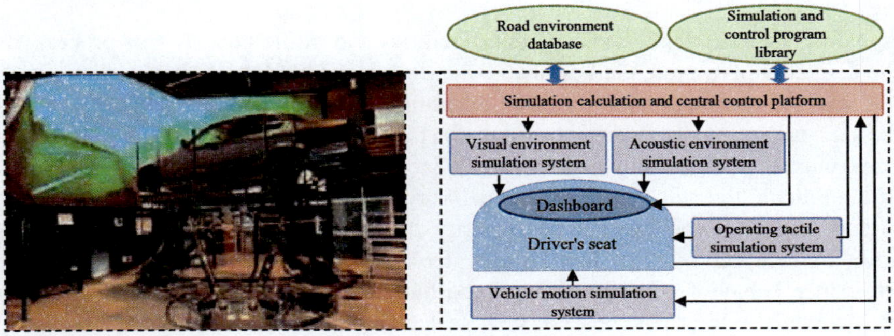

Fig. 4.41 Driving simulator on 6-DOF movable platform

with high generalization representation on learning ability, due to the singleness of simulation scenario and high experimental cost. The establishment of a testing approach for efficiently collecting driver's perception, decision-making and collision avoidance behavior is conducive to the establishment of a large-scale data set of driver's active behavior (Fig. 4.42). By identifying driver's behavior patterns (visual attention performance, alertness capability and behavior representation) in hazardous traffic scenarios, an active risk avoidance behavior model can be set up, so as to lay the theoretical foundation for the establishment of reasonable driving AIDS or anthropomorphic automatic driving decision-making system.

(3) Research on Pedestrian's Natural Behaviors for Intelligent Protection

Pedestrian's active behavior (such as collision avoidance) is a critical factor influencing the occurrence of road traffic accidents. Taking pedestrians crossing the road in the traffic environment for example, they choose a safe time node and path according to acquired experience and knowledge, as well as the surrounding environment,

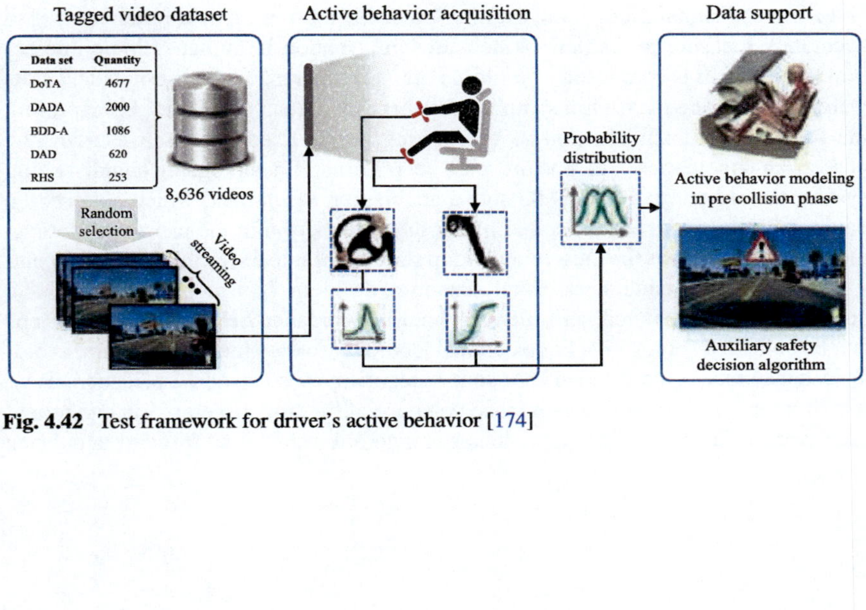

Fig. 4.42 Test framework for driver's active behavior [174]

signal lights, road feature, surrounding facilities and traffic flow; In this process, if the danger is perceived, the best avoidance way is selected through experience and judgment of external conditions. The dynamic interaction between pedestrians and the surrounding environment is based on the decision-making mechanism that mobilizes muscles to drive body movement. Due to individual difference and complexity of conditions, the active behavior of pedestrian in real traffic scenarios presents a high degree of randomness and nonlinearity. Although the process of human-vehicle collision recorded by road monitoring video and tachograph can display the reflex avoidance behavior of pedestrians in the face of danger [175, 176], it is unable to accurately extract feature data, such as speed, collision posture, collision angle, muscle activation status and other kinetic and dynamic information. In recent years, with the development of immersive Virtual Reality (VR), virtual traffic environment is more applied to the development of automatic driving technology [177], and also to the research of pedestrian's active behavior [178]. VR technology can obtain the real data of pedestrians under hazardous traffic conditions on the premise of safety assurance. With the pedestrian's active behavior research based on immersive road traffic scenario as the example, the experimental research on the characteristics of pedestrian's natural reflex behavior is introduced. The specific experiment and data flow are shown in Fig. 4.43 [178, 179].

The immersive virtual road traffic environment module is composed of virtual scenario generator, control system and test equipment. VR equipment is a virtual scenario generator provided for experimental volunteers (pedestrians) an approximately real feeling. The virtual traffic environment includes road infrastructure, buildings, trees, traffic lights, traffic noise and other elements. The control system is handled by experimenters and used to control part of information in the virtual scenario as experimental variables, including traffic light conversion, emergence of obstacle vehicles and conflicting vehicles, speed and position of conflicting vehicles, etc. During the experiment, the relative position between the volunteers and the virtual vehicle, the speed of the virtual vehicle, the conversion of traffic lights and other information are also accurately recorded.

In terms of signal acquisition, the high-precision motion capture system is used to accurately measure pedestrians' kinematics information in human-vehicle interaction. The system is composed of multiple optical cameras, marking points and software systems. The maximum sampling frequency can reach 1000 Hz. By capturing the three-dimensional coordinates of marking points fixed on the pedestrian, the pedestrian kinematics and posture can be restored. Physiological signal testing system mainly refers to the EMG signal acquisition equipment, which collects the epidermal electric signal generated by muscle contraction in the process of human movement, and shows the muscle activation state of volunteers during the experiment.

Based on the immersive virtual scenario, different typical pedestrian-vehicle conflicts can be designed, and then the natural avoidance behavior of pedestrians can be collected (Fig. 4.44). For example, according to the kinematic feature, avoidance behavior can be divided into four categories: ① backward avoidance (BA), which refers to the collision avoidance behavior of braking, deceleration and retreat after pedestrians notice the approaching dangerous vehicle; ② forward avoidance

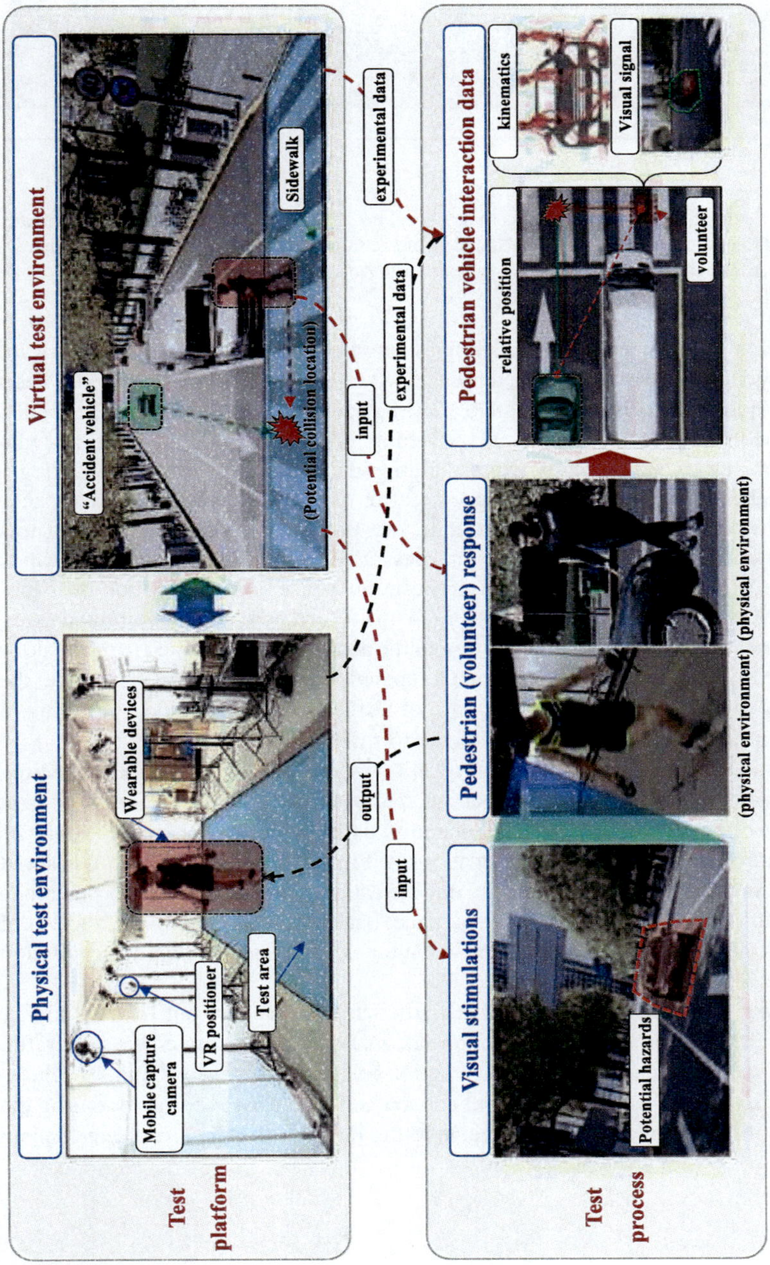

Fig. 4.43 Virtual test platform and data flow of immersive active response of pedestrian

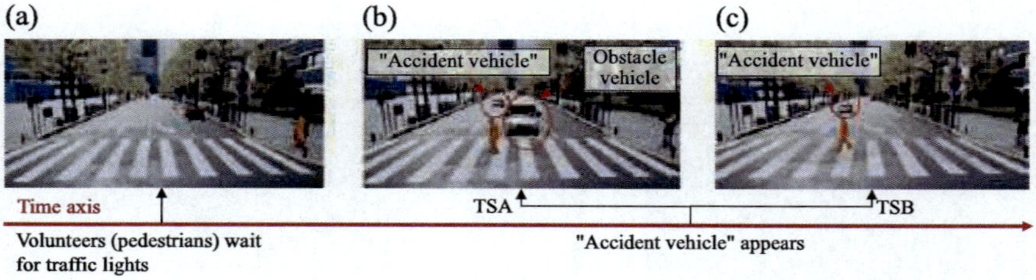

Fig. 4.44 Setting of pedestrian-vehicle conflict scenario. TSA and TSB in the figure respectively represent different traffic scenarios. In TSA, there are certain visual obstacles ("obstacle vehicles") when pedestrians cross the road. In TSB, pedestrians have a better field of vision

(FA) refers to the behavior that pedestrians avoid collision by accelerating forward when they find danger; ③ oblique stepping (OS) refers to the behavior of pedestrians stepping in the direction of the vehicle after they are aware of the danger (generally, this behavior can not effectively avoid the vehicle); ④ the behavior of pedestrians who walk through the road normally without finding danger is called Normal Walking.

Active behavior feature data include reaction time, avoidance speed, avoidance posture, lower limb muscle activation state, etc. For the cases that adopt backwards and forward avoidance behavior and successfully avoid collision, the backwards avoidance behavior generally presents a shorter response time (perception-decision-execution) than the forward behavior (Fig. 4.45). When perceiving potential danger, pedestrians whose initial position is close to the accident vehicle lane are more likely to choose forward avoidance behavior. Compared with real collision avoidance, the "decision" time of pedestrians in the case of virtual collision is longer, resulting in insufficient time for avoidance behavior performance (Fig. 4.46).

The change of pedestrian's speed is an important overall kinematic index reflecting natural avoidance behavior, which can be calculated from the acceleration data at the center of the pelvis (Fig. 4.47). Generally, the behavior of backward avoidance includes three stages: during normal walking, the initial speed of pedestrians is about 1 m/s; during braking deceleration, the uniform deceleration is about 0.73 s (average deceleration $= -2.4$ m/s^2); during avoidance, backward movement is at a speed of about -1 m/s. The forward avoidance behavior shows a clear forward acceleration (average acceleration $= 7.4$ m/s^2).

In general, the uncertainty of traffic participants' behaviors put forward higher requirements for vehicle safety protection design. Based on the immersive road traffic experimental environment, the natural reaction behavior in the face of human-vehicle conflict is extracted, as well as relevant data on human behavior before or during the conflict with vehicle, which is conducive to the integration of behavior mechanism

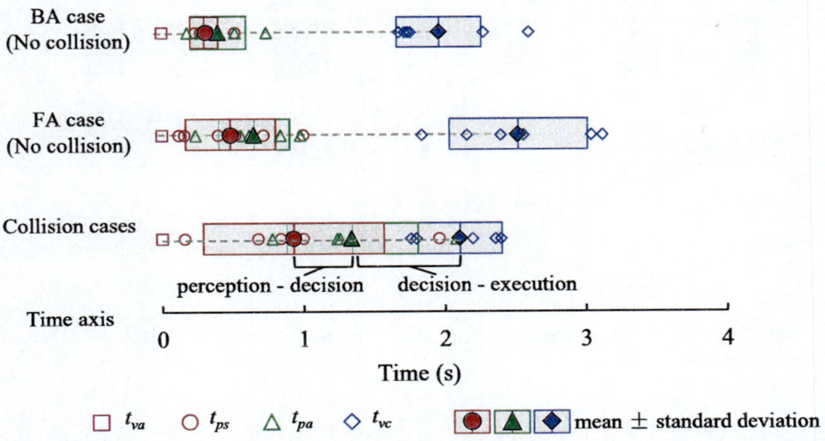

Fig. 4.45 Perception-decision-execution reaction time during pedestrian avoidance (t_{va}: the time when the "accident vehicle" appears; t_{ps}: the time when the pedestrian notices the "accident vehicle"; t_{pa}: the time when the pedestrian begins to avoid; t_{vc}: the time when the vehicle reaches the potential "collision place" (the intersection of pedestrian and vehicle trajectory)

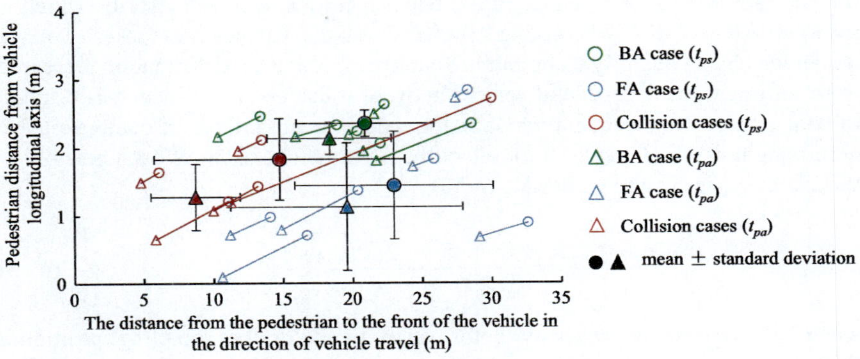

Fig. 4.46 Relative position between pedestrian and vehicle during perception and avoidance

in subsequent dynamics and risk prediction and the development of more advanced vehicle safety protection system.

2. Multi-level Quantitative Assessment on Vehicle Collision Risk

The risk of collision accident during driving can be quantitatively evaluated from different levels such as vehicle dynamics and body deformation depth. The body deformation depth can be used to evaluate the occupant injury risk and vehicle

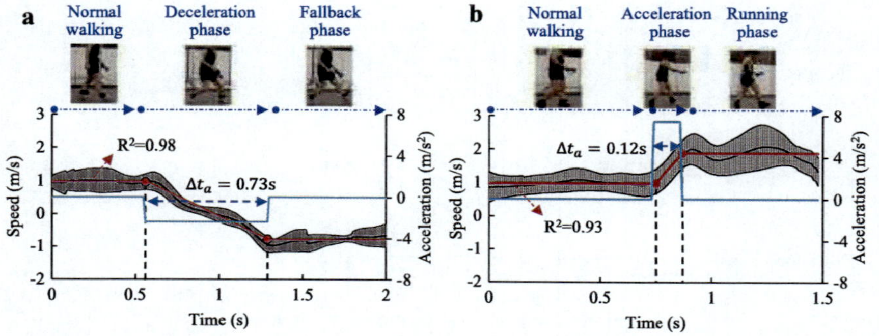

Fig. 4.47 Speed-time curve during pedestrian's avoidance: **a** backward avoidance behavior **b** forward avoidance behavior

safety after the accident, including three aspects: ① establish the functional relationship between the deformation depth and the occupant injury risk through regression analysis; ② establish the functional relationship between deformation depth and deformation energy; ③ propose the passenger injury risk calculation framework on the basis of the above two functional relationships [180].

Occupant injury risk refers to the possibility of reaching a specific injury degree (mostly take MAIS 2+ as an example) under a certain accident severity, which is recorded as P (MAIS 2+). By taking P (MAIS 2+) as the dependent variable, a damage model which can quantify its output is established. Since the deformation measured from different positions of the vehicle body is a discrete value, which is difficult to be used as an independent variable, the independent variable of comprehensive deformation depth (C) is used to reflect the deformation depth of each part of the vehicle body, as shown in Formula (4.56).

$$C = \frac{\frac{C_m}{2} + \frac{C_n}{2} + \sum_{i=m+1}^{n-1} C_i}{n - m} \tag{4.55}$$

where, C_i represents the body deformation depth at the sampling position i (Fig. 4.48b), and n and m are the sampling number.

The deformation contour of vehicle body can be simplified and approximated to triangular, rectangular or trapezoidal geometry (Fig. 4.50b). It is represented by two parameters: comprehensive deformation depth C and collision deformation length L_d. Taking them as independent variables and occupant injury risk P(MAIS 2+) as dependent variable, cubic polynomials can be selected as regression forms, as shown in Formula (4.56).

$$P(\text{MAIS 2+}) = a_0 + a_1C + a_2L_\mathrm{d} + a_3CL_\mathrm{d} + a_4C^2 + a_5C^2L_\mathrm{d} + a_6C^3 \tag{4.56}$$

where, a_0–a_6 are coefficients to be determined.

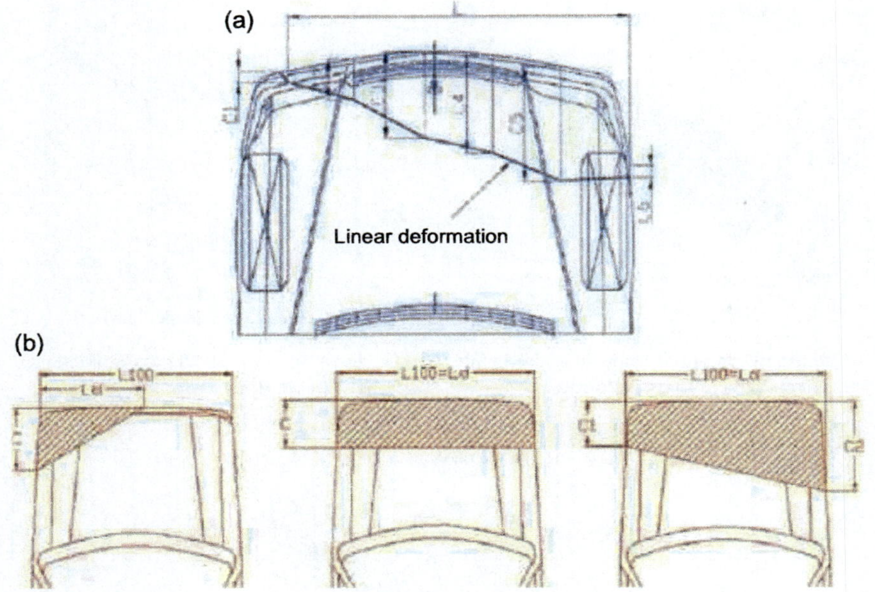

Fig. 4.48 Deformation diagram and deformation form of various parts of vehicle [180]

The real accident data points and the regression effect of regression curve surface are shown in Fig. 4.49a, with most data points distributed near the surface. However, due to the increase of fitting dimension (types of independent variables), the unreasonable distribution of original sample points is more prominent, resulting in some scattered points still far away from the surface. The degree of injury risk rises with the increase of deformation depth, indicating that the regression result can accurately represent the relationship between the degree of occupant injury risk and the depth of vehicle collision deformation [180] (Fig. 4.49b).

The energy absorbed by vehicle body deformation, usually characterized as equivalent energy speed (EES), is also an important parameter to reproduce the accident dynamic collision process. Taking the collision deformation length L_d and comprehensive deformation depth C as independent variables and the equivalent energy speed EES as dependent variable, the regression relationship between the three is established, as shown in Formula (4.57).

$$\text{EES} = ab_0 + b_1C + b_2L_d \tag{4.57}$$

where, b_0–b_2 represent the coefficients to be determined, and b_0, usually as constant, represents the speed without plastic deformation when the vehicle collides with the fixed rigid wall. The regression relationship between EES and deformation depth can be used to evaluate the probability of occupant injury. Figure 4.49c shows the comparison between the real data points and the regression surface. The results reveal that

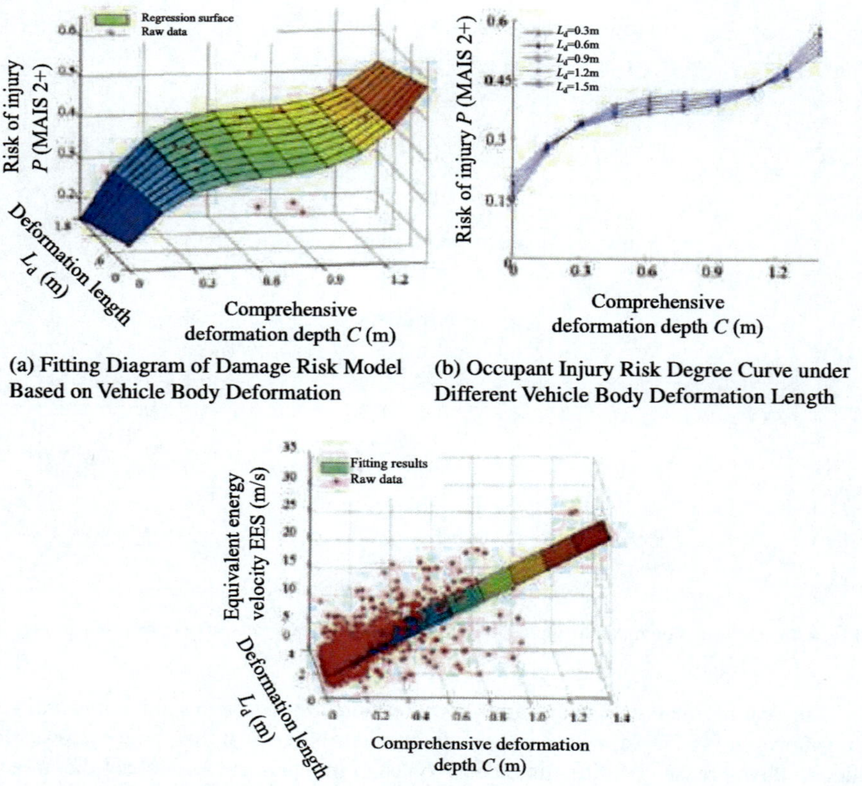

(a) Fitting Diagram of Damage Risk Model Based on Vehicle Body Deformation

(b) Occupant Injury Risk Degree Curve under Different Vehicle Body Deformation Length

(c) Fitting Diagram of Comprehensive Deformation Depth C and EES Model of Vehicle Body [180]

Fig. 4.49 Occupant injury risk model based on vehicle body deformation depth [180]

the data points are distributed on or near the surface. When the deformation depth is large, some scattered points are far from the surface, due to the small number of original accidents in the deformation depth stage and the relatively random distribution of EES [180].

4.3 Cooperative Protection Technology

The traditional vehicle active and passive safety technology is difficult to fully meet the driving safety requirements under the intensified interaction of traffic elements. It is necessary to establish a cooperative way to solve the problem of safety protection, so as to comprehensively consider the impact of various factors such as driver, vehicle and road conditions on the damage risk and realize the protection of multi-factor coupling. The specific way includes the prediction of damage risk of human-vehicle system, vehicle protection based on structural energy absorption properties and sensing, and IP technology for traffic participants.

4.3.1 Damage Risk Prediction of Human-Vehicle System

In hazardous traffic scenarios, the dynamic interaction among human, vehicle and road constitutes a complex time-varying generalized mechanical system. At the system level, a vehicle collision accident can be equivalent to a sudden change in the state of the DVE system. When the change exceeds a certain conditional threshold, it will result in the loss of system stability and the injury risk of traffic participants. Injury risk prediction is targeted at the driving scenario where collision accidents are likely to happen, inputs the relevant information of the DVE system, and takes the quantitative injury severity index (such as AIS level) and its occurrence probability as the prediction object. In addition, the whole-body injury metrics (WBIMs) can be further calculated from AIS of various body parts, for the application in single objective optimization and problem prediction (Fig. 4.50).

1. Damage Risk Prediction Based on DVE Information Perception

The characteristics of injury risk prediction problems are reflected in three aspects: (1) in terms of human, driver's decision-making behavior in hazardous traffic scenarios will affect vehicle dynamics, and physiological information and its dynamic response (muscle tension, posture change, etc.) also have a significant influence on passive injury behavior under collision conditions; (2) In terms of vehicle, the structural features of vehicle models, restraint systems and passenger seat configurations are complex and diverse, and a great amount of data is needed to fully reflect the spatial distribution of large parameters; (3) In terms of road, various environmental conditions, such as lighting, weather, and road conditions, have direct or indirect effects on dangerous driving conditions and collision forms, with a strong nonlinear correlation to damage behavior.

For risk prediction technology, the additive input information will increase the amount of calculation, which leads to the difficulty in carrying out real-time calculation under dangerous conditions. In order to make efficient risk identification, the key factors affecting the risk of DVE system can be reasonably extracted first. For instance, such factors can be abandoned as light of road sections and road conditions, which have little change for a long time and have no substantive contribution to the quickly updated damage prediction algorithm; The input features are compressed, so as to highlight the human-vehicle factors with significant effect on collision damage (critical speed at the time of collision, occupant attitude, etc.). The benefits of these adjustments include: ① prediction performance is improved. Accuracy of premature prediction on complicated systems, which might be chaotic, is relatively low, resulting in invalid prediction. Instead, the critical short-term prediction can improve

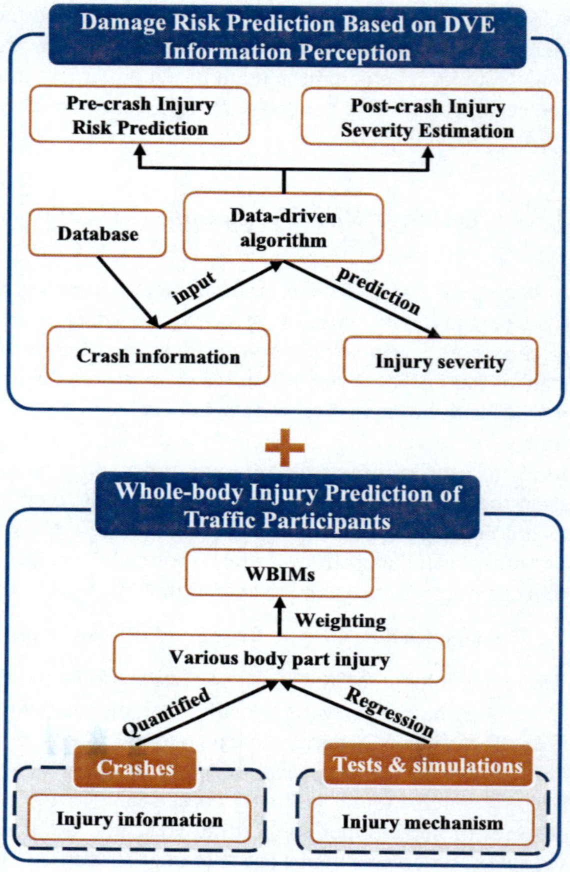

Fig. 4.50 Injury risk prediction of human-vehicle system

prediction accuracy; ② by converging the input variables and using a variety of on-board sensing devices, the feature information significantly related to damage can be collected quickly and accurately.

According to different periods and application scopes, occupant injury risk prediction technology can be divided into two categories, and its framework is shown in Fig. 4.51. Pre-collision injury risk prediction is mainly applied to the trajectory planning algorithm and adaptive restraint system of IV. Its main task is not only to identify damage risk, but also to output strategies for collision risk reduction. The collision risk is minimized by adjustment of the vehicle trajectory before the accident and application of the appropriate constraint system configuration during the accident. Therefore, the algorithm shall ensure the accuracy while taking into account the real-time performance. Post-crash injury severity estimation mainly serves the Advanced Automatic Collision Notification (AACN). After a road traffic accident, it estimates injury severity of the occupants in the vehicle and timely transmits this information

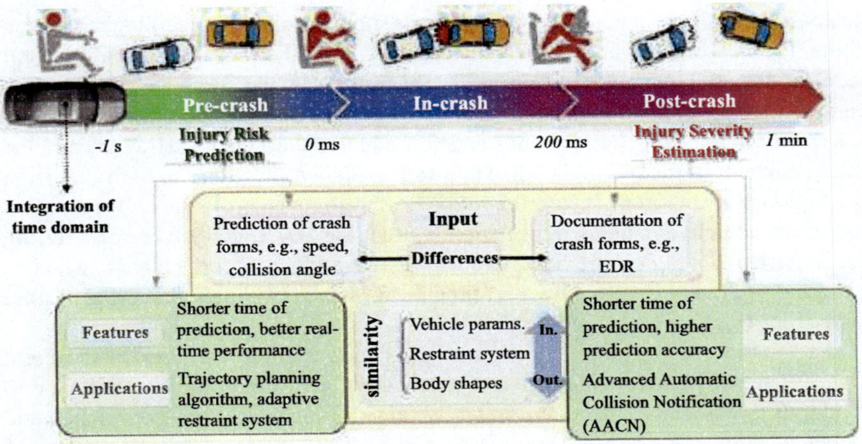

Fig. 4.51 Application framework of occupant injury risk prediction technology

to the appropriate emergency center for improving rescue efficiency and survival rate. Therefore, while assuring high-precision prediction, it aims to realize the injury prediction of different body parts and refine the judgment of injury degree.

The initial impact condition input and occupant injury output show a highly nonlinear multi-layer mapping relationship. Therefore, the injury risk prediction is essentially a multi-input prediction issue. In terms of prediction approaches, response surface or multiple regression can be used to predict mathematically. These traditional methods have strong interpretability and high stability, and are also the most widely-used models [181–183]. However, this stability shows the disadvantage in the lack of prediction capability and poor fitting capability for strong nonlinear problems.

In view of this problem, current injury risk studies begin to gradually introduce data-driven algorithms, such as decision tree, support vector machine and k-nearest neighbor. The support vector machine usually performs better in prediction performance. The decision tree uses the if–then rule set to represent the conditional probability distribution under the given feature conditions, with readable model and fast classification speed. As a typical unsupervised learning method, k-nearest neighbor can complete the classification and prediction of accident data without label information. As the generation model without iterative learning process, Naïve Bayes algorithm is simple in implementation and high in learning efficiency. However, limited by strong assumptions (i.e. the data feature conditions are independent from each other), it is usually difficult to achieve excellent prediction performance in the dangerous road traffic scenarios with highly coupled DVE information. Benefiting from the idea of ensemble learning, adaptive boosting (AdaBoost) algorithm carries out the learning task by combining multiple simple classifiers (i.e. base classifier), with better generalization capability than a single classifier model.

The deep learning model in terms of neural network is also a means for the prediction accuracy improvement. A large number of studies based on different databases and different neural network models show that the deep learning model can obtain 75–90% of the results in accident damage risk prediction, but its stability is still required for effective quantization proof. On the whole, the complicated algorithm can improve the prediction performance of the model, whereas the simple algorithm is superior in real-time performance and computational efficiency. How to combine the advantages of the two and realize real-time risk identification and high-precision prediction is an important issue to be studied. One approach is to filter and preprocess the initial vehicle dynamics information, extract the key kinematic features during the impact, and simplify the algorithm without sacrificing its performance, which is expected to greatly improve the real-time performance of the prediction algorithm [184, 185] (Fig. 4.52).

As for the training or verification of prediction models, the data in the Field Datasets are true and reliable, and are widely used [186]. With the attention paid to road traffic accidents, several representative accident databases have developed and expanded all over the world (shown in Sect. 4.1). The accident database is collected

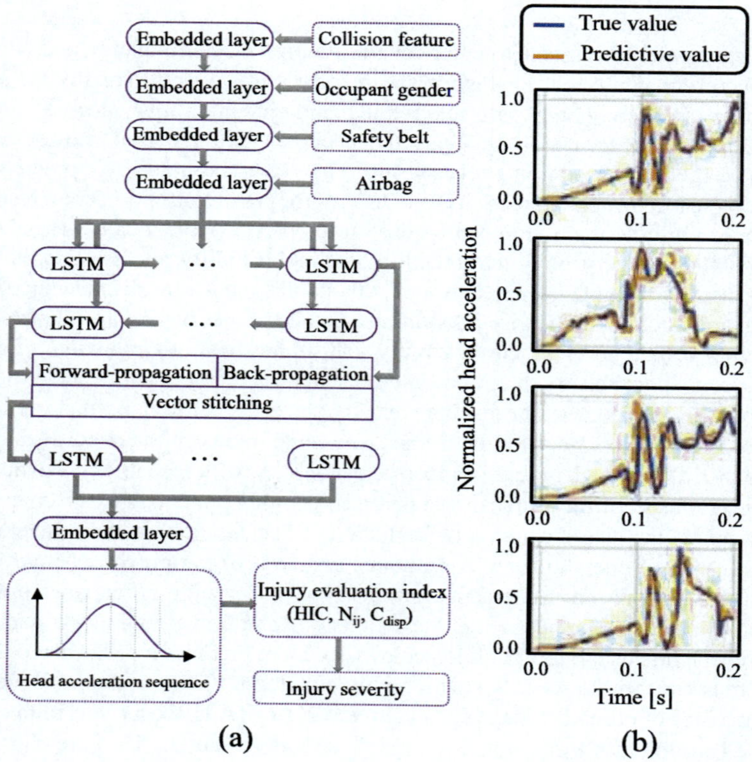

Fig. 4.52 An occupant injury prediction model based on cyclic neural network and its predicted performance [185]

by police or full-time staff from the accident scenario, marked according to the standard accident investigation, and provides a large number of optional features. For example, each accident in FARS contains 175 features, and each accident in NASS-CSD contains about 400 features. Mainly depending on extensive space–time accumulation, the huge amount of data in the accident database has large parameter space, yet without sufficient amount of data correspondingly; Due to the lack of data, the amount of effective data available for injury prediction is not large. In addition, the vast majority of accidents involve minor injuries in reality, resulting in the uneven distribution of injury severity recorded in the database [187, 188], which should be paid attention to in algorithm training. Synthetic minority oversampling Technology (SMOTE) and other methods can be adopted to preprocess data [189–191].

Due to the differences in standards, regions and time among different accident databases, there exists certain heterogeneity [192]. The coupling effect of various factors leads to different conclusions made by studies with different databases [193, 194]. Reflected in the algorithm, the algorithm trained according to the database of region A will have some problems when used in region B, and even the optimal algorithm structures of different databases are different from each other. Therefore, for each new database, the overall development process of prediction algorithm must be repeated, with difficulty, long cycle and high cost in development. To solve this problem, transfer learning between different prediction models can be conducted, as the damage essences reflected by different databases are the same. Transfer learning is a machine learning approach that reuses the knowledge learned in one task in another task. Based on existing model retraining, it requires that the features extracted by the source model are generalized, i.e., the underlying features of the source task and the target task are the same or similar [195]. This approach can reduce the amount of data required for the target task, with short development cycle and relatively low difficulty. At present, it is widely used in algorithm development in different fields. For example, as the source model, the inception-v3 model in image recognition is transferred to occupant injury risk prediction based on 4606 traffic accident samples. Under the dichotomous injury severity, ISS > 15 and ISS ≤ 15, the accuracy of the prediction model is 86.46% [196], indicating that the rapid development of damage prediction algorithm based on transfer learning has great potential in practical vehicle application.

To sum up, the occupant injury prediction in dangerous traffic scenarios is significantly influenced by quite a number of factors, including training database and prediction algorithm structure, etc. The existing prediction technology mostly depends on the accident database with unbalanced distribution and prominent heterogeneity effect, and the potential of the algorithm itself has not been fully developed. Since the existing prediction algorithms mainly focus on prediction accuracy, the further step is to improve the real-time performance of calculation and promote its practical application in on-board safety system.

2. Whole-body Injury Prediction of Traffic Participants

Traffic participants usually suffer from more than one body part injury under impact conditions. The whole-body injury is usually calculated in terms of injury degree,

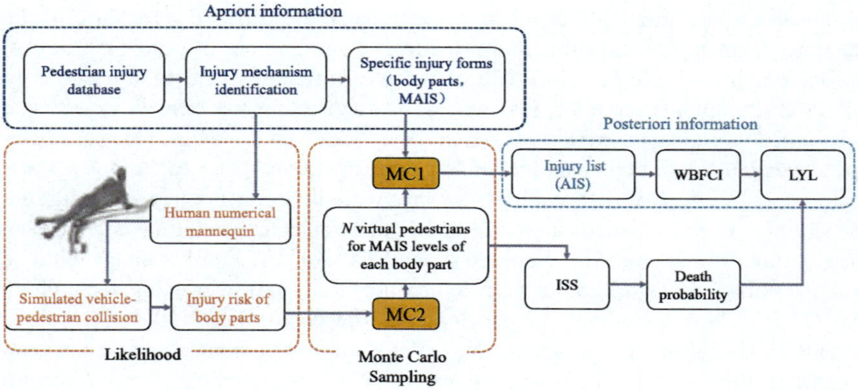

Fig. 4.53 WBIMs Estimation Framework (taking pedestrians for example) [197]

influencing factors, function, economic loss and other elements of each body part for evaluation of the overall injury severity, and can also serve the pre-accident risk prediction by reducing the calculation amount of the comprehensive injury evaluation index for rapid optimization. The following is a description on prediction and application of the Whole-Body Injury Metrics (WBIMs) for pedestrians in accordance with accident and simulation research (Fig. 4.53).

Based on the accident database within a given range, the data can be queried first to identify the form and frequency of pedestrian injury. In terms of the injury mechanism and representation capability of numerical mannequin, the whole body is divided into 17 body regions, and then the injury pattern database is constructed for MAIS levels of each body part; The two-step Monte Carlo Sampling process is adopted to generate N virtual pedestrians and calibrate injury list with AIS [197]. And the expected value $E(WBIM)$ of WBIM is estimated according to ISS, death probability [Formula (4.58)], Whole-Body Functional Capacity Index (WBFCI) [Formula (4.59)] and lost years of life (LYL) [Formula (4.60)]. The calculation is shown in Formula (4.61). The effectiveness of injury estimation can be verified according to pedestrians' injury information in the real database.

$$P(\text{death}) = 1/[1 + \exp(4.742 - 0.1172 \cdot ISS)] \quad (4.58)$$

$$WBFCI = 40 \times \prod_{d=1}^{10} \left(\frac{FCI_{\text{d}} - 60}{40} \right) + 60 \quad (4.59)$$

where, FCI is the Functional Capacity Index (FCI), which refers to the score of restricted functional capacity in ten dimensions such as audio, visual and oral dimensions one year after injury [198]. The score of each dimension ranges from 60 (total loss of function) to 100 (no loss of function), and FCI_{d} refers to the lowest score of each dimension.

$$LYL = e_X \cdot P\,(\text{death}) + e_X \cdot [1 - P(\text{ death })](1 - WBFCI/10) \tag{4.60}$$

where, e_x represents the expected value of remaining life at a given age (expected value of remaining life = expected life value of population − age of the injured); $e_x \cdot P$ (death) and $e_x \cdot [1 - P$ (death)](1 − $WBFCI/10$) represent the lost years of life due to the probability of death and the lost years of life due to the loss of functional capacity, respectively.

$$E(WBIM) = \frac{1}{N}\sum_{j=1}^{N}(WBIM)_j \tag{4.61}$$

By comparison with the probability of death obtained in epidemiological studies, the result indicates that for a given injury risk distribution or MAIS level, this approach can accurately predict comprehensive injuries, such as ISS and WBFCI. The estimated probability of death based on collision speed has a close correlation with results of epidemiological research. The approach can be applied to quickly estimate pedestrian injury probability in accordance with the collision speed.

4.3.2 Vehicle Protection Technology in Terms of Structural Energy Absorption Property and Sensing

At the moment of the collision accident, the initial kinetic energy carried by the vehicle under normal driving conditions is quickly released and transmitted to the vehicle body and occupants, resulting in deformation or destruction, as well as the risk of vehicle body damage and occupant injury. After a collision accident, the lower the damage level of the vehicle and/or its occupants and loads is, the higher the crashworthiness is. Crashworthiness can be measured by mechanical indexes such as force, acceleration and deformation. In order to meet the corresponding requirements for crashworthiness, some energy absorbing components for damage level reduction are designed and installed in the vehicle, such as bumper, front longitudinal beam and bonnet. These classical energy absorbing components shall be designed according to specific indicators. With the development of new energy vehicles, power batteries have become new protection objects, and protection design such as thermal runaway shall be carried out. Common technologies include optimization of battery pack layout, flame retardant and explosion resistance technique, collision protection structure design, etc. And timely thermal runaway warning is demanded for alleviating the consequences of accidents (Fig. 4.54).

1. Analysis and Design for Collision Energy Absorption of Vehicle Body Structure

In a collision accident, the energy absorbing components in the body structure first absorb the initial kinetic energy reasonably to the greatest extent through their own plastic deformation, so as to reduce the injury risk of traffic participants. In the front

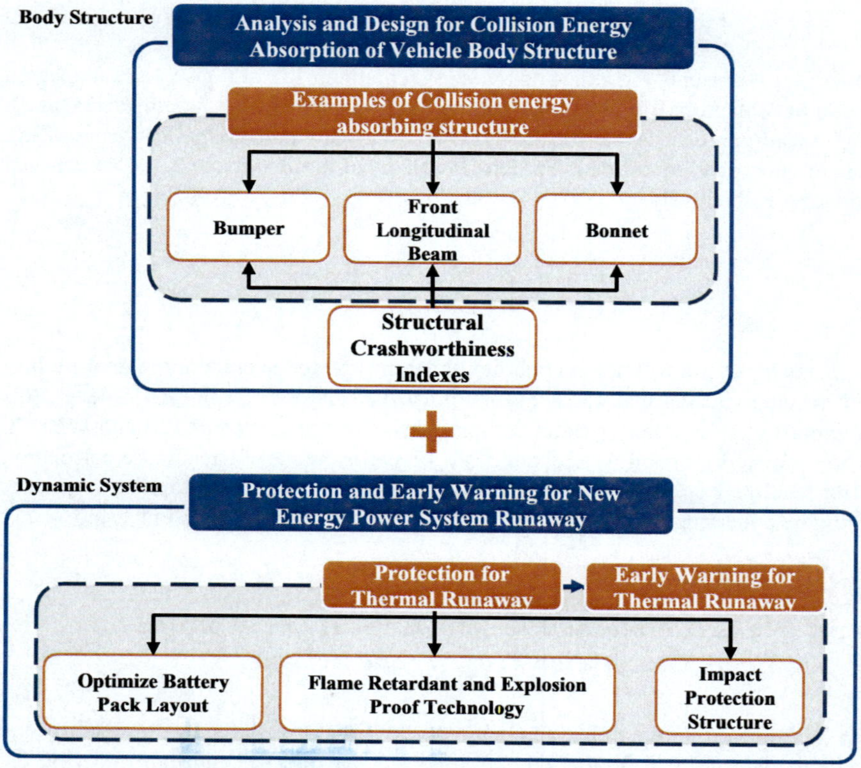

Fig. 4.54 Protection technology in terms of structural energy absorption property and sensing

collision, the main energy absorbing components are the bumper system and front longitudinal beam in the front of the vehicle body. Its design goal is to protect the engine (or power battery), chassis and other important components from damage on the premise of ensuring the safety of occupants in the vehicle, so as to optimize the safety protection performance of the vehicle. In addition, when the vehicle collides with vulnerable road users such as pedestrians outside the vehicle, the collision contact between the pedestrian's head and the front bonnet is one of the main reasons for the serious head injury. Therefore, in vehicle safety design, the design form of the energy absorption structure of the bonnet is also an important issue. Taking the three key body energy absorbing components of bumper, front longitudinal beam and bonnet for example, the analysis and design for collision energy absorbing structure are as follows.

(1) Design Scheme of Energy Absorbing Box and Front Longitudinal Beam for Occupant Protection in Front Collision

In frontal collisions with different severity, the ideal body collision force-overall deformation displacement curve is shown in Fig. 4.55. The initial kinetic energy carried by the vehicle is mainly absorbed by the energy absorbing box in the front bumper system and the compression deformation of the front longitudinal beam (Fig. 4.56). When the collision speed is low, the energy absorbing box in the front bumper system serves as the main component, which absorbs most of the energy through its own crushing deformation for avoiding damage to the front longitudinal beam, engine and chassis components [199]. When the collision speed is high, the front bumper and the front longitudinal beam participate in the deformation together, and the front longitudinal beam becomes the main energy absorbing and deformation component, prolonging collision time and decreasing peak acceleration of occupant compartment for reduction of occupant's injury risk. Therefore, the deformation mode and energy absorption capacity of the energy absorption box and the front longitudinal beam significantly affect the safety protection performance of the vehicle [200].Global(sps)2023

The design of the energy absorbing box and front longitudinal beam require that the force is uniform in the impact process, close to and cannot exceed the threshold of the maximum impact force. In addition, it is also necessary to ensure that the front longitudinal beam will not be damaged as much as possible before the energy

Fig. 4.55 Idealized overall vehicle impact force–deformation curve

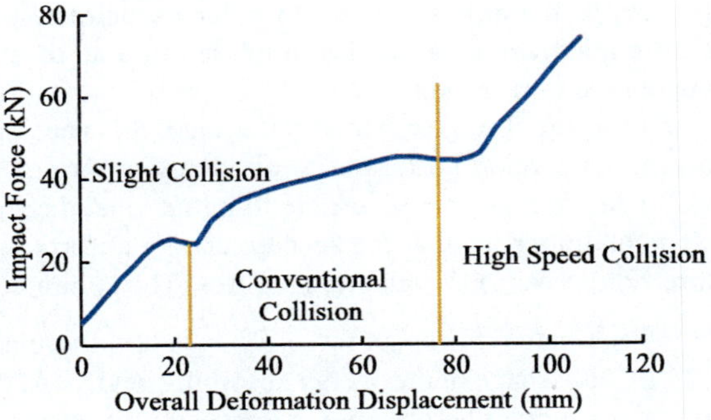

Fig. 4.56 Energy absorbing box and front longitudinal beam system

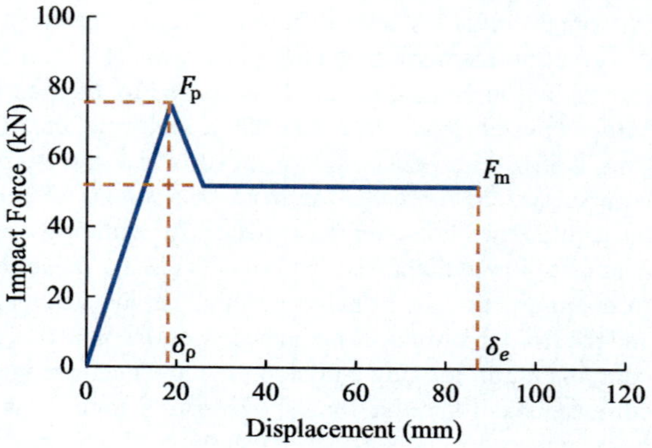

Fig. 4.57 Idealized impact force–deformation displacement curve of single energy absorbing component in front of vehicle. F_m—average impact force; F_p—peak impact force; δ_e—compression displacement of energy absorbing box; δ_p—the amount of deformation at the peak of impact force

absorbing box is completely crushed; For the front longitudinal beam, its design should ensure that the collision acceleration in the occupant compartment of the vehicle does not have too high peak value in the process of high-speed collision, so as to keep occupant away from large impact force and injury risk, as well as improve safety in collision for the whole vehicle. The deformation of the energy absorbing box and front longitudinal beam in collision can be regarded as the axial crushing process of the pipe structure. In order to facilitate expression and calculation, its deformation behavior can be simplified into an ideal collision force displacement curve as shown in Fig. 4.57.

At the initial stage of structural design, it is necessary to reasonably estimate the energy absorption performance of energy absorbing box, front longitudinal beam and other components according to initial kinetic energy of vehicle in typical collision conditions, so that the specific energy absorption design can follow the given threshold of various evaluation indexes. The common indexes are as follows:

(1) Peak value of impact force F_p: it is usually related to the material and cross-section shape of the energy absorbing device. At the initial moment of collision, the structure deformation depends on the elastic–plastic buckling threshold of structure. When the whole structure is completely compressed, the peak value of impact force F_p will rise rapidly. F_p of the energy absorbing box shall be determined by the stiffness of vehicle connector. Once the impact force exceeds the peak value of impact force that energy absorbing box can bear, the complete compression of energy absorbing box will cause damage to other parts of the vehicle. Therefore, it is necessary to carry out a reasonable design on impact peak force of the energy absorbing box for protecting other key parts of the vehicle body under the condition of low-speed collision. In high-speed collision, the energy absorbing box contributes less to energy absorption. The collision kinetic energy is mainly absorbed by the front longitudinal beam, and F_p of the front longitudinal beam is determined by the maximum acceleration peak that occupant can bear.

(2) Absorbed total energy E_d: that is, the total energy absorbed by the energy absorbing component in low-speed collision, which is calculated by Formula (4.62):

$$E_d = \int_0^{\delta_e} F(s)\mathrm{d}s \tag{4.62}$$

where, s represents the compression displacement; $F(s)$ represents the impact force.

In the same deformation displacement, the more energy the component absorbs, the better the crashworthiness is and the more favorable it is to the safety protection for occupants.

(3) Compression displacement δ_e: In order to save the front space of vehicle, the energy absorbing box shall absorb as much initial kinetic energy as possible with as little compression displacement as possible, provide more design space for the front longitudinal beam and improve the vehicle safety in high-speed collision.

(4) Average impact force F_m: measure the energy absorption capacity of energy absorbing structure. For the energy absorbing box or front longitudinal beam of a given length, F_m is calculated by Formula (4.63). The higher its value is, the more energy it absorbs. On the premise of ensuring that the impact force received by the component is less than the bearing capacity of F_p, the larger the average impact force is, the more favorable it is.

$$F_m = \frac{E_d}{\delta_e} \tag{4.63}$$

where, δ_e represents compression displacement; and E_d refers to the energy absorbed.

(5) Specific energy absorption η_c: refers to the energy that can be absorbed by a unit mass component in a certain state, calculated by Formula (4.64). Obviously, the higher the specific energy absorption is, the higher the energy absorption efficiency of the structure is.

$$F_m = \frac{E_d}{M_s} \tag{4.64}$$

where, M_s represents the mass of the structural component.

(6) Average acceleration in the process of vehicle collision $\overline{a}$: the acceleration time course in vehicle collision a(t), which is generally measured by installing an acceleration sensor on the B-pillar or the rigid structure under the driver's seat. Calculated by Formula (4.65), the average acceleration $\overline{a}$ is inversely proportional to the collision time t and directly proportional to the impact force:

$$\overline{a} = \frac{1}{T}\int_{0}^{T} a(t)\mathrm{d}t \tag{4.65}$$

(7) Energy absorption efficiency S_{te}: when the initial kinetic energy of the vehicle is completely absorbed by the body parts, the energy absorption efficiency is measured by the utilization rate of the parts and calculated by Formula (4.66):

$$S_{\text{te}} = \frac{S}{L} \tag{4.66}$$

where, L represents the length of energy absorbing part; S indicates the maximum crushing amount of energy absorbing part.

(2) Design Scheme of Energy Absorbing Bonnet for Pedestrian Protection

In human-vehicle collision accidents, the front bumper of vehicle is usually the first component to contact and collide with pedestrians. As the pedestrian's body rotates around the front of vehicle, the head and other body parts will collide with the front edge of the bonnet, the upper surface of the cover or the windshield. Head injury is the most fatal form of injury in pedestrian accidents. Its injury evaluation is based on the headform impactor (see Sect. 5.1 of Chap. 5 for details). The collision process between pedestrian headform impactor and bonnet is shown in Fig. 4.58. At a certain initial speed, the headform impactor collides with the bonnet along the collision angle. When the bonnet and the space below it are fully occupied, the headform impactor will collide with the rigid engine, resulting in large impact force and increasing risk of head injury. The safety design of vehicle front engine hood is of great significance for pedestrian's head injury protection [201]. The allowable deformation space under the hood is generally taken as the design constraint, and the HIC threshold of headform impactor is taken as the design goal, so as to further analyze the best energy absorption effect under different structures.

The acceleration time course generated during collision is defined as crash pulse. Affected by the active mass under the inertial effect and the structural stiffness of bonnet, the impact waveform of headform impactor often presents to be multi-peak. Figure 4.59 shows a typical head impact waveform when the energy absorption space

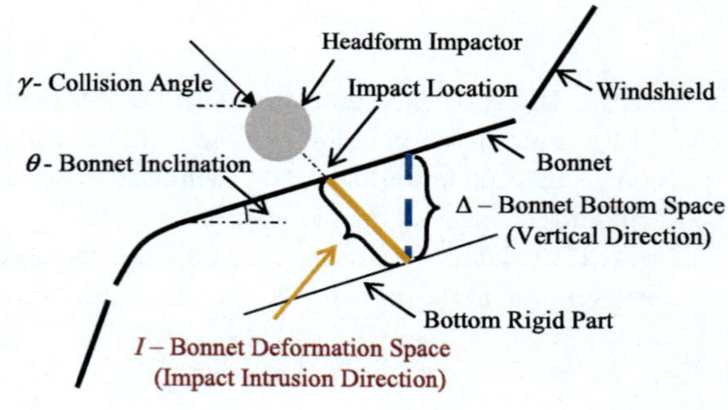

Fig. 4.58 Pedestrian's headform impactor intrusion distance and bonnet bottom space

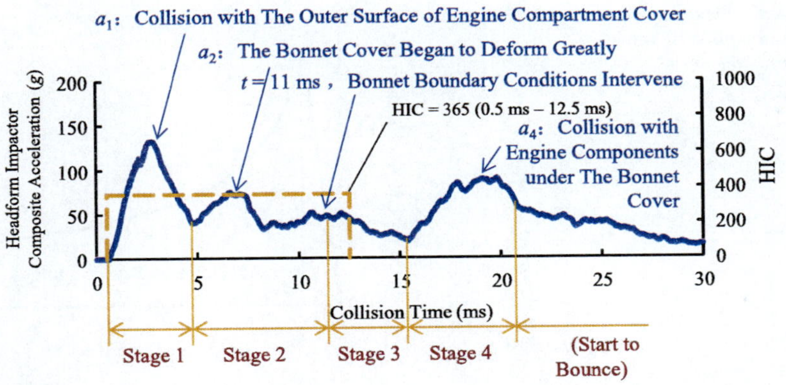

Fig. 4.59 Acceleration-time course of headform impactor

of bonnet is fully utilized. According to the waveform feature, it can be divided into different stages. In stage 1, the headform impactor collides with the bonnet, and its rubber material begins to make extrusion deformation and drives part of the mass on the bonnet to move together; In stage 2, the bonnet structure moves together under the impact, and gradually makes large deformation with the structural stiffness involved, resulting in a relatively stable acceleration; In stage 3, the impact contact reaction force on the headform impactor is mainly determined by the boundary constraint of the bonnet, and the platform area appears; In stage 4, with the further intrusion of headform impactor, the deformation space I of bonnet is completely exhausted, and the headform impactor will contact the components under the cover (such as battery and engine) at a certain residual speed.

According to the HIC calculation formula, the HIC value is sensitive to the acceleration time course and is significantly affected by the peak acceleration. Based on the acceleration time course, the theoretical minimum intrusion distance of headform impactor under certain initial collision speed and HIC value can be deduced. For the collision deformation of the actual bonnet, the idealized acceleration waveform shall be as shown in Fig. 4.60, including two curves: the first section is the rising section from the initial time $t = 0$ to t_p time, where t_p time corresponds to the acceleration peak A; The second curve decays exponentially from t_p time. Based on this ideal synthetic acceleration curve, the deformation energy absorption space of bonnet required by headform impactor under different initial speeds and inclination angles of the cover can be calculated without exceeding a certain HIC value. Accordingly, the design of bonnet shall ensure reasonable structural stiffness and control the impact waveform shape of headform impactor, so as to make it close to the ideal impact waveform as much as possible when HIC meets the requirement.

2. Protection and Early Warning for New Energy Power System Runaway

The collision safety of new energy power system represented by lithium-ion power battery is a mandatory requirement and essential technology for the development of

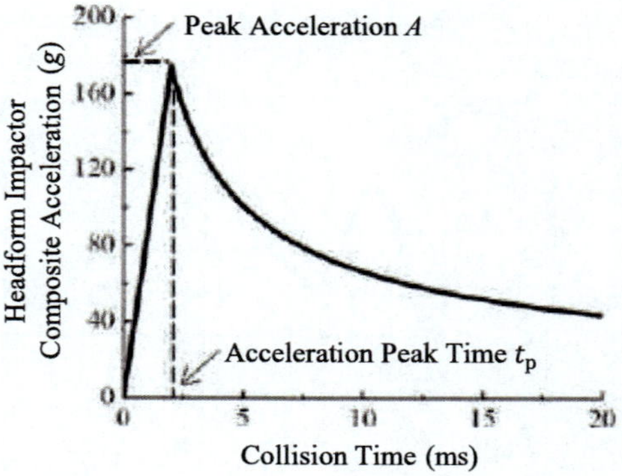

Fig. 4.60 Ideal waveform under condition of bonnet deformation

electric vehicles. Based on the extrusion experiment of power battery under multiple conditions, the correlation of response features of battery material failure, steep voltage drop and temperature rise under external load is set up, which can be applied to the lightweight and collision safety design of battery pack and electric vehicle, such as targeted protection structure and battery layout.

(1) Protection Technology for Thermal Runaway of Vehicle Power Battery

The dangerous conditions involved in the collision protection for the power battery and the occupant are different. Since the battery boxes are mostly allocated on the floor of the passenger compartment, the impact of side column, the bottom support and foreign objects at the bottom will become more hazardous conditions. Some studies have reproduced and analyzed the boundary condition, the breaking mode of the bottom cover board under impact, as well as the extrusion deformation of the local power battery in the fire accident of power battery induced by the collision with foreign objects at the bottom of the electric vehicle [202] (Fig. 4.61). The studies also compared several types of bottom collision protection structures, and proposed the sandwich bottom cover structure filled with lightweight corrugated configuration in consideration of the bottom collision protection performance [203].

The allocation of power battery influences the moment of inertia and the position of center of mass in the whole electric vehicle, and then the impact response (Fig. 4.62). The simulation analysis in terms of the regular allocation of power battery, typical collision conditions and multi-condition accident with different intensities indicates that in the full-width frontal collision, the higher the center of gravity of the vehicle is, the larger the pitch motion range in the collision is, and the more the initial kinetic energy is converted into gravitational potential energy; in the offset collision, the lower the center of gravity of the vehicle is, the smaller the pitch motion range in the collision is, the greater the contact pressure between the rear wheel and the ground is, and the larger the friction energy consumption is required for the deflection motion of the vehicle. By changing the layout of battery pack and adjusting the center of gravity of vehicle, the motion posture of electric vehicle such as pitch and deflection in collision can be controlled, so that new control factors for optimizing collision

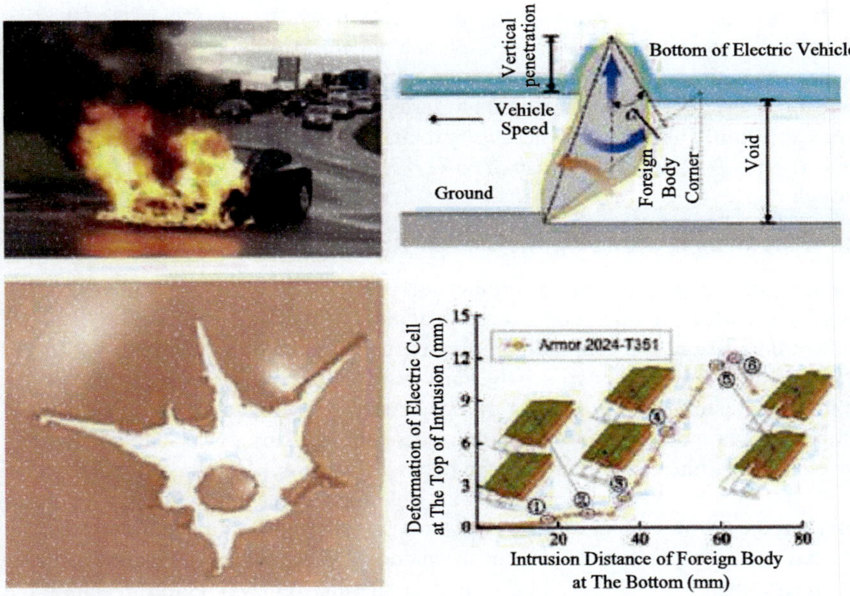

Fig. 4.61 Analysis on collision with foreign objects at bottom of electric vehicle [204]. ①–⑥—Electric core

safety performance is introduced [205]. On this basis, a collision motion posture control strategy that converts one-way collision into composite motion response is formed, which is conducive to solving the design conflict between the power battery layout space and the energy absorption displacement of collision deformation for small electric vehicles.

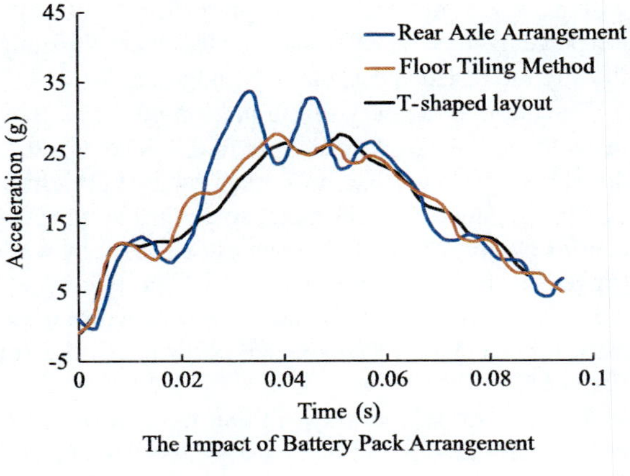

Fig. 4.62 Analysis on impact response of electric vehicle

Runaway protection technique for new energy power system can reduce the risk of thermal runaway of power battery, but can not completely avoid the risk. In case of thermal runaway of lithium-ion power battery due to mechanical abuse such as collision, there will be accompanying hazards such as smoke, fire, combustion and even explosion. Therefore, it is necessary to make judgement and provide warning in the early stage of thermal runaway of power battery system, so as to leave occupants enough escape time.

(2) Early Warning Technology for Thermal Runaway of Lithium-Ion Power Battery

The thermal runaway process of lithium-ion power battery is closely related to the temperature of the battery, and is accompanied by the internal resistance increase, voltage fluctuation, gas generation, pressure increase and other parameter changes. These parameter changes support the early warning of battery thermal runaway. As an important parameter for judging the occurrence and progress of thermal runaway, temperature is monitored by temperature detector in many power battery warning systems. When the temperature of the battery system exceeds the critical temperature, the early warning system sends an alarm. However, the biggest problem of early warning with temperature as a parameter is that there is a certain error in the internal and external temperatures of thermocouples or temperature sensors during the measurement, which may lead to battery thermal runaway before reaching the set warning temperature and eventually result in the failure of warning. To solve this problem, infrared imaging technique can be adopted to monitor the thermal runaway of power battery pack. It can better detect the temperature with high efficiency, fast response time and capability for detecting the temperature of each part of the battery [206].

As a considerable parameter of lithium-ion power battery, internal resistance will change with the state of charge (SOC), working ambient temperature and other conditions. It is often used to evaluate the battery life, state of health (SOH) and performance, as well as to detect the abnormality of the battery. When the thermal runaway of lithium-ion power battery occurs, the internal resistance of the battery will increase significantly. It is appropriate to judge the thermal runaway of lithium-ion power battery for early warning together with other parameters, instead of taking the internal resistance as the only judgment factor.

The thermal runaway of lithium-ion power battery due to mechanical abuse will make the voltage to drop sharply to 0 V. The puncture experiment on 18,650 cylindrical battery shows that the voltage value of lithium-ion power battery pack is irregular in the puncture experiment, so that the voltage can not serve as the only parameter for judging thermal runaway. However, early warning can be given based on the trigger mechanism. By detecting the battery voltage, temperature and heating rate, the early warning system model under different trigger mechanisms can be substituted for calculating the expected occurrence time of thermal runaway, as shown in Fig. 4.63.

During thermal runaway of lithium-ion power battery, the reaction of electrolyte and binder with positive and negative electrodes and their own decomposition reaction will produce a large amount of gas and smoke, resulting in the battery's changing

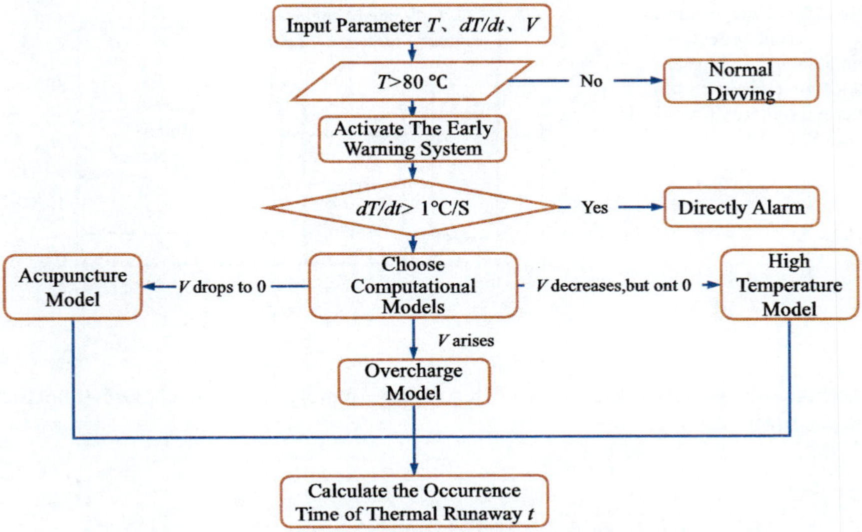

Fig. 4.63 Logic operation diagram of early warning system [210]

pressure, bulge and finally ejecting gas through the pressure reducing valve. Before the ejection of gas, the thermal runaway of the battery can also be warned by monitoring the change of the shell pressure of the lithium-ion power battery.

The gases produced in the process of thermal runaway mainly include H_2, CO, HF and various hydrocarbons. Corresponding sensors can be used to monitor these gases and smoke. In the early stage of thermal runaway of lithium-ion power battery, smoke will occur, and the concentration of small molecule gas will suddenly increase from 0 to hundreds or even thousands of milligram per cubic metre. Therefore, compared with temperature, the characteristic gas generated by battery thermal runaway is more suitable as the judgment basis of early warning. According to the sampling analysis of the gas generated in the process of thermal runaway of lithium-ion power battery, the gas concentration has been rising continuously before the pressure relief valve explodes, and the CO gas concentration has the highest change rate, which can be used as a warning signal of battery thermal runaway [207]. Application of infrared spectroscopy technology, gas sensing technology and other real-time monitoring over inorganic gases such as CO_2 and CO produced by volatile organic electrolyte or side reaction in lithium-ion power battery can effectively realize the early warning of thermal runaway of lithium-ion power battery in theory [208, 209].

Temperature detection, voltage detection, gas detection and other techniques for early warning of thermal runaway of lithium-ion power battery have different TADVs (Fig. 4.64) [211]. The gas detection technique can realize detection and early warning 7–8 min before the peak value of thermal runaway, which is 2 and 7.4 min earlier than voltage detection technique and temperature detection technique respectively. Therefore, the gas detection means can send early warning information earlier, avoid

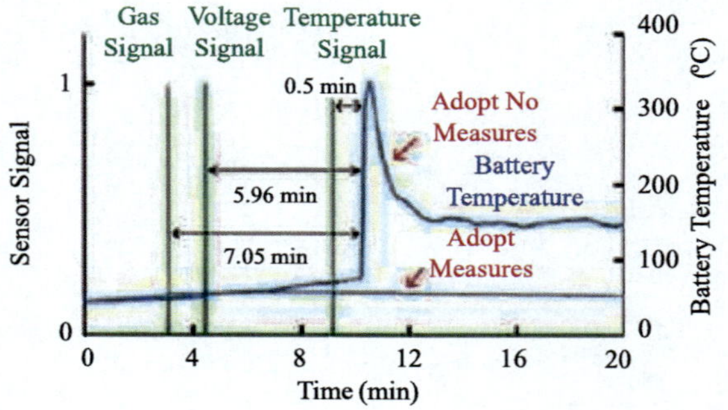

Fig. 4.64 Comparison of early warning effects of temperature, voltage and gas detection techniques on thermal runaway (modified from Ref. [211])

thermal runaway of lithium-ion power storage battery, and leave more time for evacuation.

4.3.3 IP Technology for Traffic Participants

The IP technology for traffic participants involve two types of protection objects: drivers and passengers in the vehicle and vulnerable road users outside the vehicle (pedestrians, cyclists and electric cyclists). In terms of optimizing the protection for drivers and passengers, the research of adaptive configuration technology of occupant restraint system mainly aims at adjusting parameters of restraint system to realize the adaptive adjustment for different collision conditions and signs of drivers and passengers. The ideal equilibrium restraint system makes full use of the characteristics of human anatomy to realize equilibrium in time, space and amplitude. The protection technology for vulnerable road users (taking pedestrians for example) first obtains pedestrians' information by the perception system, predicts pedestrians' behavior on this basis, generates human-vehicle interaction risk domain according to the behavior characteristics of pedestrians and vehicles, and implements early warning and active–passive protection for inevitable accidents (Fig. 4.65).

1. Adaptive Configuration of Occupant Restraint System

The current vehicle occupant restraint system is usually designed according to the given occupant body shape (such as medium-sized adult males) and collision conditions (such as regulations or standard conditions), so it is difficult to provide targeted protection for a variety of occupant bodies and different collision intensities. In view of the above problems, the industry has put forward the concept of adaptive restraint system, that is, to adjust the parameters of the restraint system and optimize the protection effect according to body shape, riding posture and collision strength in the accident. Adaptive occupant restraint system is also considered to be the predecessor and foundation of modern intelligent occupant restraint system. In 2000, the EU set up a special research team to carry out the preliminary investigation and

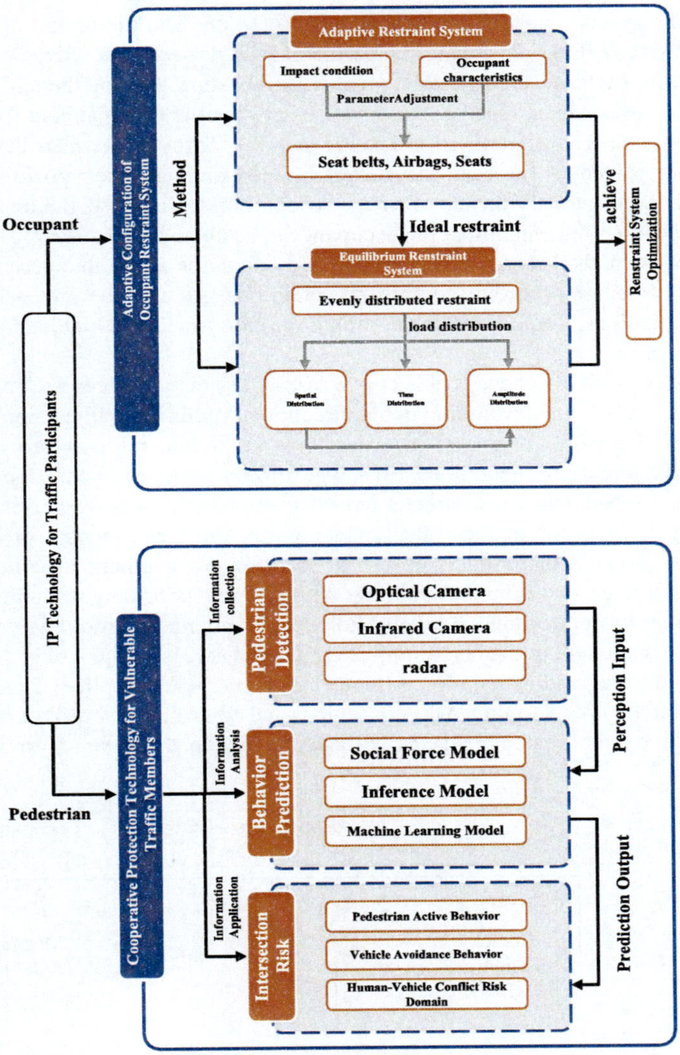

Fig. 4.65 Logic diagram of IP technology for traffic participants

research preparation for adaptive occupant restraint. The research results show that the protection potential of adaptive restraint system is huge, which is expected to reduce the injury of MAIS 2+ by 33– 41% and MAIS 3+ by 14–25% [212].

(1) Basic Concept of Adaptive Restraint System

Based on the traditional occupant restraint system hardware, the adaptive restraint system adds collision condition sensing device and adjustable occupant restraint device. The system composition is shown in Fig. 4.66. The important input information obtained in the occupant compartment includes relevant characteristics of the occupant, such as weight, height, position and posture. Among them, the spatial position of the occupant (such as the distance between the occupant and the airbag) can be detected by infrared and ultrasonic sensors; Through the distributed pressure sensor placed on the seat, the weight of occupants and their position on the seat can be detected; By means of image recognition technology, the body shape, posture and other information of the occupant can be obtained (Fig. 4.67) [213]. The restraint system shall reasonably optimize and adjust the adjustable parameters of the restraint device according to specific conditions (such as the pretension time, pretension amount, force limit value, airbag ignition time, vent size, and inflation mass flow rate).

Consistent with the concept of adaptive restraint system, devices such as vehicle airbag and seatbelt are developing in the direction of multi-level adjustment. Taking the United States for example, in order to adapt to the relevant provisions of FMVSS 208 and other regulations, two-stage airbags have been generally installed on mass-produced vehicles to protect out-of-position occupants from airbag injury (Fig. 4.68a). In order to meet the impact protection requirements of different occupants, research or products have proposed the development of continuously adjustable airbags, that is, to make airbags with different unfolding strengths or variable volumes by controlling the exhaust holes, and regulate the unfolding volume by controlling the length of the restraining strap in airbags [214] (Fig. 4.68b). The seatbelt with two-stage adjustable force limiter has been widely applied. The restraint force of seatbelt is developing in the continuous and controllable way. By controlling the restraint force of seatbelt in the early stage of collision, the forward moving space

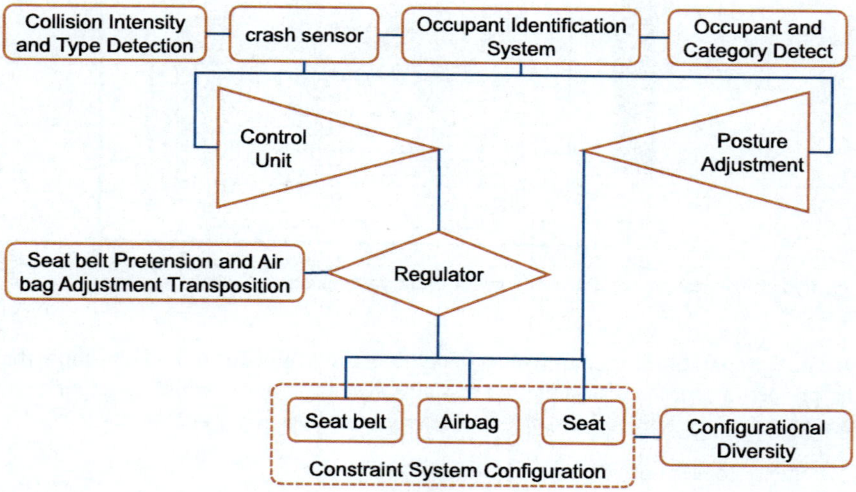

Fig. 4.66 Composition of adaptive occupant restraint system

Fig. 4.67 Example of occupant posture detection device based on image recognition [213]

of occupant relative to the occupant compartment can be fully utilized in the collision to reduce the injury of occupant [215].

(2) Configuration Technology for Adaptive Occupant Restraint System

The specific technique of adaptive occupant restraint is mainly divided into two steps. Firstly, it is to establish the relationship between the parameters of restraint system and the impact mode, occupant feature and injury results through experiment or simulation, find out the reliable and reasonable system configuration spectrum, and confirm the restraint mode to achieve the "personalized" protection by optimization calculation. Secondly, based on the feedback control strategy, such advanced restraint devices as continuously variable seatbelts with semi-active mode are used to restrain occupants in real time. Due to the strong coupling relationship between various

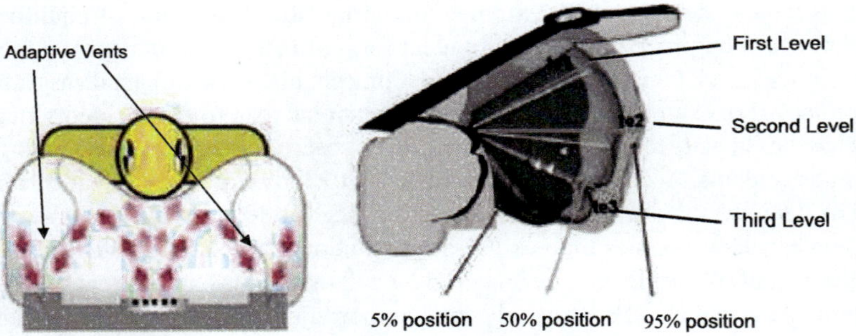

Fig. 4.68 Multi-stage adjustable occupant airbag: **a** variable exhaust property airbag; **b** variable unfolding volume [214]

parameter variables, constraint optimization is a strong nonlinear optimization issue [216]. Most of them take weighted injury index as the optimization objective. This single index is more intuitive and convenient for rapid evaluation of experimental results. Heuristic algorithm and fitting response surface show good performance in dealing with constrained system optimization. With simple application and strong robustness, the heuristic algorithm can be used to deal with problems that have multiple local optimal solutions, but it demands more times of calculation to obtain better results [217]; The response surface algorithm can acquire better results by less solving times, but the option of specific response surface types restricts the prediction accuracy of the model [218].

The optimization of adaptive occupant restraint system can be realized by building an optimization platform that can automatically carry out simulation pre-processing. The structural relationship and operation flow of the optimization platform are shown in Fig. 4.69, which mainly includes three parts: core module, pre-processing and post-processing. The core module is mainly responsible for the setting of model parameters and the output of key loss indicators in the iterative process of constraint system parameter optimization. The pre-processing mainly relies on the interactive interface provided in the kinematics calculation platform. For example, when the seat position of occupants with different body shapes changes, pre-simulation is conducted to judge whether the seat position meets the shape requirement, so as to realize the functions of model replacement, attribute adjustment and seatbelt fitting. The post-processing part extracts the output information related to the injury by the script. The next optimization iteration is mainly based on the predicted and estimated occupant injury value, which is continuously sent to the core module for optimization.

(3) Concept and Conceptual Prototype of Equilibrium Constraint

At present, the conventional vehicle occupant restraint form is "seatbelt-airbag", which still can not play the most ideal restraint protection performance in line with the characteristics of human anatomical structure. The starting point of equilibrium constraint is to realize the nearly idealized constraint form that minimizes the risk of injury under a given condition and under the optimal parameter configuration for constraint system. In some extreme conditions, the application of equilibrium restraint has been proved to be a more ideal form of restraint in occupant collision protection research (such as racing car) and biomechanical research (such as cadaver experiment). For example, compared with conventional road traffic accidents, human body has the capacity to withstand faster collision speed and acceleration in racing car collision accidents. According to the statistics of 202 Indy Car Racing accidents from 1993 to 1998, the maximum collision speed is as high as 307 km/h and the maximum collision acceleration is as high as 127 g (gravitational acceleration). The collision acceleration of 62 accidents is as high as 60 g and that of 3 accidents reaches 100 g. In these 202 accidents, there are only some minor injuries to occupants in vehicles, without any serious ones [219].

During the car racing, the included angle between the driver's seat back and the vertical direction is about 45°, and the driver wears 6-point seatbelt, helmet, head and

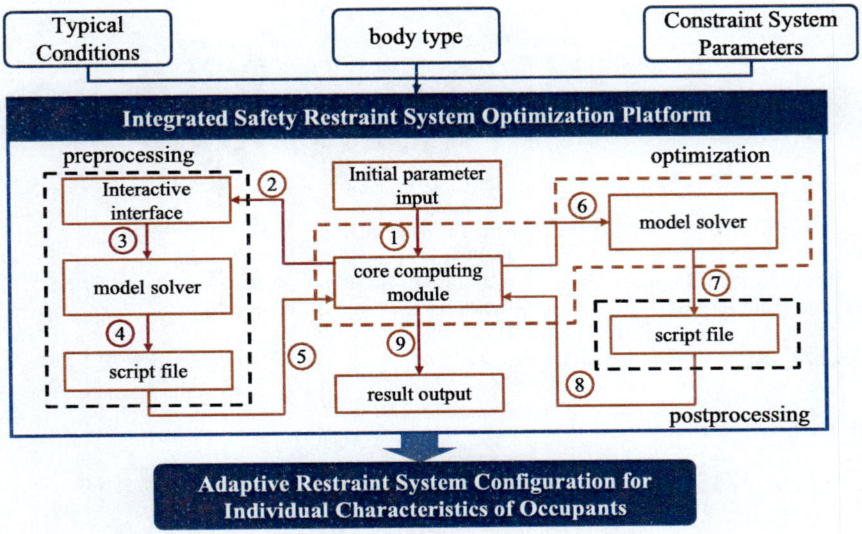

Fig. 4.69 Optimization platform of adaptive occupant restraint system. ①–⑨—Workflow sequence

neck support (HANS) and other collision protection devices [220]. Among them, the helmet is also connected to the HANS device through a sliding rope. Relevant studies show that this form of restraint can effectively reduce the displacement of the head and the stress on the neck, so as to reduce the risk of injury [221]. Meanwhile, for side collision conditions, the car also provides a side anti-collision net to reduce the risk of head and neck injury [222]. The protection mechanism of 6-point seatbelt, helmet plus HANS and side anti-collision net is to apply the binding force more evenly to human body and control the motion response of occupant in the collision, which can not only reduce the injury of direct loading parts, but also alleviate the injury due to the relative motion between parts (Fig. 4.70). Compared with the seatbelt-airbag restraint in ordinary passenger cars, the restraint device in racing cars can provide effective protection under higher collision intensity. Although with some comfort lost, the direct application of restraint devices in racing cars to ordinary passenger cars also shows that there is still room for improvement of human protection devices in ordinary passenger cars.

In view of the impact of acceleration load on human cardiovascular injury, a sandbox trolley test with uniform restraint load is carried out (Fig. 4.71). By providing uniform load to the trunk and thorax with sand, it is found that under the impact load of initial speed of 60 km/h, there is no rib fracture and cardiovascular rupture, yet only slight visceral trauma due to inertia [223]. This shows that under the uniformly distributed restraint load, the chest can bear the high-strength collision load. This conclusion is consistent with the design experience of human restraint system of the racing car mentioned before, i.e., the human body injury in the collision occurs not only in the parts directly loaded by the restraint, but also in the parts not directly

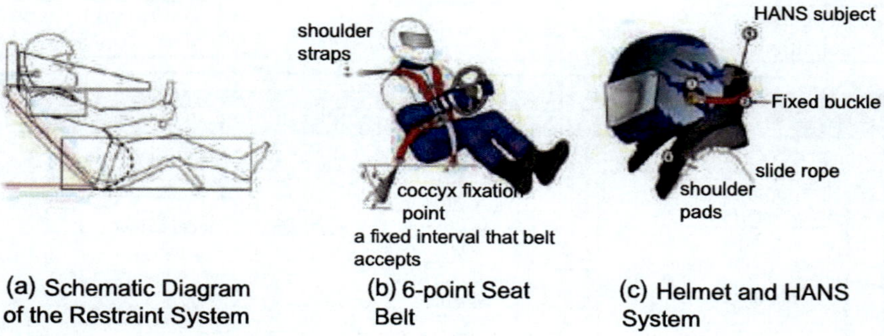

Fig. 4.70 Driver restraint system in racing car

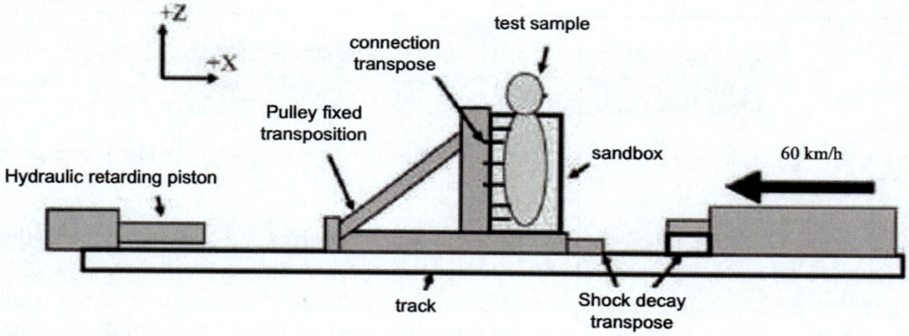

Fig. 4.71 Sandbox trolley test for vehicle occupants [223]

loaded by the restraint but with large relative movement between the connected parts. Therefore, making full use of the idea of equilibrium constraint can effectively restrict the relative motion between various parts of human body, make it bear impact load more evenly, and finally achieve the goal of no serious injury under the impact load larger than that required by the current vehicle collision regulations.

The conceptual prototype of equilibrium constraint is designed with the goal of uniformly applying the constraint load to the stronger parts of human body and reducing the overall injury risk. It shall meet the following three principles:

(1) In terms of the spatial distribution of binding force, it makes full use of the bearing capacity of human body and evenly apply the binding force to the stronger parts of the body, including shoulder, chest, hip and knee. At the same time, to take into account the injury risk to the neck due to the relative movement of adjacent parts, it is also necessary to impose the binding force on the head.
(2) The binding force should be kept as constant as possible over time. It shall make human body decelerate evenly as much as possible in a given space, so as to avoid the injury of the directly loaded part caused by the amplitude of the load.

(3) In terms of the relative amplitude of the binding force in each part, it shall reasonably distribute the binding force, control the overall movement and the relative movement between parts at the same time, so as to prevent the non-directly loaded parts from being injured due to the uncoordinated movement between adjacent parts.

The protection effect of the conceptual model based on equilibrium constraint is simulated and analyzed by using the mannequin. The results show that compared with the ordinary seatbelt restraint, the equilibrium constraint can effectively alleviate the chest deformation and rib strain; Under the high-speed frontal collision intensity of 80 km/h, the occupant injury output is still controllable; Compared with seatbelt restraint, knee baffle can effectively control the motion posture and injury output of backward occupants in the process of collision. It is preliminarily proved that equilibrium constraint is a more ideal form of restraint than seatbelt-airbag constraint. Based on the principle of equilibrium constraint system, the parameter configuration of restraint system with the lowest risk of occupant injury can be found by optimizing the restraint amplitude of each part under a given condition. The equilibrium constraint form and the corresponding constraint system parameter configuration can be taken as the objective of real constraint system design.

(2) Cooperative Protection Technology for Vulnerable Road Users

IP technology for vulnerable road users represented by pedestrians is an important part of the safety of AV. It is necessary to combine active and passive protection techniques and implement cooperative protection, so as to reduce the risk and severity of collision accidents. Firstly, pedestrian collision prevention and early warning system shall perceive, track and predict pedestrians. The sensing detection end relies on the sensor to detect the area in front of the vehicle body, detect the pedestrians in the area, obtain the pedestrian position and motion posture information, and predict the pedestrian behavior as well as the interaction risk between the pedestrian and vehicle. The system judges the risk degree according to the obtained information, and then reminds the driver or controls the vehicle to make collision avoidance operation.

(1) Pedestrian Detection

Pedestrian perception and detection is the first part of collision warning, which can be realized by vehicle optical camera, infrared camera, radar and other sensors. For instance, the charge coupled device (CCD) camera as a visual sensor has the advantages of wide detection range, large viewing angle and clear image, and collected visual information conforming to the cognitive mode of human eyes. At present, there are two kinds of pedestrian detection algorithms based on sensor information: one is the traditional detection algorithm based on machine learning training classifier, and the other is a new detection algorithm based on large-scale network model in deep learning training. The machine learning selects the feature points that can describe the characteristics of the target, and then realizes the target detection by training the classifier. For example, human body includes head, body, left and right arms, which are described by Haar wavelet features respectively, and then the corresponding classifiers are respectively trained and together describe the detection on the whole body [224]. Deformable part model (DPM) is another common pedestrian detection classifier, which adopts the feature of Histogram of Oriented Gradient (HOG) and trains the pedestrian part sub-model. In consideration of the movement and deformation of pedestrians, it can effectively detect pedestrians with different postures [225, 226].

With such disadvantages as poor robustness, low computational efficiency and bad effect, the traditional machine learning can not adapt to the variability of pedestrian postures. The emergence of deep learning has brought many new detection algorithms with better performance. The main process of deep learning is to build a multi-layer neural network model, spread the underlying features reversely layer by layer, form semantic features, and finally shape an accurate feature map depicting pedestrian targets. Currently, deep learning has become the mainstream target detection mode. Yolo detection network is better in the real-time performance [227]. Taking the detection as a regression problem, its network model learns features of the whole input image based on the convolution network, and predicts the location and category information of pedestrian target, so as to realize end-to-end target detection. Its detection speed is very fast, and the detection accuracy of small targets is also high.

(2) Pedestrian Behavior Prediction

Pedestrian behavior prediction is a difficulty in collision warning system. In the traffic environment, the pedestrian behavior has strong mobility and high randomness. It is difficult to obtain the behavior characteristics only by visual information. Common pedestrian behavior prediction models are mainly divided into three categories:

(1) The social force model based on human-to-human and human-to-objects interaction. The pedestrian motion is simplified into rigid body motion, and the pedestrian walking trajectory is predicted through the interaction planning of gravity and repulsion. The model has simple representation and few parameters, but has high requirement for the setting of model parameters. In practical application, it has limited effect for predicting individual pedestrian behavior, but more suitable for the study of human group motion behavior [228].
(2) The inference model based on transformation among different states. The transformation between different states during pedestrian movement is regarded as a Markov process. Among the behavior observation variables, the states of pedestrian standing, walking, jogging and fast running are discretized into finite variables. The transformation relationship between different state variables is studied and defined for predicting pedestrian behavior [229]. Based on Dynamic Bayesian Network (DBN), another common approach represents the degree of danger perceived by pedestrians by characterizing the visual direction of pedestrians and the dynamic relative position between pedestrians and vehicles, and further infers the future motion behavior of pedestrians in combination with their motion state [230].
(3) The machine learning model based on data-driven model. It is to first collect a large number of pedestrian trajectory data in the traffic environment, apply

the machine learning algorithm for large-scale data learning, and infer the motion states of pedestrians in the future based on the learned high-dimensional prediction model, including position, speed and other information [231].

(3) Prediction of Interaction Risk between Pedestrians and Vehicles.

In terms of the interaction risk between pedestrians and vehicles, if pedestrians and vehicles perceive each other, they may take active avoidance behavior in the interaction process, which is an important factor affecting the occurrence and severity of the collision. Based on the perception capability of vehicles and pedestrians, as well as the quantitative pedestrian avoidance behavior (refer to Sect. 4.2), four types of human-vehicle interaction can be defined:

(1) When pedestrian and vehicle notice each other, pedestrian takes the initiative to avoid danger, and vehicle (driver or on-board active safety system) takes emergency braking.
(2) Pedestrian notices the approaching vehicle and takes the initiative to avoid it, whereas vehicle does not notice the pedestrian and still drives at the initial speed.
(3) Pedestrian does not notice the approaching vehicle and still moves forward at the initial speed, whereas vehicle notices the pedestrian and takes emergency braking.
(4) Both pedestrian and vehicle does not notice each other and continue to move forward at their initial speeds.

Under any given interaction condition, the risk domain of human-vehicle interaction can be calculated and visualized by vehicle dynamics and pedestrian kinematics. Taking the typical conditions in Table 4.15 as an example, Fig. 4.72 shows the corresponding four situations in human-vehicle interaction. If pedestrians appear in the shaded area, there is a collision risk. The area where I* and I** intersect indicates that the collision cannot be avoided regardless of any avoidance behavior taken by vehicles and pedestrians in this area. The identification of unavoidable areas helps to realize the coordination of pedestrian active and passive protection. In other words, while predicting the danger before the accident, the vehicle takes necessary avoidance measures in time, and takes passive protection measures in the case of inevitable collision, so as to minimize the risk of pedestrian injury.

4.4 IP System

IP system involves safety risk warning and automatic implementation of specific protection behavior. According to different protected objects, it can be divided into vehicle body protection, power system protection and traffic participant safety protection.

Table 4.15 A typical pedestrian-vehicle interaction condition (for example), under the assumption that vehicle detection system can perceive pedestrians

Vehicle parameters	$L_{vw}(m)$	$a_{v\cdot a}(g)$	$a_{v.al}\ (g)$	v_v^0(km/h)	$D_{vr}(m)$	$D_v^0(m)$
	2	−0.7	−0.65	60	100	$(0, D_{vr})$
Pedestrian parameters	$v_p(m/s^2)$	Moving direction				
	1.0	Approach from the left front of the vehicle, and the forward direction is perpendicular to the forward direction of the vehicle				

Where, L_{vw} represents the width of the vehicle body; $a_{v\cdot a}$ indicates vehicle emergency braking deceleration; $a_{v.al}$ represents the maximum lateral acceleration of the vehicle; v_v^0 is the initial speed of the vehicle; D_{vr} represents the detection distance of on-board detection system; D_v^0 indicates the distance from the pedestrian in the forward direction of the vehicle (the distance from the potential collision point); And v_p is the initial speed of the pedestrian

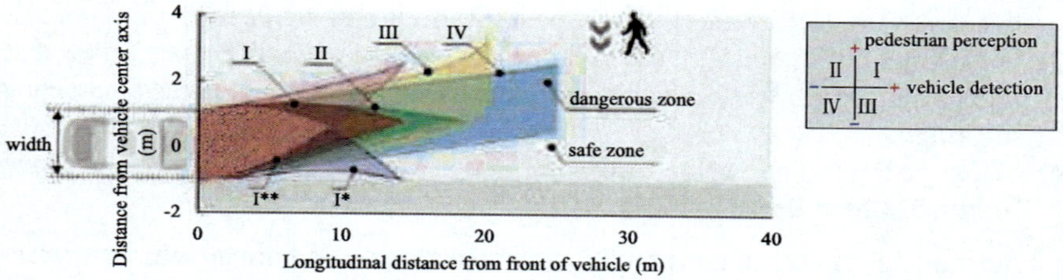

Fig. 4.72 Pedestrian-vehicle interaction risk domain (I* represents that the vehicle takes emergency braking and steering at the same time on the premise of ensuring stability of the vehicle body (I* turn left, I** turn right))

4.4.1 *Collision Risk Warning and Automatic Protection System*

In the development of automatic driving traffic scenario, the collision risk warning and automatic protection system facilitates the avoidance of collision accidents by the perception and judgment of collision risk, or provides effective on-board protection devices in inevitable accidents. In the DVE system, there is real-time interaction among traffic participants, IVs and road scenarios. Taking the collision avoidance process for example, the IV automatically senses and continuously monitors other vehicles, pedestrians or obstacles that may collide with the ego-vehicle, gives an alarm or takes braking, steering and other measures simultaneously to change its dynamics and minimize the risk of accidents (Fig. 4.73). At the same time, as the physical carrier of the occupant protection system and a new product in the future automatic driving, the intelligent cockpit is capable of such functions as state detection.

1. Vehicle Risk Identification and Safety Protection System

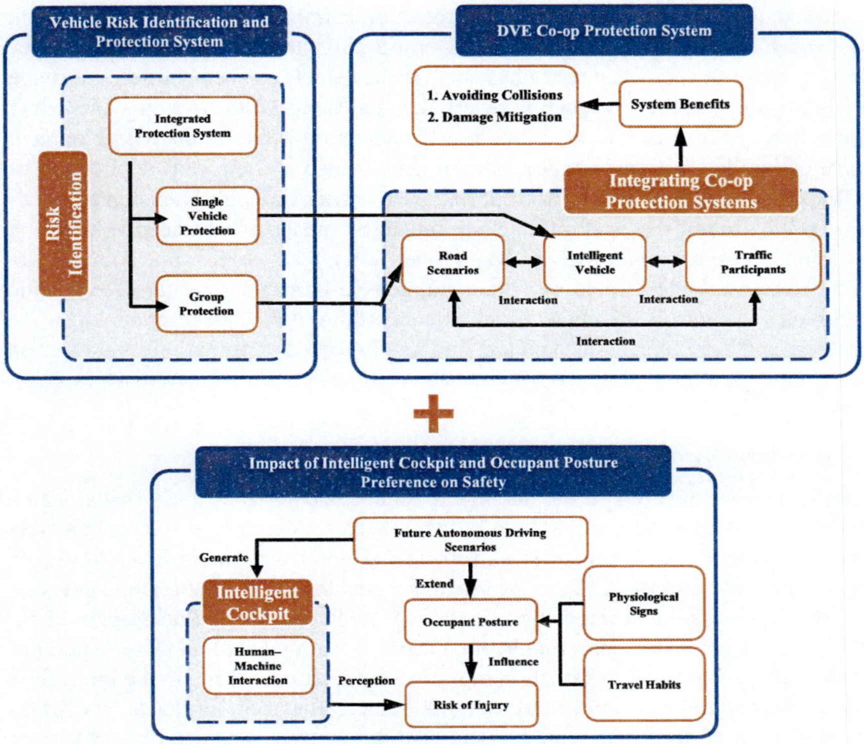

Fig. 4.73 Research framework of collision risk warning and automatic protection system

Regarding the DVE system as a whole, vehicle risk identification integrates the characteristics of vehicle motion state and driver state with road environment information, monitors the potential collision risk during driving, and ensures the in-time start of safety protection system, so as to reduce the accident risk. Taking AEB as an example, its working principle is to detect the movement state of surrounding vehicles, motorcycles, bicycles, pedestrians and other obstacles on the road through sensors such as radar, camera and lidar, and judge whether to alarm (such as reminding the driver to brake) or actively brake according to the estimated collision time for avoidance or reduction of collision injury. Its function can be generally realized in three stages: when detecting the risk of collision, the system will remind the driver of the risk and to brake for avoiding danger by means of vision (flashing of alarm light), hearing (warning tone) or touch (vibration); However, when the driver does not respond, the system will enter the pre-braking stage. If the driver starts braking at this time, the braking system will respond faster than usual, and increase the braking force to assist the driver in completing the braking process; If the driver still does not respond during the pre-braking phase of the system, the system will automatically brake for avoiding the risk.

Based on risk early warning, safety protection is implemented by corresponding systems, including single vehicle and real-time group interaction protection systems. The occupant and occupant restraint system in the vehicle can be regarded as a single vehicle system, and its interaction determines the safety of the system. The safety protection system based on real-time group interaction is the ultimate development form of intelligent transportation system, and is also a significant application of IOT technology in the field of transportation. It carries out high-precision and real-time perception of the road traffic environment by means of advanced vehicle and road sensing equipment (radar, on-board video, etc.), realizes real-time communication between vehicle-vehicle, vehicle–road, vehicle-human and vehicle–road traffic facilities and controls command execution according to the agreed communication protocol and data interaction standard, and finally form an optimal safety protection system integrating intelligent traffic management and control, dynamic information service and IOV automatic driving.

2. Intelligent Cockpit and Safety Protection

Intelligent cockpit refers to the intelligent vehicle driving and riding space, which mainly integrates intelligent products of vehicle seats, interior trim and cockpit electronics, cooperates with human, road, vehicle and cloud, and finally develops into an intelligent mobile space [232]. In the intelligent development of automobile industry, the shape, function and interaction mode of vehicle cockpit are changing without exception. In a narrow sense, intelligent cockpit is mainly based on driver's point of view, forms a variety of information input and output modes through the innovation and coordination of cockpit interior and in-vehicle electronic products, constructs an intelligent HMI system, and conducts the driving assistance and travel service functions during driving.

Current intelligent cockpit is still in a relatively primary stage of development, and its development of normal driving function takes priority over that of safety function. In fact, from the perspective of IP, intelligent cockpit can provide input information for collision risk warning by fatigue driving detection and driver's health state monitoring. At the same time, it is also the carrier of automatic protection system. Through the independent perception layer, the intelligent cockpit system perceives sufficient data, including vision (optics), voice (acoustics) in the vehicle, chassis and body data such as steering wheel, brake pedal, accelerator pedal, gear and seatbelt. It comprehensively judges the physiological state (portrait, facial features, etc.) and behavior state (driving behavior, sound, limb behavior) of the driver and passenger with biometric technique (focus on face recognition and voice recognition in the cabin). And then it sends the interaction request according to specific scenario, such as providing consultation information, providing vehicle status information, facilitating "vehicle to human" active interaction, releasing "human to vehicle" interaction burden of the driver in the driving process, enhancing interaction experience, and decreasing frequency of driver's inattention, etc., so as to improve driving safety. For the driver's condition monitoring, the common means is to monitor its psychophysiological activity process based on eye tracking technology, and monitor its physiological signal changes based on EEG, ECG, EMG and other means, so as to estimate the driver's current fatigue state and health; or to input the obtained data reflecting the driver's condition into the calculation model, accurately predict the driver's subsequent behavior state, then estimate the impact of its behavior mode on road safety, and judge the necessity of adopting corresponding intelligent technical means (such as risk early warning and emergency braking) to ensure road safety.

Intelligent cockpit can also be combined with adaptive restraint system to serve the performance improvement of passive safety technology. For example, after an accident, the system will carry out targeted protection according to the interior environment and occupant status in the vehicle, so as to minimize the injury degree of occupant. Relevant technologies play an important role in reasonably improving the layout of operating system in the vehicle, enhancing HMI experience, reducing the occurrence of accidents and ensuring road safety.

3. Occupant Posture Preference and Its Impact on Safety in Automatic Driving Scenario

With the development of automatic driving and intelligent transportation technology, when the vehicle occupants no longer need to make driving operation, their requirements for the interior environment and seating postures tend to be personalized and diversified. Therefore, it is of great significance for road traffic safety improvement by understanding the seating posture preference of occupants and the impact of different postures on safety in the future automatic driving scenario, appropriately designing seat arrangement in the vehicle and putting forward reasonable posture suggestions for occupants.

There is a significant correlation between posture preferences and body shapes and daily habits of occupants [233]. By investigation towards populations in different regions of China, different nonparametric tests are conducted according to specific data types, and the correlation between different independent variables and seating posture preference can be inferred. The results show that female, older, heavier passengers by public transport tend to choose the posture of leaning back, which in fact may lead to a relatively high safety risk in collision accidents, especially for such parameters as old age and weight are related to the high risk of injury. In terms of travel mode, passengers by public transport tend to lean back to the seat back and stretch their bodies, whereas passengers travelling by car prefer to keep away from the seat back. In terms of travel distance, most passengers choose the relatively traditional seat arrangement mode in short-distance daily travel, whereas the selection rate of "face-to-face" seat arrangement in long-distance travel reaches 42.0% (Fig. 4.74). These conclusions based on observation and investigation provide an important reference for designing reasonable and personalized seat layout schemes for different types of passengers.

4. Connotation, Objective and Benefit of DVE Cooperative System

The protection based on real-time group interaction needs to be realized by DVE cooperation, including: ① human-vehicle interaction system: refers to human-vehicle interaction state, that is, the state of interaction between pedestrians and vehicles in traffic [234]. ② vehicle-vehicle interaction system: refers to the two-way

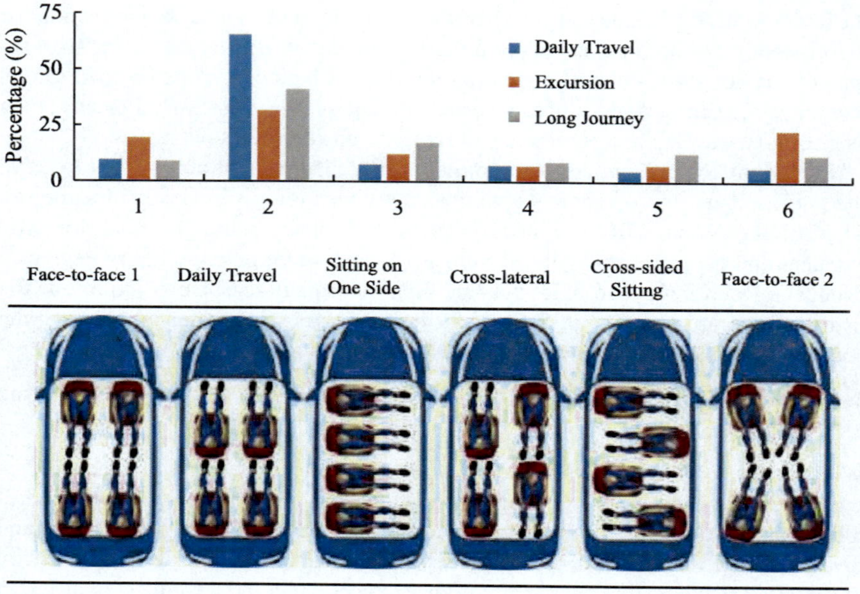

Fig. 4.74 Seat layout preferences in different travel scenarios

communication between vehicles in a limited geographical area, which facilitates the mutual transmission of basic safety information between neighbouring vehicles, such as location, speed and driving direction, and greatly reduces the probability of vehicle collision accidents. ③ vehicle–road cooperation system: aims to realize multi-faceted, real-time and dynamic information interaction and judgment between vehicle-vehicle, road-vehicle and human-vehicle, actively control the vehicle with application of more advanced wireless communication and new Internet technologies, and thus facilitate the full interaction of DVE, and consequently ensure traffic safety, improve road traffic efficiency, and form a safe, efficient and connected road traffic environment [235].

The main objectives for application of DVE cooperation system include: ① ensure reliable information transmission, enable the system to obtain the information of vehicles and road sections in real time, and provide necessary information and reliably comprehensive services for traffic participants; ② improve road traffic efficiency and ensure road safety, so as to ensure the safe and efficient operation of the whole road network and minimize congestion and traffic accidents [236]. Among them, in terms of information transmission, pedestrian and vehicle judge each other's intention by exchanging their own information, and then make decisions conducive to their own safety. The vehicle obtains the road section situation with on-board equipment and sensors, and transmits these data to the roadside equipment through information interaction. The roadside equipment analyzes and processes the data and formulates reasonable driving strategies for the vehicle, so that the vehicle can reach the destination through the optimal path. The vehicle transmits its own operation status to other adjacent vehicles through information interaction to ensure its safe operation. In terms of operation assurance, when the vehicle encounters sudden traffic conditions such as emergent steering and braking, it will inform the surrounding vehicles and pedestrians of the sudden situation through information interaction, so that other vehicle drivers and pedestrians can quickly make lane-change decisions to avoid traffic accidents.

The benefits of the DVE cooperation system are mainly reflected in the following three aspects: ① economic benefits: by controlling traffic behavior and making full use of the space–time resources of the traffic system, it can improve the traffic bearing capacity of the road network, reduce travel time and travel cost; ② safety benefits: it can break through constraints of limited human perception and response ability, and automatically perceive and pre-brake by means of relevant communication equipment to effectively avoid collision accidents; ③ environmental benefits: vehicle–road cooperation technology makes the traffic flow on the road more smooth and orderly, improves the driving speed of vehicles and the overall traffic efficiency of the road, and thus reduces the energy consumption of vehicles. However, there are also many problems in the wide promotion and application of DVE cooperation system in the current era, such as the urgent need to improve information infrastructure, low degree of interconnection and sharing, singleness of obtained external data, imperfect market mechanism, unsound standardization system, unclear definition of safety accident responsibility and high cost [235]. In conclusion, the DVE cooperation system still has large space and potential for development.

4.4.2 Monitoring System for IP-Based Vehicle Body and Power System

The on-board monitoring system for IP is targeted at three objects, i.e., vehicle body structure, new energy power system and driving system. The crashworthiness can be improved by application of the crashworthiness design, structural construction and new materials to the lightweight vehicle body structure; The real-time monitoring and early warning device for the runaway state of new energy power system represented by battery management system estimates the battery state, diagnoses the fault type and carries out early warning by monitoring the feature parameters such as overcharge and discharge, abnormal temperature, smoke, leakage and collision, so as to win time for human protection; In addition, dynamic adaptive protection systems such as electromagnetic active suspension and adaptive front-lighting system achieve protection objectives with corresponding mechanisms (Fig. 4.75).

1. Crashworthiness Design of Lightweight Vehicle Body Structure

Vehicle body structure plays an important role in absorbing deformation energy, transmitting collision load, maintaining living space and reducing the risk of occupant injury. Capable of reducing body weight, collision kinetic energy and human

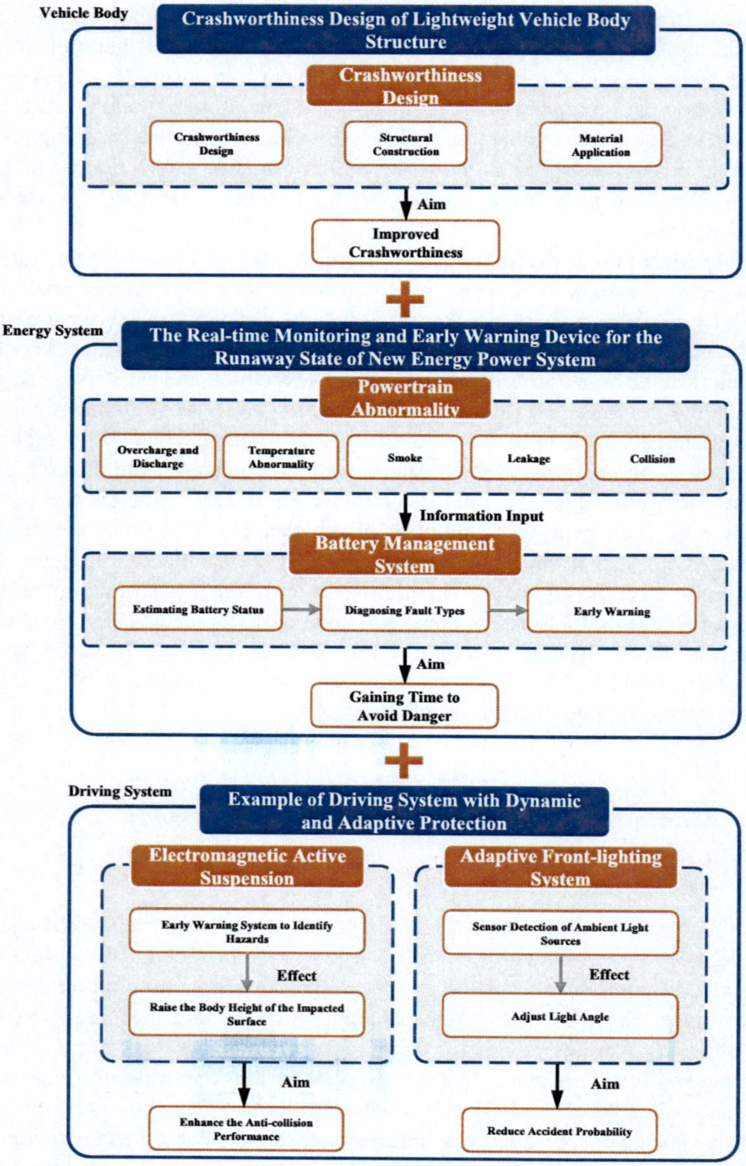

Fig. 4.75 IP-based vehicle body structure, power system monitoring system and adaptive driving system

injury risk, lightweight body design is of great significance to improve vehicle power, economy and safety [237]. Improving the crashworthiness of lightweight body structure is the key technical means to realize the three themes of vehicle energy saving, environmental protection and safety.

A specific body lightweight design technique often covers three aspects, i.e., design method selection, material selection and structural construction. The three complement each other under certain constraints to jointly realize design optimization. Among them, the design methods mainly include topology optimization, size optimization, shape optimization, single objective, multi-objective cooperative optimization, etc.; Lightweight materials mainly include high-strength steel, lightweight alloys (such as aluminum alloy, magnesium alloy and titanium alloy), engineering plastics, composites, carbon fiber reinforced plastic (CFRP), etc.; Structural construction techniques mainly include hot stamping, hydroforming, tailor welded blanks, tailor rolling blanks, structural adhesive technique, dissimilar material connection technique, etc. At present, more than 90% of the body parts are composed of thin-walled pipe fittings with different sections and shapes (Fig. 4.76). The crashworthiness of thin-walled pipe fittings is essential for the overall body crashworthiness. The thin-walled structure has such advantages for improvement of vehicle safety and lightweight as light weight, excellent crashworthiness and energy absorption performance, strong designability and easy processing and manufacturing.

Foam filling is a relatively mature lightweight design technique for the thin-walled pipe structure of vehicle body. In short, the original hollow thin-walled parts and components are filled with lightweight foam materials, such as polyurethane foam and metal foam aluminum (Fig. 4.77), so as to further improve their energy absorption performance on the premise of little change in the structural quality. The most widely used filler material is metal foam aluminum, including the following performance advantages: ① low density, light weight, and changing density with porosity, which is only 0.1–0.6 times that of solid aluminum of the same volume; ② high strength and excellent energy absorption performance. During compression, there is a high and wide stress platform, which can absorb energy under basically constant stress; ③ simple preparation, strong designability and wide application range.

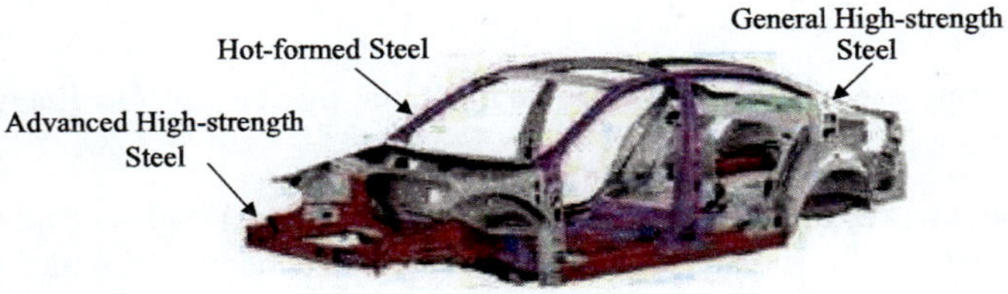

Fig. 4.76 Typical lightweight vehicle body structure (different color areas represent the use of metal materials with different strength) (Graph: 2015 Volkswagen Passat) [238]

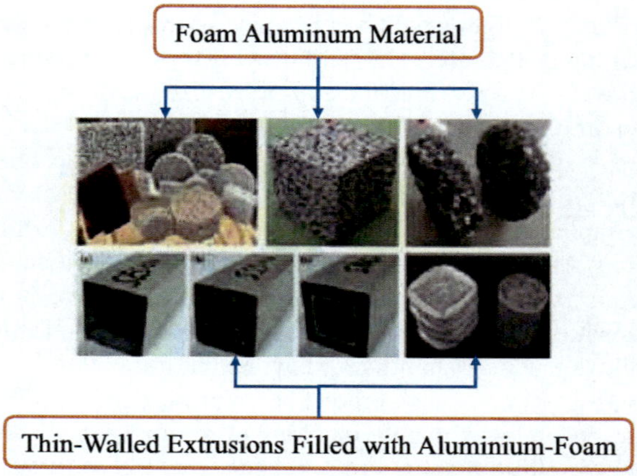

Fig. 4.77 Foam aluminum material and thin-wall structure filled with foam aluminum

Functional gradient design is to instruct the structural or material parameters of thin-walled pipe fittings to change according to a certain gradient law, so as to meet the functional or performance requirements of different positions and thus realize the overall design optimization. The idea of the design is often adopted in the coordinated design of bumper energy absorption box and front longitudinal beam. The specific design parameters include:

(1) Geometric sizes of components: design geometric size of the thin-walled structure (such as the gradual diameter of circular tube and the gradual side length of the square tube) according to a certain continuous function.
(2) Gradient wall thickness: adjust and optimize the wall thickness of thin-walled structure according to a certain law in the axial direction.
(3) Gradient change of filler material: when filling the pipe fitting structure with foam material or negative Poisson's ratio material, design the density according to a certain law of energy absorption demand.
(4) Strength of gradient materials: conduct certain heat treatment on thin-walled structural materials to make the components change differently in strength in the axial direction.
(5) Combined gradient: freely combine the above four gradient design methods according to certain requirements, so as to form a composite gradient design with the advantages of different designs.

2. Monitoring and Early Warning Device for Runaway State of New Energy Power System

The thermal runaway of lithium-ion battery due to mechanical abuse and other factors will lead to smoke, fire, combustion, explosion and other accidents, which will pose a great threat to the life safety of drivers and passengers. According to relevant regulations, the lithium-ion battery pack or system shall provide a heating accident alarm about 5 min in advance to remind the passengers to evacuate before the runaway heat of a single battery and the following heat diffusion and danger in the occupant compartment [239]. Therefore, it is particularly important to equip a set of targeted battery management system (BMS) for the power battery pack of electric vehicles, so as to realize the effective protection, monitoring and early warning of the battery pack, ensure its operation safety, and improve the work efficiency and service life. The main components of BMS include battery terminal module (mainly for data acquisition, including voltage parameters, current parameters, temperature, communication signals, etc.), intermediate control module (mainly for controlling the charger and communicating with the whole vehicle system, etc.), and display module (mainly for data presentation and HMI) (Fig. 4.78).

The operating principle of BMS is shown in Fig. 4.79. By equipped with different sensors, the power battery management system can detect the total current and total voltage, the voltage (prevent overcharge, over discharge and even reverse polarity) and temperature of single battery, smoke (monitor electrolyte leakage), leakage (insulation monitoring) and collision, etc., and then estimate the state of power battery in terms of charge, function, energy, health, fault and safety, etc. Based on the collected sensor signals, the diagnosis algorithm is adopted to diagnose the fault type, carry out early warning, and make the vehicle controller to deal with the fault effectively (when it exceeds a certain threshold, BMS can cut off the power supply of the main circuit independently), so as to prevent overcharge, over discharge, high temperature, low temperature, overcurrent, leakage and other damages to power battery or safety of occupants.

According to the topology, the power battery management system can be divided into centralized and distributed structures (Fig. 4.80), which serve different projects respectively. The centralized BMS hardware system includes high-voltage area and low-voltage area: the high-voltage area is responsible for the collection of single battery voltage and system total voltage, and the monitoring of insulation resistance; the low voltage area is comprised of power supply circuit, CPU circuit, controller CAN, communication circuit, control circuit, etc. With the advantages of low cost, compact structure and high reliability, centralized BMS is ordinarily applied in electric sightseeing vehicles, patrol vehicles and light hybrid electric vehicles with low capacity, low total voltage and small battery system volume.

The distributed BMS architecture can better realize hierarchical management over the module and the system level. Its advantages include: it can be configured efficiently according to the series and parallel design of different power battery systems; its harness distance with power battery is shorter, more uniform and more reliable; besides, it can also support the design of larger power battery system. With the continuous development of power battery system in passenger vehicle towards high capacity, high total voltage and large volume, BMS with distributed architecture is mainly used in plug-in hybrid electric vehicles and pure electric vehicles.

3. Example of Driving System with Dynamic Adaptive Protection

Driving systems with dynamic adaptive protection are mainly represented by electromagnetic active suspension and adaptive front-lighting system. In case of the

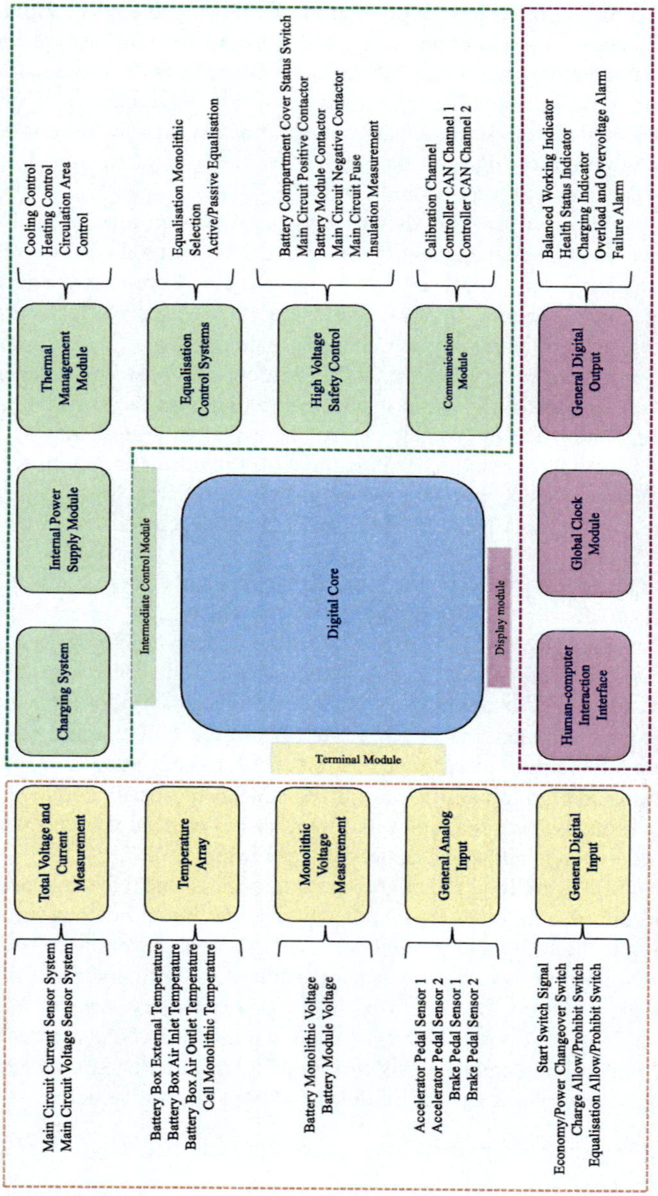

Fig. 4.78 Framework of software and hardware of power battery management system [240]

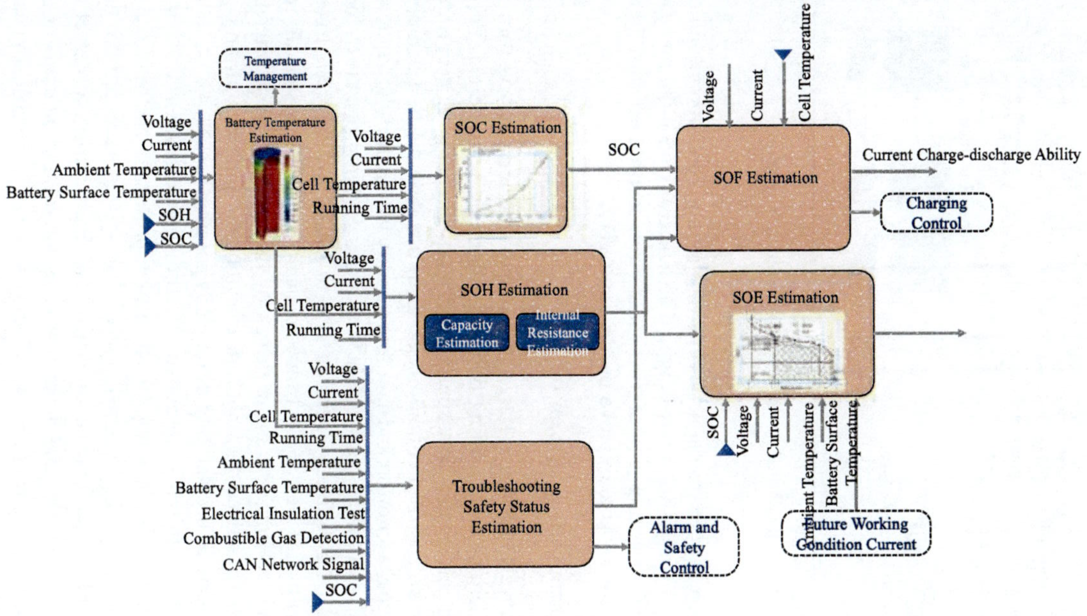

Fig. 4.79 Working principle of power battery management system

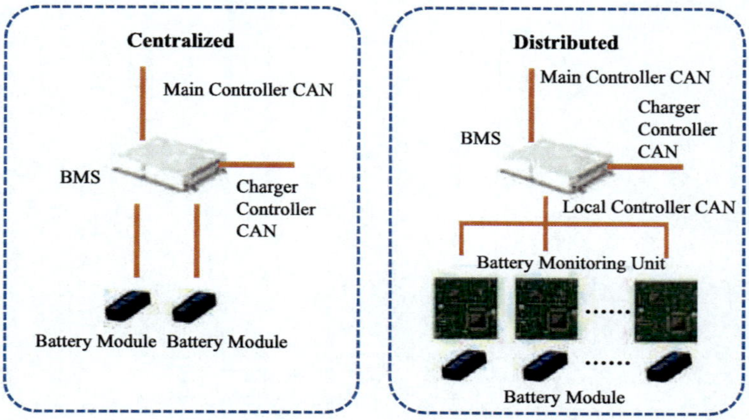

Fig. 4.80 Centralized and distributed power battery management systems

inevitable collision, the electromagnetic active suspension can improve vehicle collision performance by adjusting body height. The adaptive front-lighting system can automatically adjust lighting area of the headlamp in accordance with the lighting condition in driving, so as to improve the driving safety.

(1) Electromagnetic Active Suspension

Active suspension, also known as active guidance suspension or active form suspension, has the function of controlling body vibration and body height. Pure active suspension can timely adjust the suspension stiffness and damping according to the vehicle motion state and road conditions, so as to maintain the best damping state, balance the vehicle ride comfort and handling stability within the full frequency range of body vibration, control the body height in real time and improve the trafficability, as shown by a 2-DOF active suspension system model in Fig. 4.81. Each wheel of the vehicle with pure active electromechanical suspension system is equipped with a motor, and the four wheels are driven by the electrical system respectively (Fig. 4.82), which can reduce the amplitude of rolling motion when turning, pitching motion when braking or accelerating as much as possible [241].

The cooperation between pure active electromagnetic suspension system and safety early warning system can improve the safety protection performance of vehicles. The safety early warning system employs such devices as central driving assistant controller, built-in interconnection sensor and front camera to detect the pre-collision around the vehicle, and make real-time control and adjustment over the active suspension for the prevention of vehicle collision. If the collision cannot be avoided and the vehicle may have a side collision, the suspension system can raise the body height of the impacted surface in a short time for enhancing the anti-collision performance of the vehicle. At the same time, the control unit of the early warning system will be involved, with the capacity of reducing the vehicle speed before the

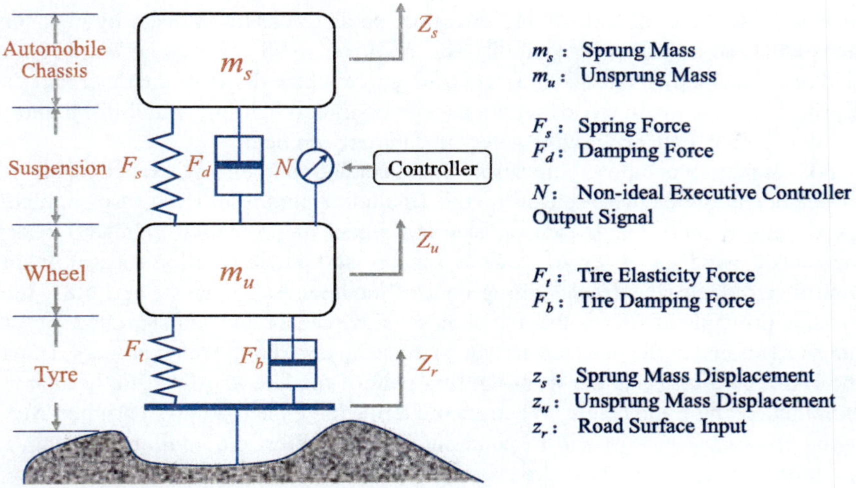

Fig. 4.81 A 2-DOF active suspension system model [241]

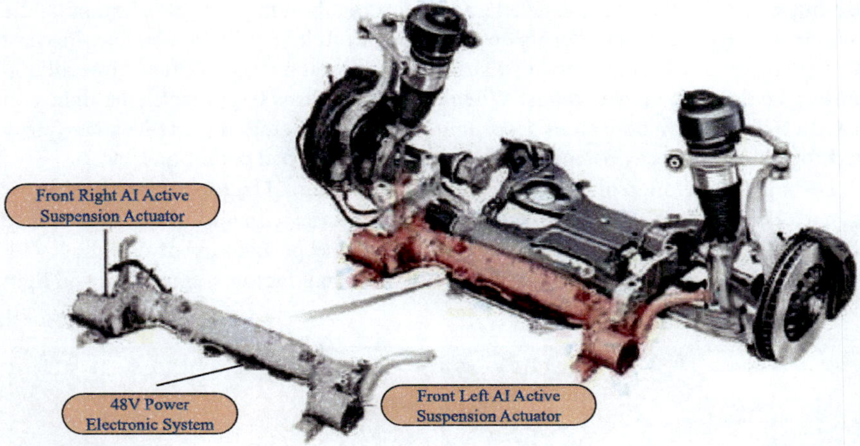

Fig. 4.82 Pure active electromechanical suspension system [242]

collision by up to 40 km/h and significantly alleviating the severity of potential collision accidents.

(2) Adaptive Front-Lighting System

Adaptive front-lighting system (AFS) can dynamically adjust the headlamp according to the vehicle deflection rate, steering wheel angle and driving speed during driving, adapt to the current steering angle, and keep the light orientation consistent with the current driving direction, so as to ensure the best lighting for the road ahead and the best visibility for the driver. AFS can expand the driver's field of vision (shown in Fig. 4.83) and timely remind the driver of the other party's approaching [243] in the road conditions with poor road lighting or multiple curves, and thus greatly enhance driving safety in dark environment.

AFS is mainly composed of sensor, signal conditioning circuit, executive, etc. It is specifically divided into sensor module (including photosensitive sensor, vehicle speed sensor, body height sensor, steering wheel angle sensor, rainfall sensor, fog sensor, wind speed sensor, particle sensor and vehicle position sensor), main controller module and front-lighting control module. As shown in Fig. 4.84, the working principle of AFS is that the photosensitive sensor and vehicle speed sensor constantly transmit the detected signals to the main controller, which processes data and signal and instructs the front-lighting control module to adjust the headlamp illumination angle and height. When encountering large fluctuations in driving, AFS signal processing system will respond quickly, and adjust the light angle within a few milliseconds to make the driver's field of vision clearer.

Under driving condition, traditional headlamp may limit the visibility of the driver or dazzle the driver in the opposite vehicle due to the internal or external factors, directly affecting driving safety. The internal factor mainly refers to the change of driving state (acceleration and deceleration, turning, braking, etc.). For example, the vehicle will lean backward during acceleration or full load driving, and lean forward during emergency braking, or show striking change in the light height of headlamp during the drive on uneven roads; When the vehicle turns left or right, the light will leave a blind spot on both sides accordingly. External factors include weather (rainy and foggy), driving environment (urban or suburban), road conditions, etc.

During the operation of AFS, these internal and external factors are used as trigger signals. For instance, when the vehicle turns, AFS turns with motion, so as to reduce the blind area of the driver's field of vision and the probability of colliding with vulnerable road users or obstacles; To identify external factors, rain sensor and light

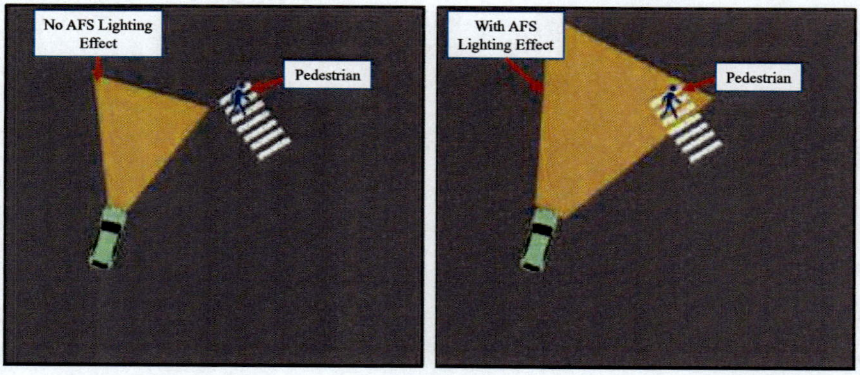

Fig. 4.83 Effect diagram of steering AFS [244]

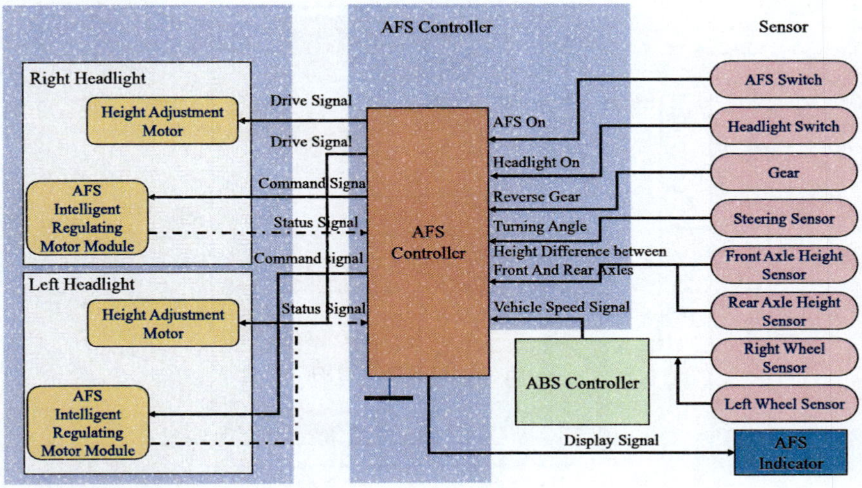

Fig. 4.84 Working principle of AFS

intensity identification sensor are required. When the vehicle is driving at dusk or entering tunnel, AFS will automatically turn on to supplement lighting; When the intensity of the outside light rises to the intensity threshold set by the system, the headlamp will turn off automatically. In addition, during the drive on urban roads, the interference led by lamp illumination to the opposite driver must be taken into consideration. The research on night traffic accidents shows that dazzle is one of the main causes of accidents. Therefore, the vehicle headlamp with adaptive steering function can deflect a certain angle in the vertical direction at this time, which can cut down the dazzling light intensity to opposite drivers or other vulnerable road users such as pedestrians, and play an important role in the prevention of traffic accidents.

4.4.3 *Safety Protection System for Traffic Participants*

Traffic participant protection system is the last barrier for protecting traffic members such as drivers and passengers in the vehicle and pedestrians outside the vehicle in the collision accident. For drivers and passengers, typical active and passive protection system includes reversible pretension seatbelt and rotatable seat, etc. For vulnerable road users, protection system involves pedestrian safety warning system, pop-up bonnet, sandwich bonnet, etc. In case of accident, all on-board active and passive systems work together to achieve the goal of reducing human injury risk (Fig. 4.85).

1. Active and Passive Protection Systems for Drivers and Passengers

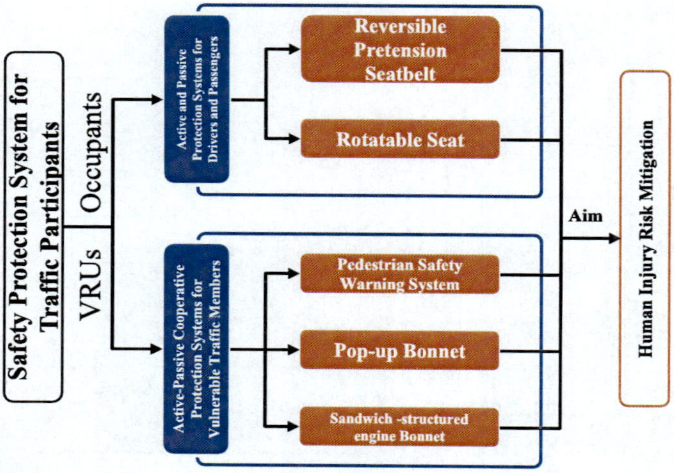

Fig. 4.85 Logic diagram of safety protection system for traffic participants

The traditional passive safety system design takes the collision time as the time zero, and fails to fully consider the impact of some key information before the accident (such as occupant posture) on human safety protection. In recent years, with the improvement of active safety technology, the shortcomings of the traditional safety protection design are gradually revealed. For instance, if active and passive safety systems are simply connected in series, the active intervention before the collision may have an adverse impact on the protection effect of the passive safety system in the collision stage. Some experimental research results show that under the frontal collision conditions of the same strength, occupants in the vehicle with emergency braking may bear more serious injury risk compared with the case without emergency braking [245]. The reason is that the emergency braking pushes drivers and passengers to make longitudinal displacement relative to the vehicle under the action of inertia and deviate from the normal seating position in the passive safety design, so that the original restraint matching scheme can not give full play to the best restraint effect, resulting in an increasing degree of injury. In other words, although the involvement of active safety technology can reduce the severity of collision accidents, the change of vehicle dynamics due to emergency braking or steering will also make occupants in the vehicle out of position and lead to a high risk of injury.

The current situation of active and passive system segmentation can not meet the safety protection requirement of IV. Integrated active and passive system design has become a new development direction of IV protection system and a bridge to effectively connect the traditional active and passive safety technology. The design purpose of active–passive cooperative protection is to minimize the risk of human injury in the whole process from the occurrence of conflict to collision. In the pre-collision stage, various sensors receive information such as traffic environment, vehicle movement and occupant's posture, and adjust the corresponding actuators. When collision approaches, the protection system is engaged in advance for restraining occupants in the vehicle and ensuring that they are in a posture conducive to the implementation of the passive safety system when the collision occurs.

(1) Reversible Pretension Seatbelt

Reversible pretension seatbelt (RPS) is currently recognized as a direct and effective technical means for avoiding occupant displacement. Especially for intelligent assisted driving vehicles with emergency braking, the RPS can effectively reduce the forward displacement of occupants due to vehicle braking. In terms of perception, the ignition trigger time of traditional restraint system still depends on the vehicle collision deceleration received by the on-board acceleration sensor, which can not realize the pretension trigger before the collision [246]. In contrast, when the conflict risk occurs (before the accident), the RPS system obtains the risk quantitative indicators such as TTC through sensing the environmental information as the trigger signal source to drive RPS timely and accurately. In terms of driving, compared with traditional explosive pretension mechanism of gunpowder, the reversible pretension is driven by DC motor and drives the retractor to tighten the safety webbing through the deceleration transmission mechanism (Fig. 4.86), where the clutch mechanism is designed for connecting the deceleration gear and the retractor mandrel. Under normal driving conditions, the clutch mechanism is in the separation state, and the usage of pretension seatbelt is no different from that of ordinary seatbelt. When pretension is required, the DC motor starts, the clutch mechanism engages and drives the retractor mandrel to rotate and recover the seatbelt webbing, and tightens the seatbelt until the motor is blocked due to excessive load. When the on-board system recognizes the collision risk, the seatbelt begins to recover the webbing before the accident, which prolongs the time of the seatbelt pretension process and reduces the risk of injury to occupant in abnormal riding posture due to the high-speed recovery of seatbelt made by the explosive retractor. In addition, the motor drive can be repeated to make the pretension reversible.

(2) Rotatable Seat

In theory, IV is capable to reduce traffic accidents caused by human factors, but difficult to achieve the goal of zero accident due to the limitation of existing technical

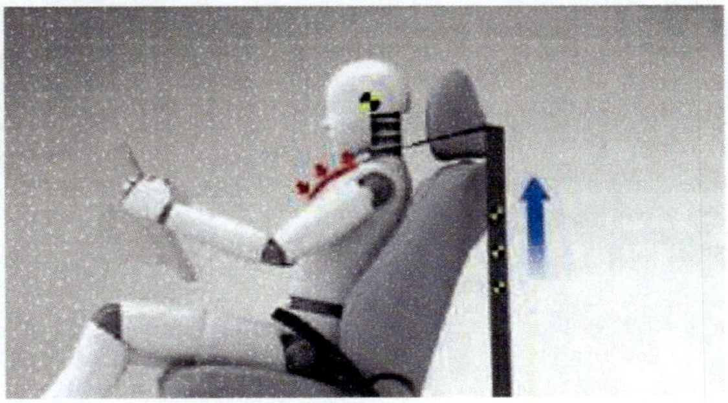

Fig. 4.86 RPS mechanism [247] (the red arrow in the figure indicates that in the case of danger, the seatbelt will be tightened actively for reduction of the gap with occupant; the blue arrow indicates that after the danger state is relieved, the webbing will be released actively for ride comfort)

conditions. For highly automatic vehicles, since the driver is no longer required to operate, the seats in the occupant compartment can be arranged with a higher degree of freedom. The concept of rotating seat with different seating orientation functions has been widely discussed (Fig. 4.87). The effectiveness of protection performance of traditional occupant restraint system is sensitive to the initial seating position of the occupant [248]. In the frontal collision simulation study, the occupant is placed in four different seating positions (seat orientation): 0° (facing the collision direction), 90°, 135° and 180° (back to the collision direction). And the corresponding injury risk and the protection performance of the restraint system can be compared accordingly. When the seat is facing 0°, the occupant's pelvis is well limited on the seat, and the chest seatbelt takes effective restraint in preventing the forward displacement of the torso. In the 90° and 135° directions, although the movement of the occupant's torso is limited by seatbelt, the neck will be seriously injured due to the extrusion of seatbelt. After the collision, the head will deflect laterally due to inertia, resulting in further stretching of the neck; Under the action of the neck, the head may hit the rigid shell of the headrest during rotating. In the seating posture with 180° back to the impact direction, the occupant has a large contact area with the seat back and headrest, which can deform and bear part of the impact load. In addition, the headrest limits the deflection of the head during the collision and reduces the severity of injury due to head rotation. In all collision angles, although the occupant in 180° direction suffers larger compression in some positions of sternum, but the least number of rib fractures and much lower value of neck ligament injury. Therefore, with the lowest comprehensive injury, the 180° orientation is a relatively safe seat orientation.

The implementation scheme of an intelligent occupant restraint system is to actively rotate the seat to 180° after detection of the danger and judgment of the inevitable collision, so as to effectively reduce the comprehensive injury of human body. However, it is also found in the experiment that too fast rotation speed will lead to direct contact between the head and the edge of rigid support structure of headrest, and probably result in brain injury. On the premise of preventing occupant from additional injury, it is feasible to rotate the seat orientation 200 ms before the collision and make it reach the orientation most conducive to occupant protection [251]. Compared with non-rotatable seat, the rotatable seat can not only fulfil the

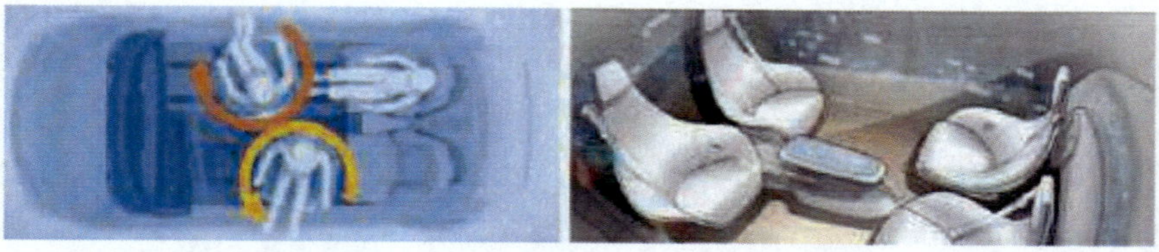

(a) Schematic Diagram of Reversible Pretension Seatbelt (Toyota boshoku)

(b) Conceptual Design of Rotatable Seat (Mercedes-Benz F105)

Fig. 4.87 Rotatable seats in highly automatic vehicle [249, 250] functions required in the occupant compartment of IV, but also effectively reduce the occupant injury in the collision.

2. Active–Passive Cooperative Protection Systems for Vulnerable Road Users

The safety protection technology for vulnerable road users involves active and passive protection systems (Fig. 4.88). Taking the active protection system for pedestrian safety as an example, the perception module senses pedestrians with cameras and radars, predicts their behavior, and prejudges the risk of human-vehicle interaction. In case of a potential danger, the vehicle movement is adjusted by the control system for collision avoidance. For inevitable collision conditions, active safety provides the passive protection system with prejudged information of collision conditions, and controls the triggering of passive protection equipment to minimize collision damage.

(1) Pedestrian Safety Warning System

Pedestrian safety early warning system is gradually applied in mass-produced vehicles. Its workflow is mainly to perceive the pedestrians in front of the vehicle in advance based on video or radar information and predict the behavior intention. When finding potentially dangerous conflicts between human and vehicles, it warns the driver or automatically control the vehicle, or takes braking or steering avoidance when necessary (especially for distracted driving by the driver) to avoid danger. Due to the high randomness of pedestrian behavior and complex traffic environment, the existing pedestrian perception and behavior prediction based on visual information still faces some difficulties. In view of it, Daimler AG established a pedestrian behavior prediction model with the dynamic Bayesian model framework based on the relative position data of human-vehicle in the process of image recognition and

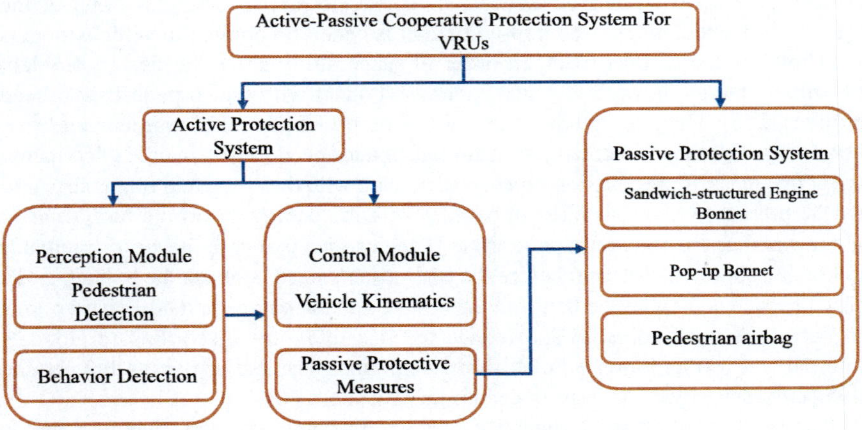

Fig. 4.88 Framework of pedestrian active–passive cooperative protection system

Fig. 4.89 Pedestrian collision risk warning system [252]

pedestrian avoidance, and developed a safety early warning system based on the model [252] (Fig. 4.89).

The SAVE-U system developed by Chrysler LLC in 2005 is also one of the more representative pedestrian detection systems. Based on various sensors such as optical camera, infrared camera and radar, it can adapt to different vehicle speeds and weather, quickly detect pedestrians and offer timely warning information. The C2-270 system developed by Mobileye company is a typical pedestrian detection and early warning system based on computer vision, which can effectively detect roads, pedestrians, vehicles and other objects, and help to reduce the occurrence of traffic accidents. The pedestrian behavior prediction algorithm can also further anticipate the future trend and pedestrian-vehicle collision point through the current movement of pedestrians [253], or estimate the risk of human-vehicle conflict in combination with pedestrian walking direction, speed, cognition on the vehicle, as well as speed and longitudinal distance of the vehicle, so as to form the multi-level warning strategy with multi-source information fusion [254].

(2) Pop-Up Bonnet

In the case of inevitable collision accidents with pedestrian, vehicle shall start the passive protection system to reduce the risk of pedestrian injury. As one of the representative techniques, the pop-up bonnet is generally applied in vehicle models with low engine compartment, in order to successfully solve the design problem of limited energy absorption space under the bonnet with which pedestrian's head collides [255]. The pop-up bonnet is mainly composed of control unit, bonnet lifter, hood hinge release mechanism, pressure sensor and speed sensor. In case of collision, the active mass of the collision object is estimated with the speed and force measured by the pressure sensor installed in parallel with the energy absorbing box, so as to judge whether the object is a pedestrian. If so, a signal is sent to the system actuator to bounce about 70–100 mm before the pedestrian's head contacts the bonnet, and a certain space is maintained between the bonnet and the engine and other hard points to buffer the head collision and reduce the risk of injury. According to Honda's internal test results based on Collision Regulations, the use of pop-up bonnet can reduce the severity of pedestrian head injury by about 40%.

The pedestrian airbag is generally used together with the pop-up bonnet and is installed under the bonnet near the windshield. Its function is mainly to reduce the head injury induced by the contact and collision between the pedestrian head and the rear edge of the bonnet and the A-pillar area during the collision. The workflow of the pedestrian airbag is similar to that of the pop-up bonnet. The pressure sensor located at the front bumper detects the collision and judges whether it collides with the pedestrian to avoid false start. After the collision with the pedestrian at a certain speed (its trigger speed is generally 20–50 km/h), the bonnet is lifted, the airbag is filled within a few milliseconds and maintained in a full state in about 300 ms, which plays a buffer role in the collision between the pedestrian's head and the vehicle, The head impact test of Euro NCAP show that the system has an obvious effect in protecting pedestrian head [256].

(3) Sandwich-Structured Bonnet Design

In vehicle–pedestrian collision accidents, the structural design of bonnet significantly affects the protection for pedestrian head. The acceleration time curve of pedestrian head and bonnet is critical for evaluating the effect of pedestrian head protection. The structural design of the bonnet with wavy sandwich structure is one of the schemes that can optimize the waveform of head impact [257]. The mechanism of the collision between pedestrian head and front-end structure of vehicle body is that the bonnet partly deforms in the initial stage and produces the first peak force under the impact; the bonnet continues to deform in the middle of the collision, emerges the membrane force and reaches the second peak force; when the energy absorption displacement under the bonnet is exhausted, the partial collapse of bonnet can absorb the residual energy. The sandwich bonnet configuration can make the collision waveform more uniform in the time course and improve the utilization efficiency of energy absorption displacement. At the same time, the head impact injury index can be optimized and controlled by structural design parameters (Fig. 4.90). Taking a mass-produced vehicle model as an example, the central area of its bonnet is designed in a sandwich plate structure. The supplementation with the sandwich plate structure can complement the traditional stiffness distribution of "soft in the middle and hard at the edge", and form a more uniform structural stiffness distribution and a more optimized pedestrian head impact waveform.

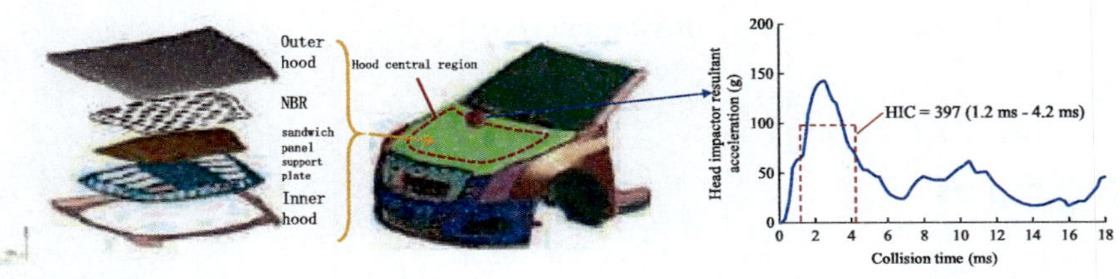

Fig. 4.90 Bonnet design with sandwich structure

4.5 Summary

Focusing on the traffic participants (including drivers, passengers and pedestrians) in the DVE micro traffic system, IP aims to reduce the risk and severity of injury in dangerous driving conditions. With the development of new energy vehicles, the runaway reaction mechanism and protection technology of vehicle power systems such as batteries also becomes an important part of IP. This chapter introduces the fundamental theory, injury quantification approach, injury suppression technique and adaptive protection system involved in the relevant research, mainly including:

(1) By means of the feature information extraction and data mining of road traffic accidents, the distribution features and evolution mechanism of vehicle collision accidents are accurately identified, and the key factors affecting the damage of human-vehicle system are confirmed, so as to provide reference for the implementation of IP. In addition, the mechanism of human body impact injury is explored from the perspective of injury biomechanics; And the deformation and failure behavior of body structure under impact load and the thermal runaway reaction mechanism of power battery under mechanical abuse are identified from the perspective of impact dynamics.

(2) The quantitative measure, injury criteria and corresponding risk probability relationship for damage or failure of traffic participants, vehicle structure and new energy power system are set. The active behavior extraction of traffic participants in hazardous traffic scenarios is helpful for establishing the active behavior model of drivers and pedestrians, and then facilitate the implementation of reasonably assisted driving or anthropomorphic safety decision-making. By quantitatively evaluating the multi-level collision damage risk of human-vehicle system, the prediction on collision risk and severity can be made.

(3) Based on such techniques as condition monitoring and risk early warning for human-vehicle system, as well as the multi-source sensing information and advanced materials, intelligent safety protection systems such as adaptive restraint are established, which can effectively ensure safety in dangerous conditions and collision process for traffic participants and new energy power system, and thus improve driving safety.

References

1. WHO., *Global launch: decade of action for road safety 2011–2020*. 2011, World Health Organization: Geneva.
2. WHO, *Global status report on road safety 2018: Summary*. 2018, World Health Organization.
3. Kamnik, R., M.N. Perc, D. Topolšek, *Using the scanners and drone for comparison of point cloud accuracy at traffic accident analysis*. Accident Analysis Prevention, 2020. **135**: p. 105391.

4. Su, S., W. Liu, K. Li, G. Yang, C. Feng, J. Ming, G. Liu, S. Liu, Z. Yin, *Developing an unmanned aerial vehicle-based rapid mapping system for traffic accident investigation.* Australian journal of forensic sciences, 2016. **48**(4): p. 454–468.
5. IRTAD. Available from: https://www.itf-oecd.org/IRTAD.
6. GIDAS. Available from: https://www.gidas.org/en/willkommen/.
7. CIDAS. Available from: http://www.c-ncap.org/cncap/yqlj.
8. NAIS. Available from: https://www.dpac.org.cn/NAIS/.
9. CISS. Available from: https://www.nhtsa.gov/crash-data-systems/crash-investigation-sampling-system.
10. NASS. Available from: https://www.nhtsa.gov/research-data/national-automotive-sampling-system-nass.
11. NTHSA. Available from: https://www.nhtsa.gov/crash-data-systems.
12. PCDS. Available from: https://catalog.data.gov/sv/dataset/the-pedestrian-crash-data-study-pcds.
13. WADB. Available from: https://www.fia.com/wadb-world-accident-database.
14. Sun Chuan, *Research on vehicle driving risk identification and control based on natural driving data 2016*. Wuhan University of Technology: Wuhan [in Chinese].
15. EuroFOT. Available from: https://www.eurofot-ip.eu/.
16. SHRP2. Available from: https://www.shrp2nds.us/.
17. Zhan, W., L. Sun, D. Wang, H. Shi, A. Clausse, M. Naumann, J. Kummerle, H. Konigshof, C. Stiller, A. de La Fortelle, *Interaction dataset: An international, adversarial and cooperative motion dataset in interactive driving scenarios with semantic maps.* arXiv preprint arXiv: 1910.03088, 2019.
18. National Bureau of Statistics of the People's Republic of China. Available from: http://www.stats.gov.cn/tjsj/ndsj/2020/indexch.htm [in Chinese].
19. Shi, L., Y. Han, H. Huang, J. Davidsson, R. Thomson, *Evaluation of injury thresholds for predicting severe head injuries in vulnerable road users resulting from ground impact via detailed accident reconstructions.* Biomechanics and Modeling in Mechanobiology, 2020. **19**(5): p. 1845–1863.
20. CIREN. Available from: https://one.nhtsa.gov/Research/Crash-Injury-Research-(CIREN).
21. Wang, S., Z. Li, Y. Guo, *Exploring the mechanism of crashes with automated vehicles using statistical modeling approaches.* PLoS ONE, 2019. **14**(3).
22. Bucsuházy, K., E. Matuchová, R. Zůvala, P. Moravcová, M. Kostíková, R. Mikulec, *Human factors contributing to the road traffic accident occurrence.* Transportation research procedia, 2020. **45**: p. 555–561.
23. Dingus, T.A., F. Guo, S. Lee, J.F. Antin, M. Perez, M. Buchanan-King, J. Hankey, *Driver crash risk factors and prevalence evaluation using naturalistic driving data.* Proceedings of the National Academy of Sciences, 2016. **113**(10): p. 2636–2641.
24. Lombardi, D.A., W.J. Horrey, T.K. Courtney, *Age-related differences in fatal intersection crashes in the United States.* Accident Analysis & Prevention, 2017. **99**: p. 20–29.
25. Martín, L., L. Baena, L. Garach, G. López, J. De Oña, *Using data mining techniques to road safety improvement in Spanish roads.* Procedia-social and behavioral sciences, 2014. **160**: p. 607–614.
26. Li, L., S. Shrestha, G. Hu. *Analysis of road traffic fatal accidents using data mining techniques.* in *2017 IEEE 15th International Conference on Software Engineering Research, Management and Applications (SERA).* 2017. IEEE.
27. Chen, C., G. Zhang, R. Tarefder, J. Ma, H. Wei, H. Guan, *A Multinomial Logit Model-Bayesian Network Hybrid Approach for Driver Injury Severity Analyses in Rear-End Crashes.* Accident Analysis & Prevention, 2015. **80**(jul.): p. 76–88.
28. Das, S., A. Dutta, I. Tsapakis, *Automated vehicle collisions in California: Applying Bayesian latent class model.* IATSS Research, 2020. **44**(4): p. 300–308.
29. Kockelman, K.M., Y.J. Kweon, *Driver injury severity: an application of ordered probit models.* Accident Analysis & Prevention, 2002. **34**(3): p. 313–321.

30. Abu-Zidan, F.M., H.O. Eid, *Factors affecting injury severity of vehicle occupants following road traffic collisions.* Injury, 2015. **46**(1): p. 136–141.
31. Kröyer, H.R., *Is 30 km/ha 'safe'speed? Injury severity of pedestrians struck by a vehicle and the relation to travel speed and age.* IATSS research, 2015. **39**(1): p. 42–50.
32. Ji Peijun, Research on occupant crash protection based on balanced constraint concept. 2016, Tsinghua University: Beijing [in Chinese].
33. Longhitano, D., B. Henary, K. Bhalla, J. Ivarsson, J. Crandall, *Influence of vehicle body type on pedestrian injury distribution.* SAE transactions, 2005: p. 2283–2288.
34. Shang, S., D. Otte, G. Li, C. Simms, *Detailed assessment of pedestrian ground contact injuries observed from in-depth accident data.* Accident Analysis & Prevention, 2018. **110**: p. 9–17.
35. Otte, D., *Influence of the fronthood length for the safety of pedestrians in car accidents and demands to the safety of small vehicles.* SAE transactions, 1994: p. 1923–1933.
36. Howson, J.E., D. Lord, D.L. Bullard. *The effect of driver height on the death rate in single-vehicle rollover accidents.* in *Transportation Research Board 91st Annual Meeting.* 2012.
37. Zhu, S., J.E. Kim, X. Ma, A. Shih, D.B. Allison, *BMI and Risk of Serious Upper Body Injury Following Motor Vehicle Crashes: Concordance of Real-World and Computer-Simulated Observations.* PLoS Medicine, 2010. **7**(3): p. e1000250.
38. Itti, L., C. Koch, *Computational modelling of visual attention.* Nature Reviews Neuroscience, 2001. **2**(3): p. 194–203.
39. Ruan, J.S., P. Prasad, *Head injury potential assessment in frontal impacts by mathematical modeling.* SAE Transactions, 1994: p. 1677–1687.
40. Nahum, A.M., J.W. Melvin, *Accidental Injury: Biomechanics and Prevention.* 2012, New York: Springer-Verlag New York.
41. Ruan, J., R. El-Jawahri, L. Chai, S. Barbat, P. Prasad, *Prediction and analysis of human thoracic impact responses and injuries in cadaver impacts using a full human body finite element model.* 2003, SAE Technical Paper.
42. Crandall, J.R., B.S. Myers, D.F. Meaney, S.Z. Schmidtke, *Pediatric injury biomechanics: archive & textbook.* 2012: Springer Science & Business Media.
43. Moran, S.G., G. McGwin Jr, J.S. Metzger, J.E. Alonso, L.W. Rue III, *Relationship between age and lower extremity fractures in frontal motor vehicle collisions.* Journal of Trauma and Acute Care Surgery, 2003. **54**(2): p. 261–265.
44. Kuppa, S., O. Fessahaie, *An overview of knee-thigh-hip injuries in frontal crashes in the United States.* National Highway Traffic Safety Administration, ISSI, 2003. **416**: p. 1–9.
45. Woo, S.L., R.E. Debski, J.D. Withrow, M.A. Janaushek, *Biomechanics of knee ligaments.* The American journal of sports medicine, 1999. **27**(4): p. 533–543.
46. Woo, S.L.-Y., S.D. Abramowitch, R. Kilger, R. Liang, *Biomechanics of knee ligaments: injury, healing, and repair.* Journal of biomechanics, 2006. **39**(1): p. 1–20.
47. Weiss, J.A., J.C. Gardiner, *Computational modeling of ligament mechanics.* Critical Reviews™ in Biomedical Engineering, 2001. **29**(3).
48. Henninger, H.B., C.J. Underwood, S.J. Romney, G.L. Davis, J.A. Weiss, *Effect of elastin digestion on the quasi-static tensile response of medial collateral ligament.* Journal of Orthopaedic Research, 2013. **31**(8): p. 1226–1233.
49. Wan, C., Z. Hao, S. Wen, *A review on research on development of ligament constitutive relations on macro, meso, and micro levels.* Acta Mechanica Solida Sinica, 2013. **26**(4): p. 331–343.
50. Fung, Y.-C., *Bio-viscoelastic solids,* in *Biomechanics.* 1981, Springer, New York: New York. p. 196–260.
51. Seeman, E., P.D. Delmas, *Bone quality—the material and structural basis of bone strength and fragility.* New England journal of medicine, 2006. **354**(21): p. 2250–2261.
52. Meachim, G., *Effect of age on the thickness of adult articular cartilage at he shoulder joint.* Annals of the rheumatic diseases, 1971. **30**(1): p. 43.
53. Accardi, M.A., S.D. McCullen, A. Callanan, S. Chung, P.M. Cann, M.M. Stevens, D. Dini, *Effects of fiber orientation on the frictional properties and damage of regenerative articular cartilage surfaces.* Tissue Engineering Part A, 2013. **19**(19–20): p. 2300–2310.

54. Burstein, A.H., J. Zika, K. Heiple, L. Klein, *Contribution of collagen and mineral to the elastic-plastic properties of bone.* JBJS, 1975. **57**(7): p. 956–961.
55. Sun Xingleng. Multi-scale experimental study on mechanical behavior and deformation mechanism of cortical bone materials under combined load. 2018, Jilin University: Changchun [in Chinese].
56. Wang, X., X. Xu, *Assessing the relationship between self-reported driving behaviors and driver risk using a naturalistic driving study.* Accident Analysis & Prevention, 2019. **128**: p. 8–16.
57. Haufe, S., J.-W. Kim, I.-H. Kim, A. Sonnleitner, M. Schrauf, G. Curio, B. Blankertz, *Electrophysiology-based detection of emergency braking intention in real-world driving.* Journal of neural engineering, 2014. **11**(5): p. 056011.
58. Deng, T., H. Yan, L. Qin, T. Ngo, B. Manjunath, *How do drivers allocate their potential attention? driving fixation prediction via convolutional neural networks.* IEEE Transactions on Intelligent Transportation Systems, 2019. **21**(5): p. 2146–2154.
59. Fang, J., D. Yan, J. Qiao, J. Xue, H. Yu, *DADA: Driver Attention Prediction in Driving Accident Scenarios.* IEEE Transactions on Intelligent Transportation Systems, 2021.
60. Choi, H.Y., S.J. Sah, B. Lee, *Experimental and numerical studies of muscular activations of bracing occupant.* 2005: p. 10–10.
61. Kallieris, D., D. Otte, R. Mattern, P. Wiedmann. *Comparison of Sled Tests with Real Traffic Accidents.* in *Stapp Car Crash Conference.* 1995.
62. Begeman, P.C., A.I. King, R.S. Levine, D.C. Viano. *Biodynamic Response of the Musculoskeletal System to Impact Acceleration.* in *Stapp Car Crash Conference.* 1980.
63. Shaw, G., D. Lessley, J. Crandall, R. Kent, L. Kitis, *Elimination of thoracic muscle tensing effects for frontal crash dummies.* SAE Transactions, 2005. **2005**(724): p. 177–191.
64. Huber, P., M. Christova, G.A. D'Addetta, E. Gallasch, S. Kirschbichler, C. Mayer, A. Prüggler, A. Rieser, W. Sinz, D. Wallner. *Muscle activation onset latencies and amplitudes during lane change in a full vehicle test.* in *Proceedings of the IRCOBI Conference.* 2013.
65. Jonas Östh, Marín Ólafsdóttir, J. Davidsson, K. Brolin. *Driver kinematic and muscle responses in braking events with standard and reversible pre-tensioned restraints: validation data for human models.* in *Stapp Car Crash Conference.* 2013. SAE Technical Paper.
66. Ejima, S., Y. Zama, F. Satou, S. Holcombe, K. Ono, K. Kaneoka, I. Shiina. *Prediction of the physical motion on human body based on muscle activities during Pre-ImpactIn.* in *IRCOBI Conference.* 2008.
67. Nie, B., Q. Zhou, *Can new passenger cars reduce pedestrian lower extremity injury? A review of geometrical changes of front-end design before and after regulatory efforts.* Traffic injury prevention, 2016. **17**(7): p. 712–719.
68. Mizuno, K., J.T. Andrish, A.J. van den Bogert, S.G. McLean, *Gender dimorphic ACL strain in response to combined dynamic 3D knee joint loading: implications for ACL injury risk.* The Knee, 2009. **16**(6): p. 432–440.
69. Huang, Y., Q. Zhou, C. Koelper, Q. Li, B. Nie, *Are riders of electric two-wheelers safer than bicyclists in collisions with motor vehicles?* Accident Analysis & Prevention, 2020. **134**: p. 105336.
70. Hu, J., K.D. Klinich, *Toward designing pedestrian-friendly vehicles.* International Journal of Vehicle Safety, 2015. **8**(1): p. 22–54.
71. Yang, J., *Review of Injury Biomechanics in Car-Pedestrian Collisions.* International Journal of Vehicle Safety, 2005. **1**.
72. Kramer, M., K. Burow, A. Heger, *Fracture Mechanism of Lower Legs Under Impact Load.* 1973.
73. Klinich, K., *Biomechanics of pedestrian injuries related to lower extremity injury assessment tools: a review of the literature and analysis of pedestrian crash database.* 2003.
74. Chou, C.C., R.J. Howell, B.Y. Chang, *A Review and Evaluation of Various HIC Algorithms.* 1988, SAE International.
75. Huang Jun, *Research on rapid optimization design of vehicle front-end structure for pedestrian lower limb collision protection.* 2013, Tsinghua University: Beijing [in Chinese].

76. Yu Tongxi, Lu Guoxing, Zhang Xiong, *Energy absorption: energy absorption of materials and structures*. 2019, Beijing: Science Press [in Chinese].
77. Forrestal, M., D. Wesenberg, *Elastic-Plastic Response of Simply Supported 1018 Steel Beams to Impulse Loads*. Journal of Applied Mechanics, 1977. **44**.
78. Forrestal, M.J., M.J. Sagartz, *Elastic-Plastic Response of 304 Stainless Steel Beams to Impulse Loads*. Journal of Applied Mechanics, 1978. **45**.
79. Symonds, P.S., *Viscoplastic behavior in response of structures to dynamic loading// HUFFINGTON N. Behavior of materials under dynamic loading*. New York:ASME: 106–124. 1965.
80. Bodner, S., W.G. Speirs, *Dynamic plasticity experiments on aluminium cantilever beams at elevated temperature*. Journal of The Mechanics and Physics of Solids - J MECH PHYS SOLIDS, 1963. **11**: p. 65–68.
81. Krempl, E., F. Khan, *Rate (time)-dependent deformation behavior: an overview of some properties of metals and solid polymers*. International Journal of Plasticity, 2003. **19**(7): p. 1069–1095.
82. Yu Tongxi, Lu Guoxing, *Energy absorption: energy absorption of materials and structures*. 2006, Beijing: Chemical Industry Press [in Chinese].
83. Andrews, K.R.F., G.L. England, E. Ghani, *Classification of the axial collapse of cylindrical tubes under quasi-static loading*. International Journal of Mechanical Sciences, 1983. **25**(9): p. 687–696.
84. Alexander, J., *An approximate analysis of the collapse of thin cylindrical shells under axial loading*. Quarterly Journal of Mechanics and Applied Mathematics - QUART J MECH APPL MATH, 1960. **13**: p. 10–15.
85. Stronge, W.J., *Impact on metal tubes: Indentation and perforation, in structural crashworthiness and failure*. International Journal of Solids and Structures, 1993: p. 165–188.
86. Singace, A., *Axial crushing analysis of tubes deforming in the multi-lobe mode*. International Journal of Mechanical Sciences - INT J MECH SCI, 1999. **41**: p. 865–890.
87. Guillow, S.R., G. Lu, R.H. Grzebieta, *Quasi-static axial compression of thin-walled circular aluminium tubes*. International Journal of Mechanical Sciences, 2001. **43**(9): p. 2103–2123.
88. Feng Xuning, *Induction and expansion mechanism, modeling and prevention and control of thermal runaway of vehicle lithium-ion power battery*. 2016, Tsinghua University: Beijing [in Chinese].
89. Wen, J., Y. Yu, C. Chen, *A review on lithium-ion batteries safety issues: existing problems and possible Solutions*. Materials Express, 2012. **2**(3): p. 197–212.
90. Crandall, J.R., D. Bose, J. Forman, C.D. Untaroiu, C. Arregui-Dalmases, C.G. Shaw, J. Kerrigan, *Human surrogates for injury biomechanics research*. Clinical anatomy, 2011. **24**(3): p. 362–371.
91. Carter, P.M., C.A. Flannagan, M.P. Reed, R.M. Cunningham, J.D. Rupp, *Comparing the effects of age, BMI and gender on severe injury (AIS 3+) in motor-vehicle crashes*. Accident Analysis & Prevention, 2014. **72**: p. 146–160.
92. Zhou, Q., S.W. Rouhana, J.W. Melvin, *Age effects on thoracic injury tolerance*. SAE transactions, 1996: p. 1819–1830.
93. Burstein, A.H., D.T. Reilly, M. Martens, *Aging of bone tissue: mechanical properties*. The Journal of bone and joint surgery. American volume, 1976. **58**(1): p. 82–86.
94. Kent, R., S.-H. Lee, K. Darvish, S. Wang, C.S. Poster, A.W. Lange, C. Brede, D. Lange, F. Matsuoka, *Structural and material changes in the aging thorax and their role in crash protection for older occupants*. 2005, SAE Technical Paper.
95. Reed, M.P., S.M. Ebert-Hamilton, J.D. Rupp, *Effects of obesity on seatbelt fit*. Traffic injury prevention, 2012. **13**(4): p. 364–372.
96. Kent, R.W., J.L. Forman, O. Bostrom, *Is there really a "cushion effect"?: a biomechanical investigation of crash injury mechanisms in the obese*. Obesity, 2010. **18**(4): p. 749–753.
97. Cormier, J.M., *The influence of body mass index on thoracic injuries in frontal impacts*. Accident Analysis & Prevention, 2008. **40**(2): p. 610–615.

98. Bose, D., S. Segui-Gomez, Maria, J.R. Crandall, *Vulnerability of female drivers involved in motor vehicle crashes: an analysis of US population at risk.* American journal of public health, 2011. **101**(12): p. 2368–2373.
99. Huang Yi, *Study on the influence parameters of restraint system considering the difference of occupant figure and collision strength.* 2014, Tsinghua University: Beijing [in Chinese].
100. Wang Cong, *Research on configuration design of occupant restraint system under various collision conditions.* 2010, Tsinghua University: Beijing [in Chinese].
101. Melvin, J.W., *Injury assessment reference values for the CRABI 6-month infant dummy in a rear-facing infant restraint with airbag deployment.* SAE transactions, 1995: p. 1553–1564.
102. Lucas, S.R., C.R. Bass, J.R. Crandall, R.W. Kent, F.H. Shen, R.S. Salzar, *Viscoelastic and failure properties of spine ligament collagen fascicles.* Biomechanics and modeling in mechanobiology, 2009. **8**(6): p. 487–498.
103. Liao, H., S.M. Belkoff, *A failure model for ligaments.* Journal of biomechanics, 1999. **32**(2): p. 183–188.
104. Shi, X., L. Cao, M.P. Reed, J.D. Rupp, C.N. Hoff, J. Hu, *A statistical human rib cage geometry model accounting for variations by age, sex, stature and body mass index.* Journal of Biomechanics, 2014. **47**(10): p. 2277–2285.
105. Yates, K.M., Y.-C. Lu, C.D. Untaroiu, *Statistical shape analysis of the human spleen geometry for probabilistic occupant models.* Journal of biomechanics, 2016. **49**(9): p. 1540–1546.
106. Bose, D., J.R. Crandall, C.D. Untaroiu, E. Maslen, *Influence of pre-collision occupant parameters on injury outcome in a frontal collision.* Accident Analysis & Prevention, 2010. **42**(4): p. 1398–1407.
107. Nie, B., J.L. Forman, H. Joodaki, T. Wu, R.W. Kent, *Scaling approach in predicting the seatbelt loading and kinematics of vulnerable occupants: How far can we go?* Traffic injury prevention, 2016. **17**(sup1): p. 93–100.
108. Nicolella, D.P., B.H. Thacker, H. Katoozian, D.T. Davy, *The effect of three-dimensional shape optimization on the probabilistic response of a cemented femoral hip prosthesis.* Journal of Biomechanics, 2006. **39**(7): p. 1265–1278.
109. Faraway, J., M.P. Reed, *Statistics for digital human motion modeling in ergonomics.* Technometrics, 2007. **49**(3): p. 277–290.
110. Liu, Y., Q. Zhou, S. Gan, B. Nie, *Influence of population variability in ligament material properties on the mechanical behavior of ankle: a computational investigation.* Computer methods in biomechanics and biomedical engineering, 2020. **23**(2): p. 43–53.
111. Nie, B., Y. Xia, Q. Zhou, J. Huang, B. Deng, M. Neal, *A response–surface–based tool for vehicle front–end design for pedestrian impact protection using human body model.* International Journal of Vehicle Design, 2014. **66**(4): p. 347–362.
112. Nie, B., J.L. Forman, M.B. Panzer, A.R. Mait, J.-P. Donlon, R.W. Kent, *Fiber-based modeling of in situ ankle ligaments with consideration of progressive failure.* Journal of biomechanics, 2017. **61**: p. 102–110.
113. Nie, B., M.B. Panzer, A. Mane, A.R. Mait, J.-P. Donlon, J.L. Forman, R.W. Kent, *A framework for parametric modeling of ankle ligaments to determine the in situ response under gross foot motion.* Computer methods in biomechanics and biomedical engineering, 2016. **19**(12): p. 1254–1265.
114. Dokko, Y., O. Ito, K. Ohashi, *Development of human lower limb and pelvis fe models for adult and the elderly.* SAE Technical Paper, 2009.
115. Klein, K.F., J. Hu, M.P. Reed, L.W. Schneider, J.D. Rupp, *Validation of a parametric finite element human femur model.* Traffic Injury Prevention, 2017. **18**(4): p. 420–426.
116. Dokko, Y., O. Ito, K. Ohashi, *Development of human lower limb and pelvis fe models for adult and the elderly.* 2009, SAE Technical Paper.
117. Burstein, A.H., D.T. Reilly, M.J.T.J.o.b. Martens, *Aging of bone tissue: mechanical properties.* The Journal of Bone Joint Surgery. American Volume, 1976. **58**(1): p. 82–86.
118. Olafsdottir, J.M., K. Brolin, J.S. Blouin, G.P. Siegmund, *Dynamic Spatial Tuning of Cervical Muscle Reflexes to Multidirectional Seated Perturbations.* Spine, 2015. **40**(4): p. E211–E219.

119. Haug, E., H.-Y. Choi, S. Robin, M. Beaugonin, *Human Models for Crash and Impact Simulation*, in *Handbook of Numerical Analysis*. 2004, Elsevier. p. 231–452.
120. Chang, C.-Y., J.D. Rupp, N. Kikuchi, L.W. Schneider. *Development of a finite element model to study the effects of muscle forces on knee-thigh-hip injuries in frontal crashes*. in *Stapp Car Crash Conference*. 2008.
121. Bloecher, H.L., J. Dickmann, M. Andres. *Automotive active safety & comfort functions using radar*. in *2009 IEEE International Conference on Ultra-Wideband*. 2009.
122. Mansouri, M., J.A. Reinbolt, *A platform for dynamic simulation and control of movement based on OpenSim and MATLAB*. Journal of Biomechanics, 2012. **45**(8): p. 1517–1521.
123. Iwamoto, M., Y. Nakahira, H. Kimpara, T. Sugiyama, K. Min. *Development of a Human Body Finite Element Model with Multiple Muscles and their Controller for Estimating Occupant Motions and Impact Responses in Frontal Crash Situations*. in *Stapp Car Crash Conference*. 2012.
124. Iwamoto, M., *Modeling Passive and Active Muscles*, in *Basic Finite Element Method as Applied to Injury Biomechanics*. 2018, Elsevier: Amsterdam. p. 447–468.
125. Chang, C.Y., *Prediction of the Effects of Lower-extremity Muscle Forces on Knee, Thigh, and Hip Injuries in Frontal Motor Vehicle Crashes*. 2009, University of Michigan: Michigan.
126. Nie, B., D. Sathyanarayan, X. Ye, J.R. Crandall, M.B. Panzer, *Active muscle response contributes to increased injury risk of lower extremity in occupant-knee airbag interaction*. Traffic Injury Prevention, 2018. **19**: p. S76–S82.
127. Happee, R., E. de Bruijn, P.A. Forbes, F.C.T. van der Helm, *Dynamic head-neck stabilization and modulation with perturbation bandwidth investigated using a multisegment neuromuscular model*. Journal of Biomechanics, 2017. **58**: p. 203–211.
128. Östh, J., E. Eliasson, R. Happee, K. Brolin, *A method to model anticipatory postural control in driver braking events*. Gait and Posture, 2014. **40**(4): p. 664–669.
129. Choi, H.Y., S.J. Sah, B. Lee. *Experimental and numerical studies of muscular activations of bracing occupant*. in *19th ESV*. 2005.
130. Rooij, L.V., *Effect of Various Pre-Crash Braking Strategies on Simulated Human Kinematic Response With Varying Levels of Driver Attention*. The 22nd International Technical Conference on the Enhanced Safety of Vehicles (ESV), 2011: p. 1–9.
131. Iwamoto, M., Y. Nakahira, H. Kimpara, *Development and Validation of the Total HUman Model for Safety (THUMS) Toward Further Understanding of Occupant Injury Mechanisms in Precrash and During Crash*. Traffic Injury Prevention, 2015. **16**: p. 36–48.
132. Behr, M., *Tonic Finite Element Model of the Lower Limb*. Journal of Biomechanical Engineering, 2005. **128**(2): p. 223–227.
133. Mo, F., F. Li, M. Behr, Z. Xiao, G. Zhang, X. Du, *A Lower Limb-Pelvis Finite Element Model with 3D Active Muscles*. Annals of Biomedical Engineering, 2018. **46**(1): p. 86–96.
134. Sugiyama, T., H. Kimpara, M. Iwamoto. *Effects of muscle tense on impact responses of lower extremity*. in *IRCOBI Conference*. 2007.
135. Li, F., R. Lu, W. Hu, H. Li, S. Hu, J. Hu, H. Wang, H. Xie, *The Influence of Neck Muscle Activation on Head and Neck Injuries of Occupants in Frontal Impacts*. Applied Bionics & Biomechanics, 2018. **2018**(4): p. 1–14.
136. Gennarelli, T.A., E. Wodzin, *AIS 2005: a contemporary injury scale*. Injury, 2006. **37**(12): p. 1083–1091.
137. Hollowell, W.T., H.C. Gabler, S.L. Stucki, S. Summers, J.R. Hackney, *Updated review of potential test procedures for FMVSS No. 208*. NHTSA Docket, 1999: p. 6407–6.
138. Kuppa, S., *Injury criteria for side impact dummies*. Washington, DC: National Transportation Biomechanics Research Center, National Highway Saftey Administration, US DOT, 2004. **67**.
139. Takhounts, E.G., M.J. Craig, K. Moorhouse, J. McFadden, V. Hasija, *Development of brain injury criteria (BrIC)*. 2013, SAE Technical Paper.
140. National Standard of the People's Republic of China. *Occupant protection in vehicle frontal collision* [in Chinese].

141. Gabler, L.F., J.R. Crandall, M.B. Panzer, *Development of a Metric for Predicting Brain Strain Responses Using Head Kinematics.* Annals of Biomedical Engineering, 2018. **46**(7): p. 972–985.
142. Tannous, R., R. Eppinger, E. Sun, F. Bandak, R. Saul, *Development of Improved Injury Criteria for the Assessment of Advanced Automotive Restraint Systems - II.* nhtsa, 1999.
143. Eichberger, A., H. Steffan, B. Geigl, M. Svensson, O. Boström, P.E. Leinzinger, M. Darok. *Evaluation of the applicability of the neck injury criterion (NIC) in rear end impacts on the basis of human subject tests.* in *Proceedings of the International Research Council on the Biomechanics of Injury conference.* 1998. International Research Council on Biomechanics of Injury.
144. Cavanaugh, J.M., Y. Zhu, Y. Huang, A. King, *Injury and response of the thorax in side impact cadaveric tests.* 1993, SAE Technical Paper.
145. Yang, J., *Review of injury biomechanics in car-pedestrian collisions.* International journal of vehicle safety, 2005. **1**(1–3): p. 100–117.
146. Kuppa, S., J. Wang, M. Haffner, R. Eppinger, *Lower extremity injuries and associated injury criteria.* 2001, SAE Technical Paper.
147. Ivarsson, J., D. Lesley, J. Kerrigan, K.S. Bhalla, D. Bose, J.R. Crandall, R.W. Kent, *Dynamic Response Corridors and Injury Thresholds of the Pedestrian Lower Extremities.* Proceedings of the International Research Council on the Biomechanics of Injury conference, 2004. **32**.
148. Quapp, K.M., J.A. Weiss, *Material characterization of human medial collateral ligament.* Journal of Biomechanical Engineering, 1998. **120**(6): p. 757.
149. Race, A., A.A. Amis, *The mechanical properties of the two bundles of the human posterior cruciate ligament.* Journal of Biomechanics, 1994. **27**(1): p. 13–24.
150. Butler, D.L., M.D. Kay, D.C. Stouffer, *Comparison of material properties in fascicle-bone units from human patellar tendon and knee ligaments.* Journal of biomechanics, 1986. **19**(6): p. 425–432.
151. Baker, S.P., B. o'Neill, W. Haddon Jr, W.B. Long, *The injury severity score: a method for describing patients with multiple injuries and evaluating emergency care.* Journal of Trauma and Acute Care Surgery, 1974. **14**(3): p. 187–196.
152. Osler, T., S.P. Baker, W. Long, *A modification of the injury severity score that both improves accuracy and simplifies scoring.* Journal of Trauma and Acute Care Surgery, 1997. **43**(6): p. 922–926.
153. Stevenson, M., M. Segui-Gomez, I. Lescohier, C. Di Scala, G. McDonald-Smith, *An overview of the injury severity score and the new injury severity score.* Injury Prevention, 2001. **7**(1): p. 10–13.
154. Tay, S.-Y., E.P. Sloan, L. Zun, P. Zaret, *Comparison of the new injury severity score and the injury severity score.* Journal of Trauma and Acute Care Surgery, 2004. **56**(1): p. 162–164.
155. Balogh, Z., P.J. Offner, E.E. Moore, W.L. Biffl, *NISS predicts postinjury multiple organ failure better than the ISS.* Journal of Trauma and Acute Care Surgery, 2000. **48**(4): p. 624–628.
156. Balogh, Z.J., E. Varga, J. Tomka, G. Süveges, L. Tóth, J.A. Simonka, *The new injury severity score is a better predictor of extended hospitalization and intensive care unit admission than the injury severity score in patients with multiple orthopaedic injuries.* Journal of orthopaedic trauma, 2003. **17**(7): p. 508–512.
157. Munhoz, D., J.A.R. Gregolin, L.I.L. de Faria, T.F. de Andrade, *Automotive Materials: current status, technology trends and challenges.* 2007, SAE International.
158. Bai, Y., T. Wierzbicki, *Application of extended Mohr-Coulomb criterion to ductile fracture.* International Journal of Fracture, 2010. **161**: p. 1–20.
159. Beese, A., M. Luo, Y. Li, Y. Bai, T. Wierzbicki, *Partially coupled anisotropic fracture model for aluminum sheets.* Engineering Fracture Mechanics - ENG FRACTURE MECH, 2010. **77**: p. 1128–1152.
160. Luo, M., T. Wierzbicki, *Numerical failure analysis of a stretch-bending test on dual-phase steel sheets using a phenomenological fracture model.* International Journal of Solids and Structures, 2010. **47**: p. 3084–3102.

161. Dørum, C., O. Hopperstad, O.-G. Lademo, M. Langseth, *Numerical Modeling of the Structural Behavior of Thin-Walled Cast Magnesium Components Using a Through-Process Approach.* Materials & Design - MATER DESIGN, 2007. **28**: p. 2619–2631.
162. Dørum, C., O. Hopperstad, T. Berstad, D. Dispinar, *Numerical modelling of magnesium die-castings using stochastic fracture parameters.* Engineering Fracture Mechanics - ENG FRACTURE MECH, 2009. **76**: p. 2232–2248.
163. Farrokh, B., A.S. Khan, *A strain rate dependent yield criterion for isotropic polymers: Low to high rates of loading.* European Journal of Mechanics - A/Solids, 2010. **29**(2): p. 274–282.
164. Holliday, L., J. Mann, G.A. Pogany, H.L.D. Pugh, D.A. Gunn, *Ductility of Polystyrene.* Nature, 1964. **202**(4930): p. 381–382.
165. Bowden, P.B., J.A. Jukes, *The plastic flow of isotropic polymers.* Journal of Materials Science, 1972. **7**(1): p. 52–63.
166. Lubineau, G., P. Ladevèze, *Construction of a micromechanics-based intralaminar meso-model, and illustrations in ABAQUS/Standard.* Computational Materials Science, 2008. **43**(1): p. 137–145.
167. Chen Guanhua, *Collision response research and structure optimization of lithium-ion battery module.* 2018, Tsinghua University: Beijing [in Chinese].
168. Li Wei, Research on deformation failure of lithium-ion battery based on fine model. 2019, Tsinghua University: Beijing [in Chinese].
169. Luo Hailing, *Research on structural failure mechanism and modeling of flexible lithium-ion battery under mechanical abuse.* 2018, Tsinghua University: Beijing [in Chinese].
170. Soni, A., T. Robert, P. Beillas. *Effects of pedestrian pre-crash reactions on crash outcomes during multi-body simulations.* in *2013 IRCOBI Conference Proceedings - International Research Council on the Biomechanics of Injury.* 2013.
171. Ding, L., M. Glazer, M. Wang, B. Mehler, B. Reimer, L. Fridman. *MIT-AVT Clustered Driving Scene Dataset: Evaluating Perception Systems in Real-World Naturalistic Driving Scenarios.* in *2020 IEEE Intelligent Vehicles Symposium (IV).* IEEE.
172. Zhou, H., Z. Zhong, M. Hu, J. Huang, *Determining the steering direction in critical situations: A decision tree–based method.* Traffic injury prevention, 2020. **21**(6): p. 395–400.
173. Hu, M., Y. Li. *Drivers' avoidance patterns in near-collision intersection conflicts.* in *2017 IEEE 20th International Conference on Intelligent Transportation Systems (ITSC).* 2017. IEEE.
174. Gan, S., Q. Li, Q. Wang, T. Cui, B. Nie. *Capturing Driver Evasive Manoeuvres In Pre-crash Phases From Large-scale Real-world Critical Scene Videos.* in *IRCOBI Asia conference.* 2021. Beijing, China.
175. Han, Y., Q. Li, F. Wang, B. Wang, K. Mizuno, Q. Zhou, *Analysis of pedestrian kinematics and ground impact in traffic accidents using video records.* International Journal of Crashworthiness, 2019. **24**(2): p. 211–220.
176. Li, Q., Y. Han, K. Mizuno, *Ground Landing Mechanisms in Vehicle-To-Pedestrian Impacts Based on Accident Video Records.* SAE Technical Paper, 2018.
177. Sportillo, D., A. Paljic, L. Ojeda, *Get ready for automated driving using Virtual Reality.* Accident Analysis and Prevention, 2018. **118**: p. 102–113.
178. Nie, B., Q. Li, S. Gan, B. Xing, Y. Huang, S.E. Li, *Safety envelope of pedestrians upon motor vehicle conflicts identified via active avoidance behaviour.* Scientific Reports, 2021. **11**(1): p. 3996.
179. Li, Q., S. Shang, X. Pei, Q. Wang, Q. Zhou, B. Nie, *Kinetic and kinematic features of pedestrian avoidance behavior in motor vehicle conflicts.* Frontiers in bioengineering and biotechnology, 2021. **9**.
180. Chen Long, *Effectiveness evaluation of intelligent vehicle driving system based on multi-source data mining.* 2017, Tsinghua University: Beijing [in Chinese].
181. Malliaris, A.C., K.H. Digges, J.H. Deblois. *Relationships between crash casualties and crash attributes.* in *SAE International Congress and Exposition.* 1997.
182. Bahouth, G., K. Digges, C. Schulman, *Influence of Injury Risk Thresholds on the Performance of an Algorithm to Predict Crashes with Serious Injuries.* Annals of Advances in Automotive Medicine, 2012. **56**: p. 223–230.

183. Stitzel, J.D., A.A. Weaver, J.W. Talton, R.T. Barnard, S.L. Schoell, A.N. Doud, R.S. Martin, J.W. Meredith, *An Injury Severity-, Time Sensitivity-, and Predictability-Based Advanced Automatic Crash Notification Algorithm Improves Motor Vehicle Crash Occupant Triage.* Journal of the American College of Surgeons, 2016. **222**(6): p. 1211–1219.e6.
184. Wang, Q., S. Gan, W. Chen, Q. Li, B. Nie, *A data-driven, kinematic feature-based, near real-time algorithm for injury severity prediction of vehicle occupants.* Accident Analysis & Prevention, 2021. **156**: p. 106149.
185. Bance, I., B. Nie. *A Framework for Near Real-Time Occupant Injury Risk Prediction using a Sequence-to-Sequence Deep Learning Approach.* in *International Research Council on Biomechanics of Injury (IRCOBI).* 2019.
186. Gabler, H.C., A.A. Weaver, J.D. Stitzel, *Automotive Field Data in Injury Biomechanics.* 2015: Springer New York.
187. Lord, D., F. Mannering, *The Statistical Analysis of Crash-Frequency Data: A Review and Assessment of Methodological Alternatives.* Transportation Research Part A, 2010. **44**(5): p. 291–305.
188. Savolainen, P.T., F.L. Mannering, D. Lord, M.A. Quddus, *The Statistical Analysis of Highway Crash-Injury Severities: A Review and Assessment of Methodological Alternatives.* Accident Analysis & Prevention, 2011. **43**(5): p. 1666–1676.
189. Jeong, H., Y. Jang, P.J. Bowman, N. Masoud, *Classification of motor vehicle crash injury severity: A hybrid approach for imbalanced data.* Accident Analysis & Prevention, 2018. **120**: p. 250–261.
190. Al-Moqri, T., X. Haijun, J.P. Namahoro, E.N. Alfalahi, I. Alwesabi, *Exploiting Machine Learning Algorithms for Predicting Crash Injury Severity in Yemen: Hospital Case Study.* Applied and Computational Mathematics, 2020. **9**(5): p. 155–164.
191. Fiorentini, N., M. Losa, *Handling imbalanced data in road crash severity prediction by machine learning algorithms.* Infrastructures, 2020. **5**(7): p. 61.
192. Delen, D., L. Tomak, K. Topuz, E. Eryarsoy, *Investigating Injury Severity Risk Factors in Automobile Crashes with Predictive Analytics and Sensitivity Analysis Methods.* Journal of Transport & Health, 2017. **4**: p. 118–131.
193. Zheng, M., T. Li, R. Zhu, J. Chen, Z. Wang, *Traffic Accident's Severity Prediction: A Deep-Learning Approach-Based CNN Network.* IEEE Access, 2019. **PP**(99): p. 1–1.
194. Geyik, B., M. Kara. *Severity Prediction with Machine Learning Methods.* in *2020 International Congress on Human-Computer Interaction, Optimization and Robotic Applications (HORA).* 2020.
195. Yosinski, J., J. Clune, Y. Bengio, H. Lipson. *How Transferable are Features in Deep Neural Networks?* in *International Conference on Neural Information Processing Systems.* 2014.
196. Song, H.S., Y. Lee, S. Park, H. Kim, S.W. Kim. *A Study on Classification of Traffic Accident Injury Grade Using CNN and NASS-CDS Data.* in *the 2018 VII International Conference.* 2018.
197. Kim, T., G. Poplin, V. Bollapragada, T. Daniel, J. Crandall, *Monte Carlo method for estimating whole-body injury metrics from pedestrian impact simulation results.* Accident Analysis & Prevention, 2020. **147**: p. 105761.
198. MacKenzie, E.J., A. Damiano, T. Miller, S. Luchter, *The development of the Functional Capacity Index.* J Trauma, 1996. **41**(5): p. 799–807.
199. Liu Yanjie, *Research on the crashworthiness of energy-absorbing components in low-speed vehicle collision.* 2012, Harbin Engineering University [in Chinese].
200. Duan Libin, *Research on lightweight and crashworthiness design of front longitudinal beams of vehicles with variable thickness.* 2017, Hunan University: Changsha [in Chinese].
201. Nie Bingbing, *Impact of vehicle front-end structure and modeling design on pedestrian collision protection.* 2013 [in Chinese].
202. Xia, Y., T. Wierzbicki, E. Sahraei, X. Zhang, *Damage of cells and battery packs due to ground impact.* Journal of Power Sources, 2014. **267**: p. 78–97.
203. Zhu, J., X. Zhang, T. Wierzbicki, Y. Xia, G. Chen, *Structural Designs for Electric Vehicle Battery Pack against Ground Impact.* 2018, SAE Technical Paper.

204. Zhou Qing, Xia Yong, Nie Bingbing, Huang Yi, Lai Xinghua, *Several challenging topics in the research and development of vehicle collision safety and lightweight*. China Journal of Highway and Transport, 2019. 32 (7): p. 1–14 [in Chinese].
205. Zhang, S., Q. Zhou, Y. Xia, *Influence of mass distribution of battery and occupant on crash response of small lightweight electric vehicle*. 2015, SAE Technical Paper.
206. Li, B., M.H. Parekh, R.A. Adams, T.E. Adams, C.T. Love, V.G. Pol, V. Tomar, *Lithium-ion Battery thermal Safety by early internal Detection, prediction and prevention*. Scientific reports, 2019. **9**(1): p. 1–11.
207. Wang Chunli, Gong Limiao, Kang Ping, Tan Yechao, Li Mingming, *Research on early warning system of lithium-ion battery storage power station*. Collection, June 2018 [in Chinese].
208. Fernandes, Y., A. Bry, S. De Persis, *Identification and quantification of gases emitted during abuse tests by overcharge of a commercial Li-ion battery*. Journal of Power Sources, 2018. **389**: p. 106–119.
209. Wang Shuyang, *Research on gas generation of lithium-ion battery*. 2012 [in Chinese].
210. Deng Yuanbing, *Experimental and simulation research on thermal runaway and its early warning mechanism of lithium-ion power battery*. 2017 [in Chinese].
211. Cummings, S.L., N. Swartz. *Off-gas monitoring for lithium ion battery health and safety [R]*.
212. Cuerden, R., J. Hill, A. Kirk, M. Mackay, *The potential effectiveness of adaptive restraints*, in *International Research Council on the Biomechanics of Impact*. 2001.
213. IOSB, F. Available from: https://www.iosb.fraunhofer.de/en/press/press-releases/2021/activity-detection-inside-the-vehicle.html.
214. Richert, J., D. Coutellier, C. Götz, W. Eberle, *Advanced smart airbags: The solution for real-life safety?* International journal of crashworthiness, 2007. **12**(2): p. 159–171.
215. Wang, C., Q. Zhou. *Concept study of adaptive seatbelt load limiter using magnetorheological fluid*. in *The 21 st International Technical Conference on the Enhanced Safety of Vehicles Conference (ESV), Stuttgart, Germany*. 2009. Citeseer.
216. Bennett, J., G. Park, *Automotive occupant dynamics optimization*. Shock and Vibration, 1995. **2**(6): p. 471–479.
217. Tang, L., M. Luo, Q. Zhou. *Optimization of a child restraint system by using a particle swarm algorithm*. in *International Conference on Intelligent Computing*. 2006. Springer.
218. Yeh, I., Y. Fu, B. Kachnowski, T. Subbian, *Optimization of a vehicle restraint system using a genetic algorithm*. SAE transactions, 2005: p. 1504–1511.
219. Melvin, J.W., K.J. Baron, W.C. Little, T.W. Gideon, J. Pierce, *Biomechanical analysis of Indy race car crashes*. SAE transactions, 1998: p. 2872–2891.
220. Hubbard, R.P., P.C. Begeman, *Biomechanical performance of a new head and neck support*. 1990, SAE Technical Paper.
221. Melvin, J.W., T. Gideon, *Biomechanical principles of racecar seat design for side impact protection*. 2004, SAE Technical Paper.
222. Gideon, T., J. Melvin, P. Begeman, *Race car nets for the control of neck forces in side impacts*. 2004, SAE Technical Paper.
223. Forman, J., S. Stacey, J. Evans, R. Kent, *Posterior acceleration as a mechanism of blunt traumatic injury of the aorta*. Journal of biomechanics, 2008. **41**(6): p. 1359–1364.
224. Mohan, A., C. Papageorgiou, T. Poggio, *Example-based object detection in images by components*. IEEE Transactions on Pattern Analysis and Machine Intelligence, 2001. **23**(4): p. 349–361.
225. Felzenszwalb, P.F., R.B. Girshick, D. McAllester, D. Ramanan, *Object detection with discriminatively trained part-based models*. IEEE Trans Pattern Anal Mach Intell, 2010. **32**(9): p. 1627–45.
226. Felzenszwalb, P., R. Girshick, D. McAllester, D. Ramanan, *Visual object detection with deformable part models*. Commun. ACM, 2013. **56**(9): p. 97–105.
227. Redmon, J., S. Divvala, R. Girshick, A. Farhadi, *You Only Look Once: Unified, Real-Time Object Detection*. 2016. 779–788.
228. Helbing, D., P. Molnar, *Social Force Model for Pedestrian Dynamics*. Physical Review E, 1998. **51**.

229. Wakim, C., J. Oksman, *A Markovian model of pedestrian behavior*. 2004. 4028–4033.
230. Kooij, J., N. Schneider, F. Flohr, D. Gavrila, *Context-Based Pedestrian Path Prediction*. 2014.
231. Bonnin, S., T. Weisswange, F. Kummert, J. Schmuedderich, *General Behavior Prediction by a Combination of Scenario-Specific Models*. Intelligent Transportation Systems, IEEE Transactions on, 2014. **15**: p. 1478–1488.
232. Berger, R., H. Robotics, *White paper on the development trend of intelligent cabin*. 2019 [in Chinese].
233. Nie, B., S. Gan, W. Chen, Q. Zhou, *Seating preferences in highly automated vehicles and occupant safety awareness: A national survey of Chinese perceptions*. Traffic injury prevention, 2020. **21**(4): p. 247–253.
234. Yao Guangpeng, *Analysis on safety and efficiency of pedestrian-vehicle interaction at signalized intersections in urban areas*. 2016 [in Chinese].
235. Wang Lin, Li Zhao, Zhan Pengyu, *Design of vehicle-road cooperation system based on IEEE802*. 11p. Industrial Control Computer, 2017 (August 2017): p. 1–2 [in Chinese].
236. Cai Zhili, Sun Fengrui, Wei Lingxiang, Wang Nan. Design of vehicle-road coooperation system based on Internet of vehicles technology. 2011 [in Chinese].
237. Xiong Feng, *Research on multi-objective cooperative optimization design for lightweight and crashworthiness of vehicle body structure*. 2018, Jilin University: Changchun [in Chinese].
238. EXTRICATION, B. Available from: http://www.boronextrication.com/2014/07/12/2015-volkswagen-passat-body-structure-extrication/.
239. State Administration for Market Regulation. *Safety requirements for traction batteries of electric vehicles*. 2020, Domestic-National Standards-State Administration for Market Regulation CN-GB [in Chinese].
240. Lu, L., X. Han, J. Li, J. Hua, M. Ouyang, *A review on the key issues for lithium-ion battery management in electric vehicles*. Journal of power sources, 2013. **226**: p. 272–288.
241. Pan, H., W. Sun, X. Jing, H. Gao, J. Yao, *Adaptive tracking control for active suspension systems with non-ideal actuators*. Journal of Sound and Vibration, 2017. **399**: p. 2–20.
242. International, S., *Audi details new A8 active suspension*. 2017.
243. Magar, S.G. *Adaptive Front Light Systems of Vehicle for Road Safety*. 2015. IEEE.
244. MAZDA. Available from: https://www.mazda.com/en/innovation/technology/safety/active_safety/afs/.
245. Mages, M., M. Seyffert, U. Class. *Analysis of the pre-crash benefit of reversible belt pre-pretensioning in different accident scenarios*. in *Proceedings of the 22nd ESV Conference*. 2011.
246. Schoeneburg, R., K.-H. Baumann, M. Fehring, D. Ag, M. Cars. *The efficiency of PRE-SAFE systems in pre-braked frontal collision situations*. in *Proceedings of the 22nd ESV Conference*. 2011.
247. Toyota. Available from: https://www.youtube.com/watch?v=0jqOOFs9-7o.
248. Wu Hequan, Zhang Jiafei, Hu Lin, *Effect of seat rotation speed on passengers in automated driving*. Automotive Engineering, 2021. 43 (2): p. 226–231 [in Chinese].
249. Mercedes-Benz. Available from: https://www.mercedes-benz.com/en/innovation/autonomous/research-vehicle-f-015-luxury-in-motion/.
250. TOYOTA-BOSHOKU. Available from: https://www.toyota-boshoku.com/global/special/ces2020/mx191.html.
251. Wu Hequan, Hou Haibin, Hu Lin, Huang Jing, *Injury risk and avoidance strategy of occupants in automated driving vehicle under different seat orientations*. China Journal of Highway and Transport, 2019. 32 (6): p. 206–215225 [in Chinese].
252. Roth, M., F. Flohr, D.M. Gavrila, Ieee, *Driver and Pedestrian Awareness-based Collision Risk Analysis*, in *2016 Ieee Intelligent Vehicles Symposium*. 2016. p. 454–459.
253. Kuo, Y.-C., C.-M. Fu, C.-T. Tsai, C.-C. Lin, G.-H. Chang, *Pedestrian Collision Warning of Advanced Driver Assistance Systems*. 2016. 740–743.
254. Hariyono, J., A. Shahbaz, L. Kurnianggoro, K.-H. Jo, *Estimation of collision risk for improving driver's safety*. 2016. 901–906.

255. Takahashi, H., H. Miyazaki, S. Narita, T. Shamoto, Y. Mase, *Development of Pop-Up Hood System for Pedestrian Protection*, in *23rd International Technical Conference on the Enhanced Safety of Vehicles (ESV)*. 2013: Seoul , South Korea.
256. Jakobsson, L., T. Broberg, H. Karlsson, A. Fredriksson, N. Gråberg, C. Gullander, M. Lindman. *Pedestrian Airbag Technology@ A Production System*. 2013.
257. Nie Bingbing, Zhou Qing, Xia Yong, *Multi-wave characteristics and structural design of pedestrian head impact on vehicle engine hood*. Journal of Automotive Safety and Energy, January 2017 (v.8): p. 69–75 [in Chinese].

Chapter 5
Testing and Evaluation for Vehicle Safety

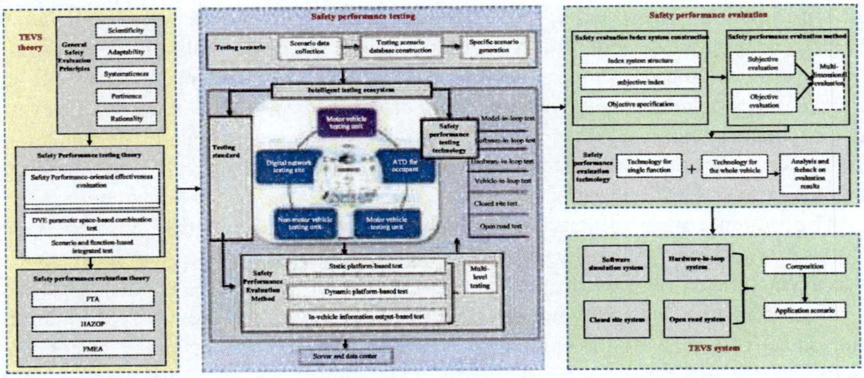

Testing and evaluation for vehicle safety (TEVS) is to measure, analyze and judge vehicle safety performance through a series of techniques and methods with reference to certain standards, and then make qualitative or quantitative evaluation conclusions.

TEVS aims to implement full and efficient test and conduct a comprehensive and reasonable evaluation on vehicles. Based on the general safety evaluation principles, safety performance testing and evaluation theory, it takes the core methods and key technologies in testing and evaluation as the research object, realizes the multi-dimensional evaluation for vehicle safety performance in each development stage, and establishes a standardized TEVS system.

The vehicle safety performance test takes the testing scenario as the input data, relies on the intelligent test ecosystem, uses the multi-level testing methods, and carries out safety test on components, systems and vehicles through simulation test, hardware and vehicle in-loop test, closed site test, open road test and other technologies from the virtual and practical perspectives. Based on the data obtained in the

J. Wang et al., *The Intelligent Safety of Automobile*, Key Technologies on New Energy Vehicles, https://doi.org/10.1007/978-981-99-6399-7_5

testing process, the vehicle safety performance evaluation is to establish the subjective evaluation, objective evaluation and multi-dimensional comprehensive evaluation approaches by constructing the subjective and objective safety evaluation index system, so as to provide the safety evaluation results ranging from single function to whole vehicle system.

5.1 Theory of TEVS

5.1.1 General Safety Evaluation Principles

Safety evaluation is essential for testing and evaluation. It refers to the application of system safety engineering principles and methods for identifying and analyzing hazards in the project or system, judging the possibility and severity of accidents, and formulating corresponding preventive measures. This process usually includes object selection, hazard identification, consequence analysis, qualitative evaluation, quantitative evaluation, and preventive measures [1]. The safety evaluation shall be carried out comprehensively and rigorously based on the actual situation of the evaluated object and the relevant national laws, regulations and technical standards. To select safety assessment methods, the principles of scientificity, adaptability, systematicness, pertinence and rationality shall be followed [2], as shown in Fig. 5.1.

The principle of scientificity means that in the process of evaluation, it is necessary to make rational analysis in the spirit of science, take qualitative and quantitative analysis as the process of scientific practice, and seek truth from facts to obtain objective evaluation results. By following the principle of scientificity, the safety evaluation can objectively analyze the evaluated objects, and ensure the accuracy of conclusions and the effectiveness of evaluation results with a standardized process, and finally put forward corresponding improvement measures to ensure the rationality of countermeasures [3].

The principle of adaptability means that the selected safety evaluation method shall adapt to the evaluated object. By following the principle of adaptability, the safety evaluation can adapt to the evaluated object, and evaluate the key points of

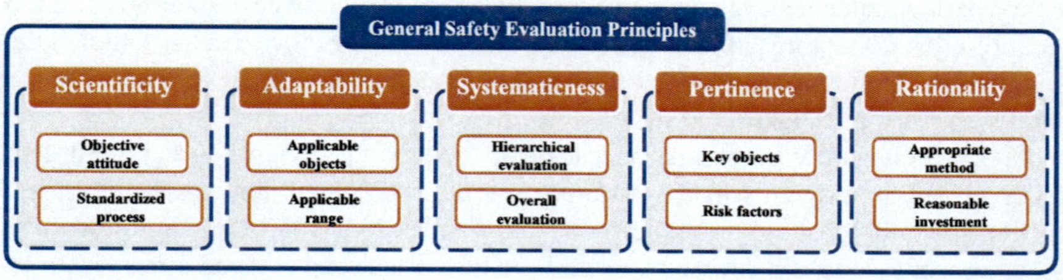

Fig. 5.1 General safety evaluation principles

each part in line with system composition of the object. When adopting different safety evaluation methods, attention shall be paid to their respective application scope and applicable objects, as well as the properties of systems and subsystems [3].

The systematic principle means that the safety evaluation and the evaluated object should be treated as a complete system. Following the systematic principle, the whole system can be evaluated from different angles and different levels. By combining hierarchical evaluation with overall evaluation, the safety evaluation can more comprehensively comprehend the evaluated object, make a true and reasonable evaluation, and thus obtain systematic data and scientific and reliable conclusions [3].

The principle of pertinence means that the safety evaluation shall find out the key points of the evaluation, and analyze the accident causes, probability and possible consequences pertinently. Following the principle, the objectives can be determined according to the actual situation and evaluation requirement of the evaluated objects. For the key or dangerous evaluated objects, the possible hazard sources and the potential hazards shall be analyzed and evaluated in particular. Similarly, in the follow-up improvement measures, countermeasures with high safety and operability should also be put forward to ensure the effectiveness of the evaluation and countermeasures [3].

The principle of rationality refers to the reasonable allocation of resources on the premise of the safety assessment. Following the principle of rationality, a relatively simple, intuitive and reasonable evaluation method can be adopted for safety evaluation, and the evaluation purpose can be achieved with low investment. In addition, it is necessary to conduct evaluation on dynamic data with a reasonable frequency, as too frequent evaluation will bring additional burden [4]. In order to prevent resource waste, it is required to reasonably carry out safety evaluation, instead of making in a complicated way [5].

5.1.2 Theory of Safety Performance Test

Before the road test, vehicle safety performance should be tested and verified in multiple aspects. The safety performance test mainly includes acceleration test based on hybrid reality, combination test based on DVE parameter space, and integrated test based on scenario and function, as shown in Fig. 5.2. Before safety performance test, it is also necessary to evaluate the effectiveness of the vehicle in terms of safety mechanism, that is, to select corresponding parameters for safety description and comprehensively conclude safety performance evaluation results in different test scenarios. Among them, the acceleration test based on hybrid reality is aimed at complex and diverse test scenarios. Tools such as automatic scenario generation are used to accelerate the test process, improve the test efficiency, and realize full coverage of traffic scenarios; The combination test based on DVE parameter space can effectively reduce the size of the parameter space while maintaining a high potential fault- detection rate; The integrated test based on scenarios and functions connects the internal relations and commonalities between scenarios and functions, with high iteration efficiency and repeatability.

1. Effectiveness Evaluation Oriented on Safety Performance

Effectiveness evaluation oriented on safety performance aims to quantitatively evaluate the extent to which the vehicle can meet the safety requirements. This section first describes the related concepts and general steps of effectiveness evaluation,

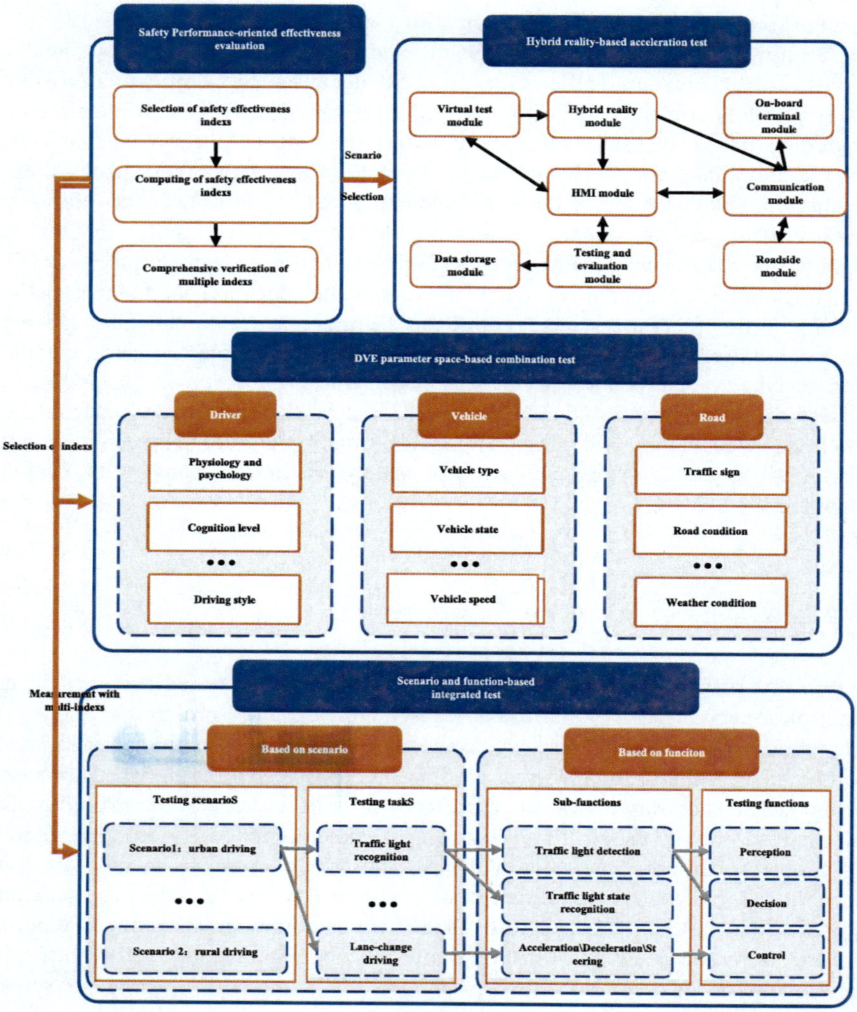

Fig. 5.2 Logic diagram of safety performance testing theory

then introduces the commonly-used ways, and puts forward the calculation method of effectiveness value.

(1) Definition of Effectiveness Evaluation

Effectiveness is generally defined as the capability of fulfilling a series of specific tasks and requirements, that is, the capability to achieve specified goals under specified conditions [6]. From the perspective of probability, effectiveness is defined as the probability that the system can meet the operation requirements under the specified working conditions and within the specified time [7]. As a complex information physical system, IV generally has a series of performance parameters, so that it is apparently not possible to evaluate vehicle safety performance with single parameter. Instead, parameters for safety description should be selected according to the test scenario. Therefore, it is necessary to integrate all indexes into one or several values as the quantitative results of safety performance. Since the dimensions and physical attributes of indexes are different, the pretreatment process of dimensional normalization should be introduced. In addition, different indexes have different representational capacities for vehicle safety performance. In the process of synthesis, different weights should be given to the indexes according to their importance, and taken to evaluate the performance of IV in meeting the application requirements, i.e. effectiveness. The above process of selecting, preprocessing, synthesizing and evaluating indexes is the general process of performance evaluation, which usually includes three steps: (1) screening effectiveness indexes and establishing evaluation index system reasonably, (2) calculating specific values of effectiveness indexes, (3) and synthesizing multiple indexes. The detailed steps are shown in Fig. 5.3.

Step 1: determine the objectives, i.e., to specify the objectives to be achieved for the effectiveness evaluation oriented on IV safety performance.

Step 2: define the system, and analyze the structure, function and performance of the evaluation system.

Step 3: select and pre-process indexes, i.e., to select indexes for describing system effectiveness in line with the criteria of comprehensiveness, definiteness and sensitivity, and carry out dimensional normalization and other pretreatment.

Step 4: establish the model, i.e., to determine the effectiveness evaluation way, and build the evaluation model.

Step 5: test and evaluate, i.e., to design and conduct evaluation experiments.

Step 6: analyze and verify the evaluation results, and make modification and improvement according to the found problems.

(2) Effectiveness Evaluation Methods

Proposed by the US Weapon System Effectiveness Industry Advisory Commission (WSEIAC) in the mid-1960s, Availability-Dependability-Capability (ADC) is one of the widely-used quantitative effectiveness evaluation methods [8]. Its basic principle can be summarized as specification of system effectiveness indexes with the function of the three performance in system availability, task reliability and system capability, which play an important role in evaluating the system performance for its target tasks. Firstly, the three performance requirements are analyzed respectively, and then their

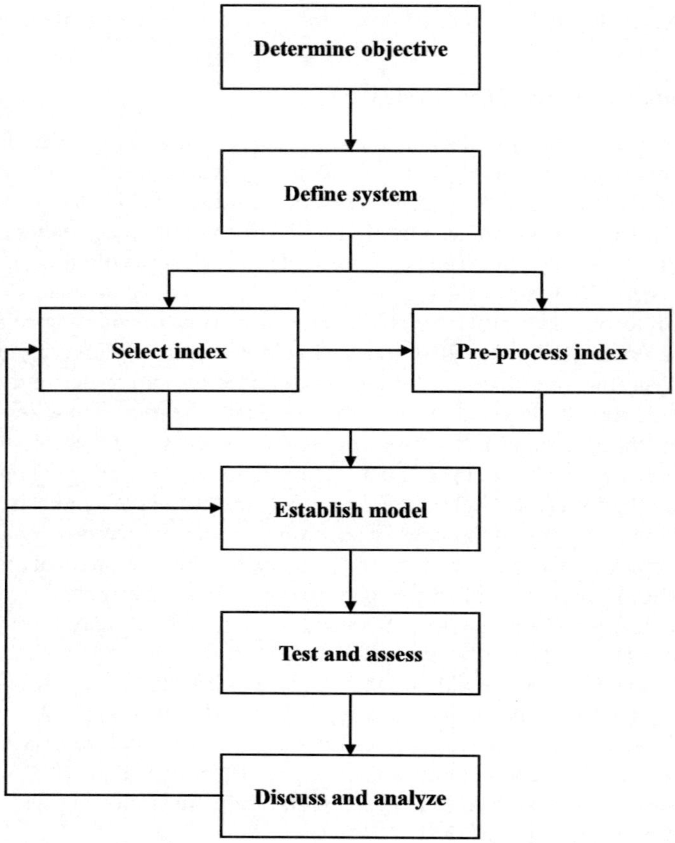

Fig. 5.3 Steps of effectiveness evaluation oriented on safety performance

coupling mode is determined according to the dependency. Finally, the capability of the system to complete the target task is calculated, that is, the system effectiveness value. The formula is as follows:

$$E^* = A \cdot D \cdot C \tag{5.1}$$

where, $E^* = (e_1, e_2, \ldots, e_m)$ represents the system effectiveness index vector; e_i is the system effectiveness index component for completion of task i.

A represents the availability vector, reflecting the level of the system state in the beginning, as shown in the following formula:

$$A = (a_1, a_2, \ldots, a_n) \tag{5.2}$$

where, a_j represents the probability that the system is in the jth state when it starts to work. The possible states of the system are the set of working state, assured working state and fault state of subsystems. Availability is related to such factors of the system as reliability, maintainability and quality [9].

D refers to the dependability (or reliability) matrix, which is related to the system reliability and repairability during operation, as shown in the following formula:

$$\mathrm{D} = \begin{bmatrix} d_{11} & d_{12} & \dots & d_{1n} \\ d_{21} & d_{22} & \dots & d_{2n} \\ \vdots & \vdots & \ddots & \vdots \\ d_{n1} & d_{n2} & \dots & d_{nn} \end{bmatrix} \quad (5.3)$$

where, d_{ij} indicates the probability that the system is in j state (effective state) on the premise of starting in i state during work.

C represents the probability that the mission objective can be achieved when the system is available and reliable, which is also called the inherent capability. c_{ij} is the characteristic value of the jth capability factor of the system in i state. The matrix form is as follows:

$$\mathrm{C} = \begin{bmatrix} c_{11} & c_{12} & \dots & c_{1m} \\ c_{21} & c_{22} & \dots & c_{2m} \\ \vdots & \vdots & \ddots & \vdots \\ c_{n1} & c_{n2} & \dots & c_{nm} \end{bmatrix} \quad (5.4)$$

2. Acceleration Test Based on Hybrid Reality

Hybrid reality-based acceleration test is a common testing way in the safety performance testing theory. It accelerates the testing process by means of the automatic scenario generation, in order to significantly improve the testing efficiency while reducing the testing cost.

(1) Hybrid Reality Test

Before the road test, the vehicle performance should be tested and verified in various aspects for safety assurance. The traditional closed site test is available to a limited number of scenarios and simpler road conditions compared with the open road environment However, it can not carry out long-distance test and other tests under ultimate conditions, so it is difficult to completely verify the vehicle performance [10]. The hybrid reality test has better solved the above defects. It combines the advantages of software simulation with the representativeness of driving test, and expands the hardware in-loop technology from the vehicle level to the traffic level at the same time, which makes the testing process more flexible and repeatable on the basis of ensuring high accuracy and safety, and facilitates the tests in key scenarios as well.

Hybrid reality test often adopts Monte Carlo method to generate virtual test environment, that is, sampling according to the distribution probabilistic density

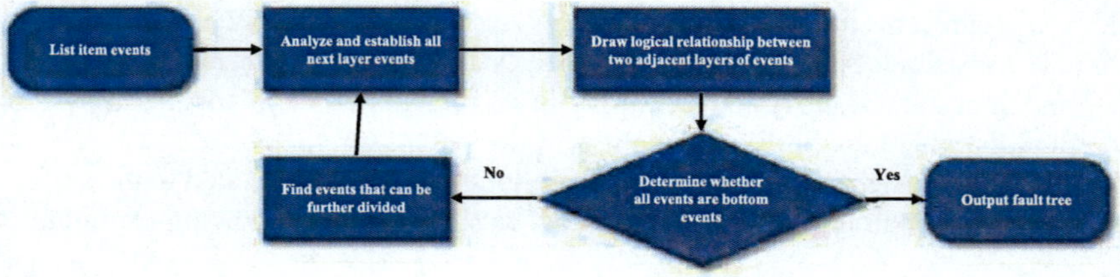

Fig. 5.4 Architecture of autonomous driving test system based on hybrid reality

of samples, and approximately estimating the mathematical expectation of sample function value, as shown in the following formula:

$$E[f(x)] = \int_x f(x)p(x)\mathrm{d}x \tag{5.5}$$

Take the driving risk for instance, $f(x)$ represents the vehicle driving risk corresponding to the scenario parameter x. $E[f(x)]$ represents the mathematical expectation of risk corresponding to the natural probabilistic distribution of parameter $p(x)$. The lower the risk expectation is, the stronger the driving ability of IV.

In typical applications, for example, the testing platform is built by involving IV and automatic intersection simulators, and the hybrid reality method can be used to realize the simulation and testing of intersection scenarios. For another example, the testing data in virtual environment is transmitted to the vehicle in real test through the cloud server, and the data of the virtual environment and the real environment are fused. Subsequently, the vehicle can select the driving scheme according to the fused data, and save driving parameters through central computing platform [11], so as to improve the integrity of vehicle performance test. Its architecture is shown in Fig. 5.4.

Hybrid reality test is a favorable transition between software in-loop test and open road test. It can significantly reduce the testing cost and effectively improve the testing efficiency.

(2) Acceleration Test

With the improvement of vehicle automation level, the application scenarios and situations of auto drive system are becoming more and more complex, and the requirements for the number of testing cases and testing scenarios are also increasing [12]. With this demand, it is necessary to accelerate the testing process. The existing research [13] proposes a hybrid reality acceleration test on the basis of hybrid reality. The specific process is as follows: firstly, to reconstruct and configure the simulation environment with the automatic scenario generation tool (Fig. 5.5), carry out the virtual test in the established simulation environment based on the real traffic scenario, then conduct the test in the corresponding actual environment, and modify the simulation models of all components, so as to solve the problem of inaccurate

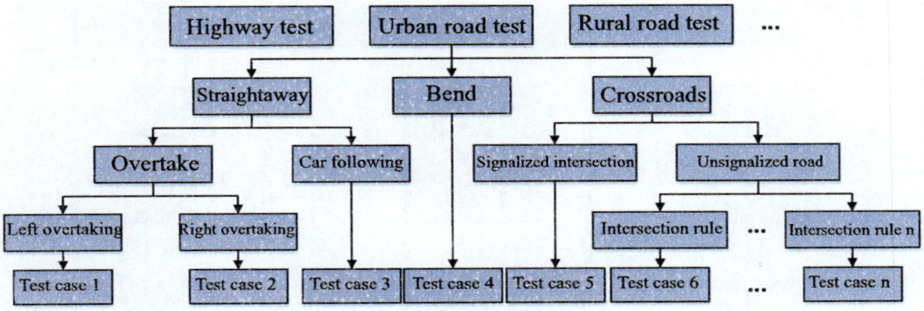

Fig. 5.5 Automatic generation of vehicle testing scenario

simulation models. This method can not only completely verify the performance of AV, but also significantly improve testing efficiency and reduce testing cost.

If only with the probabilistic distribution in the natural scenario $p(x)$ after the generation of virtual testing environment, the proportion of hazard scenarios will be relatively low, resulting in less sampling times and inaccurate mathematical expectation toward the hazard scenario. Therefore, in the process of acceleration test, sampling in accordance with the importance is usually used for special scenario, and the concept of importance weight is introduced, that is, the samples with higher sampling proportion in the real distribution should have lower weight during scenario generation, as shown in Formula (5.6):

$$E[f(x)] = \int_x f(x)p(x)\mathrm{d}x = \int_x f(x)\frac{p(x)}{q(x)}q(x)\mathrm{d}x = \int_x g(x)q(x)\mathrm{d}x \tag{5.6}$$

where, $p(x)$ is the probabilistic density distribution of each scenario when the original scenario is generated, and $q(x)$ is the probabilistic density distribution of each scenario when the importance sampling is applied. It can be seen that the importance sampling method adopts a better sampling probabilistic density distribution $q(x)$ instead of $p(x)$, which can enhance the sampling frequency of events with small probability and make expected estimation more accurate with less sampling times.

When the hybrid reality acceleration test is applied to simulation environment, it can automatically run and evaluate a large number of simulation scenarios, compared with the traditional test; The automatic scenario generation tool can significantly accelerate testing process, improve testing efficiency, and better cover more traffic scenarios. This method can also simulate and evaluate the performance of IV in traffic accident scenarios.

3. Combination Test Based on DVE Parameter Space

With the continuous improvement of intelligent levels of vehicles, the DVE traffic system involves a huge parameter space, as shown in Fig. 5.6, which is the main reason for the complexity of vehicle safety testing technology. The proposed combination test based on DVE parameter space can effectively reduce the size of parameter

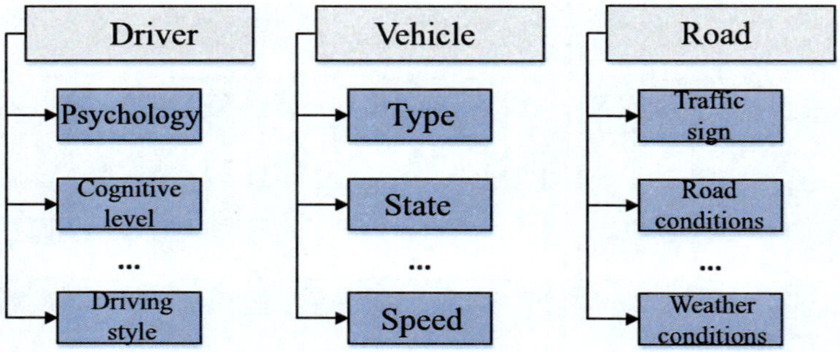

Fig. 5.6 Parameter space in DVE system

space while maintaining a high potential hazard detection rate. In the combination test, it is necessary to define the parameter space of scenario elements. There are a large amount of unlabeled information in IV and their traffic environment. It is very necessary to use clustering algorithm to learn these unlabeled training samples to reveal the inherent properties and laws of the data, and then obtain the parameter space of scenario elements. The existing clustering algorithms mainly include K-Means clustering and Gaussian mixture clustering. On the assumption that the sample set of the DVE parameter space can be divided into K clusters, K-Means clustering first initializes the division prototype, and then iteratively solves it step by step to obtain the final cluster division, as shown in Formula (5.7).

Where, E represents the sum of squares of Euclidean distances of all objects in the data set; *dist* represents the Euclidean distance between any two points; k represents the total number of data points in the parameter space; p represents data points in the parameter space; c_i represents the center of the parameter space. The smaller E is, the better the clustering effect is.

Gaussian mixture clustering uses probabilistic model to describe the partition of DVE parameter space, and finally obtains the probability of data points belonging to each category. For each data point x_i, the probability that it belongs to the kth category $\gamma(i, k)$ is recorded as:

$$\gamma(i, k) = \frac{\pi_k N(x_i | \mu_k, \Sigma_k)}{\sum_{j=1}^{K} \pi_j N(x_i | \mu_j, \Sigma_j)} \tag{5.7}$$

where, π_k represents the proportion of the kth category in the total parameter space; μ_k,Σ_k respectively represent the mean and variance of Gaussian distribution of the kth category.

How to select a small subset from the parameter space as the test case set is the critical point in the case generation of automobile test research. The widely-used combination test is to make a compromise between testing performance and cost. It selects testing cases in accordance with the principle that for any t (generally 2 or 3) parameters, all possible combinations are covered by at least one test case. This test criterion is also known as the t-combination test.

Combination test is quite effective in system testing. If the parameter space contains k parameters, the minimum number of testing cases required to complete t-combination test increases by the logarithmic order of k. The model based on the combination test is simple and effective, and can better meet the large-scale test demand [14, 15]. It is a feasible testing scheme in practice, and can significantly reduce testing cost and testing difficulty. Firstly, according to the principle of parameter combination test, the test parameters are divided into four categories: driver's cognitive level and driving style, vehicle's own configuration parameters, vehicle's motion state parameters and vehicle's surrounding environment parameters. And then the t-combination test is applied for selecting and combining the scenario parameters [16], which can effectively detect potential faults while reducing testing workload and significantly improve testing efficiency of IV.

4. Integrated Test Based on Scenario and Function

Scenario-based test means that the vehicle to be tested is required to complete the specified goals and tasks under the preset scenarios. At present, there is no recognized and clear definition of the scenario used for automatic driving test. One definition is the continuous change of the dynamic environment around the vehicle to be tested within a specific time range, including the behavior of VUT in the environment [17]. Many definitions are common in the core elements of the scenario, including the dynamic road environment that changes with time, traffic participants and vehicle driving tasks. Therefore, the scenario can be regarded as a combination of driving occasion and driving situation.

Being controllable and customizable, the scenario-based test effectively solves the problems of high testing cost and long test cycle in the open road testing. At the same time, there is no clear requirement for testing results. Without violating the given goals or tasks, the auto drive system can independently choose the way to deal with the current situation, with a high degree of freedom [10]. The typical application is the DARPA Grand Challenge and China Intelligent Vehicle Future Challenge. By simulating a variety of traffic scenarios and environments on the site, the comprehensive performance of the VUT in the real testing scenarios is investigated. However, whether for virtual simulation or scenario test in a closed field, it is based on certain assumptions and has the limitation that the scenario for test cannot cover all possible real traffic scenarios. In addition, the research on testing scenario design and complexity quantification related to IV is not sufficient.

Another kind of test is function-based, with the characteristics of high iteration efficiency and well repeatability. The function-based test is generally conducted for the single module of IV, such as perception, decision, planning and control [18]. Each module can be further decomposed into specific functions. For example, the vision-based recognition function includes the recognition of road lane, vehicles, traffic signs, traffic lights, pedestrians, obstacles and so on. From the perspective of testing, the IV that can successfully pass the function test shall have all driving capabilities similar to human beings. However, the application of function-based test

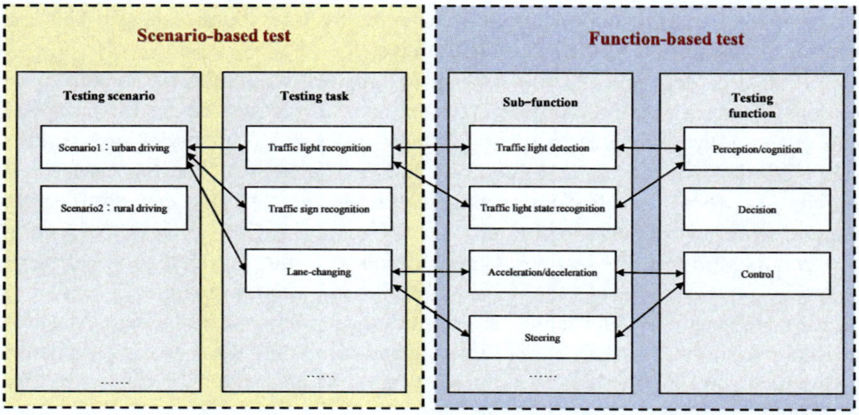

Fig. 5.7 Semantic diagram of test based on scenario and function [19]

in the relatively complex and comprehensive auto drive system shows some shortcomings. For example, quite a number of existing function-based tests are carried out independently and can not test the comprehensive performance of multiple functions.

In general, the two tests focus on either the overall or partial performance of IV in the testing process, without the capacity to realize comprehensive test. Therefore, the integrated test based on scenario and function can make up for the shortcomings of the two tests to a certain extent. The testing task is considered as the internal connection and commonness between scenarios and functions. Therefore, the semantic diagram of automatic driving test based on test task is established, as shown in Fig. 5.7 [19]. The testing design process is abstracted into four steps, i.e., testing function screening, testing task generation, task space–time allocation, and traffic scenario correlation. The performance of the VUT is quantitatively evaluated in accordance with the set evaluation criteria. The testing theory and method have been applied to the 2016 and 2017 China Intelligent Vehicle Future Challenge.

In addition, the German Pegasus project has also made explorations and attempts in the comprehensive test based on scenarios and functions. The project proposes to divide the scenario information into three categories according to the degree of abstraction: functional scenario, logical scenario and specific scenario [20]. Functional scenario is mainly used to divide driving tasks in specific scenarios, and is mainly described in words and terms; Logical scenario builds a multi-dimensional testing parameter space according to the distribution characteristics of key elements of the scenario, and expresses and refines each functional scenario for each functional scenario; Specific scenario is the smallest basic unit of virtual test, and is also the combination of multi-element parameters in logical scenario. The three scenarios are interrelated and constitute a high-coverage test scenario library. The scenario division and definition of PEGASUS has been widely accepted and applied till now [21].

5.1.3 Theory of Safety Performance Evaluation

The safety performance evaluation theory mainly includes fault tree analysis (FTA), failure mode and effect analysis (FMEA) and hazard and operability analysis (HAZOP), which logic is shown in Fig. 5.8. FTA focuses on the specific accident, and finds the direct causes and basic events of the accident from the whole to the part by means of deductive inference. FMEA focuses on the specific failure, and excavates the essential causes of failure by means of analytic induction. HAZOP comprehensively and systematically identifies design defects and event risks by means of qualitation.

FTA conducts analysis hierarchically from top to bottom according to the tree structure. In this process, the various factors that may lead to system fault (such as hardware factors, software factors, environmental factors and human factors) are analyzed layer by layer from the whole to the part, so as to determine the cause and probability of the fault, find the weak links of the system, and facilitate the follow-up improvement and optimization accordingy. The analysis process is shown in Fig. 5.9 [22, 23].

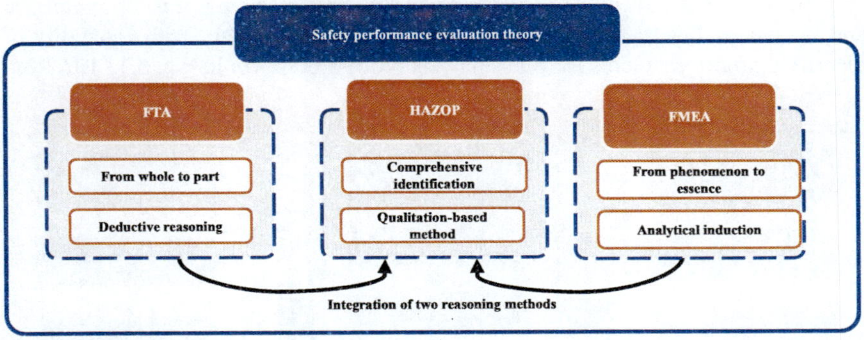

Fig. 5.8 Logical diagram of safety performance evaluation theory

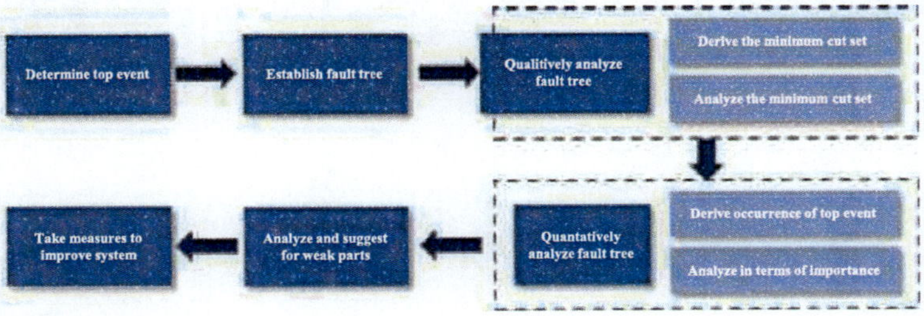

Fig. 5.9 FTA analysis process

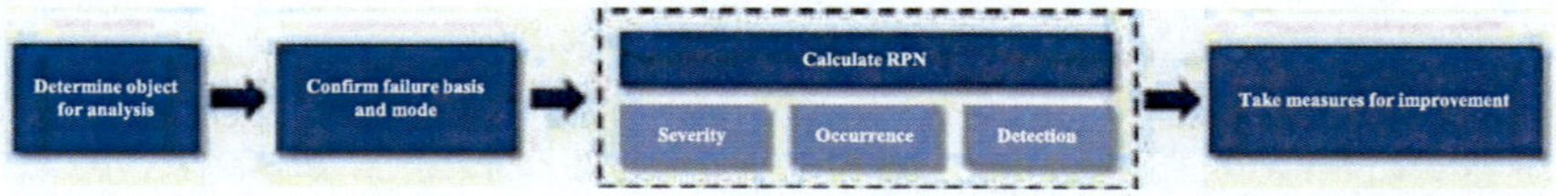

Fig. 5.10 FMEA analysis process

FMEA is a theory developed in practice and has a wide application prospect in the industrial field. It can be used in the pre-prevention stage to find the weak links possibly leading to accidents through failure mode analysis and take measures to improve in advance, or in the post improvement stage to find the problems through accident causation analysis and then make up for defects or loopholes. The analysis process is shown in Fig. 5.10 [24].

HAZOP can be used to identify potential hazards in the process of design and development, production and operation, systematically and comprehensively identify various deviations inconsistent with the design purpose, analyze the causes and evaluate corresponding consequences, and finally put forward targeted suggestions to improve system safety [25]. HAZOP is a creative process, which is generally completed by a team composed of experts with different professional knowledge backgrounds. To collect views and ideas by brainstorming, it is more creative and systematic than the independent work of individual expert, with capability of identifying more problems [26]. Its analysis process is shown in Fig. 5.11 [27, 28].

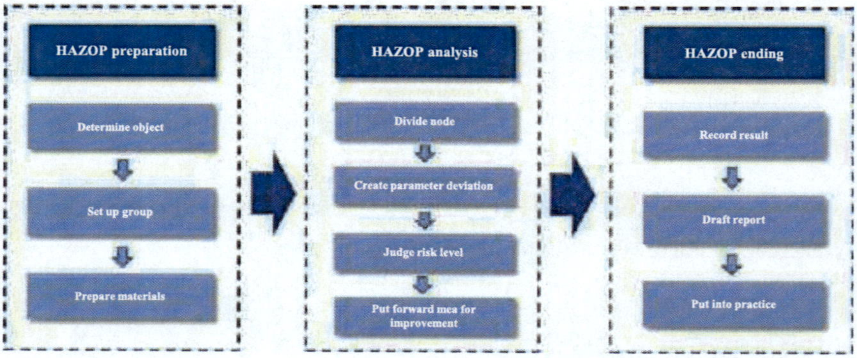

Fig. 5.11 HAZOP analysis process

5.2 Vehicle Safety Performance Testing

5.2.1 Testing Specifications and Standards

In order to regulate the production and application safety of vehicle parts, the mandatory regulations on vehicle safety were primarily put forward in Europe and the US. With the popularity of vehicles in the world, Japan and South Korea have also put forward their own mandatory regulations with reference to European and US regulations. With the rapid innovation of automotive technology in recent decades, the existing laws and regulations have been difficult to meet the demand of IV safety performance evaluation. Certification and evaluation systems for new cars in different countries came into being. In the past decade, the domestic evaluation regulations for new cars have followed the intelligent and electrified development trend of vehicles, as well as the actual road traffic situation in China, and gradually supplemented and improved the existing testing items.

1. Mandatory Regulations on Vehicle Safety

The enforcement of vehicle safety regulations is driven by increasing number of casualties in traffic accidents, and its improvement is accompanied by gradual development of automobile industry. Countries around the world have progressively established corresponding testing standards and regulations for making mandatory safety requirements, firstly led by Europe and the US. The UK began to implement the road vehicle lighting act in 1929, and the regulation of *Manufacture and Usage of Motor Vehicles* in 1931. In 1952, Germany issued the road traffic law in terms of safety regulations for vehicle and its parts. In 1958, the European Economic Commission formulated a unified European Economic Community (EEC) instructions and Economic Commission of Europe (ECE) regulations to eliminate trade barriers and promote the internationalization of European legislation. ECE regulations cover quite a number of aspects such as vehicle lighting, vehicle lock, tire, steering system, windshield, seatbelt, child seat and typical collision conditions, and involve more than 100 test items. They are accepted by most EU member states and introduced into their own legal systems.

In 1966, the U.S. Senate and House of Representatives passed the *National Traffic and Motor Vehicle Safety Act*, which also stipulated mandatory regulations for car models after 1968. As the law enforcement department, NHTSA is responsible for formulating and promoting vehicle safety standards. It implemented 17 regulations within one year, and added 9 regulations soon, all of which are the predecessor of FMVSS. After decades of development, FMVSS has formed a complete system, which covers all aspects of vehicle active and passive safety, accident avoidance, collision injury protection and other rules that can help improve vehicle collision safety.

Referring to European and US vehicle safety regulations, most countries in the world, including Japan, Canada and China, successively promulgated regulations at the end of the twentieth century. The government departments of China's vehicle industry began to introduce FMVSS series of safety regulations in the 1980s and implemented them in the late 1990s. Relevant safety regulations originally focused on frontal collision, such as collision protection for vehicle occupants. With the development of China's vehicle industry, a vehicle standard system involving mandatory national standards (GB), recommended national standards (GB/T) and vehicle industry standards (QC/T) has been established. Early GB mainly refers to ECE system, including safety, environmental protection, energy saving and anti-theft. The safety part includes active safety standards, passive safety standards and general safety standards. Active safety items mainly involve lighting of light signal devices, braking, steering, tires, etc.; Passive safety items involve seats, door locks, seatbelts, protrusions, body anti-collision and fire prevention, etc.; General safety items include vision, indicator lights, signal devices, anti-theft, etc. Mandatory national standards have made comprehensive specifications for the products manufactured and sold by vehicle enterprises, which are also the prerequisite for new cars to be on sale.

2. Foreign New Car Assessment System

For evaluating the crashworthiness and other vehicle safety levels, the New Car Assessment Program (NCAP) directly publishes testing results to the public, different from mandatory safety regulations. NCAP is generally initiated by regional government departments or vehicle industry associations to select the best-selling car models in the region and the vehicles randomly purchased from the market for the test on specified items, with testing results open to consumers [29]. In 1978, the US National Highway Traffic Safety Administration proposed and organized the establishment of the earliest NCAP system. By the late 1990s, Europe, Japan, Australia and South Korea had also established their own vehicle safety evaluation systems, including US-NCAP and IIHS in the US, Euro-NCAP in Europe, J-NACA in Japan, A-NCAP in Australia and K-NCAP in South Korea.

In spite of being not mandatory, NCAP system adopts more stringent testing standards and comprehensive quantitative evaluation system to thoroughly investigate the safety performance of new cars. Therefore, the public test results will directly affect consumers' recognition of vehicle safety. Take the most influential Euro-NCAP for example, it is led by the Federation of International Automobile (FIA), and is composed of European vehicle federations, government agencies, consumer rights organizations, vehicle clubs and other organizations. Euro-NCAP is not attached to any vehicle manufacturing enterprises, and its required funds are provided by the EU. It tests the sold or imported vehicles from time to time. Similarly, with the goal of reducing the casualty rate in traffic accidents and meeting the requirement of the insurance industry for assessment on collision damage risks of different car models, Insurance Institute for Highway Safety (IIHS) of the US is one of the most stringent third-party safety testing institutions, and its test results are directly linked to the vehicle insurance rate. As the evaluation results will affect the sale and premium of products, the new car evaluation system has been highly valued by major engine manufacturers, and excellent test results are often taken as the major selling points in model promotion. The new car evaluation system has become an important force in promoting engine manufacturers to actively improve vehicle design and safety.

3. Domestic New Car Assessment System

China New Car Assessment Program (C-NCAP) was first released by China China Automotive Technology & Research Center in 2006, and is revised every three years. Six versions have been revised in the past 14 years, and the latest 2021 version has been officially released. Its evolution process is shown in Fig. 5.12. During the development, C-NCAP not only learns from advantages of foreign new car assessment system, but also takes into account the characteristics of China's traffic accidents and market share of different car models in the setting of testing procedures and scoring items, which is conducive to the improvement of the domestic traffic safety level to a certain extent. For example, the pedestrian fatality rate in China's human-vehicle collision accidents is high. C-NCAP added the new pedestrian protection test in 2018 and further expanded the impact testing area between pedestrian head impactor and the front end of the vehicle in 2021. With the continuous increase in the share of new energy vehicles in China, collision test requirements related to the safety of new energy vehicles have been added to the assessment system in the last two editions.

Similar to Euro-NCAP, C-NCAP consists of four parts, i.e., adult (driver and passenger) protection, child protection, pedestrian protection and safety assistance system (Fig. 5.13). It covers various safety tests including in-vehicle occupant collision protection, pedestrian protection in human-vehicle collision, children restraint system devices, automatic emergency braking and electronic stability system. Each part shall be scored separately, and the total score shall be finally calculated and divided into 6 grades. For Electric Vehicles/Hybrid Electric Vehicles (EV/HEV) in accordance with the electrical safety requirements, they will be separately marked

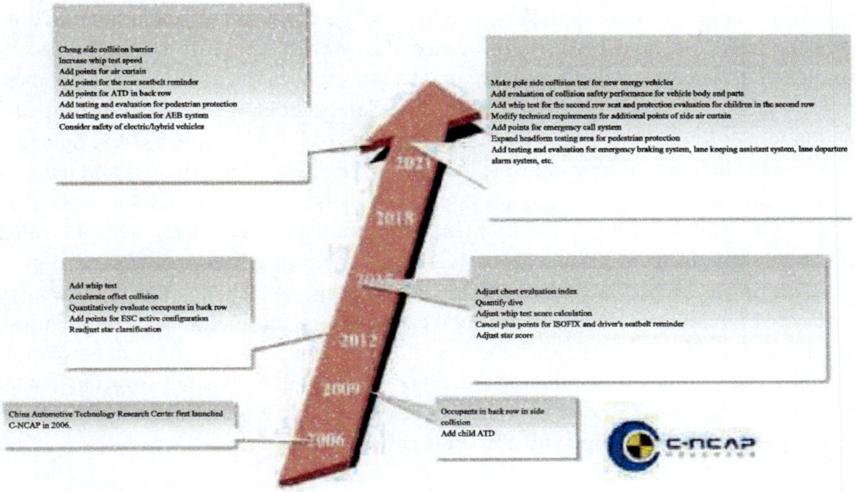

Fig. 5.12 Historical evolution of C-NCAP testing items

Fig. 5.13 Euro-NCAP test scenario [30]

with electrical safety signs, in addition to the star rating. Official data show that since 2015, C-NCAP has conducted safety evaluation on more than 200 car models in the market.

With reference to the US IIHS test standards, the Insurance Association of China took the lead to establish China Insurance Research Institute of automobile technology (CIRI) in March 2015 with financial assistance from a number of companies, and released the China Insurance Automobile Safety Index (C-IASI) for testing and evaluation system. Compared with the four aspects of C-NCAP, the C-IASI management center focuses on four indexes, including vehicle crashworthiness and maintenance economy, in-vehicle occupant safety, pedestrian safety, and vehicle assistance safety. The evaluation result of each index is published separately and presented in the intuitive form of grades: good (G), acceptable (A), marginal (M), and poor (P). And in terms of losses in traffic accidents, the center takes the maintenance economy of vehicles in low-speed frontal collision as a particular focus of attention. In addition, C-IASI puts forward more stringent requirements for the crashworthiness of vehicle body, by introducing the test condition with 25% small-overlap frontal impact. On the whole, the comparison of specific collision conditions involved in new car assessment at home and abroad is shown in Fig. 5.14.

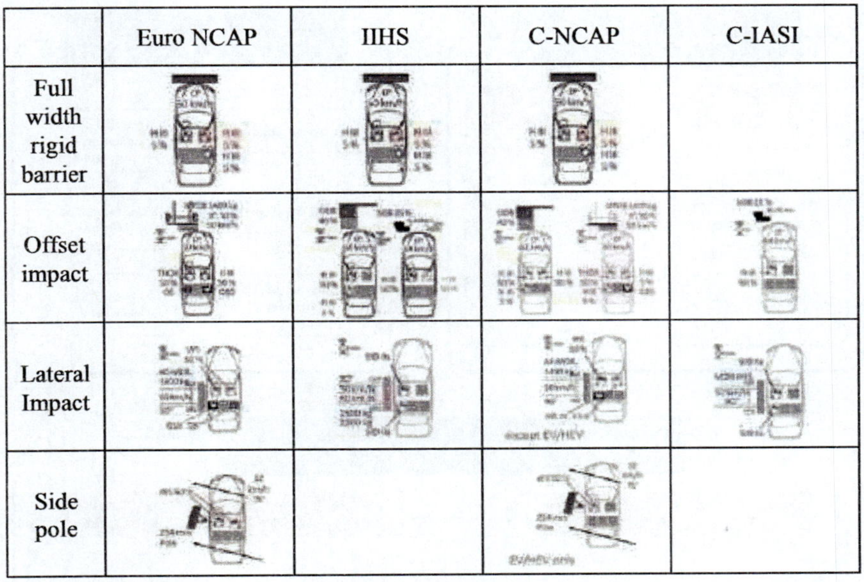

Fig. 5.14 Comparison of collision conditions for new car assessment at home and abroad

5.2.2 Testing Scenario

Testing scenarios are the input data of the testing process and the basis of testing execution. Testing scenario construction includes three steps: scenario data collection, testing scenario database construction and specific scenario generation. The specific process is shown in Fig. 5.15.

Scenario data collection is to obtain a large number of testing data with rich types and real scenarios. Based on real data, simulation data and expert experience data, the basic information of traffic environment elements and VUTs in the data are extracted. The richness and authenticity of these information determines whether the subsequently generated scenarios are sufficient and effective. The construction of testing scenario database refers to the classification and arrangement of scenario data, as well as the unified storage and management in database. During the test, logical scenario parameters can be set according to the test requirements, and then specific scenarios can be automatically generated from the testing scenario database. According to different testing requirements, the specific scenario generation process can be divided into automatic generation of typical scenarios and accelerated generation of high-risk scenarios. Automatic generation of typical scenarios refers to the generation of specific testing scenarios by random sampling based on the probabilistic density of each scenario in reality. Accelerated generation of high-risk scenarios is to increase the probabilistic density of high-risk scenarios in testing scenarios, so as

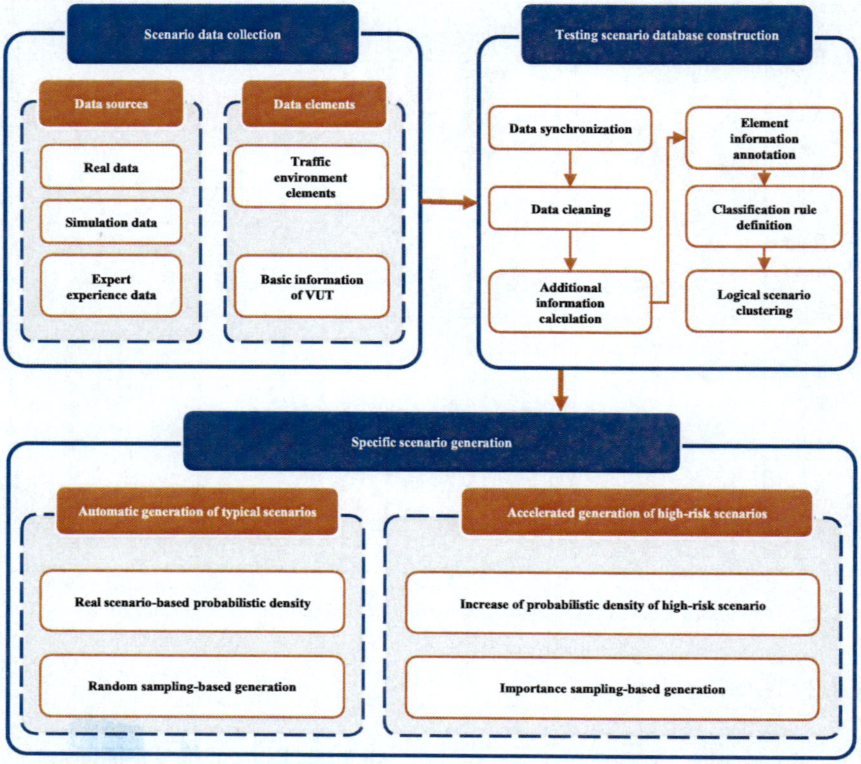

Fig. 5.15 Testing scenario construction process

to present more high-risk scenarios under the same test mileage for achieving the purpose of testing acceleration.

1. Scenario Data Collection

Scenario data is a specific description for various scenarios from three sources, i.e., real data, simulation data and expert experience data. Among them, the real data is mainly collected by the data acquisition vehicle through the installation of such multi-sensors as radar, camera and high-precision inertial navigation. Other real data come from accident, roadside unit monitoring, driver test and closed testing site. Simulation data mainly includes driving simulator data and simulated data. The driving simulator is to simulate the virtual scenario, sound effect and motion effect in the driving process by applying computer technology, and then collect the testing scenario data by placing the driver in the simulated cockpit for a real driving feeling. Simulation data refers to the data obtained from IV under test in the virtual simulation environment. At present, the mainstream simulation software includes VTD, PreScan, PanoSim, CARLA, CarMaker, etc. Expert experience data refers to the typical scenario element information from relevant experts by summarizing historical test data. The testing scenario in the relevant test standards for IV can be regarded as the expert experience data [10].

Scenario data is used to extract the key information required for test scenario generation, which can be abstracted into two major elements, namely, basic information of VUT and traffic environment elements [31]. The basic information of the vehicle mainly includes the inherent state, target information and driving behavior. The inherent state includes the geometric features, performance features and driving system of the vehicle, which will play a key role in the behavior decision of the automatic driving system; The target information is the driving task, which will affect the coverage and duration of the testing scenario; Driving behavior refers to the current motion state of the vehicle, such as longitudinal speed and lateral speed. During the test, a closed loop is formed between the vehicle and the surrounding driving environment, and environmental elements including other traffic participants will be affected by behaviors of the vehicle. Therefore, the basic information of VUT is indispensable in the testing scenario [10].

Traffic environment elements mainly include weather lighting information, static road information, dynamic road information and other traffic participant information [10]. Weather and light information will affect the perception system of IV. Static road information is the basis of scenario elements, including lane information (number of lanes, intersection shape, etc.) and road facilities information (signal lights, traffic signs, etc.). Dynamic road information includes road changes (road maintenance, blocking, collapse, etc.), dynamic changes of road facilities (roadside equipment moving, trees breaking, etc.) and dynamic changes of communication environment (signal blocking, electromagnetic interference, etc.). Dynamic road information will greatly increase the complexity of the driving scenario and the difficulty of IV decision-making. Other traffic participant information includes pedestrian information, non-motor vehicle information and motor vehicle information, which has great uncertainty in the real driving environment, as its state in the next moment is closely related to pedestrian's intention or driver's characteristics.

2. Testing Scenario Database Construction

The testing scenario database is a collection of testing scenarios in the form of database, which is built completely and hierarchically based on real data, virtual data and expert experience data. The construction process of typical testing scenario database includes data synchronization, data cleaning, additional information calculation, element information annotation, classification rule definition and logical scenario clustering, as shown in Fig. 5.16. Since different sensors have different working frequencies and different data resolutions, it is necessary to synchronize time and space for the collected data. Time synchronization can achieve nanosecond synchronization by means of global positioning system, Global Navigation Satellite System, Galileo Satellite Navigation System, etc.; Spatial synchronization can unify the data of different sensors in the same coordinate system through joint calibration of multiple sensors.

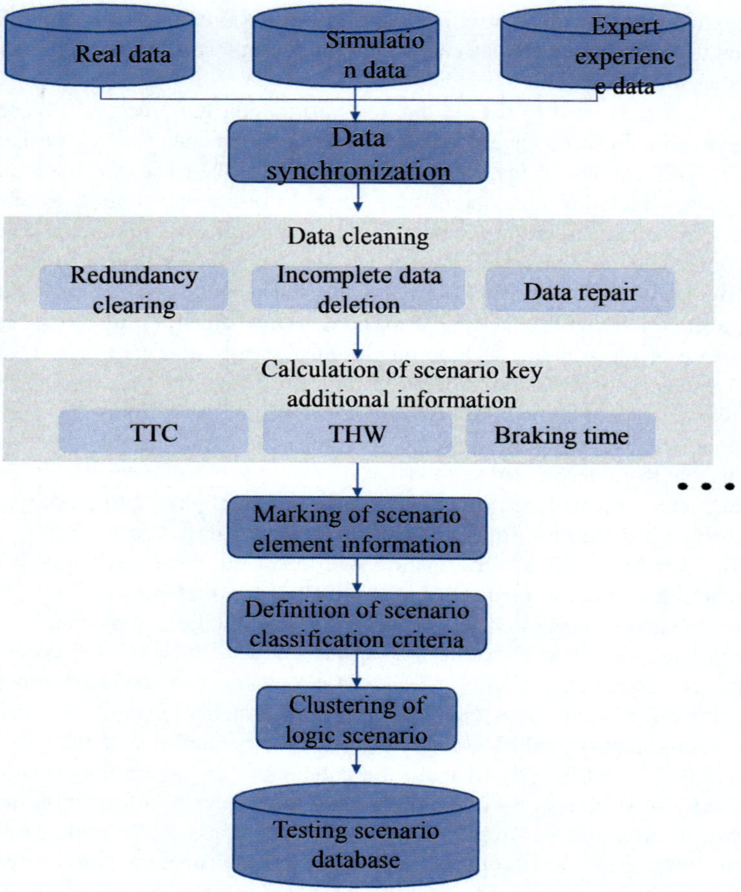

Fig. 5.16 Typical testing scenario database construction

Data cleaning generally involves redundant data clearing, incomplete data repair with statistical rules, and deletion of data that cannot be repaired. During data cleaning, it is necessary to formulate reasonable cleaning rules to ensure data integrity. At the same time, the cleaning cost should be minimized without reduction of data quality [10]. The calculation of key additional information of the scenario refers to the analysis of data that is vital to the scenario design but cannot be directly obtained from the sensors, such as TTC, THW and braking time. Scenario element information annotation is to annotate the key information in the scenario through semantic analysis, machine learning and other methods. After the element information annotation is completed, the scenario classification rules shall be set according to the annotated results and the functional requirements of the tested vehicle, and then the scenarios that meet the classification rules shall be aggregated into the same logical scenario by means of clustering. Finally, the generated logical scenario set is saved in the computer system in the form of a database. The establishment of a testing scenario database can achieve unified management of scenario data, and automatically generate test scenarios for application in test when necessary.

3. Specific Scenario Generation

The established testing scenario database contains different kinds of logical scenario data, which shall be exported to specific scenarios by specific scenario generation. Specific scenario generation usually includes automatic generation of typical scenarios and accelerated generation of high-risk scenarios. According to the testing requirements, these two kinds of generation can be integrated, so as to comprehensively generate common scenarios and efficiently generate high-risk scenarios.

(1) Automatic Generation of Typical Scenarios

Automatic generation of typical scenarios is one of the ways for generating specific scenarios. Based on the probabilistic density of each scenario in reality, the test scenarios are generated randomly according to the vehicle testing requirements, and a large number of testing scenarios are generated in a short time. Its main technical routes include the random sampling represented by Monte Carlo algorithm and rapid-exploring random tree, as well as the generation based on analytic hierarchy process and machine learning.

In order to improve the test efficiency of decision-planning system, a typical way is to automatically generate testing scenarios based on the OpenX standard series [32] developed by the Association for Standardization of Automation and Measuring Systems [33]. Firstly, it shall define the functional scenarios of decision planning system based on the hierarchical model, which consists of six layers, namely, basic road layer, infrastructure layer, temporary static obstacle layer, traffic participant layer, environment layer and electronic communication facility layer [34]. According to the testing requirements of decision-planning system, the necessary information includes road topology (basic road layer), traffic environment signal (infrastructure layer) and traffic participant information (traffic participant layer). On this basis, the combination of the positions and actions of the vehicle and other traffic participants under certain road conditions is defined as the functional scenario of the decision-planning system test.

According to the above definition, the keywords are set as road topology information, traffic signal information, the ego-vehicle information and other traffic participant information. Next, the keywords are mapped to parameters and the parameter space is constructed (Table 5.1), and finally the logic scenario is generated. It should be noted that the parameter types corresponding to keywords in the actual test should be analyzed in view of the test conditions.

Different traffic participants includes vehicle and non-vehicle participants according to their action characteristics. At present, there have been quite a number of studies on vehicle action categories [35], which can generally be divided into transverse actions including left–right turning and left–right lane changing, and

Table 5.1 Mapping relationship between keywords and key parameter types [33]

Hierarchy	Keyword	Key parameter type	Whether the value can be assigned directly or not
Basic road	Road topology	Topology type	No
		Lane ID	Yes
		Lane start coordinates	Yes
		Lane width	Yes
		Lane curvature	Yes
		Lane length	Yes
		Lane line type	Yes
Transportation infrastructure	Traffic signal	Traffic light status	Yes
		Speed limit sign	Yes
		Traffic signal location	Yes
Traffic participants	Ego-vehicle	Initial position	Yes
		Initial speed	Yes
		Action	No
	Other traffic participants	Initial relative position	No
		Initial speed	Yes
		Action	No

longitudinal actions including constant speed, acceleration and deceleration [36, 37]. Non-vehicle traffic participants have a higher degree of freedom and greater uncertainty. Table 5.2 shows the action categories and corresponding parameters of different traffic participants.

During the simulation test, in order to coordinate the start time of each action, the action parameters shall also include trigger conditions. Time-based triggering conditions are easy to be set. However, due to the uncertainty of other traffic participants' behavirs, it is necessary to integrate time, event or space-based triggering conditions. When generating a specific scenario file, the static scenario format standard Open Drive can be adopted for determining the static scenario, and then the mapping relationship between the logical scenario parameter space and Open SCENARIO format is analyzed for automatic generation of the file [33]. Open SCENARIO format can be divided into three layers: catalog, entity and event board. The logical scenario parameter space corresponding to each structure is shown in Fig. 5.17. Non-key contents such as vehicle catalog can be mapped to non-key parameters with their keywords set as default values, and finally the mapping relationship among content, keywords and parameter types (Table 5.3) can be obtained. Random sampling is carried out in the parameter space according to the scenario keywords, and the content, keywords and parameter values are expressed according to the format requirements. Consequently, a complete scenario file is made for the automatic generation of the testing scenario.

Table 5.2 Action categories and corresponding parameters of traffic participants [33]

Traffic participant	Action category		Corresponding parameter
IV	Automatic driving		Destination coordinates
Manual driving vehicle	Transverse action	Left/right turn	Turning duration/ target lane
		Left/right lane change	Lane change duration
	Longitudinal action	Constant speed	Speed
		Acceleration	Acceleration/ acceleration duration
		Deceleration	Deceleration/ deceleration duration
Non vehicle	Constant speed		Speed
	Acceleration		Acceleration/ acceleration duration
	Deceleration		Deceleration/ deceleration duration

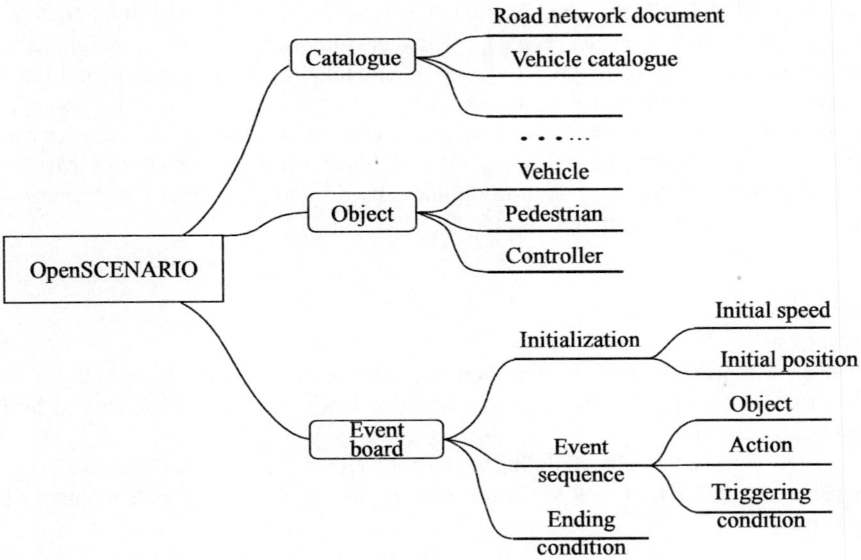

Fig. 5.17 OpenSCENARIO format analysis [33]

Table 5.3 Mapping relationship among OpenSCENARIO, keywords and parameter types [33]

OpenSCENARIO content	Keyword	Parameter type
Road network file	Road topology traffic signal	Key parameter
Vehicle catalog	Non	Non-key parameter
Controller	Non	Non-key parameter
Initialization	Ego-vehicle and other traffic participants	Initial speed initial position
Event sequence	Ego-vehicle and other traffic participants	Action
Ending condition	Non	Non-key parameter

(2) Accelerated Generation of High-risk Scenarios

If the safety performance of IV in high-risk scenarios is favorable, their performance in general scenarios shall be guaranteed as well in theory [38]. Therefore, IV test in high-risk scenarios has gradually become the research focus of researchers [10]. According to the test requirements, the required high-risk scenarios can be screened out by setting specific parameter values such as the expected braking time, the expected TTC, the expected braking deceleration, traffic flow, speed fluctuation, average speed and acceleration change. Then, the importance sampling is conducted to replace the probabilistic density function generated by the original scenario and increase the generation probability of high-risk scenarios, so as to accelerate the generation of high-risk scenarios. Accelerated generation of high-risk scenarios can test a large number of high-risk scenarios with fewer test times, which improves the test efficiency. When the random sampling is used to generate high-risk scenarios, the number of tests required n_1 is at least:

$$n_1 = \frac{\gamma^2}{\alpha^2}\frac{1-\beta}{\beta}$$

where, α represents a given constant; β indicates the occurrence of high-risk scenarios; γ It is related to the inverse cumulative distribution function of the standard normal distribution.

When using importance sampling to generate high-risk scenarios, the probabilistic density function of high-risk scenarios is $g^*(x)$, and the number of tests required n_2 is at least:

$$n_2 = \frac{\gamma^2}{\alpha^2}\left\{\frac{E_{g^*}[I^2(x)L^2(x)]}{\beta^2-1} - 1\right\} \tag{5.8}$$

$$I(x) = \begin{cases} 1, & x \in \varepsilon \\ 0, & \text{others} \end{cases} \tag{5.9}$$

$$L(x) = \frac{g(x)}{g^*(x)} \tag{5.10}$$

where, $I(x)$ refers to the indicating function of high-risk scenario ε, with a value of 0 or 1; $g(x)$ represents the probabilistic density function generated by the high-risk scenario when the random sampling is conducted; $g^*(x)$ represents the probabilistic density function generated by high-risk scenarios when importance sampling is conducted; $E_{g^*}[I^2(x)L^2(x)]$ represents the occurrence of high-risk scenarios after the probabilistic density function changes from $g(x)$ to $g^*(x)$ [39].

The test shows that when the accelerated generation of high-risk scenario is carried out to test the front vehicle cut-in, front vehicle braking and other typical scenarios, the testing speed can reach 7000 times that of the random scenario generation.

5.2.3 Intelligent Testing Ecosystem

The intelligent testing ecosystem refers to the closed-loop system that takes the digital network testing site as the core and uses classical testing units to build a testing scenario with high degree of reduction, so as to conduct the functional safety test for the whole unit and components of motor vehicle, as shown in Fig. 5.18. The digital network testing site involves hardware equipment such as computing, storage, communication, perception and traffic scenario infrastructure, and realizes efficient monitoring and management for the testing process through the testing system parameter configuration, information transmission and storage, vehicle and environment information collection, and test information processing system. Classic testing units include ATD representing occupant, collision impactor representing pedestrian, non-motor vehicle and motor vehicle testing units. Their physical features are similar to those of real objects and can be reused, which not only ensures the authenticity of data during high-risk scenario test, but also ensures the safety of test. The motor vehicle unit under test is the tested object in the testing process. By collecting the perception, decision and control data of the equipment in the VUT, the function and performance of each component is tested.

1. Digital Network Testing Site

In order to ensure the safety, comfort, intelligence and economy of IV, it is necessary to carry out relevant tests on the functions and performance of the whole vehicle. Since 2018, relevant departments have issued policies to regulate and guide the testing of IV. In this context, it is imperative to standardize the construction of digital network testing environment.

Digital network testing site refers to a closed and semi-closed intelligent connected traffic test site that can efficiently facilitate test scenario generation, simulation of real scenario construction, test data collection and processing by database and communication network technology and digital management. Its hardware system includes

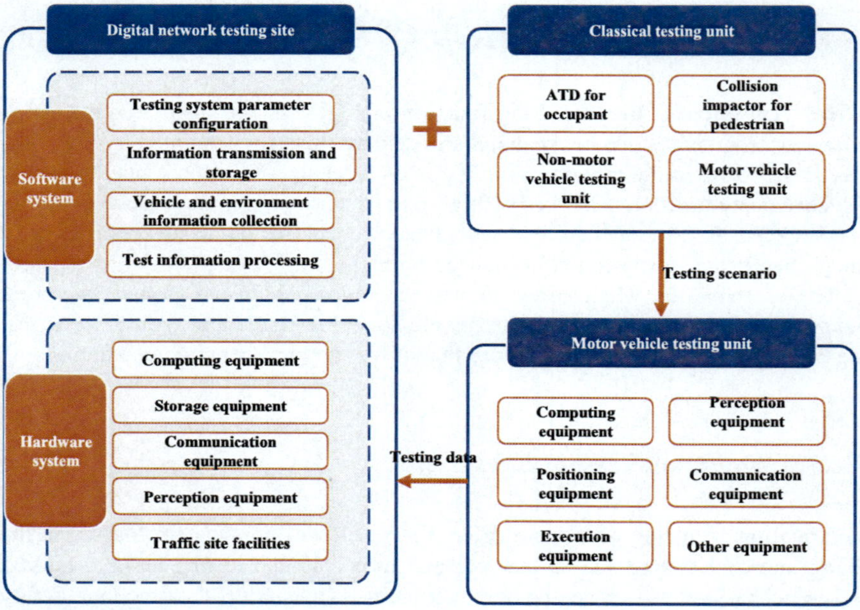

Fig. 5.18 Structure of intelligent testing ecosystem

computing equipment composed of servers, etc.; storage devices composed of disk arrays; communication equipment composed of base stations, switches, routers, etc.; sensing equipment composed of camera, laser radar, ground induction coil, etc.; and traffic site infrastructure composed of signal lights and signboards, etc. The software system includes four parts, i.e., testing system parameter configuration, information transmission and storage, vehicle and environment information collection, and testing information processing. Figure 5.19 shows a way to build the digital network testing site [40], which uses modular design ideas to divide the testing system of the site into four levels, including application scenario layer, perception release layer, network link layer and management service layer. The application scenario layer is responsible for simulating the real traffic scenarios, including meteorological conditions, road conditions and network communication conditions. The perception release layer relies on the information release platform, cameras, lidars and other sensors for collecting environmental information and releasing corresponding control and service information. The network link layer is composed of LTE-V (long term evolution technology-vehicle communication), DSRC (dedicated short-range communication technology), 5G (5th generation mobile communication technology), Wi-Fi (wireless fidelity), EUHT (enhanced ultra high throughput) and other wireless communication technology devices, which are applied to provide information transmission services for the perception release layer and application scenario layer. The management service layer is responsible for the processing, storage and

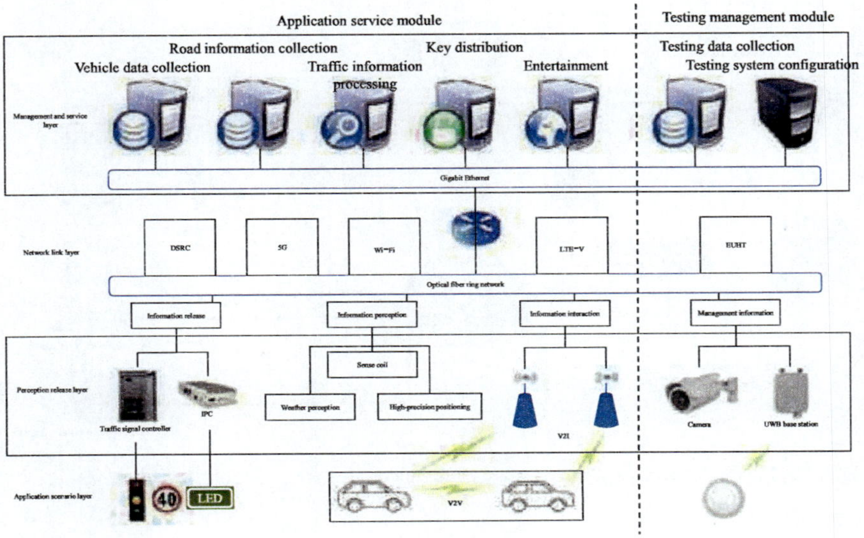

Fig. 5.19 Testing environment of intelligent connected transportation system [40]

backup of data, as well as the management and maintenance of equipment at other layers.

In the above testing system, the functional modules of each layer operate independently without mutual dependence and interference. All modules in the system are connected to Ethernet through all IP network and uniformly assigned addresses. When a new device or module is added, it is only necessary to locate it in a specific functional level, assign IP address to it and configure corresponding network services to realize device interconnection.

2. Motor Vehicle Unit Under Test

Figure 5.20 shows the physical architecture of the motor vehicle unit under test refitted by the Intelligent Driving Laboratory of Tsinghua University, which is characterized by the fact that all the computing equipment, sensing equipment, positioning equipment and communication equipment related to the external intelligent connected driving realize the Ethernet communication of the star topology through a Gigabit management switcher. The external system conducts highly reliable communication with the original vehicle controller by the single chip microcomputer (i.e. the bottom controller in the figure) through the gateway, and completes the reading of vehicle bottom information and the transmission of control instructions. At the software level, the platform builds a virtual bus between external devices based on ZeroMQ suite to realize flexible information exchange between devices with low delay and high bandwidth, and also facilitate the mounting of new devices.

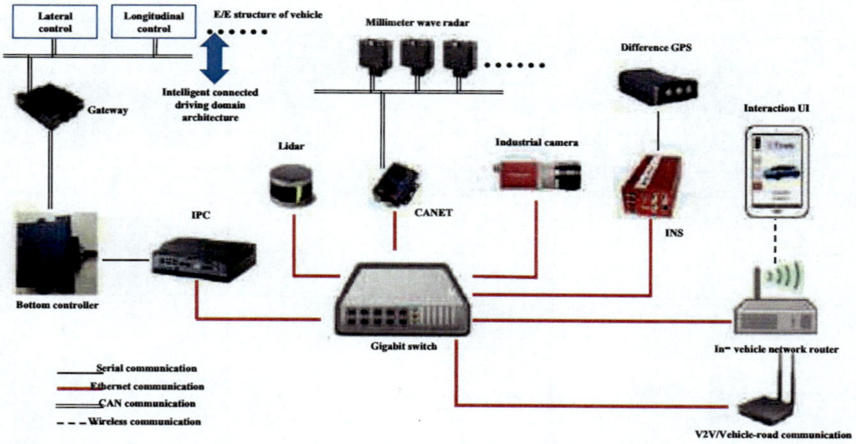

Fig. 5.20 Physical structure of motor vehicle unit under test

(1) Computing Equipment

Computing equipment mainly includes AI unit, computing unit and control unit. The AI unit is mainly responsible for processing and fusing the perceived data of each sensor with GPU, FPGA, ASIC and other chips; The computing unit is composed of several multi-core CPUs, mainly responsible for loading the vehicle operation system, completing task scheduling, and executing the algorithms related to automatic driving; The control unit is composed of multiple ECUs and DCUs, mainly responsible for safety-related vehicle bottom control, such as vehicle drive assembly control, and dynamics transverse and longitudinal control.

(2) Sensing Equipment

Sensing equipment mainly includes camera, millimeter wave radar, laser radar, ultrasonic radar, etc., which detection range, advantages and disadvantages are shown in Table 5.4. These equipment can realize cooperative sensing through sensor information fusion, so as to achieve a larger sensing range and higher recognition accuracy.

Cameras can be classified into monocular and binocular, etc. Monocular cameras have low cost and are widely used. After obtaining the image, it can recognize the target through image matching and other algorithms, and then obtain the information of road vehicles, lane lines, traffic signs, traffic lights and so on. The cost of binocular camera is relatively high. Based on the triangular ranging principle, the ranging accuracy can be much higher than that of monocular camera, and the calibration and calculation is relatively more complicated.

Millimeter wave radar senses the environment by transmitting electromagnetic waves with the wavelength of 1–10 mm. Compared with the camera, it can more thoroughly penetrate smoke and dust, and is less affected by the weather. But at the same time, due to its fan-shaped coverage area, it has large blind area. And since it is

Table 5.4 Comparison of advantages and disadvantages of sensing equipment

Sensor type	Maximum detection range (m)	Advantage	Disadvantage
Camera	About 200	Direct perception mode, and able to identify traffic signals	More affected by the weather, and large distance error
Millimeter wave radar	150–250	Accurate position and speed detection, and high resolution	Low accuracy in long-distance detection, and less perceived information
Lidar	100–200	Rich perceived information, and able to obtain 3D point cloud information	High cost, and difficult to penetrate signals in bad weather
Ultrasonic radar	About 5	Low cost, miniaturization and mature technology	Limited measurement accuracy, and not applicable in high-speed scenarios

not capable to obtain the color features of the environment, it is difficult to identify road signs and traffic signs.

Lidar works in the optical frequency band. With principle similar to millimeter wave radar, it uses laser beam to detect targets. Owing to the capacity of more accurate acquisition for information such as speed, position and angle, its object detection effect is better than millimeter wave radar, and the cost is also higher. Ultrasonic radar senses the environment by transmitting ultrasonic waves, and is mainly used for low-speed and short-range ranging, such as assisting detection of surrounding obstacles in parking scenarios.

(3) Positioning Equipment

Positioning equipment mainly includes global satellite navigation and positioning equipment, and inertial navigation and positioning equipment. The global satellite navigation and positioning equipment uses radio signals transmitted by satellites for navigation and positioning. It is characterized by global, all-weather, high precision and fast real-time. Its composition is shown in Fig. 5.21. Representative applications mainly include the Global Positioning System (GPS) of the US, the Global Navigation Satellite System (GLONASS) of Russia and Beidou Navigation Satellite System (BDS) of China and the Galileo Satellite Navigation System (Galileo) of EU. The inertial navigation and positioning equipment uses gyroscopes and accelerometers to measure the position and direction of the vehicle. The equipment is not affected by external factors such as signal occlusion, and has the problem of cumulative error in long-time operation. It is generally applied in combination with satellite navigation and positioning equipment.

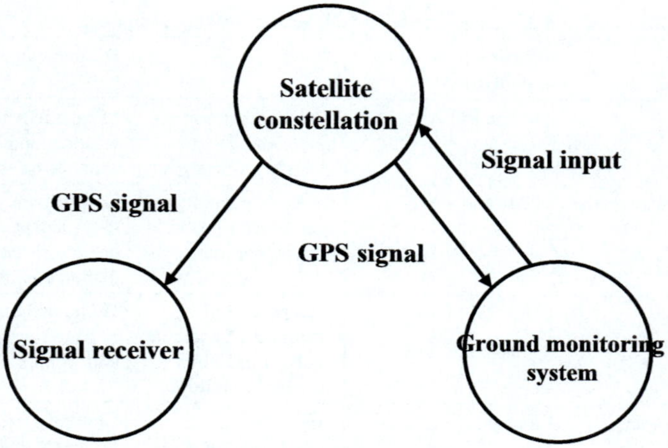

Fig. 5.21 Composition of satellite navigation and positioning system

(4) Communication Equipment

Communication equipment mainly includes on-board communication equipment and roadside communication equipment. The on-board communication equipment integrates mobile communication module, wireless communication module, satellite positioning module, data acquisition, storage module, etc. Under the unified dispatching of the central processing unit, it can facilitate in-vehicle communication, V2V communication and vehicle–road communication. Installed on both sides of the road, roadside communication equipment obtains vehicle speed, position and other status information by wireless communication technology, sends the information to the cloud traffic control center for analysis and processing, and finally returns the processing results to the on-board communication equipment.

(5) Execution Equipment

The execution equipment mainly includes three types of equipment, i.e., drive by wire, steering by wire and power by wire. The drive-by-wire equipment is mainly composed of accelerator pedal, pedal position sensor, electronic control unit, data bus, servo motor and throttle actuator, as shown in Fig. 5.22. It changes the traditional mechanical connection structure of pedal-controlled throttle opening into electronic connection. The control mode driven by electric signal can facilitate the system or driver in more proper control of the vehicle.

Steering-by-wire equipment includes steering wheel assembly (steering wheel, turning angle sensor of steering wheel, return torque motor of steering wheel), ECU, steering actuator motor, power supply, etc., as shown in Fig. 5.23. The steering-by-wire equipment changes the traditional steering shaft, universal joint and other connection structures to the form of electric signal driving motor for the steering gear control, and adds the road sense analog motor for generation of the steering

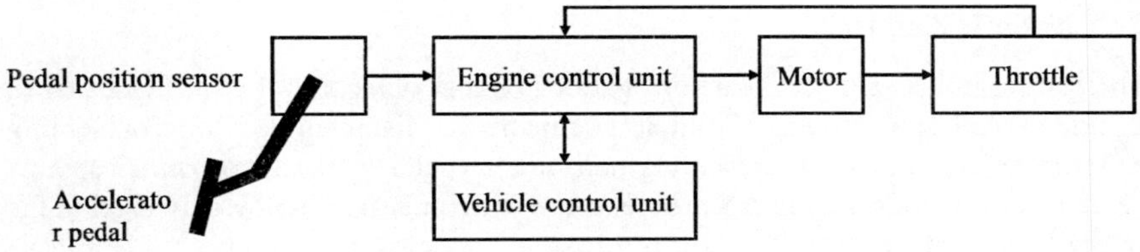

Fig. 5.22 Structure of drive-by-wire system

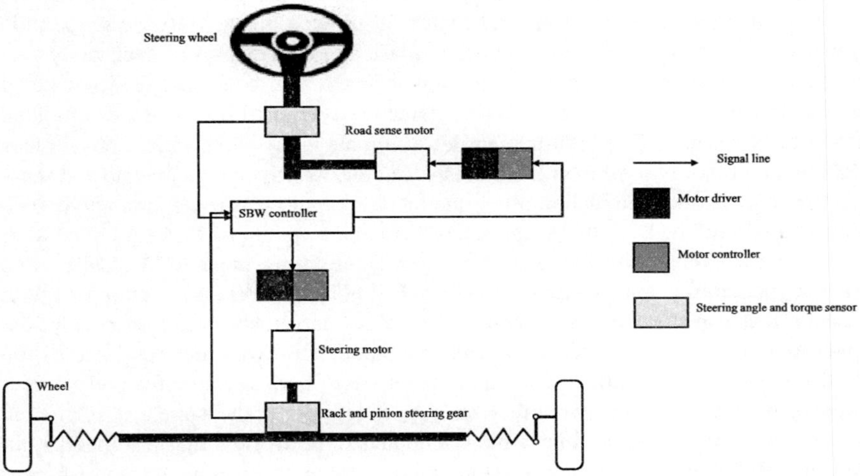

Fig. 5.23 Structure of steering-by-wire system

wheel return torque, so as to provide the driver with the corresponding road sense information.

Brake-by-wire equipment consists of brake pedal, pedal travel sensor, brake controller, brake actuator, etc. The brake controller is responsible for receiving the brake signal and controlling the actuator for braking, as well as identifying abnormal states such as wheel locking and slipping, so as to realize the functions of wheel anti-lock and drive anti-skid.

(6) Other Equipment

Other equipment such as intelligent cockpit equipment interacts with drivers and passengers through liquid crystal instruments, head-up display, central control screen, interior and exterior rear-view mirrors and other carriers, and facilitates the improvement of driving and riding experience based on touch control, voice control, gesture control and other intelligent interaction.

3. Classic Testing Unit

The testing unit is defined as a surrogate of typical objects in the traffic scenario, including ATDs representing vehicle occupants, collision impactors representing pedestrians, and models representing non-motor vehicles and motor vehicles. As a key tool in vehicle safety performance test, testing unit has been widely used in the safety development of road vehicles.

(1) ATD

As a human surrogate in the collision test, the Anthropomorphic Test Device (ATD) is a test instrument made of special materials in accordance with the ergonomic principle (Fig. 5.24). The geometric size, mass and distribution of each body part of the ATD are similar to those of the human body. The bones are made of metal or plastic, and contain joint parts. The outside is covered with soft tissue simulated with plastic or foam. The 50th-percentile adult male ATD with standard physique is the most commonly used in collision tests. In other words, the geometric and mass features can be obtained by compiling the median of all measured human body data into a "medium" ATD, with a height of 175 cm and a weight of 77.5 kg. Other adult sizes used in collision tests are the 95th-percentile larger male ATD (AM95) and the 5th- percentile smaller female ATD (AF05) [41]. The ATDs are equipped with sensors that can measure the acceleration, force, deformation and other physical quantities related to human injury. The kinematics and dynamics response of the ATD can basically reproduce the impact response of human, and effectively output injury parameters, with repeatability and reproducibility in the same test.

In 1949, Sierra Corporation of the US launched the world's first test mannequin Sierra Sam, which was mainly developed to test the seat ejection system of fighter aircraft. In 1966, ARL company of the US developed VIP (very important people)-series ATD. The geometric size and mass distribution of Sierra Sam and VIP-series ATD are basically proper. Since the 1950s, they have been used in vehicle safety performance evaluation experiments. However, due to low biological fidelity and lack of modern sensors, they are difficult to meet the requirement of vehicle collision test. In 1974, the head manufactured by Sierra, the rubber neck manufactured by General Motors and the torso VIP50A manufactured by the Army Research Laboratory (ARL) of the US were combined to form a hybrid II ATD. The word "hybrid" in the name means that the components are mixed from different manufacturers. Hybrid II is the first ATD used in vehicle collision test. In 1975, General Motors developed a new generation of Hybrid III ATD, which was specially improved in head and chest and possibly capable to measure the biomechanical parameters such as neck load and chest deformation, with higher biological fidelity and test performance. GM transferred the intellectual property right of Hybrid III to NHTSA. And Hybrid III was adopted as standard test equipment by the US frontal collision regulation (FMVSS 208) [42].

Since Hybrid III is only verified in terms of its biological fidelity in frontal collision test, it is not applicable to side collision and other conditions. Since the 1970s, researchers have carried out research and development of ATD applicable to side

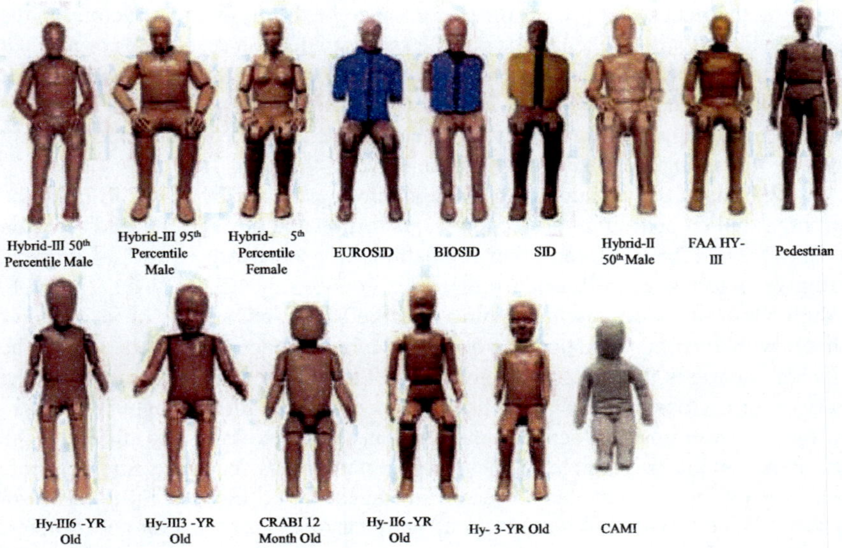

Fig. 5.24 ATD family

collision conditions based on the results of a large number of biomechanical tests for side collision. The EuroSID ATD prototype originally used for side collision was designed and manufactured by the EU research institutes with TNO (the Netherlands Organization) as the center under EEVC. The EuroSID-1 for mass production was completed in 1989 and was adopted by the European side collision regulation ECE R95 in 1998, as well as by the US side impact regulation FMVSS 214 [43]. However, EuroSID-1 has a large oblique stiffness in rib. During its application in lateral test conditions specified by FMVSS 214, the deformation of the rib does not increase significantly when the oblique load input increases. Moreover, due to the mutual interference between the seat back and the ATD back plate in the side collision test, the motion response of the EuroSID-1 is not realistic. In the SID-2000 project, after improving these shortcomings, the ES-2 was developed by EEVC WG1 and is now applied in the European ECE R95. ES-2re, the modified version of ES-2, is currently applied in US FMVSS 214, with the rib length increased on the basis of ES-2.

(2) Pedestrian Collision Impactor

Compared with the full-size ATD used in the occupant test, the impactor representing different body parts is mainly used as the test unit for pedestrian collision safety. The European Experimental Vehicles Committee (EEVC) began to study the testing ways for pedestrian collision protection in 1987, and successively established working groups WG 10 and WG 17. Based on the full investigation of the research results in accident statistics, biomechanics, test analysis, etc., it has formulated WG 10 and then WG 17 standards [44], and proposed the pedestrian impactor for testing and evaluating the pedestrian protection performance of the front-end structure of the vehicle, which is still applied to the vehicle safety evaluation regulations or new car assessment projects in various countries. Pedestrian impactor test includes the test of adult and child headform impactor impacting the bonnet and windshield, the test of lower legform impactor impacting the vehicle bumper, and the test of upper legform impactor impacting the leading edge of bonnet.

On the basis of pedestrian protection regulations proposed by EEVC in 1994, the currently-applied pedestrian protection test is formulated through IHRA discussion by UN R127 (Table 5.5) and other regulations in accordance with the European testing regulations for pedestrian protection proposed by EEVC WG17 in 1998. In the pedestrian safety test procedures specified by C-NCAP, the head impactor collides with the designated position on the top of the bonnet after accelerated by the launcher, measures the average acceleration at the head centroid and calculates the injury criteria. Considering that the height of people with different ages is different and the collision position between the head and the vehicle is also different, the regulations divide the vehicle bonnet into collision areas for adults and children. In the corresponding areas, adult headform impactor (4.5 kg) and child headform impactor (3.5 kg) with different mass and geometry sizes are used for collision test respectively. The lower limb impactor is the flexible FlexPLI developed by JAMA/JARI with bending characteristics similar to human body, and can measure impact force, tibial moment, femur moment and elongation of equivalent ligament. Some test methods and corresponding pedestrian injury criteria are shown in Table 5.5.

(3) Non-motor Vehicle Unit under Test

The non-motor vehicle unit under test mainly includes bicycles and electric scooters, etc., as shown in Fig. 5.25. These models are made of specific materials, and are capable to simulate the real shape and reflection features of non-motor vehicles for the detection of radar and camera from all angles and heights. The wheels of the test model can rotate to form a specific micro-Doppler effect, so as to simulate the real driving state of non-motor vehicle.

(4) Motor Vehicle Unit under Test

The motor vehicle unit under test mainly includes motorcycles and cars, etc., as shown in Fig. 5.26. These models are also made of specific reflective materials, with each weight about dozens of kilograms, which are convenient for operators to disassemble, transport and assemble quickly. The motor vehicle unit under test can also be used together with the corresponding skateboard platform, and can achieve the required motion speed under the cont.

5.2.4 *Safety Performance Testing Methods*

The safety performance test refers to the collection of testing data and completion of the whole testing process based on the testing environment and scenario. Specifically,

Table 5.5 Injury criteria for pedestrian protection in some regulations and standards

Position	Project		UN R127 (2015)	C-CNAP (2021)
Head	*Front of the bonnet (child head)*			
	Impactor mass		3.5 kg	3.5 kg
Head	Impact speed		35 km/h	40 km/h
	Impact angle		50°	50°
	Test area		WAD1000 or BLE[b] + 82.5–1700 mm	If BLE[b] $\leq$ 1700, WAD1000–BLE[b] If BLE[b] $\geq$ 1700, WAD1000–1700 mm
	Judgement criteria		HIC $\leq$ 1000 1/2 area HIC $\leq$ 1700 remaining area	–
	Back of the bonnet (adult head)			
	Impactor mass		4.5 kg	4.5 kg
	Impact speed		35 km/h	40 km/h
	Impact angle		65°	65°
	Test area		WAD1700–2100 mm or BRR[c]	If BLE[b] $\leq$ 1700, BLE[b]–WAD 2300 mm If BLE[b] $\geq$ 1700, WAD1700–2300 mm
	Judgement criteria		Total with children's area HIC $\leq$ 1000 2/3 area HIC $\leq$ 1700 remaining area	Total with children's area HIC $\leq$ 1000 2/3 area HIC $\leq$ 1700 remaining area
Leg	Impactor		FlePLI (LBRL[a] < 500 mm)	aPLI
	Impact speed		40 km/h	40 km/h
	Judgement criteria	Tibial fracture	340 N · m (mitigation area 380 N · m)	400 N · m
		Knee flexion degree	22 mm (MCL elongation)	38 mm
		Shear deformation degree of knee joint	13 mm (ACL, PCL elongation)	–
	Imapctor		Upper legform impactor (LBRL $\geq$ 425)	–
	Imapct speed		40 km/h	–

(continued)

Table 5.5 (continued)

Position	Project		UN R127 (2015)	C-CNAP (2021)
	Judgement criteria	Load	7.5 kN	–
		Moment	510 N · m	600 N · m

Note
[a](LRBL)—lower bumper reference line
[b](BLE)—bonnet leading edge
[c](BRR)—bonnet rear reference

Fig. 5.25 Models of non-motor vehicle unit under test [45]

Fig. 5.26 Model of motor vehicle unit under test [45]

it includes three methods based on static testing platform, dynamic testing platform and in-vehicle information output. In the testing process, the appropriate testing method shall be selected in order to quickly and accurately collect the motion state information of each target in the testing scenario, ensure the accuracy of testing data and improve testing efficiency.

1. Test Based on Static Testing Platform

The test based on the static testing platform is conducted to collect the specific behavior data of IV by setting up static testing equipment in the corresponding places in various road testing environment. The static test equipment includes wired and wireless communication equipment, which can return testing information to testing monitoring center in real time [46]. In order to ensure the accuracy of the test, the applied test equipment shall be practical and mature products and their combinations, such as the speed detection equipment for road traffic management, and the infrared testing equipment for parking test in the driving school [47].

Due to the wide driving range and complex driving environment in the test, it is difficult to meet the data acquisition requirements only with a single sensor. Although the current video detection technology is capable to automatically detect state parameters and specific behaviors (such as reversing and turning) of moving vehicles to a certain extent, it is greatly affected by factors such as occlusion, shadow and light changes.

The static test platform integrates multiple ladars to collect vehicle driving state data (such as position and speed), fusion data, detect and track the motion state of the vehicle, and output its driving trajectory. The vehicle trajectory data includes speed, direction, position, etc., which can be used to estimate the tested vehicle state at specific detection points (such as intersections, parking lots, highway entrances and exits) [48].

Figure 5.27 is the diagram of static system-based test with integration of multiple ladars. The system places multiple ladars in monitoring areas such as intersections, in order to obtain the appearance of vehicles, pedestrians, cyclists and other targets in the coverage area. Each lidar sends the monitoring data to the server through the client for synchronization of data space and time as well as data fusion. Since lidars are placed in different positions for different monitoring perspectives, the fused data can monitor a wider range and avoid some measurement problems induced by occlusion [48].

The test based on the static platform can accurately measure the detailed behavior data of vehicles passing through the local detection area. It has the advantage in high test accuracy, but the disadvantage in small test coverage area. Although the coverage of the test can be further improved by increasing the number of static test equipment, more manpower and material resources are required accordingly for the full coverage of the entire testing area.

2. Test Based on Dynamic Testing Platform

The sensor equipment used in the test based on the dynamic testing platform is installed on the mobile platform, which follows the vehicle during the testing process.

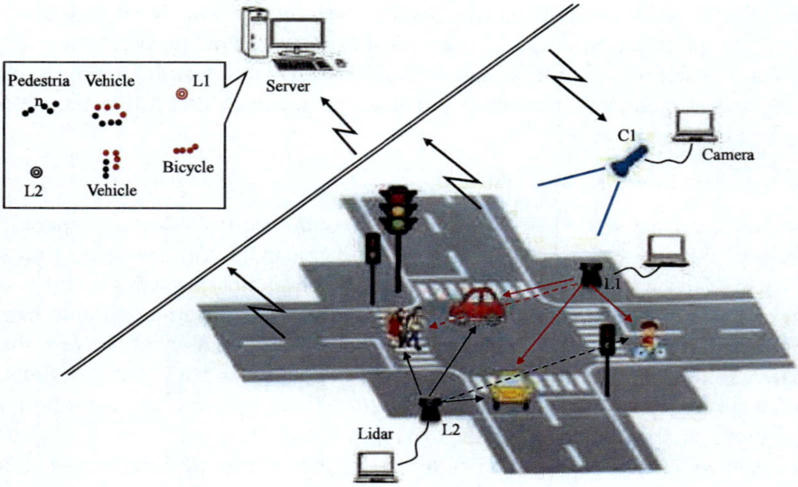

Fig. 5.27 Diagram of static system-based test with multi-lidar integration [47]

It detects the change of relative position between the vehicle and the mobile platform as well as the driving state of the mobile platform itself, and obtains the driving state information of the vehicle through data fusion solution. The test sensors usually include cameras, lidars and GPS/inertial navigation integrated positioning devices. Camera is used to capture the driving state of the VUT in the whole process. Lidar is used to measure the distance between the mobile platform and the vehicle, and then calculate the relative speed. The GPS /inertial navigation integrated positioning device can record the motion trajectory of the mobile platform. Based on the motion trajectory of the mobile platform and the relative position between the platform and the VUT, the trajectory and driving state of the VUT can be obtained [46]. In order to obtain the vehicle driving state information in real time, a wireless communication device can be set on the dynamic mobile platform for sending the captured images and calculated vehicle driving state information to the test monitoring center.

Figure 5.28 shows the framework of vehicle evaluation system based on dynamic testing platform, which can be divided into two parts.

(1) Acquisition of vehicle driving parameters (dotted line area on the right side of Fig. 5.28): is to take the on-board lidar and GPS/inertial unit integrated navigation system as sensor data input, detect and track moving targets around the main vehicle through simultaneous positioning/map creation and moving target detection and tracking. An interactive method is established to manually assign a specific tracked moving target and input its relevant information, such as serial number, so as to accurately estimate and output driving parameters of the specific target.
(2) 3D visualization of the vehicle and its driving environment (dotted line area on the left of Fig. 5.28): is to take other on-board ladars, non-horizontally set single

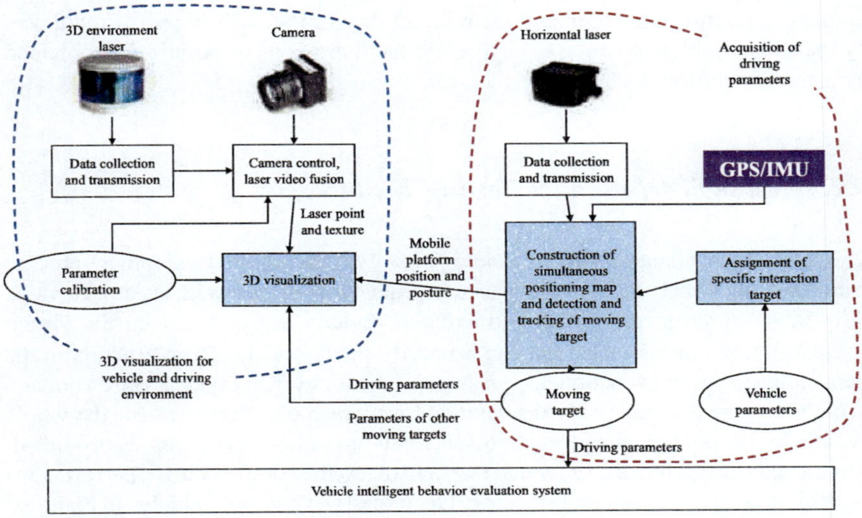

Fig. 5.28 Framework of vehicle evaluation system based on dynamic testing platform [47]

or multi-line lasers and cameras as sensor data input, and use the vehicle position, state and sensor coordinate parameters for coordinate conversion through the fusion of data obtained by lasers and cameras, so as to establish a 3D model for the driving environment of AV; The dynamically three-dimensional visualization for AV and its driving environment is realized in real time by the input of AV driving parameters module.

3. Test Based on Vehicle Internal Information Output

The test based on the internal information output of the vehicle refers to the real-time collection of all process data and intermediate output of the perception, decision, control and other systems of the VUT. According to the testing requirements, the data can be directly stored to the on-board data acquisition device of the VUT, or be transmitted to the testing mobile platform or monitoring center by wireless communication. Taking the parking test at stop line as an example, the vehicle is required to slow down after detecting the stop line and stop within 1 m away from the stop line. Once detecting the stop line during the actual test, the VUT shall collect such data as the distance from the stop line, vehicle speed and the distance from the stop line after the stop, and send them to the monitoring center [48].

In order to avoid misleading information output from the evaluated vehicles, this testing method can also be used together with the test based on static sensors or based on dynamic sensors. It integrates the internal data and external behavior data of vehicles, so as to lay an important basis for objective evaluation.

For vehicle intelligent behavior test in complex driving environment, the testing methods based on static and dynamic platforms and vehicle internal information output can be integrated to obtain the relevant data of the VUT in an all-round way by means of multiple sensors, so as to avoid the limitations of vehicle data obtained by a single platform.

5.2.5 Safety Performance Testing Technologies

The safety performance testing technology involves the whole development process from the functional design to the actual production of the vehicle, as shown in Fig. 5.29. Within it, the model and software in-loop test is based on the virtual simulation test environment, and can primarily verify the algorithm function in the demand analysis and function design stage without physical objects. Hardware and vehicle in-loop test can verify the safety performance of components or the whole vehicle by involving some physical objects into the virtual simulation environment. The closed site test relies on a special test field to test the comprehensive performance of vehicles in a simulated real scenario. Open road test can test vehicles in real road environment without affecting the overall traffic safety by appropriately selecting real road segments.

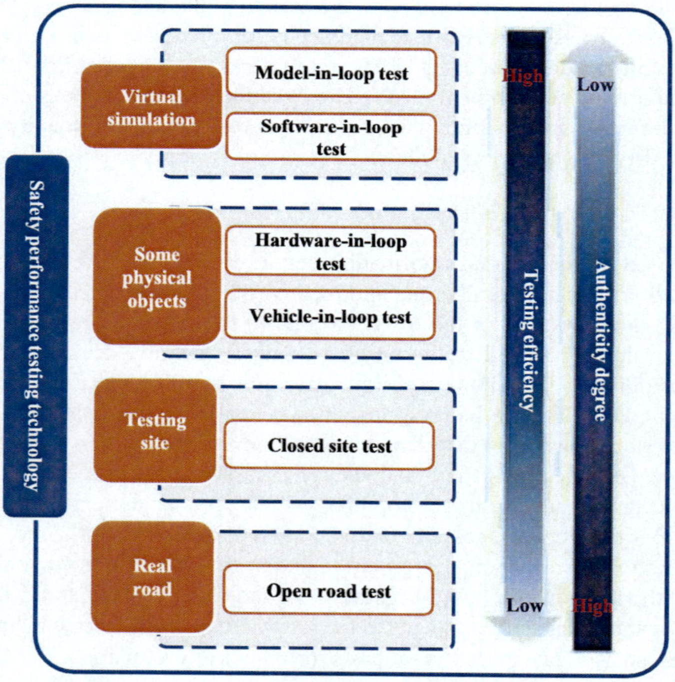

Fig. 5.29 Safety performance testing technology

From the model in-loop test to the open road test, the testing scenarios are becoming more authentic continuously, with more reliable testing results. However, the complexity of scenario construction is also increasing, and the testing efficiency is gradually decreasing. By reasonable application of various testing technologies, such as testing a large number of random scenarios in the virtual environment and testing typical scenarios in the real environment, both the efficiency and effectiveness of the test can be taken into account for reliable testing results.

1. Model In-Loop Test

In the development process of various control algorithms, it is usually necessary to carry out corresponding tests in the design stage, in order to reduce the research and development cost and find the problems and errors in the algorithms earlier. Model in-loop test refers to the technology of testing-related algorithms in combination with the virtual simulation environment before the actual controller is developed [1]. Through the model in-loop test, the correctness of the algorithm can be verified in the early stage of system development without physical hardware objects. The complete model in-loop test usually includes four parts: decision planning algorithm, sensor model, vehicle dynamics model and simulation scenario [10, 49]. The decision planning algorithm is the tested object of the model in-loop test. Sensor models mainly include monocular camera model based on aperture imaging [50], binocular camera model based on extended Kalman filter [51], radar signal reflection model based on target geometric and physical characteristics [52], etc. In terms of vehicle dynamics modeling, in addition to the more mature model based on theory, there are also data-based, system identification-based and object-oriented modeling methods [53–55]. The simulation scenario is designed and built by developers according to the testing requirements.

The process of model in-loop test generally involves five steps: testing requirement analysis, testing environment configuration, testing case design, testing execution, and testing result analysis [56]. The testing requirement analysis is targeted at the function and performance of the tested auto drive system, the requirements for the simulation platform performance as well as the requirements for the simulation result output. Testing environment configuration refers to the parameter configuration for the simulation test platform according to the test requirements, including vehicle model, static scenario, dynamic scenario and sensor model configuration. Testing case design refers to the design of specific testing scenarios in accordance with the requirements for functions and performance of the system under test, which should take into account the principles of sufficiency, efficiency and repeatability. Testing execution refers to the process of formulating test outline according to the requirements of the tested system, then carrying out specific simulation scenario test through software operation, and finally obtaining the response data of the tested object on the input signal of the simulation platform. Testing execution specifically involves initial state setting, testing vehicle operation, vehicle target addition, testing vehicle decision, testing process monitoring, testing process automation, data storage, etc. Testing result analysis refers to data processing of simulation results, including data classification, statistics, screening, visualization, etc. If the testing results show that the system under test has completed the scheduled test as required, and there is a reasonable explanation or a correct and effective treatment way for the abnormal phenomena in the test, the testing process can be ended.

2. Software In-Loop Test

Software in-loop test generally refers to the evaluation of the generated code or handwritten code in the simulation on the host, that is, the change of the tested object from algorithm to algorithm code. The purpose of software in-loop test is to verify the code in the early stage, that is, whether the code and the model used for code generation are consistent in behavior, so as to prevent errors in the code generation process. Therefore, the software in-loop test can be regarded as the equivalence of the model in-loop test, as shown in Fig. 5.30. The "software" in the software in-loop test refers to the software after the "model" is converted into code (the "C code" in the figure is taken for example) and compiled.

3. Hardware In-Loop Test

Hardware in-loop (HIL) test, which combines physical components with software models, is a technical form widely used in component or control system test. Some components or systems of the auto drive system are real, while the environment is virtual. The hardware in-loop test can be carried out in environment perception system and control execution system of IV, etc. [57]. Figure 5.31 shows a typical hardware in-loop test platform for perception system [10], which integrates radar, camera, V2X and other real sensing equipment, and simultaneously builds radar echo simulation system, camera field of view simulation system and V2X channel simulation system. The real sensor can obtain the actual perceived information and transmit it to the real-time controller. The real-time controller is responsible for

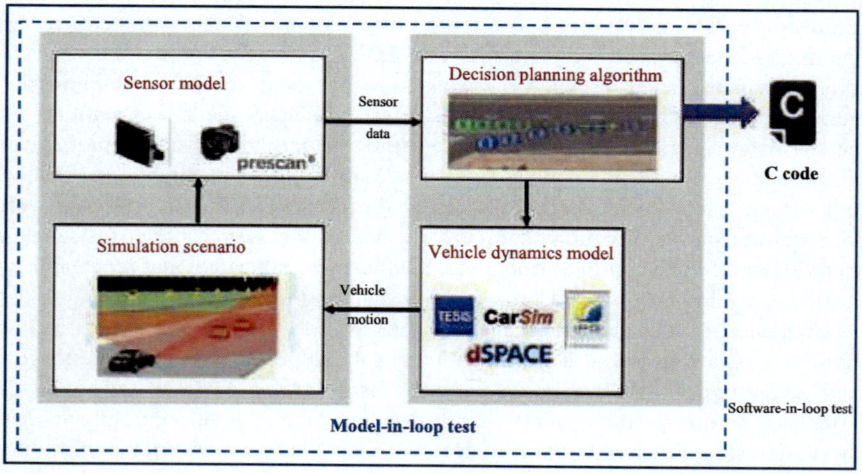

Fig. 5.30 Model in-loop test and software in-loop test

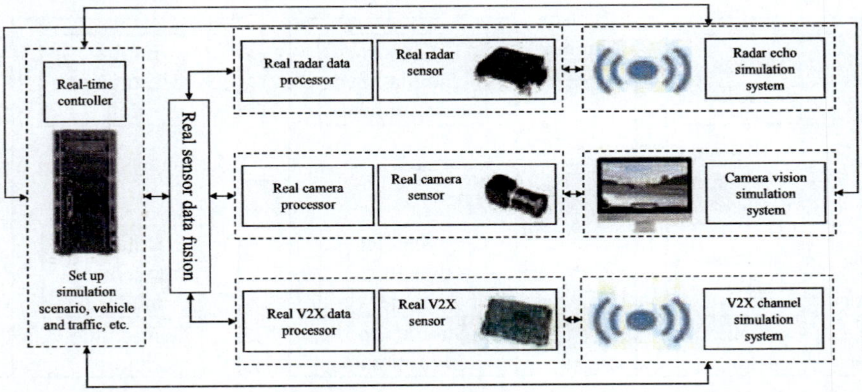

Fig. 5.31 Framework of typical hardware in-loop test platform for perception system [10]

analyzing the accuracy of the perceived results and timely updating the scenarios in each simulation system.

The in-loop test of control execution system usually includes the test of braking, steering and other systems. Taking the braking system for instance, the hardware in-loop testing system can be composed of four parts, i.e., upper computing unit, lower controller unit, driver unit and actuator unit. The upper computing unit is connected with the controller through Ethernet, and the control model is loaded into the controller. The controller collects the current and voltage signals of the driving unit and controls the voltage output of the driving unit to the actuator. The actuator drives the wheel cylinder to generate pressure and target braking force by controlling the on–off of the solenoid valve [58]. The above electromechanical process is usually verified by hardware in-loop test.

4. Vehicle In-Loop Test

Vehicle in-loop test means that all parts of the hardware and systems are combined into a complete vehicle, which is connected to the virtual simulation environment as physical hardware for unified test, so as to detect the possible deviation in the process of hardware combination and make the test results more accurate. It includes the vehicle in-loop test based on closed site and the vehicle in-loop test based on rotating hub platform [11]. The typical test scheme is shown in Fig. 5.32.

Vehicle in-loop test based on closed site refers to the test of connecting the real vehicle in the closed open field to the virtual simulation environment. It is to generate the virtual environment (including road, weather, other traffic participants, etc.) by the simulation software, and input the electronic signal of the virtual environment to the vehicle control unit. The vehicle will make decisions and control its movement according to the environmental information. At the same time, the simulation software reads the vehicle position, heading and other information in real time for the update of virtual environment. And the above cycle is repeated again and again.

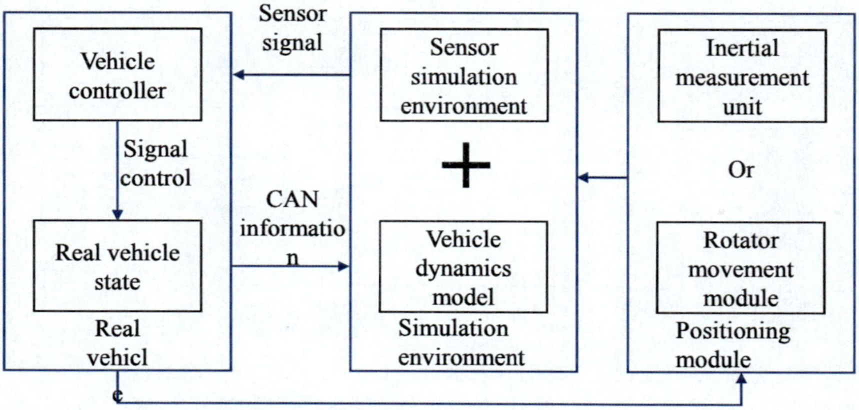

Fig. 5.32 Solution of vehicle in-loop test [10]

As shown in Fig. 5.33, the vehicle in-loop test based on rotating drum platform refers to the test of placing the vehicle on the rotating drum platform and generating the surrounding environment with virtual or simulated equipment. During the test, the absolute position of the vehicle remains unchanged, and the driving state of the vehicle is simulated by changing the relative position between other target and the vehicle. Compared with the vehicle in-loop test based on closed site, the vehicle in-loop test based on rotating drum platform overcomes the site restriction and is applicable for testing large-scale traffic scenarios. At the same time, since the environment information is perceived by real sensors, the test results of the perception system are more reliable. However, due to the limited dynamic performance of the rotating drum platform, it is difficult to accurately test the performance of control and execution systems, etc.

5. Closed Site Test

The closed site test can simulate or restore the real scenario in the limited site by building a special closed testing site with high authenticity props or objects for test of comprehensive performance [11]. At present, many countries and regions have converted or established some closed testing sites. MCity, located in Michigan of the US, is a typical closed testing site for AV. The testing site is composed of a variety of pavement and road elements. At the same time, the testing site is also equipped with a wealth of road elements such as traffic signs and signal lights. In the urban scenario, various simulated buildings, urban auxiliary facilities and other road elements that AV may encounter in the real world are built.

The closed site test is different from the vehicle in-loop test based on closed site. In the latter, the vehicle runs in an open field, but the environment around the vehicle is virtual, which is the analog signal generated by the sensor model. In the former, the vehicle is driving in a professional closed site, which is the physical environment

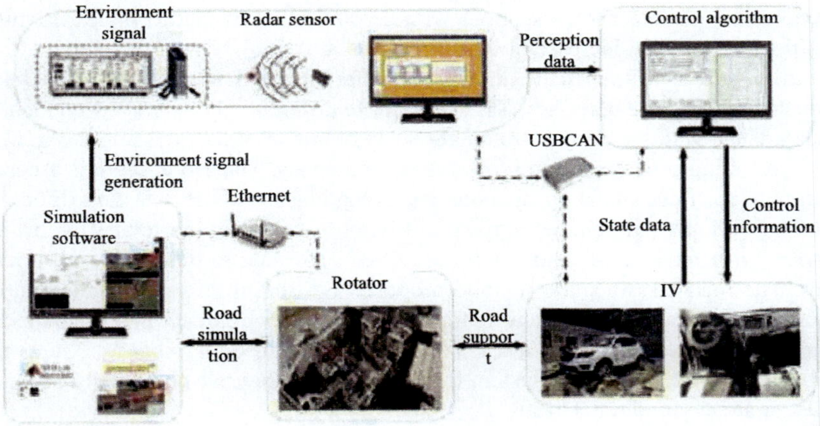

Fig. 5.33 Vehicle in-loop test based on rotating drum platform [59]

rather than electronic signals, and other traffic participants can be simulated by the wheeled mobile robot platform.

The advantage of closed site test is that it is convenient to test a specific scenario by setting up the traffic element props and operating conditions required by the scenario in the site. The test can be repeated for many times under the same working conditions. At the same time, the props used in closed site test generally have high authenticity, and their physical characteristics are similar to physical objects, so the testing results are highly reliable. The disadvantages of closed site test include low testing efficiency and certain test risks. Before the test, it shall spend a lot of manpower and material resources on the design and arrangement of the scenario. In extreme cases, there is a risk of injury for the vehicle or driver. Therefore, in order to improve testing efficiency and reduce testing risk, it is usually necessary to carry out sufficient software and hardware in- loop tests first, and select the most valuable scenario for closed site test.

6. Open Road Test

Open road test refers to driving in an open environment and recording the driving status of an AV with a safety officer. Currently, not all open road areas are applicable for IV test. How to scientifically and reasonably select test open roads and coordinate the relationship between test demand and traffic safety are the practical problems faced by cities in carrying out IV testing [60]. According to the existing open road test experience, at the initial stage when the vehicle function is not perfect, it is best to select the road segment with relatively simple road structure and small traffic flow, which can not only ensure testing safety, but also avoid affection on traffic efficiency of urban roads. However, in order to promote the development and progress of IV technology and ensure the safety and comfortable driving in complex scenarios, it is also necessary to select some sections with dense traffic flow and complex driving conditions for gradual increase of the testing difficulty [60].

With regard to testing roads, due to the different automation levels of vehicle products of different enterprises, it is necessary to differentially open road segments in accordance with the product automation capability, in order to ensure the traffic safety in testing process and improve testing efficiency. Therefore, there is a need to propose road classification methods corresponding to different testing difficulties. The difficulty of vehicle automatic driving is positively correlated with the complexity of road traffic, and can be classified according to the features of open roads. The road classification method proposed by Beijing takes the three major factors of road traffic infrastructure, traffic operation environment and road social environment as the primary indexes, and sets different secondary indexes, so as to establish the evaluation index system for testing comprehensive complexity of road traffic, as shown in Fig. 5.34 [60].

According to the features of IV technology and road traffic, the weight coefficients of the primary and secondary indexes in the index system for comprehensive complexity of road traffic can be further determined (Table 5.6). Combined with the above index system, the calculation formula of road comprehensive complexity is as follows:

$$\text{Score}_l = \sum_{i=0}^{m} \left(\alpha_i \sum_{j=0}^{n_i} \gamma_{ijl} \beta_{ij} \right) \tag{5.11}$$

where, Score_l is the total score of the lth road segment; α_i represents the weight of the ith primary index; β_{ij} represents the weight of the jth secondary index under the ith primary index; γ_{ijl} refers to the scoring value of the jth secondary index under

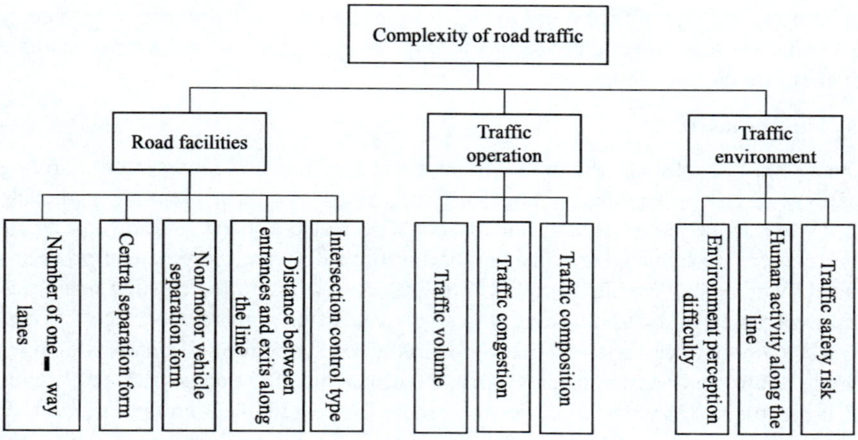

Fig. 5.34 Evaluation index system for testing comprehensive complexity of road traffic [60]

Table 5.6 Weight coefficients in index system for testing comprehensive complexity of road traffic

Primary indexes	Weight	Secondary indexes	Weight
Road facilities	0.4	Number of one-way lanes	0.20
		Motor-non-motor classification	0.20
		Central classification	0.05
		Intersection control	0.25
		Distance between entrance and exit along the road	0.30
Traffic operation	0.30	Traffic volume	0.35
		Traffic congestion	0.35
		Traffic composition	0.30
Traffic environment	0.30	Environmental perception difficulty	0.30
		Population activity along the road	0.35
		Traffic safety risk	0.35

the ith primary index of the lth road segment; m represents the number of primary indexes,; and n_i represents the number of secondary index under the ith primary index [60].

The traffic complexity score of each section can be calculated by Formula (5.11). The higher the score is, the more complex the road is. According to the scoring and testing requirements, the grading classification scheme for different test open sections is obtained, and the grading test can be conducted finally.

5.3 Vehicle Safety Performance Evaluation

5.3.1 Construction of Safety Evaluation Index System

Safety is a key element in many public concerns about IV [61]. Therefore, scientific and accurate safety performance evaluation is vital for the technical breakthrough and application of IV. Evaluation indexes are often used as the quantitative expression of evaluation results to reflect the performance of evaluated subjects. However, a single vehicle safety evaluation index can only reflect the vehicle performance in a certain test section, yet difficult to scientifically and reasonably represent the overall safety level. In consideration of the comprehensiveness of vehicle safety performance evaluation, it should be carried out from multiple levels and multiple angles. Its system logic is shown in Fig. 5.35. Guided by the general safety evaluation principles, an evaluation index system with multi-level tree structure is established from top to bottom, through extensive preliminary selection and fine screening and optimization of indexes. In line with different definitions and characteristics, the evaluation indexes can be divided into two categories, i.e., subjective and objective. They are different

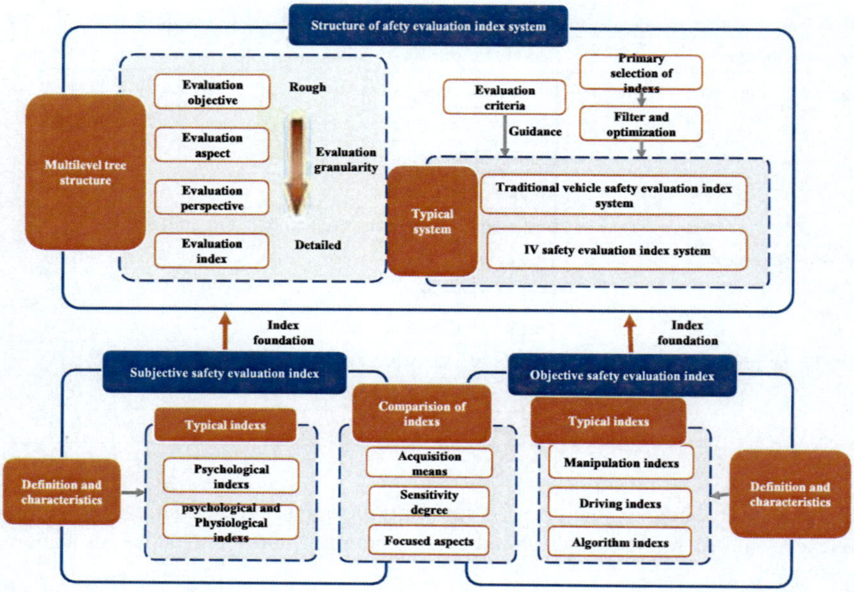

Fig. 5.35 Logic diagram of safety evaluation index system

in terms of acquisition means, sensitivity and emphasis, but all are indispensable for the safety evaluation process.

The vehicle safety evaluation index system can be divided into four levels according to the tree structure, which are evaluation objective, evaluation aspect, evaluation perspective, and evaluation index. The evaluation objective of vehicle safety level is the highest level in the index system. Taking it as the starting point, the evaluation can be carried out from different evaluation aspects of safety performance, and reflect the safety levels of vehicle in different sections in terms of different evaluation dimensions. The next level of evaluation is the evaluation perspective, which expands the content to be evaluated more finely, and directly affects the selection of follow-up evaluation indexes. The evaluation index directly corresponds to each evaluation perspective, and its selection process involves primary selection, screening and optimization. The primary selection is generally conducted by means of comprehensive method and analytical method, etc. Based on the principle of comprehensiveness, indexes that are repetitive or difficult to operate are included, so as to provide a set of all possible indexes for subsequent screening and optimization. For example, the expert decision is adopted to assist in selecting appropriate evaluation indexes and eliminating improper or redundant ones.

During the evaluation for safety performance of traditional vehicles, the safety performance index system is often built according to two evaluation aspects of active safety and passive safety. A typical tree structure is shown in Fig. 5.36. At present, the research on the construction of IV performance evaluation index system is not sufficient, and there is a lack of unified index system with both comprehensiveness and scientificity. According to a mainstream evaluation system, the safety performance evaluation of IV is divided into three aspects, i.e., FuSa, SOTIF and cybersecurity [62], as shown in Fig. 5.37.

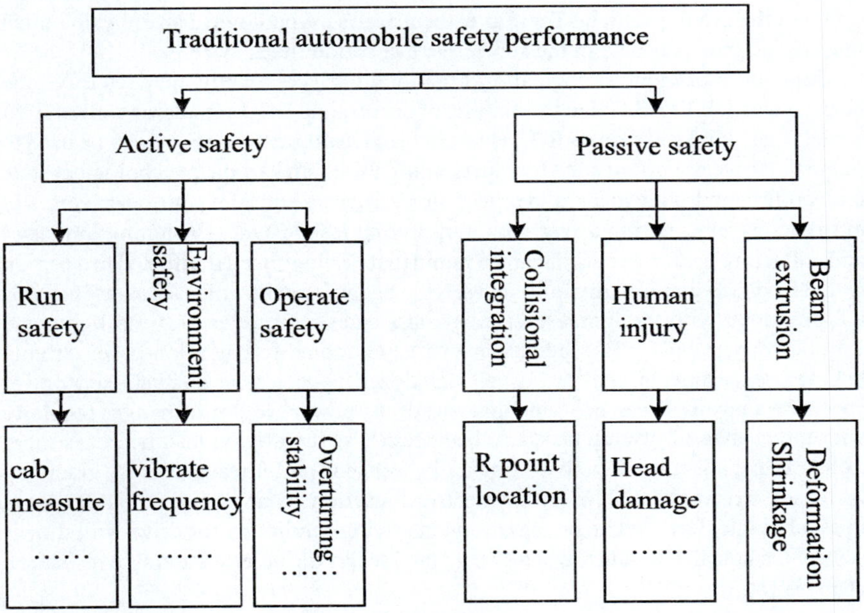

Fig. 5.36 Evaluation index system for traditional vehicle safety performance

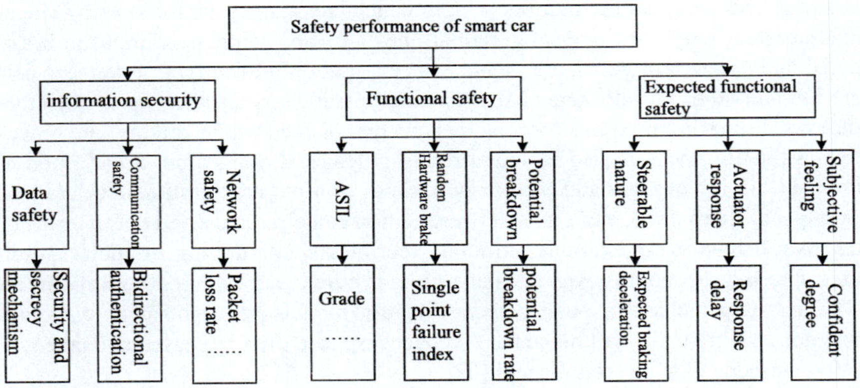

Fig. 5.37 Evaluation index system for iv safety performance

1. Subjective Safety Evaluation Index

The indexes in the IV safety evaluation index system include objective ones (such as the longitudinal and lateral acceleration during vehicle driving) or subjective ones (which are the determined attribute values of indexes, such as the sense of safety during driving). As a vital part in the DVE system, the driver shows psychological and physiological changes in the traffic environment, which can directly reflect the safety of the driving vehicle. Therefore, the driver's psychological and physiological indexes are indispensable in the subjective evaluation index system.

Commonly-used physiological information of drivers mainly includes ECG, the heart rate variability obtained by analysis of the original ECG signals, as well as EEG signals and skin level signals [63]. Heart rate and heart rate variability can be used to judge the driver's current state by representing the change of the psychological state under different driving loads, and reflect subjective sense of safety during driving. For example, by interrupting driver with non-driving tasks (such as listening to music) and collecting heart rate data, it can be found that driving speed is affected to a certain extent by different music rhythms, and driving behaviors with potential safety hazards are presented with the increase of heart rate, such as ignorance of red lights and zebra crossings [64]. EEG indexes are widely adopted in cognition-based driving behavior experiments, and are usually analyzed by potential analysis and power spectrum analysis. It can be found that part of the power spectrum changes regularly with the change of driving tasks. To be specific, as the driving task becomes more difficult, driving α wave activity gradually weakens and θ wave activity gradually increases in driver's EEG [65]. Skin electrical reaction is one of the common indexes in psychological test. When facing emergent driving conditions, the driver will show a state of tension and pressure, when sweat gland secretion increases and skin resistance changes [66].

2. Objective Safety Evaluation Index

Objective evaluation indexes for vehicle operation behavior mainly include lateral and longitudinal operation indexes [63], in which lateral ones are for steering wheel and steering lamp, etc., and longitudinal ones are for quantities of input to brake pedal and accelerator pedal, etc. Vehicle speed, motion trajectory, transverse acceleration and other vehicle state characterization parameters also belong to objective indexes. From the more macroscopic perspective of road traffic driving, the objective evaluation indexes also include driving mileage, driving time, driving period (such as driving at night and peak hours), times of rapid acceleration/deceleration, overspeed, sharp turns, etc. The safety evaluation can also be carried out in terms of takeover behavior. For example, through experiments with driving simulators, such objective indexes as takeover response time, synthetic acceleration and minimum TTC are extracted, and takeover safety evaluation models are established. It is found that non-driving tasks will make the takeover response time increased and takeover safety descended to a certain extent [67].

For the advanced auto drive system, the algorithm plays a vital role in ensuring the IV safe driving. Therefore, selecting scientific and proper objective evaluation indexes for evaluation of the algorithm is also essential in the safety performance evaluation. For example, for the environment awareness algorithm, the perceived result is the main data source of the decision control. Therefore, the environment awareness algorithm is often evaluated in terms of accuracy, real-time, robustness, etc., by the specific indexes such as missed detection rate, false detection rate, single frame processing time, and stability in different scenarios.

In general, the subjective and objective evaluation indexes have their own focuses as well as advantages and disadvantages in the evaluation process. The specific comparison is shown in Table 5.7. Subjective evaluation indexes are easy to implement and closer to human's real feelings, but are different in sensitivity and poor in reproducibility due to individual differences. Objective evaluation indexes can represent the vehicle performance level more specifically and accurately, but can not truly reflect the actual driving experience.

3. Virtual Safety Evaluation Based on Numerical Model

Compared with the hardware test in the real world, virtual evaluation based on numerical model has the advantages of low cost, high efficiency, repeatability, controllable parameters and easy optimization. It is widely applied in the development and design of industrial products. Virtual evaluation can be traced back to the concept of Cyber-Physical Systems (CPS) proposed by Helen Gill of the National Science Foundation (NSF) of the US in 2006. In 2011, based on this concept, Germany proposed Industrial 4.0 (Industrie 4.0), which is characterized by "intelligence" and "Networking". In 2016, Siemens began trying to use digital twins for improving Industrial 4.0 applications, and officially released a complete Digital Twin application model at the end of 2017 (Fig. 5.38).

In terms of virtual safety evaluation system, the digital model is capable to quickly predict the response of various virtual scenarios and working conditions, and serve for the IV safety evaluation system in the face of real scenarios. Taking collision safety for instance, the refined vehicle numerical model can not only precisely simulate the mechanical behavior of the vehicle in various test procedures, but also accurately output the crashworthiness indexes. At the same time, human body numerical models (such as THUMS and GHBMC), as digital substitutes for traffic participants, have also been widely used in academic research and safety system development.

Table 5.7 Comparison of subjective and objective evaluation index

Compare dimension	Subjective evaluation index	Objective evaluation index
Acquisition means	Human (diver, passenger, etc.)	Various sensors
Evaluation process	Physiology, psychology	Physics
Output	Driver subjective parameters and description	System and vehicle measurement values
Sensitivity	Low	High
Reproducibility	Low	High
Focus	Driving experience	Vehicle status

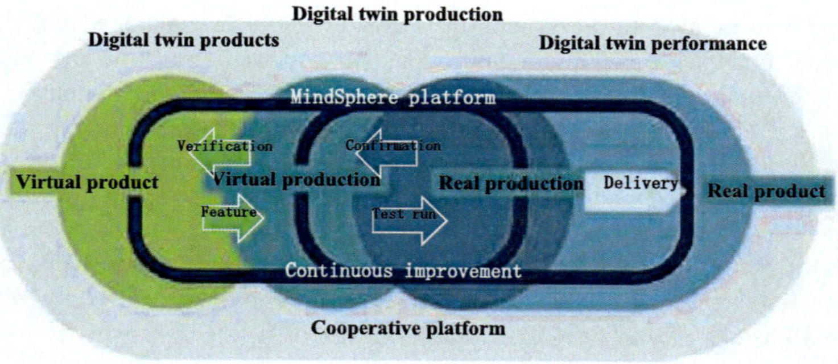

Fig. 5.38 Digital twin application model of Siemens

Compared with the physical ATD of the classical test unit, the human body numerical model has the following advantages. It has a high degree of biological simulation, that is, can characterize the impact mechanical response of human tissues or organs at different scales (see Sect. 4.1 of Chap. 4 for details); It can parameterize the physical signs of human body. Compared with the ATD family that can only characterize the individuals with typical physical signs in the population distribution, the parameterized human body model can personalize the parameters including age, height and weight in the virtual environment (see Sect. 4.2 of Chap. 4 for details); It has the characteristics of active passive behavior coupling. The ATD is unable to express active stress behavior, while the active human body model can test and evaluate the passive injury behavior under the influence of active behavior (such as muscle tension) (see Sect. 4.1 of Chap. 4 for details).

In general, virtual evaluation is becoming an effective supplement to test evaluation, and is expected to replace hardware test to a great extent for integrated safety evaluation. Some current regulations and test standards have put forward the requirements for virtual test standardization based on human body numerical model and other means. For example, the latest version of Euro NCAP stipulates that the numerical model for pedestrian head safety system evaluation is provided by the original equipment manufacturer. For the designated impact points in the head impact area of children and adults, the test agency randomly selects several test points for physical test, and compares the results with the numerical model results at the same impact point, so as to determine whether to expand the number of test samples. The safety evaluation specifications, centered in virtual test and supplemented by physical test, not only indicate that virtual test has been gradually accepted by official institutions, but also reflect the importance and reliability of digital models in safety testing.

Accordingly, the virtual evaluation system involving the future intelligent vehicle active and passive cooperative safety systems in the full collision process shall also be established. The relevant system shall be based on the intelligent traffic driving scenario, and take into account the characteristics of different regions and population, as well as the requirements of multiple working conditions and full stages (Fig. 5.39).

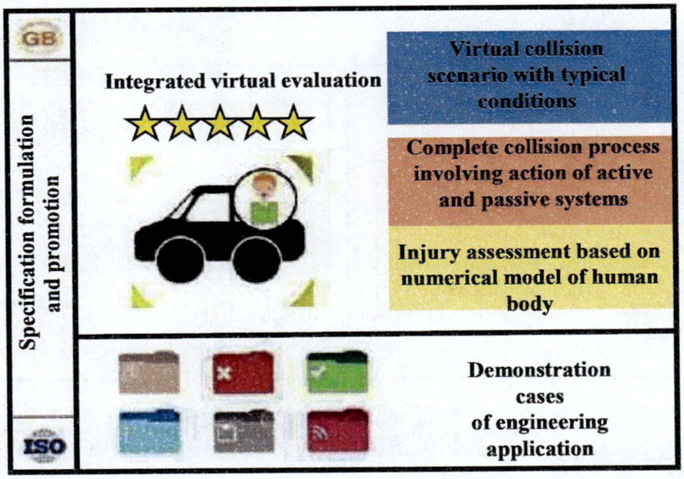

Fig. 5.39 Research and establishment of integrated safety evaluation system

The specific process shall include the development and demonstration of repeatable test methods and typical test scenarios, establishment of standardized test method based on numerical model by integrating the traditional hardware test for verifying and applying the refined body numerical model, establishment of a general damage risk judgment and assessment method, and construction of the corresponding safety protection strategy and product development criteria. In view of the differences in body shape characteristics, physiological parameters and injury tolerance in different countries, as well as the characteristics of China's road traffic, an integrated safety evaluation system shall be established, so as to provide truly applicable industry standards for the safety evaluation of IV and further provide drafts or technical suggestions for international standards.

5.3.2 Safety Performance Evaluation Methods

After determination of the safety evaluation index system, it is necessary to apply specific evaluation methods to conclude the overall evaluation results based on the index values at all levels in the comprehensive system. The method logic is shown in Fig. 5.40. Based on the subjective driving experience of human drivers, the subjective evaluation is to observe and evaluate the performance of the vehicle through the senses in the process of driving in accordance with subjective evaluation specifications, and record the results for subsequent analysis. The objective evaluation based on the measured data starts from the actual performance value of the VUT, and uses quantitative index to describe the objective performance in a certain task. The

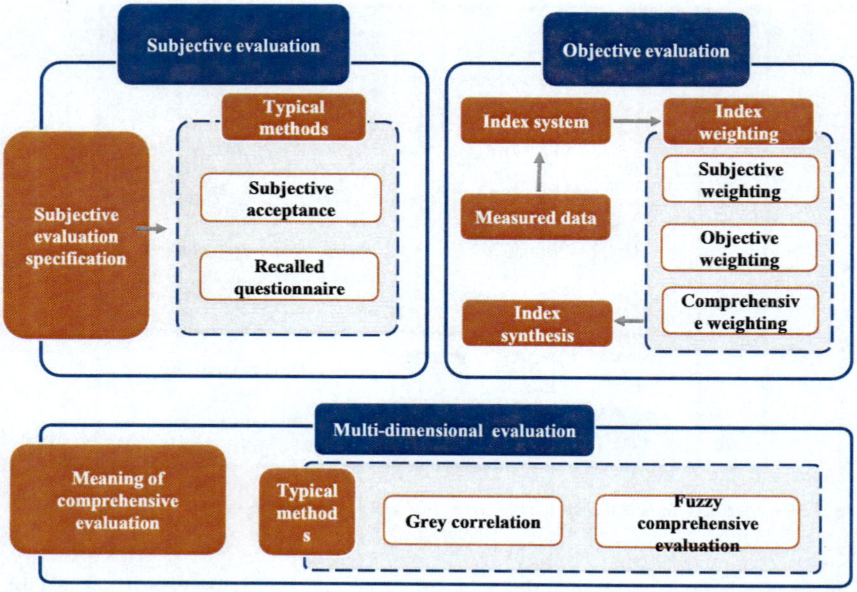

Fig. 5.40 Logical diagram of safety performance evaluation

multi-dimensional comprehensive evaluation combined with subjective and objective evaluation methods can facilitate more comprehensive and reasonable evaluation by combining the actual driving experience with the quantitative results of measured performance.

1. Subjective Evaluation Based on Driver's Acceptance

Whether for traditional vehicles or IV, the commonly-used evaluation methods are generally divided into subjective evaluation and objective evaluation. The subjective evaluation refers to the evaluation process in which the trained evaluators observe, operate, record and analyze the evaluation results of a certain performance of the vehicle through their senses in a typical road scenario or test environment according to certain subjective evaluation procedures, and then conclude the driving experience. It is an indispensable and important part in the vehicle evaluation system.

Researchers from Brunel University London have conducted a subjective evaluation on ACC system from the perspective of psychology. The research is based on the driving simulator. The subjects are required to perform manual driving and ACC assisted driving on the simulator respectively. The variance analysis is used to analyze the psychological variables. It is found that the ACC system will lead to the decline in driver's sense of control, trust for the vehicle as well as environmental perception, whereas will reduce workload and pressure [66].

NHTSA and the University of Michigan Transportation Research Institute (UMTRI) have carried out Intelligent Cruise Control Field Operational Test

(ICCFOT) [68], during which the subjective evaluation of ACC system is conducted in the form of a subjective questionnaire. The questionnaire includes questions designed and classified specifically for drivers with different ages, driving experience and experience in using ACC system in terms of such subjective feelings as comfort, safety and trust towards the system. And the 7-point scoring is applied to evaluate the ACC system according to the average value and standard deviation of the evaluation scores of various questions.

2. Objective Evaluation Based on Measured Data

Since the subjective evaluation is based on human drivers, who have different characteristics and sensitivities and also vulnerable to external environment, the subjective evaluation has the disadvantages of strong fuzziness and poor reproducibility. The objective evaluation based on the measured data starts from the actual performance value of the VUT, and uses quantitative index to describe the specific performance of a certain event. In the objective evaluation, it is necessary to establish an evaluation index system by combining the vehicle under test, the testing methods and the evaluation objectives, then assign weights to each index, and finally conclude the performance evaluation results of the system or the whole vehicle through comprehensive evaluation.

As a vital part of the evaluation process, the weighting of evaluation index greatly influences the final evaluation score. Therefore, this section first introduces the commonly-used index weighting methods, which can be divided into subjective weighting and objective weighting.

(1) Subjective Weighting

Subjective weighting is based on the subjective judgment of experts according to their experience, and is conducted more frequently by means of Analytic Hierarchy Process (AHP), which is a typical representative of eigenvalue. Its basic idea is to establish a hierarchical structure to describe the system functions or features according to the requirements. By comparing the relative importance of the evaluation indexes in pairs, the corresponding scale is given to form the judgment matrix of the upper and lower indexes, and the important sequence of the relative importance between the upper and lower-level indexes is given. The following describes in detail the steps to determine the index weight with this method [47].

The first step is to construct a reasonable and consistent judgment matrix required by AHP. The evaluation indexes of IV can be multi-level. For each evaluation index of each level, the following ways can be used to build a judgment matrix. The evaluator shall weigh the relative importance of the two evaluation indexes, and assign a certain value to the importance by the 1–9 scale, as shown in Table 5.8.

According to the above definition table, experts and evaluators weigh and compare the relative importance of various evaluation indexes, and then determine the results of compared importance, as shown in Table 5.9.

According to the above compared importance, the judgment matrix A^* can be obtained, as shown in the following formula:

Table 5.8 Importance/less importance and definition of scale

Order	Importance/less importance	Scale
1	Both indexes are equally important	1
2	One index is slightly more important than the other	3
3	One index is more important than the other	5
4	One index is much more important than the other	7
5	One index is extremely more important than the other	9
6	One index is slightly less important than the other	1/3
7	One index is less important than the other	1/5
8	One index is much less important than the other	1/7
9	One index is extremely less importance than the other	1/9

Table 5.9 Results of compared importance between indexes

Indexes at the same level	A_1	A_2	……	A_n
A_1	a_{11}	a_{12}	……	a_{1n}
A_2	a_{21}	a_{22}	……	a_{2n}
……	……	……	……	……
A_n	a_{n1}	a_{n2}	……	a_{nn}

$$A^* = [a_{ij}]_{n \times n} \tag{5.12}$$

where, a_{ij} represents the importance of the ith element relative to the jth element. According to the construction definition and method of judgment matrix, it is clear that matrix A^* has the following properties:

Any element is a positive number: $a_{ij} > 0$;

The value of diagonal element is 1: $a_{ii} = 1$;

Symmetrical elements are reciprocal to each other: $a_{ij} = 1/a_{ji}$.

The second step is to calculate the weight of each evaluation index.

Assuming that there are n evaluation indexes at the same level, the weight of each evaluation index can be calculated according to the following steps.

First, the judgment matrix A^* of the evaluation index is normalized by column, that is, the normalization of each column in the judgment matrix A^*, as shown in the following formula:

$$\overline{a}_{ij} = \frac{a_{ij}}{\sum_{i=1}^{n} a_{ij}}, \quad \forall_{i,j} = 1, 2, \ldots, n \tag{5.13}$$

Secondly, the normalized judgment matrix is added by rows to obtain the sum vector W, and its elements W_i are as follows:

$$W_i = \sum_{j=1}^{n} \overline{a}_{ij}, \quad \forall_i = 1, 2, \ldots, n \tag{5.14}$$

Thirdly, the obtained sum vector is normalized to obtain the weight vector $\overline{W}$, and its elements are as follows:

$$\overline{W}_i = \frac{W_i}{\sum_{i=1}^{n} W_i}, \quad \forall_i = 1, 2, \ldots, n \tag{5.15}$$

Finally, before determination of the weight vector, the consistency test shall also be conducted for the reliability of evaluation results. For consistency test, it is necessary to first obtain the maximum eigenvalue $\gamma_{\max}$, of the judgment matrix A^*, which is calculated by means of ANC, as shown in the following formula:

$$\gamma_{\max} = \sum_{i=1}^{n} \frac{\left[A^*\overline{W}_i\right]_{\mathrm{i}}}{n(\overline{W}_i)_i} \tag{5.16}$$

where, $[A^*\overline{W}_i]_i$ represents the ith element of vector $A^*\overline{W}_i$; $(\overline{W}_i)_i$ represents the ith element of the vector $\overline{W}_i$. The meaning of each symbol is consistent with the above.

According to the above formula, the consistency index $C.I.$ can be calculated as follows:

$$C.I. = \frac{\gamma_{\max} - n}{n - 1} \tag{5.17}$$

Then, based on the quantity n of the evaluation indexes, the average random consistency index $C.R.$ (Table 5.10) is obtained in accordance with the table. As long as the relationship of the following formula is satisfied, the consistency test can be considered to be passed and the weight vector $\overline{W}_i$ is valid. Otherwise, the judgment matrix A^* shall be modified and recalculated and tested until it is passed.

$$\frac{C.I.}{C.R.} < 0.1(5 - 19)$$

(2) Objective Weighting

The objective weighting calculates and analyzes the actually obtained attribute values of the indexes, and then derives the weight coefficients. Entropy weight is widely used for objective weight, and its basic idea comes from Claude Shannon's information

Table 5.10 Values of average random consistency index

n	3	4	5	6	7	8	9	10	11
C.R	0.58	0.9	1.12	1.24	1.32	1.41	1.45	1.49	1.51

theory. For the weight coefficient of the evaluation index, the smaller the information entropy of an index is, the greater the amount of information it provides, and the greater the impact on the results of the comprehensive evaluation is. The weight corresponding to the index should be larger, and vice versa. The following is a brief introduction to the steps of determining the weight of IV evaluation index by means of entropy weight.

The first step is to calculate the information entropy of each evaluation index according to the test data. The test data of IV shall be standardized. Here, it is assumed that there are k IV to be evaluated, and each has n evaluation indexes at the same level. After the obtained test data is normalized, the information entropy of a group of test data corresponding to any index can be calculated according to the definition of information entropy in information theory by the following formula:

$$e_j = -\frac{1}{\ln k}\sum_{i=1}^{k} \mathrm{x}_{\mathrm{ij}}^{*} \cdot \ln \mathrm{x}_{\mathrm{ij}}^{*}, \quad j = 1, 2, \ldots, n \tag{5.18}$$

where, x_{ij}^{*} represents the test data of the jth evaluation index of the ith vehicle under test; e_j represents the information entropy of the jth evaluation index.

The second step is to calculate the weight of each evaluation index. According to the above formula for calculating the information entropy of the evaluation index, the difference factor g_j, of the index is defined as follows:

$$g_j = 1 - e_j \tag{5.19}$$

Then the weight coefficient w_j, of the index can be calculated, as shown in the following formula:

$$w_j = \frac{g_j}{\sum_{j=1}^{n} \mathrm{g_j}}, \quad j = 1, 2, \ldots, n \tag{5.20}$$

The weight vector W of each evaluation index is obtained as follows:

$$W = (w_1, w_2, \ldots, w_n)^{\mathrm{T}} \tag{5.21}$$

(3) Comprehensive Weighting

The comprehensive weighting combines the characteristics of the subjective and objective weighting. According to the specific evaluation situation, the proportion of the subjective and objective weighting in the process of index weighting is determined, and finally the weight coefficient of the comprehensive weighting is obtained. Suppose that for any evaluation index, the corresponding weight coefficient is determined by subjective and objective weighting as α_i and β_i respectively. Then determination of the weight coefficient by the comprehensive weighting is as follows [1]:

$$w_i = \varepsilon\alpha_i + (1-\varepsilon)\beta_i, i = 1, 2, \ldots, n \tag{5.22}$$

where, w_i represents the weight coefficient corresponding to the ith evaluation index; ε is the compromise coefficient of subjective and objective weighting, and represents the relative importance of the two weighting methods, with value range [0, 1].

By adjusting the contribution of subjective and objective weighting in the process of weighting, the comprehensive weighting can adjust the degree of influence of subjective and objective components on the evaluation results according to the specific evaluation task, which not only reflects the subjective evaluation information of the evaluator, but also makes use of the objective measured data, so that the final evaluation conclusion takes into account the feeling of operation and horizontal comparability of the evaluator.

3. Multi-dimensional Comprehensive Evaluation

After determining the weight of each subjective and objective evaluation index, multiple evaluation indexes are integrated into a comprehensive evaluation value through the comprehensive evaluation model for reflecting the multi-dimensional comprehensive evaluation results of the vehicle to be evaluated. The specific methods of multi-dimensional comprehensive evaluation include weighted average, fuzzy comprehensive evaluation, grey correlation and TOPSIS. The following mainly introduces the gray correlation and fuzzy comprehensive evaluation.

According to the measured data of each evaluation index, GRA sets the optimal attribute value of each evaluation index as the reference data, compares the measured data with the optimal attribute value, obtains the corresponding value of each evaluation index for weighting, and then combines the weights of all evaluation index in the weighting process to obtain the quantitative evaluation results. For example, in practical research and application, GRA can be applied to the evaluation of U-shaped steering performance of IV [69]. The evaluation result of GRA depends on the optimal attribute value and each IV to be evaluated, which change will be likely to lead to the change of evaluation results. For another example, for m IV to be evaluated and n evaluation indexes, the comprehensive evaluation results of each vehicle after the combination of indexes can be calculated as follows:

$$S_i = \sum_{j=1}^{n} \mathrm{w_j} \cdot \xi_{\mathrm{i}}(j), \quad i = 1, 2, \ldots, m \tag{5.23}$$

where, S_i represents the comprehensive evaluation result of the ith vehicle to be evaluated; ξ_i refers to the correlation coefficient of the ith vehicle to be evaluated; w_j represents the weight coefficient corresponding to the jth evaluation index, which is obtained by the weighting.

For the determination of the correlation coefficient of each evaluation index, it is necessary to first set the optimal attribute value of each index and carry out standardization, and then calculate the correlation coefficient ξ of each index based on the Grey Systems Theory as follows:

$$\xi_i(j) = \frac{\min_i \min_j \left| x_o^*(j) - x_i^*(j) \right| + \rho \max_i \max_j \left| x_o^*(j) - x_i^*(j) \right|}{\left| x_o^*(j) - x_i^*(j) \right| + \rho \max_i \max_j \left| x_o^*(j) - x_i^*(j) \right|} \tag{5.24}$$

where, x_o^* represents the optimal attribute value of each evaluation index after standardization; x_i^* represents the value of indexes of the ith vehicle to be evaluated after standardization; ρ represents the resolution coefficient and $0 < \rho < 1$. The smaller the resolution coefficient is, the higher the resolution is. In practical research, the resolution coefficient can be taken as 0.5.

Fuzzy comprehensive evaluation transforms the factors that are unclear in boundary and difficult in quantification into quantitative evaluation in accordance with the membership degree theory of fuzzy mathematics. Owing to its simple mathematical model and clear structure, it has been widely adopted in many scientific fields [70]. Firstly, the fuzzy comprehensive evaluation shall determine the set of evaluation indexes and the set of evaluation grades. The former constitutes the evaluation framework, and the latter specifies the selection range of evaluation results. Then, it determines the membership degree of each evaluation index to each evaluation grade and constructs the decision matrix, synthesizes the fuzzy comprehensive evaluation result vector by combining the decision matrix with the weight vector, calculates with the corresponding score vector of each evaluation grade and obtains the final comprehensive evaluation results. In addition, it is also applicable to the multi-level index system, and can better quantify the subjective indexes. In practical research and application, fuzzy comprehensive evaluation and analytic hierarchy process can be combined to determine the weight of each index through analytic hierarchy process, and then comprehensively evaluate the safety performance of IV in multiple dimensions by fuzzy comprehensive evaluation [47].

For the evaluation index set and evaluation grade set composed of n evaluation indexes and m evaluation grades, the valuation matrix for membership degree can be expressed as:

$$R = [r_{ij}]_{n \times m} \tag{5.25}$$

where, R represents the evaluation matrix; r_{ij} refers to the fuzzy membership degree of the ith evaluation index to the jth evaluation grade, $0 \leq r_{ij} \leq 1$. The membership degree in the evaluation matrix usually depends on the knowledge, experience and judgment of experts.

The calculation of the result vector of fuzzy comprehensive evaluation can be obtained by combining the weight vector and the evaluation matrix. Generally, matrix multiplication can be used, as shown in the following formula:

$$B = W \cdot R \tag{5.26}$$

where, B represents the result vector of fuzzy comprehensive evaluation; and W represents the weight vector.

The comprehensive evaluation result is calculated by the fuzzy comprehensive evaluation result vector and the corresponding score vector of each evaluation grade, as shown in the following formula:

$$G = B \cdot \mu \tag{5.27}$$

where, G represents the comprehensive evaluation result; μ represents the corresponding score vector of each evaluation grade. For example, for the five evaluation grades (such as excellent, good, average, poor and bad), the corresponding score vector can be $\mu = [1.0, 0.8, 0.6, 0.4, 0.2]$.

5.3.3 Safety Performance Evaluation Technologies

The vehicle safety performance evaluation technology provides the partial and overall evaluation results in terms of the safety of the single function and the whole vehicle system, as shown in Fig. 5.41. Taking the vehicle ADAS function evaluation for example, the safety evaluation technology for single function is combined with the multi-scenario evaluation specifications for the evaluation score calculation. The evaluation technology for the whole vehicle system inspects the safety performance evaluation results in the evaluation environment from the perspective of takeover degree and scenario trafficability. The analysis on the safety performance evaluation results can infer the high-risk scenarios prone to accidents and define of the safety driving boundary; The diagnostic analysis can confirm the existing problems of the vehicle and provide targeted suggestions for improvement; Finally, on the basis of reducing the risk and cost of evaluation, it can facilitate the improvement of test and evaluation methods for more efficient and accurate evaluation conclusions.

(1) Safety Evaluation Technology for Single Function

Low-level IV usually integrates multiple driving assistance functions. Quantitative evaluation of its FuSa is essential for the safety of the whole vehicle. Taking AEB as an example, Euro-NCAP has included AEB in the evaluation specifications since 2014. At first, the evaluation of AEB was conducted by two types of tests in urban driving and suburban driving, and each had different scoring methods [71]. In the 2020 Euro-NCAP, the safety assistance part refined the evaluation of AEB in car-to-car collision. The total evaluation score is mainly composed of three parts: car-to-car rear (CCR) score, car-to-car front turn across path (CCFtap) score, and human–machine interaction (HMI) score [72]. The vehicles in the CCR scenario test include the vehicle under test (VUT) and the target vehicle (Euro-NCAP vehicle target, EVT) which is internally padded by gas filler and externally covered by PVC material with vehicle mark. The specific CCR test scenario is shown in Fig. 5.42 [71]. Since FCW is the pre-trigger function in AEB system, its evaluation standards are also included in Euro-NCAP's evaluation specifications for AEB system. Under the CCR

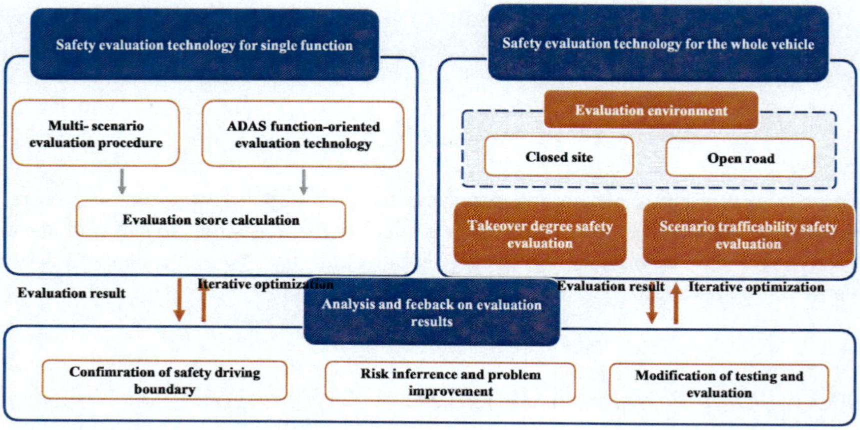

Fig. 5.41 Logic diagram of safety performance evaluation technology

scenario specified above, the scoring criteria for AEB and FCW functions are shown in Table 5.11 [72].

In the CCFtap scenario, VUT turns left, and the target vehicle drives at a constant speed and makes conflict, as shown in Fig. 5.43. When VUT successfully avoids collision at a specific speed, it can be scored, as shown in Table 5.12.

The total score of HMI is 2 points. When the system meets the following requirements, one point can be obtained for each:

(1) Supplementary warning for FCW system. In addition to the required audio-visual warning, head-up display, seatbelt jitter, brake jitter or any other tactile feedback can be sent when TTC > 1.2 s.

Scenario description	Scenario diagram
Car to Car Rear Stationary (CCRs)	10-50km/h 30-80km/h → Static
Car to Car Rear Moving (CCRm)	30-80km/h → 20km/h
Car to Car Rear Braking (CCRb)	50km/h ← 12 or 40m → 50km/h -2 or -6m/s^2

Fig. 5.42 CCR testing scenario [71]

Table 5.11 Speed scoring under CCR scenario [72]

Test speed (km/h)	AEB			FCW		
	CCRs	CCRm	CCRb	CCRs	CCRm	CCRb
10	1					
15	2					
20	2					
25	2					
30	2	1		2		
35	2	1		2		
40	1	1		2		
45	1	1		2		
50	1	1	4	3	1	4
55		1		2	1	
60		1		1	1	
65		2		1	2	
70		2		1	2	
75		2		1	2	
80		2		1	2	
Total score	14	15	4	18	11	4

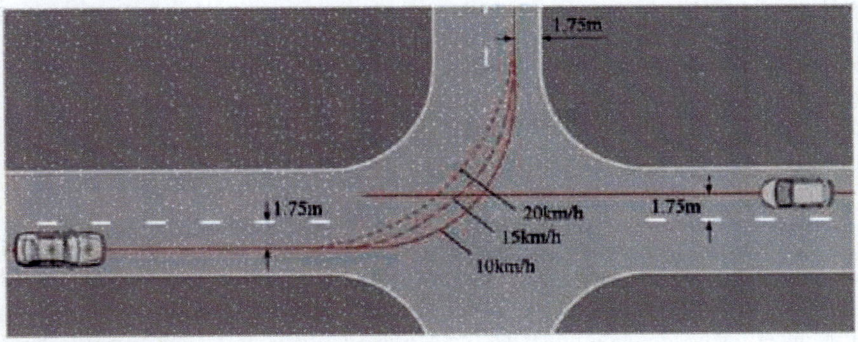

Fig. 5.43 CCFtap scenario [73]

Table 5.12 CCFtap scoring [72]

Test speed (km/h)	CCFtap		
	GVT: 30 km/h	GVT: 45 km/h	GVT: 55 km/h
10	1	1	1
15	1	1	1
20	1	1	1
Total score	9		

(2) Pretension function of seatbelt. When an emergent situation with a potential collision risk is detected, the seatbelt can be tightened in advance in the face of collision.

By test of the above three sub-items, their evaluation scores are obtained, and then the scores are integrated into total score of AEB in car-to-car scenario. The total score is calculated as [72]:

$$S_{\text{total}} = 2 \cdot S_{\text{CCR}}^{\text{AEB}} \cdot f_{\text{correction}}^{\text{AEB}} + 1.5 \cdot S_{\text{CCR}}^{\text{FCW}} \cdot f_{\text{corretion}}^{\text{FCW}} + 2 \cdot S_{\text{CCFtap}} + 0.5 \cdot S_{\text{HMI}} \tag{5.28}$$

where, $S_{\text{CCR}}^{\text{AEB}}$ and $S_{\text{CCR}}^{\text{FCW}}$ represent the scores of AEB and FCW in CCR scenario respectively; S_{CCFtap} represents the score of AEB in CCFtap scenario; S_{HMI} indicates HMI score; $f_{\text{correction}}^{\text{AEB}}$ and $f_{\text{correction}}^{\text{FCW}}$ represent the correction factors of AEB and FCW respectively; f is calculated with the actual test score S_{actual} and the expected test score $S_{\text{predicted}}$, as shown in the following formula:

$$f = \frac{S_{\text{actual}}}{S_{\text{predicted}}} \tag{5.29}$$

(2) Safety Evaluation Technology for Whole Vehicle System

In addition to evaluating the functions of IV, it is also necessary to evaluate the overall safety performance of vehicle. At present, IV shall usually be tested on open roads or closed sites. According to the test data, the quantitative evaluation results of the VUT safety are obtained by means of the vehicle safety evaluation technology. There is no unified standard for selecting vehicle safety evaluation indexes, and more application-oriented indexes such as vehicle collision probability can be adopted for evaluation [74]. Some studies have taken the number of manual takeover and the trafficability of scenario test as the main evaluation index [75]. The number of manual takeovers, that is, the average number of times a driver takes over an IV every 1000 miles (about 1600 km), reflects the frequency at which the auto drive system reaches the specified system boundary during the test, and also shows the coverage of the IV on the specified test tasks and scenarios. However, this index has high risk in test and low efficiency in obtaining data. Index of scenario test trafficability refers to the capability of the whole vehicle system to safely pass the specifically designed test scenarios. However, since it is the only one-sided evaluation for specific scenarios, its application is quite limited. Due to the complex and changeable driving environment, the safety test and evaluation must be oriented towards the multi-dimensional scenario for comprehensive assessment. The safety clustering evaluation based on multi-dimensional scenario further couples the two indexes of hazard area dispersion and hazard area range through the clustering of the multi-dimensional scenario and the analysis of the hazard scenario, and obtains the safety clustering evaluation index in terms of the scenario risk rate. More comprehensively quantitative evaluation results of system safety can be obtained by testing and evaluating the automatic driving algorithm with this method [76].

(3) Analysis and Feedback for Evaluation Results

In accordance with the safety evaluation results of single function and whole vehicle system, it is necessary to further carry out safety diagnosis, analysis and other work for guiding the continuous optimization of vehicle function and system development, or improve the evaluation technology and methods for obtaining more comprehensive and objective TEVS results on the premise of reducing the evaluation risk and cost. The details are as follows:

(1) According to the safety evaluation results, the inference for the causes of IV high-risk behaviors is studied, the functional modules of the coupled vehicle system are decoupled, and the causal relationship between high-risk behaviors and algorithms of functional modules is established, so as to reduce or even avoid the high-risk behaviors and hazard scenarios. On the other hand, the HMI logic is optimized in terms of HMI safety, so as to realize the timely takeover by the human driver before the risk.
(2) Due to the large number, the deep learning algorithms adopted by IV are complicated and poorly deconstructable. Only the intuitive evaluation results can not directly facilitate the improvement of algorithms. Therefore, it is necessary to study the diagnostic analysis on IV algorithms, classify the algorithms to be improved (such as structural problems and parametric problems), and then put forward targeted improvement suggestions [21].
(3) In addition to providing guidance for the improvement of vehicle safety, the evaluation results can also facilitate the improvement of test and evaluation methods. The test and evaluation of IV can be designed more reasonably through iterative optimization in terms of key element and parameter design of test scenarios, test boundary conditions, index type selection and weight distribution.

5.4 TEVS System

According to the vehicle performance and different interactions required by surrounding vehicle environment, the TEVS system is divided into four types, i.e., software simulation, hardware in-loop, closed site and open road evaluation system. TEVS is carried out comprehensively in such stages as early function development, component or control system testing, closed site testing and open road comprehensive testing respectively. The TEVS system plays an important role in ensuring the intelligent safety of vehicles.

5.4.1 Software Simulation Testing and Evaluation System

The software simulation TEVS system is a pure digital simulation testing tool, and applied in the early stage of function development, when there is no physical hardware. Based on the specific simulation requirements, it provides environmental information to the vehicle model through static environment simulation and dynamic scenario simulation, so as to verify the algorithms of related system functions [11].

The structure of software simulation TEVS system includes four parts, i.e., static and dynamic scenario simulation (also known as external environment and scenario module), environment perception, decision planning and execution control. The system logic is shown in Fig. 5.44. According to the simulation requirements, environment perception modules in terms of sensor simulation, communication simulation, positioning simulation, perception fusion, etc., first perceives data from the simulation environment; The decision planning modules include decision algorithm, behavior prediction, task decision, trajectory planning, etc. According to the preset algorithm, it is input with the perception data, output the decision behavior and planning trajectory, and transmit the instructions to the control execution module; The control execution modules include state tracking and control execution in vehicle dynamics, and controls the steering, braking and other behaviors of the vehicle model according to the decision-planning results.

In terms of typical system applications, the existing software-based simulation tests are mainly conducted with virtual simulation software [1] (Table 5.13). Taking the widely-used PreScan for instance [77], its core function is sensor perception system modeling, which is mainly used for the development of ADAS system and automatic driving function. However, its vehicle dynamics model has limited accuracy and low modeling efficiency, and can only deal with simple traffic scenarios. SiVIC software [78] has similar function to PreScan, but its sensor model is more real

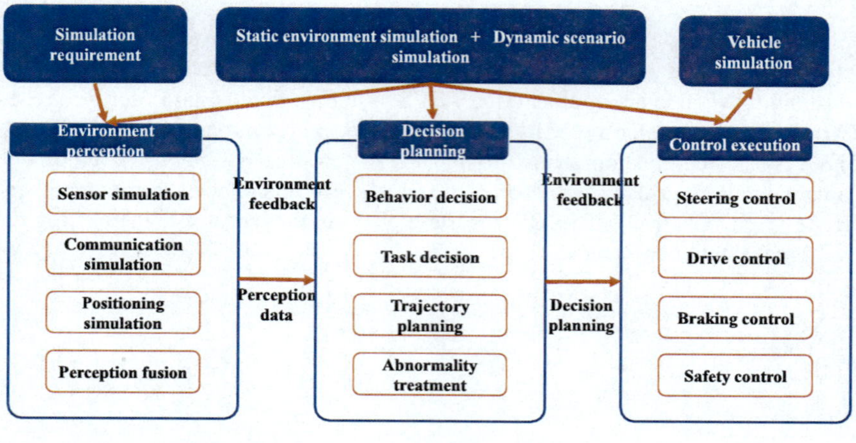

Fig. 5.44 Logic diagram of software simulation testing system

and complete, and can record the data generated by the sensor through third-party software such as RTMap (real timemap). Virt Test Drive (VTD) is used to create, configure, present and evaluate virtual environment based on highway and railway simulation, ranging from the generation of 3D content to the simulation of complex traffic scenarios. Its interface is simple and easy to integrate. As a simulation software based on macro traffic environment, Simulation of Urban Mobility (SUMO) [79] has the functions of building complex traffic environment and reproducing real traffic scenarios. It is often used for joint simulation with USARSim (unified system for automation and robot simulation) software. It is applicable to traffic flow simulation under network connection as well as simulation test of multiple IV under large-scale traffic network. However, it has disadvantages of simple vehicle model, low reduction degree, and only a few on-board sensor information. The Carcraft virtual testing environment built by Google, based on the data collected by the sensor system of Waymo IV and high-precision map information, has realized a relatively complete reproduction of the real road traffic environment, including driver model, traffic environment model, and vehicle dynamics model, etc. It is commonly used in the development and testing of advanced driver assistance system (ADAS) and vehicle dynamics-related control systems.

The virtual simulation test based on software simulation evaluation system can get rid of the demand for hardware and real testing environment, which has the advantages of high testing efficiency [11] and low cost and risk. However, it faces two main problems: one is that the testing results depend on the correctness of sensor model and vehicle model; the other is that it is difficult to quickly restore the real scenario in the simulation environment, which will affect the overall evaluation efficiency.

Table 5.13 Comparison of virtual simulation software for different types of vehicles

Software	Application scenario	Limitation
PreScan and SiVIC	Development of ADAS system and automatic driving function; Applicable to micro traffic scenarios with small spatial scale and a small number of traffic participants	Low modeling efficiency for large spatial scale and complex traffic scenarios
SUMO	Build complex traffic environment, simulate traffic flow and large-scale traffic network	Low reduction degree of the vehicle model, and poor dynamics properties
Carcraft	Accurately restore real and complex traffic scenarios	High development cost
VTD	Comprehensive scenario coverage and simple interface	Simplified physical engine, and poor dynamics properties

5.4.2 Hardware In-Loop Testing and Evaluation System

Hardware in-loop test combines physical components and software models, and is widely used in component or control system testing. It has the three test requirements, i.e., continuous test based on test purpose; test on different combinations of performance indexes such as safety and comfort; and certain expansibility of testing results on SOTIF. It consists of three parts: environment awareness system, decision planning system and in-loop test of control execution system, as shown in Fig. 5.45 [10].

Vehicle hardware in-loop testing and evaluation system is an important tool for parallel development, and has many typical applications. For example, the in-loop tests of environment perception system include the in-loop tests of camera, radar, V2X and multi-source sensor fusion system. The in-loop fusion environment perception framework based on the environment perception system is also capable to test the data interaction in the system [80].

The in-loop test of the decision planning system is conducted in virtual environment, where the real vehicle dynamics model is placed, and the vehicle data is transmitted in real time through CAN and I/O interface, etc. In this way, the real vehicle model is synchronously combined with the simulation environment. By integration of different components or systems, a hardware in-loop integrated test system can be established [81]. Ford conducted CarSim-Simulink joint simulation for testing the path following algorithm of IV based on the power train and chassis hardware in-loop platform. In view of the complexity and safety of IV testing, the hardware in-loop test system will further verify the availability of the decision planning and other functions.

The hardware in-loop test of control and execution system has been developed earlier and mature at present. Based on the hardware in-loop test platform [82], the relevant vehicle control execution algorithms can be effectively verified through the combination of hardware platform and execution control system.

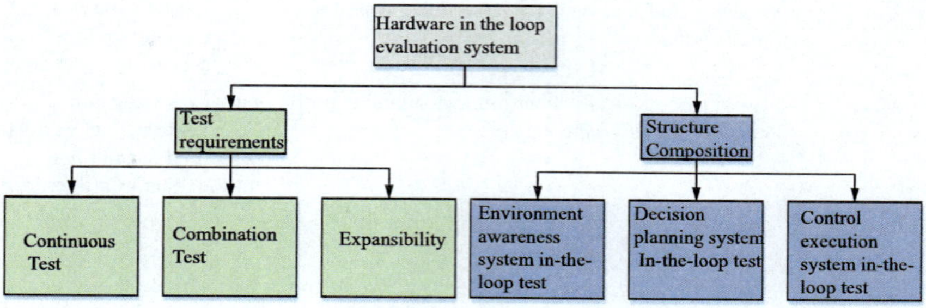

Fig. 5.45 Structure composition and test requirements of hardware in-loop testing and evaluation system

5.4.3 Closed Site Testing and Evaluation System

Based on the special closed test site, as well as the real environment and real vehicle system, the closed site testing and evaluation system focuses on the reduction capability of the environment and scenario. The system mostly adopts flexible design to ensure that the vehicle can be tested in different environment and scenarios as much as possible under limited site conditions. It is specifically composed of five parts, as shown in Fig. 5.46:

(1) Basic test site: mainly includes ADAS testing site, simulated expressway (fast road), urban road network, highway, low-grade road, tunnel, parking lot, street view, etc., as well as relevant facilities.
(2) Natural environment simulation system: simulates the surrounding natural environment during driving, including rainy days, fog days, snowy days and natural lighting.
(3) Traffic element system: mainly includes motor vehicles, non-motor vehicles, pedestrians, traffic control facilities and abnormal objects (such as animals).
(4) Communication system: mainly includes wired communication system, wireless communication system, traffic environment detection system, edge computing system and high-precision positioning system.
(5) Control center: mainly includes site management system, cloud storage system, traffic signal control system and interaction system, which can realize real-time transmission of information and real-time control of the system.

The closed site evaluation system can test single or multiple specific functions. In terms of typical applications, the US, the EU, Japan and China have built and transformed some special closed test sites according to the specific test demand of vehicle functions in the R & D process (Fig. 5.47). As introduced above, MCity [83] in Michigan of the US is a world-famous driverless demonstration and test site. It can represent road features of ordinary US cities and towns, with a small area but rich test environment and facilities, and is mainly used for testing automatic driving technology, IOV technology and electric safety system. AstaZero [84], located near Gothenburg of Sweden, is also a typical closed test site. It has four test environment, including rural road segment, urban area, expressway area and multi-vehicle road. It is equipped with testing facilities such as differential GPS, simulation test system and control center, which can present various customized test environment, such as

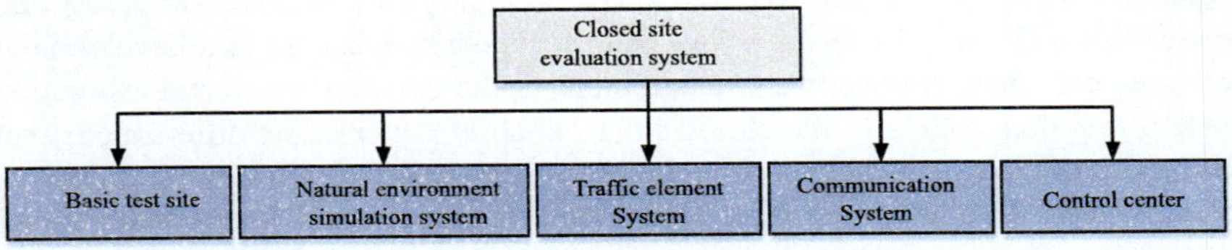

Fig. 5.46 Structure of closed site testing and evaluation system

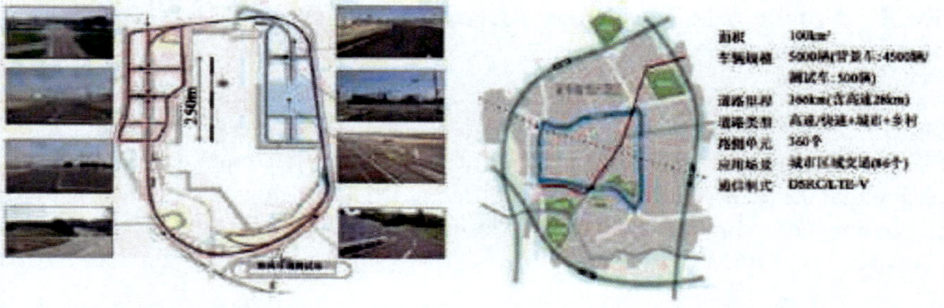

a) 日本J-town测试场[77] b) 中国上海智能网联汽车示范区

Fig. 5.47 Typical application of closed site testing and evaluation system

crowded urban roads, expressways, multi-lane parallel road conditions, roundabouts and intersections, as well as some hazardous road conditions in particular. It cooperates with universities and industry institutions, and is a future-oriented research and development platform for vehicle safety technology, where Volvo and other enterprises have carried out research, development and testing of intelligent safety assistance system. The J-town [85] automatic driving test site in Japan is a closed test site with early establishment, complete functions and mature development, mainly including urban area-based V2X test, multi-purpose test and special environment test areas. The IV pilot demonstration zone [85] located in Shanghai is the first IOV open test site of China. The scenario planning covers a variety of typical road traffic features, as well as four fields, including safety, efficiency, information service and new energy vehicle (NEV) application.

5.4.4 Open Road Testing and Evaluation System

The open road testing and evaluation system serves for the interactive function evaluation of auto drive system on the open road by obtaining the information of the vehicle and surrounding environment in real time, so as to facilitate the landing of corresponding on-board systems and vehicle technologies.

1. System Structure Composition

Open road testing and evaluation system is a necessary way to verify the safety and reliability of IV, and is the only way from technology research and development to large-scale mass production. The structure of the existing open road evaluation system is similar to that of the closed site evaluation system. The difference is that the road traffic elements change from being closed to open, and the real traffic flow is introduced. In addition, it shall take such factors as signal lights and weather conditions into consideration. In general, the system consists of five parts, as shown in Fig. 5.48:

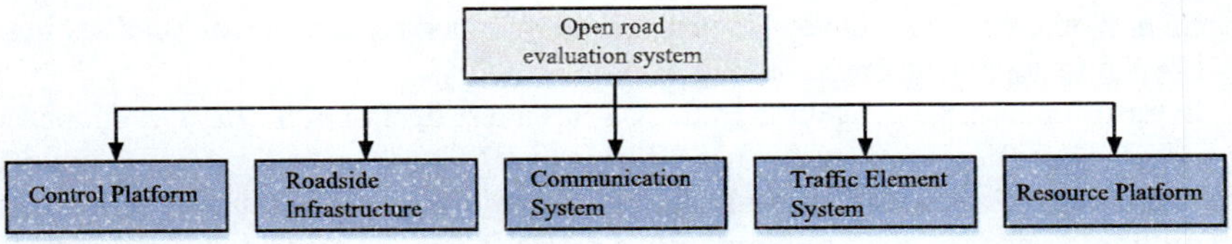

Fig. 5.48 Structure of open road testing and evaluation system

(1) Control platform: provides communication link, real-time data of all traffic elements and real-time operation environment of applications for vehicle testing, and can realize real-time transmission of information and real-time control of system.
(2) Roadside infrastructure: realizes the perception fusion at the roadside and the real-time transmission of information among human, vehicle and road.
(3) Communication system: realizes the extensive interconnection communication among human, vehicle, road and cloud in the system based on the standardized communication mechanism, and achieves high performance and high controllability based on 5G, software-defined network and other advanced communication technologies.
(4) Traffic element system: includes random motor vehicle flow, non-motor vehicle flow, pedestrians, traffic facilities and abnormal objects in open road.
(5) Resource platform: provides professional cooperative application services and involves other data, such as high-precision map, ground-based argumentation positioning, meteorology, traffic management and public security.

2. Typical Application of System

At present, the planning of open testing roads at home and abroad has been accelerated, and a highly intelligent connected demonstration zone has been built for promoting the development of open road testing. In the initial testing stage of IOV, the open road evaluation system should be applied to road segments with simple traffic conditions and small traffic flow, with the goal of giving priority to urban commuting traffic and reducing traffic safety risks. In order to promote the development of IV technology and ensure the safe passage on various roads in the future, it shall also be applied to roads with complex road conditions and gradually increase the difficulty of testing, so as to finally provide a more comprehensive IV test scenario.

(1) Status of Foreign Open Road Testing and Evaluation

Quite a few governments and relevant institutions have begun to vigorously develop the R & D and testing for IV. In terms of government and policy, developed countries led by the US have begun to carry out relevant research. The US has always been supporting and advocating the open road test of autonomous driving. By 2020, more than 20 states have promulgated laws and regulations related to autonomous driving, and gradually gave AV access to open roads. Germany, France, Sweden, South Korea,

Japan and othe countries have also progressively issued regulations on road testing, and began to support the road testing of AV.

In terms of enterprise, Waymo of the US has a test fleet of 600 Chrysler Pacificas models, and conducts road tests in 25 cities in the US at the same time. It can collect 25,000 miles (about 40,000 km) of data in one day, and has accumulated more than 20 million miles (about 32 million km) of real AV data for algorithm development and improvement by 2020.

In terms of testing site, most of open roads used for testing abroad are expressways, and gradually expanding to the central urban area. Typical open test sites include US GoMentum Station [86] testing site, Texas automated vehicle testing alliance, etc., which are shown in Table 5.14 for basic overview.

(2) Status of Domestic Open Road Testing and Evaluation

In the context of the national policy for promoting ICV development, local governments have also actively responded to the layout and began to focus on building intelligent transportation and IV demonstration areas. Sites for IV testing have been open in several cities, among which Beijing, Shanghai and Shenzhen are the earliest ones [85], as shown in Table 5.15. In addition, there are cities under layout or rapid development, such as Chongqing and Changchun. In general, the open test sites and open roads built in China are mainly distributed around cities and suburban areas, with favorable traffic environment and relatively simple road conditions, which improves the flexibility and safety of open road testing.

The Specifications for Road Test Management of ICV (Trial) [87] jointly issued by the Ministry of Industry and Information Technology, the Ministry of Public Security and the Ministry of Transport of China clearly specifies 14 aspects of test contents

Table 5.14 Typical overseas open sites for ICV testing

Name	Testing and evaluation types and features of open roads
US GoMentum station testing site	It is mainly a highway corridor and supports the test of AV in commercial areas in an open environment
Texas automated vehicle testing alliance	It involves diversified urban traffic test environment, such as parks, expressways, urban low-speed roads, border ports and other real test environment

Table 5.15 Main domestic open roads for ICV testing

City	Traffic profile of testing road
Beijing	More than 200 automatic driving test roads are open, with a total length of 700 km. There are various environment for all-weather test, weather test and high-speed test
Shanghai	There are diversified urban traffic test environment, including urban highways and parks, etc.
Shenzhen	There are favorable road conditions, mainly including secondary trunk road and general trunk road, and relatively simple traffic scenario

and 34 scenarios, specifically including the identification and response of traffic signs and markings, the identification and response of traffic lights, the identification and response of driving state of vehicles in front (including opposite vehicles), the identification and response of obstacles, the identification and response of pedestrians and non-motor vehicles, car-following (including parking and starting), stopping by the roadside, overtaking, parallel driving, intersection traffic, roundabout traffic, automatic emergency braking, manual takeover, network communication, etc., all of which can be of guidance to the closed site and open road testing and evaluation system.

5.5 Summary

Accurate, efficient and comprehensive TEVS is a considerable demand in the development towards intelligence. In this chapter, introductions are made in terms of theory, method and system.

(1) The safety evaluation methods should be selected in accordance with the five basic principles of scientificity, adaptability, systematicness, pertinence and rationality. In the theory of safety performance test, the effectiveness evaluation based on safety mechanism is to quantitatively or qualitatively evaluate the degree to which the IV or system reaches the safety application goal under the specified conditions. Hybrid reality-based acceleration test combines software simulation with driving test, which improves the flexibility and repeatability of the test process. DVE parameter space-based combination test can effectively reduce the size of parameter space by combination testing technology. By exploration of the internal relationship between scenario test and function test based on their characteristics, a comprehensive testing method with more all-round dimensions can be established. The theory of safety performance evaluation can be applied to find out the causes of system failure and its effects, and then improve and optimize the design defects of the system.

(2) Establishment of a testing scenario database for generating typical scenarios or high-risk scenarios can accelerate the testing process and improve the testing efficiency. In the intelligent testing ecosystem, various testing scenarios can be presented with the network environment and rich testing units for efficient testing process and driving state monitoring. Based on the static and dynamic testing platforms and other testing methods, driving data can be accurately obtained in the testing process. Various in-loop testing technologies can be applied for testing various components or systems in the whole cycle of vehicle research and development, so as to find problems as soon as possible and solve them in time.

(3) The safety evaluation index system is established through screening, optimizing and grading various safety indexes in accordance with the safety evaluation criteria. The subjective evaluation method is for obtaining the actual feelings of

drivers and passengers, and the objective evaluation is for the quantitative value of vehicle safety. Through the combination of subjective and objective evaluation, the comprehensive evaluation of safety performance can be made from multiple dimensions. Single function and vehicle system evaluation technologies can explore the response of vehicle functions and the safety degree of the whole vehicle in the scenario-based road test. The evaluation results are thoroughly analyzed, which can not only lay a theoretical basis for the optimization of the software and hardware of each vehicle module, reduce high-risk behaviors and scenarios, but also facilitate the improvement of the testing and evaluation methods for more accurate and reasonable evaluation results.

(4) TEVS system can be generally divided into four categories, namely software simulation, hardware in-loop, closed site and open road testing and evaluation systems. Software simulation system can effectively reduce the development cost; Hardware in-loop test is capable to restore the working conditions of real components; Closed site system can simulate the real traffic environment; And open road evaluation system can evaluate the relevant technologies through the real traffic environment. All of them ensures the intelligent safety performance of vehicles in a significant way.

References

1. China Center for Information Industry Development. Testing and evaluation technology of ICVs [M]. Beijing: People's Posts and Telecom Press, 2019. [in Chinese]
2. Cao Qinggui. Safety assessment [M]. Beijing: China Machine Press, 2017. [in Chinese]
3. Zhu Dongjie. Research on building fire safety evaluation based on analytic hierarchy process [D]. China University of Science and Technology, 2005. [in Chinese]
4. Liao Jun, An Yisheng, Zhang Shaoyang, Zhao Xiangmo, Ma Tianshan. Comprehensive evaluation model for dynamic traffic safety of road sections [J]. Journal of Traffic and Transportation Engineering, 2009, 9 (04): 79–84. [in Chinese]
5. Wang Hai. Research on safety assessment of dangerous chemical terminal based on FSA [D]. Dalian: Dalian Maritime University, 2007. [in Chinese]
6. Guo Qisheng. Introduction to equipment effectiveness evaluation [M]. Beijing: National Defense Industry Press, 2005. [in Chinese]
7. Liu Gaoliang. Research on effectiveness evaluation of integrated avionics system [D]. Xi'an: Northwest University of Technology, 2007. [in Chinese]
8. Liu Shilei, Li Hao. Improved ADC method and its application in effectiveness evaluation of weapon equipment system [J]. Journal of National University of Defense Technology, 2017, 39 (03): 130–135. [in Chinese]
9. Chen Lei, Yao Weizhao, Guo Quankui. Theory, method and application of effectiveness evaluation [M]. Beijing: Beijing University of Posts and Telecommunications Press, 2016. [in Chinese]
10. Zhu Bing, Zhang Peixing, Zhao Jian, et al. Research progress in scenario-based virtual testing of automated driving vehicle [J]. China Journal of Highway and Transport, 2019, 32 (06): 1–19. [in Chinese]
11. Zhang Zihui, Li Yanqiang, Wang Jianqiang, et al. Automated driving test method and system based on hybrid reality [P]. Shandong Province: CN109781431A, 2019–05–21. [in Chinese]

12. Yu Zhuoping, Xing Xingyu, Chen Junyi. Progress in testing technology and application of automated driving vehicles [J]. Journal of Tongji University (Natural Science Edition), 2019, 47 (04): 540–547. [in Chinese]
13. HUANG W L,WANG K F. Autonomous vehicles testing methods review [C]|| International Conference on Intelligent Transportation Systems (ITSC). 2016 IEEE 19th.
14. KUHN D R, REILLY M J. An investigation of the applicability of design of experiments to software testing. In: Caulfield M, ed. Proc. of the Annual NASA/IEEE Software Engineering Workshop (SEW). Los Alamitos: IEEE Press, 2009.5.91
15. GRINDAL M. Handling combinatorial explosion in software testing [D]. Linköpings: Linköpings Universitet, 2007.
16. WOTAWA F. Testing Autonomous and Highly Configurable Systems: Challenges and Feasible Solutions [M]|| Automated Driving. Springer International Publishing, 2017.
17. ELROFAI H, WORM D, CAMP O O D. Scenario Identification for Validation of Automated Driving Functions [M]|| Advanced Microsystems for Automotive Applications 2016. Springer International Publishing, 2016.
18. HUANG W L, LV Y, CHEN L, et al. Accelerate the autonomous vehicles reliability testing in parallel paradigm [C]|| IEEE 20th International Conference on Intelligent Transportation Systems (ITSC), 2017: 922–927.
19. LI L, HUANG W L, LIU Y, et al. Intelligence Testing for Autonomous Vehicles: A New Approach [J]. IEEE Transactions on Intelligent Vehicles, 2017, 1(2):158–166.
20. MENZEL T, BAGSCHIK G, MAURER M. Scenarios for Development, Test and Validation of Automated Vehicles [J]. IEEE Intelligent Vehicles Symposium (IV), 2018: 1821–1827.
21. Yu Rongjie, Tian Ye, Sun Jian. Virtual testing of high-grade automated driving vehicle: research progress and frontier [J]. China Journal of Highway and Transport, 2020, 33 (11): 125–138. [in Chinese]
22. Zhang Genbao, Yang Xingyong. Reliability analysis technology [J]. Manufacturing Technology and Machine Tools, 2015 (1): 7–14. [in Chinese]
23. Zeng Shengkui. Course of system reliability design analysis [M]. Beijing: Beijing University of Aeronautics and Astronautics Press, 2001. [in Chinese]
24. Sun Haiyun. Research on the application of FMEA in automobile axle [D]. Qingdao: Qingdao University of Technology, 2014. [in Chinese]
25. Dunjo J, Fthenakis V, Vilchez J A, et al. Hazard and operability (HAZOP) analysis. A literature review [J]. Journal of Hazardous Materials, 2010, 173(1–3):19–32.
26. Li Na, Sun Wenyong, Ning Xindao. Application analysis of HAZOP, LOPA and SIL [J]. Journal of Safety Science and Technology, 2012. [in Chinese]
27. Standardization Administration of the P.R.C. Application guide for hazard and operability analysis (HAZOP analysis): GB/T 35320-2017 [S]. Beijing: China Quality and Standards Publishing and Media Co., Ltd., 2018. [in Chinese]
28. Zhang Yanhui, Chen Xiaochun. Improved HAZOP risk assessment [J]. Journal of Safety Science and Technology, 2011, 07 (007): 174–178. [in Chinese]
29. Xu Yingbo, Zhang Jinhuan, Ma Chunsheng, et al. Research on NCAP in foreign countries and some reflections on NCAP in China [J]. Automobile Technology, 2006 (S1): 1–4. [in Chinese]
30. THE RATINGS EXPLAINED[EB/OL]. [2022-02-15]. https://www.euroncap.com/en/vehicle-safety/the-ratings-explained/.
31. Li Keqiang. Electric vehicle engineering manual Volume 6th: Intelligent network connection [M]. Beijing: China Machine Press, 2019. [in Chinese]
32. BEST A, NARANG S, BARBER D, et al. AutonoVi: Autonomous Vehicle Planning with Dynamic Maneuvers and Traffic Constraints [C]|| 2017 IEEE/RSJ International Conference on Intelligent Robots and Systems (IROS). IEEE, 2017: 2629–2636.
33. Chen Junyi, Feng Tianyue, Liu Lihao, et al. Automatic generation way of specific scenarios for decision-planning system test [J]. Automobile Technology, 2020 (10): 45–50. [in Chinese]
34. SCHULDT F. Ein Beitrag für den methodischen Test von automatisierten Fahrfunktionen mit Hilfe von virtuellen Umgebungen [D]. Braunschweig: Technische Universitat Braunschweig, 2017.

35. Geng Xinli. Research on decision-making of driverless vehicles in uncertain urban environment [D]. Hefei: University of Science and Technology of China, 2017. [in Chinese]
36. Li Wei, Gao Dezhi, Duan Jianmin, et al. Research on free lane-changing model of intelligent vehicles [J]. Journal of Highway and Transportation Research and Development, 2010, 27 (2): 119–123. [in Chinese]
37. RIGOLLI M, BRADY M. Towards a Behavioural Traffic Monitoring System [C]|| International Joint Conference on Autonomous Agents & Multiagent Systems. ACM, 2005: 449–454.
38. ECKSTEIN L, ZLOCKI A. Safety potential of ADAS-combined methods for an effective evaluation [C]|| 23rd International Technical Conference on the Enhanced Safety of Vehicles (ESV) Seoul, South Korea. 2013: 15–25.
39. ROYDEN H L. Real Analysis [M]. Englewood Cliffs: Prentice Hall, 1948.
40. Li Xiaochi, Zhao Xiangmo, Xu Zhigang, et al. Modular flexible testing field for intelligent connected transport system [J]. China Journal of Highway and Transport, 2019, 32 (06): 137–146. [in Chinese]
41. WISMANS J, JANSSEN E, BEUSENBERG M, et al. Injury Biomechanics (4J610): Third printing,Eindhoven University of Technology, Eindhoven, The Netherlands, 2000.
42. FOSTER J K, KORTGE J O, WOLANIN M J. Hybrid III-a biomechanically-based crash test dummy [J]. SAE Transactions, 1977: 3268–3283.
43. NEILSON L, LOWNE R, TARRIERE C, et al. The EUROSID side impact dummy [R]. SAE Technical Paper, 1985.
44. EEVC. Improved test methods to evaluate pedestrian protection afforded by passenger cars (December 1998 with September 2002 updates) [R]. EEVC Working Group 17, 2002.
45. 4activesystems [EB/OL]. [2022-02-15]. https://www.4activesystems.at/downloads. [in Chinese]
46. Sun Yang, Chen Huiyan. Research on the evaluation system of unmanned ground vehicles [J]. Acta Armamentarii, 2015, 36 (06): 978–986. [in Chinese]
47. Xiong Guangming, Gao Li, Wu Shaobin, et al. Intelligent behavior of driverless vehicles and its testing and evaluation [M]. Beijing: Beijing University of Technology Press, 2015. [in Chinese]
48. Sha Jie, Zhao Huijing, Cui Jinshi, Zha Hongbin. Intersection traffic data collection and analysis based on distributed laser scanner [J]. Journal of Highway and Transportation Research and Development, 2009, 26 (S1): 30–34. [in Chinese]
49. KIROVSKII O. Determination of Validation Testing Scenarios for an ADAS Functionality: Case Study [J]. SAE Paper 2019-01-0137.
50. DA VISION A J, REID I D, MOLTON N D, et al. MonoSLAM: Real-time Single Camera SLAM[J]. IEEE Transactions on Pattern Analysis and Machine Intelligence, 2007, 29 (6): 1052–1067.
51. Xia Houkaishun, Chen Shanxing, Wu Yilin. Modeling and simulation of target tracking system based on binocular PTZ camera [J]. Journal of System Simulation, 2015, 27 (2): 362–368. [in Chinese]
52. ETINGER A, LITVAK B, PINHASI Y. Multi Ray Model for Near-ground Millimeter Wave Radar [J]. Sensors, 2017, 17 (9):1983.
53. Wang Huili, Yang Haizhong. Vehicle dynamics modeling based on system identification [J]. Chinese Journal of Scientific Instrument, 2015, 36 (6): 1275–1282. [in Chinese]
54. Nie Yinyu. Research on data-driven vehicle dynamics modeling and simulation [D]. Xi'an: Xi'an Jiaotong University, 2016. [in Chinese]
55. Xu Xuejin. Research on vehicle dynamics modeling based on driving simulator [D]. Wuhan: Wuhan University of Technology, 2007. [in Chinese]
56. Yan Zhi. Virtual simulation technology of automated driving (IV): simulation testing process and requirements [EB/OL]. [2021-12-27] https://auto.vogel.com.cn/c/2021-05-24/1110694.shtml. [in Chinese]
57. OTTEN S, BACH J, WOHLFAHRT C, et al. Automated Assessment and Evaluation of Digital Test Drives[M] || Springer. Advanced Microsystems for Automotive Applications 2017. Berlin: Springer, 2017:189–199.

58. Chen Zhicheng, Wu Jian, Zhao Jian, et al. Control strategy for precise adjustment of braking force of hybrid-line control dynamic system [J]. Automotive Engineering, 2018, 40 (4): 86–93. [in Chinese]
59. Gao Y, Xu Z, Zhao X, et al. Hardware-in-loop Simulation Platform for Autonomous Vehicle AEB Prototyping and Validation [C]// 2020 IEEE 23rd International Conference on Intelligent Transportation Systems (ITSC). IEEE, 2020: 1–6.
60. Li Xiaoqing, Liu Qi, Li Binglin, et al. Research on route selection and classification of open roads for testing ICVs [J]. Journal of Traffic and Transportation Engineering, 2019, 19 (05): 74–78. [in Chinese]
61. Tang Li, Qing Sandong, Xu Zhigang, Zhou Houqing. Summary of research on public acceptance of automated driving [J]. Journal of Traffic and Transportation Engineering, 2020, 20 (02): 131–146. [in Chinese]
62. Li Jun. Challenges and solutions to China's SOTIF [J]. ICV, 2021 (05): 12–13. [in Chinese]
63. Li Yang. Research on multi-attribute evaluation method and application of driving behavior safety [D]. Changchun: Jilin University, 2016. [in Chinese]
64. BRODSKY W, SLOR Z. Background music as a risk factor for distraction among young-novice drivers [J]. Accid Anal Prev, 2013, 59(oct.):382–393.
65. SMITH M E, GEVINS A, BROWN H, et al. Monitoring task loading with multivariate EEG measures during complex forms of human-computer interaction [J]. Human Factors: The Journal of the Human Factors and Ergonomics Society, 2001, 43(3): 366–380.
66. Guo Zhangyong. Research on evaluation of driver acceptance of adaptive cruise control system [D]. Changchun: Jilin University, 2014. [in Chinese]
67. Lin Qingfeng, Wang Zhaojie, Lu Guangquan. Takeover safety evaluation model of L3 automated driving vehicle [J]. Automotive Engineering, 2019, 41 (11): 1258–1264. [in Chinese]
68. FANCHER, P. Intelligent cruise control field operational test.No.UMTRI-98-17[R]. 1998.
69. DONG F, ZHAO Y N, GAO L. Application of gray correlation and improved AHP to evaluation on intelligent U-Turn behavior of unmanned vehicles [C]ll 2015 8th International Symposium on Computational Intelligence and Design (ISCID). IEEE, 2015, 1: 25–29.
70. XIONG G, HAO L, DING Z, et al. Subjective evaluation of vehicle active safety using PreScan and Simulink: Lane departure warning system as an example [C]ll 2017 IEEE International Conference on Vehicular Electronics and Safety (ICVES). IEEE, 2017.
71. Sun Yong, Guo Kuiyuan, Gao Mingqiu. Status and development of autonomous emergency baking system in the new vehicle evaluation procedure [J]. Automobile Technology, 2016 (02): 1–6. [in Chinese]
72. European new car assessment programme assessment protocol: Safety Assist Version 9.0.4. [S/OL]. Euro NCAP, 2021.
73. European new car assessment programme test protocol: AEB Car-to-Car systems Version 3.0.3. [S/OL]. Euro NCAP, 2021.
74. Wang Runmin, Deng Xiaofeng, Xu Zhigang, Zhao Xiangmo. Overview of research on simulation testing and evaluation technology for Internet of vehicles [J]. Application Research of Computers, 2019, 36 (07): 1921–1926+1939. [in Chinese]
75. GOOGLE. On the road to fully self-driving-waymo safety report [DB/OL]. https://storage.googleapis.com/sdc-prod/v1/safetyreport/waymo-safety-report-2017-10.pdf.
76. Zhu Bing, Zhang Peixing, Zhao Jian. Cluster evaluation of automated driving safety for multi-dimensional logical scenarios [J]. Automotive Engineering, 2020, 42 (11): 1458–1463+1505. [in Chinese]
77. GIETELINK O J. Design and validation of advanced driver assistance systems [D]. Delft: Delft University of Technology, 2007.
78. GRUYER D, PERCHBERTI S, GLASER S. Development of full speed range ACC with SIVIC, a virtual platform for ADAS prototyping, test and evaluation [C]ll Intelligent Vehicles Symposium. Gold Coast: IEEE, 2013:100–105.
79. ROSSETTI R J F. An integrated architecture for autonomous vehicles simulation [C]ll ACM Symposium on Applied Computing. Riva del Garda: ACM, 2012:286–292.

80. HAGER B, ALLEN J. Raw Data Injection and Failure Testing of Camera, Radar, and Lidar for Highly Automated Systems [J]. SAE Paper 2019-01-1378.
81. DROLIA, WANG Z, VEMURI, et al. Demo Abstract: AutoPlug -An Automotive Test-bed for ECU Testing,Validation and Verification [C]ll IEEE.International Conference on Information Processing in Sensor Networks. New York: IEEE, 2011: 131–132.
82. WE K S, LEE C G, LEE J, et al. ECU-in-loop Real-time Simulation Technique for Developing Integrated Vehicle Safety System [C]ll IEEE. International Conference on Control, Automation and Systems. New York: IEEE, 2014: 1432–1437.
83. BRIEFS U. M-city grand opening [J]. The UMTRI Research Review, 2015, 46(3):1.
84. JACOBSON J, JANEVIK P, WALLIN P. Challenges in creating AstaZero, the active safety test area [C]ll Transport Research Arena (TRA) 5th Conference. Paris: Transport Research Arena (TRA), 2014: 14–20.
85. Chen Guihua, Yu Shengbo, Li Qiao, et al. Investigation and research on the development of China ICV testing demonstration zone [J]. Chinese Journal of Automotive Engineering, 2020, 10 (02): 79–87. [in Chinese]
86. American Honda; GoMentum Station And Honda Offer Demo Of Autonomous Vehicle TechnologyJ [J]. Journal of Transportation, 2016.
87. Science and Technology Department of the Ministry of Transport of the People's Republic of China. Management specification for road testing of ICVs (Trial) [EB/OL]. [2021-07-15]. https://xxgk.mot.gov.cn/2020/jigou/kjs/2006/t20200623_3317132.html. [in Chinese]

Chapter 6
Prospects for ISV Technology

6.1 Vision and Development Requirement of ISV

6.1.1 Safety Vision

The inherent attributes and connotations of vehicles have been expanded in the new round of scientific revolution and disruptive industry innovation. Vehicles are developing rapidly toward "new four modernizations", namely electrification, intelligence, networking and sharing. IV not only offers a diversified travel experience, but also significantly improves road traffic safety and efficiency. With the development goal of reducing accident casualties and people's property losses, the active and passive integration of intelligent safety technology will identify potential risks in the driving process, improve driving safety and lessen accident losses, to provide the essential and fundamental guarantee for the industrial application of IV.

6.1.2 Development Requirement

As a pillar of the nation, the vehicle serves as a significant driving force of national economy and national defense modernization, as well as a vital field for great power competitions. Safe and efficient transportation is not only the expectation of the public, but also the primary motive of vehicle design, development and application. The safe operation and intelligent development of vehicles have been attached great importance to the government and industry. And new demand for its more efficient, convenient, intelligent and safe development has also been put forward by the state, society and people.

J. Wang et al., *The Intelligent Safety of Automobile*, Key Technologies on New Energy Vehicles, https://doi.org/10.1007/978-981-99-6399-7_6

(1) National Demand

The CPC Central Committee and the State Council pointed out in the outline of *Building a Leading Transportation Nation* [1] that we should build China's strength in transportation by 2035. With the large volume, numerous personnel and high output value, China's vehicle industry is a pillar for the national economy and social development, taking a strategic position in the construction of a transportation power. In 2020, 11 ministries and commissions, including the National Development and Reform Commission, the Ministry of Industry and Information Technology and the Ministry of Science and Technology, jointly issued the *IV Innovation and Development Strategy* [2], which focuses on the strategic planning on six IV systems including cooperative and open technology innovation, cross-border integration of industrial ecology, advanced and complete infrastructure, systematic and perfect regulations and standards, scientific and standardized product supervision, and comprehensive and efficient cybersecurity systems. The strategy will play an important role in accelerating the innovation and development of IV and enhancing the comprehensive national strength in the new era.

The development paradigm, demonstration application and standardization of IV have been specified by relevant laws, regulations and standards. In 2019, the Ministry of Industry and Information Technology proposed in the *Key Points for Standardization of Intelligent and Connected Vehicles* [3], to orderly promote the formulation of vehicle information security standards, and deeply engage in the formulation of key standards such as expected function security and information security. And in 2021, the Ministry issued the *Management Specification for Road Test and Demonstration Application of Intelligent and Connected Vehicles networked vehicles (for Trial Implementation)* [4], together with the Ministry of Public Security and the Ministry of Transport, to standardize the road test and demonstration application of intelligent and connected vehicles. The *Road Traffic Safety Law (Revised Proposal)*, drafted and issued by the Ministry of Public Security in 2021, further pushed forward the solution to new situations and problems from the perspective of road traffic safety and management. With the principal ideology of ensuring people's safe and efficient travel, the state has implemented various measures in recent years for the safety development of IV in terms of strategic guidance, standard specification, legal and regulatory constrains, etc., as shown in Fig. 6.1.

(2) Social Demand

Road traffic accident is a major public health concern all over the world. According to the data of *China Statistical Yearbook* [5–9], the number of vehicle traffic accidents in China has been remaining high. Taking the years of 2015–2019 for example, it is as high as 140,000 every year, which not only result in huge direct or indirect economic losses, but also serious physical and mental harm to individuals and families; not only a huge waste on social resources, but also an adverse impact on social harmony and stability. Preventing and reducing traffic accidents and improving traffic efficiency are important social demands, and vehicle intelligent safety is the key to meet them. It takes DVE system as the research object, studies the safety development theory

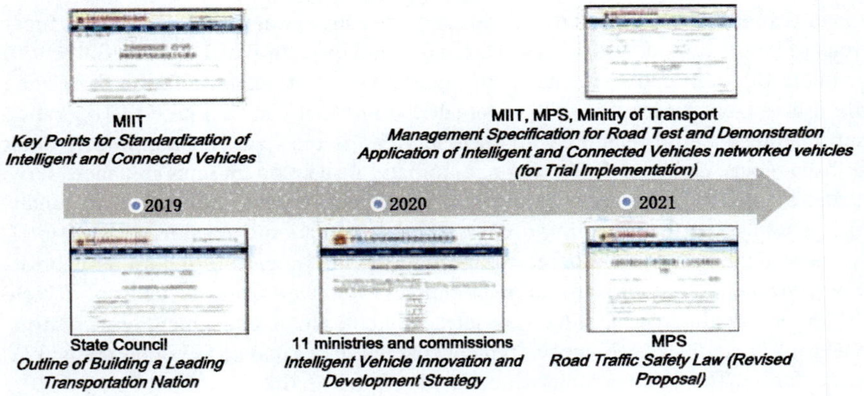

Fig. 6.1 National policies in terms of IV safety

of systems engineering, and improves the capability of environmental awareness, situation assessment, cooperative decision-making and safety control based on FuSa, SOTIF and cybersecurity, which is expected to prevent accidents in the whole process, ensure safe driving and significantly reduce personal injury and property loss.

(3) People's Demand

The scale of civil vehicles in China has been growing rapidly. According to *China Statistical Yearbook* [5–9], the number of civil vehicles in China was about 254 million by 2019, an increase of 9.23% over 2018, with most of them used in large and medium-sized cities. It was indicated in the research of *2019 China Urban Traffic Report* [10] that most cities have traffic congestion with different degrees, especially during peak hours. Traffic accident is one of the main causes of road congestion, and leads to problems such as energy consumption and environmental pollution, resulting in huge losses. Safe, efficient, green and convenient travel is the expectation of people, and the development of vehicle intelligent safety technology can present more possibilities to meet such demands. Employing IOV technologies, information exchange between vehicles and multiple platforms such as intelligent transportation system and cloud coordination system can be realized. And the algorithms such as multi-vehicle cooperative path planning can be applied to optimize travel schemes, improve traffic efficiency and ensure travel safety.

6.2 Future IV and Its Safety

Vehicle is one of the best carriers for integrated application of new technologies. The "new four modernizations" push forward far-reaching changes in the element attributes and coupling relationship in the DVE system. Accordingly, the intelligent systems (collectively referred to intelligent driving system) that realize the functions of perception, decision-making, control and execution of IV in the future will promote the continuous extension of vehicle connotation. In addition to serving the drivers and passengers with automated driving, it can facilitate the operation and maintenance of vehicles with such services as the state self-monitoring, fault self-diagnosis, safety self-protection, automatic inspection and maintenance, serve traffic managers and other road users as intelligent mobile information terminals, and guarantee performance improvement and connotation extension with the intelligent evaluation system. Therefore, through the intelligent driving system and its advanced technology, driver can be completely liberated from the tasks of vehicle driving and maintenance, and enjoy safe, efficient, green, convenient and comfortable travel, which has become the typical characteristics and basic connotation of IV in the future. The system composition is shown in Fig. 6.2.

With the development of new energy, electric wire control, vehicle design and other technologies, the attributes of elements such as vehicle power system, chassis configuration and body structure, together with the coupling relationship among them, are bound to change. Vehicle is endowed with new scientific connotation, by transformation from "manned tool" to "intelligent mobile space". For example, the low-carbon driving power system has made a breakthrough. Through multi-domain integration and integrated control, it optimizes and matches the controllable area of vehicle stability, promotes its development towards high-dimensional, multi-domain, cooperative and integrated development, and realizes the matching of DVE systems; The adaptive innovation of electric drive chassis configuration realizes

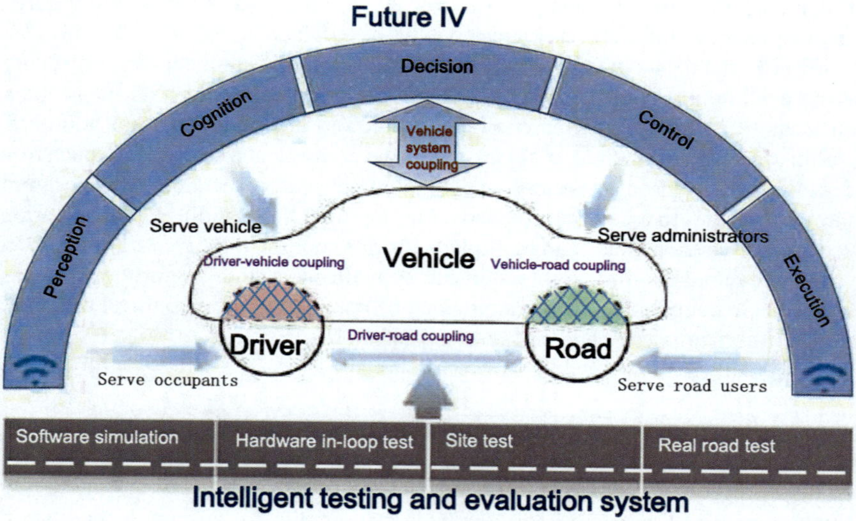

Fig. 6.2 Composition of future IV system

the deep immersion of distribution, wire control and generalization by simplifying the chassis structure and increasing the control freedom. With a short transmission chain, high efficiency and flexible spatial layout, it can improve the vehicle power and safety; Subversive change of lightweight structure realizes new body design and optimized health management, by establishing personalized, modular and digital multi-dimensional material-structure-architecture-evaluation cooperation by in-depth construction of cooperative design theory and approach.

At the same time, IV is a multi-domain cross-research carrier. The development of big data, artificial intelligence and communication technology gives impetus to the gradual evolution of the coupling relationship among different elements such as vehicles, traffic participants and road environment, as well as the change of the properties and operation law of the DVE system. The gradual application of technologies such as vehicle–road cooperation, IOV and cloud-end control has continuously enhanced the synergy among elements, and can better improve the safety, economy and transportation efficiency of vehicles from the perspective of the overall system, so as to realize autonomous driving and rapid mobility, and ensure the convenient usage, safe drive and green efficiency under the complex regional environment such as plateau mountains and urban roadways, as well as severe weather conditions such as rain, snow and fog.

In view of its characteristics of multi-component, nonlinear and strong coupling, the future IV will carry out real-time data detection, fault autonomous analysis and diagnosis for itself in terms of such aspects as operation status monitoring, performance failure mechanism, reliable communication mechanism and cloud information management, so as to share vehicle–road-cloud data, realize the automatic upgrading and maintenance of vehicle system, and improve its operation and maintenance safety.

In view of time-varying intention, personalized behavior and cognitive differentiation and other characteristics possessed by different drivers, the future IV will make the human–machine switching function design and function application mode of the carrier platform for serving drivers and passengers, identify the dynamic driving behavior and its physiological signals, implement high-precision prediction, build a safety decision-making control algorithm with self-adaptation to different drivers' behaviors, so as to provide overall safety strategies for drivers and passengers, as well as the safe and efficient driving and riding experience.

In view of the characteristics of different types of off-vehicle road users, such as high randomness of behaviors, mixed operation, interaction and mutual restriction with drivers, the future IV will focus on the development of all-round intelligent assistance technology for serving road users, comprehend environmental situation and driving scenario employing active perception and collaborative perception, etc., and identify the safety requirements of other road users, so as to ensure two-way safety for multiple types of road users and drivers with the optimal allocation and cooperative utilization of road resources.

In view of the characteristics of large randomness, strong coupling and low reproducibility of different traffic elements, the future IV will conduct comprehensive and efficient testing and verification on perception and cognition, decision-making and planning, and control and implementation for serving administrators, in terms of their compliance with road laws and regulations, conformity with standards and specifications, and adaptability to the applied environment, so as to promote the accurate formulation of traffic situation response strategies and completely guarantee the driving safety in any scenarios.

In the process of IV development, the change of its attribute and connotation also defines the development direction of intelligent safety technology in the future. With complex and time-varying coupling relationship among its diverse elements, the future IV system will have increasing occurrence of sudden fault of its core part, and be vulnerable to cyber attack, communication interference and other environmental effects. Hence, it is urgent to carry out all-around systematic improvement in terms of FuSa, SOTIF and cybersecurity in IV system safety; At the same time, due to the complex and changeable real environment, it is difficult to make cognitive decision-making for multiple types of traffic elements. The existing safety technology has a single applied scenario and independent function, mostly available for specific scenarios and working conditions. Accordingly, IV in the future shall have self-adaptability based on the research by learning from people, simulating people and surpassing people, to ensure the safe operation in a complex driving environment; When accidents are unavoidable, it is also necessary to adaptively take the behavior laws of different driving system elements into consideration, comprehensively utilize the system perception, cognition, decision-making, control and execution capabilities, and realize multi-dimensional intelligent protection with assistance safety technology. The above intelligent safety technologies can support the organic integration of originally independent system elements, coordinate the impacts of DVE elements, ensure the safe and efficient operation of the system from a macro perspective, make the interconnection and resource sharing of various elements, and facilitate the optimization of cooperative operation and the improvement of safety performance.

6.3 Cutting Edge Technology of System Safety

IV is a comprehensive system integrating a number of functions, including perception and positioning, decision-making and planning, control and execution. The complexity and diversity of the system itself and the dynamic changes in environmental conditions are likely to pose a wide range of challenges to system safety, including various types of potential faults, insufficient functions and cybersecurity vulnerabilities. The development goal of vehicle system safety is to ensure the overall safety of IV system, and thoroughly reduce the residual risks by safety analysis, function improvement, verification and confirmation in the design and development stage, so as to realize the real-time monitoring and prevention of potential system risks in the process of operation. The mature system safety technology shall aim at different intelligent levels of autopilot vehicles and their operation design domains, and prevent system risks caused by potential faults, insufficient functions and cyber attacks in real time, so as to give full play to its intelligent potential and minimize system risk. And the technology should be guided by system thinking, as well as interdisciplinary technologies such as machinery, electronics, artificial intelligence and statistics. The development trend mainly includes the following three aspects:

(1) **FuSa solutions in accordance with the characteristics of IV**. Based on the existing electronic technology, the whole vehicle and its core devices will cover more technical fields, such as machinery, electrochemistry, and optics, to meet the overall safety requirements; incorporate open-source operating systems and uncertain algorithms at the system level, and improve the strategy and system of FuSa design, testing, evaluation and safety demonstration; introduce the design concept of service-oriented architecture (SOA) in IT field, to adapt to the complex and cross coupling feature of IV supply chain, and form design criteria in consideration of multiple dimensions, modular reuse and multi-factor architecture; break through the limitation that the current FuSa standard is only applicable to single vehicle system, and build a design, analysis, risk assessment and test evaluation system involving multiple vehicles and vehicle–road-cloud-network integration.
(2) **Risk propagation mechanism and prediction and pre-control mechanism for SOTIF of IV**. The potential triggering conditions and insufficient functions of system risks shall be further analyzed in order to build a qualitative or quantitative risk triggering mechanism related to SOTIF, and explore the internal generation mechanism and dynamic communication theory of SOTIF in combination with the function realization logic of IV; Since the SOTIF risks are influenced by external triggering conditions and insufficient interior functions, the unified model for risk quantification should be built, by the research in consideration of cognition and quantification for environmental risk and system interior risk, together with driving environment cognition and system performance evaluation; Based on the risk quantification model, it is necessary to further explore the prediction and pre-control technology of the SOTIF risk, monitor and predict the risk during driving, take into account the commonness and specialty of different risk sources, and combine the functional logic and structural framework of IV system to build a targeted system risk pre-control mechanism and adjust the system strategy adaptively, so as to ensure the safe and stable operation of IV in complex and dynamic traffic scenarios.
(3) **A comprehensive defense system for cybersecurity running through the whole chain of "end-pipe-cloud" of IV shall be established**, as follows: foster a trusted environment for IV, and gradually improve the trusted environment for IOV and enhance the defense capability and efficiency against unknown threats, with the help of cryptography and trusted computing system; breakthrough dynamic and active cybersecurity protection technology, carry out researches including automatic threat identification, active risk management and attack traceability, so as to realize the transformation from single point or passive defense to comprehensive defense in combination of dynamic perception safety detection and active safety management; build a in-depth defense

system with clear layers in the whole life cycle. Around the whole life cycle of IV in design, R&D, production, maintenance and scrapping, a hierarchical and multi-domain protection system covering on-board intelligent terminal, mobile intelligent terminal, IOV service platform and multi-mode network communication protocol shall be built, based on key technologies such as safety classification, access control, encryption safety, intrusion detection and security audit assurance.

In view of the development trends of vehicle system safety technology in the above-mentioned three aspects, the corresponding scientific problems, main difficulties and challenges are summarized as follows:

(1) **In terms of FuSa**, the core sensors of IV are short of FuSa design. Especially for IV of L3 and above, the driving system is responsible for safety. Based on the quality and reliability required by vehicle- specification-level devices, the sensors also need to have safety functions such as self-inspection, fault diagnosis and reminder alarm. At present, the main sensors for automated driving such as camera, millimeter wave radar, ultrasonic radar and lidar are still insufficient in FuSa design, which has become a major obstacle to IV safety; IV have both computational diversity and safety requirements for the main control chip. Currently, the main control chip of automated driving widely adopts CPU, GPU, FPGA and other processor structures, resulting in the high complexity of the chip, as well as the increasing difficulty and challenge in FuSa design; IV widely adopt open source operating systems (such as Linux), which can not conduct comprehensive safety analysis, control flow and data flow analysis, and difficult to carry out coverage test to meet the high FuSa level; Machine learning and other algorithms have black-box nature, which shows a high degree of uncertainty for the identification of highly-correlated safety targets such as lane lines and traffic participants. It is difficult to carry out sufficient safety analysis and software testing, and can not ensure the safety requirement at the high FuSa level.
(2) **In terms of SOTIF**, the current research on the generation, quantification and protection of SOTIF risks is far from enough, which is mainly reflected in the difficulty of comprehensively and deeply revealing the generation and propagation mechanism of SOTIF risks of complex systems. The problem of SOTIF involves the complex interaction among driver, vehicle and road environment. The complexity, dynamics, uncertainty and variability of scenarios pose a great challenge to the safe operation of IV. The complexity, coupling, diversity and black box characteristics of the system make it difficult to explore in depth the mechanism of its functional deficiencies; It is difficult to objectively and accurately characterize and evaluate the comprehensive risk jointly caused by external environmental factors and insufficient system functions, as well as to make clear of system risks caused by them. There is a lack of an effective risk quantification model; It is difficult to timely and efficiently predict the safety-related risks of SOTIF of the system in complex scenarios, and formulate safety decision-making approaches and cooperative protection mechanisms. In the process of safety decision-making, the existing hierarchical automated driving

system mainly estimates the external driving risks by means of motion or collision prediction for collision avoidance and other goals, without sufficient consideration of system function. Although the end-to-end automated driving system can implicitly compensate the insufficient functions of sensors and perception and decision-making models, its reliability and poor interpretability lead to constrained application. In addition, the monitoring indexes and degradation strategies adopted by some system monitoring and degradation technologies are relatively simple, and the risk protection capability under complex environments and system conditions is very limited.

(3) **In terms of cybersecurity**, the current research mostly focuses on using cybersecurity means from Internet to solve the cybersecurity problem of IV. However, due to different application objects and scenarios of IV, the relevant research still faces many challenges, mainly including: the cybersecurity schemes are mostly made in the above-mentioned way, and their effects shall be further tested and verified; Compared with data encryption and message authentication technology, how to establish efficient and accurate intrusion detection approaches for precise identification of intrusion behavior is a difficult issue to be solved at present; The increasing complexity of heterogeneous software and hardware components of IV brings difficulties to the testing and evaluation of cybersecurity. It is urgent to make a breakthrough in testing and evaluation technology and establish a fundamental security testing platform supporting unified standards and services.

In order to overcome the above problems and challenges, it is necessary to conduct a more comprehensive and in-depth study on the system safety issues caused by the DVE system as well as its internal and external functions. There are typical cutting-edge technologies in the system safety as follows:

(1) **In the field of FuSa**, cutting-edge technologies mainly include: ① vehicle specification qualification and FuSa design technology for IV's core devices. It can improve the quality and reliability of such vehicle-specification-level devices as main sensors, including camera, millimeter wave radar, ultrasonic radar, lidar, etc. On this basis, it improves the self-inspection, fault diagnosis, warning and other safety functions of single sensor and sensor link; To meet the computing power diversity and safety requirements of IV for the main control chip, in addition to providing parallel computing power, it enhances the comprehensive safety design and processing technology of multiprocessor architecture chip, and meets the FuSa requirements of main control chip with high complexity. ② FuSa technology for complex software and system with large scale. Based on systematic thinking, standardized development process and safety culture construction, the comprehensive safety of highly complicated automated driving software is improved through a variety of technical means, such as white box-based open source system, algorithm interpretability design, antagonism training, independent redundant channel increase, as well as comprehensive test and evaluation system establishment. ③ Multi-dimensional modular reuse and architecture multi-factor trade-off technology. In order to meet the complex

and cross coupling characteristics of IV supply chain, a modular reuse safety architecture topology design scheme suitable for service-oriented architecture (SOA) concept is constructed, so that the derived alternative for logical function architecture conforms with the service design principles of "reuse-abstraction-encapsulation-granularity-coordination", and then a variety of alternative architectures are obtained through the above topology design. The trade-off elements are selected from such perspectives as the number of ECUs, intra-domain bandwidth, FuSa, incremental development, system cost and software development difficulty, and then the multi-factor trade-off and architecture optimization are conducted to support the FuSa concept of IV and road-cloud-network integration system.

(2) **In the field of SOTIF**, self-monitoring and adaptive SOTIF protection technology is a typical cutting-edge technology, which specifically refers to exploring self-monitoring and risk protection technology from multiple dimensions by in-depth research on the generation mechanism and risk quantification approach of SOTIF: ① Internal condition monitoring aims to study the direct quantification of the risks induced by an insufficient function of vehicle, in particular, evaluate the on-line state of the model towards anomaly detection and uncertainty estimation of on-board artificial intelligence algorithm, so as to provide a risk criterion for adaptive safety decision. ② Environmental condition monitoring is mainly based on the impact of external factors such as weather conditions, road conditions and network communication conditions on the SOTIF of the system. The corresponding relationship between the two is established by means of theoretical analysis, test comparison and data statistics, so as to map the external factors leading to insufficient functions of the system into quantifiable and monitorable environmental condition indexes. ③ Compliance monitoring and restraint can ensure the adherence of IV to traffic rules by digitizing road traffic laws and regulations and transforming them into system safety monitoring indexes. The ego-vehicle adheres to laws and regulations in driving for responsibility avoidance, which is conducive to the prediction and avoidance of surrounding traffic participants, as well as the establishment of a safe road ecological environment. ④ The detection on drivers and passengers' states involves fatigue state, behavior and takeover ability. It is necessary to refine the common operation mechanism and individual behavior characteristics of drivers and passengers, systematically identify their potential misoperation and state performance, and extract the states with driver monitoring technology. ⑤ Based on the above information, it can comprehensively evaluate the SOTIF risks faced by IV, then design the safety decision-making approach sensitive to risks, and build a system strategy adaptive regulation mechanism to respond to the overall system risks, so as to effectively ensure the SOTIF. Figure 6.3 shows a typical implementation framework of this technology, which includes three monitoring layers: the first layer monitors the state of algorithms such as perception, prediction and positioning; the second layer monitors the ODD status, such as weather, road conditions, V2X communication conditions; the third layer is the digital layer of traffic regulations for monitoring whether the vehicle operates

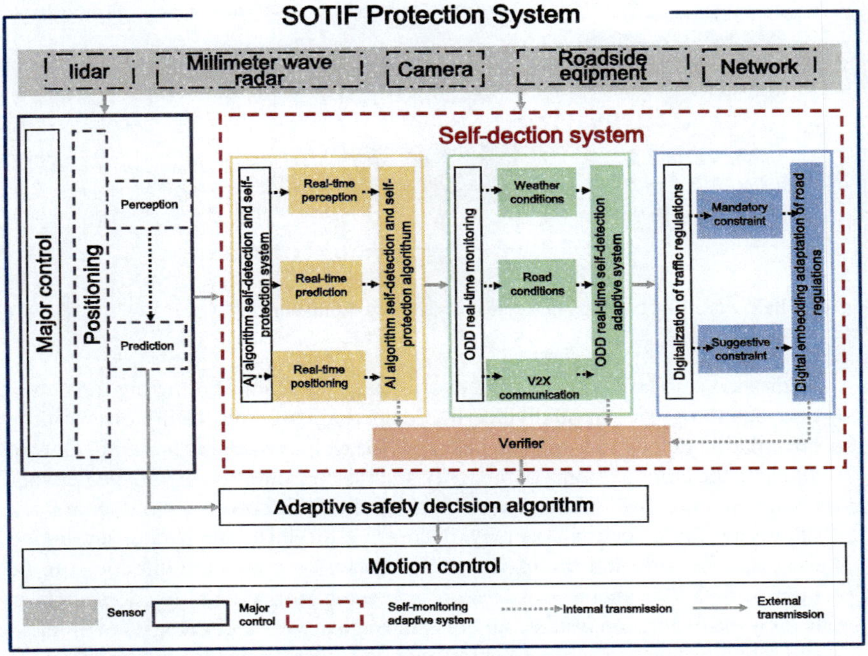

Fig. 6.3 SOTIF protection system based on Self-monitoring and Self-adaptation

in compliance. The monitored information from the three layers is simultaneously transferred to the checker, and the safety decision-making algorithms are adaptively chosen to match the monitored results, so that the safety of motion control is guaranteed.

As shown in Fig. 6.4, the SOTIF protection technology for insufficient perceptual function adopts deep integration [11] and other means to estimate the cognitive uncertainty of the perceptual model, and constructs an artificial potential field sensitive to the perceptual uncertainty as one of the optimization objectives of the controller for decision planning based on model prediction, to dynamically adjust the safety decision to deal with the possible error detection of the perceptual model, and improve decision safety in the case of insufficient perceptual function.

(3) **In the field of cybersecurity**, cutting-edge technologies mainly include: ① lightweight data encryption and authentication of IV on-board networks. To reduce the additional computing power, network bandwidth and storage space cost due to encryption operation, such means as lightweight algorithms and hardware acceleration can be adopted for message authentication and encryption in vehicle network to improve the operation efficiency. ② IV intrusion detection technology. In accordance with the data source, intrusion detection can

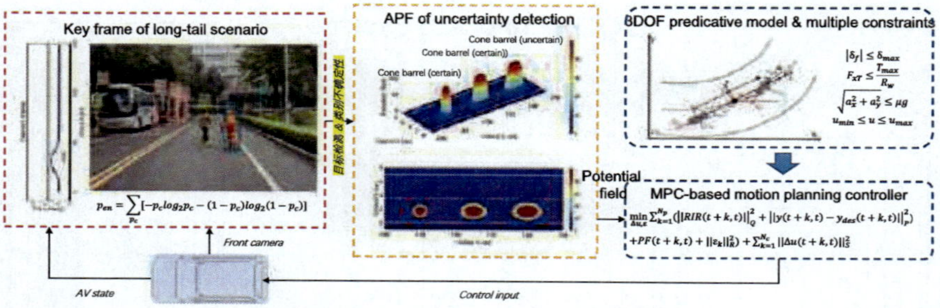

Fig. 6.4 Safety decision based on perceptual uncertainty estimation

be divided into host-based and network-based detection; Following the detection technology, it can be divided into detection based on feature observation, information theory and statistical analysis, as well as machine learning. How to alleviate the computational complexity and the consumption of vehicle network communication bandwidth is a problem to be further solved. At the same time, improving the detection accuracy, decreasing the false alarm rate, shortening the response time and improving the robustness are also the directions to be improved. ③ IV cybersecurity testing and evaluation technology. Research on testing and evaluation is made, to break through the key technologies of efficient and reliable cybersecurity evaluation, and develop relevant evaluation tools; On this basis, the cybersecurity testing platform is established in four directions, i.e., IV key components, operating system, communication environment and information service system.

In addition, it is necessary to build a comprehensive technical system for the three types of system safety problems, such as exploring the hazard /threat mechanism analysis and risk quantitative assessment technology for IV and multi-domain electrical/electronic information (EEI) architecture of vehicle–road-cloud-network in complex environment, to break through the limitation of hazard and threat separation and mere applicability to single vehicle system in the current safety standards; studying the multi-domain and multi-dimensional hybrid modeling technology of vehicle–road-cloud-network and the comprehensive mechanism analysis of hazard/threat, to fully cover the risk sources such as software and hardware faults, functional deficiency and cyber attack in EEI architecture; and by further quantitative evaluation on of FuSa factors "severity/exposure rate/controllability" and cybersecurity factors "security/property/operation/privacy" affecting the risk level, as well as the analysis on operation scenarios of vehicle–road-cloud-network, establishing the quantitative risk assessment theory of FuSa, SOTIF and cybersecurity under the new generation of centralized EEI architecture, to form a new breed of integrated solutions to vehicle–road-cloud-network system safety of IOV.

6.4 Cutting Edge Technology of Driving Safety

To implement the concept of "human-centered" and "serve for human", the development goal of DSV is to promote the adaptive and personalized development of intelligent driving safety systems, so as to realize the safe and efficient operation of IV at different stages, especially the high-grade AV. Accordingly, the DSV system shall be aimed at different levels of IV as well as drivers and passengers with different characteristics, to ensure driving safety of traffic participants. To achieve the above goals, the development trend of DSV technology mainly includes the following three aspects:

(1) **DVE generalized system Modeling technology based on properties of interactive coupling of elements.** Precise and objective technology concerning generalized dynamics modeling of DVE is vital to accurately predict behavior of driver, vehicle and environment and ensure operation safety of systems. The dynamic changes and coupling characteristics of various elements of the DVE system are complicated. Therefore, it is of great significance to explore the interactive relationship of various elements, clarify interactive law among various parts, and then establish a generalized dynamics model of DVE.
(2) **Comprehensive cognition and situation assessment technology for adaptive complex driving environment.** In the driving process, IV closely interacts with its surrounding vehicles, dynamic and static traffic obstacles, as well as pedestrians. In order to ensure DSV, it is necessary to explore the cognitive technology on complex driving environment, accurately distinguish behaviors of interactive objects and driving environment status, and precisely evaluate and predict risk situations.
(3) **Vehicle cooperative control technology based on multi-objective, high real-time and human-likeness.** Driving safety is the main factor to be considered during vehicle driving, which shall be optimized in coordination with multi-dimensional indexes such as mobility, fuel economy and driving comfort. In addition, it is also a considerable development trend to explore the decision-making mechanism of human driver and design human-like vehicle decision and control strategies, and make the vehicle evolved beyond driver in the process of self-learning.

Concerning the above three critical DSV technologies, the corresponding scientific problems, main difficulties and challenges are summarized as follows:

(1) **In terms of the construction of a generalized DVE system model, it is difficult for the current research to fully describe the multi-factor coupling and cooperative operation mechanism.** Most of the existing studies extend from a single object to the perspective of driver-vehicle, driver-road and vehicle–road interaction, and less from the overall perspective of DVE system. In specific, although the current study on vehicle–road coupling has been put forth to some extent, it is mainly based on the interaction mechanism between tire dynamics and road dynamics, yet hard to extend and establish a closed-loop system model

from an overall perspective of DVE. The research of driver-vehicle interaction focuses on the safety technology for advanced driving assistance systems, but rarely considers driver's acceptance of the system, with poor maneuverability. The research of driver-road interaction mainly concentrates on the intention of single vehicle driving, without in-depth study on interactive intention identification. On the whole, relevant studies are limited to the scope of narrow concepts, oversimplify some factors, and lack a unified and complete description on the system despite having covered such traffic elements as driver, vehicle and road. There is an urgent need to accurately analyze the characteristics of driving cognition, build a cognitive model, solve the trust problem in the driver-vehicle interaction, and realize cooperative driving of driver and vehicle; to build a driver-road interaction mechanism model for exploring the interaction between the vehicle driver and surrounding traffic participants, so as to facilitate the active control of collision risk avoidance in theory; and finally to establish a novel and scalable description and unified modeling on the complex DVE system, and obtain the law of influence of traffic environment change on DSV on this basis, so as to guide vehicle motion control or traffic management in theory.

(2) **In terms of environmental cognition and situation assessment, it is difficult for the current research to reveal the risk generation mechanism of a complex DVE system and make unified quantitative assessment**. In the existing research, the algorithm model based on vehicle kinematics is simple and easy to understand, but difficult to adapt to complex road conditions; The algorithm based on collision probability can take the uncertainty of environmental perception and behavior into account, yet with limited accuracy on uncertainty modeling; The risk assessment algorithm based on data-driven approaches such as machine learning has acceptable accuracy, but requires high data completeness and lacks interpretability; The APF model describing system risk can consider the impact of road environment on driving risk, but is complex and difficult to calibrate parameters. Generally speaking, it is easy to obtain and understand quite a number of existing evaluation indexes, yet hard to fully describe the joint impact of driver behavior, vehicle state, road condition and environmental characteristics on driving risk situation according to these indexes, and hard to fully reflect the internal transformation law among driving states and analyze the dynamic evolution mechanism of system operation risk. Therefore, towards the closed-loop DVE system, some technical bottlenecks are to be solved urgently in the existing research, as follows: by starting with the analysis of the coupling mechanism of multiple elements of driver, vehicle and road environment, how to reveal the interaction of driver-vehicle, vehicle-vehicle and vehicle-environment and causation mechanism of collision accidents; how to describe the uncertainty of closed-loop system, excavate the unified dynamic evolution mechanism of system operation risk, and build situation assessment approach.

(3) **In terms of the research progress of control theory, it is difficult for the current research to analyze the driver-vehicle interaction mechanism and self-learning cooperative decision mechanism in complex scenarios.**

In essence, the correlation among multiple optimization objectives is affected by the coupling characteristics of the research object. Current multi-objective decision control approaches usually weight multiple optimization objectives to form a single optimization objective for calculation, or use it as an independent index to explore the multi-dimensional performance of the system. Due to a lack of analysis and research on internal relationship and coupling mechanism of multiple optimization objectives, it is difficult to analyze the interactive coupling mechanism of DVE in complex scenarios. Hence, its application is constrained. In addition, the current research on the law of driver's cognition and handling behavior mostly focuses on the relationship between the overall driving behavior and cognitive process, which is difficult to summarize the common driving features and establish a self-learning cooperative decision mechanism. In the existing decision-making research, the rule-based method is difficult to traverse all real traffic scenarios, while the deep learning method tends to consume a lot of computing resources in the research and development process, with poor interpretability. Therefore, it is necessary to accurately identify driver behavior and analyze individual differences of behavior decision-making, so as to formulate the decision and control strategy of simulating people and surpassing people, which is of far-reaching significance to realize the development of vehicle intelligence.

In general, in order to deal with various problems in the development of DSV technology and its essential scientific issues, it is necessary to conduct more in-depth research on the DVE system and its internal action mechanism, especially the role of "driver" in the system. Therefore, intelligent-based idea of learning from driver, simulating driver and surpassing driver is the key to solve the technical problems of DSV [12]. The establishment of human-like decision and control driven by driver cognitive mechanism and oriented to self-learning cooperation in complex scenarios is one of the typical cutting-edge technologies in the field of driving safety. A typical solution is to analyze the common laws of driver behavior and control mechanism (Fig. 6.5). For example, in accordance with the law of natural biological behaviors (such as heat avoidance of fruit flies), it is to summarize driver's "pursuing benefit and avoiding harm" toward safety, efficiency, comfort and energy saving, explore driver's lateral and longitudinal control behaviors at the same time, refine the online dynamic identification of driver's personality on this basis and put its key parameters into the design of automated driving decision strategy.

To study the human-like decision control of IV, it is necessary to refine the general decision-making mechanism of drivers and establish individual driving characteristics. A typical research idea is to design driver's general decision-making mechanism based on risk identification results in mixed traffic and the common driving objective model considering efficiency, safety and comfort [13]; Secondly, to design driver's personalized driving characteristics based on driver risk cognition and modeling of personalized driving objectives, establish a virtual mechanical system for unified

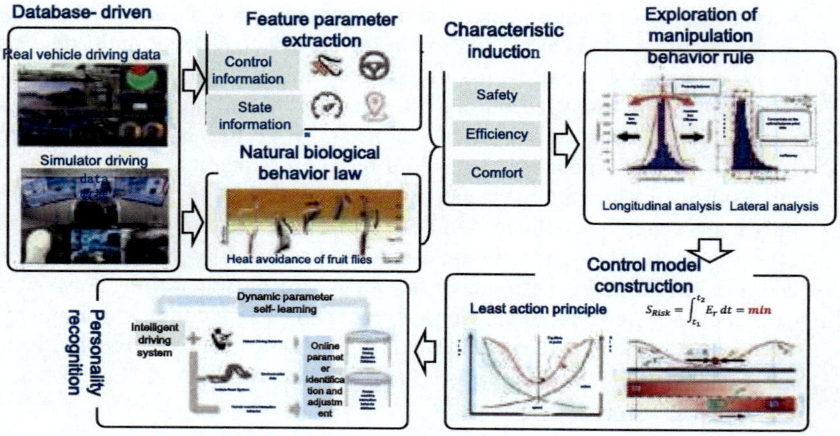

Fig. 6.5 Common laws of driver behaviors and analysis of control mechanism

description of traffic elements, carry out unified planning of critical driving parameters such as path and speed with the principle of least action, and finally build a human-like decision control way of IV [14], as shown in Fig. 6.6.

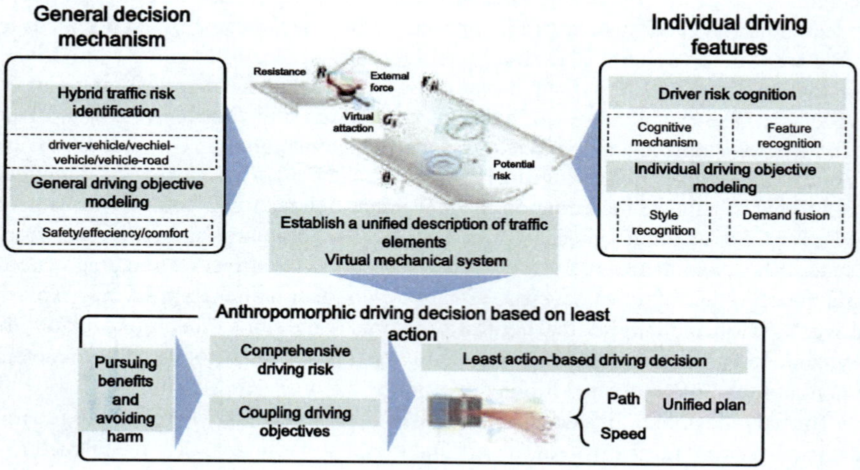

Fig. 6.6 Human-like decision control way of IV

6.5 Cutting Edge Technology of Intelligent Protection

In the face of hazardous road traffic scenarios such as inevitable collision accidents, the existing assistance safety technology usually takes emergency braking for mitigating the collision intensity, and fails to realize the optimal protection, without due consideration to driver-vehicle dynamic interaction and damage coupling mechanism in the collision process. With the help of perception, decision-making and execution capability of ICV, accurately predicting injury risks of traffic participants (including occupants in the vehicle and pedestrians outside the vehicle) and implementing an adaptive protection strategy with the lowest injury risk as the optimization goal are considerable ways to establish innovative vehicle intelligent safety protection strategies. To achieve the above objectives, the development trend of ISV technology mainly includes the following three aspects:

(1) **Real-time damage risk prediction technology based on DVE state**. The dynamic changes and coupling properties of various elements within DVE system are complex, which is reflected in the multi-dimensional and highly dynamic load transfer path among the deformation of vehicle structure, the action of restraint system and the causation of human injury in the process of accidents. By identifying the basic damage mechanism under impact conditions and comprehensively using the environmental perception information of IV, human injury risk in a certain time before the accident can be quantified accurately in real time, which will provide important data reference for safety decision-making of IV.
(2) **Algorithm fast migration technology based on vehicle model specificity**. Vehicle design has a significant impact on the risk and severity of traffic participant injury. Among various vehicle models, there are big differences in material, structure, occupant restraint system, etc. The establishment of a injury prediction algorithm taking the specificity of vehicle models into account can realize the rapid migration prediction among different models, which is of great significance to shorten the safe development period and reduce the cost, and also one of the keys to the application and promotion of relevant technologies in the industry.
(3) **IP technology for traffic participants based on active and passive cooperation**. Most of the current protection technologies take the collision time as the dividing line, with the division into two protection stages: active safety and passive safety. The design is separated from each other, without effective coordination. Correspondingly, the active and passive cooperated safety protection strategy implements the integrated development of risk warning, risk prediction and protection strategy, which can jointly give full play to the potential of pre-accident and post-accident protection technology and realize the "closed-loop" optimal protection for traffic participants.

In the development of vehicle IP technology, the essential scientific issues behind it shall be explored urgently. The scientific problems, main difficulties and challenges corresponding to the above three key IP technologies are as follows:

(1) **In face of the complex and changeable road traffic hazard scenarios, the current research is difficult to leverage the accuracy and real-time damage risk prediction**. The dynamic interaction of DVE in road traffic systems constitutes a dynamic, strongly-coupled and nonlinear generalized dynamics unit. In terms of human, drivers' active behaviors in hazardous scenarios significantly change vehicle dynamics state, and their physiological signs and dynamic responses (such as posture change) will affect their passive injuries during collisions; In terms of vehicle, it is the main medium for load transmission to driver. The structural stiffness and restraint system state of vehicle body will significantly affect the impact load borne by occupants in the vehicle; In terms of road, the obstacles in contact with the vehicle will directly determine collision intensity, and various environmental conditions, directly and indirectly, affect the initial working conditions. These complexities lead to the fact that high-precision risk prediction models often rely on deep learning algorithms with complex structures. With quite a number of network parameters, computation tends to be highly complicated, resulting in the difficulty of real-time prediction in hazardous scenarios [15]. How to explore the key characteristic parameters affecting accident injury risk in the road traffic system, improve the prediction in case of low computational complexity, and take into account both the accuracy and real-time injury risk prediction technology are the technical bottlenecks to be resolved for the present research.
(2) **In view of vehicle type specificity, the current research has not developed the damage risk prediction technology for rapid migration among different vehicle models**. The risk of collision accidents is significantly related to the characteristics of vehicle models. The damage risk prediction model for a specific vehicle model is hard to be directly applied to another vehicle model. In addition, to establish the risk prediction algorithm, it is necessary to establish a large-scale collision simulation database for a specific vehicle model, carry out numerical simulation verification based on real vehicle collision test, and train and verify the data-driven prediction algorithm or model. If the above process is completely repeated for each new vehicle model, it will lead to such problems as difficult development, long cycle, high cost and hard application of algorithms. Therefore, the rapid migration of damage risk prediction model among different vehicle models is still undeveloped, which is not conducive to the application and popularization of damage risk prediction technology.
(3) **Concerning the development of IP strategy, the current research has not organically combined active and passive safety technologies**. The active safety technology focuses on the driving period before the accident, to decrease accident rate and enhance occupants' experience; while the passive technology focuses on the damage initiation and evolution period after the accident, to improve the crashworthiness of vehicles and reduce the risks of human injuries. The root of the time segmentation between the two is that it is difficult to accurately evaluate and predict damage risks in the pre-collision stage, leading to the difficulty of damage mitigation by actively intervening in vehicle dynamics or adaptively adjusting restraint system in advance; Besides, there is no accurate

definition of on-board safety criteria in the industry, leading to the difficulty in defining the optimization direction without concise and clear guiding objectives for vehicle safety decision. Therefore, the establishment of on-board safety criteria under hazardous scenarios of road traffic as well as the combination of active safety and passive safety technologies are of far-reaching significance for the realization of IP safety, and also the essential guarantee for IV on the road.

In order to tackle various problems in the development of IP technology and its essential scientific issues, it is necessary to grasp the core of problems and establish the hazardous scenario damage prediction technology. It contributes to obtaining multiple source information related to the severity of accident, and implementing an accurate assessment of injury severity of relevant traffic participants (such as occupants in the vehicle or pedestrians outside the vehicle), with the assistance of on-board perception system. The key variables that have a noteworthy impact on injury involve three aspects related to DVE: as for driver, physiological characteristics (age, gender, height, weight, etc.) and occupants' postures (seat orientation, seat back angle, body posture, etc.); for vehicle, physical parameters (mass, size, stiffness, etc.) and restraint system information (seatbelt, airbag, etc.); for road, types of obstacles (vehicles, pedestrians, buildings, signs, etc.), collision speed, collision location, etc. The acquisition of the above key accident information is vital to accurate prediction of damage. The external factors including obstacle types and collision speed can be captured and identified by the external sensing equipment such as on-board camera, radar and lidar. And the internal factors including physiological features and body postures can be identified by the internal sensing equipment such as the in-vehicle camera, pressure sensor and inclination sensor.

The damage severity prediction technology can be divided into two categories: pre-accident damage prediction and post-accident damage estimation (Fig. 6.7). Pre-accident prediction mainly serves the trajectory planning algorithm and adaptive constraint system. It is necessary to ensure the accuracy as well as the real-time performance of the algorithm, update damage prediction results in real time for optimizing the configuration of constraint system when inevitable collision accident is judged, and assist the trajectory planning algorithm to complete the trajectory operation for minimizing damage [15]. Post-accident damage estimation mainly facilitates advanced systems such as automatic accident alarm, to estimate the damage severity of road traffic accident according to collision information of the on-board accident data recording system (e.g., collision speed and vehicle collision waveform) and transmit to the appropriate emergency center in time. Taking into account both the efficiency and survival rate of rescue work, it has high requirements for prediction accuracy. The full-time damage prediction in the collision process can optimize the control strategy and improve the IV safety protection.

The active and passive protection strategy for traffic participants is not only critical to IV safety protection, but also the basis of safety system design and development. Taking occupants in the vehicle for example, to build a "closed-loop" safety protection with risk warning, risk prediction as well as active and passive cooperation, it

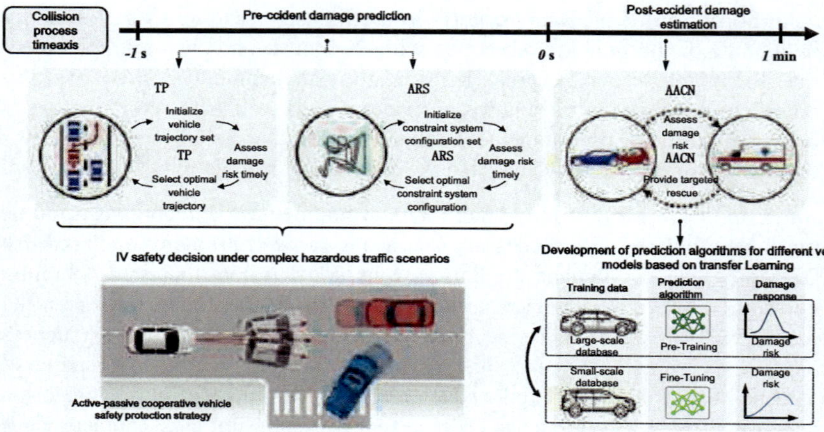

Fig. 6.7 Damage risk prediction and protection strategy in hazardous scenarios

is necessary to focus on the inevitable road traffic hazard scenarios related to collision accidents, and determine vehicle kinematics state and occupant posture within a certain time of the future, by making full use of vehicle perception information, together with occupant posture, vehicle and scenario conditions. And it is to build on-board safety criteria that comprehensively consider potential damage of all road traffic participants in hazardous scenarios based on the hierarchical prediction algorithm for injury risk with mechanism fusion, actively intervene dynamics state of IV according to the criteria and vehicle perception information, integrate potential damage risk into the vehicle trajectory planning algorithm and adaptive restraint system together with the traditional passive safety protection strategy, implement the optimal decision-making for occupant safety protection to minimize potential injury risk, and eventually build an on-board protection strategy with active and passive cooperation [15–17].

In hazardous road traffic scenarios such as collisions, based on damage risk prediction algorithm and vehicle safety criteria, the implementation of active and passive cooperative protection strategy with the goal of minimizing potential damage risk is a significant path to further improvement of the IV safety protection for traffic participants.

6.6 Cutting Edge Technology of Cooperative Safety

The development goal of cooperative safety is to build an interconnected cloud-control driving traffic mode for the overall transportation system with ICV as the carrier, based on adhering to national laws and regulations and ensuring complete FuSa. Traditional traffic facilities and vehicles are independent from each other. In other words, vehicles are offline in the whole traffic control network, and there is no cooperative system between vehicles and environment. The current technology separates the extensive communication among driver, vehicle, road and cloud, as well as the physical layer, information layer and application layer formed by them, leading to two problems in the automated driving framework based on single vehicle intelligence as follows. The first one is at the perception level. Restrained by the perception range and cooperative capability of a single vehicle, it can not meet the requirement of long-distance, large-scale and real-time traffic dispatching and control, and is difficult to ensure the safety and efficiency of the overall traffic; Second is at the computing level. Functional modules, such as perception, decision-making and control, are placed on the vehicle-end computing platform in single vehicle intelligent technology. Due to limited computing power, it can not meet the requirement of real-time vehicle control when carrying out large-scale optimization calculations.

Employing the IOV technology, ICV realizes the interconnection, data sharing and unified allocation of resources among driver, vehicle, road, cloud, etc., providing global traffic element data for the cooperative operation optimization and the overall traffic performance improvement of road network, as well as the support for multi-group, multi-objective fusion perception, intelligent decision and cooperative control of complex traffic system. In response to scientific issues and key technologies related to intelligent safety including intelligent decision-making and cooperative control, cooperative safety mainly has the following three development trends:

(1) **Environment fusion perception technology based on multi-source cooperation**. Under the existing technical conditions, the detection and recognition accuracy of a single vehicle environment perception algorithm is constrained by a large number of factors including perception range, weather environment, lighting and so on. Under extreme environmental conditions, the reliability of a single sensor cannot be guaranteed, which endangers driving safety. Visual sensor, millimeter wave radar, lidar, ultrasonic radar, on-board multi-sensor information fusion and multi-source cooperative sensing based on C-V2X can fuse multiple types of perceived information at the vehicle end and road end, improve the detection under such conditions as visual blind area and bad weather, and greatly facilitate vehicle's comprehensive perception over environment.
(2) **Vehicle safety control technology based on real-time cooperation between vehicle and environment**. Based on environmental perception, vehicle-environment cooperative technology can further play a role in advancing vehicle-environment fusion perception, intention judgment and cooperative decision, as well as vehicle safety control technology. Specifically speaking, the first step is to build the multi-vehicle cooperative communication topology and formation optimization approach for ICV, and then apply the multi-vehicle cooperative control algorithm suitable for various scenarios under different market penetration of ICV through interaction of hierarchically- connected auxiliary information, so as to further improve vehicle safety and driving comfort.

(3) **Traffic safety management and control technology based on cloud control platform**. The traffic big data cloud control platform is an intelligent connected driving infrastructure providing basic services such as real-time information fusion and sharing, computing, application scheduling, data analysis and cybersecurity. For traffic control, a large quantity of traffic data collection, cooperative decision, dispatching assignment and other work shall be carried out, and the existing roadside processor can not meet this real-time demand. Cloud control platform can effectively solve this problem and provide real-time dynamic data such as vehicle operation, road infrastructure, traffic environment and traffic management, as well as real-time cooperative computing environment of large-scale connected applications for ICV, their users and regulators, etc. Facilitated by this platform, the overall cost of AV in the future traffic control will be reduced, and the goal of optimizing system resource utilization and improving road traffic safety can be ultimately achieved.

The development of cooperative safety technology also faces a series of difficulties and challenges, as follows:

(1) **The existing vehicle-environment cooperative technology is capable of point-to-point communication, but is still insufficient in cooperative decision at the system level and difficult to guarantee system safety during multi-vehicle cooperative decision**. Traditional vehicle-environment cooperation mainly stresses the coordination between vehicle and roadside equipment. Although it can solve some problems related to intelligent safety of single vehicle, its application scenarios are limited. Hence, it is difficult to realize a wide range of intelligent connected application and multi-vehicle cooperation regarding safety decisions for road network, and unable to meet technical requirements for global optimization towards vehicle–road-cloud in the development process of the transportation system composed of ICV, as well as high-level application for safety control and traffic data interaction.
(2) **Massive computing such as perception and decision-making are carried out at the vehicle end, and the on-board computing platform is difficult to realize safe and effective computing**. At present, most of the perceptual decision algorithms of commercialized autonomous driving technology are still carried out at the vehicle end, and cooperative safety is constrained to a small amount of information interaction and cooperation with roadside facilities. According to the report of China Academy of Information and Communications Technology, ICV will generate real-time information data of up to 4 TB per hour, including road conditions, weather, surrounding objects, traffic and street signs. As these information shall be used for real-time vehicle control within extremely low time delay, it is difficult to ensure vehicle safety with the current computing capacity. Therefore, only by a single onboard computing platform with high calculation pressure, the efficient and safe computing power can not be guaranteed.
(3) **The industry has not established the infrastructure system with a unified standard, and conducted the research and development of critical common**

technologies, resulting in less effective implementation of key cooperative safety algorithms in the equipment of different manufacturers. At present, in the various infrastructure of vehicle enterprises and roads, the equipment with different platforms and interfaces lacks unified and standardized infrastructure or relevant middleware products as well as common key common technologies. How to effectively use the available road infrastructure to guarantee key algorithms of cooperative safety is an important issue for ISV technology development in the future. Currently, various enterprises develop cooperative safety algorithms with obvious value-added effects by cooperation with Internet enterprises, yet pay less attention to the critical common technologies supporting cloud control platforms.

To sum up, the ISV technology based on a single vehicle is difficult to ensure the safe and efficient operation of the future transportation system. To deal with the above problems, it is necessary to build a unified interconnection system of vehicle–road-cloud, that is, ICV cloud control system. By using the new generation of information and communication technology, the system connects the physical layer, information layer and application layer of DVE-cloud, carries out perception, decision and control under multi-source integration, and further improves ISV with multi-vehicle cooperation technology [18]. As shown in Fig. 6.8, the cloud control system establishes digital mapping of a physical system through integration of ICV, roadside infrastructure and resource platform, and implements hierarchically integrated decision and cooperative control over ICV and roadside infrastructure.

The application of multi-vehicle cooperative intelligent safety is divided into three categories in the cloud control system, which are respectively conducted in the three-level cloud computing platform: the edge cloud serving the high reliability and low delay for ICV, the regional cloud serving the traffic control application for traffic management and control departments, and the central cloud serving the traffic big data-based business for industrial chain users. Among them, edge cloud is the basic operation platform of multi-vehicle cooperative intelligent safety algorithm, with main objectives: (1) facilitate cooperative decision and control based on edge computing for multi-vehicle cooperative intelligent safety applications at different levels, and improve vehicle driving safety, including risk warning in blind area and over the horizon, optimal speed planning, cooperative lane-change planning, formation control, etc.; (2) provide all kinds of real-time, quasi real-time and non real-time basic data related to vehicles and their driving environment for various cooperative safety application platforms, and offer common services in terms of safety, efficiency, comfort and energy saving for vehicle enterprises or external systems on demand by means of container management and unified interface.

In the edge cloud system, the requirements for millisecond delay and ultra-high reliability of information transmission can be met, which can facilitate a variety of cooperative safety technology applications based on ICV cloud control system. The typical technologies include over-the-horizon perception and trajectory re-planning algorithm, and multi-vehicle traffic algorithm in intersection with vehicle–road cooperation. On the cloud control platform with over-the-horizon perception, operational

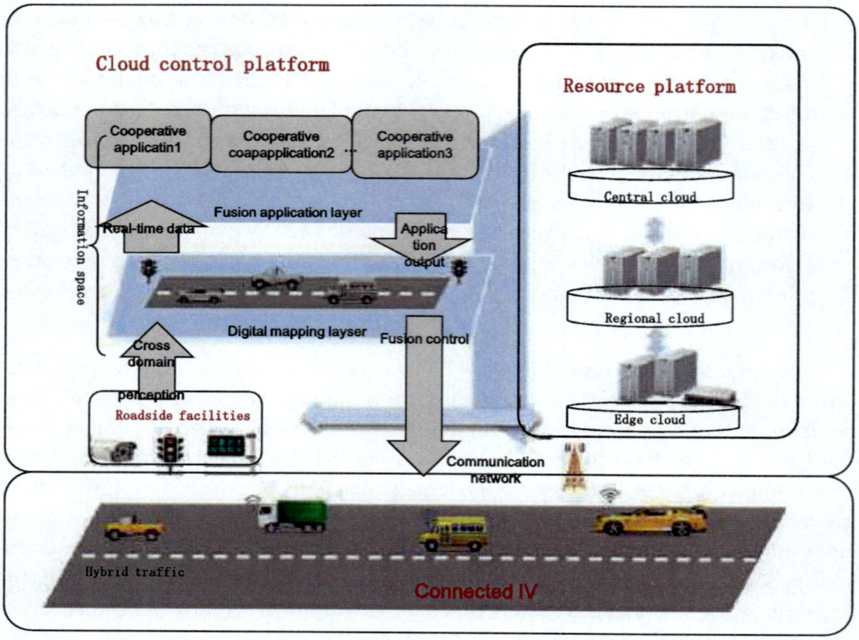

Fig. 6.8 Basic structure of ICV cloud control system

information of all connected vehicles is shared through standardized communication protocol and standardized interconnection services for different vehicles, and the four-dimensional perception including time and space is realized (Fig. 6.9). And thus ICV is endowed with the capability of over-the-horizon perception and dynamic update over road state, which greatly secures travel decision-making and behavior control.

The cloud control driving platform interconnects traffic and vehicle through standardized protocols, so that the decision and control of vehicle movement is cooperated with traffic regulation system on the platform. And it introduces the decision and control of vehicle movement into traffic optimization, to ensure the cooperation between vehicle and traffic signals based on common objectives for optimization, and substantially resolve bottlenecks of traffic travel. In Fig. 6.10, the algorithm for multi-vehicle cooperation at the intersection is shown. By the centralized decision-making on such basis as first-come-first-served [19] and virtual platoon [20], the optimal vehicle arrival sequence and signal light timing are calculated at the edge roadside end, and the distributed vehicle control is carried out at the vehicle end after dispatching. The multi-vehicle cooperation can effectively improve traffic safety at the intersection.

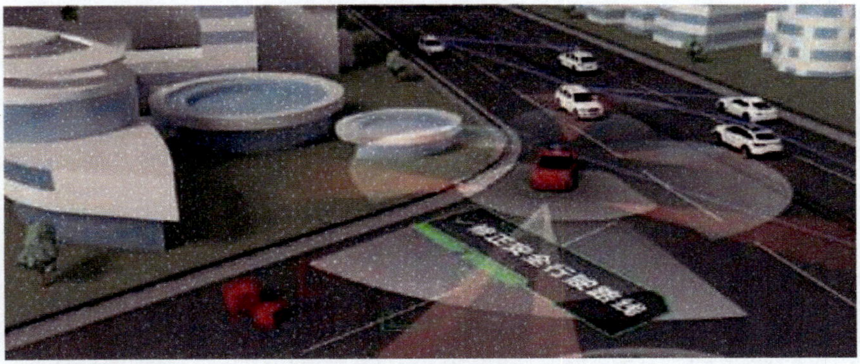

Fig. 6.9 Over-the-horizon perception and Trajectory Re-planning algorithm

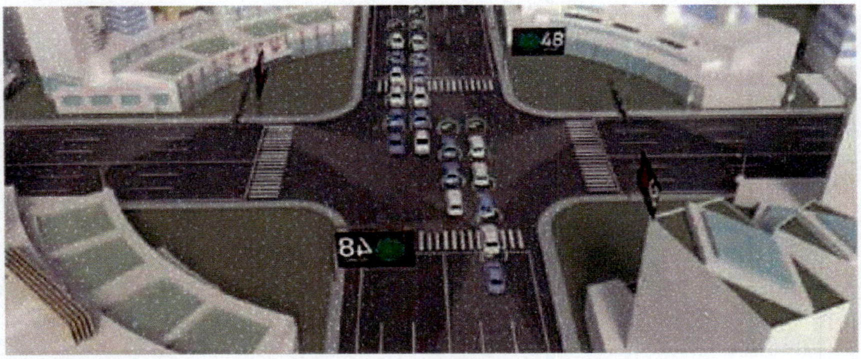

Fig. 6.10 Algorithm for Multi-Vehicle cooperation at intersection

6.7 Summary

Based on the DVE system, a comprehensive ISV technology has been established in terms of system safety, driving safety, intelligent protection, and safety testing and evaluation. With the enhancement of intelligent and connected proficiency of vehicles, cutting-edge technologies such as safety protection system with self-monitoring and adaptive SOTIF, human-like decision control, damage risk prediction, intelligent protection, and cooperative safety based on ICV cloud control system will be developed continuously, to improve driving safety in an all-around way and better secure traffic participants and road traffic.

References

1. The State Council of CPC Central Committee. Outline of building a leading transportation nation [EB/OL]. (2019–9–19)[2022–2–21]. http://www.gov.cn/gongbao/content/2019/content_5437132.htm. [in Chinese]
2. National Development and Reform Commission. IV innovation and Development strategy [EB/OL](2020–02–24)[2022–2–21]. https://www.ndrc.gov.cn/xxgk/zcfb/tz/202002/t20200224_1221077.html?code=&state=123. [in Chinese]
3. Ministry of Industry and Information Technology. Key points for standardization of ICVs [EB/OL].(2019–05–18)[2021–07–08]. http://www.gov.cn/xinwen/2019-05/18/content_5392700.htm. [in Chinese]
4. Ministry of Industry and Information Technology. Management specification for road test and demonstration application of ICVs (for Trial Implementation) [EB/OL](2021–07–27)[2022–2–21]. http://www.gov.cn/zhengce/zhengceku/2021-08/03/content_5629199.htm. [in Chinese]
5. National Bureau of Statistics of the P.R.C. China Statistical Yearbook 2020 [M]. Beijing: China Statistics Press, 2020. [in Chinese]
6. National Bureau of Statistics of the P.R.C. China Statistical Yearbook 2019 [M]. Beijing: China Statistics Press, 2019. [in Chinese]
7. National Bureau of Statistics of the P.R.C. China Statistical Yearbook 2018 [M]. Beijing: China Statistics Press, 2018. [in Chinese]
8. National Bureau of Statistics of the P.R.C. China Statistical Yearbook 2017 [M]. Beijing: China Statistics Press, 2017. [in Chinese]
9. National Bureau of Statistics of the P.R.C. China Statistical Yearbook 2016 [M]. Beijing: China Statistics Press, 2016. [in Chinese]
10. Baidu Map. 2019 China urban traffic report. [2021–07–08]. http://jiaotong.baidu.com/landings/landing?id=50. [in Chinese]
11. Lakshminarayanan B, Pritzel A, Blundell C. Simple and scalable predictive uncertainty estimation using deep ensembles[J]. Advances in neural information processing systems, 2017, 30
12. Wang J, Wu J, Zheng X, Ni D, Li K. Driving safety field theory modeling and its application in pre-collision warning system[J]. Transportation Research Part C: Emerging Technologies, 2016, 72: 306–324
13. Huang H, Zheng X, Yang Y, et al. An integrated architecture for intelligence evaluation of automated vehicles[J]. Accident Analysis & Prevention, 2020, 145: 105681
14. Zheng X, Huang H, Wang J, et al. Behavioral decision-making model of the intelligent vehicle based on driving risk assessment[J]. Computer-Aided Civil and Infrastructure Engineering, 2021, 36(7): 820-837
15. Wang Q, Chen W, Sun Z, et al. A Safety-focused Vehicle Driving Strategy Towards Minimal Injury Severity In Imminent Collision Scenarios[J]. Proceedings of the Internation Research Council on Biomechanics of Injury Asia (IRCOBI Aisa). Online, 2021
16. Parseh M, Asplund F, Svensson L, et al. A data-driven approach towards minimizing collision severity for highly automated vehicles[J]. IEEE Transactions on Intelligent Vehicles, 2021, 6(4): 723-735
17. Wang Q, Gan S, Chen W, et al. A data-driven, kinematic feature-based, near real-time algorithm for injury severity prediction of vehicle occupants[J]. Accident Analysis & Prevention, 2021, 156: 106149
18. Li Keqiang, Chang Xueyang, Li Jiawen, et al. Cloud control system of ICVs and its implementation [J] Automotive Engineering, 2021, 42 (12): 1595–1605. [in Chinese]
19. Xu B, Ban X J, Bian Y, et al. Cooperative approach of traffic signal optimization and speed control of connected vehicles at isolated intersections[J]. IEEE Transactions on Intelligent Transportation Systems, 2018, 20(4): 1390-1403
20. Xu B, Li S E, Bian Y, et al. Distributed conflict-free cooperation for multiple connected vehicles at unsignalized intersections[J]. Transportation Research Part C: Emerging Technologies, 2018, 93: 322-334